INFORMATION THEORY AND CODING BY EXAMPLE

This fundamental monograph introduces both the probabilistic and the algebraic aspects of information theory and coding. It has evolved from the authors' years of experience teaching at the undergraduate level, including several Cambridge Mathematical Tripos courses. The book provides relevant background material, a wide range of worked examples and clear solutions to problems from real exam papers. It is a valuable teaching aid for undergraduate and graduate students, or for researchers and engineers who want to grasp the basic principles.

MARK KELBERT is a Reader in Statistics in the Department of Mathematics at Swansea University. For many years he has also been associated with the Moscow Institute of Information Transmission Problems and the International Institute of Earthquake Prediction Theory and Mathematical Geophysics (Moscow).

YURI SUHOV is a Professor of Applied Probability in the Department of Pure Mathematics and Mathematical Statistics at the University of Cambridge (Emeritus). He is also affiliated to the University of São Paulo in Brazil and to the Moscow Institute of Information Transmission Problems.

INFORMATION THEORY
AND CODING BY EXAMPLE

MARK KELBERT

Swansea University, and Universidade de São Paulo

YURI SUHOV

University of Cambridge, and Universidade de São Paulo

CAMBRIDGE
UNIVERSITY PRESS

CAMBRIDGE
UNIVERSITY PRESS

University Printing House, Cambridge CB2 8BS, United Kingdom

One Liberty Plaza, 20th Floor, New York, NY 10006, USA

477 Williamstown Road, Port Melbourne, VIC 3207, Australia

314-321, 3rd Floor, Plot 3, Splendor Forum, Jasola District Centre, New Delhi - 110025, India

103 Penang Road, #05-06/07, Visioncrest Commercial, Singapore 238467

Cambridge University Press is part of the University of Cambridge.

It furthers the University's mission by disseminating knowledge in the pursuit of education, learning and research at the highest international levels of excellence.

www.cambridge.org
Information on this title: www.cambridge.org/9780521769358

First published 2013

A catalogue record for this publication is available from the British Library

ISBN 978-0-521-76935-8 Hardback
ISBN 978-0-521-13988-5 Paperback

Contents

Preface

This book is partially based on the material covered in several Cambridge Mathematical Tripos courses: the third-year undergraduate courses *Information Theory* (which existed and evolved over the last four decades under slightly varied titles) and *Coding and Cryptography* (a much younger and simplified course avoiding cumbersome technicalities), and a number of more advanced Part III courses (Part III is a Cambridge equivalent to an MSc in Mathematics). The presentation revolves, essentially, around the following core concepts: (a) the entropy of a probability distribution as a measure of 'uncertainty' (and the entropy rate of a random process as a measure of 'variability' of its sample trajectories), and (b) coding as a means to measure and use redundancy in information generated by the process.

Thus, the contents of this book includes a more or less standard package of information-theoretical material which can be found nowadays in courses taught across the world, mainly at Computer Science and Electrical Engineering Departments and sometimes at Probability and/or Statistics Departments. What makes this book different is, first of all, a wide range of examples (a pattern that we followed from the onset of the series of textbooks *Probability and Statistics by Example* by the present authors, published by Cambridge University Press). Most of these examples are of a particular level adopted in Cambridge Mathematical Tripos exams. Therefore, our readers can make their own judgement about what level they have reached or want to reach.

The second difference between this book and the majority of other books on information theory or coding theory is that it covers both possible directions: probabilistic and algebraic. Typically, these lines of inquiry are presented in different monographs, textbooks and courses, often by people who work in different departments. It helped that the present authors had a long-time association with the Institute for Information Transmission Problems, a section of the Russian Academy of Sciences, Moscow, where the tradition of embracing a broad spectrum of problems was strongly encouraged. It suffices to list, among others, the names of Roland Dobrushin, Raphail Khas'minsky, Mark Pinsker, Vladimir Blinovsky, Vyacheslav Prelov, Boris Tsybakov, Kamil Zigangirov (probability and statistics), Valentin Afanasiev, Leonid Bassalygo, Serguei Gelfand, Valery Goppa, Inna Grushko, Grigorii Kabatyansky, Grigorii Margulis, Yuri Sagalovich, Alexei Skorobogatov, Mikhail Tsfasman, Victor Zinov'yev, Victor Zyablov (algebra, combinatorics, geometry, number theory), who worked or continue to work there (at one time, all these were placed in a five-room floor of a converted building in the centre of Moscow). Importantly, the Cambridge mathematical tradition of teaching information-theoretical and coding-theoretical topics was developed along similar lines, initially by Peter Whittle (Probability and Optimisation) and later on by Charles Goldie (Probability), Richard Pinch (Algebra and Geometry), Tom Körner and Keith Carne (Analysis) and Tom Fisher (Number Theory).

We also would like to add that this book has been written by authors trained as mathematicians (and who remain still mathematicians to their bones), who nevertheless have a strong background in applications, with all the frustration that comes with such work: vagueness, imprecision, disputability (involving, inevitably, personal factors) and last – but by no means least – the costs of putting any mathematical idea – however beautiful – into practice. Still, they firmly believe that mathematisation is the mainstream road to survival and perfection in the modern competitive world, and therefore that Mathematics should be taken and studied seriously (but perhaps not beyond reason).

Both aforementioned concepts (entropy and codes) forming the base of the information-theoretical approach to random processes were introduced by Shannon in the 1940s, in a rather accomplished form, in his publications [139], [141]. Of course, entropy already existed in thermodynamics and was understood pretty well by Boltzmann and Gibbs more than a century ago, and codes have been in practical (and efficient) use for a very long time. But it was Shannon who fully recognised the role of these concepts and put them into a modern mathematical framework, although, not having the training of a professional mathematician, he did not always provide complete proofs of his constructions. [Maybe he did not bother.] In relevant sections we comment on some rather bizarre moments in the development of Shannon's relations with the mathematical community. Fortunately, it seems that this did not bother him much. [Unlike Boltzmann, who was particularly sensitive to outside comments and took them perhaps too close to his heart.] Shannon definitely understood the full value of his discoveries; in our view it puts him on equal footing with such towering figures in mathematics as Wiener and von Neumann.

It is fair to say that Shannon's name still dominates both the probabilistic and the algebraic direction in contemporary information and coding theory. This is quite extraordinary, given that we are talking of the contribution made by a person who was active in this area more than 40 years ago. [Although on several advanced topics Shannon, probably, could have thought, re-phrasing Einstein's words: "Since mathematicians have invaded the theory of communication, I do not understand it myself anymore."]

During the years that passed after Shannon's inceptions and inventions, mathematics changed drastically, and so did electrical engineering, let alone computer science. Who could have foreseen such a development back in the 1940s and 1950s, as the great rivalry between Shannon's information-theoretical and Wiener's cybernetical approaches was emerging? In fact, the latter promised huge (even fantastic) benefits for the whole of humanity while the former only asserted that a modest goal of correcting transmission errors could be achieved within certain limits. Wiener's book [171] captivated the minds of 1950s and 1960s thinkers in practically all domains of intellectual activity. In particular, cybernetics became a serious political issue in the Soviet Union and its satellite countries: first it was declared "a bourgeois anti-scientific theory", then it was over-enthusiastically embraced. [A quotation from a 1953 critical review of cybernetics in a leading Soviet ideology journal *Problems of Philosophy* reads: "Imperialists are unable to resolve the controversies destroying the capitalist society. They can't prevent the imminent economical crisis. And so they try to find a solution not only in the frenzied arms race but also in ideological warfare. In their profound despair they resort to the help of pseudo-sciences that give them some glimmer of hope to prolong their survival." The 1954 edition of the Soviet *Concise Dictionary of Philosophy* printed in hundreds of thousands of copies defined cybernetics as a "reactionary pseudo-science which appeared in the USA after World War II and later spread across other capitalist countries: a kind of modern mechanicism." However, under pressure from top Soviet physicists who gained authority after successes of the Soviet nuclear programme, the same journal, *Problems of Philosophy*, had to print in 1955 an article proclaiming positive views on cybernetics. The authors of this article included Alexei Lyapunov and Sergei Sobolev, prominent Soviet mathematicians.]

Curiously, as was discovered in a recent biography on Wiener [35], there exist "secret [US] government documents that show how the FBI and the CIA pursued Wiener at the height of the Cold War to thwart his social activism and the growing influence of cybernetics at home and abroad." Interesting comparisons can be found in [65].

However, history went its own way. As Freeman Dyson put it in his review [41] of [35]: "[Shannon's theory] was mathematically elegant, clear, and easy to apply to practical problems of communication. It was far more user-friendly than cybernetics. It became the basis of a new discipline called 'information theory' … [In modern times] electronic engineers learned information theory, the gospel according to Shannon, as part of their basic training, and cybernetics was forgotten."

Not quite forgotten, however: in the former Soviet Union there still exist at least seven functioning institutes or departments named after cybernetics: two in Moscow and two in Minsk, and one in each of Tallinn, Tbilisi, Tashkent and Kiev (the latter being a renowned centre of computer science in the whole of the former USSR). In the UK there are at least four departments, at the Universities of Bolton, Bradford, Hull and Reading, not counting various associations and societies. Across the world, cybernetics-related societies seem to flourish, displaying an assortment of names, from concise ones such as the Institute of the Method (Switzerland) or the Cybernetics Academy (Italy) to the Argentinian Association of the General Theory of Systems and Cybernetics, Buenos Aires. And we were delighted to discover the existence of the Cambridge Cybernetics Society (Belmont, CA, USA). By contrast, information theory figures only in a handful of institutions' names. Apparently, the old Shannon *vs.* Wiener dispute may not be over yet.

In any case, Wiener's personal reputation in mathematics remains rock solid: it suffices to name a few gems such as the Paley–Wiener theorem (created on Wiener's numerous visits to Cambridge), the Wiener–Hopf method and, of course, the Wiener process, particularly close to our hearts, to understand his true role in scientific research and applications. However, existing recollections of this giant of science depict an image of a complex and often troubled personality. (The title of the biography [35] is quite revealing but such views are disputed, e.g., in the review [107]. In this book we attempt to adopt a more tempered tone from the chapter on Wiener in [75], pp. 386–391.) On the other hand, available accounts of Shannon's life (as well as other fathers of information and coding theory, notably, Richard Hamming) give a consistent picture of a quiet, intelligent and humorous person. It is our hope that this fact will not present a hindrance for writing Shannon's biographies and that in future we will see as many books on Shannon as we see on Wiener.

As was said before, the purpose of this book is twofold: to provide a synthetic introduction both to probabilistic and algebraic aspects of the theory supported by a significant number of problems and examples, and to discuss a number of topics rarely presented in most mainstream books. Chapters 1–3 give an introduction into the basics of information theory and coding with some discussion spilling over to more modern topics. We concentrate on typical problems and examples [many of them originated in Cambridge courses] more than on providing a detailed presentation of the theory behind them. Chapter 4 gives a brief introduction into a variety of topics from information theory. Here the presentation is more concise and some important results are given without proofs.

Because the large part of the text stemmed from lecture notes and various solutions to class and exam problems, there are inevitable repetitions, multitudes of

notation and examples of pigeon English. We left many of them deliberately, feeling that they convey a live atmosphere during the teaching and examination process.

Two excellent books [52] and [36] had a particularly strong impact on our presentation. We feel that our long-term friendship with Charles Goldie played a role here, as well as YS's amicable acquaintance with Tom Cover. We also benefited from reading (and borrowing from) the books [18], [110], [130] and [98]. The warm hospitality at a number of programmes at the Isaac Newton Institute, University of Cambridge, in 2002–2010 should be acknowledged, particularly Stochastic Processes in Communication Sciences (January–July 2010). Various parts of the material have been discussed with colleagues in various institutions, first and foremost, the Institute for Information Transmission Problems and the Institute of Mathematical Geophysics and Earthquake Predictions, Moscow (where the authors have been loyal staff members for a long time). We would like to thank James Lawrence, from Statslab, University of Cambridge, for his kind help with figures.

References to PSE I and PSE II mean the books by the present authors *Probability and Statistics by Example*, Cambridge University Press, Volumes I and II. We adopted the style used in PSE II, presenting a large portion of the material through 'Worked Examples'. Most of these Worked Examples are stated as problems (and many of them originated from Cambridge Tripos Exam papers and keep their specific style and spirit).

1

Essentials of Information Theory

Throughout the book, the symbol $\mathbb{P}$ denotes various probability distributions. In particular, in Chapter 1, $\mathbb{P}$ refers to the probabilities for sequences of random variables characterising sources of information. As a rule, these are sequences of independent and identically distributed random variables or discrete-time Markov chains; namely, $\mathbb{P}(U_1 = u_1, \ldots, U_n = u_n)$ is the joint probability that random variables $U_1, \ldots, U_n$ take values $u_1, \ldots, u_n$, and $\mathbb{P}(V = v \,|\, U = u, W = w)$ is the conditional probability that a random variable V takes value v, given that random variables U and W take values u and w, respectively. Likewise, $\mathbb{E}$ denotes the expectation with respect to $\mathbb{P}$.

The symbols p and P are used to denote various probabilities (and probability-related objects) loosely. The symbol $\sharp A$ denotes the cardinality of a finite set A. The symbol $\mathbf{1}$ stands for an indicator function. We adopt the following notation and formal rules for logarithms: $\ln - \log_e$, $\log - \log_2$, and for all $b > 1$: $0 \cdot \log_b 0 = 0 \cdot \log_b \infty = 0$. Next, given $x > 0$, $\lfloor x \rfloor$ and $\lceil x \rceil$ denote the maximal integer that is no larger than x and the minimal integer that is no less than x, respectively. Thus, $\lfloor x \rfloor \le x \le \lceil x \rceil$; equalities hold here when x is a positive integer ($\lfloor x \rfloor$ is called the integer part of x.)

The abbreviations LHS and RHS stand, respectively, for the left-hand side and the right-hand side of an equation.

1.1 Basic concepts. The Kraft inequality. Huffman's encoding

A typical scheme used in information transmission is as follows:

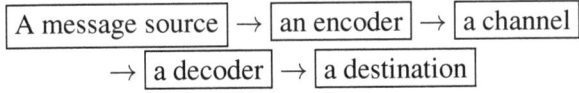

A message source $\to$ an encoder $\to$ a channel $\to$ a decoder $\to$ a destination

Example 1.1.1   (a) A message source: a Cambridge college choir.
(b) An encoder: a BBC recording unit. It translates the sound to a binary array and writes it to a CD track. The CD is then produced and put on the market.
(c) A channel: a customer buying a CD in England and mailing it to Australia. The channel is subject to 'noise': possible damage (mechanical, electrical, chemical, etc.) incurred during transmission (transportation).
(d) A decoder: a CD player in Australia.
(e) A destination: an audience in Australia.
(f) The goal: to ensure a high-quality sound despite damage.

   In fact, a CD can sustain damage done by a needle while making a neat hole in it, or by a tiny drop of acid (you are not encouraged to make such an experiment!). In technical terms, typical goals of information transmission are:

  (i) fast encoding of information,
 (ii) easy transmission of encoded messages,
(iii) effective use of the channel available (i.e. maximum transfer of information per unit time),
 (iv) fast decoding,
  (v) correcting errors (as many as possible) introduced by noise in the channel.

   As usual, these goals contradict each other, and one has to find an optimal solution. This is what the chapter is about. However, do not expect perfect solutions: the theory that follows aims mainly at providing knowledge of the basic principles. A final decision is always up to the individual (or group) responsible.

   A large part of this section (and the whole of Chapter 1) will deal with *encoding* problems. The aims of encoding are:

(1) compressing data to reduce redundant information contained in a message,
(2) protecting the text from unauthorised users,
(3) enabling errors to be corrected.

We start by studying *sources* and *encoders*. A source emits a sequence of letters (or symbols),

$$u_1 \, u_2 \, \ldots \, u_n \ldots, \qquad (1.1.1)$$

where $u_j \in I$, and $I(= I_m)$ is an m-element set often identified as $\{1, \ldots, m\}$ (a source alphabet). In the case of literary English, $m = 26 + 7$, 26 letters plus 7 punctuation symbols: . , : ; – ( ). (Sometimes one adds ? ! ' ' and "). Telegraph English corresponds to $m = 27$.

   A common approach is to consider (1.1.1) as a *sample* from a random source, i.e. a sequence of random variables

$$U_1, U_2, \ldots, U_n, \ldots \qquad (1.1.2)$$

and try to develop a theory for a reasonable class of such sequences.

Example 1.1.2 (a) The simplest example of a random source is a sequence of independent and identically distributed random variables (IID random variables):

$$\mathbb{P}(U_1 = u_1,\ U_2 = u_2, \ldots, U_k = u_k) = \prod_{j=1}^{k} p(u_j), \qquad (1.1.3a)$$

where $p(u) = \mathbb{P}(U_j = u)$, $u \in I$, is the marginal distribution of a single variable. A random source with IID symbols is often called a *Bernoulli source*.

A particular case where $p(u)$ does not depend on $u \in U$ (and hence equals $1/m$) corresponds to the equiprobable Bernoulli source.

(b) A more general example is a Markov source where the symbols form a discrete-time Markov chain (DTMC):

$$\mathbb{P}(U_1 = u_1,\ U_2 = u_2, \ldots,\ U_k = u_k) = \lambda(u_1) \prod_{j=1}^{k-1} P(u_j, u_{j+1}), \qquad (1.1.3b)$$

where $\lambda(u) = \mathbb{P}(U_1 = u)$, $u \in I$, are the initial probabilities and $P(u, u') = \mathbb{P}(U_{j+1} = u' | U_j = u)$, $u, u' \in I$, are transition probabilities. A Markov source is called *stationary* if $\mathbb{P}(U_j = u) = \lambda(u)$, $j \geq 1$, i.e. $\lambda = \{\lambda(u), u = 1, \ldots, m\}$ is an invariant row-vector for matrix $P = \{P(u, v)\}$: $\sum_{u \in I} \lambda(u) P(u, v) = \lambda(v)$, $v \in I$, or, shortly, $\lambda P = \lambda$.

(c) A 'degenerated' example of a Markov source is where a source emits repeated symbols. Here,

$$\begin{aligned} \mathbb{P}(U_1 = U_2 = \cdots = U_k = u) &= p(u),\ u \in I, \\ \mathbb{P}(U_k \neq U_{k'}) &= 0,\ 1 \leq k < k', \end{aligned} \qquad (1.1.3c)$$

where $0 \leq p(u) \leq 1$ and $\sum_{u \in I} p(u) = 1$.

An initial piece of sequence (1.1.1)

$$\mathbf{u}^{(n)} = (u_1, u_2, \ldots, u_n) \quad \text{or, more briefly,} \quad \mathbf{u}^{(n)} = u_1 u_2 \ldots u_n$$

is called a (source) sample *n-string*, or *n-word* (in short, a string or a word), with digits from I, and is treated as a 'message'. Correspondingly, one considers a random *n*-string (a random message)

$$\mathbf{U}^{(n)} = (U_1, U_2, \ldots, U_n) \quad \text{or, briefly,} \quad \mathbf{U}^{(n)} = U_1 U_2 \ldots U_n.$$

An encoder (or coder) uses an alphabet $J(= J_q)$ which we typically write as $\{0, 1, \ldots, q-1\}$; usually the number of encoding symbols $q < m$ (or even $q \ll m$); in many cases $q = 2$ with $J = \{0, 1\}$ (a binary coder). A *code* (also *coding*, or *encoding*) is a map, f, that takes a symbol $u \in I$ into a finite string, $f(u) = x_1 \ldots x_s$, with digits from J. In other words, f maps I into the set J^* of all possible strings:

$$f : I \to J^* = \bigcup_{s \geq 1} \left(J \times \cdots \text{ (s times) } \times J \right).$$

Strings $f(u)$ that are images, under f, of symbols $u \in I$ are called *codewords* (in code f). A code has (constant) length N if the value s (the length of a code-word) equals N for all codewords. A message $\mathbf{u}^{(n)} = u_1 u_2 \ldots u_n$ is represented as a concatenation of codewords

$$f(\mathbf{u}^{(n)}) = f(u_1) f(u_2) \ldots f(u_n);$$

it is again a string from J^*.

Definition 1.1.3   We say that a code is *lossless* if $u \neq u'$ implies that $f(u) \neq f(u')$. (That is, the map $f : I \to J^*$ is one-to-one.) A code is called *decipherable* if any string from J^* is the image of at most one message. A string x is a *prefix* in another string y if $y = xz$, i.e. y may be represented as a result of a concatenation of x and z. A code is *prefix-free* if no codeword is a prefix in any other codeword (e.g. a code of constant length is prefix-free).

A prefix-free code is decipherable, but not vice versa:

Example 1.1.4   A code with three source letters $1, 2, 3$ and the binary encoder alphabet $J = \{0, 1\}$ given by

$$f(1) = 0, \quad f(2) = 01, \quad f(3) = 011$$

is decipherable, but not prefix-free.

Theorem 1.1.5   (The Kraft inequality) *Given positive integers* $s_1, \ldots, s_m$, *there exists a decipherable code* $f : I \to J^*$, *with codewords of lengths* $s_1, \ldots, s_m$, *iff*

$$\sum_{i=1}^{m} q^{-s_i} \leq 1. \qquad (1.1.4)$$

Furthermore, under condition (1.1.4) *there exists a prefix-free code with codewords of lengths* $s_1, \ldots, s_m$.

Proof   (I) Sufficiency. Let (1.1.4) hold. Our goal is to construct a prefix-free code with codewords of lengths $s_1, \ldots, s_m$. Rewrite (1.1.4) as

$$\sum_{l=1}^{s} n_l q^{-l} \leq 1, \qquad (1.1.5)$$

or

$$n_s q^{-s} \le 1 - \sum_{l=1}^{s-1} n_l q^{-l},$$

where n_l is the number of codewords of length l and $s = \max s_i$. Equivalently,

$$n_s \le q^s - n_1 q^{s-1} - \cdots - n_{s-1} q. \tag{1.1.6a}$$

Since $n_s \ge 0$, deduce that

$$n_{s-1} q \le q^s - n_1 q^{s-1} - \cdots - n_{s-2} q^2,$$

or

$$n_{s-1} \le q^{s-1} - n_1 q^{s-2} - \cdots - n_{s-2} q. \tag{1.1.6b}$$

Repeating this argument yields subsequently

$$\begin{aligned} n_{s-2} &\le q^{s-2} - n_1 q^{s-3} - \cdots - n_{s-3} q \\ \vdots \qquad & \qquad \vdots \qquad\qquad\qquad \vdots \\ n_2 &\le q^2 - n_1 q \end{aligned} \tag{1.1.6.s-1}$$

$$n_1 \le q. \tag{1.1.6.s}$$

Observe that actually either $n_{i+1} = 0$ or n_i is less than the RHS of the inequality, for all $i = 1, \ldots, s-1$ (by definition, $n_s \ge 1$ so that for $i = s-1$ the second possibility occurs). We can perform the following construction. First choose n_1 words of length 1, using distinct symbols from J: this is possible in view of (1.1.6.s). It leaves $(q - n_1)$ symbols unused; we can form $(q - n_1)q$ words of length 2 by appending a symbol to each. Choose n_2 codewords from these: we can do so in view of (1.1.6.s−1). We still have $q^2 - n_1 q - n_2$ words unused: form n_3 codewords, etc. In the course of the construction, no new word contains a previous codeword as a prefix. Hence, the code constructed is prefix-free.

(II) Necessity. Suppose there exists a decipherable code in J^* with codeword lengths $s_1, \ldots, s_m$. Set $s = \max s_i$ and observe that for any positive integer r

$$\left(q^{-s_1} + \cdots + q^{-s_m} \right)^r = \sum_{l=1}^{rs} b_l q^{-l}$$

where b_l is the number of ways r codewords can be put together to form a string of length l.

Because of decipherability, these strings must be distinct. Hence, we must have $b_l \leq q^l$, as q^l is the total number of l-strings. Then

$$\left(q^{-s_1} + \cdots + q^{-s_m} \right)^r \leq rs,$$

and

$$q^{-s_1} + \cdots + q^{-s_m} \leq r^{1/r} s^{1/r} = \exp \left[\frac{1}{r} (\log r + \log s) \right].$$

This is true for any r, so take $r \to \infty$. The RHS goes to 1. $\qquad\square$

Remark 1.1.6 A given code obeying (1.1.4) is not necessarily decipherable.

Leon G. Kraft introduced inequality (1.1.4) in his MIT PhD thesis in 1949.

One of the principal aims of the theory is to find the 'best' (that is, the shortest) decipherable (or prefix-free) code. We now adopt a probabilistic point of view and assume that symbol $u \in I$ is emitted by a source with probability $p(u)$:

$$\mathbb{P}(U_k = u) = p(u).$$

[At this point, there is no need to specify a joint probability of more than one subsequently emitted symbol.]

Recall, given a code $f : I \mapsto J^*$, we encode a letter $i \in I$ by a prescribed codeword $f(i) = x_1 \ldots x_{s(i)}$ of length $s(i)$. For a random symbol, the generated codeword becomes a random string from J^*. When f is lossless, the probability of generating a given string as a codeword for a symbol is precisely $p(i)$ if the string coincides with $f(i)$ and 0 if there is no letter $i \in I$ with this property. If f is not one-to-one, the probability of a string equals the sum of terms $p(i)$ for which the codeword $f(i)$ equals this string. Then the length of a codeword becomes a *random variable*, S, with the probability distribution

$$\mathbb{P}(S = s) = \sum_{1 \leq i \leq m} \mathbf{1}(s(i) = s) p(i). \qquad (1.1.7)$$

We are looking for a decipherable code that minimises the expected word-length:

$$\mathbb{E} S = \sum_{s \geq 1} s \mathbb{P}(S = s) = \sum_{i=1}^{m} s(i) p(i).$$

The following problem therefore arises:

$$\text{minimise } g(s(1), \ldots, s(m)) = \mathbb{E} S$$

$$\text{subject to } \sum_i q^{-s(i)} \leq 1 \text{ (Kraft)} \qquad (1.1.8)$$

with $s(i)$ positive integers.

Theorem 1.1.7   *The optimal value for problem* (1.1.8) *is lower-bounded as follows:*

$$\min \mathbb{E}S \geq h_q(p(1),\ldots,p(m)), \tag{1.1.9}$$

where

$$h_q(p(1),\ldots,p(m)) = -\sum_i p(i) \log_q p(i). \tag{1.1.10}$$

Proof   The algorithm (1.1.8) is an integer-valued optimisation problem. If we drop the condition that $s(1),\ldots,s(m) \in \{1,2,\ldots\}$, replacing it with a 'relaxed' constraint $s(i) > 0$, $1 \leq i \leq m$, the Lagrange sufficiency theorem could be used. The Lagrangian reads

$$\mathscr{L}(s(1),\ldots,s(m),z;\lambda) = \sum_i s(i)p(i) + \lambda\left(1 - \sum_i q^{-s(i)} - z\right)$$

(here, $z \geq 0$ is a slack variable). Minimising $\mathscr{L}$ in $s_1,\ldots,s_m$ and z yields

$$\lambda < 0, \quad z = 0, \quad \text{and} \quad \frac{\partial \mathscr{L}}{\partial s(i)} = p(i) + q^{-s(i)}\lambda \ln q = 0,$$

whence

$$-\frac{p(i)}{\lambda \ln q} = q^{-s(i)}, \quad \text{i.e. } s(i) = -\log_q p(i) + \log_q(-\lambda \ln q), \ 1 \leq i \leq m.$$

Adjusting the constraint $\sum_i q^{-s(i)} = 1$ (the slack variable $z = 0$) gives

$$\sum_i p(i)/(-\lambda \ln q) = 1, \quad \text{i.e. } -\lambda \ln q = 1.$$

Hence,

$$s(i) = -\log_q p(i), \quad 1 \leq i \leq m,$$

is the (unique) optimiser for the relaxed problem, giving the value h_q from (1.1.10). The relaxed problem is solved on a larger set of variables $s(i)$; hence, its minimal value does not exceed that in the original one. $\square$

Remark 1.1.8   The quantity h_q defined in (1.1.10) plays a central role in the whole of information theory. It is called the q-ary entropy of the probability distribution $(p(x), x \in I)$ and will emerge in a great number of situations. Here we note that the dependence on q is captured in the formula

$$h_q(p(1),\ldots,p(m)) = \frac{1}{\log q} h_2(p(1),\ldots,p(m))$$

where h_2 stands for the binary entropy:

$$h_2(p(1),\ldots,p(m)) = -\sum_i p(i) \log p(i). \tag{1.1.11}$$

Worked Example 1.1.9   (a) *Give an example of a lossless code with alphabet* J_q *which does not satisfy the Kraft inequality. Give an example of a lossless code with the expected code-length strictly less than* $h_q(X)$.

(b) *Show that the 'Kraft sum'* $\sum_i q^{-s(i)}$ *associated with a lossless code may be arbitrarily large (for sufficiently large source alphabet).*

Solution (a) Consider the alphabet $I = \{0, 1, 2\}$ and a lossless code f with $f(0) = 0, f(1) = 1, f(2) = 00$ and codeword-lengths $s(0) = s(1) = 1, s(2) = 2$. Obviously, $\sum_{x \in I} 2^{-s(x)} = 5/4$, violating the Kraft inequality. For a random variable X with $p(0) = p(1) = p(2) = 1/3$ the expected codeword-length $\mathbb{E}s(X) = 4/3 < h(X) = \log 3 = 1.585$.

(b) Assume that the alphabet size $m = \sharp I = 2(2^L - 1)$ for some positive integer L. Consider the lossless code assigning to the letters $x \in I$ the codewords $0, 1, 00, 01, 10, 11, 000, \ldots$, with the maximum codeword-length L. The Kraft sum is

$$\sum_{x \in I} 2^{-s(x)} = \sum_{l \leq L} \sum_{x : s(x) = l} 2^{-s(x)} = \sum_{l \leq L} 2^l \times 2^{-l} = L,$$

which can be made arbitrarily large.   □

The assertion of Theorem 1.1.7 is further elaborated in

Theorem 1.1.10   (Shannon's noiseless coding theorem (NLCT)) *For a random source emitting symbols with probabilities* $p(i) > 0$, *the minimal expected codeword-length for a decipherable encoding in alphabet* J_q *obeys*

$$h_q \leq \min \mathbb{E}S < h_q + 1, \tag{1.1.12}$$

where $h_q = -\sum_i p(i) \log_q p(i)$ *is the* q-*ary entropy of the source; see (1.1.10).*

Proof The LHS inequality is established in (1.1.9). For the RHS inequality, let $s(i)$ be a positive integer such that

$$q^{-s(i)} \leq p(i) < q^{-s(i)+1}.$$

The non-strict bound here implies $\sum_i q^{-s(i)} \leq \sum_i p(i) = 1$, i.e. the Kraft inequality. Hence, there exists a decipherable code with codeword-lengths $s(1), \ldots, s(m)$. The strict bound implies

$$s(i) < -\frac{\log p(i)}{\log q} + 1,$$

and thus

$$\mathbb{E}S < -\frac{\sum\limits_{i} p(i)\log p(i)}{\log q} + \sum_{i} p(i) = \frac{h}{\log q} + 1.$$

□

Example 1.1.11 An instructive application of Shannon's NLCT is as follows. Let the size m of the source alphabet equal 2^k and assume that the letters $i = 1, \ldots, m$ are emitted equiprobably: $p(i) = 2^{-k}$. Suppose we use the code alphabet $J_2 = \{0, 1\}$ (binary encoding). With the binary entropy $h_2 = -\log 2^{-k} \sum\limits_{1 \le i \le 2^k} 2^{-k} = k$, we need, on average, at least k binary digits for decipherable encoding. Using a term *bit* for a unit of entropy, we say that on average the encoding requires at least k bits.

Moreover, the NLCT leads to a *Shannon–Fano encoding* procedure: we fix positive integer codeword-lengths $s(1), \ldots, s(m)$ such that $q^{-s(i)} \le p(i) < q^{-s(i)+1}$, or, equivalently,

$$-\log_q p(i) \le s(i) < -\log_q p(i) + 1; \text{ that is, } s(i) = \left\lceil -\log_q p(i) \right\rceil. \quad (1.1.13)$$

Then construct a prefix-free code, from the shortest $s(i)$ upwards, ensuring that the previous codewords are not prefixes. The Kraft inequality guarantees enough room. The obtained code may not be optimal but has the mean codeword-length satisfying the same inequalities (1.1.13) as an optimal code.

Optimality is achieved by *Huffman's encoding* $f_m^H \colon I_m \mapsto J_q^*$. We first discuss it for binary encodings, when $q = 2$ (i.e. $J = \{0, 1\}$). The algorithm constructs a binary tree, as follows.

(i) First, order the letters $i \in I$ so that $p(1) \ge p(2) \ge \cdots \ge p(m)$.

(ii) Assign symbol 0 to letter $m - 1$ and 1 to letter m.

(iii) Construct a reduced alphabet $I_{m-1} = \{1, \ldots, m-2, (m-1, m)\}$, with probabilities

$$p(1), \ldots, p(m-2), p(m-1) + p(m).$$

Repeat steps (i) and (ii) with the reduced alphabet, etc. We obtain a binary tree. For an example of Huffman's encoding for $m = 7$ see Figure 1.1.

The number of branches we must pass through in order to reach a root i of the tree equals $s(i)$. The tree structure, together with the identification of the roots as source letters, guarantees that encoding is prefix-free. The optimality of binary Huffman encoding follows from the following two simple lemmas.

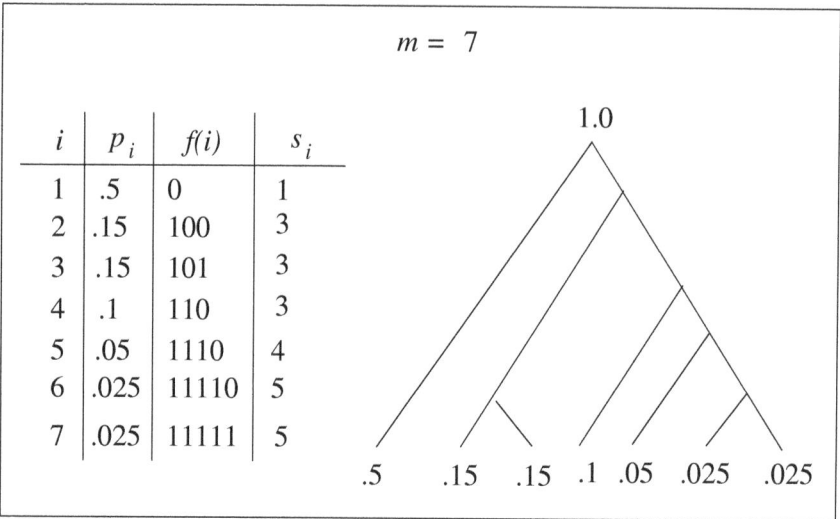

i	p_i	$f(i)$	s_i
1	.5	0	1
2	.15	100	3
3	.15	101	3
4	.1	110	3
5	.05	1110	4
6	.025	11110	5
7	.025	11111	5

Figure 1.1

Lemma 1.1.12   *Any optimal prefix-free binary code has the codeword-lengths reverse-ordered versus probabilities:*

$$p(i) \geq p(i') \quad \text{implies} \quad s(i) \leq s(i'). \tag{1.1.14}$$

Proof   If not, we can form a new code, by swapping the codewords for i and i'. This shortens the expected codeword-length and preserves the prefix-free property. □

Lemma 1.1.13   *In any optimal prefix-free binary code there exist, among the codewords of maximum length, precisely two agreeing in all but the last digit.*

Proof   If not, then either (i) there exists a single codeword of maximum length, or (ii) there exist two or more codewords of maximum length, and they all differ before the last digit. In both cases we can drop the last digit from some word of maximum length, without affecting the prefix-free property. □

Theorem 1.1.14   *Huffman's encoding is optimal among the prefix-free binary codes.*

Proof   The proof proceeds with induction in m. For $m = 2$, the Huffman code f_2^{H} has $f_2^{\mathrm{H}}(1) = 0$, $f_2^{\mathrm{H}}(2) = 1$, or vice versa, and is optimal. Assume the Huffman code f_{m-1}^{H} is optimal for I_{m-1}, whatever the probability distribution. Suppose further that the Huffman code f_m^H is not optimal for I_m for some probability distribution. That is, there is another prefix-free code, f_m^*, for I_m with a shorter expected word-length:

$$\mathbb{E}S_m^* < \mathbb{E}S_m^H. \tag{1.1.15}$$

The probability distribution under consideration may be assumed to obey

$$p(1) \geq \cdots \geq p(m).$$

By Lemmas 1.1.12 and 1.1.13, in both codes we can shuffle codewords so that the words corresponding to $m-1$ and m have maximum length and differ only in the last digit. This allows us to reduce both codes to I_{m-1}. Namely, in the Huffman code f_m^H we remove the final digit from $f_m^H(m)$ and $f_m^H(m-1)$, 'glueing' these codewords. This leads to Huffman encoding f_{m-1}^H. In f_m^* we do the same, and obtain a new prefix-free code f_{m-1}^*.

Observe that in Huffman code f_m^H the contribution to $\mathbb{E}S_m^H$ from $f_m^H(m-1)$ and $f_m^H(m)$ is $s^H(m)(p(m-1)+p(m))$; after reduction it becomes $(s^H(m)-1)$ $(p(m-1)+p(m))$. That is, $\mathbb{E}S$ is reduced by $p(m-1)+p(m)$. In code f_m^* the similar contribution is reduced from $s^*(m)(p(m-1)+p(m))$ to $(s^*(m)-1)(p(m-1)$ $+p(m))$; the difference is again $p(m-1)+p(m)$. All other contributions to $\mathbb{E}S_{m-1}^H$ and $\mathbb{E}S_{m-1}^*$ are the same as the corresponding contributions to $\mathbb{E}S_m^H$ and $\mathbb{E}S_m^*$, respectively. Therefore, f_{m-1}^* is better than f_{m-1}^H: $\mathbb{E}S_{m-1}^* < \mathbb{E}S_{m-1}^H$, which contradicts the assumption. $\qquad\square$

In view of Theorem 1.1.14, we obtain

Corollary 1.1.15 *Huffman's encoding is optimal among the decipherable binary codes.*

The generalisation of the Huffman procedure to q-ary codes (with the code alphabet $J_q = \{0,1,\ldots,q-1\}$) is straightforward: instead of merging two symbols, $m-1, m \in I_m$, having lowest probabilities, you merge q of them (again with the smallest probabilities), repeating the above argument. In fact, Huffman's original 1952 paper was written for a general encoding alphabet. There are numerous modifications of the Huffman code covering unequal coding costs (where some of the encoding digits $j \in J_q$ are more expensive than others) and other factors; we will not discuss them in this book.

Worked Example 1.1.16   *A drawback of Huffman encoding is that the codeword-lengths are complicated functions of the symbol probabilities $p(1)$, ..., $p(m)$. However, some bounds are available. Suppose that $p(1) \geq p(2) \geq \cdots \geq p(m)$. Prove that in any binary Huffman encoding:*

(a) *if $p(1) < 1/3$ then letter 1 must be encoded by a codeword of length ≥ 2;*
(b) *if $p(1) > 2/5$ then letter 1 must be encoded by a codeword of length 1.*

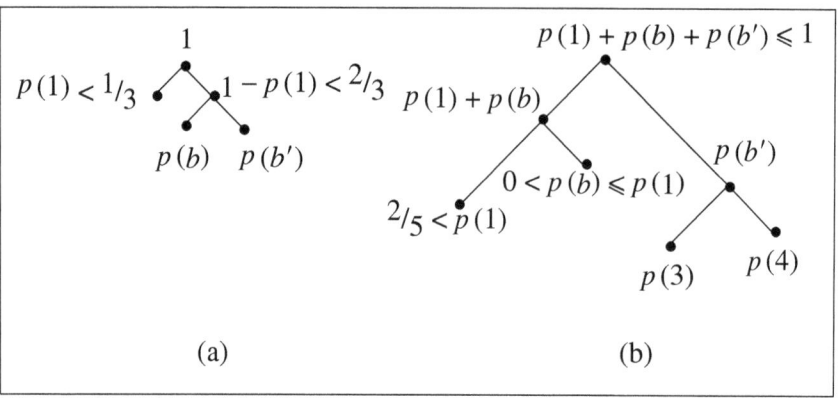

Figure 1.2

Solution (a) Two cases are possible: the letter 1 either was, or was not merged with other letters before *two* last steps in constructing a Huffman code. In the first case, $s(1) \geq 2$. Otherwise we have symbols 1, b and b', with

$$p(1) < 1/3, \quad p(1) + p(b) + p(b') = 1 \quad \text{and hence} \quad \max[p(b), p(b')] > 1/3.$$

Then letter 1 is to be merged, at the last but one step, with one of b, b', and hence $s(1) \geq 2$. Indeed, suppose that at least one codeword has length 1, and this codeword is assigned to letter 1 with $p(1) < 1/3$. Hence, the top of the Huffman tree is as in Figure 1.2(a) with $0 \leq p(b), p(b') \leq 1 - p(1)$ and $p(b) + p(b') = 1 - p(1)$.

But then $\max[p(b), p(b')] > 1/3$, and hence $p(1)$ should be merged with $\min[p(b), p(b')]$. Hence, Figure 1.2(a) is impossible, and letter 1 has codeword-length ≥ 2.

The bound is sharp as both codes

$$\{0, 01, 110, 111\} \quad \text{and} \quad \{00, 01, 10, 11\}$$

are binary Huffman codes, e.g. for a probability distribution $1/3$, $1/3$, $1/4$, $1/12$.

(b) Now let $p(1) > 2/5$ and assume that letter 1 has a codeword-length $s(1) \geq 2$ in a Huffman code. Thus, letter 1 was merged with other letters before the last step. That is, at a certain stage, we had symbols 1, b and b' say, with

(A) $p(b') \geq p(1) > 2/5$,

(B) $p(b') \geq p(b)$,

(C) $p(1) + p(b) + p(b') \leq 1$

(D) $p(1), p(b) \geq 1/2 \, p(b')$.

Indeed, if, say, $p(b) < 1/2p(b')$ then b should be selected instead of $p(3)$ or $p(4)$ on the previous step when $p(b')$ was formed. By virtue of (D), $p(b) \geq 1/5$ which makes (A)+(C) impossible.

A piece of the Huffman tree over $p(1)$ is then as in Figure 1.2(b), with $p(3) + p(4) = p(b')$ and $p(1) + p(b') + p(b) \leq 1$. Write

$$p(1) = 2/5 + \varepsilon, \quad p(b') = 2/5 + \varepsilon + \delta, \quad p(b) = 2/5 + \varepsilon + \delta - \eta,$$

with $\varepsilon > 0$, $\delta, \eta \geq 0$. Then

$$p(1) + p(b') + p(b) = 6/5 + 3\varepsilon + 2\delta - \eta \leq 1, \quad \text{and} \quad \eta \geq 1/5 + 3\varepsilon + 2\delta.$$

This yields

$$p(b) \leq 1/5 - 2\varepsilon - \delta < 1/5.$$

However, since

$$\max\left[p(3), p(4)\right] \geq p(b')/2 \geq p(1)/2 > 1/5,$$

probability $p(b)$ should be merged with $\min\left[p(3), p(4)\right]$, i.e. diagram (b) is impossible. Hence, the letter 1 has codeword-length $s(1) = 1$. ☐

Worked Example 1.1.17 *Suppose that letters $i_1, \ldots, i_5$ are emitted with probabilities 0.45, 0.25, 0.2, 0.05, 0.05. Compute the expected word-length for Shannon–Fano and Huffman coding. Illustrate both methods by finding decipherable binary codings in each case.*

Solution In this case $q = 2$, and

	$p(i)$	$\lceil -\log_2 p(i) \rceil$	codeword
	.45	2	00
	.25	2	01
Shannon–Fano:	.2	3	100
	.05	5	11100
	.05	5	11111

with $\mathbb{E}(\text{codeword-length}) = .9 + .5 + .6 + .25 + .25 = 2.5$, and

	p_i	codeword
	.45	1
	.25	01
Huffman:	.2	000
	.05	0010
	.05	0011

with $\mathbb{E}(\text{codeword-length}) = 0.45 + 0.5 + 0.6 + 0.2 + 0.2 = 1.95$. ☐

Worked Example 1.1.18   *A Shannon–Fano code is in general not optimal. However, it is 'not much' longer than Huffman's. Prove that, if S_{SF} is the Shannon–Fano codeword-length, then for any $r = 1, 2, \ldots$ and any decipherable code f^* with codeword-length S^*,*

$$\mathbb{P}\left(S^* \leq S_{SF} - r\right) \leq q^{1-r}.$$

Solution   Write

$$\mathbb{P}\left(S^* \leq S_{SF} - r\right) = \sum_{i \in I:\, s^*(i) \leq s_{SF}(i) - r} p(i).$$

Note that $s_{SF}(i) < -\log_q p(i) + 1$, hence

$$\sum_{i \in I:\, s^*(i) \leq s_{SF}(i) - r} p(i) \leq \sum_{i \in I:\, s^*(i) \leq -\log_q p(i) + 1 - r} p(i)$$

$$= \sum_{i \in I:\, s^*(i) - 1 + r \leq -\log_q p(i)} p(i)$$

$$= \sum_{i \in I:\, p(i) \leq q^{-s^*(i) + 1 - r}} p(i)$$

$$\leq \sum_{i \in I} q^{-s^*(i) + 1 - r}$$

$$= q^{1-r} \sum_{i \in I} q^{-s^*(i)}$$

$$\leq q^{1-r};$$

the last inequality is due to Kraft.   □

A common modern practice is not to encode each letter $u \in I$ separately, but to divide a source message into 'segments' or 'blocks', of a fixed length n, and encode these as 'letters'. It obviously increases the nominal number of letters in the alphabet: the blocks are from the Cartesian product $I^{\times n} = I \times \cdots$ (n times) $\times I$. But what matters is the entropy

$$h_q^{(n)} = -\sum_{i_1, \ldots, i_n} \mathbb{P}(U_1 = i_1, \ldots, U_n = i_n) \log_q \mathbb{P}(U_1 = i_1, \ldots, U_n = i_n) \qquad (1.1.16)$$

of the probability distribution for the blocks in a typical message. [Obviously, we need to know the joint distribution of the subsequently emitted source letters.] Denote by $S^{(n)}$ the random codeword-length in a decipherable segmented code. The minimal expected codeword-length per source letter is defined by $e_n := \min \dfrac{1}{n} \, \mathbb{E} S^{(n)}$; by Shannon's NLCT, it obeys

$$\frac{h_q^{(n)}}{n} \leq e_n \leq \frac{h_q^{(n)}}{n} + \frac{1}{n}. \qquad (1.1.17)$$

We see that, for large n, $e_n \sim h_q^{(n)}/n$.

Example 1.1.19   For a Bernoulli source emitting letter i with probability $p(i)$ (cf. Example 1.1.2), equation (1.1.16) yields

$$h_q^{(n)} = - \sum_{i_1,\dots,i_n} p(i_1)\cdots p(i_n) \log_q \left(p(i_1)\cdots p(i_n) \right)$$

$$= - \sum_{j=1}^{n} \sum_{i_1,\dots,i_n} p(i_1)\cdots p(i_n) \log_q p(i_j) = n h_q, \qquad (1.1.18)$$

where $h_q = -\sum p(i) \log_q p(i)$. Here, $e_n \sim h_q$. Thus, for n large, the minimal expected codeword-length per source letter, in a segmented code, eventually attains the lower bound in (1.1.13), and hence does not exceed $\min \mathbb{E} S$, the minimal expected codeword-length for letter-by-letter encodings. This phenomenon is much more striking in the situation where the subsequent source letters are dependent. In many cases $h_q^{(n)} \ll n\, h_q$, i.e. $e_n \ll h_q$. This is the gist of data compression.

Therefore, statistics of long strings becomes an important property of a source. Nominally, the strings $\mathbf{u}^{(n)} = u_1 \dots u_n$ of length n 'fill' the Cartesian power $I^{\times n}$; the total number of such strings is m^n, and to encode them all we need $m^n = 2^{n \log m}$ distinct codewords. If the codewords have a fixed length (which guarantees the prefix-free property), this length is between $\lfloor n \log m \rfloor$, and $\lceil n \log m \rceil$, and the rate of encoding, for large n, is $\sim \log m$ bits/source letter. But if some strings are rare, we can disregard them, reducing the number of codewords used. This leads to the following definitions.

Definition 1.1.20   A source is said to be (reliably) *encodable at rate $R > 0$* if, for any n, we can find a set $\Lambda_n \subset I^{\times n}$ such that

$$\sharp \Lambda_n \leq 2^{nR} \quad \text{and} \quad \lim_{n \to \infty} \mathbb{P}(\mathbf{U}^{(n)} \in \Lambda_n) = 1. \qquad (1.1.19)$$

In other words, we can encode messages at rate R with a negligible error for long source strings.

Definition 1.1.21   The *information rate H* of a given source is the infimum of the reliable encoding rates:

$$H = \inf[R \colon R \text{ is reliable}]. \qquad (1.1.20)$$

Theorem 1.1.22   *For a source with alphabet I_m,*

$$0 \leq H \leq \log m, \qquad (1.1.21)$$

both bounds being attainable.

Proof The LHS inequality is trivial. It is attained for a degenerate source (cf. Example 1.1.2c); here A_n contains $\leq m$ constant strings, which is eventually beaten by 2^{nR} for any $R > 0$. On the other hand, $\sharp I^{\times n} = m^n = 2^{n \log m}$, hence the RHS inequality. It is attained for a source with IID letters and $p(u) = 1/m$: in this case $P(A_n) = (1/m^n) \, \sharp A_n$, which goes to zero when $\sharp A_n \leq 2^{nR}$ and $R < \log m$. $\square$

Example 1.1.23   (a) For telegraph English, $m = 27 \simeq 2^{4.76}$, i.e. $H \leq 4.76$. Fortunately, $H \ll 4.76$, and this makes possible: (i) data compression, (ii) error-correcting, (iii) code-breaking, (iv) crosswords. The precise value of H for telegraph English (not to mention literary English) is not known: it is a challenging task to assess it accurately. Nevertheless, modern theoretical tools and computing facilities make it possible to assess the information rate of a given (long) text, *assuming* that it comes from a source that operates by allowing a fair amount of 'randomness' and 'homogeneity' (see Section 6.3 of [36].)

Some results of numerical analysis can be found in [136] analysing three texts: (a) the collected works of Shakespeare; (b) a mixed text from various newspapers; and (c) the King James Bible. The texts were stripped of punctuation and the spaces between words were removed. Texts (a) and (b) give values 1.7 and 1.25 respectively (which is rather flattering to modern journalism). In case (c) the results were inconclusive; apparently the above assumptions are not appropriate in this case. (For example, the genealogical enumerations of Genesis are hard to compare with the philosophical discussions of Paul's letters, so the homogeneity of the source is obviously not maintained.)

Even more challenging is to compare different languages: which one is more appropriate for intercommunication? Also, it would be interesting to repeat the above experiment with the collected works of Tolstoy or Dostoyevsky.

For illustration, we give below the original table by Samuel Morse (1791–1872), creator of the Morse code, providing the frequency count of different letters in telegraph English which is dominated by a relatively small number of common words.

$$
\begin{pmatrix}
E & T & A & I & N & O & S & H & R \\
12000 & 9000 & 8000 & 8000 & 8000 & 8000 & 8000 & 6400 & 6200 \\
D & L & U & C & M & F & W & Y & G \\
4400 & 4000 & 3400 & 3000 & 3000 & 2500 & 2000 & 2000 & 1700 \\
P & B & V & K & Q & J & X & Z & \\
1700 & 1600 & 1200 & 800 & 500 & 400 & 400 & 200 &
\end{pmatrix}
$$

(b) A similar idea was applied to the decimal and binary decomposition of a given number. For example, take number π. If the information rate for its binary

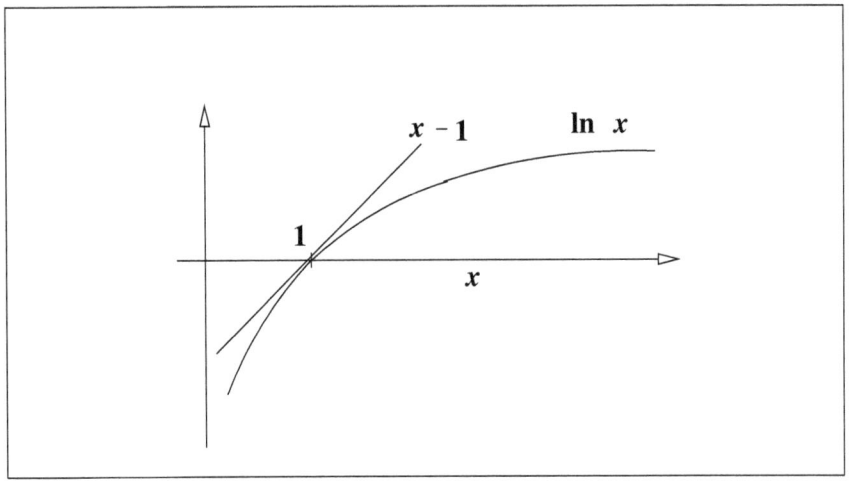

Figure 1.3

decomposition approaches value 1 (which is the information rate of a randomly chosen sequence), we may think that π behaves like a completely random number; otherwise we could imagine that π was a 'Specially Chosen One'. The same question may be asked about e, $\sqrt{2}$ or the Euler–Mascheroni constant $\gamma = \lim\limits_{N \to \infty} \left(\sum\limits_{1 \le n \le N} \frac{1}{n} - \ln N \right)$. (An open part of one of Hilbert's problems is to prove or disprove that γ is a transcendental number, and transcendental numbers form a set of probability one under the Bernoulli source of subsequent digits.) As the results of numerical experiments show, for the number of digits $N \sim 500,000$ all the above-mentioned numbers display the same pattern of behaviour as a completely random number; see [26]. In Section 1.3 we will calculate the information rates of Bernoulli and Markov sources.

We conclude this section with the following simple but fundamental fact.

Theorem 1.1.24 (The Gibbs inequality: cf. PSE II, p. 421) *Let $\{p(i)\}$ and $\{p'(i)\}$ be two probability distributions (on a finite or countable set I). Then, for any $b > 1$,*

$$\sum_i p(i) \log_b \frac{p'(i)}{p(i)} \le 0, \text{ i.e. } -\sum_i p(i) \log_b p(i) \le -\sum_i p(i) \log_b p'(i), \quad (1.1.22)$$

and equality is attained iff $p(i) = p'(i)$, $1 \in I$.

Proof The bound

$$\log_b x \le \frac{x-1}{\ln b}$$

holds for each $x > 0$, with equality iff $x = 1$. Setting $I' = \{i : p(i) > 0\}$, we have

$$
\begin{aligned}
\sum_i p(i) \log_b \frac{p'(i)}{p(i)} &= \sum_{i \in I'} p(i) \log_b \frac{p'(i)}{p(i)} \leq \frac{1}{\ln b} \sum_{i \in I'} p(i) \left(\frac{p'(i)}{p(i)} - 1 \right) \\
&= \frac{1}{\ln b} \left(\sum_{i \in I'} p'(i) - \sum_{i \in I'} p(i) \right) = \frac{1}{\ln b} \left(\sum_{i \in I'} p'(i) - 1 \right) \leq 0.
\end{aligned}
$$

For equality we need: (a) $\sum_{i \in I'} p'(i) = 1$, i.e. $p'(i) = 0$ when $p(i) = 0$; and (b) $p'(i)/p(i) = 1$ for $i \in I'$. $\qquad \square$

1.2 Entropy: an introduction

Only entropy comes easy.
Anton Chekhov (1860–1904), Russian writer and playwright

This section is entirely devoted to properties of entropy. For simplicity, we work with the binary entropy, where the logarithms are taken at base 2. Consequently, subscript 2 in the notation h_2 is omitted. We begin with a formal repetition of the basic definition, putting a slightly different emphasis.

Definition 1.2.1 Given an event A with probability $p(A)$, the *information* gained from the fact that A has occurred is defined as

$$
i(A) = -\log p(A).
$$

Further, let X be a random variable taking a finite number of distinct values $\{x_1, \ldots, x_m\}$, with probabilities $p_i = p_X(x_i) = \mathbb{P}(X = x_i)$. The binary *entropy* $h(X)$ is defined as the expected amount of information gained from observing X:

$$
h(X) = -\sum_{x_i} p_X(x_i) \log \, p_X(x_i) = -\sum_i p_i \log p_i = \mathbb{E}\left[-\log p_X(X) \right]. \qquad (1.2.1)
$$

[In view of the adopted equality $0 \cdot \log 0 = 0$, the sum may be reduced to those x_i for which $p_X(x_i) > 0$.]

Sometimes an alternative view is useful: $i(A)$ represents the amount of information needed to specify event A and $h(X)$ gives the expected amount of information required to specify a random variable X.

Clearly, the entropy $h(X)$ depends on the probability distribution, but not on the values $x_1, \ldots, x_m$: $h(X) = h(p_1, \ldots, p_m)$. For $m = 2$ (a two-point probability distribution), it is convenient to consider the function $\eta(p)(= \eta_2(p))$ of a single variable $p \in [0, 1]$:

$$
\eta(p) = -p \log p - (1 - p) \log(1 - p). \qquad (1.2.2a)
$$

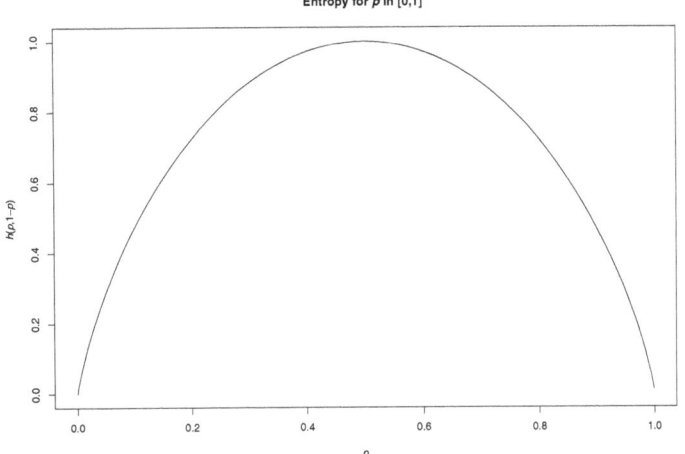

Figure 1.4

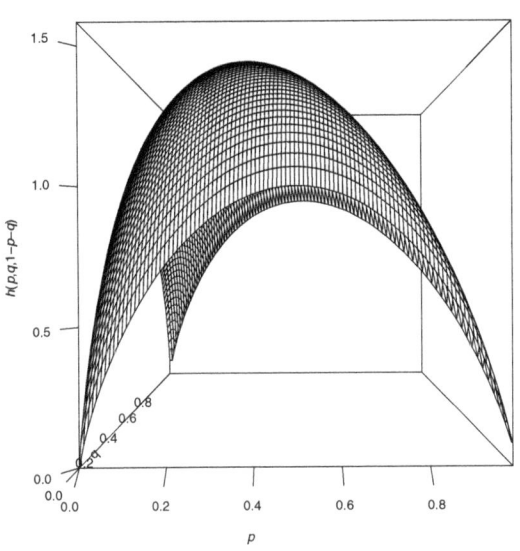

Figure 1.5

The graph of $\eta(p)$ is plotted in Figure 1.4. Observe that the graph is concave as

$$\frac{d^2}{dp^2}\eta(p) = -\left(\log e\right)/\left[p(1-p)\right] < 0.$$ See Figure 1.4.

The graph of the entropy of a three-point distribution

$$\eta_3(p,q) = -p\log p - q\log q - (1-p-q)\log(1-p-q)$$

is plotted in Figure 1.5 as a function of variables $p, q \in [0, 1]$ with $p + q \leq 1$: it also shows the concavity property.

Definition 1.2.1 implies that for independent events, A_1 and A_2,

$$i(A_1 \cap A_2) = i(A_1) + i(A_2), \tag{1.2.2b}$$

and $i(A) = 1$ for event A with $p(A) = 1/2$.

A justification of Definition 1.2.1 comes from the fact that any function $i^*(A)$ which (i) depends on probability $p(A)$ (i.e. obeys $i^*(A) = i^*(A')$ if $p(A) = p(A')$), (ii) is continuous in $p(A)$, and (iii) satisfies (1.2.2b), coincides with $i(A)$ (for axiomatic definitions of entropy, cf. Worked Example 1.2.24 below).

Definition 1.2.2   Given a pair of random variables, X, Y, with values x_i and y_j, the *joint entropy* $h(X, Y)$ is defined by

$$h(X, Y) = - \sum_{x_i, y_j} p_{X,Y}(x_i, y_j) \log p_{X,Y}(x_i, y_j) = \mathbb{E}\left[- \log p_{X,Y}(X, Y) \right], \tag{1.2.3}$$

where $p_{X,Y}(x_i, y_j) = \mathbb{P}(X = x_i, Y = y_j)$ is the joint probability distribution. In other words, $h(X, Y)$ is the entropy of the random vector (X, Y) with values (x_i, y_j).

The *conditional entropy*, $h(X|Y)$, of X given Y is defined as the expected amount of information gained from observing X given that a value of Y is known:

$$h(X|Y) = - \sum_{x_i, y_j} p_{X,Y}(x_i, y_j) \log p_{X|Y}(x_i|y_j) = \mathbb{E}\left[- \log p_{X|Y}(X|Y) \right]. \tag{1.2.4}$$

Here, $p_{X,Y}(i, j)$ is the joint probability $\mathbb{P}(X = x_i, Y = y_j)$ and $p_{X|Y}(x_i|y_j)$ the conditional probability $\mathbb{P}(X = x_i|Y = y_j)$. Clearly, (1.2.3) and (1.2.4) imply

$$h(X|Y) = h(X, Y) - h(Y). \tag{1.2.5}$$

Note that in general $h(X|Y) \neq h(Y|X)$.

For random variables X and Y taking values in the same set I, and such that $p_Y(x) > 0$ for all $x \in I$, the *relative entropy* $h(X||Y)$ (also known as the entropy of X relative to Y or Kullback–Leibler distance $D(p_X||p_Y)$) is defined by

$$h(X||Y) = \sum_x p_X(x) \log \frac{p_X(x)}{p_Y(x)} = \mathbb{E}_X\left[- \log \frac{p_Y(X)}{p_X(X)} \right], \tag{1.2.6}$$

with $p_X(x) = \mathbb{P}(X = x)$ and $p_Y(x) = \mathbb{P}(Y = x)$, $x \in I$.

Straightforward properties of entropy are given below.

Theorem 1.2.3

(a) *If a random variable X takes at most m values, then*

$$0 \leq h(X) \leq \log m; \tag{1.2.7}$$

the LHS equality occurring iff X takes a single value, and the RHS equality occurring iff X takes m values with equal probabilities.

(b) *The joint entropy obeys*

$$h(X,Y) \leq h(X) + h(Y), \qquad (1.2.8)$$

with equality iff X and Y are independent, i.e. $\mathbb{P}(X = x, Y = y) = \mathbb{P}(X = x)$ $\mathbb{P}(Y = y)$ for all $x, y \in I$.

(c) *The relative entropy is always non-negative:*

$$h(X||Y) \geq 0, \qquad (1.2.9)$$

with equality iff X and Y are identically distributed: $p_X(x) \equiv p_Y(x)$, $x \in I$.

Proof Assertion (c) is equivalent to Gibbs' inequality from Theorem 1.1.24. Next, (a) follows from (c), with $\{p(i)\}$ being the distribution of X and $p'(i) \equiv 1/m$, $1 \leq i \leq m$. Similarly, (b) follows from (c), with i being a pair (i_1, i_2) of values of X and Y, $p(i) = p_{X,Y}(i_1, i_2)$ being the joint distribution of X and Y and $p'(i) = p_X(i_1)p_Y(i_2)$ representing the product of their marginal distributions. Formally:

(a) $h(X) = -\sum_i p(i) \log p(i) \leq \sum_i p(i) \log m = \log m,$

(b) $h(X,Y) = -\sum_{(i_1,i_2)} p_{X,Y}(i_1,i_2) \log p_{X,Y}(i_1,i_2)$

$\leq -\sum_{(i_1,i_2)} p_{X,Y}(i_1,i_2) \times \log \left(p_X(i_1) p_Y(i_2) \right)$

$= -\sum_{i_1} p_X(i_1) \log p_X(i_1) - \sum_{i_2} p_Y(i_2) \log p_Y(i_2)$

$= h(X) + h(Y).$

We used here the identities $\sum_{i_2} p_{X,Y}(i_1,i_2) = p_X(i_1)$, $\sum_{i_1} p_{X,Y}(i_1,i_2) = p_Y(i_2)$. □

Worked Example 1.2.4

(a) *Show that the geometric random variable Y with $p_j = \mathbb{P}(Y = j) = (1-p)p^j$, $j = 0,1,2,\ldots$, yields maximum entropy amongst all distributions on $\mathbb{Z}_+ = \{0,1,2,\ldots\}$ with the same mean.*

(b) *Let Z be a random variable with values from a finite set K and f be a given real function $f : K \to \mathbb{R}$, with $f_* = \min \left[f(k) : k \in K \right]$ and $f^* = \max \left[f(k) : k \in K \right]$. Set $E(f) = \sum_{k \in K} f(k)/(\sharp K)$ and consider the problem of maximising the entropy $h(Z)$ of the random variable Z subject to a constraint*

$$\mathbb{E}f(Z) \leq \alpha. \qquad (1.2.10)$$

Show that:

(bi) *when $f^* \geq \alpha \geq E(f)$ then the maximising probability distribution is uniform on K, with $\mathbb{P}(Z = k) = 1/(\sharp K)$, $k \in K$;*

(bii) *when $f_* \leq \alpha < E(f)$ and f is not constant then the maximising probability distribution has*

$$\mathbb{P}(Z = k) = p_k = e^{\lambda f(k)} \Big/ \sum_i e^{\lambda f(i)}, \quad k \in K, \qquad (1.2.11)$$

where $\lambda = \lambda(\alpha) < 0$ is chosen so as to satisfy

$$\sum_k p_k f(k) = \alpha. \qquad (1.2.12)$$

Moreover, suppose that Z takes countably many values, but $f \geq 0$ and for a given α there exists a $\lambda < 0$ such that $\sum_i e^{\lambda f(i)} < \infty$ and $\sum_k p_k f(k) = \alpha$ where p_k has form (1.2.11). Then:

(biii) *the probability distribution in (1.2.11) still maximises $h(Z)$ under (1.2.10).*

Deduce assertion (a) from (biii).

(c) *Prove that $h_Y(X) \geq 0$, with equality iff $\mathbb{P}(X = x) = \mathbb{P}(Y = x)$ for all x. By considering Y, a geometric random variable on $\mathbb{Z}_+$ with parameter chosen appropriately, show that if the mean $\mathbb{E}X = \mu < \infty$, then*

$$h(X) \leq (\mu + 1)\log(\mu + 1) - \mu \log \mu, \qquad (1.2.13)$$

with equality iff X is geometric.

Solution   (a) By the Gibbs inequality, for all probability distribution $(q_0, q_1, \ldots)$ with mean $\sum_{i \geq 0} i q_i \leq \mu$,

$$h(q) = -\sum_i q_i \log q_i \leq -\sum_i q_i \log p_i = -\sum_i q_i (\log(1 - p) + i \log p)$$
$$\leq -\log(1 - p) - \mu \log p = h(Y)$$

as $\mu = p/(1 - p)$, and equality holds iff q is geometric with mean μ.

(b) First, observe that the uniform distribution, with $p_k = 1/(\sharp K)$, which renders the 'global' maximum of $h(Z)$ is obtained for $\lambda = 0$ in (1.2.11). In part (bi), this distribution satisfies (1.2.10) and hence maximises $h(Z)$ under this constraint.

Passing to (bii), let $p_k^* = e^{\lambda f(k)} \Big/ \sum_i e^{\lambda f(i)}$, $k \in K$, where λ is chosen to satisfy $\mathbb{E}^* f(Z) = \sum_k p_k^* f(k) = \alpha$. Let $q = \{q_k\}$ be any probability distribution satisfying

$\mathbb{E}_q f = \sum\limits_k q_k f(k) \le \alpha$. Next, observe that the mean value (1.2.12) calculated for the probability distribution from (1.2.11) is a non-decreasing function of λ. In fact, the derivative

$$\frac{d\alpha}{d\lambda} = \frac{\sum\limits_k [f(k)]^2 \, e^{\lambda f(k)}}{\sum\limits_i e^{\lambda f(i)}} - \frac{\left(\sum\limits_k f(k) e^{\lambda f(k)}\right)^2}{\left(\sum\limits_i e^{\lambda f(i)}\right)^2} = \mathbb{E}[f(Z)]^2 - [\mathbb{E}f(Z)]^2$$

is positive (it yields the variance of the random variable $f(Z)$); for a non-constant f the RHS is actually non-negative. Therefore, for non-constant f (i.e. with $f_* < E(f) < f^*$), for all α from the interval $[f_*, f^*]$ there exists exactly one probability distribution of form (1.2.11) satisfying (1.2.12), and for $f_* \le \alpha < E(f)$ the corresponding $\lambda(\alpha)$ is < 0.

Next, we use the fact that the Kullback–Leibler distance $D(q\|p^*)$ (cf. (1.2.6)) satisfies $D(q\|p^*) = \sum\limits_k q_k \log{(q_k/p_k^*)} \ge 0$ (Gibbs' inequality) and that $\sum\limits_k q_k f(k) \le \alpha$ and $\lambda < 0$ to obtain that

$$h(q) = -\sum_k q_k \log q_k = -D(q\|p^*) - \sum_k q_k \log p_k^*$$

$$\le -\sum_k q_k \log p_k^* = -\sum_k q_k \left(-\log \sum_i e^{\lambda f(i)} + \lambda f(k) \right)$$

$$\le -\sum_k q_k \left(-\log \sum_i e^{\lambda f(i)} \right) - \lambda \alpha$$

$$= -\sum_k p_k^* \left(-\log \sum_i e^{\lambda f(i)} + \lambda f(k) \right)$$

$$= -\sum_k p_k^* \log p_k^* = h(p^*).$$

For part (biii): the above argument still works for an infinite countable set K provided that the value $\lambda(\alpha)$ determined from (1.2.12) is < 0.

(c) By the Gibbs inequality $h_Y(X) \ge 0$. Next, we use part (b) by taking $f(k) = k$, $\alpha = \mu$ and $\lambda = \ln q$. The maximum-entropy distribution can be written as $p_j^* = (1-p)p^j$, $j = 0, 1, 2, \ldots$, with $\sum\limits_k k p_k^* = \mu$, or $\mu = p/(1-p)$. The entropy of this distribution equals

$$h(p^*) = -\sum_j (1-p)p^j \log\left((1-p)p^j\right)$$

$$= -\frac{p}{1-p} \log p - \log(1-p) = (\mu+1)\log(\mu+1) - \mu \log \mu,$$

where $\mu = p/(1-p)$.

Alternatively:

$$0 \leq h_Y(X) = \sum_i p(i) \log \frac{p(i)}{(1-p)p^i}$$

$$= -h(X) - \log(1-p)\sum_i p(i) - (\log p)\left[\sum_i ip(i)\right]$$

$$= -h(X) - \log(1-p) - \mu \log p.$$

The optimal choice of p is $p = \mu/(\mu+1)$. Then

$$h(X) \leq -\log \frac{1}{\mu+1} - \mu \log \frac{\mu}{\mu+1} = (\mu+1)\log(\mu+1) - \mu\log\mu.$$

The RHS is the entropy $h(Y)$ of the geometric random variable Y. Equality holds iff $X \sim Y$, i.e. X is geometric. $\qquad\square$

A simple but instructive corollary of the Gibbs inequality is

Lemma 1.2.5   (The pooling inequalities) *For any $q_1, q_2 \geq 0$, with $q_1 + q_2 > 0$,*

$$-(q_1+q_2)\log(q_1+q_2) \leq -q_1\log q_1 - q_2\log q_2$$

$$\leq -(q_1+q_2)\log\frac{q_1+q_2}{2}; \qquad (1.2.14)$$

the first equality occurs iff $q_1 q_2 = 0$ (i.e. either q_1 or q_2 vanishes), and the second equality iff $q_1 = q_2$.

Proof   Indeed, (1.2.14) is equivalent to

$$0 \leq h\left(\frac{q_1}{q_1+q_2}, \frac{q_2}{q_1+q_2}\right) \leq \log 2\, (= 1).$$

$\qquad\square$

By Lemma 1.2.5, 'glueing' together values of a random variable could diminish the corresponding contribution to the entropy. On the other hand, the 're-distribution' of probabilities making them equal increases the contribution. An immediate corollary of Lemma 1.2.5 is the following.

Theorem 1.2.6   *Suppose that a discrete random variable X is a function of discrete random variable $Y\colon X = \phi(Y)$. Then*

$$h(X) \leq h(Y), \qquad (1.2.15)$$

with equality iff ϕ is invertible.

Proof   Indeed, if ϕ is invertible then the probability distributions of X and Y differ only in the order of probabilities, which does not change the entropy. If ϕ 'glues' some values y_j then we can repeatedly use the LHS pooling inequality. $\qquad\square$

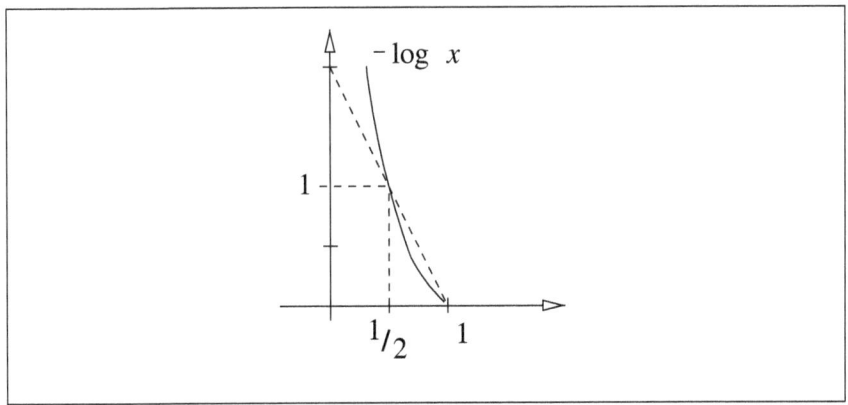

Figure 1.6

Worked Example 1.2.7 *Let $p_1,\ldots,p_n$ be a probability distribution, with $p^* = \max[p_i]$. Prove the following lower bounds for the entropy $h = -\sum_i p_i \log p_i$:*

(i) $h \geq -p^* \log p^* - (1-p^*) \log(1-p^*) = \eta(p^*)$;
(ii) $h \geq -\log p^*$;
(iii) $h \geq 2(1-p^*)$.

Solution Part (i) follows from the pooling inequality, and (ii) holds as

$$h \geq -\sum_i p_i \log p^* = -\log p^*.$$

To check (iii), assume first that $p^* \geq 1/2$. Since the function $p \mapsto \eta(p)$, $0 \leq p \leq 1$, is concave (see (1.2.3)), its graph on $[1/2,1]$ lies above the line $x \mapsto 2(1-p)$. Then, by (i),

$$h \geq \eta(p^*) \geq 2(1-p^*). \qquad (1.2.16)$$

On the other hand, if $p^* \leq 1/2$, we use (ii):

$$h \geq -\log p^*,$$

and apply the inequality $-\log p \geq 2(1-p)$ for $0 \leq p \leq 1/2$. $\qquad\qquad \square$

Theorem 1.2.8 (The Fano inequality) *Suppose a random variable X takes $m > 1$ values, and one of them has probability $(1-\varepsilon)$. Then*

$$h(X) \leq \eta(\varepsilon) + \varepsilon \log(m-1) \qquad (1.2.17)$$

where η is the function from (1.2.2a).

Proof Suppose that $p_1 = p(x_1) = 1 - \varepsilon$. Then

$$h(X) = h(p_1, \ldots, p_m) = -\sum_{i=1}^{m} p_i \log p_i$$

$$= -p_1 \log p_1 - (1 - p_1) \log(1 - p_1) + (1 - p_1) \log(1 - p_1)$$
$$- \sum_{2 \le i \le m} p_i \log p_i$$

$$= h(p_1, 1 - p_1) + (1 - p_1) h \left(\frac{p_2}{1 - p_1}, \ldots, \frac{p_m}{1 - p_1} \right);$$

in the RHS the first term is $\eta(\varepsilon)$ and the second one does not exceed $\varepsilon \log(m-1)$.

$\square$

Definition 1.2.9   Given random variables X, Y, Z, we say that X and Y are *conditionally independent* given Z if, for all x and y and for all z with $\mathbb{P}(Z = z) > 0$,

$$\mathbb{P}(X = x, Y = y | Z = z) = \mathbb{P}(X = x | Z = z) \mathbb{P}(Y = y | Z = z). \qquad (1.2.18)$$

For the conditional entropy we immediately obtain

Theorem 1.2.10   (a) *For all random variables X, Y,*

$$0 \le h(X|Y) \le h(X), \qquad (1.2.19)$$

the first equality occurring iff X is a function of Y and the second equality holding iff X and Y are independent.

(b) *For all random variables X, Y, Z,*

$$h(X|Y, Z) \le h(X|Y) \le h(X|\phi(Y)), \qquad (1.2.20)$$

the first equality occurring iff X and Z are conditionally independent given Y and the second equality holding iff X and Z are conditionally independent given $\phi(Y)$.

Proof (a) The LHS bound in (1.2.19) follows from definition (1.2.4) (since $h(X|Y)$ is a sum of non-negative terms). The RHS bound follows from representation (1.2.5) and bound (1.2.8). The LHS quality in (1.2.19) is equivalent to the equation $h(X, Y) = h(Y)$ or $h(X, Y) = h(\phi(X, Y))$ with $\phi(X, Y) = Y$. In view of Theorem 1.2.6, this occurs iff, with probability 1, the map $(X, Y) \mapsto Y$ is invertible, i.e. X is a function of Y. The RHS equality in (1.2.19) occurs iff $h(X, Y) = h(X) + h(Y)$, i.e. X and Y are independent.

(b) For the lower bound, use a formula analogous to (1.2.5):

$$h(X|Y, Z) = h(X, Z|Y) - h(Z|Y) \qquad (1.2.21)$$

and an inequality analogous to (1.2.10):

$$h(X, Z|Y) \le h(X|Y) + h(Z|Y), \qquad (1.2.22)$$

with equality iff X and Z are conditionally independent given Y. For the RHS bound, use:

(i) a formula that is a particular case of (1.2.21): $h(X|Y,\phi(Y)) = h(X,Y|\phi(Y)) - h(Y|\phi(Y))$, together with the remark that $h(X|Y,\phi(Y)) = h(X|Y)$;

(ii) an inequality which is a particular case of (1.2.22): $h(X,Y|\phi(Y)) \leq h(X|\phi(Y)) + h(Y|\phi(Y))$, with equality iff X and Y are conditionally independent given $\phi(Y)$. □

Theorems 1.2.8 above and 1.2.11 below show how the entropy $h(X)$ and conditional entropy $h(X|Y)$ are controlled when X is 'nearly' a constant (respectively, 'nearly' a function of Y).

Theorem 1.2.11 (The generalised Fano inequality) *For a pair of random variables, X and Y taking values $x_1,\ldots,x_m$ and $y_1,\ldots,y_m$, if*

$$\sum_{j=1}^{m} \mathbb{P}(X = x_j, Y = y_j) = 1 - \varepsilon, \tag{1.2.23}$$

then

$$h(X|Y) \leq \eta(\varepsilon) + \varepsilon \log(m-1), \tag{1.2.24}$$

where $\eta(\varepsilon)$ is defined in (1.2.3).

Proof Denoting $\varepsilon_j = \mathbb{P}(X \neq x_j | Y = y_j)$, we write

$$\sum_j p_Y(y_j)\varepsilon_j = \sum_j \mathbb{P}(X \neq x_j, Y = y_j) = \varepsilon. \tag{1.2.25}$$

By definition of the conditional entropy, the Fano inequality and concavity of the function $\eta(\cdot)$,

$$h(X|Y) \leq \sum_j p_Y(y_j)\Big(\eta(\varepsilon_j) + \varepsilon_j \log(m-1)\Big)$$
$$\leq \sum_j p_Y(y_j)\eta(\varepsilon_j) + \varepsilon \log(m-1) \leq \eta(\varepsilon) + \varepsilon \log(m-1).$$

□

If the random variable X takes countably many values $\{x_1, x_2, \ldots\}$, the above definitions may be repeated, as well as most of the statements; notable exceptions are the RHS bound in (1.2.7) and inequalities (1.2.17) and (1.2.24).

Many properties of entropy listed so far are extended to the case of random strings.

Theorem 1.2.12 *For a pair of random strings, $\mathbf{X}^{(n)} = (X_1,\ldots,X_n)$ and $\mathbf{Y}^{(n)} = (Y_1,\ldots,Y_n)$,*

(a) *the joint entropy, given by*

$$h(\mathbf{X}^{(n)}) = -\sum_{\mathbf{x}^{(n)}} \mathbb{P}(\mathbf{X}^{(n)} = \mathbf{x}^{(n)}) \log \mathbb{P}(\mathbf{X}^{(n)} = \mathbf{x}^{(n)}),$$

obeys

$$h(\mathbf{X}^{(n)}) = \sum_{i=1}^{n} h(X_i | \mathbf{X}^{(i-1)}) \le \sum_{i=1}^{n} h(X_i), \qquad (1.2.26)$$

with equality iff components $X_1, \ldots, X_n$ are independent;

(b) *the conditional entropy, given by*

$$h(\mathbf{X}^{(n)} | \mathbf{Y}^{(n)})$$
$$= -\sum_{\mathbf{x}^{(n)}, \mathbf{y}^{(n)}} \mathbb{P}(\mathbf{X}^{(n)} = \mathbf{x}^{(n)}, \mathbf{Y}^{(n)} = \mathbf{y}^{(n)}) \log \mathbb{P}(\mathbf{X}^{(n)} = \mathbf{x}^{(n)} | \mathbf{Y}^{(n)} = \mathbf{y}^{(n)}),$$

satisfies

$$h(\mathbf{X}^{(n)} | \mathbf{Y}^{(n)}) \le \sum_{i=1}^{n} h(X_i | \mathbf{Y}^{(n)}) \le \sum_{i=1}^{n} h(X_i | Y_i), \qquad (1.2.27)$$

with the LHS equality holding iff $X_1, \ldots, X_n$ are conditionally independent, given $\mathbf{Y}^{(n)}$, and the RHS equality holding iff, for each $i = 1, \ldots, n$, X_i and $\{Y_r : 1 \le r \le n, \ r \ne i\}$ are conditionally independent, given Y_i.

Proof The proof repeats the arguments used previously in the scalar case. ☐

Definition 1.2.13 The *mutual information* or *mutual entropy*, $I(X : Y)$, between X and Y is defined as

$$I(X : Y) := \sum_{x,y} p_{X,Y}(x,y) \log \frac{p_{X,Y}(x,y)}{p_X(x) p_Y(y)} = \mathbb{E} \log \frac{p_{X,Y}(X,Y)}{p_X(X) p_Y(Y)}$$
$$= h(X) + h(Y) - h(X,Y) = h(X) - h(X|Y)$$
$$= h(Y) - h(Y|X). \qquad (1.2.28)$$

As can be seen from this definition, $I(X : Y) = I(Y : X)$.

Intuitively, $I(X : Y)$ measures the amount of information about X conveyed by Y (and vice versa). Theorem 1.2.10(b) implies

Theorem 1.2.14 *If a random variable $\phi(Y)$ is a function of Y then*

$$0 \le I(X : \phi(Y)) \le I(X : Y), \qquad (1.2.29)$$

the first equality occurring iff X and $\phi(Y)$ are independent, and the second iff X and Y are conditionally independent, given $\phi(Y)$.

Worked Example 1.2.15   *Suppose that two non-negative random variables X and Y are related by Y = X + N, where N is a geometric random variable taking values in $\mathbb{Z}_+$ and is independent of X. Determine the distribution of Y which maximises the mutual entropy between X and Y under the constraint that the mean $\mathbb{E}X \leq K$ and show that this distribution can be realised by assigning to X the value zero with a certain probability and letting it follow a geometrical distribution with a complementary probability.*

Solution Because $Y = X + N$ where X and N are independent, we have

$$I(X:Y) = h(Y) - h(Y|X) = h(Y) - h(N).$$

Also $\mathbb{E}(Y) = \mathbb{E}(X) + \mathbb{E}(N) \leq K + \mathbb{E}(N)$. Therefore, if we can guarantee that Y may be taken geometrically distributed with mean $K + \mathbb{E}(N)$ then it gives the maximal value of $I(X:Y)$. To this end, write an equation for probability-generating functions:

$$\mathbb{E}(z^Y) = \mathbb{E}(z^X)\mathbb{E}(z^N), \ z > 0,$$

with $\mathbb{E}(z^N) = (1-p)/(1-zp), 0 < z < 1/p$, and

$$\mathbb{E}(z^Y) = \frac{1-p^*}{1-zp^*}, \ \ 0 < z < \frac{1}{p^*},$$

where p^* is to be found from an equation

$$\mu_Y = \frac{p^*}{1-p^*} = K + \frac{p}{1-p} = \frac{K(1-p)+p}{1-p}.$$

This yields

$$p^* = \frac{K(1-p)+p}{1+K(1-p)}, \ \ E(z^Y) = \frac{1-p}{1+K(1-p)-z(p+K(1-p))},$$

and

$$\mathbb{E}(z^X) = \frac{1-zp}{1+K(1-p)-z(p+K(1-p))}. \tag{1.2.30}$$

The form of the distribution of X suggested in the example leads to

$$\mathbb{E}(z^X) = \kappa_0 + (1-\kappa_0)\frac{1-p_X}{1-zp_X}, \tag{1.2.31}$$

where $\kappa_0 + (1-\kappa_0)(1-p_X) = \mathbb{P}(X=0)$. Selecting

$$p_X = \frac{p+K(1-p)}{1+K(1-p)}, \kappa_0 = \frac{p}{p+K(1-p)},$$

we see that (1.2.30) and (1.2.31) coincide. $\square$

> *I only ask for information...*
> Charles Dickens (1812–1870), English writer,
> from *David Copperfield*

In Definition 1.2.13 and Theorem 1.2.14, random variables X and Y may be replaced by random strings. In addition, by repeating the above arguments for strings $\mathbf{X}^{(n)}$ and $\mathbf{Y}^{(n)}$, we obtain

Theorem 1.2.16 (a) *The mutual entropy between random strings obeys*

$$I(\mathbf{X}^{(n)} : \mathbf{Y}^{(n)}) \geq h(\mathbf{X}^{(n)}) - \sum_{i=1}^{n} h(X_i|\mathbf{Y}^{(n)}) \geq h(\mathbf{X}^{(n)}) - \sum_{i=1}^{n} h(X_i|Y_i). \qquad (1.2.32)$$

(b) *If $X_1, \ldots, X_n$ are independent then*

$$I(\mathbf{X}^{(n)} : \mathbf{Y}^{(n)}) \geq \sum_{i=1}^{n} I(X_i : \mathbf{Y}^{(n)}). \qquad (1.2.33)$$

Observe that

$$\sum_{i=1}^{n} I(X_i : \mathbf{Y}^{(n)}) \geq \sum_{i=1}^{n} I(X_i : Y_i). \qquad (1.2.34)$$

Worked Example 1.2.17 Let X, Z be random variables and $\mathbf{Y}^{(n)} = (Y_1, \ldots, Y_n)$ be a random string.

(a) *Prove the inequality*

$$0 \leq I(X : Z) \leq \min\{h(X), h(Z)\}.$$

(b) *Prove or disprove by producing a counter-example the inequality*

$$I(X : \mathbf{Y}^{(n)}) \leq \sum_{j=1}^{n} I(X : Y_j), \qquad (1.2.35)$$

first under the assumption that $Y_1, \ldots, Y_n$ are independent random variables, and then under the assumption that $Y_1, \ldots, Y_n$ are conditionally independent given X.

(c) *Prove or disprove by producing a counter-example the inequality*

$$I(X : \mathbf{Y}^{(n)}) \geq \sum_{j=1}^{n} I(X : Y_j), \qquad (1.2.36)$$

first under the assumption that $Y_1, \ldots, Y_n$ are independent random variables, and then under the assumption that $Y_1, \ldots, Y_n$ are conditionally independent given X.

Solution (a) By the Gibbs inequality, $I(X:Z) \geq 0$, and

$$I(X:Z) := -\sum_{x,z} \mathbb{P}(X=x, Z=z) \log \frac{\mathbb{P}(X=x, Z=z)}{\mathbb{P}(X=x)\mathbb{P}(Z=z)}$$

$$= h(X) - h(X|Z) = h(Z) - h(Z|X).$$

Here $h(X|Z) \geq 0$ and $h(Z|X) \geq 0$. Hence $I(X:Z) \leq h(X)$ and $I(X:Z) \leq h(Z)$, so $I(X:Z) \leq \min \big[h(X), h(Z)\big]$.

(b) Write

$$I(X:\mathbf{Y}^{(n)}) = h(\mathbf{Y}^{(n)}) - h(\mathbf{Y}^{(n)}|X). \tag{1.2.37}$$

Then, if $Y_1, \ldots, Y_n$ are conditionally independent given X, the RHS of (1.2.37) equals

$$h(\mathbf{Y}^{(n)}) - \sum_{j=1}^{n} h(Y_j|X) \leq \sum_{j=1}^{n} \big[h(Y_j) - h(Y_j|X)\big] = \sum_{j=1}^{n} I(X:Y_j),$$

giving that of (1.2.35).

(c) Next, if $Y_1, \ldots, Y_n$ are independent, the RHS of (1.2.37) equals

$$\sum_{j=1}^{n} h(Y_j) - h(\mathbf{Y}^{(n)}|X) \geq \sum_{j=1}^{n} \big[h(Y_j) - h(Y_j|X)\big] = \sum_{j=1}^{n} I(X:Y_j),$$

giving the RHS of (1.2.36).

On the other hand, property (b) fails under the independence condition. Indeed, set $n=2$, with $\mathbf{Y}^{(2)} = (Y_1, Y_2)$, and let Y_1 and Y_2 take values 0 or 1 with probabilities $1/2$, $j=1,2$, independently, and set $X = (Y_1 + Y_2) \bmod 2$. Then

$$h(X) = h(X|Y_j) = 1, \quad \text{so } I(X:Y_j) \equiv 0, \ j=1,2,$$

but

$$h(X|\mathbf{Y}^{(2)}) = 0, \quad \text{so } I(X:\mathbf{Y}^{(2)}) = 1.$$

Also, (c) fails under the conditional independence condition. Indeed, take a DTMC $(U_1, U_2, \ldots)$ with states ± 1, the initial probability distribution $\{1/2, 1/2\}$ and the transition matrix $\begin{pmatrix} 0 & 1 \\ 1 & 0 \end{pmatrix}$. Set

$$Y_1 = U_1, \ X = U_2, \ Y_2 = U_3.$$

Then Y_1, Y_2 are conditionally independent given X: $Y_1 = Y_2 = -X$. On the other hand,

$$1 = I(X:\mathbf{Y}^{(2)}) = h(\mathbf{Y}^{(2)}) = h(Y_1) = h(Y_2)$$
$$< h(Y_1) + h(Y_2) = I(X:Y_1) + I(X:Y_2) = 2.$$

$\square$

Recall that a real function $f(\mathbf{y})$ defined on a convex set $\mathbb{V} \subseteq \mathbb{R}^m$ is called *concave* if

$$f(\lambda_0 \mathbf{y}^{(0)} + \lambda_1 \mathbf{y}^{(1)}) \geq \lambda_0 f(\mathbf{y}^{(0)}) + \lambda_1 f(\mathbf{y}^{(1)})$$

for any $\mathbf{y}^{(0)}, \mathbf{y}^{(1)} \in \mathbb{V}$ and $\lambda_0, \lambda_1 \in [0,1]$ with $\lambda_0 + \lambda_1 = 1$. It is called *strictly concave* if the equality is attained only when either $\mathbf{y}^{(0)} = \mathbf{y}^{(1)}$ or $\lambda_0 \lambda_1 = 0$. We treat $h(X)$ as a function of variables $\mathbf{p} = (p_1, \ldots, p_m)$; set $\mathbb{V}$ in this case is $\{\mathbf{y} = (y_1, \ldots, y_m) \in \mathbb{R}^m : y_i \geq 0, \ 1 \leq i \leq m, \ y_1 + \cdots + y_m = 1\}$.

Theorem 1.2.18 *Entropy is a strictly concave function of the probability distribution.*

Proof Let the random variables $X^{(i)}$ have probability distributions $\mathbf{p}^{(i)}$, $i = 0, 1$, and assume that the random variable Λ takes values 0 and 1 with probabilities λ_0 and λ_1, respectively, and is independent of $X^{(0)}$, $X^{(1)}$. Set $X = X^{(\Lambda)}$; then the inequality $h(\lambda_0 \mathbf{p}^{(0)} + \lambda_1 \mathbf{p}^{(1)}) \geq \lambda_0 h(\mathbf{p}^{(0)}) + \lambda_1 h(\mathbf{p}^{(1)})$ is equivalent to

$$h(X) \geq h(X|\Lambda) \tag{1.2.38}$$

which follows from (1.2.19). If we assume equality in (1.2.38), X and Λ must be independent. Assume in addition that $\lambda_0 > 0$ and write, by using independence,

$$\mathbb{P}(X = i, \Lambda = 0) = \mathbb{P}(X = i)\mathbb{P}(\Lambda = 0) = \lambda_0 \mathbb{P}(X = i).$$

The LHS equals $\lambda_0 \mathbb{P}(X = i|\Lambda = 0) = \lambda_0 p_i^{(0)}$ and the RHS equals $\lambda_0 \big(\lambda_0 p_i^{(0)} + \lambda_1 p_i^{(1)}\big)$. We may cancel λ_0 obtaining

$$(1 - \lambda_0) p_i^{(0)} = \lambda_1 p_i^{(1)},$$

i.e. the probability distributions $\mathbf{p}^{(0)}$ and $\mathbf{p}^{(1)}$ are proportional. Then either they are equal or $\lambda_1 = 0$, $\lambda_0 = 1$. The assumption $\lambda_1 > 0$ leads to a similar conclusion. $\square$

Worked Example 1.2.19 *Show that the quantity*

$$\rho(X, Y) = h(X|Y) + h(Y|X)$$

obeys

$$\rho(X, Y) = h(X) + h(Y) - 2I(X:Y)$$
$$= h(X, Y) - I(X:Y) = 2h(X, Y) - h(X) - h(Y).$$

Prove that ρ is symmetric, i.e. $\rho(X, Y) = \rho(Y, X) \geq 0$, and satisfies the triangle inequality, i.e. $\rho(X, Y) + \rho(Y, Z) \geq \rho(X, Z)$. Show that $\rho(X, Y) = 0$ iff X and Y are functions of each other. Also show that if X' and X are functions of each other then $\rho(X, Y) = \rho(X', Y)$. Hence, ρ may be considered as a metric on the set of the random variables X, considered up to equivalence: $X \sim X'$ iff X and X' are functions of each other.

Solution Check the triangle inequality

$$h(X|Z) + h(Z|X) \le h(X|Y) + h(Y|X) + h(Y|Z) + h(Z|Y),$$

or

$$h(X,Z) \le h(X,Y) + h(Y,Z) - h(Y).$$

To this end, write $h(X,Z) \le h(X,Y,Z)$ and note that $h(X,Y,Z)$ equals

$$h(X,Z|Y) + h(Y) \le h(X|Y) + h(Z|Y) + h(Y)$$
$$= h(X,Y) + h(Y,Z) - h(Y).$$

Equality holds iff (i) $Y = \phi(X,Z)$ and (ii) X, Z are conditionally independent given Y. □

Remark 1.2.20 The property that $\rho(X,Z) = \rho(X,Y) + \rho(Y,Z)$ means that 'point' Y lies on a 'line' through X and Z; in other words, that all three points X, Y, Z lie on a straight line. Conditional independence of X and Z given Y can be stated in an alternative (and elegant) way: the triple $X \to Y \to Z$ satisfies the Markov property (in short: is Markov). Then suppose we have four random variables X_1, X_2, X_3, X_4 such that, for all $1 \le i_1 < i_2 < i_3 \le 4$, the random variables X_{i_1} and X_{i_3} are conditionally independent given X_{i_2}; this property means that the quadruple $X_1 \to X_2 \to X_3 \to X_4$ is Markov, or, geometrically, that all four points lie on a line. The following fact holds: if $X_1 \to X_2 \to X_3 \to X_4$ is Markov then the mutual entropies satisfy

$$I(X_1 : X_3) + I(X_2 : X_4) = I(X_1 : X_4) + I(X_2 : X_3). \tag{1.2.39}$$

Equivalently, for the joint entropies,

$$h(X_1, X_3) + h(X_2, X_4) = h(X_1, X_4) + h(X_2, X_3). \tag{1.2.40}$$

In fact, for all triples X_{i_1}, X_{i_2}, X_{i_3} as above, in the metric ρ we have that

$$\rho(X_{i_1}, X_{i_3}) = \rho(X_{i_1}, X_{i_2}) + \rho(X_{i_2}, X_{i_3}),$$

which in terms of the joint and individual entropies is rewritten as

$$h(X_{i_1}, X_{i_3}) = h(X_{i_1}, X_{i_2}) + h(X_{i_2}, X_{i_3}) - h(X_{i_2}).$$

Then (1.2.39) takes the form

$$h(X_1, X_2) + h(X_2, X_3) - h(X_2) + h(X_2, X_3) + h(X_3, X_4) - h(X_3)$$
$$= h(X_1, X_2) + h(X_2, X_3) - h(X_2) + h(X_3, X_4) + h(X_2, X_3) - h(X_3)$$

which is a trivial identity.

Worked Example 1.2.21   *Consider the following inequality. Let a triple $X \to$*
$Y \to Z$ be Markov where $\mathbf{Z}$ is a random string $(Z_1, \ldots, Z_n)$. Then

$$\sum_{1 \leq i \leq n} I(X : Z_i) \leq I(X, Y) + I(\mathbf{Z}) \quad \text{where } I(\mathbf{Z}) := \sum_{1 \leq i \leq n} h(Z_i) - h(\mathbf{Z}).$$

Solution The Markov property for $X \to Y \to Z$ leads to the bound

$$I(X : \mathbf{Z}) \leq I(X : Y).$$

Therefore, it suffices to verify that

$$\sum_{1 \leq i \leq n} I(X : Z_i) - I(\mathbf{Z}) \leq I(X : \mathbf{Z}). \tag{1.2.41}$$

As we show below, bound (1.2.41) holds for any X and $\mathbf{Z}$ (without referring to a
Markov property). Indeed, (1.2.41) is equivalent to

$$nh(X) - \sum_{1 \leq i \leq n} h(X, Z_i) + h(\mathbf{Z}) \leq h(X) + h(\mathbf{Z}) - h(X, \mathbf{Z})$$

or

$$h(X, \mathbf{Z}) - h(X) \leq \sum_{1 \leq i \leq n} h(X, Z_i) - nh(X)$$

which in turn is nothing but the inequality $h(\mathbf{Z}|X) \leq \sum_{1 \leq i \leq n} h(Z_i|X)$.   □

Worked Example 1.2.22   *Write $h(\mathbf{p}) := -\sum_1^m p_j \log p_j$ for a probability 'vector'*

$$\mathbf{p} = \begin{pmatrix} p_1 \\ \vdots \\ p_m \end{pmatrix}, \text{ with entries } p_j \geq 0 \text{ and } p_1 + \cdots + p_m = 1.$$

(a) *Show that $h(P\mathbf{p}) \geq h(\mathbf{p})$ if $P = (P_{ij})$ is a doubly stochastic matrix (i.e. a square*
   matrix with elements $P_{ij} \geq 0$ for which all row and column sums are unity).
   Moreover, $h(P\mathbf{p}) \equiv h(\mathbf{p})$ iff P is a permutation matrix.
(b) *Show that $h(\mathbf{p}) \geq - \sum_{j=1}^m \sum_{k=1}^m p_j P_{jk} \log P_{jk}$ if P is a stochastic matrix and $\mathbf{p}$ is an*
   invariant vector of P: $P\mathbf{p} = \mathbf{p}$.

Solution (a) By concavity of the log-function $x \mapsto \log x$, for all $\lambda_i, c_i \geq 0$ such
that $\sum_1^m \lambda_i = 1$, we have $\log(\lambda_1 c_1 + \cdots + \lambda_m c_m) \geq \sum_1^m \lambda_i \log c_i$. Apply this to $h(P\mathbf{p}) =$

$$-\sum_{i,j} P_{ij} p_j \log \left(\sum_k P_{ik} p_k \right) \geq -\sum_j p_j \log \left(\sum_{i,k} P_{ij} P_{ik} p_k \right) = -\sum_j p_j \log \left((P^{\mathsf{T}} P \mathbf{p})_j \right). \text{ By}$$

the Gibbs inequality the RHS $\geq h(\mathbf{p})$. The equality holds iff $P^{\mathsf{T}} P \mathbf{p} \equiv \mathbf{p}$, i.e. $P^{\mathsf{T}} P =$
$\mathbf{I}$, the unit matrix. This happens iff P is a permutation matrix.

(b) The LHS equals $h(U_n)$ for the stationary Markov source $(U_1, U_2, \ldots)$ with equilibrium distribution $\mathbf{p}$, whereas the RHS is $h(U_n | U_{n-1})$. The general inequality $h(U_n | U_{n-1}) \leq h(U_n)$ gives the result.   □

Worked Example 1.2.23   *The sequence of random variables $\{X_j : j = 1, 2, \ldots\}$ forms a DTMC with a finite state space.*

(a) *Quoting standard properties of conditional entropy, show that $h(X_j | X_{j-1}) \leq h(X_j | X_{j-2})$ and, in the case of a stationary DTMC, $h(X_j | X_{j-2}) \leq 2h(X_j | X_{j-1})$.*
(b) *Show that the mutual information $I(X_m : X_n)$ is non-decreasing in m and non-increasing in n, $1 \leq m \leq n$.*

Solution (a) By the Markov property and stationarity

$$
\begin{aligned}
h(X_j | X_{j-1}) &= h(X_j | X_{j-1}, X_{j-2}) \\
&\leq h(X_j | X_{j-2}) \leq h(X_j, X_{j-1} | X_{j-2}) \\
&= h(X_j | X_{j-1}, X_{j-2}) + h(X_{j-1} | X_{j-2}) = 2h(X_j | X_{j-1}).
\end{aligned}
$$

(b) Write

$$
\begin{aligned}
I(X_m : X_n) - I(X_m : X_{n+1}) &= h(X_m | X_{n+1}) - h(X_m | X_n) \\
&= h(X_m | X_{n+1}) - h(X_m | X_n, X_{n+1}) \text{ (because } X_m \text{ and} \\
&\qquad\qquad X_{n+1} \text{ are conditionally independent, given } X_n)
\end{aligned}
$$

which is ≥ 0. Thus, $I(X_m : X_n)$ does not increase with n.

Similarly,

$$
I(X_{m-1} : X_n) - I(X_m : X_n) = h(X_n | X_{m-1}) - h(X_n | X_m, X_{m-1}) \geq 0.
$$

Thus, $I(X_m : X_n)$ does not decrease with m.

Here, no assumption of stationarity has been used. The DTMC may not even be time-homogeneous (i.e. the transition probabilities may depend not only on i and j but also on the time of transition).   □

Worked Example 1.2.24   *Given random variables Y_1, Y_2, Y_3, define*

$$
I(Y_1 : Y_2 | Y_3) = h(Y_1 | Y_3) + h(Y_2 | Y_3) - h(Y_1, Y_2 | Y_3).
$$

Now let the sequence X_n, $n = 0, 1, \ldots$ be a DTMC. Show that

$$
I(X_{n-1} : X_{n+1} | X_n) = 0 \text{ and hence } I(X_{n-1} : X_{n+1}) \leq I(X_n : X_{n+1}).
$$

Show also that $I(X_n : X_{n+m})$ is non-increasing in m, for $m = 0, 1, 2, \ldots$.

Solution By the Markov property, X_{n-1} and X_{n+1} are conditionally independent, given X_n. Hence,

$$h(X_{n-1},X_{n+1}|X_n) = h(X_{n+1}|X_n) + h(X_{n-1}|X_n)$$

and $I(X_{n-1} : X_{n+1}|X_n) = 0$. Also,

$$
\begin{aligned}
I(X_n &: X_{n+m}) - I(X_n : X_{n+m+1})\\
&= h(X_{n+m}) - h(X_{n+m+1}) - h(X_n,X_{n+m+1}) + h(X_n,X_{n+m})\\
&= h(X_n|X_{n+m+1}) - h(X_n|X_{n+m})\\
&= h(X_n|X_{n+m+1}) - h(X_n|X_{n+m},X_{n+m+1}) \geq 0,
\end{aligned}
$$

the final equality holding because of the conditional independence and the last inequality following from (1.2.21). □

Worked Example 1.2.25 (An axiomatic definition of entropy)

(a) *Consider a probability distribution $(p_1,\dots,p_m)$ and an associated measure of uncertainty (entropy) such that*

$$h(p_1q_1,p_1q_2,\dots,p_1q_n,p_2,p_3,\dots,p_m) = h(p_1,\dots,p_m) + p_1h(q_1,\dots,q_n),$$
(1.2.42)

if $(q_1,\dots,q_n)$ is another distribution. That is, if one of the contingencies (of probability p_1) is divided into sub-contingencies of conditional probabilities $q_1,\dots,q_n$, then the total uncertainty breaks up additively as shown. The functional h is assumed to be symmetric in its arguments, so that analogous relations holds if contingencies $2,3,\dots,m$ are subdivided.

Suppose that $F(m) := h(1/m,\dots,1/m)$ is monotone increasing in m. Show that, as a consequence of (1.2.42), $F(m^k) = kF(m)$ and hence that $F(m) = c\log m$ for some constant c. Hence show that

$$h(p_1,\dots,p_m) = -c\sum_j p_j \log p_j$$
(1.2.43)

if p_j are rational. The validity of (1.2.43) for an arbitrary collection $\{p_j\}$ then follows by a continuity assumption.

(b) *An alternative axiomatic characterisation of entropy is as follows. If a symmetric function h obeys for any $k < m$*

$$
\begin{aligned}
h(p_1,\dots,p_m) &= h(p_1 + \dots + p_k, p_{k+1},\dots,p_m)\\
&+ (p_1 + \dots + p_k)h\left(\frac{p_1}{p_1+\dots+p_k},\dots,\frac{p_k}{p_1+\dots+p_k}\right),
\end{aligned}
$$
(1.2.44)

$h(1/2,1/2) = 1$, and $h(p, 1-p)$ is a continuous function of $p \in [0,1]$, then

$$h(p_1, \ldots, p_m) = -\sum_j p_j \log p_j.$$

Solution (a) Using (1.2.42), we obtain for the function $F(m) = h(1/m, \ldots, 1/m)$ the following identity:

$$F(m^2) = h\left(\frac{1}{m} \times \frac{1}{m}, \ldots, \frac{1}{m} \times \frac{1}{m}, \frac{1}{m^2}, \ldots, \frac{1}{m^2}\right)$$

$$= h\left(\frac{1}{m}, \frac{1}{m^2}, \ldots, \frac{1}{m^2}\right) + \frac{1}{m} F(m)$$

$$\vdots$$

$$= h\left(\frac{1}{m}, \ldots, \frac{1}{m}\right) + \frac{m}{m} F(m) = 2F(m).$$

The induction hypothesis is $F(m^{k-1}) = (k-1)F(m)$. Then

$$(m^k) = h\left(\frac{1}{m} \times \frac{1}{m^{k-1}}, \ldots, \frac{1}{m} \times \frac{1}{m^{k-1}}, \frac{1}{m^k}, \ldots, \frac{1}{m^k}\right)$$

$$= h\left(\frac{1}{m^{k-1}}, \frac{1}{m^k}, \ldots, \frac{1}{m^k}\right) + \frac{1}{m} F(m)$$

$$\vdots$$

$$= h\left(\frac{1}{m^{k-1}}, \ldots, \frac{1}{m^{k-1}}\right) + \frac{m}{m} F(m)$$

$$= (k-1)F(m) + F(m) = kF(m).$$

Now, for given positive integers $b > 2$ and m, we can find a positive integer n such that $2^n \le b^m \le 2^{n+1}$, i.e.

$$\frac{n}{m} \le \log_2 b \le \frac{n}{m} + \frac{1}{m}.$$

By monotonicity of $F(m)$, we obtain $nF(2) \le mF(b) \le (n+1)F(2)$, or

$$\frac{n}{m} \le \frac{F(b)}{F(2)} \le \frac{n}{m} + \frac{1}{m}.$$

We conclude that $\left|\log_2 b - \dfrac{F(b)}{F(2)}\right| \le \dfrac{1}{m}$, and letting $m \to \infty$, $F(b) = c \log b$ with $c = F(2)$.

Now take rational numbers $p_1 = \dfrac{r_1}{r}, \ldots, p_m = \dfrac{r_m}{r}$ and obtain

$$h\left(\frac{r_1}{r}, \ldots, \frac{r_m}{r}\right) = h\left(\frac{r_1}{r} \times \frac{1}{r_1}, \ldots, \frac{r_1}{r} \times \frac{1}{r_1}, \frac{r_2}{r}, \ldots, \frac{r_m}{r}\right) - \frac{r_1}{r}F(r_1)$$

$$\vdots$$

$$= h\left(\frac{1}{r}, \ldots, \frac{1}{r}\right) - c \sum_{1 \le i \le m} \frac{r_i}{r} \log r_i$$

$$= c \log r - c \sum_{1 \le i \le m} \frac{r_i}{r} \log r_i = -c \sum_{1 \le i \le m} \frac{r_i}{r} \log \frac{r_i}{r}.$$

(b) For the second definition the point is that we do *not* assume the monotonicity of $F(m) = h(1/m, \ldots, 1/m)$ in m. Still, using (1.2.44), it is easy to check the additivity property

$$F(mn) = F(m) + F(n)$$

for any positive integers m, n. Hence, for a canonical prime number decomposition $m = q_1^{\alpha_1} \ldots q_s^{\alpha_s}$ we obtain

$$F(m) = \alpha_1 F(q_1) + \cdots + \alpha_s F(q_s).$$

Next, we prove that

$$\frac{F(m)}{m} \to 0, \ F(m) - F(m-1) \to 0 \tag{1.2.45}$$

as $m \to \infty$. Indeed,

$$F(m) = h\left(\frac{1}{m}, \ldots, \frac{1}{m}\right)$$

$$= h\left(\frac{1}{m}, \frac{m-1}{m}\right) + \frac{m-1}{m}h\left(\frac{1}{m-1}, \ldots, \frac{1}{m-1}\right),$$

i.e.

$$h\left(\frac{1}{m}, \frac{m-1}{m}\right) = F(m) - \frac{m-1}{m}F(m-1).$$

By continuity and symmetry of $h(p, 1-p)$,

$$\lim_{m \to \infty} h\left(\frac{1}{m}, \frac{m-1}{m}\right) = h(0,1) = h(1,0).$$

But from the representations

$$h\left(\frac{1}{2}, \frac{1}{2}, 0\right) = h\left(\frac{1}{2}, \frac{1}{2}\right) + \frac{1}{2}h(1,0)$$

and (the symmetry again)

$$h\left(\frac{1}{2},\frac{1}{2},0\right) = h\left(0,\frac{1}{2},\frac{1}{2}\right) = h(1,0) + h\left(\frac{1}{2},\frac{1}{2}\right)$$

we obtain $h(1,0) = 0$. Hence,

$$\lim_{m\to\infty}\left(F(m) - \frac{m-1}{m}F(m-1)\right) = 0. \tag{1.2.46}$$

Next, we write

$$mF(m) = \sum_{k=1}^{m} k\left(F(k) - \frac{k-1}{k}F(k-1)\right)$$

or, equivalently,

$$\frac{F(m)}{m} = \frac{m+1}{2m}\left[\frac{2}{m(m+1)}\sum_{k=1}^{m} k\left(F(k) - \frac{k-1}{k}F(k-1)\right)\right].$$

The quantity in the square brackets is the arithmetic mean of $m(m+1)/2$ terms of a sequence

$$F(1), F(2) - F(1), F(2) - F(1), F(3) - \frac{2}{3}F(2), F(3) - \frac{2}{3}F(2),$$

$$F(3) - \frac{2}{3}F(2), \dots, F(k) - \frac{k-1}{k}F(k-1), \dots,$$

$$F(k) - \frac{k-1}{k}F(k-1), \dots$$

that tends to 0. Hence, it goes to 0 and $F(m)/m \to 0$. Furthermore,

$$F(m) - F(m-1) = \left(F(m) - \frac{m-1}{m}F(m-1)\right) - \frac{1}{m}F(m-1) \to 0,$$

and (1.2.46) holds. Now define

$$c(m) = \frac{F(m)}{\log m},$$

and prove that $c(m) = $ const. It suffices to prove that $c(p) = $ const for any prime number p. First, let us prove that a sequence $(c(p))$ is bounded. Indeed, suppose the numbers $c(p)$ are not bounded from above. Then, we can find an infinite sequence of primes $p_1, p_2, \dots, p_n, \dots$ such that p_n is the minimal prime such that $p_n > p_{n-1}$ and $c(p_n) > c(p_{n-1})$. By construction, if a prime $q < p_n$ then $c(q) < c(p_n)$.

Consider the canonical decomposition into prime factors of the number $p_n - 1 = q_1^{\alpha_1} \ldots q_s^{\alpha_s}$ with $q_1 = 2$. Then we write the difference $F(p_n) - F(p_n - 1)$ as

$$F(p_n) - \frac{F(p_n)}{\log p_n} \log(p_n - 1) + c(p_n) \log(p_n - 1) - F(p_n - 1)$$

$$= \frac{F(p_n)}{p_n} \frac{p_n}{\log p_n} \log \frac{p_n}{p_n - 1} + \sum_{j=1}^{s} \alpha_j (c(p_n) - c(q_j)) \log q_j.$$

The previous remark implies that

$$\sum_{j=1}^{s} \alpha_j (c(p_n) - c(q_j)) \log q_j \geq (c(p_n) - c(2)) \log 2 = (c(p_n) - c(2)). \quad (1.2.47)$$

Moreover, as $\lim_{p \to \infty} \frac{p}{\log p} \log \frac{p}{p-1} = 0$, equations (1.2.46) and (1.2.47) imply that $c(p_n) - c(2) \leq 0$ which contradicts with the construction of $c(p)$. Hence, $c(p)$ is bounded from above. Similarly, we check that $c(p)$ is bounded from below. Moreover, the above proof yields that $\sup_p c(p)$ and $\inf_p c(p)$ are both attained.

Now assume that $c(\widehat{p}) = \sup_p c(p) > c(2)$. Given a positive integer m, decompose into prime factors $\widehat{p}^m - 1 = q_1^{\alpha_1} \ldots q_s^{\alpha_s}$ with $q_1 = 2$. Arguing as before, we write the difference $F(\widehat{p}^m) - F(\widehat{p}^m - 1)$ as

$$F(\widehat{p}^m) - \frac{F(\widehat{p}^m)}{\log \widehat{p}^m} \log(\widehat{p}^m - 1) + c(\widehat{p}) \log(\widehat{p}^m - 1) - F(\widehat{p}^m - 1)$$

$$= \frac{F(\widehat{p}^m)}{\widehat{p}^m} \frac{\widehat{p}^m}{\log \widehat{p}^m} \log \frac{\widehat{p}^m}{\widehat{p}^m - 1} + \sum_{j=1}^{s} \alpha_j (c(\widehat{p}) - c(q_j)) \log q_j$$

$$\geq \frac{c(\widehat{p}^m)}{\widehat{p}^m} \frac{\widehat{p}^m}{\log \widehat{p}^m} \log \frac{\widehat{p}^m}{\widehat{p}^m - 1} + (c(\widehat{p}) - c(2)).$$

As before, the limit $m \to \infty$ yields $c(\widehat{p}) - c(2) \leq 0$ which gives a contradiction. Similarly, we can prove that $\inf_p c(p) = c(2)$. Hence, $c(p) = c$ is a constant, and $F(m) = c \log m$. From the condition $F(2) = h\left(\frac{1}{2}, \frac{1}{2}\right) = 1$ we get $c = 1$. Finally, as in (a), we obtain

$$h(p_1, \ldots, p_m) = -\sum_{i=1}^{m} p_i \log p_i \quad (1.2.48)$$

for any rational $p_1, \ldots, p_m \geq 0$ with $\sum_{i=1}^{m} p_i = 1$. By continuity argument (1.2.48) is extended to the case of irrational probabilities. $\qquad \square$

Worked Example 1.2.26 *Show that 'more homogeneous' distributions have a greater entropy. That is, if $\mathbf{p} = (p_1, \ldots, p_n)$ and $\mathbf{q} = (q_1, \ldots, q_n)$ are two probability distributions on the set $\{1, \ldots, n\}$, then $\mathbf{p}$ is called more homogeneous than $\mathbf{q}$*

($\mathbf{p} \preceq \mathbf{q}$, cf. [108]) if, after rearranging values $p_1, \ldots, p_n$ and $q_1, \ldots, q_n$ in decreasing order:

$$p_1 \geq \cdots \geq p_n, \quad q_1 \geq \cdots \geq q_n,$$

one has

$$\sum_{i=1}^{k} p_i \leq \sum_{i=1}^{k} q_i, \quad \text{for all } k = 1, \ldots, n.$$

Then

$$h(\mathbf{p}) \geq h(\mathbf{q}) \quad \text{whenever } \mathbf{p} \preceq \mathbf{q}.$$

Solution We write the probability distributions $\mathbf{p}$ and $\mathbf{q}$ as non-increasing functions of a discrete argument

$$\mathbf{p} \sim p^{(1)} \geq \cdots \geq p^{(n)} \geq 0, \mathbf{q} \sim q^{(1)} \geq \cdots \geq q^{(n)} \geq 0,$$

$$\text{with } \sum_{i} p^{(i)} = \sum_{i} q^{(i)} = 1.$$

Condition $\mathbf{p} \preceq \mathbf{q}$ means that if $\mathbf{p} \neq \mathbf{q}$ then there exist i_1 and i_2 such that (a) $1 \leq i_1 \leq i_2 \leq n$, (b) $q^{(i_1)} > p^{(i_1)} \geq p^{(i_2)} > q^{(i_2)}$ and (c) $q^{(i)} \geq p^{(i)}$ for $1 \leq i \leq i_1$, $q^{(i)} \leq p^{(i)}$ for $i \geq i_2$.

Now apply induction in s, the number of values $i = 1, \ldots, n$ for which $q^{(i)} \neq p^{(i)}$. If $s = 0$ we have $\mathbf{p} = \mathbf{q}$ and the entropies coincide. Make the induction hypothesis and then increase s by 1. Take a pair i_1, i_2 as above. Increase $q^{(i_2)}$ and decrease $q^{(i_1)}$ so that the sum $q^{(i_1)} + q^{(i_2)}$ is preserved, until either $q^{(i_1)}$ reaches $p^{(i_1)}$ or $q^{(i_2)}$ reaches $p^{(i_2)}$ (see Figure 1.7). Property (c) guarantees that the modified distributions $\mathbf{p} \preceq \mathbf{q}$. As the function $x \to \eta(x) = -x \log x - (1-x) \log(1-x)$ strictly increases on $[0, 1/2]$. Hence, the entropy of the modified distribution strictly increases. At the end of this process we diminish s. Then we use our induction hypothesis. $\square$

1.3 Shannon's first coding theorem. The entropy rate of a Markov source

A useful meaning of the information rate of a source is that it specifies the minimal rates of growth for the set of sample strings carrying, asymptotically, the full probability.

Lemma 1.3.1 *Let H be the information rate of a source (see (1.1.20)). Define*

$$D_n(R) := \max \left[\mathbb{P}(\mathbf{U}^{(n)} \in A) : A \subset I^{\times n}, \, \sharp A \leq 2^{nR} \right]. \tag{1.3.1}$$

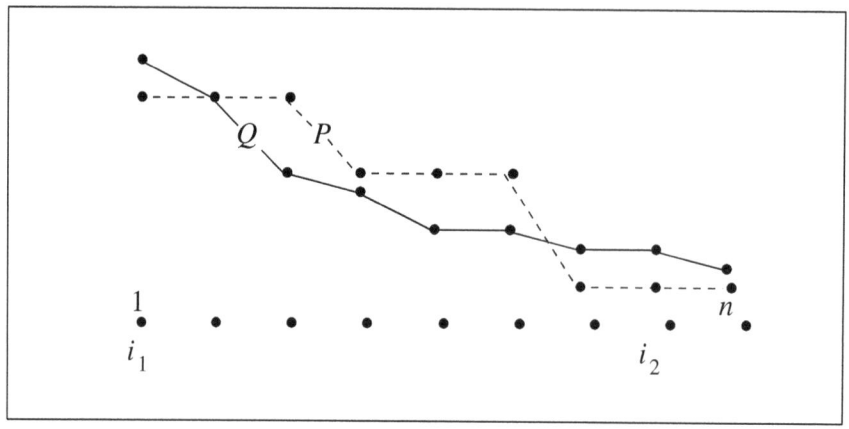

Figure 1.7

Then for any $\varepsilon > 0$, as $n \to \infty$,

$$\lim D_n(H+\varepsilon) = 1, \quad and, \ if \ H > 0, \ D_n(H-\varepsilon) \nrightarrow 1. \tag{1.3.2}$$

Proof   By definition, $R := H + \varepsilon$ is a reliable encoding rate. Hence, there exists a sequence of sets $A_n \subset I^{\times n}$, with $\sharp A_n \leq 2^{nR}$ and $\mathbb{P}(\mathbf{U}^{(n)} \in A_n) \to 1$, as $n \to \infty$. Since $D_n(R) \geq \mathbb{P}(\mathbf{U}^{(n)} \in A_n)$, then $D_n(R) \to 1$.

Now suppose that $H > 0$, and take $R := H - \varepsilon$; for ε small enough, $R > 0$. However, R is not a reliable rate. That is, there is no sequence A_n with the above properties. Take a set C_n where the maximum in (1.3.1) is attained. Then $\sharp C_n \leq 2^{nR}$, but $\mathbb{P}(C_n) \nrightarrow 1$. $\qquad\qquad\qquad\qquad\qquad\qquad\qquad\qquad\qquad\qquad\qquad\qquad\qquad\square$

Given a string $\mathbf{u}^{(n)} = u_1 \ldots u_n$, consider its 'log-likelihood' value per source-letter:

$$\xi_n(\mathbf{u}^{(n)}) = -\frac{1}{n} \log_+ p_n(\mathbf{u}^{(n)}), \ \mathbf{u}^{(n)} \in I^{\times n}, \tag{1.3.3a}$$

where $p_n(\mathbf{u}^{(n)}) := \mathbb{P}(\mathbf{U}^{(n)} = \mathbf{u}^{(n)})$ is the probability assigned to string $\mathbf{u}^{(n)}$. Here and below, $\log_+ x = \log x$ if $x > 0$, and is 0 if $x = 0$. For a random string, $\mathbf{U}^{(n)} = u_1, \ldots, u_n$,

$$\xi_n(\mathbf{U}^{(n)}) = -\frac{1}{n} \log_+ p_n(\mathbf{U}^{(n)}) \tag{1.3.3b}$$

is a random variable.

Lemma 1.3.2   *For all $R, \varepsilon > 0$,*

$$\mathbb{P}(\xi_n \leq R) \leq D_n(R) \leq \mathbb{P}(\xi_n \leq R + \varepsilon) + 2^{-n\varepsilon}. \tag{1.3.4}$$

Proof For brevity, omit the upper index (n) in the notation $\mathbf{u}^{(n)}$ and $\mathbf{U}^{(n)}$. Set

$$
\begin{aligned}
B_n &:= \{\mathbf{u} \in I^{\times n} : p_n(\mathbf{u}) \geq 2^{-nR}\} \\
&= \{\mathbf{u} \in I^{\times n} : -\log p_n(\mathbf{u}) \leq nR\} \\
&= \{\mathbf{u} \in I^{\times n} : \xi_n(\mathbf{u}) \leq R\}.
\end{aligned}
$$

Then

$$
1 \geq \mathbb{P}(\mathbf{U} \in B_n) = \sum_{\mathbf{u} \in B_n} p_n(\mathbf{u}) \geq 2^{-nR} \, \sharp B_n, \text{ whence } \sharp B_n \leq 2^{nR}.
$$

Thus,

$$
\begin{aligned}
D_n(R) &= \max \left[\mathbb{P}(\mathbf{U} \in A_n) : A_n \subseteq I^{\times n}, \, \sharp A \leq 2^{nR} \right] \\
&\geq \mathbb{P}(\mathbf{U} \in B_n) = \mathbb{P}(\xi_n \leq R),
\end{aligned}
$$

which proves the LHS in (1.3.4).

On the other hand, there exists a set $C_n \subseteq I^{\times n}$ where the maximum in (1.3.1) is attained. For such a set, $D_n(R) = \mathbb{P}(\mathbf{U} \in C_n)$ is decomposed as follows:

$$
\begin{aligned}
D_n(R) &= \mathbb{P}(\mathbf{U} \in C_n, \xi_n \leq R+\varepsilon) + \mathbb{P}(\mathbf{U} \in C_n, \, \xi_n > R+\varepsilon) \\
&\leq \mathbb{P}(\xi_n \leq R+\varepsilon) + \sum_{\mathbf{u} \in C_n} p_n(u)\mathbf{1}\left(p_n(\mathbf{u}) < 2^{-n(R+\varepsilon)}\right) \\
&< \mathbb{P}(\xi_n \leq R+\varepsilon) + 2^{-n(R+\varepsilon)} \sharp C_n \\
&= \mathbb{P}(\xi_n \leq R+\varepsilon) + 2^{-n(R+\varepsilon)} 2^{nR} \\
&= \mathbb{P}(\xi_n \leq R+\varepsilon) + 2^{-n\varepsilon}.
\end{aligned}
$$

$\square$

Definition 1.3.3   (See PSE II, p. 367.) A sequence of random variables $\{\eta_n\}$ converges in probability to a constant r if, for all $\varepsilon > 0$,

$$
\lim_{n \to \infty} \mathbb{P}\left(|\eta_n - r| \geq \varepsilon\right) = 0. \tag{1.3.5}
$$

Replacing, in this definition, r by a random variable η, we obtain a more general definition of convergence in probability to a random variable.

Convergence in probability is denoted henceforth as $\eta_n \xrightarrow{\mathrm{P}} r$ (respectively, $\eta_n \xrightarrow{\mathrm{P}} \eta$).

Remark 1.3.4   It is precisely the convergence in probability (to an expected value) that figures in the so-called law of large numbers (cf. (1.3.8) below). See PSE I, p. 78.

Theorem 1.3.5   (Shannon's first coding theorem (FCT)) *If ξ_n converges in probability to a constant γ then $\gamma = H$, the information rate of a source.*

Proof   Let $\xi_n \xrightarrow{\mathbb{P}} \gamma$. Since $\xi_n \geq 0$, $\gamma \geq 0$. By Lemma 1.3.2, for any $\varepsilon > 0$,

$$D_n(\gamma+\varepsilon) \geq \mathbb{P}(\xi_n \leq \gamma+\varepsilon) \geq \mathbb{P}(\gamma-\varepsilon \leq \xi_n \leq \gamma+\varepsilon)$$
$$= \mathbb{P}\left(|\xi_n - \gamma| \leq \varepsilon\right) = 1 - \mathbb{P}(|\xi_n - \gamma| > \varepsilon) \to 1 \quad (n \to \infty).$$

Hence, $H \leq \gamma$. In particular, if $\gamma = 0$ then $H = 0$. If $\gamma > 0$, we have, again by Lemma 1.3.2, that

$$D_n(\gamma-\varepsilon) \leq \mathbb{P}(\xi_n \leq \gamma-\varepsilon/2) + 2^{-n\varepsilon/2} \leq \mathbb{P}(|\xi_n - \gamma| \geq \varepsilon/2) + 2^{-n\varepsilon/2} \to 0.$$

By Lemma 1.3.1, $H \geq \gamma$. Hence, $H = \gamma$.   □

Remark 1.3.6   (a) Convergence $\xi_n \xrightarrow{\mathbb{P}} \gamma = H$ is equivalent to the following *asymptotic equipartition* property: for any $\varepsilon > 0$,

$$\lim_{n \to \infty} \mathbb{P}\left(2^{-n(H+\varepsilon)} \leq p_n(\mathbf{U}^{(n)}) \leq 2^{-n(H-\varepsilon)}\right) = 1. \qquad (1.3.6)$$

In fact,

$$\mathbb{P}\left(2^{-n(H+\varepsilon)} \leq p_n(\mathbf{U}^{(n)}) \leq 2^{-n(H-\varepsilon)}\right)$$
$$= \mathbb{P}\left(H - \varepsilon \leq -\frac{1}{n}\log p_n(\mathbf{U}^{(n)}) \leq H+\varepsilon\right)$$
$$= \mathbb{P}\left(|\xi_n - H| \leq \varepsilon\right) = 1 - \mathbb{P}(|\xi_n - H| > \varepsilon).$$

In other words, for all $\varepsilon > 0$ there exists $n_0 = n_0(\varepsilon)$ such that, for any $n > n_0$, the set $I^{\times n}$ decomposes into disjoint subsets, Π_n and T_n, with

(i) $\mathbb{P}(\mathbf{U}^{(n)} \in \Pi_n) < \varepsilon$,
(ii) $2^{-n(H+\varepsilon)} \leq \mathbb{P}(\mathbf{U}^{(n)} = \mathbf{u}^{(n)}) \leq 2^{-n(H-\varepsilon)}$ for all $\mathbf{u}^{(n)} \in T_n$.

Pictorially speaking, T_n is a set of 'typical' strings and Π_n is the residual set. We conclude that, for a source with the asymptotic equipartition property, it is worthwhile to encode the typical strings with codewords of the same length, and the rest anyhow. Then we have the effective encoding rate $H + o(1)$ bits/source-letter, though the source emits $\log m$ bits/source-letter.

(b) Observe that

$$\mathbb{E}\xi_n = -\frac{1}{n}\sum_{\mathbf{u}^{(n)} \in I^{\times n}} p_n(\mathbf{u}^{(n)})\log p_n(\mathbf{u}^{(n)}) = \frac{1}{n}h^{(n)}. \qquad (1.3.7)$$

The simplest example of an information source (and one among the most instructive) is a Bernoulli source.

Theorem 1.3.7   *For a Bernoulli source $U_1, U_2, \ldots$, with $\mathbb{P}(U_i = x) = p(x)$,*

$$H = -\sum_x p(x)\log p(x).$$

Proof   For an IID sequence $U_1, U_2, \ldots$, the probability of a string is

$$p_n(\mathbf{u}^{(n)}) = \prod_{i=1}^{n} p(u_i), \quad \mathbf{u}^{(n)} = u_1 \ldots u_n.$$

Hence, $-\log p_n(u) = \sum_i -\log p(u_i)$. Denoting $\sigma_i = -\log p(U_i)$, $i = 1, 2, \ldots$, we see that $\sigma_1, \sigma_2, \ldots$ form a sequence of IID random variables. For a random string $\mathbf{U}^{(n)} = U_1 \ldots U_n$, $-\log p_n(\mathbf{U}^{(n)}) = \sum_{i=1}^{n} \sigma_i$, where the random variables $\sigma_i = -\log p(U_i)$ are IID.

Next, write $\xi_n = \dfrac{1}{n} \sum_{i=1}^{n} \sigma_i$. Observe that $\mathbb{E}\sigma_i = -\sum_j p(j) \log p(j) = h$ and

$$\mathbb{E}\xi_n = \mathbb{E}\left(\frac{1}{n}\sum_{i=1}^{n}\sigma_i\right) = \frac{1}{n}\sum_{i=1}^{n}\mathbb{E}\sigma_i = \frac{1}{n}\sum_{i=1}^{n}h = h,$$

the final equality being in agreement with (1.3.7), since, for the Bernoulli source, $h^{(n)} = nh$ (see (1.1.18)), and hence $\mathbb{E}\xi_n = h$. We immediately see that $\xi_n \xrightarrow{\mathrm{P}} h$ by the law of large numbers. So $H = h$ by Theorem 1.3.5 (FCT). $\qquad\square$

Theorem 1.3.8   (The law of large numbers for IID random variables) *For any sequence of IID random variables $\eta_1, \eta_2, \ldots$ with finite variance and mean $\mathbb{E}\eta_i = r$, and for any $\varepsilon > 0$,*

$$\lim_{n\to\infty} \mathbb{P}\left(|\frac{1}{n}\sum_{i=1}^{n}\eta_i - r| \geq \varepsilon\right) = 0. \qquad (1.3.8)$$

Proof   The proof of Theorem 1.3.8 is based on the famous Chebyshev inequality; see PSE II, p. 368. $\qquad\square$

Lemma 1.3.9   *For any random variable η and any $\varepsilon > 0$,*

$$\mathbb{P}(\eta \geq \varepsilon) \leq \frac{1}{\varepsilon^2}\mathbb{E}\eta^2.$$

Proof   See PSE I, p. 75. $\qquad\square$

Next, consider a Markov source $U_1 \, U_2 \, \ldots$ with letters from alphabet $I_m = \{1, \ldots, m\}$ and assume that the transition matrix $(P(u, v))$ (or rather its power) obeys

$$\min_{u,v} P^{(r)}(u, v) = \rho > 0 \text{ for some } r \geq 1. \qquad (1.3.9)$$

This condition means that the DTMC is irreducible and aperiodic. Then (see PSE II, p. 71), the DTMC has a unique invariant (equilibrium) distribution $\pi(1), \ldots, \pi(m)$:

$$0 \le \pi(u) \le 1, \ \sum_{u=1}^{m} \pi(u) = 1, \ \pi(v) = \sum_{u=1}^{m} \pi(u)P(u,v), \qquad (1.3.10)$$

and the n-step transition probabilities $P^{(n)}(u,v)$ converge to $\pi(v)$ as well as the probabilities $\left(\lambda P^{n-1}\right)(v) = \mathbb{P}(U_n = v)$:

$$\lim_{n \to \infty} P^{(n)}(u,v) = \lim_{n \to \infty} \mathbb{P}(U_n = v) = \lim_{n \to \infty} \sum_{u} \lambda(u)P^{(n)}(u,v) = \pi(v), \qquad (1.3.11)$$

for all initial distribution $\{\lambda(u), u \in I\}$. Moreover, the convergence in (1.3.11) is exponentially (geometrically) fast.

Theorem 1.3.10   *Assume that condition* (1.3.9) *holds with* $r = 1$. *Then the DTMC* $U_1, U_2, \ldots$ *possesses a unique invariant distribution* (1.3.10), *and for any* $u, v \in I$ *and any initial distribution* λ *on* I,

$$|P^{(n)}(u,v) - \pi(v)| \le (1-\rho)^n \ \text{ and } \ |\mathbb{P}(U_n = v) - \pi(v)| \le (1-\rho)^{n-1}. \quad (1.3.12)$$

In the case of a general $r \ge 1$, *we replace, in the RHS of* (1.3.12), $(1-\rho)^n$ *by* $(1-\rho)^{\lfloor n/r \rfloor}$ *and* $(1-\rho)^{n-1}$ *by* $(1-\rho)^{\lfloor (n-1)/r \rfloor}$.

Proof   See Worked Example 1.3.13.   □

Now we introduce an information rate H of a Markov source.

Theorem 1.3.11   *For a Markov source, under condition* (1.3.9),

$$H = - \sum_{1 \le u,v \le m} \pi(u)P(u,v) \log P(u,v) = \lim_{n \to \infty} h(U_{n+1}|U_n); \qquad (1.3.13)$$

if the source is stationary then $H = h(U_{n+1}|U_n)$.

Proof   We again use the Shannon FCT to check that $\xi_n \xrightarrow{\mathbb{P}} H$ where H is given by (1.3.13), and $\xi_n = -\dfrac{1}{n} \log p_n(\mathbf{U}^{(n)})$, cf. (1.3.3b). In other words, condition (1.3.9) implies the asymptotic equipartition property for a Markov source.

The Markov property means that, for all string $\mathbf{u}^{(n)} = u_1 \ldots u_n$,

$$p_n(\mathbf{u}^{(n)}) = \lambda(u_1)P(u_1, u_2) \cdots P(u_{n-1}, u_n), \qquad (1.3.14a)$$

and $-\log p_n(\mathbf{u}^{(n)})$ is written as the sum

$$-\log \lambda(u_1) - \log P(u_1, u_2) - \cdots - \log P(u_{n-1}, u_n). \qquad (1.3.14b)$$

For a random string, $\mathbf{U}^{(n)} = U_1 \ldots U_n$, the random variable $-\log p_n(\mathbf{U}^{(n)})$ has a similar form:

$$-\log \lambda(U_1) - \log P(U_1, U_2) - \cdots - \log P(U_{n-1}, U_n). \qquad (1.3.15)$$

As in the case of a Bernoulli source, we denote

$$\sigma_1(U_1) := -\log \lambda(U_1), \quad \sigma_i(U_{i-1}, U_i) := -\log P(U_{i-1}, U_i), \quad i \geq 2, \qquad (1.3.16)$$

and write

$$\xi_n = \frac{1}{n}\left(\sigma_1 + \sum_{i=1}^{n-1} \sigma_{i+1}\right). \qquad (1.3.17)$$

The expected value of σ is

$$\mathbb{E}\,\sigma_1 = -\sum_u \lambda(u) \log \lambda(u) \qquad (1.3.18a)$$

and, as $\mathbb{P}(U_i = v) = \lambda P^{i-1}(v) = \sum_u \lambda(u) P^{(i-1)}(u, v)$,

$$\mathbb{E}\sigma_{i+1} = -\sum_{u,u'} \mathbb{P}(U_i = u, U_{i+1} = u') \log P(u, u')$$

$$= -\sum_{u,u'} (\lambda P^{i-1})(u) P(u, u') \log P(u, u'), \quad i \geq 1. \qquad (1.3.18b)$$

Theorem 1.3.10 implies that $\lim_{i \to \infty} \mathbb{E}\sigma_i = H$. Hence,

$$\lim_{n \to \infty} \mathbb{E}\xi_n = \lim_{n \to \infty} \frac{1}{n} \sum_{i=1}^{n} \mathbb{E}\sigma_i = H,$$

and the convergence $\xi_n \xrightarrow{\text{P}} H$ is again a law of large numbers, for the sequence (σ_i):

$$\lim_{n \to \infty} \mathbb{P}\left(\left|\frac{1}{n} \sum_{i=1}^{n} \sigma_i - H\right| \geq \varepsilon\right) = 0. \qquad (1.3.19)$$

However, the situation here is not as simple as in the case of a Bernoulli source. There are two difficulties to overcome: (i) $\mathbb{E}\sigma_i$ equals H only in the limit $i \to \infty$; (ii) $\sigma_1, \sigma_2, \ldots$ are no longer independent. Even worse, they do not form a DTMC, or even a Markov chain of a higher order. [A sequence $U_1, U_2, \ldots$ is said to form a DTMC of order k, if, for all $n \geq 1$,

$$\mathbb{P}(U_{n+k+1} = u' | U_{n+k} = u_k, \ldots, U_{n+1} = u_1, \ldots)$$
$$= \mathbb{P}(U_{n+k+1} = u' | U_{n+k} = u_k, \ldots, U_{n+1} = u_1).$$

An obvious remark is that, in a DTMC of order k, the vectors $\bar{U}_n = (U_n, U_{n+1}, \ldots, U_{n+k-1})$, $n \geq 1$, form an ordinary DTMC.] In a sense, the 'memory' in a sequence $\sigma_1, \sigma_2, \ldots$ is infinitely long. However, it decays exponentially: the precise meaning of this is provided in Worked Example 1.3.14.

Anyway, by using the Chebyshev inequality, we obtain

$$\mathbb{P}\left(\left|\frac{1}{n}\sum_{i=1}^{n}\sigma_i - H\right| \geq \varepsilon\right) \leq \frac{1}{n^2\varepsilon^2}\mathbb{E}\left(\sum_{i=1}^{n}(\sigma_i - H)\right)^2. \tag{1.3.20}$$

Theorem 1.3.11 immediately follows from Lemma 1.3.12 below.   □

Lemma 1.3.12   *The expectation value in the RHS of* (1.3.20) *satisfies the bound*

$$\mathbb{E}\left(\sum_{i=1}^{n}(\sigma_i - H)\right)^2 \leq Cn, \tag{1.3.21}$$

where $C > 0$ is a constant that does not depend on n.

Proof   See Worked Example 1.3.14.   □

By (1.3.21), the RHS of (1.3.20) becomes $\leq \dfrac{C}{n\varepsilon^2}$ and goes to zero as $n \to \infty$.

Worked Example 1.3.13   *Prove the following bound (cf.* (1.3.12)):

$$|P^{(n)}(u,v) - \pi(v)| \leq (1-\rho)^n. \tag{1.3.22}$$

Solution   (Compare with PSE II, p. 72.) First, observe that (1.3.12) implies the second bound in Theorem 1.3.10 as well as (1.3.10). Indeed, $\pi(v)$ is identified as the limit

$$\lim_{n\to\infty} P^{(n)}(u,v) = \lim_{n\to\infty}\sum_{\tilde{u}} P^{(n-1)}(u,\tilde{u})P(\tilde{u},v) = \sum_{\tilde{u}}\pi(\tilde{u})P(\tilde{u},v), \tag{1.3.23}$$

which yields (1.3.10). If $\pi'(1), \pi'(2), \ldots, \pi'(m)$ is another invariant probability vector, i.e.

$$0 \leq \pi'(u) \leq 1, \;\; \sum_{u=1}^{m}\pi'(u) = 1, \;\; \pi'(v) = \sum_{u}\pi'(u)P(u,v),$$

then $\pi'(v) = \sum_{u}\pi'(u)P^{(n)}(u,v)$ for all $n \geq 1$. The limit $n \to \infty$ gives then

$$\pi'(v) = \sum_{u}\pi'(u)\lim_{n\to\infty}P^{(n)}(u,v) = \sum_{u}\pi'(u)\pi(v) = \pi(v),$$

i.e. the invariant probability vector is unique.

To prove (1.3.22) denote

$$m_n(v) = \min_{u} P^{(n)}(u,v), \quad M_n(v) = \max_{u} P^{(n)}(u,v). \tag{1.3.24}$$

Then

$$m_{n+1}(v) = \min_{u} P^{(n+1)}(u,v) = \min_{u} \sum_{\tilde{u}} P(u,\tilde{u})P^{(n)}(\tilde{u},v)$$

$$\geq \min_{u} P^{(n)}(u,v) \sum_{\tilde{u}} P(u,\tilde{u}) = m_n(v).$$

Similarly,

$$M_{n+1}(v) = \max_{u} P^{(n+1)}(u,v) = \max_{u} \sum_{\tilde{u}} P(u,\tilde{u})P^{(n)}(\tilde{u},v)$$

$$\leq \max_{u} P^{(n)}(u,v) \sum_{\tilde{u}} P(u,\tilde{u}) = M_n(v).$$

Since $0 \leq m_n(v) \leq M_n(v) \leq 1$, both $m_n(v)$ and $M_n(v)$ have the limits

$$m(v) = \lim_{n\to\infty} m_n(v) \leq \lim_{n\to\infty} M_n(v) = M(v).$$

Furthermore, the difference $M(v) - m(v)$ is written as the limit

$$\lim_{n\to\infty} (M_n(v) - m_n(v)) = \lim_{n\to\infty} \max_{u,u'} (P^{(n)}(u,v) - P^{(n)}(u',v)).$$

So, if we manage to prove that

$$\max_{u,u',v} |P^{(n)}(u,v) - P^{(n)}(u',v)| \leq (1-\rho)^n, \tag{1.3.25}$$

then $M(v) = m(v)$ for each v. Furthermore, denoting the common value $M(v) = m(v)$ by $\pi(v)$, we obtain (1.3.22)

$$|P^{(n)}(u,v) - \pi(v)| \leq M_n(v) - m_n(v) \leq (1-\rho)^n.$$

    To prove (1.3.25), consider a DTMC on $I \times I$, with states (u_1, u_2), and transition probabilities

$$\mathbf{P}\big((u_1,u_2),(v_1,v_2)\big) = \begin{cases} P(u_1,v_1)P(u_2,v_2), & \text{if } u_1 \neq u_2, \\ P(u,v), & \text{if } u_1 = u_2 = u; \ v_1 = v_2 = v, \\ 0, & \text{if } u_1 = u_2 \text{ and } v_1 \neq v_2. \end{cases}$$

$$\tag{1.3.26}$$

It is easy to check that $\mathbf{P}\big((u_1,u_2),(v_1,v_2)\big)$ is indeed a transition probability matrix (of size $m^2 \times m^2$): if $u_1 = u_2 = u$ then

$$\sum_{v_1,v_2} \mathbf{P}\big((u_1,u_2),(v_1,v_2)\big) = \sum_{v} P(u,v) = 1$$

whereas if $u_1 \neq u_2$ then

$$\sum_{v_1,v_2} \mathbf{P}\big((u_1,u_2),(v_1,v_2)\big) = \sum_{v_1} P(u_1,v_1) \sum_{v_2} P(u_2,v_2) = 1$$

(the inequalities $0 \leq \mathbf{P}\big((u_1, u_2), (v_1, v_2)\big) \leq 1$ follow directly from the definition (1.3.26)).

This is the so-called *coupled* DTMC on $I \times I$; we denote it by (V_n, W_n), $n \geq 1$. Observe that both components V_n and W_n are DTMCs with transition probabilities $P(u, v)$. More precisely, the components V_n and W_n move independently, until the first (random) time τ when they coincide; we call it the coupling time. After time τ the components V_n and W_n 'stick' together and move synchronously, again with transition probabilities $P(u, v)$.

Suppose we start the coupled chain from a state (u, u'). Then

$$|P^{(n)}(u, v) - P^{(n)}(u', v)|$$
$$= |\mathbb{P}(V_n = v | V_1 = u, W_1 = u') - \mathbb{P}(W_n = v | V_1 = u, W_1 = u')|$$

(because each component of (V_n, W_n) moves with the same transition probabilities)

$$= |\mathbb{P}(V_n = v, W_n \neq v | V_1 = u, W_1 = u')$$
$$\qquad - \mathbb{P}(V_n \neq v, W_n = v | V_1 = u, W_1 = u')|$$
$$\leq \mathbb{P}(V_n \neq W_n | V_1 = u, W_1 = u')$$
$$= \mathbb{P}(\tau > n | V_1 = u, W_1 = u'). \tag{1.3.27}$$

Now, the probability obeys

$$\mathbb{P}(\tau = 1 | V_1 = u, W_1 = u') \geq \sum_v P(u, v) P(u', v) \geq \rho \sum_v P(u', v) = \rho,$$

i.e. the complementary probability satisfies

$$\mathbb{P}(\tau > 1 | V_1 = u, W_1 = u') \leq 1 - \rho.$$

By the strong Markov property (of the coupled chain),

$$\mathbb{P}(\tau > n | V_1 = u, W_1 = u') \leq (1 - \rho)^n. \tag{1.3.28}$$

Bounds (1.3.28) and (1.3.27) together give (1.3.25). $\qquad\qquad\square$

Worked Example 1.3.14 *Under condition (1.3.9) with $r = 1$ prove the following bound:*

$$\left|\mathbb{E}\big[(\sigma_i - H)(\sigma_{i+k} - H)\big]\right| \leq \big(H + |\log \rho|\big)^2 (1 - \rho)^{k-1}. \tag{1.3.29}$$

Solution For brevity, we assume $i > 1$; the case $i = 1$ requires minor changes. Returning to the definition of random variables σ_i, $i > 1$, write

$$\mathbb{E}\left[(\sigma_i - H)(\sigma_{i+k} - H)\right]$$
$$= \sum_{u,u'}\sum_{v,v'} \mathbb{P}(U_i = u, U_{i+1} = u'; U_{i+k} = v, U_{i+k+1} = v')$$
$$\times \left(-\log P(u, u') - H\right)\left(-\log P(v, v') - H\right). \quad (1.3.30)$$

Our goal is to compare this expression with

$$\sum_{u,u'}\sum_{v,v'}\left(\lambda P^{i-1}\right)(u)P(u, u')\left[-\log P(u, u') - H\right]$$
$$\times \pi(v)P(v, v')\left[-\log P(v, v') - H\right]. \quad (1.3.31)$$

Observe that (1.3.31) in fact vanishes because the sum $\sum\limits_{v,v'}$ vanishes due to the definition (1.3.13) of H.

The difference between sums (1.3.30) and (1.3.31) comes from the fact that the probabilities

$$\mathbb{P}(U_i = u, U_{i+1} = u'; U_{i+k} = v, U_{i+k+1} = v')$$
$$= \left(\lambda P^{i-1}\right)(u)P(u, u')P^{(k-1)}(u', v)P(v, v')$$

and

$$\left(\lambda P^{i-1}\right)(u)P(u, u')\pi(v)P(v, v')$$

do not coincide. However, the difference of these probabilities in absolute value does not exceed

$$|P^{(k-1)}(u', v) - \pi(v)| \le (1 - \rho)^{k-1}.$$

As $|-\log P(\cdot, \cdot) - H| \le H + |\log \rho|$, we obtain (1.3.29). $\qquad \square$

Proof of Theorem 1.3.11. This is now easy to complete. To prove (1.3.21), expand the square and use the additivity of the expectation:

$$\mathbb{E}\left[\sum_{i=1}^{n}(\sigma_i - H)\right]^2 = \sum_{1 \le i \le n} \mathbb{E}\left[(\sigma_i - H)^2\right]$$
$$+ 2\sum_{1 \le i < j \le n} \mathbb{E}\left[(\sigma_i - H)(\sigma_j - H)\right]. \quad (1.3.32)$$

The first sum in (1.3.32) is OK: it contains n terms $\mathbb{E}(\sigma_i - H)^2$ each bounded by a constant (say, C' may be taken to be $(H + |\log \rho|)^2$). Thus this sum is at most $C'n$.

It is the second sum that causes problems: it contains $n(n-1)/2$ terms. We bound it as follows:

$$\left| \sum_{1 \leq i < j \leq n} \mathbb{E}\left[(\sigma_i - H)(\sigma_j - H)\right] \right| \leq \sum_{i=1}^{n} \left(\sum_{k=1}^{\infty} \left| \mathbb{E}\left[(\sigma_i - H)(\sigma_{i+k} - H)\right] \right| \right), \quad (1.3.33)$$

and use (1.3.29) to finish the proof. $\qquad\square$

Our next theorem shows the role of the (relative) entropy in the asymptotic analysis of probabilities; see PSE I, p. 82.

Theorem 1.3.15   Let $\zeta_1, \zeta_2, \ldots$ *be a sequence of IID random variables taking values 0 and 1 with probabilities* $1-p$ *and* p, *respectively*, $0 < p < 1$. *Then, for any sequence* k_n *of positive integers such that* $k_n \to \infty$ *and* $n - k_n \to \infty$ *as* $n \to \infty$,

$$\mathbb{P}\left(\sum_{i=1}^{n} \zeta_i = k_n \right) \sim \left(2\pi np^*(1-p^*)\right)^{-1/2} \exp\left(-nD(p\|p^*)\right). \quad (1.3.34)$$

Here, $\sim$ *means that the ratio of the left- and right-hand sides tends to 1 as* $n \to \infty$, $p^*(= p_n^*)$ *denotes the ratio* $\dfrac{k_n}{n}$, *and* $D(p\|p^*)$ *stands for the relative entropy* $h(X\|Y)$ *where* X *is distributed as* ζ_i *(i.e. it takes values 0 and 1 with probabilities* $1-p$ *and* p), *while* Y *takes the same values with probabilities* $1 - p^*$ *and* p^*.

Proof   Use Stirling's formula (see PSE I, p.72):

$$n! \sim \sqrt{2\pi n} n^n e^{-n}. \quad (1.3.35)$$

[In fact, this formula admits a more precise form: $n! = \sqrt{2\pi n} n^n e^{-n+\theta(n)}$, where $\dfrac{1}{12n+1} < \theta(n) < \dfrac{1}{12n}$, but for our purposes (1.3.35) is enough.] Then the probability in the LHS of (1.3.34) is (for brevity, the subscript n in k_n is omitted)

$$\binom{n}{k} p^k(1-p)^{n-k} \sim \left(\frac{n}{2\pi k(n-k)}\right)^{1/2} \frac{n^n}{k^k(n-k)^{n-k}} p^k(1-p)^{n-k}$$

$$= \left(2\pi np^*(1-p^*)\right)^{-1/2}$$
$$\times \exp\left[-k\ln k/n - (n-k)\ln(n-k)/n + k\ln p + (n-k)\ln(1-p)\right].$$

But the RHS of the last formula coincides with the RHS of (1.3.34). $\qquad\square$

If p^* is close to p, we can write

$$D(p\|p^*) = \frac{1}{2}\left(\frac{1}{p} + \frac{1}{1-p}\right)(p^* - p)^2 + O(|p^* - p|^3), \quad (1.3.36)$$

as $D(p\|p^*)|_{p^*=p} = \left(\dfrac{d}{dp^*} D(p\|p^*)\right)\Big|_{p^*=p} = 0$, and immediately obtain

Corollary 1.3.16   (The local De Moivre–Laplace theorem; cf. PSE I, p. 81) *If* $n(p^* - p) = k_n - np = o(n^{2/3})$ *then*

$$\mathbb{P}\left(\sum_{i=1}^{n} \zeta_i = k_n\right) \sim \frac{1}{\sqrt{2\pi np(1-p)}} \exp\left(-\frac{n}{2p(1-p)}(p^* - p)^2\right). \qquad (1.3.37)$$

Worked Example 1.3.17   *At each time unit a device reads the current version of a string of N characters each of which may be either 0 or 1. It then transmits the number of characters which are equal to 1. Between each reading the string is perturbed by changing one of the characters at random (from 0 to 1 or vice versa, with each character being equally likely to be changed). Determine an expression for the information rate of this source.*

Solution The source is Markov, with the state space $\{0, 1, \ldots, N\}$ and the transition probability matrix

$$\begin{pmatrix} 0 & 1 & 0 & 0 & \cdots & 0 & 0 \\ 1/N & 0 & (N-1)/N & 0 & \cdots & 0 & 0 \\ 0 & 2/N & 0 & (N-2)/N & \cdots & 0 & 0 \\ & & \cdots & & \cdots & & \\ 0 & 0 & 0 & 0 & \cdots & 0 & 1/N \\ 0 & 0 & 0 & 0 & \cdots & 1 & 0 \end{pmatrix}.$$

The DTMC is irreducible and periodic. It possesses a unique invariant distribution

$$\pi_i = 2^{-N}\binom{N}{i}, \quad 0 \le i \le N.$$

By Theorem 1.3.11,

$$H = -\sum_{i,j} \pi_i P(i,j) \log P(i,j) = 2^{1-N} \frac{1}{N} \sum_{j=1}^{N-1} \binom{N}{j} j \log \frac{N}{j}.$$

$\square$

Worked Example 1.3.18   *A stationary source emits symbols $0, 1, \ldots, m$ ($m \ge 4$ is an even number), according to a DTMC, with the following transition probabilities $p_{jk} = P(U_{n+1} = k \mid U_n = j)$:*

$$p_{jj+2} = 1/3, \ 0 \le j \le m-2, \ \ p_{jj-2} = 1/3, \ 2 \le j \le m,$$

$$p_{jj} = 1/3, \ 2 \le j \le m-2, \ \ p_{00} = p_{11} = p_{m-1m-1} = p_{mm} = 2/3.$$

The distribution of the first symbol is equiprobable. Find the information rate of the source. Does the result contradict Shannon's FCT?

How does the answer change if m is odd? How can you use, for m odd, Shannon's FCT to derive the information rate of the above source?

Solution For m even, the DTMC is *reducible*: there are two communicating classes, $I_1 = \{0, 2, \ldots, m\}$ with $m/2 + 1$ states, and $I_2 = \{1, 3, \ldots, m-1\}$ with $m/2$ states. Correspondingly, for any set A_n of n-strings,

$$\mathbb{P}(A_n) = q\mathbb{P}_1(A_{n1}) + (1-q)\mathbb{P}_2(A_{n2}), \qquad (1.3.38)$$

where $A_{n1} = A_n \cap I_1$ and $A_{n1} = A_n \cap I_2$; $\mathbb{P}_i$ refers to the DTMC on class I_i, $i = 1, 2$, and $q = \mathbb{P}(U_1 \in I_1)$.

The random variable from (1.3.3b) is $\xi_n = -\frac{1}{n}\log p_n(\mathbf{U}^{(n)})$; according to (1.3.38),

$$\xi_n = -\frac{1}{n}\log p_{n1}(\mathbf{U}^{(n)}) \text{ with probability } q,$$
$$= -\frac{1}{n}\log p_{n2}(\mathbf{U}^{(n)}) \text{ with probability } 1-q. \qquad (1.3.39)$$

Both DTMCs are irreducible and aperiodic on their communicating classes and their invariant distributions are uniform:

$$\pi_i^{(1)} = \frac{2}{m+2}, \ i \in I_1, \ \pi_i^{(2)} = \frac{2}{m}, \ i \in I_2.$$

Their information rates equal, respectively,

$$H^{(1)} = \log 3 - \frac{8}{3(m+2)} \text{ and } H^{(2)} = \log 3 - \frac{8}{3m}. \qquad (1.3.40)$$

As follows from (1.3.38), the information rate of the whole DTMC equals

$$H_{\text{odd}} = \begin{cases} H^{(1)} = \max\left[H^{(1)}, H^{(2)}\right], & \text{if } 0 < q \leq 1, \\ H^{(2)}, & \text{if } q = 0. \end{cases} \qquad (1.3.41)$$

For $0 < q < 1$ Shannon's FCT is not applicable:

$$-\frac{1}{n}\log p_{n1}(\mathbf{U}^{(n)}) \xrightarrow{\mathbb{P}_1} H^{(1)} \text{ whereas } -\frac{1}{n}\log p_{n2}(\mathbf{U}^{(n)}) \xrightarrow{\mathbb{P}_2} H^{(2)},$$

i.e. ξ_n converges to a non-constant limit. However, if $q(1-q) = 0$, then (1.3.41) is reduced to a single line, and Shannon's FCT is applicable: ξ_n converges to the corresponding constant $H^{(i)}$.

If m is odd, again there are two communicating classes, $I_1 = \{0, 2, \ldots, m-1\}$ and $I_2 = \{1, 3, \ldots, m\}$, each of which now contains $(m+1)/2$ states. As before,

DTMCs $\mathbb{P}_1$ and $\mathbb{P}_2$ are irreducible and aperiodic and their invariant distributions are uniform:

$$\pi_i^{(1)} = \frac{2}{m+1}, \quad i \in I_1, \quad \pi_i^{(2)} = \frac{2}{m+1}, \quad i \in I_2.$$

Their common information rate equals

$$H_{\text{odd}} = \log 3 - \frac{8}{3(m+1)}, \tag{1.3.42}$$

which also gives the information rate of the whole DTMC. It agrees with Shannon's FCT, because now

$$\xi_n = -\frac{1}{n} \log p_n(\mathbf{U}^{(n)}) \xrightarrow{\mathbb{P}} H_{\text{odd}}. \tag{1.3.43}$$

$\square$

Worked Example 1.3.19 *Let a be the size of A and b the size of the alphabet B. Consider a source with letters chosen from an alphabet $A + B$, with the constraint that no two letters of A should ever occur consecutively.*

(a) *Suppose the message follows a DTMC, all characters which are permitted at a given place being equally likely. Show that this source has information rate*

$$H = \frac{a \log b + (a+b) \log(a+b)}{2a+b}. \tag{1.3.44}$$

(b) *By solving a recurrence relation, or otherwise, find how many strings of length n satisfy the constraint that no two letters of A occur consecutively. Suppose these strings are equally likely and let $n \to \infty$. Show that the limiting information rate becomes*

$$H = \log \left(\frac{b + \sqrt{b^2 + 4ab}}{2} \right).$$

Why are the answers different?

Solution (a) The transition probabilities of the DTMC are given by

$$P(x,y) = \begin{cases} 0, & \text{if } x,y \in \{1,\ldots,a\}, \\ 1/b, & \text{if } x \in \{1,\ldots,a\}, \ y \in \{a+1,\ldots,a+b\}, \\ 1/(a+b), & \text{if } x \in \{a+1,\ldots,a+b\}, y \in \{1,\ldots,a+b\}. \end{cases}$$

The chain is irreducible and aperiodic. Moreover, $\min P^{(2)}(x,y) > 0$; hence, an invariant distribution $\pi = (\pi(x), x \in \{1,\ldots,a+b\})$ is unique. We can find π from the *detailed balance* equations (DBEs) $\pi(x)P(x,y) = \pi(y)P(y,x)$ (cf. PSE II, p. 82), which yields

$$\pi(x) = \begin{cases} 1/(2a+b), & x \in \{1,\ldots,a\}, \\ (a+b)/[b(2a+b)], & x \in \{a+1,\ldots,a+b\}. \end{cases}$$

The DBEs imply that π is invariant: $\pi(y) = \sum_x \pi(x)P(x,y)$, but not vice versa. Thus, we obtain (1.3.44).

(b) Let M_n denote the number of allowed n-strings, A_n the number of allowed n-strings ending with a letter from A, and B_n the number of allowed n-strings ending with a letter from B. Then

$$M_n = A_n + B_n, \quad A_{n+1} = aB_n, \quad \text{and} \quad B_{n+1} = b(A_n + B_n),$$

which yields

$$B_{n+1} = bB_n + abB_{n-1}.$$

The last recursion is solved by

$$B_n = c_+\lambda_+^n + c_-\lambda_-^n,$$

where $\lambda_\pm$ are the eigenvalues of the matrix

$$\begin{pmatrix} 0 & ab \\ 1 & b \end{pmatrix},$$

i.e.

$$\lambda_\pm = \frac{b \pm \sqrt{b^2 + 4ab}}{2},$$

and $c_\pm$ are constants, $c_+ > 0$. Hence,

$$\begin{aligned} M_n &= a\left(c_+\lambda_+^{n-1} + c_-\lambda_-^{n-1}\right) + \left(c_+\lambda_+^n + c_-\lambda_-^n\right) \\ &= \lambda_+^n\left(c_-\left(a\frac{\lambda_-^{n-1}}{\lambda_+^n} + \frac{\lambda_-^n}{\lambda_+^n}\right) + c_+\left(a\frac{1}{\lambda_+} + 1\right)\right), \end{aligned}$$

and $\frac{1}{n}\log M_n$ is represented as the sum

$$\log\lambda_+ + \frac{1}{n}\log\left(c_-\left(a\frac{\lambda_-^{n-1}}{\lambda_+^n} + \frac{\lambda_-^n}{\lambda_+^n}\right) + c_+\left(a\frac{1}{\lambda_+} + 1\right)\right).$$

Note that $\left|\frac{\lambda_-}{\lambda_+}\right| < 1$. Thus, the limiting information rate equals

$$\lim_{n\to\infty} \frac{1}{n}\log M_n = \log\lambda_+.$$

The answers are different since the conditional equidistribution results in a strong dependence between subsequent letters: they do *not* form a DTMC. ☐

Worked Example 1.3.20 Let $\{U_j: j = 1, 2, \ldots\}$ be an irreducible and aperiodic DTMC with a finite state space. Given $n \geq 1$ and $\alpha \in (0, 1)$, order the strings $\mathbf{u}^{(n)}$ according to their probabilities $\left(\mathbb{P}(\mathbf{U}^{(n)} = \mathbf{u}_1^{(n)}) \geq \mathbb{P}(\mathbf{U}^{(n)} = \mathbf{u}_2^{(n)}) \geq \cdots \right)$ and select them in this order until the probability of the remaining set becomes $\leq 1 - \alpha$. Let $M_n(\alpha)$ denote the number of the selected strings. Prove that $\lim_{n \to \infty} \frac{1}{n} \log M_n(\alpha) = H$, the information rate of the source,

(a) *in the case where the rows of the transition probability matrix P are all equal (i.e. $\{U_j\}$ is a Bernoulli sequence),*

(b) *in the case where the rows of P are permutations of each other, and in a general case. Comment on the significance of this result for coding theory.*

Solution (a) Let $\mathbb{P}$ stand for the probability distribution of the IID sequence (U_n) and set $H = -\sum_{j=1}^{m} p_j \log p_j$ (the binary entropy of the source). Fix $\varepsilon > 0$ and partition the set $I^{\times n}$ of all n-strings into three disjoint subsets:

$$\mathscr{K}_+ = \{\mathbf{u}^{(n)} : p(\mathbf{u}^{(n)}) \geq 2^{-n(H-\varepsilon)}\}, \quad \mathscr{K}_- = \{\mathbf{u}^{(n)} : p(\mathbf{u}^{(n)}) \leq 2^{-n(H+\varepsilon)}\},$$

and

$$\mathscr{K} = \{\mathbf{u}^{(n)} : 2^{-n(H+\varepsilon)} < p(\mathbf{u}^{(n)}) < 2^{-n(H-\varepsilon)}\}.$$

By the law of large numbers (or asymptotic equipartition property), $-\frac{1}{n} \log \mathbb{P}(\mathbf{U}^{(n)})$ converges to $H(= h)$, i.e. $\lim_{n \to \infty} \mathbb{P}(\mathscr{K}_+ \cup \mathscr{K}_-) = 0$, and $\lim_{n \to \infty} \mathbb{P}(\mathscr{K}) = 1$. Thus, to obtain probability $\geq \alpha$, for n large enough, you (i) cannot restrict yourself to $\mathscr{K}_+$ and have to borrow strings from $\mathscr{K}$, (ii) don't need strings from $\mathscr{K}_-$, i.e. will have the last selected string from $\mathscr{K}$. Denote by $\mathscr{M}_n(\alpha)$ the set of selected strings, and $\sharp \mathscr{M}_n(\alpha)$ by M_n. You have two two-side bounds

$$\alpha \leq \mathbb{P}(\mathscr{M}_n(\alpha)) \leq \alpha + 2^{-n(H-\varepsilon)}$$

and

$$2^{-n(H+\varepsilon)} M_n(\alpha) \leq \mathbb{P}(\mathscr{M}_n(\alpha)) \leq \mathbb{P}(\mathscr{K}_+) + 2^{-n(H-\varepsilon)} M_n(\alpha).$$

Excluding $\mathbb{P}(\mathscr{M}_n(\alpha))$ yields

$$2^{-n(H+\varepsilon)} M_n(\alpha) \leq \alpha + 2^{-n(H-\varepsilon)} \text{ and } 2^{-n(H-\varepsilon)} M_n(\alpha) \geq \alpha - \mathbb{P}(\mathscr{K}_+).$$

These inequalities imply, respectively,

$$\limsup_{n\to\infty} \frac{1}{n}\log M_n(\alpha) \leq H + \varepsilon \quad \text{and} \quad \liminf_{n\to\infty} \frac{1}{n}\log M_n(\alpha) \geq H - \varepsilon.$$

As ε is arbitrary, the limit is H.

(b) The argument may be repeated without any change in the case of permutations because the ordered probabilities form the same set as in case (a), and in a general case by applying the law of large numbers to $(1/n)\xi_n$; cf. (1.3.3b) and (1.3.19). Finally, the significance for coding theory: if we are prepared to deal with the error-probability $\leq \alpha$, we do not need to encode all m^n string $\mathbf{u}^{(n)}$ but only $\sim 2^{nH}$ most frequent ones. As $H \leq \log m$ (and in many cases $\ll \log m$), it yields a significant economy in storage space (data-compression). $\qquad\square$

Worked Example 1.3.21   *A binary source emits digits 0 or 1 according to the rule*

$$P(X_n = k | X_{n-1} = j, X_{n-2} = i) = q_r,$$

where k, j, i and r take values 0 or 1, $r = k - j - i \mod 2$, and $q_0 + q_1 = 1$. Determine the information rate of the source.

Also derive the information rate of a binary Bernoulli source, emitting digits 0 and 1 with probabilities q_0 and q_1. Explain the relationship between these two results.

Solution The source is a DTMC of the second order. That is, the pairs (X_n, X_{n+1}) form a four-state DTMC, with

$$P(00,00) = q_0, \; P(00,01) = q_1, \; P(01,10) = q_0, \; P(01,11) = q_1,$$
$$P(10,00) = q_0, \; P(10,01) = q_1, \; P(11,10) = q_0, \; P(11,11) = q_1;$$

the remaining eight entries of the transition probability matrix vanish. This gives

$$H = -q_0 \log q_0 - q_1 \log q_1.$$

For a Bernoulli source the answer is the same. $\qquad\square$

Worked Example 1.3.22   *Find an entropy rate of a DTMC associated with a random walk on the 3×3 chessboard:*

$$\begin{pmatrix} 1 & 2 & 3 \\ 4 & 5 & 6 \\ 7 & 8 & 9 \end{pmatrix}. \qquad\qquad (1.3.45)$$

Find the entropy rate for a rook, bishop (both kinds), queen and king.

Solution We consider the king's DTMC only; other cases are similar. The transition probability matrix is

$$
\begin{pmatrix}
0 & 1/3 & 0 & 1/3 & 1/3 & 0 & 0 & 0 & 0 \\
1/5 & 0 & 1/5 & 1/5 & 1/5 & 1/5 & 0 & 0 & 0 \\
0 & 1/3 & 0 & 0 & 1/3 & 1/3 & 0 & 0 & 0 \\
1/5 & 1/5 & 0 & 0 & 1/5 & 0 & 1/5 & 1/5 & 0 \\
1/8 & 1/8 & 1/8 & 1/8 & 0 & 1/8 & 1/8 & 1/8 & 1/8 \\
0 & 1/5 & 1/5 & 0 & 1/5 & 0 & 0 & 1/5 & 1/5 \\
0 & 0 & 0 & 1/3 & 1/3 & 0 & 0 & 0 & 1/3 \\
0 & 0 & 0 & 1/5 & 1/5 & 1/5 & 1/5 & 0 & 1/5 \\
0 & 0 & 0 & 0 & 1/3 & 1/3 & 0 & 1/3 & 0
\end{pmatrix}
$$

By symmetry the invariant distribution is $\pi_1 = \pi_3 = \pi_9 = \pi_7 = \lambda$, $\pi_4 = \pi_2 = \pi_6 = \pi_8 = \mu$, $\pi_5 = \nu$, and by the DBEs

$$
\lambda/3 = \mu/5, \ \lambda/3 = \nu/8, \ 4\lambda + 4\mu + \nu = 1
$$

implies $\lambda = \frac{3}{40}$, $\mu = \frac{1}{8}$, $\nu = \frac{1}{5}$. Now

$$
H = -4\lambda \frac{1}{3}\log\frac{1}{3} - 4\mu\frac{1}{5}\log\frac{1}{5} - \nu\frac{1}{8}\log\frac{1}{8} = \frac{1}{10}\log 15 + \frac{3}{40}.
$$

□

1.4 Channels of information transmission. Decoding rules. Shannon's second coding theorem

In this section we prove a core statement of Shannon's theory: the second coding theorem (SCT), also known as the noisy coding theorem (NCT). Shannon stated its assertion and gave a sketch of its proof in his papers and books in the 1940s. His argument was subject to (not entirely unjustified) criticism by professional mathematicians. It took the mathematical community about a decade to produce a rigorous and complete proof of the SCT. However, with hindsight, one cannot stop admiring Shannon's intuition and his firm grasp of fundamental notions such as entropy and coding as well their relation to statistics of long random strings. We point at various aspects of this topic, not avoiding a personal touch palpable in writings of the main players in this area.

So far, we have considered a source emitting a random text $U_1 U_2 \ldots$, and an encoding of a message $\mathbf{u}^{(n)}$ by a binary codeword $\mathbf{x}^{(N)}$ using a code $f_n : I^{\times n} \to J^{\times N}$, $J = \{0,1\}$. Now we focus upon the relation between the length of a message n and the codeword-length N: it is determined by properties of the *channel* through which the information is sent. It is important to remember that the code f_n is supposed to be known to the receiver.

Typically, a channel is subject to 'noise' which distorts the messages transmitted: a message at the output differs in general from the message at the input. Formally, a channel is characterised by a conditional distribution

$$\mathbf{P}_{\text{ch}}\left(\text{receive word } \mathbf{y}^{(N)}|\text{codeword } \mathbf{x}^{(N)} \text{ sent}\right); \qquad (1.4.1)$$

we again suppose that it is known to both sender and receiver. (We use a distinct symbol $\mathbf{P}_{\text{ch}}\left(\,\cdot\,|\text{codeword } \mathbf{x}^{(N)} \text{ sent}\right)$ or, briefly, $\mathbf{P}_{\text{ch}}\left(\,\cdot\,|\mathbf{x}^{(N)}\right)$, to stress that this probability distribution is generated by the channel, conditional on the event that codeword $\mathbf{x}^{(N)}$ has been sent.) Speaking below of a channel, we refer to a conditional probability (1.4.1) (or rather a family of conditional probabilities, depending on N). Consequently, we use the symbol $\mathbf{Y}^{(N)}$ for a random string representing the output of the channel; given that a word $\mathbf{x}^{(N)}$ was sent,

$$\mathbf{P}_{\text{ch}}(\mathbf{Y}^{(N)} = \mathbf{y}^{(N)}|\mathbf{x}^{(N)}) = \mathbf{P}_{\text{ch}}(\mathbf{y}^{(N)}|\mathbf{x}^{(N)}).$$

An important example is the so-called *memoryless binary channels* (MBCs) where

$$\mathbf{P}_{\text{ch}}\left(\mathbf{y}^{(N)}|\mathbf{x}^{(N)}\right) = \prod_{i=1}^{N} P(y_i|x_i), \qquad (1.4.2)$$

if $\mathbf{y}^{(N)} = y_1 \ldots y_N$, $\mathbf{x}^{(N)} = x_1 \ldots x_N$. Here, $P(y|x)$, $x, y = 0, 1$, is a symbol-to-symbol channel probability (i.e. the conditional probability to have symbol y at the output of the channel given that symbol x has been sent). Clearly, $\{P(y|x)\}$ is a 2×2 stochastic matrix (often called the channel matrix). In particular, if $P(1|0) = P(0|1) = p$, the channel is called *symmetric* (MBSC). The channel matrix then has the form

$$\Pi = \begin{pmatrix} 1-p & p \\ p & 1-p \end{pmatrix}$$

and p is called the row error-probability (or the symbol-error-probability).

Example 1.4.1   Consider the memoryless channel, where $Y = X + Z$, and an additive noise Z takes values 0 and a with probability $1/2$; a is a given real number. The input alphabet is $\{0, 1\}$ and Z is independent of X.

Properties of this channel depend on the value of a. Indeed, if $a \neq \pm 1$, the channel is uniquely decodable. In other words, if we have to use the channel for transmitting messages (strings) of length n (there are 2^n of them altogether) then any message can be sent straightaway, and the receiver will be able to recover it. But if $a = \pm 1$, there are errors possible, and to make sure that the receiver can recover our message we have to encode it, which, typically, results in increasing the length of the string sent into the channel, from n to N, say.

In other words, strings of length N sent to the channel will be codewords representing source messages of a shorter length n. The maximal ratio n/N which still
allows the receiver to recover the original message is an important characteristic of
the channel, called the *capacity*. As we will see, passing from $a \neq \pm 1$ to $a = \pm 1$
changes the capacity from 1 (no encoding needed) to $1/2$ (where the codeword-
length is twice as long as the length of the source message).

So, we need to introduce a decoding rule $\widehat{f}_N : J^{\times N} \to I^{\times n}$ such that the overall
probability of error $\varepsilon (= \varepsilon(f_n, \widehat{f}_N, \mathbb{P}))$ defined by

$$\varepsilon = \sum_{\mathbf{u}^{(n)}} \mathbb{P}\big(\widehat{f}_N(\mathbf{Y}^{(N)}) \neq \mathbf{u}^{(n)}, \mathbf{u}^{(n)} \text{ emitted}\big)$$

$$= \sum_{\mathbf{u}^{(n)}} \mathbb{P}\big(\mathbf{U}^{(n)} = \mathbf{u}^{(n)}\big) \mathbf{P}_{\mathrm{ch}}\big(\widehat{f}_N(\mathbf{Y}^{(N)}) \neq \mathbf{u}^{(n)} \mid f_n(\mathbf{u}^{(n)}) \text{ sent}\big) \qquad (1.4.3)$$

is small. We will try (and under certain conditions succeed) to have the error-
probability (1.4.3) tending to zero as $n \to \infty$.

The idea which is behind the construction is based on the following facts:

(1) For a source with the asymptotic equipartition property the number of distinct n-strings emitted is $2^{n(H+o(1))}$ where $H \leq \log m$ is the information rate of
the source. Therefore, we have to encode not $m^n = 2^{n \log m}$ messages, but only
$2^{n(H+o(1))}$ which may be considerably less. That is, the code f_n may be defined
on a subset of $I^{\times n}$ only, with the codeword-length $N = \lceil nH \rceil$.

(2) We may try even a larger N: $N = \lceil \overline{R}^{-1} nH \rceil$, where $\overline{R}$ is a constant with $0 < \overline{R} <$
1. In other words, the increasing length of the codewords used from $\lceil nH \rceil$ to
$\lceil \overline{R}^{-1} nH \rceil$ will allow us to introduce a redundancy in the code f_n, and we may
hope to be able to use this redundancy to diminish the overall error-probability
(1.4.3) (provided that in addition a decoding rule is 'good'). It is of course
desirable to minimise $\overline{R}^{-1}$, i.e. maximise $\overline{R}$: it will give the codes with optimal
parameters. The question of how large $\overline{R}$ is allowed to be depends of course on
the channel.

It is instrumental to introduce a notational convention. As the codeword-length is
a crucial parameter, we write N instead of $\overline{R}^{-1} Hn$ and $\overline{R}N$ instead of Hn: the number of distinct strings emitted by the source becomes $2^{N(\overline{R}+o(1))}$. In future, the index
$n \sim \dfrac{N\overline{R}}{H}$ will be omitted wherever possible (and replaced by N otherwise). It is convenient to consider a 'typical' set $\mathcal{U}_N$ of distinct strings emitted by the source, with
$\sharp \mathcal{U}_N = 2^{N(\overline{R}+o(1))}$. Formally, $\mathcal{U}_N$ can include strings of different length; it is only
the log-asymptotics of $\sharp \mathcal{U}_N$ that matter. Accordingly, we will omit the superscript
(n) in the notation $\mathbf{u}^{(n)}$.

Definition 1.4.2 A value $\overline{R} \in (0,1)$ is called a *reliable transmission rate* (for a given channel) if, given that the source strings take *equiprobable* values from a set $\mathcal{U}_N$ with $\sharp\,\mathcal{U}_N = 2^{N(\overline{R}+o(1))}$, there exist an encoding rule $f_N : \mathcal{U}_N \to \mathcal{X}_N \subseteq J^{\times N}$ and a decoding rule $\widehat{f}_N : J^{\times N} \to \mathcal{U}_N$ with the error-probability

$$\sum_{\mathbf{u}\in\mathcal{U}_N} \frac{1}{\sharp\,\mathcal{U}_N} \mathbf{P}_{\mathrm{ch}}\left(\widehat{f}_N(\mathbf{Y}^{(N)}) \neq \mathbf{u}\,\big|\,f_N(\mathbf{u})\ \text{sent}\right) \tag{1.4.4}$$

tending to zero as $N \to \infty$. That is, for each sequence $\mathcal{U}_N$ with $\lim\limits_{N\to\infty} \frac{1}{N}\log\sharp\,\mathcal{U}_N = \overline{R}$, there exist a sequence of encoding rules $f_N : \mathcal{U}_N \to \mathcal{X}_N$, $\mathcal{X}_N \subseteq J^{\times N}$, and a sequence of decoding rules $\widehat{f}_N : J^{\times N} \to \mathcal{U}_N$ such that

$$\lim_{N\to\infty} \frac{1}{\sharp\,\mathcal{U}_N} \sum_{\mathbf{u}\in\mathcal{U}_N}\sum_{\mathbf{Y}^{(N)}} \mathbf{1}\big(\widehat{f}_N(\mathbf{Y}^{(N)}) \neq \mathbf{u}\big)\mathbf{P}_{\mathrm{ch}}\left(\mathbf{Y}^{(N)}\,\big|\,f_N(\mathbf{u})\right) = 0. \tag{1.4.5}$$

Definition 1.4.3 The *channel capacity* is defined as the supremum

$$C = \sup\left[\overline{R} \in (0,1) : \overline{R}\ \text{is a reliable transmission rate}\,\right]. \tag{1.4.6}$$

Remark 1.4.4 (a) Physically speaking, the channel capacity can be thought of as a limit $\lim\limits_{N\to\infty} \frac{1}{N}\log n(N)$ where $n(N)$ is the maximal number of strings of length N which can be sent through the channel with a vanishing probability of erroneous decoding.

(b) The reason for the equiprobable distribution on $\mathcal{U}_N$ is that it yields the worst-case scenario. See Theorem 1.4.6 below.

(c) If encoding rule f_N used is one-to-one (lossless) then it suffices to treat the decoding rules as maps $J^{\times N} \to \mathcal{X}_N$ rather than $J^{\times N} \to \mathcal{U}_N$: if we guess correctly what codeword $\mathbf{x}^{(N)}$ has been sent, we simply set $\mathbf{u} = f_N^{-1}(\mathbf{x}^{(N)})$. If, in addition, the source distribution is equiprobable over $\mathcal{U}$ then the error-probability ε can be written as an average over the set of codewords $\mathcal{X}_N$:

$$\varepsilon = \frac{1}{\sharp\,\mathcal{X}} \sum_{\mathbf{x}\in\mathcal{X}_N} \left[1 - \mathbf{P}_{\mathrm{ch}}\left(\widehat{f}_N(\mathbf{Y}^{(N)}) = \mathbf{x}\,\big|\,\mathbf{x}\ \text{sent}\right)\right].$$

Accordingly, it makes sense to write $\varepsilon = \varepsilon^{\mathrm{ave}}$ and speak about the average probability of error. Another form is the maximum error-probability

$$\varepsilon^{\mathrm{max}} = \max\left[1 - \mathbf{P}_{\mathrm{ch}}\left(\widehat{f}_N(\mathbf{Y}^{(N)}) = \mathbf{x}\,\big|\,\mathbf{x}\ \text{sent}\right) : \mathbf{x} \in \mathcal{X}_N\right];$$

obviously, $\varepsilon^{\mathrm{ave}} \leq \varepsilon^{\mathrm{max}}$. In this section we work with $\varepsilon^{\mathrm{ave}} \to 0$ leaving the question of whether $\varepsilon^{\mathrm{max}} \to 0$. However, in Section 2.2 we reduce the problem of assessing $\varepsilon^{\mathrm{max}}$ to that with $\varepsilon^{\mathrm{ave}}$, and as a result, the formulas for the channel capacity deduced in this section will remain valid if $\varepsilon^{\mathrm{ave}}$ is replaced by $\varepsilon^{\mathrm{max}}$.

Remark 1.4.5 (a) By Theorem 1.4.17 below, the channel capacity of an MBC is given by

$$C = \sup_{p_{X_k}} I(X_k : Y_k). \tag{1.4.7}$$

Here, $I(X_k : Y_k)$ is the mutual information between a single pair of input and output letters X_k and Y_k (the index k may be omitted), with the joint distribution

$$\mathbb{P}(X = x, Y = y) = p_X(x)P(y|x), \quad x, y = 0, 1, \tag{1.4.8}$$

where $p_X(x) = \mathbb{P}(X = x)$. The supremum in (1.4.7) is over all possible distributions $p_X = (p_X(0), p_X(1))$. A useful formula is $I(X : Y) = h(Y) - h(Y|X)$ (see (1.3.12)). In fact, for the MBSC

$$
h(Y|X) = - \sum_{x=0,1} p_X(x) \sum_{y=0,1} P(y|x) \log P(y|x)
$$
$$
- \sum_{y=0,1} P(y|x) \, \log P(y|x) = h_2(p, 1-p) = \eta_2(p); \tag{1.4.9}
$$

the lower index 2 will be omitted for brevity.

Hence $h(Y|X) = \eta(p)$ does not depend on input distribution p_X, and for the MBSC

$$C = \sup_{p_X} h(Y) - \eta(p). \tag{1.4.10}$$

But $\sup_{p_X} h(Y)$ is equal to $\log 2 = 1$: it is attained at $p_X(0) = p_X(1) = 1/2$, and $p_Y(0) = p_Y(1) = 1/2(p + 1 - p) = 1/2$. Therefore, for an MBSC, with the row error-probability p,

$$C = 1 - \eta(p). \tag{1.4.11}$$

(b) Suppose we have a source $U_1 U_2 \ldots$ with the asymptotic equipartition property and information rate H. To send a text emitted by the source through a channel of capacity C we need to encode messages of length n by codewords of length $\dfrac{n(H + \varepsilon)}{C}$ in order to have the overall error-probability tending to zero as $n \to \infty$. The value $\varepsilon > 0$ may be chosen arbitrarily small. Hence, if $H/C < 1$, a text can be encoded with a higher speed than it is produced: in this case the channel is used reliably for transmitting information from the source. On the contrary, if $H/C > 1$, the text will be produced with a higher speed than we can encode it and send reliably through a channel. In this case reliable transmission is impossible. For a Bernoulli or stationary Markov source and an MBSC, condition $H/C < 1$ is equivalent to $h(U) + \eta(p) < 1$ or $h(U_{n+1}|U_n) + \eta(p) < 1$ respectively.

In fact, Shannon's ideas have not been easily accepted by leading contemporary mathematicians. It would be interesting to see the opinions of the leading scientists who could be considered as 'creators' of information theory.

Theorem 1.4.6 *Fix a channel (i.e. conditional probabilities $\mathbf{P}_{ch}$ in (1.4.1)) and a set $\mathscr{U}$ of the source strings and denote by $\varepsilon(\mathbb{P})$ the overall error-probability (1.4.3) for $\mathbf{U}^{(n)}$ having a probability distribution $\mathbb{P}$ over $\mathscr{U}$, minimised over all encoding and decoding rules. Then*

$$\varepsilon(\mathbb{P}) \leq \varepsilon(\mathbb{P}^0), \tag{1.4.12}$$

where $\mathbb{P}^0$ is the equidistribution over $\mathscr{U}$.

Proof Fix encoding and decoding rules f and $\widehat{f}$, and let a string $\mathbf{u} \in \mathscr{U}$ have probability $\mathbb{P}(\mathbf{u})$. Define the error-probability when $\mathbf{u}$ is emitted as

$$\beta(\mathbf{u}) := \sum_{\mathbf{y}:\widehat{f}(\mathbf{y})\neq\mathbf{u}} \mathbf{P}_{ch}(\mathbf{y}|f(\mathbf{u})).$$

The overall error-probability equals

$$\varepsilon(=\varepsilon(\mathbb{P},f,\widehat{f})) = \sum_{\mathbf{u}\in\mathscr{U}} \mathbb{P}(\mathbf{u})\beta(\mathbf{u}).$$

If we permute the allocation of codewords (i.e. encode $\mathbf{u}$ by $f(\mathbf{u}')$ where $\mathbf{u}' = \lambda(\mathbf{u})$ and λ is a permutation of degree $\sharp\mathscr{U}$), we get the overall error-probability $\varepsilon(\lambda)$ $= \sum_{\mathbf{u}\in\mathscr{U}} \mathbb{P}(\mathbf{u})\beta(\lambda(\mathbf{u}))$. In the case $\mathbb{P}(\mathbf{u}) = \left(\sharp\mathscr{U}\right)^{-1}$ (equidistribution), $\varepsilon(\lambda)$ does not depend on λ and is given by

$$\overline{\varepsilon} = \frac{1}{\sharp\mathscr{U}} \sum_{\mathbf{u}\in\mathscr{U}} \beta(\mathbf{u}) = \varepsilon(\mathbb{P}^0,f,\widehat{f}).$$

It is claimed that for each probability distribution $\{\mathbb{P}(\mathbf{u}), \mathbf{u} \in \mathscr{U}\}$, there exists λ such that $\varepsilon(\lambda) \leq \overline{\varepsilon}$. In fact, take a *random* permutation, Λ, equidistributed among all $(\sharp\mathscr{U})!$ permutations of degree $\sharp\mathscr{U}$. Then

$$\min_{\lambda}\varepsilon(\lambda) \leq \mathbb{E}\varepsilon(\Lambda) = \mathbb{E}\sum_{\mathbf{u}\in\mathscr{U}} \mathbb{P}(\mathbf{u})\beta(\Lambda\mathbf{u})$$

$$= \sum_{\mathbf{u}\in\mathscr{U}} \mathbb{P}(\mathbf{u})\mathbb{E}\beta(\Lambda\mathbf{u}) = \sum_{\mathbf{u}\in\mathscr{U}} \mathbb{P}(\mathbf{u})\frac{1}{\sharp\mathscr{U}} \sum_{\widetilde{\mathbf{u}}\in\mathscr{U}} \beta(\widetilde{\mathbf{u}}) = \overline{\varepsilon}.$$

Hence, given any f and $\widehat{f}$, we can find new encoding and decoding rules with overall error-probability $\leq \varepsilon(\mathbb{P}^0,f,\widehat{f})$. Minimising over f and $\widehat{f}$ leads to (1.4.12). $\square$

Worked Example 1.4.7   *Let the random variables X and Y, with values from finite 'alphabets' I and J, represent, respectively, the input and output of a transmission channel, with the conditional probability $P(x \mid y) = \mathbb{P}(X = x \mid Y = y)$. Let $h(P(\cdot \mid y))$ denote the entropy of the conditional distribution $P(\cdot \mid y)$, $y \in J$:*

$$h(P(\cdot \mid y)) = -\sum_x P(X \mid y) \log P(x \mid y).$$

Let $h(X \mid Y)$ denote the conditional entropy of X given Y Define the ideal observer decoding rule as a map $f^{\mathrm{IO}} : J \to I$ such that $P(f(y) \mid y) = \max_{x \in I} P(x \mid y)$ for all $y \in J$. Show that

(a) *under this rule the error-probability*

$$\pi_{\mathrm{er}}^{\mathrm{IO}}(y) = \sum_{x \in I} \mathbf{1}(x \neq f(y)) P(x \mid y)$$

satisfies $\pi_{\mathrm{er}}^{\mathrm{IO}}(y) \leq \frac{1}{2} h(P(\cdot \mid y))$;

(b) *the expected value of the error-probability obeys $\mathbb{E}\pi_{\mathrm{er}}^{\mathrm{IO}}(Y) \leq \frac{1}{2} h(X \mid Y)$.*

Solution Indeed, (a) follows from (iii) in Worked Example 1.2.7, as

$$\pi_{\mathrm{err}}^{\mathrm{IO}} = 1 - P(f(y) \mid y) = 1 - P_{\max}(\cdot \mid y),$$

which is less than or equal to $\frac{1}{2} h(P(\cdot \mid y))$. Finally, (b) follows from (a) by taking expectations, as $h(X|Y) = \mathbb{E}h(P(\cdot|Y))$. □

As was noted before, a general *decoding rule* (or a *decoder*) is a map $\widehat{f}_N : J^{\times N} \to \mathcal{U}_N$; in the case of a lossless encoding rule f_N, $\widehat{f}_N$ is a map $J^{\times N} \to \mathcal{X}_N$. Here $\mathcal{X}$ is a set of codewords. Sometimes it is convenient to identify the decoding rule by fixing, for each codeword $\mathbf{x}^{(N)}$, a set $A(\mathbf{x}^{(N)}) \subset J^{\times N}$, so that $A(x_1^{(N)})$ and $A(x_2^{(N)})$ are disjoint for $x_1^{(N)} \neq x_2^{(N)}$, and the union $\cup_{\mathbf{x}^{(N)} \in \mathcal{X}_N} A(\mathbf{x}^{(N)})$ gives the whole $J^{\times N}$. Given that $\mathbf{y}^{(N)} \in A(\mathbf{x}^{(N)})$, we decode it as $\widehat{f}_N(\mathbf{y}^{(N)}) = \mathbf{x}^{(N)}$.

Although in the definition of the channel capacity we assume that the source messages are equidistributed (as was mentioned, it gives the worst case in the sense of Theorem 1.4.6), in reality of course the source does not always follow this assumption. To this end, we need to distinguish between two situations: (i) the receiver knows the probabilities

$$p(\mathbf{u}) = \mathbb{P}(U = \mathbf{u}) \tag{1.4.13}$$

of the source strings (and hence the probability distribution $p_N(\mathbf{x}^{(N)})$ of the codewords $\mathbf{x}^{(N)} \in \mathcal{X}_N$), and (ii) he does not know $p_N(\mathbf{x}^{(N)})$. Two natural decoding rules are, respectively,

(i) the *ideal observer* (IO) rule decodes a received word $\mathbf{y}^{(N)}$ by a codeword $\mathbf{x}^{(N)\star}$ that maximises the posterior probability

$$\mathbb{P}\left(\mathbf{x}^{(N)} \text{ sent } |\mathbf{y}^{(N)} \text{ received }\right) = \frac{p_N(\mathbf{x}^{(N)})\mathbf{P}_{\text{ch}}(\mathbf{y}^{(N)}|\mathbf{x}^{(N)})}{p_{\mathbf{Y}^{(N)}}(\mathbf{y}^{(N)})}, \qquad (1.4.14)$$

where

$$p_{\mathbf{Y}^{(N)}}(\mathbf{y}^{(N)}) = \sum_{\widetilde{\mathbf{x}}^{(N)} \in \mathcal{X}_N} p_N(\widetilde{\mathbf{x}}^{(N)})\mathbf{P}_{\text{ch}}(\mathbf{y}^{(N)}|\widetilde{\mathbf{x}}^{(N)}),$$

and

(ii) the *maximum likelihood* (ML) rule decodes a received word $\mathbf{y}^{(N)}$ by a codeword $x_{\star}^{(N)}$ that maximises the prior probability

$$\mathbf{P}_{\text{ch}}(\mathbf{y}^{(N)}|\mathbf{x}^{(N)}). \qquad (1.4.15)$$

Theorem 1.4.8   *Suppose that an encoding rule f is defined for all messages that occur with positive probability and is one-to-one. Then:*

(a) *For any such encoding rule, the IO decoder minimises the overall error-probability among all decoders.*
(b) *If the source message $\mathbf{U}$ is equiprobable on a set $\mathcal{U}$, then for any encoding rule $f : \mathcal{U} \to \mathcal{X}_N$ as above, the random codeword $\mathbf{X}^{(N)} = f(\mathbf{U})$ is equiprobable on $\mathcal{X}_N$, and the IO and ML decoders coincide.*

Proof   Again, for simplicity let us omit the upper index (N).

(a) Note that, given a received word $\mathbf{y}$, the IO obviously maximises the joint probability $p(\mathbf{x})\mathbf{P}_{\text{ch}}(\mathbf{y}|\mathbf{x})$ (the denominator in (1.4.14) is fixed when word $\mathbf{y}$ is fixed). If we use an encoding rule f and decoding rule $\widehat{f}$, the overall error-probability (see (1.4.3)) is

$$\sum_{\mathbf{u}} \mathbb{P}(\mathbf{U} = \mathbf{u})\mathbf{P}_{\text{ch}}\left(\widehat{f}(\mathbf{y}) \neq \mathbf{u}|f(\mathbf{u}) \text{ sent}\right)$$
$$= \sum_{\mathbf{x}} p(\mathbf{x})\sum_{\mathbf{y}} \mathbf{1}\left(\widehat{f}(\mathbf{y}) \neq \mathbf{x}\right)\mathbf{P}_{\text{ch}}(\mathbf{y}|\mathbf{x})$$
$$= \sum_{\mathbf{y}}\sum_{\mathbf{x}} \mathbf{1}\left(\mathbf{x} \neq \widehat{f}(\mathbf{y})\right) p(\mathbf{x})\mathbf{P}_{\text{ch}}(\mathbf{y}|\mathbf{x})$$
$$= \sum_{\mathbf{y}}\sum_{\mathbf{x}} p(\mathbf{x})\mathbf{P}_{\text{ch}}(\mathbf{y}|\mathbf{x}) - \sum_{\mathbf{y}} p\left(\widehat{f}(\mathbf{y})\right)\mathbf{P}_{\text{ch}}\left(\mathbf{y}|\widehat{f}(\mathbf{y})\right)$$
$$= 1 - \sum_{\mathbf{y}} p\left(\widehat{f}(\mathbf{y})\right)\mathbf{P}_{\text{ch}}\left(\mathbf{y}|\widehat{f}(\mathbf{y})\right).$$

It remains to note that each term in the sum $\sum_{y} p\left(\widehat{f}(\mathbf{y})\right)\mathbf{P}_{\text{ch}}\left(\mathbf{y}|\widehat{f}(\mathbf{y})\right)$ is maximised when $\widehat{f}$ coincides with the IO rule. Hence, the whole sum is maximised, and the overall error-probability minimised.

(b) The first statement is obvious, as, indeed is the second. $\square$

Assuming in the definition of the channel capacity that the source messages are equidistributed, it is natural to explore further the ML decoder. While using the ML decoder, an error can occur because either the decoder chooses a wrong codeword x or an encoding rule f used is not one-to-one. The probability of this is assessed in Theorem 1.4.8. For further simplification, we write $\mathbf{P}$ instead of $\mathbf{P}_{\mathrm{ch}}$; symbol $\mathbb{P}$ is used mainly for the joint input/output distribution.

Lemma 1.4.9 *If the source messages are equidistributed over a set $\mathscr{U}$ then, while using the ML decoder and an encoding rule f, the overall error-probability satisfies*

$$\varepsilon(f) \leq \frac{1}{\sharp \mathscr{U}} \sum_{\mathbf{u} \in \mathscr{U}} \sum_{\mathbf{u}' \in \mathscr{U} \,:\, \mathbf{u}' \neq \mathbf{u}} \mathbb{P}\left(\mathbf{P}\left(\mathbf{Y}|f(\mathbf{u}')\right) \geq \mathbf{P}\left(\mathbf{Y}|f(\mathbf{u})\right) | \mathbf{U} = \mathbf{u}\right). \qquad (1.4.16)$$

Proof If the source emits $\mathbf{u}$ and the ML decoder is used, we get

(a) an error when $\mathbf{P}(\mathbf{Y} \mid f(\mathbf{u}')) > \mathbf{P}(\mathbf{Y} \mid f(\mathbf{u}))$ for some $\mathbf{u}' \neq \mathbf{u}$,
(b) possibly an error when $\mathbf{P}(\mathbf{Y}|f(\mathbf{u}')) = \mathbf{P}(\mathbf{Y} \mid f(\mathbf{u}))$ for some $\mathbf{u}' \neq \mathbf{u}$ (this includes the case when $f(\mathbf{u}) = f(\mathbf{u}')$), and finally
(c) no error when $\mathbf{P}(\mathbf{Y} \mid f(\mathbf{u}')) < \mathbf{P}(\mathbf{Y} \mid f(\mathbf{u}))$ for any $\mathbf{u}' \neq \mathbf{u}$.

Thus, the probability is bounded as follows:

$$\begin{aligned}
\mathbb{P}&(\text{error} \mid \mathbf{U} = \mathbf{u}) \\
&\leq \mathbb{P}\left(\mathbf{P}\left(\mathbf{Y} \mid f(\mathbf{u}')\right) \geq \mathbf{P}(\mathbf{Y} \mid f(\mathbf{u})) \text{ for some } \mathbf{u}' \neq u \mid \mathbf{U} = \mathbf{u}\right) \\
&\leq \sum_{\mathbf{u}' \in \mathscr{U}} \mathbf{1}(\mathbf{u}' \neq \mathbf{u}) \mathbb{P}\left(\mathbf{P}\left(\mathbf{Y} \mid f(\mathbf{u}')\right) \geq \mathbf{P}(\mathbf{Y} \mid f(\mathbf{u})) \mid \mathbf{U} = \mathbf{u}\right).
\end{aligned}$$

Multiplying by $\dfrac{1}{\sharp \mathscr{U}}$ and summing up over $\mathbf{u}$ yields the result. $\quad\square$

Remark 1.4.10 Bound (1.4.16) of course holds for any probability distribution $p(\mathbf{u}) = \mathbb{P}(\mathbf{U} = \mathbf{u})$, provided $\dfrac{1}{\sharp \mathscr{U}}$ is replaced by $p(\mathbf{u})$.

As was already noted, a *random* coding is a useful tool alongside with deterministic encoding rules. A deterministic encoding rule is a map $f \colon \mathscr{U} \to J^{\times N}$; if $\sharp \mathscr{U} = r$ then f is given as a collection of codewords $\{f(\mathbf{u}_1), \ldots, f(\mathbf{u}_r)\}$ or, equivalently, as a concatenated 'megastring' (or codebook)

$$f(\mathbf{u}_1) \ldots f(\mathbf{u}_r) \in \left(J^{\times N}\right)^{\times r} = \{0,1\}^{\times Nr}.$$

Here, $u_1, \ldots, u_r$ are the source strings (not letters!) constituting set $\mathscr{U}$. If f is lossless then $f(\mathbf{u}_i) \neq f(\mathbf{u}_j)$ whenever $i \neq j$. A random encoding rule is a random element F of $\left(J^{\times N}\right)^r$, with probabilities $\mathbb{P}(F = f)$, $f \in \left(J^{\times N}\right)^r$. Equivalently, F may be regarded as a collection of random codewords $F(\mathbf{u}_i)$, $i = 1, \ldots, r$, or, equivalently, as a random codebook

$$F(\mathbf{u}_1)F(\mathbf{u}_2)\ldots F(\mathbf{u}_r) \in \{0,1\}^{Nr}.$$

A typical example is where codewords $F(\mathbf{u}_1)$, $F(\mathbf{u}_2)$, $\ldots$, $F(\mathbf{u}_r)$ are independent, and (random) symbols $W_{i1}, \ldots, W_{iN}$ constituting word $F(\mathbf{u}_i)$ are independent too.

The reasons for considering random encoding rules are:

(1) the existence of a 'good' deterministic code frequently follows from the existence of a good random code;

(2) the calculations for random codes are usually simpler than for optimal deterministic codes, because a discrete optimisation is replaced by an optimisation over probability distributions.

A drawback of random coding is that it is not always one-to-one ($F(\mathbf{u})$ may coincide with $F(\mathbf{u}')$ for $\mathbf{u} \neq \mathbf{u}'$). However, this occurs, for large N, with negligible probability.

The idea of random coding goes back to Shannon. As often happened in the history of mathematics, a brilliant idea solves one problem but opens a Pandora box of other questions. In this case, a particular problem that emerged from the aftermath of random coding was the problem of *finding* 'good' non-random codes. A major part of modern information and coding theory revolves around this problem, and so far no general satisfactory solution has been found. However, a number of remarkable partial results have been achieved, some of which are discussed in this book.

Continuing with random coding, write the expected error-probability for a random encoding rule F:

$$E := \mathscr{E}\varepsilon(F) = \sum_f \varepsilon(f)\mathscr{P}(F = f). \tag{1.4.17}$$

Theorem 1.4.11

(i) *There exists a deterministic encoding rule f with $\varepsilon(f) \leq E$.*

(ii) $\quad \mathscr{P}\left(\varepsilon(F) < \dfrac{E}{1-\rho}\right) \geq \rho$ *for any $\rho \in (0,1)$.*

Proof Part (i) is obvious. For (ii), use the Chebyshev inequality (see PSE I, p. 75):

$$\mathscr{P}\left(\varepsilon(F) \geq \frac{E}{1-\rho}\right) \leq \frac{1-\rho}{E}E = 1 - \rho. \qquad \square$$

Definition 1.4.12 For random words $\mathbf{X}^{(N)} = X_1 \ldots X_N$ and $\mathbf{Y}^{(N)} = Y_1 \ldots Y_N$ define

$$C_N := \sup \left[\frac{1}{N} I\left(\mathbf{X}^{(N)} : \mathbf{Y}^{(N)}\right), \quad \text{over input} \right.$$

$$\left. \text{probability distributions } P_{\mathbf{X}^{(N)}} \right]. \tag{1.4.18}$$

Recall that $I\left(\mathbf{X}^{(N)} : \mathbf{Y}^{(N)}\right)$ is the mutual entropy given by

$$h\left(\mathbf{X}^{(N)}\right) - h\left(\mathbf{X}^{(N)}|\mathbf{Y}^{(N)}\right) = h\left(\mathbf{Y}^{(N)}\right) - h\left(\mathbf{Y}^{(N)}|\mathbf{X}^{(N)}\right).$$

Remark 1.4.13 A simple heuristic argument (which will be made rigorous in Section 2.2) shows that the capacity of the channel cannot exceed the mutual information between its input and output. Indeed, for each typical input N-sequence, there are

$$\text{approximately } 2^{h(\mathbf{Y}^{(N)}|\mathbf{X}^{(N)})} \text{ possible } \mathbf{Y}^{(N)} \text{ sequences,}$$

all of them equally likely. We will not be able to detect which sequence $\mathbf{X}$ was sent unless no two $\mathbf{X}^{(N)}$ sequences produce the same $\mathbf{Y}^{(N)}$ output sequence. The total number of typical $\mathbf{Y}^{(N)}$ sequences is $2^{h(\mathbf{Y}^{(N)})}$. This set has to be divided into subsets of size $2^{h(\mathbf{Y}^{(N)}|\mathbf{X}^{(N)})}$ corresponding to the different input $\mathbf{X}^{(N)}$ sequences. The total number of disjoint sets is

$$\leq 2^{h(\mathbf{Y}^{(N)}) - h(\mathbf{Y}^{(N)}|\mathbf{X}^{(N)})} = 2^{I\left(\mathbf{X}^{(N)} : \mathbf{Y}^{(N)}\right)}.$$

Hence, the total number of distinguishable signals of the length N could not be bigger than $2^{I\left(\mathbf{X}^{(N)} : \mathbf{Y}^{(N)}\right)}$. Putting the same argument slightly differently, the number of typical sequences $\mathbf{X}^{(N)}$ is $2^{Nh(\mathbf{X}^{(N)})}$. However, there are only $2^{Nh(\mathbf{X}^{(N)}, \mathbf{Y}^{(N)})}$ *jointly typical* sequences $(\mathbf{X}^{(N)}, \mathbf{Y}^{(N)})$. So, the probability that any randomly chosen pair is jointly typical is about $2^{-I\left(\mathbf{X}^{(N)} : \mathbf{Y}^{(N)}\right)}$. So, the number of distinguished signals is bounded by $2^{h(\mathbf{X}^{(N)}) + h(\mathbf{Y}^{(N)}) - h(\mathbf{X}^{(N)}|\mathbf{Y}^{(N)})}$.

Theorem 1.4.14 (Shannon's SCT: converse part) *The channel capacity C obeys*

$$C \leq \limsup_{N \to \infty} C_N. \tag{1.4.19}$$

Proof Consider a code $f = f_N : \mathscr{U}_N \to \mathscr{X}_N \subseteq J^{\times N}$, where $\sharp \mathscr{U}_N = 2^{N(\overline{R} + o(1))}$, $\overline{R} \in (0,1)$. We want to prove that for any decoding rule

$$\varepsilon(f) \geq 1 - \frac{C_N + o(1)}{\overline{R} + o(1)}. \tag{1.4.20}$$

The assertion of the theorem immediately follows from (1.4.20) and the definition of the channel capacity because

$$\liminf_{N\to\infty} \varepsilon(f) \geq 1 - \frac{1}{\overline{R}} \limsup_{N\to\infty} C_N$$

which is > 0 when $\overline{R} > \limsup_{N\to\infty} C_N$.

Let us check (1.4.20) for one-to-one f (otherwise $\varepsilon(f)$ is even bigger). Then a codeword $\mathbf{X}^{(N)} = f(\mathbf{U})$ is equidistributed when string U is, and, if a decoding rule is $\widehat{f}: J^{\times N} \to \mathscr{X}$, we have, for N large enough,

$$NC_N \geq I\left(\mathbf{X}^{(N)} : \mathbf{Y}^{(N)}\right) \geq I\left(\mathbf{X}^{(N)} : \widehat{f}(\mathbf{Y}^{(N)})\right) \quad \text{(cf. Theorem 1.2.6)}$$

$$= h\left(\mathbf{X}^{(N)}\right) - h\left(\mathbf{X}^{(N)}|\widehat{f}(\mathbf{Y}^{(N)})\right)$$

$$= \log r - h\left(\mathbf{X}^{(N)}|\widehat{f}(\mathbf{Y}^{(N)})\right) \text{ (by equidistribution)}$$

$$\geq \log r - \varepsilon(f)\log(r-1) - 1.$$

Here and below $r = \sharp \mathscr{U}$. The last bound follows by the generalised Fano inequality (1.2.25). Indeed, observe that the (random) codeword $\mathbf{X}^{(N)} = f(\mathbf{U})$ takes r values $\mathbf{x}_1^{(N)}, \ldots, \mathbf{x}_r^{(N)}$ from the codeword set $\mathscr{X}(= \mathscr{X}_N)$, and the error-probability is

$$\varepsilon(f) = \sum_{i=1}^{r} \mathbb{P}(\mathbf{X}^{(N)} = \mathbf{x}_i^{(N)}, \widehat{f}(\mathbf{Y}^{(N)}) \neq \mathbf{x}_i^{(N)}).$$

So, (1.2.25) implies

$$h\left(\mathbf{X}^{(N)}|\widehat{f}(\mathbf{Y}^{(N)})\right) \leq h_2(\varepsilon) + \varepsilon \log(r-1) \leq 1 + \varepsilon(f)\log(r-1),$$

and we obtain $NC_N \geq \log r - \varepsilon(f)\log(r-1) - 1$. Finally, $r = 2^{N(\overline{R}+o(1))}$ and

$$NC_N \geq N(\overline{R}+o(1)) - \varepsilon(f)\log\left(2^{N(\overline{R}+o(1))} - 1\right),$$

i.e.

$$\varepsilon(f) \geq \frac{N(\overline{R}+o(1)) - NC_N}{\log\left(2^{N(\overline{R}+o(1))} - 1\right)} = 1 - \frac{C_N + o(1)}{\overline{R}+o(1)}.$$

$\square$

Let $p(\mathbf{X}^{(N)}, \mathbf{Y}^{(N)})$ be the random variable that assigns, to random words $\mathbf{X}^{(N)}$ and $\mathbf{Y}^{(N)}$, the joint probability of having these words at the input and output of a channel, respectively. Similarly, $p_X(\mathbf{X}^{(N)})$ and $p_Y(\mathbf{Y}^{(N)})$ denote the random variables that give the marginal probabilities of words $\mathbf{X}^{(N)}$ and $\mathbf{Y}^{(N)}$, respectively.

Theorem 1.4.15 (Shannon's SCT: direct part) *Suppose we can find a constant $c \in (0,1)$ such that for any $\overline{R} \in (0,c)$ and $N \geq 1$ there exists a random coding $F(\mathbf{u}_1),\ldots,F(\mathbf{u}_r)$, where $r = 2^{N(\overline{R}+o(1))}$, with IID codewords $F(\mathbf{u}_i) \in J^{\times N}$, such that the (random) input/output mutual information*

$$\Theta_N := \frac{1}{N} \log \frac{p(\mathbf{X}^{(N)}, \mathbf{Y}^{(N)})}{p_X(\mathbf{X}^{(N)}) p_Y(\mathbf{Y}^{(N)})} \qquad (1.4.21)$$

converges in probability to c as $N \to \infty$. Then the channel capacity $C \geq c$.

The proof of Theorem 1.4.15 is given after Worked Examples 1.4.24 and 1.4.25 (the latter is technically rather involved). To start with, we explain the strategy of the proof outline by Shannon in his original 1948 paper. (It took about 10 years before this idea was transformed into a formal argument.)

First, one generates a random codebook $\mathscr{X}$ consisting of $r = 2^{\lceil NR \rceil}$ words, $\mathbf{X}^{(N)}(1),\ldots,\mathbf{X}^{(N)}(r)$. The codewords $\mathbf{X}^{(N)}(1),\ldots,\mathbf{X}^{(N)}(r)$ are assumed to be known to both the sender and the receiver, as well as the channel transition matrix $\mathbf{P}_{ch}(y|x)$. Next, the message is chosen according to a uniform distribution, and the corresponding codeword is sent over a channel. The receiver uses the maximum likelihood (ML) decoding, i.e. choose the *a posteriori* most likely message. But this procedure is difficult to analyse. Instead, a suboptimal but straightforward typical set decoding is used. The receiver declares that the message w is sent if there is only one input such that the codeword for w and the output of the channel are jointly typical. If no such word exists or it is non-unique then an error is declared. Surprisingly, this procedure is asymptotically optimal. Finally, the existence of a good random codebook implies the existence of a good non-random coding.

In other words, channel capacity C is no less than the supremum of the values c for which the convergence in probability in (1.4.21) holds for an appropriate random coding.

Corollary 1.4.16 *With c as in the assumptions of Theorem 1.4.15, we have that*

$$\sup c \leq C \leq \limsup_{N \to \infty} C_N. \qquad (1.4.22)$$

So, if the LHS and RHS sides of (1.4.22) coincide, then their common value gives the channel capacity.

Next, we use Shannon's SCT for calculating the capacity of an MBC. Recall (cf. (1.4.2)), for an MBC,

$$\mathbf{P}\left(\mathbf{y}^{(N)}|\mathbf{x}^{(N)}\right) = \prod_{i=1}^{N} P(y_i|x_i). \qquad (1.4.23)$$

Theorem 1.4.17   *For an MBC,*

$$I\left(\mathbf{X}^{(N)} : \mathbf{Y}^{(N)}\right) \le \sum_{j=1}^{N} I(X_j : Y_j), \tag{1.4.24}$$

with equality if the input symbols $X_1, \ldots, X_N$ are independent.

Proof   Since $\mathbf{P}\left(\mathbf{y}^{(N)} | \mathbf{x}^{(N)}\right) = \prod_{j=1}^{N} P(y_j | x_j)$, the conditional entropy $h\left(\mathbf{Y}^{(N)} | \mathbf{X}^{(N)}\right)$

equals the sum $\sum_{j}^{N} h(Y_j | X_j)$. Then the mutual information

$$\begin{aligned}
I\left(\mathbf{X}^{(N)} : \mathbf{Y}^{(N)}\right) &= h\left(\mathbf{Y}^{(N)}\right) - h\left(\mathbf{Y}^{(N)} | \mathbf{X}^{(N)}\right) \\
&= h\left(\mathbf{Y}^{(N)}\right) - \sum_{1 \le j \le N} h(Y_j | X_j) \\
&\le \sum_{j} \left(h(Y_j) - h(Y_j | X_j)\right) = \sum_{j} I(X_j : Y_j).
\end{aligned}$$

The equality holds iff $Y_1, \ldots, Y_N$ are independent. But $Y_1, \ldots, Y_N$ are independent if $X_1, \ldots, X_N$ are.   □

Remark 1.4.18   Compare with inequalities (1.4.24) and (1.2.27). Note the opposite inequalities in the bounds.

Theorem 1.4.19   *The capacity of an MBC is*

$$C = \sup_{p_{X_1}} I(X_1 : Y_1). \tag{1.4.25}$$

The supremum is over all possible distributions p_{X_1} of the symbol X_1.

Proof   By the definition of C_N, $N C_N$ does not exceed

$$\sup_{p_X} I(\mathbf{X}^{(N)} : \mathbf{Y}^{(N)}) \le \sum_{j} \sup_{p_{X_j}} I(X_j : Y_j) = N \sup_{p_{X_1}} I(X_1 : Y_1).$$

So, by Shannon's SCT (converse part),

$$C \le \limsup_{N \to \infty} C_N \le \sup_{p_{X_1}} I(X_1 : Y_1).$$

On the other hand, take a random coding F, with codewords $F(\mathbf{u}_l) = V_{l1} \cdots V_{lN}$, $1 \le l \le r$, containing IID symbols V_{lj} that are distributed according to p^*, a probability distribution that maximises $I(X_1 : Y_1)$. [Such random coding is defined for *any r*, i.e. for any $\overline{R}$ (even $\overline{R} > 1$!).] For this random coding, the (random) mutual entropy Θ_N equals

$$\frac{1}{N} \log \frac{p\left(\mathbf{X}^{(N)}, \mathbf{Y}^{(N)}\right)}{p_{\mathbf{X}}\left(\mathbf{X}^{(N)}\right) p_{\mathbf{Y}}\left(\mathbf{Y}^{(N)}\right)}$$

$$= \frac{1}{N} \sum_{j=1}^{N} \log \frac{p(X_j, Y_j)}{p^*(X_j) p_Y(Y_j)} = \frac{1}{N} \sum_{j=1}^{N} \zeta_j,$$

where $\zeta_j := \log \dfrac{p(X_j, Y_j)}{p^*(X_j) p_Y(Y_j)}$.

The random variables ζ_j are IID, and

$$\mathbb{E} \zeta_j = \mathbb{E} \log \frac{p(X_j, Y_j)}{p^*(X_j) p_Y(Y_j)} = I_{p^*}(X_1 : Y_1).$$

By the law of large numbers for IID random variables (see Theorem 1.3.5), for the random coding as suggested,

$$\Theta_N \xrightarrow{\mathbb{P}} I_{p^*}(X_1 : Y_1) = \sup_{p_{X_1}} I(X_1 : Y_1).$$

By Shannon's SCT (direct part),

$$C \geq \sup_{p_{X_1}} I(X_1 : Y_1).$$

Thus, $C = \sup_{p_{X_1}} I(X_1 : Y_1)$. $\qquad\qquad\square$

Remark 1.4.20 (a) The pair (X_1, Y_1) may be replaced by any (X_j, Y_j), $j \geq 1$.

(b) Recall that the joint distribution of X_1 and Y_1 is defined by $\mathbb{P}(X_1 = x, Y_1 = y) = p_{X_1}(x) P(y|x)$ where $(P(y|x))$ is the channel matrix.

(c) Although, as was noted, the construction holds for each r (that is, for each $\overline{R} \geq 0$) only $\overline{R} \leq C$ are reliable.

Example 1.4.21 A helpful statistician preprocesses the output of a memoryless channel (MBC) with transition probabilities $P(y|x)$ and channel capacity $C = \max_{p_X} I(X : Y)$ by forming $Y' = g(Y)$: he claims that this will strictly improve the capacity. Is he right? Surely not, as preprocessing (or doctoring) does not increase the capacity. Indeed,

$$I(X : Y) = h(X) - h(X|Y) \geq h(X) - h(X|g(Y)) = I(X : g(Y)). \qquad (1.4.26)$$

Under what condition does he not strictly decrease the capacity? Equality in (1.4.26) holds iff, under the distribution p_X that maximises $I(X : Y)$, the random variables X and Y are conditionally independent given $g(Y)$. [For example, $g(y_1) = g(y_2)$ iff for any x, $P_{X|Y}(x|y_1) = P_{X|Y}(x|y_2)$; that is, g glues together only those values of y for which the conditional probability $P_{X|Y}(\cdot|y)$ is the same.] For an MBC, equality holds iff g is one-to-one, or $p = P(1|0) = P(0|1) = 1/2$.

Formula (1.4.25) admits a further simplification when the channel is symmetric (MBSC), i.e. $P(1|0) = P(0|1) = p$. More precisely, in accordance with Remark 1.4.5(a) (see (1.4.11)) we obtain

Theorem 1.4.22   For an MBSC, with the row error-probability p,

$$C = 1 - h(p, 1 - p) = 1 - \eta(p) \qquad (1.4.27)$$

(see (1.4.11)). *The channel capacity is realised by a random coding with the IID symbols V_{1j} taking values 0 and 1 with probability 1/2.*

Worked Example 1.4.23

(a) *Consider a memoryless channel with two input symbols A and B, and three output symbols, $A, B, *$. Suppose each input symbol is left intact with probability 1/2, and transformed into a $*$ with probability 1/2. Write down the channel matrix and calculate the capacity.*

(b) *Now calculate the new capacity of the channel if the output is further processed by someone who cannot distinguish A and $*$, so that the matrix becomes*

$$\begin{pmatrix} 1 & 0 \\ 1/2 & 1/2 \end{pmatrix}.$$

Solution (a) The channel has the matrix

$$\begin{pmatrix} 1/2 & 0 & 1/2 \\ 0 & 1/2 & 1/2 \end{pmatrix}$$

and is symmetric (the rows are permutations of each other). So, $h(Y|X = x) = -2 \times \dfrac{1}{2} \log \dfrac{1}{2} = 1$ does not depend on the value of $x = A, B$. Then $h(Y|X) = 1$, and

$$I(X : Y) = h(Y) - 1. \qquad (1.4.28)$$

If $\mathbb{P}(X = A) = \alpha$ then Y has the output distribution

$$\left(\frac{1}{2}\alpha, \frac{1}{2}(1 - \alpha), \frac{1}{2} \right)$$

and $h(Y|X)$ is maximised at $\alpha = 1/2$. Then the capacity equals

$$h(1/4, 1/4, 1/2) - 1 = \frac{1}{2}. \qquad (1.4.29)$$

(b) Here, the channel is not symmetric. If $\mathbb{P}(X = A) = \alpha$ then the conditional entropy is decomposed as

$$\begin{aligned} h(Y|X) &= \alpha h(Y|X = A) + (1 - \alpha)h(Y|X = B) \\ &= \alpha \times 0 + (1 - \alpha) \times 1 = (1 - \alpha). \end{aligned}$$

Then

$$h(Y) = -\frac{1+\alpha}{2}\log\frac{1+\alpha}{2} - \frac{1-\alpha}{2}\log\frac{1-\alpha}{2}$$

and

$$I(X:Y) = -\frac{1+\alpha}{2}\log\frac{1+\alpha}{2} - \frac{1-\alpha}{2}\log\frac{1-\alpha}{2} - 1 + \alpha$$

which is maximised at $\alpha = 3/5$, with the capacity given by

$$\left(\log 5\right) - 2 = 0.321928.$$

$\square$

Our next goal is to prove the direct part of Shannon's SCT (Theorem 1.4.15). As was demonstrated earlier, the proof is based on two consecutive Worked Examples below.

Worked Example 1.4.24 *Let F be a random coding, independent of the source string $\mathbf{U}$, such that the codewords $F(\mathbf{u}_1), \ldots, F(\mathbf{u}_r)$ are IID, with a probability distribution p_F:*

$$p_F(\mathbf{v}) = \mathbb{P}(F(\mathbf{u}) = \mathbf{v}), \quad \mathbf{v}\, (= \mathbf{v}^{(N)}) \in J^{\times N}.$$

Here, $\mathbf{u}_j$, $j = 1, \ldots, r$, are source strings, and $r = 2^{N(\bar{R}+o(1))}$. Define random codewords $\mathbf{V}_1, \ldots, \mathbf{V}_{r-1}$ by

$$
\begin{aligned}
\text{if } \mathbf{U} = \mathbf{u}_j \quad &\text{then } \mathbf{V}_i := F(\mathbf{u}_i) &&\text{for } i < j \text{ (if any)},\\
&\text{and } \mathbf{V}_i := F(\mathbf{u}_{i+1}) &&\text{for } i \geq j \text{ (if any)}, &&(1.4.30)\\
&&& 1 \leq j \leq r,\ 1 \leq i \leq r-1.
\end{aligned}
$$

Then $\mathbf{U}$ (the message string), $\mathbf{X} = F(\mathbf{U})$ (the random codeword) and $\mathbf{V}_1, \ldots, \mathbf{V}_{r-1}$ are independent words, and each of $\mathbf{X}, \mathbf{V}_1, \ldots, \mathbf{V}_{r-1}$ has distribution p_F.

Solution This is straightforward and follows from the formula for the joint probability,

$$
\begin{aligned}
\mathbb{P}(\mathbf{U} = \mathbf{u}_j, \mathbf{X} = \mathbf{x}, \mathbf{V}_1 = \mathbf{v}_1, \ldots, \mathbf{V}_{r-1} = \mathbf{v}_{r-1})\\
= \mathbb{P}(\mathbf{U} = \mathbf{u}_j)\, p_F(\mathbf{x})\, p_F(\mathbf{v}_1) \ldots p_F(\mathbf{v}_{r-1}). \quad (1.4.31)
\end{aligned}
$$

$\square$

Worked Example 1.4.25 *Check that for the random coding as in Worked Example 1.4.24, for any $\kappa > 0$,*

$$E = \mathbb{E}\varepsilon(F) \leq \mathbb{P}(\Theta_N \leq \kappa) + r2^{-N\kappa}. \quad (1.4.32)$$

Here, the random variable Θ_N is defined in (1.4.21), with $\mathbb{E}\Theta_N = \frac{1}{N}I\left(\mathbf{X}^{(N)} : \mathbf{Y}^{(N)}\right).$

Solution For given words $\mathbf{x}(=\mathbf{x}^{(N)})$ and $\mathbf{y}(=\mathbf{y}^{(N)}) \in J^{\times N}$, denote

$$S_{\mathbf{y}}(\mathbf{x}) := \{\mathbf{x}' \in J^{\times N} : \mathbf{P}(\mathbf{y} \mid \mathbf{x}') \geq \mathbf{P}(\mathbf{y} \mid x)\}. \tag{1.4.33}$$

That is, $S_{\mathbf{y}}(\mathbf{x})$ includes all words the ML decoder may produce in the situation where $\mathbf{x}$ was sent and $\mathbf{y}$ received. Set, for a given non-random encoding rule f and a source string $\mathbf{u}$, $\delta(f, \mathbf{u}, \mathbf{y}) = 1$ if $f(\mathbf{u}') \in S_{\mathbf{y}}(f(\mathbf{u}))$ for some $\mathbf{u}' \neq \mathbf{u}$, and $\delta(f, \mathbf{u}, \mathbf{y}) = 0$ otherwise. Clearly, $\delta(f, \mathbf{u}, \mathbf{y})$ equals

$$1 - \prod_{\mathbf{u}': \mathbf{u}' \neq \mathbf{u}} \mathbf{1}\left(f(\mathbf{u}') \notin S_{\mathbf{y}}(f(\mathbf{u}))\right)$$
$$= 1 - \prod_{\mathbf{u}': \mathbf{u}' \neq \mathbf{u}} \left[1 - \mathbf{1}\left(f(\mathbf{u}') \in S_{\mathbf{y}}(f(\mathbf{u}))\right)\right].$$

It is plain that, for all non-random encoding f, $\varepsilon(f) \leq \mathbb{E}\delta(f, \mathbf{U}, \mathbf{Y})$, and for all random encoding F, $E = \mathbb{E}\varepsilon(F) \leq \mathbb{E}\delta(F, \mathbf{U}, \mathbf{Y})$. Furthermore, for the random encoding as in Worked Example 1.4.24, the expected value $\mathbb{E}\delta(F, \mathbf{U}, \mathbf{Y})$ does not exceed

$$\mathbb{E}\left(1 - \prod_{i=1}^{r-1}\left[1 - \mathbf{1}\left(\mathbf{V}_i \in S_{\mathbf{Y}}(\mathbf{X})\right)\right]\right) = \sum_x p_X(\mathbf{x}) \sum_y \mathbf{P}(\mathbf{y}|\mathbf{x})$$
$$\times \mathbb{E}\left[\left(1 - \prod_{i=1}^{r-1}\left[1 - \mathbf{1}\left(\mathbf{V}_i \in S_{\mathbf{Y}}(\mathbf{X})\right)\right]\right)|\mathbf{X} = \mathbf{x}, \mathbf{Y} = \mathbf{y}\right],$$

which, owing to independence, equals

$$\sum_x p_X(\mathbf{x}) \sum_y \mathbf{P}(\mathbf{y}|\mathbf{x})\left(1 - \prod_{i=1}^{r-1}\mathbb{E}\left(1 - \mathbf{1}_{\{\mathbf{V}_i \in S_{\mathbf{y}}(\mathbf{x})\}}\right)\right).$$

Furthermore, due to the IID property (as explained in Worked Example 1.4.24),

$$\prod_{i=1}^{r-1}\mathbb{E}\left(1 - \mathbf{1}_{\{\mathbf{V}_i \in S_{\mathbf{y}}(\mathbf{x})\}}\right) = (1 - Q_{\mathbf{y}}(\mathbf{x}))^{r-1},$$

where

$$Q_{\mathbf{y}}(\mathbf{x}) := \sum_{\mathbf{x}'} \mathbf{1}\left(\mathbf{x}' \in S_{\mathbf{y}}(\mathbf{x})\right) p_X(\mathbf{x}').$$

Hence, the expected error-probability $E \leq 1 - \mathbb{E}(1 - Q_{\mathbf{Y}}(\mathbf{X}))^{r-1}$.

Denote by $\mathbb{T} = \mathbb{T}(\kappa)$ the set of pairs of words $\mathbf{x}, \mathbf{y}$ for which

$$\Theta_N = \frac{1}{N}\log\frac{p(\mathbf{x}, \mathbf{y})}{p_X(\mathbf{x})p_Y(\mathbf{y})} > \kappa$$

and use the identity

$$1 - (1 - Q_{\mathbf{y}}(\mathbf{x}))^{r-1} = \sum_{j=0}^{r-2}(1 - Q_{\mathbf{y}}(\mathbf{x}))^j Q_{\mathbf{y}}(\mathbf{x}). \tag{1.4.34}$$

Next observe that

$$1 - (1 - Q_{\mathbf{y}}(\mathbf{x}))^{r-1} \leq 1, \quad \text{when } (x, \mathbf{y}) \notin \mathbb{T}. \tag{1.4.35}$$

Owing to the fact that when $(x, \mathbf{y}) \in \mathbb{T}$,

$$\left(1 - (1 - Q_{\mathbf{y}}(\mathbf{x}))^r\right) = \sum_{j=1}^{r-1} \left((1 - Q_{\mathbf{y}}(\mathbf{x}))^j\right) Q_{\mathbf{y}}(\mathbf{x}) \leq (r-1)Q_{\mathbf{y}}(\mathbf{x}),$$

this yields

$$E \leq \mathbb{P}\big((X,Y) \notin \mathbb{T}\big) + (r-1) \sum_{(x,\mathbf{y}) \in \mathbb{T}} p_X(\mathbf{x})\mathbf{P}(\mathbf{y}|\mathbf{x})Q_{\mathbf{y}}(\mathbf{x}). \tag{1.4.36}$$

Now observe that

$$\mathbb{P}\big((X,Y) \notin \mathbb{T}\big) = \mathbb{P}\big(\Theta_N \leq \kappa\big). \tag{1.4.37}$$

Finally, for $(x, \mathbf{y}) \in \mathbb{T}$ and $\mathbf{x}' \in S_{\mathbf{y}}(\mathbf{x})$,

$$\mathbf{P}(\mathbf{y}|\mathbf{x}') \geq \mathbf{P}(\mathbf{y}|\mathbf{x}) \geq p_Y(\mathbf{y})2^{N\kappa}.$$

Multiplying by $\dfrac{p_X(\mathbf{x}')}{p_Y(\mathbf{y})}$ gives $\mathbb{P}\left(X = \mathbf{x}'|Y = \mathbf{y}\right) \geq p_X(\mathbf{x}')2^{N\kappa}$. Then summing over $\mathbf{x}' \in S_{\mathbf{y}}(\mathbf{x})$ gives $1 \geq \mathbb{P}\left(S_Y(\mathbf{x})|Y = \mathbf{y}\right) \geq Q_{\mathbf{y}}(\mathbf{x})2^{N\kappa}$, or

$$Q_{\mathbf{y}}(\mathbf{x}) \leq 2^{-N\kappa}. \tag{1.4.38}$$

Substituting (1.4.37) and (1.4.38) into (1.4.36) yields (1.4.32). □

Proof of Theorem 1.4.15 The proof of Theorem 1.4.15 can now be easily completed. Take $\bar{R} = c - 2\varepsilon$ and $\kappa = c - \varepsilon$. Then, as $r = 2^{N(\bar{R}+o(1))}$, we have that E does not exceed

$$\mathbb{P}(\Theta_N \leq c - \varepsilon) + 2^{N(c - 2\varepsilon - c + \varepsilon + o(1))} = \mathbb{P}(\Theta_N \leq c - \varepsilon) + 2^{-N\varepsilon}.$$

This quantity tends to zero as $N \to \infty$, because $\mathbb{P}(\Theta_N \leq c - \varepsilon) \to 0$ owing to the condition $\Theta_N \xrightarrow{\mathbb{P}} c$. Therefore, the random coding F gives the expected error probability that vanishes as $N \to \infty$.

By Theorem 1.4.11(i), for any $N \geq 1$ there exists a deterministic encoding $f = f_N$ such that, for $\bar{R} = c - 2\varepsilon$, $\lim_{N \to \infty} \varepsilon(f) = 0$. Hence, $\bar{R}$ is a reliable transmission rate. This is true for any $\varepsilon > 0$, thus $C \geq c$. □

The form of the argument used in the above proof was proposed by P. Whittle (who used it in his lectures at the University of Cambridge) and appeared in [52], pp. 114–117. We thank C. Goldie for this information. An alternative approach is based on the concept of joint typicality; this approach is used in Section 2.2 where we discuss channels with continuously distributed noise.

Theorems 1.4.17 and 1.4.19 may be extended to the case of a memoryless channel with an arbitrary (finite) output alphabet, $J_q = \{0, \ldots, q-1\}$. That is, at the input of the channel we now have a word $\mathbf{Y}^{(N)} = Y_1 \ldots Y_N$ where each Y_j takes a (random) value from J_q. The memoryless property means, as before, that

$$\mathbf{P}_{\text{ch}}\left(\mathbf{y}^{(N)}|\mathbf{x}^{(N)}\right) = \prod_{i=1}^{N} P(y_i \mid x_i), \qquad (1.4.39)$$

and the symbol-to-symbol channel probabilities $P(y|x)$ now form a $2 \times q$ stochastic matrix (the channel matrix). A memoryless channel is called symmetric if the rows of the channel matrix are permutations of each other and double symmetric if in addition the columns of the channel matrix are permutations of each other. The definitions of the reliable transmission rate and the channel capacity are carried through without change. The capacity of a memoryless binary channel is depicted in Figure 1.8.

Theorem 1.4.26   *The capacity of a memoryless symmetric channel with an output alphabet J_q is*

$$C \le \log q - h(p_0, \ldots, p_{q-1}) \qquad (1.4.40)$$

where $(p_0, \ldots, p_{q-1})$ is a row of the channel matrix. The equality is realised in the case of a double-symmetric channel, and the maximising random coding has IID symbols V_i taking values from J_q with probability $1/q$.

Proof   The proof is carried out as in the binary case, by using the fact that $I(X_1 : Y_1) = h(Y_1) - h(Y_1|X_1) \le \log q - h(Y_1|X_1)$. But in the symmetric case

$$h(Y_1 \mid X_1) = -\sum_{x,y} \mathbb{P}(X_1 = x) P(y \mid x) \log P(y \mid x)$$

$$= -\sum_x \mathbb{P}(X_1 = x) \sum_k p_k \log p_k = h(p_0, \ldots, p_{q-1}). \qquad (1.4.41)$$

If, in addition, the columns of the channel matrix are permutations of each other, then $h(Y_1)$ attains $\log q$. Indeed, take a random coding as suggested. Then $\mathbb{P}(Y = y)$
$= \sum_{x=0}^{q-1} \mathbb{P}(X_1 = x) P(y|x) = \frac{1}{q} \sum_x P(y|x)$. The sum $\sum_x P(y|x)$ is along a column of the
channel matrix, and it does not depend on y. Hence, $\mathbb{P}(Y = y)$ does not depend on $y \in I_q$, which means equidistribution. $\square$

Remark 1.4.27   (a) In the random coding F used in Worked Examples 1.4.24 and 1.4.25 and Theorems 1.4.6, 1.4.15 and 1.4.17, the expected error-probability $E \to 0$ with $N \to \infty$. This guarantees not only the existence of a 'good' non-random coding for which the error-probability E vanishes as $N \to \infty$ (see Theorem 1.4.11(i)), but also that 'almost' all codes are asymptotically good. In fact, by Theorem 1.4.11(ii),

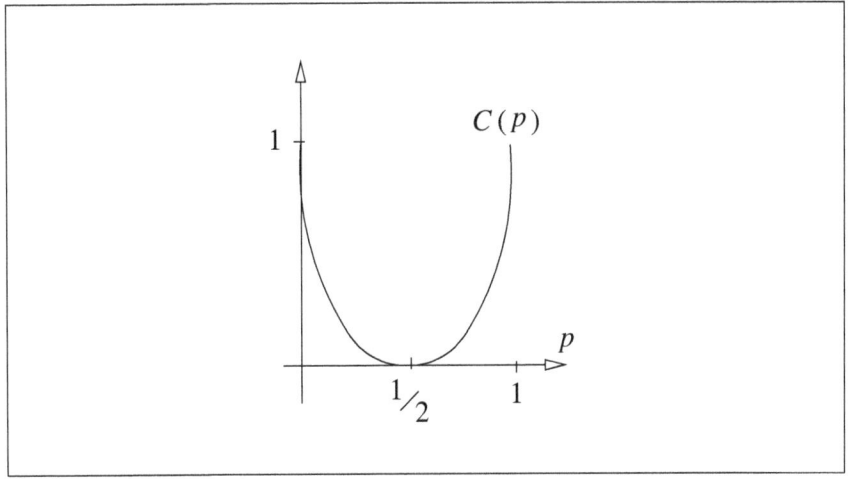

Figure 1.8

with $\rho = 1 - \sqrt{E}$, $\mathbb{P}\left(\varepsilon(F) < \sqrt{E}\right) \geq 1 - \sqrt{E} \to 1$, as $N \to \infty$. However, this does not help to *find* a good code: constructing good codes remains a challenging task in information theory, and we will return to this problem later.

Worked Example 1.4.28 *Bits are transmitted along a communication channel. With probability λ a bit may be inverted and with probability μ it may be rendered illegible. The fates of successive bits are independent. Determine the optimal coding for, and the capacity of, the channel.*

Solution The channel matrix is 2×3:

$$\Pi = \left(\begin{array}{ccc} 1-\lambda-\mu & \lambda & \mu \\ \lambda & 1-\lambda-\mu & \mu \end{array} \right);$$

the rows are permutations of each other, and hence have equal entropies. Therefore, the conditional entropy $h(Y|X)$ equals

$$h(1-\lambda-\mu, \lambda, \mu) = -(1-\lambda-\mu)\log(1-\lambda-\mu) - \lambda\log\lambda - \mu\log\mu,$$

which does not depend on the distribution of the input symbol X.

Thus, $I(X:Y)$ is maximised when $h(Y)$ is. If $p_Y(0) = p$ and $p_Y(1) = q$, then

$$h(Y) = -\mu\log\mu - p\log p - q\log q,$$

which is maximised when $p = q = (1 - \mu)/2$ (by pooling), i.e. $p_X(0) = p_X(1) = 1/2$. This gives the following expression for the capacity:

$$-(1 - \mu)\log\frac{1 - \mu}{2} + (1 - \lambda - \mu)\log(1 - \lambda - \mu) + \lambda\log\lambda$$

$$= (1 - \mu)\left(1 - h\left(\frac{1 - \lambda - \mu}{1 - \mu}, \frac{\lambda}{1 - \mu}\right)\right).$$

$\square$

Worked Example 1.4.29

(a) (Data-processing inequality) *Consider two independent channels in series. A random variable X is sent through channel 1 and received as Y. Then it is sent through channel 2 and received as Z. Prove that*

$$I(X : Z) \leq I(X : Y),$$

so the further processing of the second channel can only reduce the mutual information.

The independence of the channels means that given Y, the random variables X and Z are conditionally independent. Deduce that

$$h(X, Z | Y) = h(X | Y) + h(Z | Y)$$

and

$$h(X, Y, Z) + h(Z) = h(X, Z) + h(Y, Z).$$

Define $I(X : Z | Y)$ as $h(X | Y) + h(Z | Y) - h(X, Z | Y)$ and show that

$$I(X : Z | Y) = I(X : Y) - I(X : Z).$$

Does the equality hold in the data processing inequality

$$I(X : Z) \neq I(X : Y)?$$

(b) *The input and output of a discrete-time channel are both expressed in an alphabet whose letters are the residue classes of integers mod r, where r is fixed. The transmitted letter $[x]$ is received as $[j + x]$ with probability p_j, where x and j are integers and $[c]$ denotes the residue class of c mod r. Calculate the capacity of the channel.*

Solution (a) Given Y, the random variables X and Z are conditionally independent. Hence,

$$h(X \mid Y) = h(X \mid Y, Z) \leq h(X \mid Z),$$

and

$$I(X : Y) = h(X) - h(X | Y) \geq h(X) - h(X \mid Z) = I(X : Z).$$

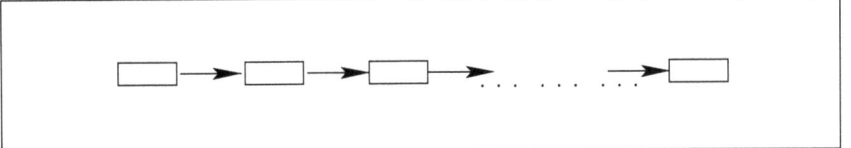

Figure 1.9

The equality holds iff X and Y are conditionally independent given Z, e.g. if the second channel is error-free $(Y,Z) \mapsto Z$ is one-to-one, or the first channel is fully noisy, i.e. X and Y are independent.

(b) The rows of the channel matrix are permutations of each other. Hence $h(Y|X) = h(p_0, \ldots, p_{r-1})$ does not depend on p_X. The quantity $h(Y)$ is maximised when $p_X(i) = 1/r$, which gives

$$C = \log r - h(p_0, \ldots, p_{r-1}).$$

$\square$

Worked Example 1.4.30 *Find the error-probability of a cascade of n identical independent binary symmetric channels (MBSCs), each with the error-probability $0 < p < 1$ (see Figure 1.9).*

Show that the capacity of the cascade tends to zero as $n \to \infty$.

Solution The channel matrix of a combined n-cascade channel is Π^n where

$$\Pi = \begin{pmatrix} 1-p & p \\ p & 1-p \end{pmatrix}.$$

Calculating the eigenvectors/values yields

$$\Pi^n = \frac{1}{2} \begin{pmatrix} 1+(1-2p)^n & 1-(1-2p)^n \\ 1-(1-2p)^n & 1+(1-2p)^n \end{pmatrix},$$

which gives the error-probability $1/2 \, (1-(1-2p)^n)$. If $0 < p < 1$, Π^n converges to

$$\begin{pmatrix} 1/2 & 1/2 \\ 1/2 & 1/2 \end{pmatrix},$$

and the capacity of the channel approaches

$$1 - h(1/2, 1/2) = 1 - 1 = 0.$$

If $p = 0$ or 1, the channel is error-free, and $C \equiv 1$.

$\square$

Worked Example 1.4.31   *Consider two independent MBCs, with capacities* C_1, C_2 *bits per second. Prove, or provide a counter-example to, each of the following claims about the capacity C of a compound channel formed as stated.*

(a) *If the channels are in series, with the output from one being fed into the other with no further coding, then* $C = \min[C_1, C_2]$.
(b) *Suppose the channels are used in parallel in the sense that at every second a symbol (from its input alphabet) is transmitted through channel 1 and the next symbol through channel 2; each channel thus emits one symbol each second. Then* $C = C_1 + C_2$.
(c) *If the channels have the same input alphabet and at each second a symbol is chosen and sent simultaneously down both channels, then* $C = \max[C_1, C_2]$.
(d) *If channel $i = 1, 2$ has matrix Π_i and the compound channel has*

$$\Pi = \begin{pmatrix} \Pi_1 & 0 \\ 0 & \Pi_2 \end{pmatrix},$$

then C is given by $2^C = 2^{C_1} + 2^{C_2}$. *To what mode of operation does this correspond?*

Solution (a)

$$X \qquad\qquad Y \qquad\qquad Z$$
$$\longrightarrow \quad \boxed{\text{channel 1}} \quad \longrightarrow \quad \boxed{\text{channel 2}} \quad \longrightarrow$$

As in Worked Example 1.4.29a,

$$I(X:Z) \leq I(X:Y), \quad I(X:Z) \leq I(Y:Z).$$

Hence,

$$C = \sup_{p_X} I(X:Z) \leq \sup_{p_X} I(X:Y) = C_1$$

and similarly

$$C \leq \sup_{p_Y} I(Y:Z) = C_2,$$

i.e. $C \leq \min[C_1, C_2]$. A strict inequality may occur: take $\delta \in (0, 1/2)$ and the matrices

$$\text{ch 1} \sim \begin{pmatrix} 1-\delta & \delta \\ \delta & 1-\delta \end{pmatrix}, \quad \text{ch 2} \sim \begin{pmatrix} 1-\delta & \delta \\ \delta & 1-\delta \end{pmatrix},$$

and

$$\text{ch } [1+2] \sim \frac{1}{2} \begin{pmatrix} (1-\delta)^2 + \delta^2 & 2\delta(1-\delta) \\ 2\delta(1-\delta) & (1-\delta)^2 + \delta^2 \end{pmatrix}.$$

Here, $1/2 > 2\delta(1-\delta) > \delta$,

$$C_1 = C_2 = 1 - h(\delta, 1 - \delta),$$

and

$$C = 1 - h(2\delta(1-\delta), 1 - 2\delta(1-\delta)) < C_i$$

because $h(\varepsilon, 1 - \varepsilon)$ strictly increases in $\varepsilon \in [0, 1/2]$.

(b)

$$X_1 \longrightarrow \boxed{\text{channel 1}} \longrightarrow Y_1$$

$$X_2 \longrightarrow \boxed{\text{channel 2}} \longrightarrow Y_2$$

The capacity of the combined channel

$$C = \sup_{p_{(X_1, X_2)}} I\big((X_1, X_2) : (Y_1, Y_2)\big).$$

But

$$\begin{aligned}
I\big((X_1, X_2) : (Y_1, Y_2)\big) &= h(Y_1, Y_2) - h\big(Y_1, Y_2 | X_1, X_2\big) \\
&\leq h(Y_1) + h(Y_2) - h(Y_1 | X_1) - h(Y_2 | X_2) \\
&= I(X_1 : Y_1) + I(X_2 : Y_2);
\end{aligned}$$

equality applies iff X_1 and X_2 are independent. Thus, $C = C_1 + C_2$ and the maximising $p_{(X_1, X_2)}$ is $p_{X_1} \times p_{X_2}$ where p_{X_1} and p_{X_2} are maximisers for $I(X_1 : Y_1)$ and $I(X_2 : Y_2)$.

(c)

$$\boxed{\text{channel 1}} \longrightarrow Y_1$$

$$X \nearrow$$

$$\searrow$$

$$\boxed{\text{channel 2}} \longrightarrow Y_2$$

Here,

$$C = \sup_{p_X} I\big(X : (Y_1 : Y_2)\big)$$

and

$$\begin{aligned}
I\big((Y_1 : Y_2) : X\big) &= h(X) - h\big(X | Y_1, Y_2\big) \\
&\geq h(X) - \min_{j=1,2} h(X | Y_j) = \min_{j=1,2} I(X : Y_j).
\end{aligned}$$

Thus, $\mathbb{C} \geq \max[C_1, C_2]$. A strict inequality may occur: take an example from part (a). Here, $C_i = 1 - h(\delta, 1 - \delta)$. Also,

$$I\big((Y_1, Y_2) : X\big) = h(Y_1, Y_2) - h\big(Y_1, Y_2 | X\big)$$
$$= h(Y_1, Y_2) - h(Y_1 | X) - h(Y_2 | X)$$
$$= h(Y_1, Y_2) - 2h(\delta, 1 - \delta).$$

If we set $p_X(0) = p_X(1) = 1/2$ then

$$(Y_1, Y_2) = (0, 0) \quad \text{with probability} \quad \big[(1 - \delta)^2 + \delta^2\big]/2,$$
$$(Y_1, Y_2) = (1, 1) \quad \text{with probability} \quad \big[(1 - \delta)^2 + \delta^2\big]/2,$$
$$(Y_1, Y_2) = (1, 0) \quad \text{with probability} \quad \delta(1 - \delta),$$
$$(Y_1, Y_2) = (0, 1) \quad \text{with probability} \quad \delta(1 - \delta),$$

with

$$h(Y_1, Y_2) = 1 + h\big(2\delta(1 - \delta), \ 1 - 2\delta(1 - \delta)\big),$$

and

$$I\big((Y_1, Y_2) : X\big) = 1 + h\big(2\delta(1 - \delta), \ 1 - 2\delta(1 - \delta)\big) - 2h(\delta, 1 - \delta)$$
$$> 1 - h(\delta, 1 - \delta) = C_i.$$

Hence, $C > C_i$, $i = 1, 2$.

(d)

The difference with part (c) is that every second only one symbol is sent, either to channel 1 or 2. If we fix probabilities α and $1 - \alpha$ that a given symbol is sent through a particular channel then

$$I(X : Y) = h(\alpha, 1 - \alpha) + \alpha I(X_1 : Y_1) + (1 - \alpha) I(X_2 : Y_2). \tag{1.4.42}$$

Indeed, $I(X : Y) = h(Y) - h(Y | X)$, where

$$h(Y) = -\sum_y \alpha p_{Y_1}(y) \log \alpha p_{Y_1}(y) - \sum_y (1 - \alpha) p_{Y_2}(y) \log (1 - \alpha) p_{Y_2}(y)$$
$$= -\alpha \log \alpha - (1 - \alpha) \log(1 - \alpha) + \alpha h(Y_1) + (1 - \alpha) h(Y_2)$$

and

$$h(Y|X) = -\sum_{x,y} \alpha p_{X_1,Y_1}(x,y) \log p_{Y_1|X_1}(y|x)$$
$$- \sum_{x,y}(1-\alpha)p_{X_2,Y_2}(y|x) \log p_{Y_2|X_2}(y|x)$$
$$= \alpha h(Y_1|X_1) + (1-\alpha)h(Y_2|X_2)$$

proving (1.4.42). This yields

$$C = \max_{0 \leq \alpha \leq 1} \left[h(\alpha, 1-\alpha) + \alpha C_1 + (1-\alpha)C_2 \right];$$

the maximum is given by

$$\alpha = 2^{C_1}/(2^{C_1}+2^{C_2}), \quad 1-\alpha = 2^{C_2}/(2^{C_1}+2^{C_2}),$$

and $C = \log\left(2^{C_1}+2^{C_2}\right)$.   □

Worked Example 1.4.32   *A spy sends messages to his contact as follows. Each hour either he does not telephone, or he telephones and allows the telephone to ring a certain number of times – not more than N, for fear of detection. His contact does not answer, but merely notes whether or not the telephone rings, and, if so, how many times. Because of deficiencies in the telephone system, calls may fail to be properly connected; the correct connection has probability p, where $0 < p < 1$, and is independent for distinct calls, but the spy has no means of knowing which calls reach his contact. If connection is made, then the number of rings is transmitted correctly. The probability of a false connection from another subscriber at a time when no call is made may be neglected. Write down the channel matrix for this channel and calculate the capacity explicitly. Determine a condition on N in terms of p which will imply, with optimal coding, that the spy will always telephone.*

Solution The channel alphabet is $\{0,1,\ldots,N\}$: $0 \sim$ non-call (in a given hour), and $j \geq 1 \sim j$ rings. The channel matrix is $P(0|0) = 1$, $P(0|j) = 1 - p$ and $P(j|j) = p$, $1 \leq j \leq N$, and $h(Y|X) = -q(p \log p + (1-p) \log(1-p))$, where $q = p_X(X \geq 1)$. Furthermore, given q, $h(Y)$ attains its maximum when

$$p_Y(0) = 1 - pq, \quad p_Y(k) = \frac{pq}{N}, \quad 1 \leq k \leq N.$$

Maximising $I(X:Y) = h(Y) - h(Y|X)$ in q yields $p(1-p)^{(1-p)/p} \times (1-pq) = pq/N$ or

$$q = \min\left[\frac{1}{p}\left(1 + \frac{1}{Np}\left(\frac{1}{1-p}\right)^{(1-p)/p}\right)^{-1}, 1\right].$$

The condition $q = 1$ is equivalent to $\log N \geq -\dfrac{1}{p}\log(1-p)$, i.e.

$$N \geq \frac{1}{(1-p)^{1/p}}.$$ $\qquad\qquad\qquad\qquad\qquad\qquad\qquad\qquad\qquad\Box$

1.5 Differential entropy and its properties

Definition 1.5.1   Suppose that the random variable X has a probability density (PDF) $p(\mathbf{x})$, $\mathbf{x} \in \mathbb{R}^n$:

$$\mathbb{P}\{X \in A\} = \int_A p(\mathbf{x})d\mathbf{x}$$

for any (measurable) set $A \subseteq \mathbb{R}^n$, where $p(\mathbf{x}) \geq 0$, $\mathbf{x} \in \mathbb{R}^n$, and $\int_{\mathbb{R}^n} d\mathbf{x}\, p(\mathbf{x}) = 1$. The *differential* entropy $h_{\mathrm{diff}}(X)$ is defined as

$$h_{\mathrm{diff}}(X) = -\int p(\mathbf{x}) \log p(\mathbf{x})d\mathbf{x}, \qquad\qquad (1.5.1)$$

under the assumption that the integral is absolutely convergent. As in the discrete case, $h_{\mathrm{diff}}(X)$ may be considered as a functional of the density $p : \mathbf{x} \in \mathbb{R}^n \mapsto \mathbb{R}_+ = [0, \infty)$. The difference is however that $h_{\mathrm{diff}}(X)$ may be negative, e.g. for a uniform distribution on $[0, a]$, $h_{\mathrm{diff}}(X) = -\int_0^a dx(1/a)\log(1/a) = \log a < 0$ for $a < 1$. [We write x instead of $\mathbf{x}$ when $x \in \mathbb{R}$.] The relative, joint and conditional differential entropy are defined similarly to the discrete case:

$$h_{\mathrm{diff}}(X||Y) = D_{\mathrm{diff}}(p||p') = -\int p(\mathbf{x}) \log \frac{p'(\mathbf{x})}{p(\mathbf{x})} d\mathbf{x}, \qquad\qquad (1.5.2)$$

$$h_{\mathrm{diff}}(X,Y) = -\int p_{X,Y}(\mathbf{x},\mathbf{y}) \log p_{X,Y}(\mathbf{x},\mathbf{y})d\mathbf{x}d\mathbf{y}, \qquad\qquad (1.5.3)$$

$$\begin{aligned} h_{\mathrm{diff}}(X|Y) &= -\int p_{X,Y}(\mathbf{x},\mathbf{y}) \log p_{X|Y}(\mathbf{x}|\mathbf{y})d\mathbf{x}d\mathbf{y} \\ &= h_{\mathrm{diff}}(X,Y) - h_{\mathrm{diff}}(Y), \end{aligned} \qquad\qquad (1.5.4)$$

again under the assumption that the integrals are absolutely convergent. Here, $p_{X,Y}$ is the joint probability density and $p_{X|Y}$ the conditional density (the PDF of the conditional distribution). Henceforth we will omit the subscript diff when it is clear what entropy is being addressed. The assertions of Theorems 1.2.3(b),(c), 1.2.12, and 1.2.18 are carried through for the differential entropies: the proofs are completely similar and will not be repeated.

Remark 1.5.2   Let $0 \leq x \leq 1$. Then x can be written as a sum $\sum_{n \geq 1} \alpha_n 2^{-n}$ where $\alpha_n (= \alpha_n(x))$ equals 0 or 1. For 'most' of the numbers x the series is not reduced to a finite sum (that is, there are infinitely many n such that $\alpha_n = 1$; the formal statement is that the (Lebesgue) measure of the set of numbers $x \in (0,1)$ with infinitely many $\alpha_n(x) = 1$ equals one). Thus, if we want to 'encode' x by means of binary digits we would need, typically, a codeword of an infinite length. In other words, a typical value for a uniform random variable X with $0 \le X \le 1$ requires infinitely many bits for its 'exact' description. It is easy to make a similar conclusion in a general case when X has a PDF $f_X(x)$.

However, if we wish to represent the outcome of the random variable X with an accuracy of first n binary digits then we need, on average, $n + h(X)$ bits where $h(X)$ is the differential entropy of X. Differential entropies can be both positive and negative, and can even be $-\infty$. Since $h(X)$ can be of either sign, $n + h(X)$ can be greater or less than n. In the discrete case the entropy is both shift and scale invariant since it depends only on probabilities $p_1, \ldots, p_m$, not on the values of the random variable. However, the differential entropy is shift but not scale invariant as is evident from the identity (cf. Theorem 1.5.7)

$$h(aX + b) = h(X) + \log |a|.$$

However, the relative entropy, i.e. Kullback–Leibler distance $D(p\|q)$, is scale invariant.

Worked Example 1.5.3   *Consider a PDF on $0 \le x \le e^{-1}$,*

$$f_r(x) = C_r \frac{1}{x(-\ln x)^{r+1}}, \quad 0 < r < 1.$$

Then the differential entropy $h(X) = -\infty$.

Solution After the substitution $y = -\ln x$ we obtain

$$\int_0^{e^{-1}} \frac{1}{x(-\ln x)^{r+1}} dx = \int_1^\infty \frac{1}{y^{r+1}} dy = \frac{1}{r}.$$

Thus, $C_r = r$. Further, using $z = \ln(-\ln x)$

$$\int_0^{e^{-1}} \frac{\ln(-\ln x)}{x(-\ln x)^{r+1}} dx = \int_0^\infty z e^{-rz} dz = \frac{1}{r^2}.$$

Hence,

$$\begin{aligned} h(X) &= -\int f_r(x) \ln f_r(x) dx \\ &= \int f_r(x) \left(-\ln r + \ln x + (r+1)\ln(-\ln x) \right) dx \\ &= -\ln r - \int_0^{e^{-1}} \left[\frac{r}{x(-\ln x)^r} - r(r+1)\frac{\ln(-\ln x)}{x(-\ln x)^{r+1}} \right] dx, \end{aligned}$$

so that for $0 < r < 1$, the second term is infinite, and two others are finite. $\quad\square$

Theorem 1.5.4   *Let* $X = (X_1,\dots,X_d) \sim N(\mu,C)$ *be a multivariate normal random vector, of mean* $\mu = (\mu_1,\dots,\mu_d)$ *and covariance matrix* $C = (c_{ij})$, *i.e.* $\mathbb{E}X_i = \mu_i$, $\mathbb{E}(X_i - \mu_i)(X_j - \mu_j) = c_{ij} = c_{ji}$, $1 \le i,j \le d$. *Then*

$$h(X) = \frac{1}{2}\log\left[(2\pi e)^d \det C\right]. \tag{1.5.5}$$

Proof   The PDF $p_X(\mathbf{x})$ is

$$p(\mathbf{x}) = \frac{1}{\left((2\pi)^d \det C\right)^{1/2}} \exp\left[-\frac{1}{2}\left(\mathbf{x} - \mu, C^{-1}(\mathbf{x} - \mu)\right)\right], \quad \mathbf{x} \in \mathbb{R}^d.$$

Then $h(X)$ takes the form

$$
\begin{aligned}
&-\int_{\mathbb{R}^d} p(\mathbf{x})\left[-\frac{1}{2}\log\left((2\pi)^d \det C\right) - \frac{\log e}{2}\left(\mathbf{x} - \mu, C^{-1}(\mathbf{x} - \mu)\right)\right]\mathrm{d}\mathbf{x}\\
&= \frac{\log e}{2}\mathbb{E}\left[\sum_{i,j}(x_i - \mu_i)(x_j - \mu_j)\left(C^{-1}\right)_{ij}\right] + \frac{1}{2}\log\left((2\pi)^d \det C\right)\\
&= \frac{\log e}{2}\sum_{i,j}\left(C^{-1}\right)_{ij}\mathbb{E}(x_i - \mu_i)(x_j - \mu_j) + \frac{1}{2}\log\left((2\pi)^d \det C\right)\\
&= \frac{\log e}{2}\sum_{i,j}\left(C^{-1}\right)_{ij}C_{ji} + \frac{1}{2}\log\left((2\pi)^d \det C\right)\\
&= \frac{d\log e}{2} + \frac{1}{2}\log\left((2\pi)^d \det C\right) = \frac{1}{2}\log\left((2\pi e)^d \det C\right).
\end{aligned}
$$

□

Theorem 1.5.5   *For a random vector* $X = (X_1,\dots,X_d)$ *with mean* μ *and covariance matrix* $C = (C_{ij})$ *(i.e.* $C_{ij} = \mathbb{E}\left[(X_i - \mu_i)(X_j - \mu_j)\right] = C_{ji}$),

$$h(X) \le \frac{1}{2}\log\left((2\pi e)^d \det C\right), \tag{1.5.6}$$

with the equality iff X *is multivariate normal.*

Proof   Let $p(\mathbf{x})$ be the PDF of X and $p^0(\mathbf{x})$ the normal density with mean μ and covariance matrix C. Without loss of generality assume $\mu = 0$. Observe that $\log p^0(\mathbf{x})$ is, up to an additive constant term, a quadratic form in x_k. Furthermore, for each monomial $x_i x_j$, $\int d\mathbf{x} p^0(\mathbf{x}) x_i x_j = \int d\mathbf{x} p(\mathbf{x}) x_i x_j = C_{ij} = C_{ji}$, and the moment of quadratic form $\log p^0(\mathbf{x})$ are equal. We have

$$
\begin{aligned}
0 \;\le\; D(p\|p^0) \;\text{(by Gibbs)} \;&=\; \int p(\mathbf{x})\log\frac{p(\mathbf{x})}{p^0(\mathbf{x})}d\mathbf{x}\\
&=\; -h(p) - \int p(\mathbf{x})\log p^0(\mathbf{x})d\mathbf{x}\\
&=\; -h(p) - \int p^0(\mathbf{x})\log p^0(\mathbf{x})d\mathbf{x}\\
\text{(by the above remark)} \;&=\; -h(p) + h(p^0).
\end{aligned}
$$

The equality holds iff $p = p^0$. $\qquad\square$

Worked Example 1.5.6

(a) *Show that the exponential density maximises the differential entropy among the PDFs on $[0,\infty)$ with given mean, and the normal density maximises the differential entropy among the PDFs on $\mathbb{R}$ with a given variance.*

Moreover, let $\mathbf{X} = (X_1,\dots,X_d)^T$ be a random vector with $\mathbb{E}\mathbf{X} = 0$ and $\mathbb{E}X_i X_j = C_{ij}, 1 \le i, j \le d$. Then $h_{\mathrm{diff}}(\mathbf{X}) \le \frac{1}{2}\log\big((2\pi e)^d \det(C_{ij})\big)$, with equality iff $\mathbf{X} \sim N(0,C)$.

(b) *Prove that the bound $h(X) \le \log m$ (cf. (1.2.7)) for a random variable X taking not more than m values admits the following generalisation for a discrete random variable with infinitely many values in $\mathbb{Z}_+$:*

$$
h(X) \le \frac{1}{2}\log\left(2\pi e\big(\mathrm{Var}\,X + \frac{1}{12}\big)\right).
$$

Solution (a) For the Gaussian case, see Theorem 1.5.5. In the exponential case, by the Gibbs inequality, for any random variable Y with PDF $f(y)$, $\int f(y)\log\big[f(y)e^{\lambda y}/\lambda\big]dy \ge 0$ or

$$
h(Y) \le (\lambda\mathbb{E}Y\log e - \log\lambda) = h(\mathrm{Exp}(\lambda)),
$$

with equality iff $Y \sim \mathrm{Exp}(\lambda)$, $\lambda = (\mathbb{E}Y)^{-1}$.

(b) Let X_0 be a discrete random variable with $\mathbb{P}(X_0 = i) = p_i$, $i = 1,2,\dots$, and the random variable U be independent of X_0 and uniform on $[0,1]$. Set $X = X_0 + U$. For a normal random variable Y with $\mathrm{Var}\,X = \mathrm{Var}\,Y$,

$$
h_{\mathrm{diff}}(X) \le h_{\mathrm{diff}}(Y) = \frac{1}{2}\log\left(2\pi e \mathrm{Var}\,Y\right) = \frac{1}{2}\log\left(2\pi e(\mathrm{Var}\,X + \frac{1}{12})\right).
$$

$\qquad\square$

The value of $\mathbb{E}X$ is not essential for $h(X)$ as the following theorem shows.

Theorem 1.5.7

(a) *The differential entropy is not changed under the shift: for all* $\mathbf{y} \in \mathbb{R}^d$,

$$h(X + \mathbf{y}) = h(X).$$

(b) *The differential entropy changes additively under multiplication:*

$$h(aX) = h(X) + \log|a|, \quad \text{for all } a \in \mathbb{R}.$$

Furthermore, if $A = (A_{ij})$ *is a* $d \times d$ *non-degenerate matrix, consider the affine transformation* $\mathbf{x} \in \mathbb{R}^d \to A\mathbf{x} + \mathbf{y} \in \mathbb{R}^d$.

(c) *Then*

$$h(AX + \mathbf{y}) = h(X) + \log|\det A|. \tag{1.5.7}$$

Proof   The proof is straightforward and left as an exercise   □

Worked Example 1.5.8   (The data-processing inequality for the relative entropy) *Let S be a finite set, and $\Pi = (\Pi(x,y), x, y \in S)$ be a stochastic kernel (that is, for all $x, y \in S$, $\Pi(x,y) \geq 0$ and $\sum_{y \in S} \Pi(x,y) = 1$; in other words, $\Pi(x,y)$ is a transition probability in a Markov chain). Prove that $D(p_1\Pi\|p_2\Pi) \leq D(p_1\|p_2)$ where $p_i\Pi(y) = \sum_{x \in S} p_i(x)\Pi(x,y)$, $y \in S$ (that is, applying a Markov operator to both probability distributions cannot increase the relative entropy).*

Extend this fact to the case of the differential entropy.

Solution   In the discrete case Π is defined by a stochastic matrix $(\Pi(x,y))$. By the log-sum inequality (cf. PSE II, p. 426), for all y

$$\sum_{x} p_1(x)\Pi(x,y) \log \frac{\sum_{w} p_1(w)\Pi(w,y)}{\sum_{z} p_2(z)\Pi(z,y)}$$

$$\leq \sum_{x} p_1(x)\Pi(x,y) \log \frac{p_1(x)\Pi(x,y)}{p_2(x)\Pi(x,y)}$$

$$= \sum_{x} p_1(x)\Pi(x,y) \log \frac{p_1(x)}{p_2(x)}.$$

Taking summation over y we obtain

$$D(p_1\Pi\|p_2\Pi) = \sum_{x}\sum_{y} p_1(x)\Pi(x,y) \log \frac{\sum_{w} p_1(w)\Pi(w,y)}{\sum_{z} p_2(z)\Pi(z,y)}$$

$$\leq \sum_{x}\sum_{y} p_1(x)\Pi(x,y) \log \frac{p_1(x)}{p_2(x)} = D(p_1\|p_2).$$

In the continuous case a similar inequality holds if we replace summation by integration.   □

The concept of differential entropy has proved to be useful in a great variety of situations, very often quite unexpectedly. We consider here inequalities for determinants and ratios of determinants of positive definite matrices (cf. [39], [36]). Recall that the covariance matrix $C = (C_{ij})$ of a random vector $\mathbf{X} = (X_1, \ldots, X_d)$ is positive definite, i.e. for any complex vector $\mathbf{y} = (y_1, \ldots, y_d)$, the scalar product $(\mathbf{y}, C\mathbf{y}) = \sum_{i,j} C_{ij} y_i \overline{y_j}$ is written as

$$\sum_{i,j} \mathbb{E}(X_i - \mu_i)(X_j - \mu_j) y_i \overline{y_j} = \mathbb{E} \left| \sum_i (X_i - \mu_i) y_i \right|^2 \geq 0.$$

Conversely, for any positive definite matrix C there exists a PDF for which C is a covariance matrix, e.g. a multivariate normal distribution (if C is not strictly positive definite, the distribution is degenerate).

Worked Example 1.5.9   *If C is positive definite then $\log[\det C]$ is concave in C.*

Solution Take two positive definite matrices $C^{(0)}$ and $C^{(1)}$ and $\lambda \in [0,1]$. Let $\mathbf{X}^{(0)}$ and $\mathbf{X}^{(1)}$ be two multivariate normal vectors, $\mathbf{X}^{(i)} \sim N(0, C^{(i)})$. Set, as in the proof of Theorem 1.2.18, $X = \mathbf{X}^{(\Lambda)}$, where the random variable Λ takes two values, 0 and 1, with probabilities λ and $1 - \lambda$, respectively, and is independent of $\mathbf{X}^{(0)}$ and $\mathbf{X}^{(1)}$. Then the random variable X has covariance $C = \lambda C^{(0)} + (1 - \lambda) C^{(1)}$, although X need not be normal. Thus,

$$
\begin{aligned}
&\frac{1}{2} \log \left(2\pi e\right)^d + \frac{1}{2} \log \left[\det \left(\lambda C^{(0)} + (1 - \lambda) C^{(1)} \right) \right] \\
&= \frac{1}{2} \log \left((2\pi e)^d \det C \right) \geq h(X) \quad \text{(by Theorem 1.5.5)} \\
&\geq h(X|\Lambda) \quad \text{(by Theorem 1.2.11)} \\
&= \frac{\lambda}{2} \log \left((2\pi e)^d \det C^{(0)} \right) + \frac{1 - \lambda}{2} \log \left((2\pi e)^d \det C^{(1)} \right) \\
&= \frac{1}{2} \left[\log \left(2\pi e\right)^d + \lambda \log \left(\det C^{(0)} \right) + (1 - \lambda) \log \left(\det C^{(1)} \right) \right].
\end{aligned}
$$

□

This property is often called the Ky Fan inequality and was proved initially in 1950 by using much more involved methods. Another famous inequality is due to Hadamard:

Worked Example 1.5.10   *For a positive definite matrix $C = (C_{ij})$,*

$$\det C \leq \prod_i C_{ii}, \tag{1.5.8}$$

and the equality holds iff C is diagonal.

Solution If $X = (X_1, \ldots, X_n) \sim N(0, C)$ then

$$\frac{1}{2} \log \left[(2\pi e)^d \det C \right] = h(X) \le \sum_i h(X_i) = \sum_i \frac{1}{2} \log(2\pi e C_{ii}),$$

with equality iff $X_1, \ldots, X_n$ are independent, i.e. C is diagonal.   □

Next we discuss the so-called entropy–power inequality (EPI). The situation with the EPI is quite intriguing: it is considered one of the 'mysterious' facts of information theory, lacking a straightforward interpretation. It was proposed by Shannon; the book [141] contains a sketch of an argument supporting this inequality. However, the first rigorous proof of the EPI only appeared nearly 20 years later, under some rather restrictive conditions that are still the subject of painstaking improvement. Shannon used the EPI in order to bound the capacity of an additive channel with continuous noise by that of a Gaussian channel; see Chapter 4. The EPI is also related to important properties of monotonicity of entropy; an example is Theorem 1.5.15 below.

The existing proofs of the EPI are not completely elementary; see [82] for one of the more transparent proofs.

Theorem 1.5.11 (Entropy–power inequality). *For two independent random variables X and Y with PDFs $f_X(x)$ and $f_Y(x)$, $x \in \mathbf{R}^1$,*

$$h(X + Y) \ge h(X' + Y'), \tag{1.5.9}$$

where X' and Y' are independent normal random variables with $h(X) = h(X')$ and $h(Y) = h(Y')$.

In the d-dimensional case the entropy–power inequality is as follows.

For two independent random variables X and Y with PDFs $f_X(x)$ and $f_Y(x)$, $x \in \mathbf{R}^d$,

$$e^{2h(X+Y)/d} \ge e^{2h(X)/d} + e^{2h(Y)/d}. \tag{1.5.10}$$

It is easy to see that for $d = 1$ (1.5.9) and (1.5.10) are equivalent. In general, inequality (1.5.9) implies (1.5.10) via (1.5.13) below which can be established independently. Note that inequality (1.5.10) may be true or false for discrete random variables. Consider the following example: let $X \sim Y$ be independent with $P_X(0) = 1/6, P_X(1) = 2/3, P_X(2) = 1/6$. Then

$$h(X) = h(Y) = \ln 6 - \frac{2}{3} \ln 4, \ h(X + Y) = \ln 36 - \frac{16}{36} \ln 8 - \frac{18}{36} \ln 18.$$

By inspection, $e^{2h(X+Y)} = e^{2h(X)} + e^{2h(Y)}$. If X and Y are non-random constants then $h(X) = h(Y) = h(X + Y) = 0$, and the EPI is obviously violated. We conclude that the existence of PDFs is an essential condition that cannot be omitted. In a different form EPI could be extended to discrete random variables, but we do not discuss this theory here.

Sometimes the differential entropy is defined as $h(X) = -\mathbb{E}\log_2 p(X)$; then (1.5.10) takes the form $2^{h(X+Y)/d} \geq 2^{h(X)/d} + 2^{h(Y)/d}$.

The entropy–power inequality plays a very important role not only in information theory and probability but in geometry and analysis as well. For illustration we present below the famous Brunn–Minkowski theorem that is a particular case of the EPI. Define the set sum of two sets as

$$A_1 + A_2 = \{x_1 + x_2 : x_1 \in A_1, x_2 \in A_2\}.$$

By definition $A + \emptyset = A$.

Theorem 1.5.12 (Brunn–Minkowski)

(a) *Let A_1 and A_2 be measurable sets. Then the volume*

$$V(A_1 + A_2)^{1/d} \geq V(A_1)^{1/d} + V(A_2)^{1/d}. \tag{1.5.11}$$

(b) *The volume of the set sum of two sets A_1 and A_2 is greater than the volume of the set sum of two balls B_1 and B_2 with the same volume as A_1 and A_2, respectively:*

$$V(A_1 + A_2) \geq V(B_1 + B_2), \tag{1.5.12}$$

where B_1 and B_2 are spheres with $V(A_1) = V(B_1)$ and $V(A_2) = V(B_2)$.

Worked Example 1.5.13 *Let C_1, C_2 be positive-definite $d \times d$ matrices. Then*

$$[\det(C_1 + C_2)]^{1/d} \geq [\det C_1]^{1/d} + [\det C_2]^{1/d}. \tag{1.5.13}$$

Solution Let $X_1 \sim N(0, C_1), X_2 \sim N(0, C_2)$, then $X_1 + X_2 \sim N(0, C_1 + C_2)$. The entropy–power inequality yields

$$(2\pi e)\left(\det(C_1 + C_2)\right)^{1/d} = e^{2h(X_1+X_2)/d}$$
$$\geq e^{2h(X_1)/d} + e^{2h(X_2)/d} = (2\pi e)\left(\det C_1\right)^{1/d} + (2\pi e)\left(\det C_2\right)^{1/d}.$$

$\square$

Worked Example 1.5.14 *A Töplitz $n \times n$ matrix C is characterised by the property that $C_{ij} = C_{rs}$ if $|i - j| = |r - s|$. Let $C_k = C(1, 2, \ldots, k)$ denote the principal minor of the Töplitz positive-definite matrix formed by the rows and columns $1, \ldots, k$. Prove that for $|C| = \det C$,*

$$|C_1| \geq |C_2|^{1/2} \geq \cdots \geq |C_n|^{1/n}, \tag{1.5.14}$$

$|C_n|/|C_{n-1}|$ is decreasing in n, and

$$\lim_{n\to\infty} \frac{|C_n|}{|C_{n-1}|} = \lim_{n\to\infty} |C_n|^{1/n}. \qquad (1.5.15)$$

Solution Let $(X_1, X_2, \ldots, X_n) \sim N(0, C_n)$. Then the quantities $h(X_k|X_{k-1}, \ldots, X_1)$ are decreasing in k, since

$$h(X_k|X_{k-1}, \ldots, X_1) = h(X_{k+1}|X_k, \ldots, X_2) \geq h(X_{k+1}|X_k, \ldots, X_1),$$

where the equality follows from the Töplitz assumption and the inequality from the fact that the conditioning reduces the entropy. Next, we use the result of Problem 1.8b from Section 1.6 that the running averages

$$\frac{1}{k} h(X_1, \ldots X_k) = \frac{1}{k} \sum_{i=1}^{k} h(X_i|X_{i-1}, \ldots X_1)$$

are decreasing in k. Then (1.5.14) follows from

$$h(X_1, \ldots X_k) = \frac{1}{2} \log[(2\pi e)^k |C_k|].$$

Since $h(X_n|X_{n-1}, \ldots, X_1)$ is a decreasing sequence, it has a limit. Hence, by the Cesáro mean theorem

$$\lim_{n\to\infty} \frac{h(X_1, X_2, \ldots X_n)}{n} = \lim_{n\to\infty} \frac{1}{n} \sum_{i=1}^{n} h(X_k|X_{k-1}, \ldots, X_1)$$

$$= \lim_{n\to\infty} h(X_n|X_{n-1}, \ldots, X_1).$$

Translating this to determinants, we obtain (1.5.15). $\qquad\square$

The entropy–power inequality could be immediately extended to the case of several summands

$$e^{2h\left(X_1 + \cdots + X_n\right)/d} \geq \sum_{i=1}^{n} e^{2h(X_i)/d}.$$

But, more interestingly, the following intermediate inequality holds true. Let $X_1, X_2, \ldots, X_{n+1}$ be IID square-integrable random variables. Then

$$e^{2h\left(X_1 + \cdots + X_n\right)/d} \geq \frac{1}{n} \sum_{j=1}^{n+1} e^{2h\left(\sum_{i\neq j} X_i\right)/d}. \qquad (1.5.16)$$

As was established, the differential entropy is maximised by a Gaussian distribution, under the constraint that the variance of the random variable under consideration is bounded from above. We will state without proof the following important result showing that the entropy increases on every summation step in the central limit theorem.

Theorem 1.5.15 *Let $X_1, X_2, \ldots$ be IID square-integrable random variables with* $\mathbb{E}X_i = 0$, *and* $\mathrm{Var}X_i = 1$. *Then*

$$h\left(\frac{X_1 + \cdots + X_n}{\sqrt{n}}\right) \le h\left(\frac{X_1 + \cdots + X_{n+1}}{\sqrt{n+1}}\right). \tag{1.5.17}$$

1.6 Additional problems for Chapter 1

Problem 1.1 *Let Σ_1 and Σ_2 be alphabets of sizes m and q. What does it mean to say that $f : \Sigma_1 \to \Sigma_2^*$ is a decipherable code? Deduce from the inequalities of Kraft and Gibbs that if letters are drawn from Σ_1 with probabilities $p_1, \ldots, p_m$ then the expected word length is at least $h(p_1, \ldots, p_m)/\log q$.*

Find a decipherable binary code consisting of codewords 011, 0111, 01111, 11111, *and three further codewords of length* 2. *How do you check that the code you have obtained is decipherable?*

Solution Introduce $\Sigma^* = \bigcup_{n \ge 0} \Sigma^n$, the set of all strings with digits from Σ. We send a message $x_1 x_2 \ldots x_n \in \Sigma_1^*$ as the concatenation $f(x_1)f(x_2)\ldots f(x_n) \in \Sigma_2^*$, i.e. f extends to a function $f^* : \Sigma_1^* \to \Sigma_2^*$. We say a code is decipherable if f^* is injective.

Kraft's inequality states that a prefix-free code $f : \Sigma_1 \to \Sigma_2^*$ with codeword-lengths $s_1, \ldots, s_m$ exists iff

$$\sum_{i=1}^{m} q^{-s_i} \le 1. \tag{1.6.1}$$

In fact, every decipherable code satisfies this inequality.

Gibbs' inequality states that if $p_1, \ldots, p_n$ and $\widehat{p}_1, \ldots, \widehat{p}_n$ are two probability distributions then

$$h(p_1, \ldots, p_n) = -\sum_{i=1}^{n} p_i \log p_i \le -\sum_{i=1}^{n} p_i \log \widehat{p}_i, \tag{1.6.2}$$

with equality iff $p_i \equiv \widehat{p}_i$.

Suppose that f is decipherable with codeword-lengths $s_1, \ldots, s_m$. Put $\widehat{p}_i = q^{-s_i}/c$ where $c = \sum_{i=1}^{m} q^{-s_i}$. Then, by Gibbs' inequality,

$$\begin{aligned}
h(p_1, \ldots, p_n) &\le -\sum_{i=1}^{n} p_i \log \widehat{p}_i \\
&= -\sum_{i=1}^{n} p_i(-s_i \log q - \log c) \\
&= \left(\sum_i p_i s_i\right) \log q + \left(\sum_i p_i\right) \log c.
\end{aligned}$$

By Kraft's inequality, $c \leq 1$, i.e. $\log c \leq 0$. We obtain that

$$\text{expected codeword-length } \sum_i p_i s_i \geq h(p_1, \ldots, p_n)/\log q.$$

In the example, the three extra codewords must be 00, 01, 10 (we cannot take 11, as then a sequence of ten 1s is not decodable). Reversing the order in every codeword gives a prefix-free code. But prefix-free codes are decipherable. Hence, the code is decipherable.

In conclusion, we present an alternative proof of necessity of Kraft's inequality. Denote $s = \max s_i$; let us agree to extend any word in $\mathscr{X}$ to the length s, say by adding some fixed symbol. If $x = x_1 x_2 \ldots x_{s_i} \in \mathscr{X}$, then any word of the form $x_1 x_2 \ldots x_{s_i} y_{s_i+1} \ldots y_s \notin \mathscr{X}$ because x is a prefix. But there are at most q^{s-s_i} of such words. Summing up on i, we obtain that the total number of excluded words is $\sum_{i=1}^m q^{s-s_i}$. But it cannot exceed the total number of words q^s. Hence, (1.6.1) follows:

$$q^s \sum_{i=1}^m q^{-s_i} \leq q^s.$$

□

Problem 1.2   *Consider an alphabet with m letters each of which appears with probability $1/m$. A binary Huffman code is used to encode the letters, in order to minimise the expected codeword-length $(s_1 + \cdots + s_m)/m$ where s_i is the length of a codeword assigned to letter i. Set $s = \max[s_i : 1 \leq i \leq m]$, and let n_ℓ be the number of codewords of length ℓ.*

(a) Show that $2 \leq n_s \leq m$.

(b) For what values of m is $n_s = m$?

(c) Determine s in terms of m.

(d) Prove that $n_{s-1} + n_s = m$, i.e. any two codeword-lengths differ by at most 1.

(e) Determine n_{s-1} and n_s.

(f) Describe the codeword-lengths for an idealised model of English (with $m = 27$) where all the symbols are equiprobable.

(g) Let now a binary Huffman code be used for encoding symbols $1, \ldots, m$ occurring with probabilities $p_1 \geq \cdots \geq p_m > 0$ where $\sum_{1 \leq j \leq m} p_j = 1$. Let s_1 be the length of a shortest codeword and s_m of a longest codeword. Determine the maximal and minimal values of s_m and s_1, and find binary trees for which they are attained.

Solution (a) Bound $n_s \geq 2$ follows from the tree-like structure of Huffman codes. More precisely, suppose $n_s = 1$, i.e. a maximum-length codeword is unique and corresponds to say letter i. Then the branch of length s leading to i can be pruned at the end, without violating the prefix-free condition. But this contradicts minimality.

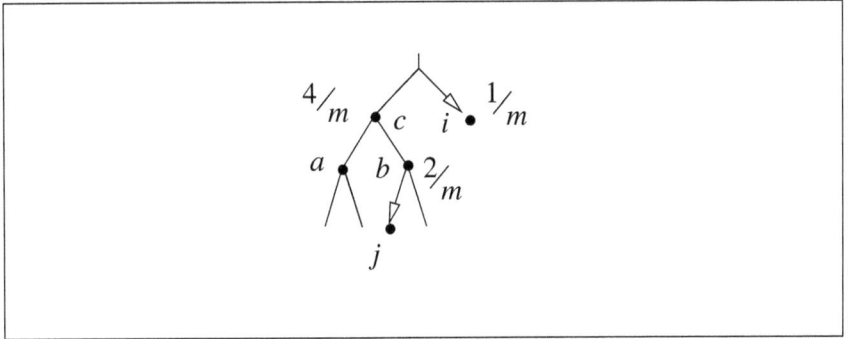

Figure 1.10

Bound $n_s \le m$ is obvious. (From what is said below it will follow that n_s is always even.)

(b) $n_s = m$ means all codewords are of equal length. This, obviously, happens iff $m = 2^k$, in which case $s = k$ (a perfect binary tree $\mathbb{T}_k$ with 2^k leaves).

(c) In general,

$$s = \begin{cases} \log m, & \text{if } m = 2^k, \\ \lceil \log m \rceil, & \text{if } m \ne 2^k. \end{cases}$$

The case $m = 2^k$ was discussed in (b), so let us assume that $m \ne 2^k$. Then $2^k < m < 2^{k+1}$ where $k = \lfloor \log m \rfloor$. This is clear from the observation that the binary tree for probabilities $1/m$ (we will call it a binary m-tree $\mathbb{B}_m$) contains the perfect binary tree $\mathbb{T}_k$ but is contained in $\mathbb{T}_{k+1}$. Hence, s is as above.

(d) Indeed, in the case of an equidistribution $1/m, \ldots, 1/m$ it is impossible to have a branch of the tree whose length differs from the maximal value s by two or more. In fact, suppose there is such a branch, B_i, of the binary tree leading to some letter i and choose a branch M_j of maximal length s leading to a letter j. In a conventional terminology, letter j was engaged in s merges and i in $t \le s-2$ merges. Ultimately, the branches B_i and M_j must merge, and this creates a contradiction. For example, the 'least controversial' picture is still 'illegal'; see Figure 1.10. Here, vertex i carrying probability $1/m$ should have been joined with vertex a or b carrying each probability $2/m$, instead of joining a and b (as in the figure), as it creates vertex c carrying probability $4/m$.

(e) We conclude that (i) for $m = 2^k$, the m-tree $\mathbb{B}_m$ coincides with $\mathbb{T}_k$, (ii) for $m \ne 2^k$ we obtain $\mathbb{B}_m$ in the following way. First, take a binary tree $\mathbb{T}_k$ where $k = \lfloor \log m \rfloor$, with $1 \le m - 2^k < 2^k$. Then $m - 2^k$ leaves of $\mathbb{T}_k$ are allowed to branch one step

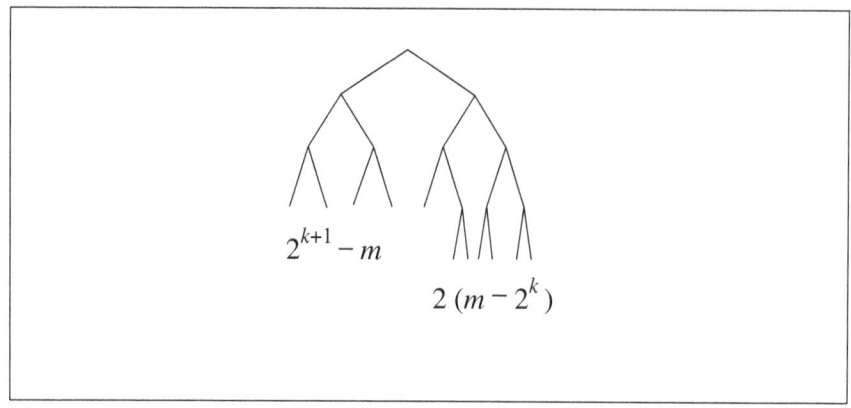

Figure 1.11

further: this generates $2(m - 2^k) = 2m - 2^{k+1}$ leaves of tree $\mathbb{T}_{k+1}$. The remaining $2^k - (m - 2^k) = 2^{k+1} - m$ leaves of T_k are left intact. See Figure 1.11. So,

$$n_{s-1} = 2^{k+1} - m, \quad n_s = 2m - 2^{k+1}, \quad \text{where} \quad k = \lceil \log m \rceil.$$

(f) In the example of English, with equidistribution among $m = 27 = 16 + 11$ symbols, we have 5 codewords of length 4 and 22 codewords of length 5. The average codeword-length is

$$\frac{5 \times 4 + 22 \times 5}{27} = \frac{130}{27} \approx 4.8.$$

(g) The minimal value for s_1 is 1 (obviously). The maximal value is $\lceil \log_2 m \rceil$, i.e. the positive integer l with $2^l < m \le 2^{l+1}$. The maximal value for s_m is $m - 1$ (obviously). The minimal value is $\lfloor \log_2 m \rfloor$, i.e. the natural l such that $2^{l-1} < m \le 2^l$.

The tree that yields $s_1 = 1$ and $s_m = m - 1$ is given in Figure 1.12.

It is characterised by

i	$f(i)$	s_i
1	0	1
2	10	2
$\vdots$	$\vdots$	$\vdots$
$m - 1$	11...10	$m - 1$
m	11...11	$m - 1$

and is generated when

$$p_1 > p_2 + \cdots + p_m > 2(p_3 + \cdots + p_m) > \cdots > 2^{m-1} p_m.$$

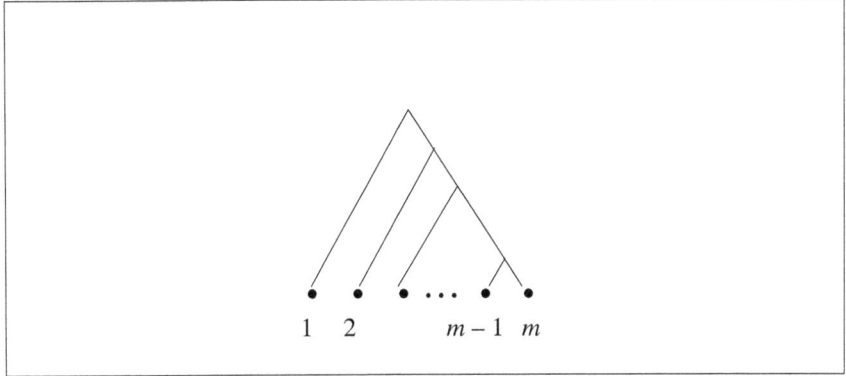

Figure 1.12

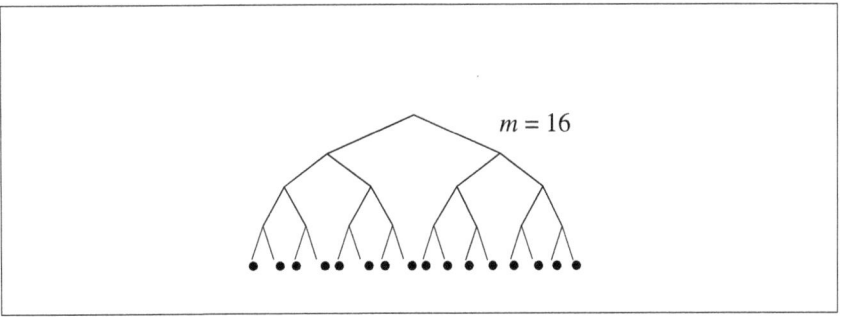

Figure 1.13

A tree that maximises s_1 and minimises s_m corresponds to uniform probabilities where $p_1 = \cdots = p_m = 1/m$. When $m = 2^l$, the branches of the tree have the same length $l = \log_2 m$ (a perfect binary tree); see Figure 1.13.

Otherwise, i.e. if $2^l < m < 2^{l+1}$, the tree has $2^{l+1} - m$ leaves at level l and $2(m - 2^l)$ leaves at level $l + 1$; see Figure 1.14.

Indeed, by the Huffman construction, the shortest branch cannot be larger than $\lceil \log_2 m \rceil$ and the longest shorter than $\lfloor \log_2 m \rfloor$, as the tree is always a subtree of a perfect binary tree.   □

Problem 1.3   *A binary erasure channel with erasure probability p is a discrete memoryless binary channel (MBC) with channel matrix*

$$\begin{pmatrix} 1-p & p & 0 \\ 0 & p & 1-p \end{pmatrix}.$$

State Shannon's second coding theorem (SCT) and use it to compute the capacity of this channel.

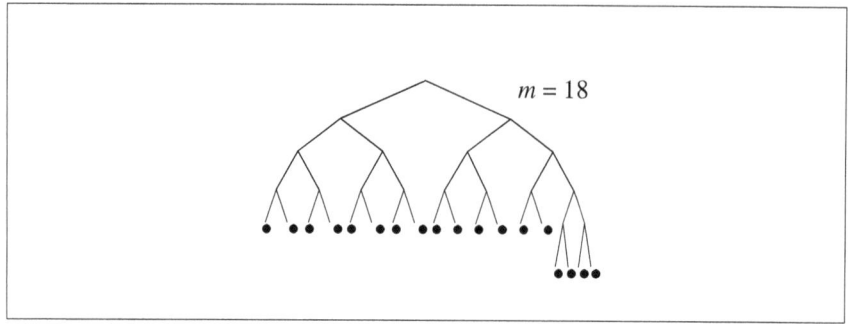

Figure 1.14

Solution The SCT states that for an MBC

$$(\text{capacity}) = (\text{maximum information transmitted per letter}).$$

Here the capacity is understood as the supremum over all reliable information rates while the RHS is defined as

$$\max_X I(X:Y)$$

where the random variables X and Y represent an input and the corresponding output.

The binary erasure channel keeps an input letter 0 or 1 intact with probability $1 - p$ and turns it to a splodge $*$ with probability p. An input random variable X is 0 with probability α and 1 with probability $1 - \alpha$. Then the output random variable Y takes three values:

$$\mathbb{P}(Y = 0) = (1 - p)\alpha,$$
$$\mathbb{P}(Y = 1) = (1 - p)(1 - \alpha),$$
$$\mathbb{P}(Y = *) = p.$$

Thus, conditional on the value of Y, we have

$$\left.\begin{array}{l} h(X|Y = 0) = 0, \\ h(X|Y = 1) = 0, \\ h(X|Y = *) = h(\alpha), \end{array}\right\} \text{ implying that } h(X|Y) = ph(\alpha).$$

Therefore,

$$\begin{aligned} \text{capacity} \ &= \max_\alpha I(X:Y) \\ &= \max_\alpha \left[h(X) - h(X|Y) \right] \\ &= \max_\alpha \left[h(\alpha) - ph(\alpha) \right] \\ &= (1 - p)\max_\alpha h(\alpha) = 1 - p, \end{aligned}$$

because $h(\alpha) = -\alpha \log \alpha - (1-\alpha) \log(1-\alpha)$ attains it maximum value 1 at $\alpha = 1/2$. $\qquad\square$

Problem 1.4 Let X and Y be two discrete random variables with corresponding cumulative distribution functions (CDF) $\mathbb{P}_X$ and $\mathbb{P}_Y$.
(a) Define the conditional entropy $h(X|Y)$, and show that it satisfies

$$h(X|Y) \le h(X),$$

giving necessary and sufficient conditions for equality.
(b) For each $\alpha \in [0,1]$, the mixture random variable $W(\alpha)$ has PDF of the form

$$\mathbb{P}_{W(\alpha)}(x) = \alpha \mathbb{P}_X(x) + (1-\alpha)\mathbb{P}_Y(x).$$

Prove that for all α the entropy of $W(\alpha)$ satisfies:

$$h(W(\alpha)) \ge \alpha h(X) + (1-\alpha)h(Y).$$

(c) Let $h_{\mathrm{Po}}(\lambda)$ be the entropy of a Poisson random variable $\mathrm{Po}(\lambda)$. Show that $h_{\mathrm{Po}}(\lambda)$ is a non-decreasing function of $\lambda > 0$.

Solution (a) By definition,

$$
\begin{aligned}
h(X|Y) &= h(X,Y) - h(Y)\\
&= -\sum_{x,y} \mathbb{P}(X=x, Y=y) \log \mathbb{P}(X=x, Y=y)\\
&\quad + \sum_{y} \mathbb{P}(Y=y) \log \mathbb{P}(Y=y).
\end{aligned}
$$

The inequality $h(X|Y) \le h(X)$ is equivalent to

$$h(X,Y) \le h(X) + h(Y),$$

and follows from the Gibbs inequality $\sum_i p_i \log \dfrac{p_i}{q_i} \ge 0$. In fact, take $i = (x,y)$ and

$$p_i = \mathbb{P}(X=x, Y=y), \quad q_i = \mathbb{P}(X=x)\mathbb{P}(Y=y).$$

Then

$$
\begin{aligned}
0 &\le \sum_{x,y} \mathbb{P}(X=x, Y=y) \log \frac{\mathbb{P}(X=x, Y=y)}{\mathbb{P}(X=x)\mathbb{P}(Y=y)}\\
&= \sum_{x,y} \mathbb{P}(X=x, Y=y) \log \mathbb{P}(X=x, Y=y)\\
&\quad - \sum_{x,y} \mathbb{P}(X=x, Y=y)\big[\log \mathbb{P}(X=x) + \log \mathbb{P}(Y=y)\big]\\
&= -h(X,Y) + h(X) + h(Y).
\end{aligned}
$$

Equality here occurs iff X and Y are independent.

(b) Define a random variable T equal to 0 with probability α and 1 with probability $1-\alpha$. Then the random variable Z has the distribution $W(\alpha)$ where

$$Z = \begin{cases} X, & \text{if } T = 0, \\ Y, & \text{if } T = 1. \end{cases}$$

By part (a),

$$h(Z|T) \leq h(Z),$$

with the LHS $= \alpha h(X) + (1-\alpha)h(Y)$, and the RHS $= h(W(\alpha))$.

(c) Observe that for independent random variables X and Y, $h(X+Y|X) = h(Y|X) = h(Y)$. Hence, again by part (a),

$$h(X+Y) \geq h(X+Y|X) = h(Y).$$

Using this fact, for all $\lambda_1 < \lambda_2$, take $X \sim \mathrm{Po}(\lambda_1)$, $Y \sim \mathrm{Po}(\lambda_2 - \lambda_1)$, independently. Then

$$h(X+Y) \geq h(X) \text{ implies } h_{\mathrm{Po}}(\lambda_2) \geq h_{\mathrm{Po}}(\lambda_1).$$

□

Problem 1.5   *What does it mean to transmit reliably at rate R through a binary symmetric channel (MBSC) with error-probability p? Assuming Shannon's second coding theorem (SCT), compute the supremum of all possible reliable transmission rates of an MBSC. What happens if: (i) p is very small; (ii) $p = 1/2$; or (iii) $p > 1/2$?*

Solution An MBSC can transmit reliably at rate R if there is a sequence of codes $\mathscr{X}_N$, $N = 1,2,\ldots$, with $\lfloor 2^{NR} \rfloor$ codewords such that

$$\widehat{e}(\mathscr{X}_N) = \max_{x \in \mathscr{X}_N} \mathbb{P}\left(\text{error}|x \text{ sent}\right) \to 0 \text{ as } N \to \infty.$$

By the SCT, the so-called operational channel capacity is $\sup R = \max_\alpha I(X:Y)$, the maximum information transmitted per input symbol. Here X is a Bernoulli random variable taking values 0 and 1 with probabilities $\alpha \in [0,1]$ and $1-\alpha$, and Y is the output random variable for the given input X. Next, $I(X:Y)$ is the mutual entropy (information):

$$I(X:Y) = h(X) - h(X|Y) = h(Y) - h(Y|X).$$

Observe that the binary entropy function $h(x) \leq 1$ with equality for $x = 1/2$. Selecting $\alpha = 1/2$ conclude that the MBSC with error probability p has the capacity

$$\max_{\alpha} I(X : Y) = \max_{\alpha} \left[h(Y) - h(Y|X) \right]$$
$$= \max_{\alpha} \left[h(\alpha p + (1-\alpha)(1-p)) - \eta(p) \right]$$
$$= 1 + p \log p + (1-p) \log(1-p).$$

(i) If p is small, the capacity is only slightly less than 1 (the capacity of a noiseless channel).

(ii) If $p = 1/2$, the capacity is zero (the channel is useless).

(iii) If $p > 1/2$, we may swap the labels on the output alphabet, replacing p by $1 - p$, and the channel capacity is non-zero. □

Problem 1.6   (i) *What is bigger π^e or e^π ?*

(ii) *Prove the log-sum inequality: for non-negative numbers $a_1, a_2, \ldots, a_n$ and $b_1, b_2, \ldots, b_n$,*

$$\sum_i a_i \log \frac{a_i}{b_i} \geq \left(\sum_i a_i \right) \log \left(\frac{\sum_i a_i}{\sum_i b_i} \right) \tag{1.6.3}$$

with equality iff $a_i/b_i = const$.

(iii) *Consider two discrete probability distributions $p(x)$ and $q(x)$. Define the relative entropy (or Kullback–Leibler distance) and prove the Gibbs inequality,*

$$D(p\|q) = \sum_x p(x) \log \left(\frac{p(x)}{q(x)} \right) \geq 0, \tag{1.6.4}$$

with equality iff $p(x) = q(x)$ for all x.

Using (1.6.4), show that for any positive functions $f(x)$ and $g(x)$, and for any finite set A,

$$\sum_{x \in A} f(x) \log \left(\frac{f(x)}{g(x)} \right) \geq \left(\sum_{x \in A} f(x) \right) \log \left(\frac{\sum_{x \in A} f(x)}{\sum_{x \in A} g(x)} \right).$$

Check that that for any $0 \leq p, q \leq 1$,

$$p \log \left(\frac{p}{q} \right) + (1-p) \log \left(\frac{1-p}{1-q} \right) \geq (2 \log_2 e)(q - p)^2, \tag{1.6.5}$$

and show that for any probability distributions $p = (p(x))$ and $q = (q(x))$,

$$D(p\|q) \geq \frac{\log_2 e}{2} \left(\sum_x |p(x) - q(x)| \right)^2. \tag{1.6.6}$$

Solution (i) Denote $x = \ln \pi$, and taking the logarithm twice obtain the inequality $x - 1 > \ln x$. This is true as $x > 1$, hence $e^\pi > \pi^e$.

(ii) Assume without loss of generality that $a_i > 0$ and $b_i > 0$. The function $g(x) = x \log x$ is strictly convex. Hence, by the Jensen inequality for any coefficients $\sum c_i = 1, c_i \geq 0$,

$$\sum c_i g(x_i) \geq g\left(\sum c_i x_i\right).$$

Selecting $c_i = b_i \left(\sum_j b_j\right)^{-1}$ and $x_i = a_i/b_i$, we obtain

$$\sum_i \frac{a_i}{\sum b_j} \log \frac{a_i}{b_i} \geq \left(\sum_i \frac{a_i}{\sum b_j}\right) \log \left(\frac{\sum_i a_i}{\sum_i b_j}\right)$$

which is the log-sum inequality.

(iii) There exists a constant $c > 0$ such that

$$\log y \geq c\left(1 - \frac{1}{y}\right), \quad \text{with equality iff } y = 1.$$

Writing $B = \{x : p(x) > 0\}$,

$$\begin{aligned} D(p\|q) &= \sum_{x \in B} p(x) \log \frac{p(x)}{q(x)} \\ &\geq c \sum_{x \in B} p(x)\left(1 - \frac{q(x)}{p(x)}\right) = c\left[1 - q(B)\right] \geq 0. \end{aligned}$$

Equality holds iff $q(x) \equiv p(x)$. Next, write

$$f(A) = \sum_{x \in A} f(x), \quad p(x) = \frac{f(x)}{f(A)} \mathbf{1}(x \in A),$$

$$g(A) = \sum_{x \in A} g(x), \quad q(x) = \frac{g(x)}{g(A)} \mathbf{1}(x \in A).$$

Then

$$\sum_{x \in A} f(x) \log \frac{f(x)}{g(x)} = f(A) \sum_{x \in A} p(x) \log \frac{f(A)p(x)}{g(A)q(x)}$$

$$= f(A) \underbrace{\sum_{x \in A} p(x) \log \frac{p(x)}{q(x)}}_{\geq \text{ by the previous part}} + f(A) \log \frac{f(A)}{g(A)}$$

$$\geq f(A) \log \frac{f(A)}{g(A)}.$$

Inequality (1.6.5) could be easily established by inspection. Finally, consider $A = \{x : p(x) \le q(x)\}$. Since

$$\sum_x |p(x) - q(x)| = 2[q(A) - p(A)] = 2[p(A^c) - q(A^c)],$$

then

$$
\begin{aligned}
D(p\|q) &= \sum_{x \in A} p(x) \log \frac{p(x)}{q(x)} + \sum_{x \in A^c} p(x) \log \frac{p(x)}{q(x)} \\
&\ge p(A) \log \frac{p(A)}{q(A)} + p(A^c) \log \frac{p(A^c)}{q(A^c)} \\
&\ge (2 \log_2 e) [p(A) - q(A)]^2 = \frac{\log_2 e}{2} \left[\sum_x |p(x) - q(x)| \right]^2.
\end{aligned}
$$

□

Problem 1.7   (a) *Define the conditional entropy, and show that for random variables U and V the joint entropy satisfies*

$$h(U, V) = h(V|U) + h(U).$$

Given random variables $X_1, \ldots, X_n$, by induction or otherwise prove the chain rule

$$h(X_1, \ldots X_n) = \sum_{i=1}^{n} h(X_i | X_1, \ldots, X_{i-1}). \tag{1.6.7}$$

(b) *Define the subset average over subsets of size k to be*

$$h_k^{(n)} = \sum_{S:|S|=k} \frac{h(X_S)}{k} \Big/ \binom{n}{k},$$

where $h(X_S) = h(X_{s_1}, \ldots, X_{s_k})$ for $S = \{s_1, \ldots, s_k\}$. Assume that, for any i, $h(X_i|X_S) \le h(X_i|X_T)$ when $T \subseteq S$, and $i \notin S$.
   By considering terms of the form

$$h(X_1, \ldots, X_n) - h(X_1, \ldots X_{i-1}, X_{i+1}, \ldots, X_n)$$

show that $h_n^{(n)} \le h_{n-1}^{(n)}$.
   Using the fact that $h_k^{(k)} \le h_{k-1}^{(k)}$, show that $h_k^{(n)} \le h_{k-1}^{(n)}$, for $k = 2, \ldots, n$.
(c) *Let $\beta > 0$, and define*

$$t_k^{(n)} = \sum_{S:|S|=k} e^{\beta h(X_S)/k} \Big/ \binom{n}{k}.$$

Prove that

$$t_1^{(n)} \ge t_2^{(n)} \ge \cdots \ge t_n^{(n)}.$$

Solution (a) By definition, the conditional entropy

$$h(V|U) = h(U,V) - h(U)$$
$$= \sum_u \mathbb{P}(U=u)h(V|U=u),$$

where $h(V|U=u)$ is the entropy of the conditional distribution:

$$h(V|U=u) = -\sum_v \mathbb{P}(V=v|U=u)\log\mathbb{P}(V=v|U=u).$$

The chain rule (1.6.7) is established by induction in n.

(b) By the chain rule

$$h(X_1,\ldots,X_n) = h(X_1,\ldots,X_{n-1}) + h(X_n|X_1,\ldots,X_{n-1}) \qquad (1.6.8)$$

and, in general,

$$\begin{aligned}
&h(X_1,\ldots,X_n)\\
&= h(X_1,\ldots,X_{i-1},X_{i+1},\ldots,X_n) + h(X_i|X_1,\ldots,X_{i-1},X_{i+1},\ldots,X_n)\\
&\leq h(X_1,\ldots,X_{i-1},X_{i+1},\ldots,X_n) + h(X_i|X_1,\ldots,X_{i-1}), \qquad (1.6.9)
\end{aligned}$$

because

$$h(X_i|X_1,\ldots,X_{i-1},X_{i+1},\ldots,X_n) \leq h(X_i|X_1,\ldots,X_{i-1}).$$

Then adding equations (1.6.9) from $i=1$ to n:

$$\begin{aligned}
nh(X_1,\ldots,X_n) &\leq \sum_{i=1}^{n} h(X_1,\ldots,X_{i-1},X_{i+1},\ldots,X_n)\\
&\quad + \sum_{i=1}^{n} h(X_i|X_1,\ldots,X_{i-1}).
\end{aligned}$$

The second sum in the RHS equals $h(X_1,\ldots,X_n)$ by the chain rule (1.6.7). So,

$$(n-1)h(X_1,\ldots,X_n) \leq \sum_{i=1}^{n} h(X_1,\ldots,X_{i-1},X_{i+1},\ldots,X_n).$$

This implies that $h_n^{(n)} \leq h_{n-1}^{(n)}$, since

$$\frac{1}{n}h(X_1,\ldots,X_n) \leq \frac{1}{n}\sum_{i=1}^{n} \frac{h(X_1,\ldots,X_{i-1},X_{i+1},\ldots,X_n)}{n-1}. \qquad (1.6.10)$$

In general, fix a subset S of size k in $\{1,\ldots,n\}$. Writing $S(i)$ for $S\setminus\{i\}$, we obtain

$$\frac{1}{k}h[X(S)] \leq \frac{1}{k}\sum_{i\in S} \frac{h(X[S(i)])}{k-1},$$

by the above argument. This yields

$$\binom{n}{k} h_k^{(n)} = \sum_{S \subset \{1,...,n\}:\, |S|=k} \frac{h[X(S)]}{k} \leq \sum_{S \subset \{1,...,n\}:\, |S|=k} \sum_{i \in S} \frac{h(X[S(i)])}{k(k-1)}. \qquad (1.6.11)$$

Finally, each subset of size $k-1$, $S(i)$, appears $[n-(k-1)]$ times in the sum (1.6.11). So, we can write $h_k^{(n)}$ as

$$\left[\sum_{T \subset \{1,...,n\}:\, |T|=k-1} \frac{h[X(T)]}{k-1} \right] \frac{n-(k-1)}{k} \Big/ \binom{n}{k}$$

$$= \sum_{T \subset \{1,...,n\}:\, |T|=k-1} \frac{h[X(T)]}{k-1} \Big/ \binom{n}{k-1} = h_{k-1}^n.$$

(c) Starting from (1.6.11), exponentiate and then apply the arithmetic mean/geometric mean inequality, to obtain for $S_0 = \{1, 2, \ldots, n\}$

$$e^{\beta h(X(S_0))/n} \leq e^{\beta [h(S_0(1)) + \cdots + h(S_0(n))]/(n(n-1))} \leq \frac{1}{n} \sum_{i=1}^n e^{\beta h(S_0(i))/(n-1)}$$

which is equivalent to $t_n^{(n)} \leq t_{n-1}^{(n)}$. Now we use the same argument as in (b), taking the average over all subsets to prove that for all $k \leq n, t_k^{(n)} \leq t_{k-1}^{(n)}$. □

Problem 1.8 *Let $p_1, \ldots, p_n$ be a probability distribution, with $p^* = \max_i [p_i]$. Prove that*

(i) $-\sum_i p_i \log_2 p_i \geq -p^* \log_2 p^* - (1-p^*) \log_2 (1-p^*)$;

(ii) $-\sum_i p_i \log_2 p_i \geq \log_2 (1/p^*)$;

(iii) $-\sum_i p_i \log_2 p_i \geq 2(1-p^*)$.

The random variables X and Y with values x and y from finite 'alphabets' I and J represent the input and output of a transmission channel, with the conditional probability $P(x \mid y) = \mathbb{P}(X = x \mid Y = y)$. Let $h(P(\cdot \mid y))$ denote the entropy of the conditional distribution $P(\cdot \mid y), y \in J$, and $h(X \mid Y)$ denote the conditional entropy of X given Y. Define the ideal observer decoding rule as a map $f : J \to I$ such that $P(f(y) \mid y) = \max_{x \in I} P(x \mid y)$ for all $y \in J$. Show that under this rule the error-probability

$$\pi_{er}(y) = \sum_{x \in I:\, x \neq f(y)} P(x \mid y)$$

satisfies $\pi_{er}(y) \leq \frac{1}{2} h(P(\cdot \mid y))$, and the expected error satisfies

$$\mathbb{E}\pi_{er}(Y) \leq \frac{1}{2} h(X \mid Y).$$

Solution Bound (i) follows from the pooling inequality. Bound (ii) holds as

$$-\sum_i p_i \log p_i \geq \sum_i p_i \log \frac{1}{p^*} = \log \frac{1}{p^*}.$$

To check (iii), it is convenient to use (i) for $p^* \geq 1/2$ and (ii) for $p^* \leq 1/2$. Assume first that $p^* \geq 1/2$. Then, by (i),

$$h(p_1,\ldots,p_n) \geq h(p^*, 1-p^*).$$

The function $x \in (0,1) \mapsto h(x, 1-x)$ is concave, and its graph on $(1/2,1)$ lies strictly above the line $x \mapsto 2(1-x)$. Hence,

$$h(p_1,\ldots,p_n) \geq 2(1-p^*).$$

On the other hand, if $p^* \leq 1/2$, we use (ii):

$$h(p_1,\ldots,p_n) \geq \log \frac{1}{p^*}.$$

Further, for $0 \leq x \leq 1/2$,

$$\log \frac{1}{x} \geq 2(1-x); \quad \text{equality iff } x = \frac{1}{2}.$$

For the concluding part, we use (iii). Write

$$\pi_{\text{er}}(y) = 1 - \mathbb{P}_{\text{ch}}(f(y)|y) = 1 - p_{\max}(\cdot|y)$$

which is $\leq h(P(\cdot|y))/2$. Finally, the mean $\mathbb{E}\pi_{\text{er}}(Y)$ is bounded by taking expectations, since $h(X|Y) = \mathbb{E}h(P(\cdot|Y))$. $\qquad\square$

Problem 1.9   *Define the information rate H and the asymptotic equipartition property of a source. Calculate the information rate of a Bernoulli source. Given a memoryless binary channel, define the channel capacity C. Assuming the statement of Shannon's second coding theorem (SCT), deduce that $C = \sup_{p_X} I(X:Y)$.*

An erasure channel keeps a symbol intact with probability $1-p$ and turns it into an unreadable splodge with probability p. Find the capacity of the erasure channel.

Solution The information rate H of a source $U_1, U_2, \ldots$ with a finite alphabet I is the supremum of all values $R > 0$ such that there exists a sequence of sets $A_n \in I \times \cdots \times I$ (n times) such that $|A_n| \leq 2^{nR}$ and $\lim_{n\to\infty} \mathbb{P}(U_1^n \in A_n) = 1$.

The asymptotic equipartition property means that, as $n \to \infty$,

$$-\frac{1}{n}\log p_n(U_1^n) \to H,$$

in one sense or another (here we mean convergence in probability). Here $U_1^n =$ $U_1 \ldots U_n$ and $p_n(u_1^n) = \mathbb{P}(U_1^n = u_1^n)$. The SCT states that if the random variable $-\log p_n(U_1^n)/n$ converges to a limit then the limit equals H.

A memoryless binary channel (MBC) has the conditional probability

$$\mathbb{P}_{ch}\left(\mathbf{Y}^{(N)}|\mathbf{X}^{(N)} \text{ sent}\right) = \prod_{1 \leq i \leq N} P(y_i|x_i)$$

and produces an error with probability

$$\varepsilon^{(N)} = \sum_u \mathbb{P}_{source}\left(U = u\right)\mathbb{P}_{ch}\left(\widehat{f}_N(\mathbf{Y}^{(N)}) \neq u \mid f_N(u) \text{ sent}\right),$$

where $\mathbb{P}_{source}$ stands for the source probability distribution, and one uses a code f_N and a decoding rule $\widehat{f}_N$. A value $\overline{R} \in (0, 1)$ is said to be a reliable transmission rate if, given that $\mathbb{P}_{source}$ is an equidistribution over a set $\mathscr{U}_N$ of source strings u with $\sharp \mathscr{U}_N = 2^{N[\overline{R}+o(1)]}$, there exist f_N and $\widehat{f}_N$ such that

$$\lim_{N\to\infty} \frac{1}{\sharp \mathscr{U}_N} \sum_{u \in \mathscr{U}_N} \mathbb{P}_{ch}\left(\widehat{f}_N(\mathbf{Y}^{(N)}) \neq u \mid f_N(u) \text{ sent}\right) = 0.$$

The channel capacity is the supremum of all reliable transmission rates.

For an erasure channel, the matrix is

$$\begin{array}{c} 0 \\ 1 \end{array} \begin{pmatrix} 1-p & 0 & p \\ 0 & 1-p & p \\ 0 & 1 & \star \end{pmatrix}$$

The conditional entropy $h(Y|X) = h(p, 1-p)$ does not depend on p_X. Thus,

$$C = \sup_{p_X} I(X : Y) = \sup_{p_X} h(Y) - h(Y|X)$$

is achieved at $p_X(0) = p_X(1) = 1/2$ with

$$h(Y) = -(1-p)\log[(1-p)/2] - p\log p = h(p, 1-p) + (1-p).$$

Hence, the capacity $C = 1 - p$. ☐

Problem 1.10 *Define Huffman's encoding rule and prove its optimality among decipherable codes. Calculate the codeword lengths for the symbol-probabilities* $\frac{1}{5}, \frac{1}{5}, \frac{1}{6}, \frac{1}{10}, \frac{1}{10}, \frac{1}{10}, \frac{1}{10}, \frac{1}{30}.$

Prove, or provide a counter-example to, the assertion that if the length of a code-word from a Huffman code equals l then, in the same code, there exists another codeword of length l' such that $|l - l'| \leq 1$.

Solution An answer to the first part:

probability	codeword	length
1/5	00	2
1/5	100	3
1/6	101	3
1/10	110	3
1/10	010	3
1/10	011	3
1/10	1110	4
1/30	1111	4

For the second part: a counter-example:

probability	codeword	length
1/2	0	1
1/8	100	3
1/8	101	3
1/8	110	3
1/8	111	3

□

Problem 1.11   *A memoryless channel with the input alphabet $\{0,1\}$ repro-
duces the symbol correctly with probability $(n-1)/n^2$ and reverses it with probability $1/n^2$. [Thus, for $n = 1$ the channel is binary and noiseless.] For $n \geq 2$ it
also produces $2(n-1)$ sorts of 'splodges', conventionally denoted by α_i and β_i,
$i = 1,\ldots,n-1$, with similar probabilities: $P(\alpha_i|0) = (n-1)/n^2$, $P(\beta_i|0) = 1/n^2$,
$P(\beta_i|1) = (n-1)/n^2$, $P(\alpha_i|1) = 1/n^2$. Prove that the capacity C_n of the channel
increases monotonically with n, and $\lim_{n\to\infty} C_n = \infty$. How is the capacity affected
if we simply treat splodges α_i as 0 and β_i as 1?*

Solution The channel matrix is

$$
\begin{array}{c}
0 \\[4pt]
1 \\[4pt]
{}
\end{array}
\left(
\begin{array}{cccccc}
\dfrac{n-1}{n^2} & \dfrac{1}{n^2} & \dfrac{n-1}{n^2} & \dfrac{1}{n^2} & \cdots & \dfrac{n-1}{n^2} & \dfrac{1}{n^2} \\[10pt]
\dfrac{1}{n^2} & \dfrac{n-1}{n^2} & \dfrac{1}{n^2} & \dfrac{n-1}{n^2} & \cdots & \dfrac{1}{n^2} & \dfrac{n-1}{n^2} \\[10pt]
0 & 1 & \alpha_1 & \beta_1 & \cdots & \alpha_{n-1} & \beta_{n-1}
\end{array}
\right).
$$

The channel is double-symmetric (the rows and columns are permutations of each
other), hence the capacity-achieving input distribution is

$$
p_X(0) = p_X(1) = \frac{1}{2},
$$

and the capacity C_n is given by

$$C_n = \log(2n) + n\,\frac{1}{n^2}\log(n^2) + n\,\frac{n-1}{n^2}\log\frac{n^2}{n-1}$$

$$= 1 + 3\log n - \frac{n-1}{n}\log(n-1) \;\to\; +\infty, \quad \text{as } n \to \infty.$$

Furthermore, extrapolating

$$C(x) = 1 + 3\log x - \left(1 - \frac{1}{x}\right)\log(x-1), \quad x \ge 1,$$

we find

$$\frac{dC(x)}{dx} = \frac{3}{x} - \frac{1}{x-1} + \frac{1}{x(x-1)} - \frac{1}{x^2}\log(x-1)$$

$$= \frac{2}{x} - \frac{1}{x^2}\log(x-1) = \frac{1}{x^2}[\,2x - \log(x-1)\,] > 0, \; x > 1.$$

Thus, C_n increases with n for $n \ge 1$. When α_i and β_i are treated as 0 or 1, the capacity does not change. $\qquad\square$

Problem 1.12 Let X_i, $i = 1, 2, \ldots$, be IID random variables, taking values $1, 0$ with probabilities p and $(1-p)$. Prove the local De Moivre–Laplace theorem with a remainder term:

$$\mathbb{P}(S_n = k) = \frac{1}{\sqrt{2\pi y(1-y)n}}\exp[-nh_p(y) + \theta_n(k)], \quad k = 1, \ldots, n-1; \quad (1.6.12)$$

here

$$S_n = \sum_{1 \le i \le n} X_i, \quad y = k/n, \quad h_p(y) = y\ln\left(\frac{y}{p}\right) + (1-y)\ln\left(\frac{1-y}{1-p}\right)$$

and the remainder $\theta_n(k)$ obeys

$$|\theta_n(k)| < \frac{1}{6ny(1-y)}, \quad y = k/n.$$

Hint: Use the Stirling formula with the remainder term

$$n! = \sqrt{2\pi n}, \quad n^n e^{-n + \vartheta(n)},$$

where

$$\frac{1}{12n+1} < \vartheta(n) < \frac{1}{12n}.$$

Find values k^+ and k^-, $0 \le k^+, k^- \le n$ (depending on n), such that $\mathbb{P}(S_n = k^+)$ is asymptotically maximal and $\mathbb{P}(S_n = k^-)$ is asymptotically minimal, as $n \to \infty$, and write the corresponding asymptotics.

Solution Write

$$\mathbb{P}(S_n = k) = \binom{n}{k}(1-p)^{n-k}p^k = \frac{n!}{k!(n-k)!}(1-p)^{n-k}p^k$$

$$= \sqrt{\frac{n}{2\pi k(n-k)}}\,\frac{n^n}{k^k(n-k)^{n-k}}(1-p)^{n-k}p^k$$

$$\times \exp\left[\vartheta(n) - \vartheta(k) - \vartheta(n-k)\right]$$

$$= \frac{1}{\sqrt{2\pi ny(1-y)}}\exp\left[-k\ln y - (n-k)\ln(1-y)\right.$$

$$\left. +k\ln p + (n-k)\ln(1-p)\right]\exp\left[\vartheta(n) - \vartheta(k) - \vartheta(n-k)\right]$$

$$= \frac{1}{\sqrt{2\pi ny(1-y)}}\exp\left[-nh_p(y)\right]$$

$$\times \exp\left[\vartheta(n) - \vartheta(k) - \vartheta(n-k)\right].$$

Now, as

$$|\vartheta(n) - \vartheta(k) - \vartheta(n-k)| < \frac{1}{12n} + \frac{1}{12k} + \frac{1}{12(n-k)} < \frac{2n^2}{12nk(n-k)},$$

(1.6.12) follows, with $\theta_n(k) = \vartheta(n) - \vartheta(k) - \vartheta(n-k)$. By the Gibbs inequality, $h_p(y) \geq 0$ and $h_p(y) = 0$ iff $y = p$. Furthermore,

$$\frac{dh_p(y)}{dy} = \ln\frac{y}{p} - \ln\frac{1-y}{1-p}, \quad \text{and} \quad \frac{d^2h_p(y)}{dy^2} = \frac{1}{y} + \frac{1}{1-y} > 0,$$

which yields

$$\frac{dh_p(y)}{dy}\bigg|_{y=p} = 0, \quad \frac{dh_p(y)}{dy} < 0, \ 0 < y < p, \quad \text{and} \quad \frac{dh_p(y)}{dy} > 0, \ p < y < 1.$$

Hence,

$$\underline{h}_p = \min h_p(y) = 0, \text{ attained at } y = p,$$

$$\overline{h}_p = \max h_p(y) = \min\left[\ln\frac{1}{p}, \ \ln\frac{1}{1-p}\right], \text{ attained at } y = 0 \text{ or } y = 1.$$

Thus, the maximal probability for $n \gg 1$ is for $y^* = p$, i.e. $k^+ = \lfloor np \rfloor$:

$$\mathbb{P}\left(S_n = \lfloor np \rfloor\right) \simeq \frac{1}{\sqrt{2\pi np(1-p)}}\exp\left(\theta_n(\lfloor np \rfloor)\right),$$

where

$$|\theta_n(\lfloor np \rfloor)| \leq \frac{1}{6np(1-p)}.$$

Similarly, the minimal probability is

$$\mathbb{P}(S_n = 0) = p^n, \qquad \text{if } 0 < p \leq 1/2,$$
$$\mathbb{P}(S_n = n) = (1-p)^n, \quad \text{if } 1/2 \leq p < 1.$$

$\square$

Problem 1.13 (a) *Prove that the entropy* $h(X) = -\sum_{i=1}^{n} p(i) \log p(i)$ *of a discrete random variable X with probability distribution* $\mathbf{p} = (p(1), \ldots, p(n))$ *is a concave function of the vector* $\mathbf{p}$.

Prove that the mutual entropy $I(X:Y) = h(Y) - h(Y \mid X)$ *between random variables X and Y, with* $\mathbb{P}(X = i, Y = k) = p_X(i)P_{Y|X}(k \mid i)$, $i, k = 1, \ldots, n$, *is a concave function of the vector* $p_X = (p_X(1), \ldots, p_X(n))$ *for fixed conditional probabilities* $\{P_{Y|X}(k \mid i)\}$.

(b) *Show that*

$$h(X) \geq -p^* \log_2 p^* - (1-p^*) \log_2(1-p^*),$$

where $p^* = \max_x \mathbb{P}(X = x)$, *and deduce that, when* $p^* \geq 1/2$,

$$h(X) \geq 2(1-p^*). \tag{1.6.13}$$

Show also that inequality (1.6.13) remains true even when $p^* < 1/2$.

Solution (a) Concavity of $h(\mathbf{p})$ means that

$$h(\lambda_1 \mathbf{p}_1 + \lambda_2 \mathbf{p}_2) \geq \lambda_1 h(\mathbf{p}_1) + \lambda_2 h(\mathbf{p}_2) \tag{1.6.14}$$

for any probability vectors $\mathbf{p}_j = (p_j(1), \ldots, p_j(n))$, $j = 1, 2$, and any $\lambda_1, \lambda_2 \in (0,1)$ with $\lambda_1 + \lambda_2 = 1$. Let X_1 have distribution $\mathbf{p}_1$ and X_2 have distribution $\mathbf{p}_2$. Let also

$$Z = 1, \text{with probability } \lambda_1 \text{ or } 2, \text{with probability } \lambda_2,$$

and $Y = X_Z$. Then the distribution of Y is $\lambda_1 \mathbf{p}_1 + \lambda_2 \mathbf{p}_2$. By Theorem 1.2.11(a),

$$h(Y) \geq h(Y|Z),$$

and by the definition of the conditional entropy

$$h(Y|Z) = \lambda_1 h(X_1) + \lambda_2 h(X_2).$$

This yields (1.6.14). Now

$$I(X:Y) = h(Y) - h(Y|X) = h(Y) - \sum p_X(i)h(P_{Y|X}(\cdot|i)). \tag{1.6.15}$$

If $P_{Y|X}(\cdot|\cdot)$ are fixed, the second term is a linear function of p_X, hence concave. The first term, $h(Y)$, is a concave function of p_Y which in turn is a linear function of p_X. Thus, $h(Y)$ is concave in p_X, and so is $I(X:Y)$.

(b) Consider two cases, (i) $p^* \geq 1/2$ and (ii) $p^* \leq 1/2$. In case (i), by pooling inequality,

$$h(X) \geq h(p^*, 1-p^*) \geq (1-p^*) \log \frac{1}{p^*(1-p^*)} \geq (1-p^*) \log \frac{1}{4} = 2(1-p^*)$$

as $p^* \geq 1/2$. In case (ii) we use induction in n, the number of values taken by X. The initial step is $n = 3$: without loss of generality, assume that $p^* = p_1 \geq p_2 \geq p_3$. Then $1/3 \leq p_1 < 1/2$ and $(1 - p_1)/2 \leq p_2 \leq p_1$. Write

$$h(p_1, p_2, p_3) = h(p_1, 1 - p_1) + (1 - p_1)h(q, 1 - q), \text{ where } q = \frac{p_2}{1 - p_1}.$$

As $1/2 \leq q \leq p_1(1 - p_1) \leq 1$,

$$h(q, 1 - q) \geq h(p_1/(1 - p_1), (1 - 2p_1)/(1 - p_1)),$$

i.e.

$$h(p_1, p_2, p_3) \geq h(p_1, p_1, 1 - 2p_1) = 2p_1 + h(2p_1, 1 - 2p_1).$$

The inequality $2p_1 + h(2p_1, 1 - 2p_1) \geq 2(1 - p_1)$ is equivalent to

$$h(2p_1, 1 - 2p_1) > 2 - 4p_1, \quad 1/3 \leq p_1 < 1/2,$$

or to

$$h(p, 1 - p) > 2 - 2p, \quad 2/3 \leq p \leq 1,$$

which follows from (a). Thus, for $n = 3$, $h(p_1, p_2, p_3) \geq 2(1 - p^*)$ regardless of the value of p^*. The initial induction step is completed.

Make the induction hypothesis $h(X) \geq 2(1 - p^*)$ for the number of values of X which is $\leq n - 1$. Then take $\mathbf{p} = (p_1, \ldots, p_n)$ and assume without loss of generality that $p^* = p_1 \geq \cdots \geq p_n$. Write $\mathbf{q} = (p_2/(1 - p_1), \ldots, p_{n-1}/(1 - p_1))$ and

$$h(\mathbf{p}) = h(p_1, 1 - p_1) + (1 - p_1)h(\mathbf{q}) \geq h(p_1, 1 - p_1) + (1 - p_1)2(1 - q_1). \quad (1.6.16)$$

The inequality $h(\mathbf{p}) \geq 2(1 - p^*)$ will follow from

$$h(p_1, 1 - p_1) + (1 - p_1)(2 - 2q_1) \geq (2 - 2p_1)$$

which is equivalent to

$$h(p_1, 1 - p_1) \geq 2(1 - p_1)(1 - 1 + q_1) = 2(1 - p_1)q_1 = 2p_2,$$

for $1/n \leq p_1 < 1/2$, $(1 - p_1)/(n - 1) \leq p_2 < p_1$. But obviously

$$h(p_1, 1 - p_1) \geq 2(1 - p_1) \geq 2p_2$$

(with equality at $p_1 = 0, 1/2$). Inequality (1.6.16) follows from the induction hypothesis.   □

Problem 1.14   *Let a probability distribution p_i, $i \in I = \{1, 2, \ldots, n\}$, be such that $\log_2(1/p_i)$ is an integer for all i with $p_i > 0$. Interpret I as an alphabet whose letters are to be encoded by binary words. A Shannon–Fano (SF) code assigns to letter i a word of length $\ell_i = \lceil \log_2(1/p_i) \rceil$; by the Kraft inequality it may be constructed* to be uniquely decodable. Prove the *competitive optimality* of the SF codes: if ℓ_i', $i \in I$, are the codeword-lengths of any uniquely decodable binary code then

$$\mathbb{P}\left(\ell_i < \ell_i'\right) \geq \mathbb{P}\left(\ell_i' < \ell_i\right), \qquad (1.6.17)$$

with equality iff $\ell_i \equiv \ell_i'$.

Hint: You may find useful the inequality $\text{sgn}(\ell - \ell') \leq 2^{\ell-\ell'} - 1$, $\ell, \ell' = 1, \ldots, n$.

Solution Write

$$
\begin{aligned}
\mathbb{P}(\ell_i' < \ell_i) - \mathbb{P}(\ell_i' > \ell_i) &= \sum_{i:\ell_i' < \ell_i} p_i - \sum_{i:\ell_i' > \ell_i} p_i \\
&= \sum_i p_i \, \text{sign}\,(\ell_i - \ell_i') \\
&= \mathbb{E}\,\text{sgn}(\ell - \ell') \leq \mathbb{E}\left(2^{\ell-\ell'} - 1\right),
\end{aligned}
$$

as $\text{sign}\, x \leq 2^x - 1$ for integer x. Continue the argument with

$$
\begin{aligned}
\mathbb{E}\left(2^{\ell-\ell'} - 1\right) &= \sum_i p_i \left(2^{\ell_i - \ell_i'} - 1\right) \\
&= \sum_i 2^{-\ell_i}\left(2^{\ell_i - \ell_i'} - 1\right) = \sum_i 2^{-\ell_i'} - \sum_i 2^{-\ell_i} \\
&\leq 1 - \sum_i 2^{-\ell_i} = 1 - 1 = 0
\end{aligned}
$$

by the Kraft inequality. This yields the inequality

$$\mathbb{P}\left(\ell_i < \ell_i'\right) \geq \mathbb{P}\left(\ell_i' < \ell_i\right).$$

To have equality, we must have (a) $2^{\ell_i - \ell_i'} - 1 = 0$ or $1, i \in I$ (because $\text{sign}\, x = 2^x - 1$ only for $x = 0$ or 1), and (b) $\sum_i 2^{-\ell_i'} = 1$. As $\sum_i 2^{-\ell_i} = 1$, the only possibility is $2^{\ell_i - \ell_i'} \equiv 1$, i.e. $\ell_i = \ell_i'$. $\qquad\qquad\square$

Problem 1.15   Define the capacity C of a binary channel. Let $C_N = (1/N)\sup I(\mathbf{X}^{(N)} : \mathbf{Y}^{(N)})$, where $I(\mathbf{X}^{(N)} : \mathbf{Y}^{(N)})$ denotes the mutual entropy between $\mathbf{X}^{(N)}$, the random word of length N sent through the channel, and $\mathbf{Y}^{(N)}$, the received word, and where the supremum is over the probability distribution of $\mathbf{X}^{(N)}$. Prove that $C \leq \limsup_{N \to \infty} C_N$.

Solution A binary channel is defined as a sequence of conditional probability distributions

$$\mathbb{P}_{\text{ch}}^{(N)}\left(\mathbf{y}^{(N)}|\mathbf{x}^{(N)}\right), \quad N = 1, 2, \ldots,$$

where $\mathbf{x}^{(N)} = x_1 \ldots x_N$ is a binary word (string) at the input and $\mathbf{y}^{(N)} = y_1 \ldots y_N$ a binary word (string) at the output port. The channel capacity C is an asymptotic parameter of the family $\{\mathbb{P}_{\mathrm{ch}}^{(N)}(\cdot \mid \cdot)\}$ defined by

$$C = \sup \left[\overline{R} \in (0,1) : \overline{R} \text{ is a reliable transmission rate} \right]. \tag{1.6.18}$$

Here, a number $\overline{R} \in (0,1)$ is called a reliable transmission rate (for a given channel) if, given that the random source string is equiprobably distributed over a set $\mathscr{U}^{(N)}$ with $\sharp \mathscr{U}^{(N)} = 2^{N[\overline{R}+O(1)]}$, there exist an encoding rule $f^{(N)} : \mathscr{U}^{(N)} \to \mathscr{X}_N \subseteq \{0,1\}^N$ and a decoding rule $\widehat{f}^{(N)} : \{0,1\}^N \to \mathscr{U}^{(N)}$ such that the error probability $e^{(N)} \to 0$ as $N \to \infty$ is given by

$$e^{(N)} := \sum_{u \in \mathscr{U}^{(N)}} \frac{1}{\sharp \mathscr{U}^{(N)}} \mathbb{P}_{\mathrm{ch}}^{(N)} \left(\left\{ \mathbf{y}^{(N)} : \widehat{f}^{(N)} \left(\mathbf{y}^{(N)} \right) \neq u \right\} \mid f^{(N)}(u) \right); \tag{1.6.19}$$

note

$$e^{(N)} = e^{(N)} \left(f^{(N)}, \widehat{f}^{(N)} \right).$$

The converse part of Shannon's second coding theorem (SCT) states that

$$C \leq \limsup_{N \to \infty} \frac{1}{N} \sup_{\mathbb{P}_{\mathbf{X}^{(N)}}} I\left(\mathbf{X}^{(N)} : \mathbf{Y}^{(N)} \right), \tag{1.6.20}$$

where $I\left(\mathbf{X}^{(N)} : \mathbf{Y}^{(N)} \right)$ is the mutual entropy between the random input and output strings $\mathbf{X}^{(N)}$ and $\mathbf{Y}^{(N)}$ and $\mathbb{P}_{\mathbf{X}^{(N)}}$ is a distribution of $\mathbf{X}^{(N)}$.

For the proof, it suffices to check that if $\sharp \mathscr{U}^{(N)} = 2^{N[\overline{R}+O(1)]}$ then, for all $f^{(N)}$ and $\widehat{f}^{(N)}$,

$$e^{(N)} \geq 1 - \frac{C_N + o(1)}{\overline{R} + o(1)} \tag{1.6.21}$$

where

$$C_N = \frac{1}{N} \sup_{\mathbb{P}_{\mathbf{X}^{(N)}}} I\left(\mathbf{X}^{(N)} : \mathbf{Y}^{(N)} \right).$$

Indeed,   if $\overline{R} > \limsup_{N \to \infty} C_N$ then,   according   to   (1.6.21) $\liminf_{N \to \infty} \inf_{f^{(N)}, \widehat{f}^{(N)}} e^{(N)} > 0$ and $\overline{R}$ is not reliable.

To prove (1.6.21), assume, without loss of generality, that $f^{(N)}$ is lossless. Then the input word $\mathbf{x}^{(N)}$ is equidistributed, with probability $1 / (\sharp \mathscr{U}^{(N)})$. For all decoding rules $\widehat{f}^{(N)}$ and any N large enough,

$$clNC_N \geq I\left(\mathbf{X}^{(N)} : \mathbf{Y}^{(N)} \right) \geq I\left(\mathbf{X}^{(N)} : \widehat{f}\left(\mathbf{Y}^{(N)} \right) \right)$$
$$= h\left(\mathbf{X}^{(N)} \right) - h\left(\mathbf{X}^{(N)} \mid \widehat{f}\left(\left(\mathbf{Y}^{(N)} \right) \right) \right)$$
$$= \log \left(\sharp \mathscr{U}^{(N)} \right) - h\left(\mathbf{X}^{(N)} \mid \widehat{f}\left(\left(\mathbf{Y}^{(N)} \right) \right) \right)$$
$$\geq \log \left(\sharp \mathscr{U}^{(N)} \right) - 1 - \varepsilon^{(N)} \log \left(\sharp \mathscr{U}^{(N)} - 1 \right). \tag{1.6.22}$$

The last bound here follows from the generalised Fano inequality

$$h\left(\mathbf{X}^{(N)}|\widehat{f}\left((\mathbf{Y}^{(N)})\right)\right) \leq -e^{(N)}\log e^{(N)} - \left(1 - e^{(N)}\right)\log\left(1 - e^{(N)}\right)$$
$$+ e^{(N)}\log\left(\sharp \mathcal{U}^{(N)} - 1\right)$$
$$\leq 1 + e^{(N)}\log\left(\sharp \mathcal{U}^{(N)} - 1\right).$$

Now, from (1.6.22),

$$NC_N \geq N\left[\overline{R} + o(1)\right] - 1 - e^{(N)}\log\left(2^{N[\overline{R} + o(1)]} - 1\right),$$

i.e.

$$e^{(N)} \geq \frac{N\left[\overline{R} + o(1)\right] - NC_N - 1}{\log\left(2^{N[\overline{R} + o(1)]} - 1\right)} = 1 - \frac{C_N + o(1)}{\overline{R} + o(1)},$$

as required. $\qquad\qquad\qquad\qquad\qquad\qquad\qquad\qquad\qquad\qquad\square$

Problem 1.16   *A memoryless channel has input 0 and 1, and output 0, 1 and $*$ (illegible). The channel matrix is given by*

$$\mathbf{P}(0|0) = 1, \mathbf{P}(0|1) = \mathbf{P}(1|1) = \mathbf{P}(*|1) = 1/3.$$

Calculate the capacity of the channel and the input probabilities $p_X(0)$ and $p_X(1)$ for which the capacity is achieved.

Someone suggests that, as the symbol $*$ may occur only from 1, it is to your advantage to treat $*$ as 1: you gain more information from the output sequence, and it improves the channel capacity. Do you agree? Justify your answer.

Solution Use the formula

$$C = \sup_{p_X} I(X:Y) = \sup_{p_X} \left[h(Y) - h(Y|X)\right],$$

where p_X is the distribution of the input symbol:

$$p_X(0) = p, \quad p_X(1) = 1 - p, \quad 0 \leq p \leq 1.$$

So, calculate $I(X:Y)$ as a function of p:

$$h(Y) = -p_Y(0)\log p_Y(0) - p_Y(1)\log p_Y(1) - p_Y(*)\log p_Y(*).$$

Here

$$p_Y(0) = p + (1 - p)/3 = (1 + 2p)/3,$$
$$p_Y(1) = p_Y(*) = (1 - p)/3,$$

and

$$h(Y) = -\frac{1 + 2p}{3}\log\frac{1 + 2p}{3} - \frac{2(1 - p)}{3}\log\frac{1 - p}{3}.$$

Also,

$$h(Y|X) = - \sum_{x=0,1} p_X(x) \sum_y P(y|x) \log P(y|x)$$
$$= -p_X(1)\log 1/3 = (1-p)\log 3.$$

Thus,

$$I(X:Y) = -\frac{1+2p}{3}\log\frac{1+2p}{3} - \frac{2(1-p)}{3}\log\frac{1-p}{3} - (1-p)\log 3.$$

Differentiating yields

$$\frac{d}{dp}I(X:Y) = -2/3(\log(1/3+2p/3)) + 2/3\log(1/3-p/3) + \log 3.$$

Hence, the maximum max $I(X:Y)$ is found from relation

$$\frac{2}{3}\log\frac{1-p}{1+2p} + \log 3 = 0.$$

This yields

$$\log\frac{1-p}{1+2p} = -\frac{3}{2}\log 3 := b,$$

and

$$\frac{1-p}{1+2p} = 2^b, \text{ i.e. } 1-2^b = p(1+2^{b+1}).$$

The answer is

$$p = \frac{1-2^b}{1+2^{b+1}}.$$

For the last part, write

$$I(X:Y) = h(X) - h(X|Y) \le h(X) - h(X|Y') = I(X:Y')$$

for any Y' that is a function of Y; the equality holds iff Y and X are conditionally independent, given Y'. It is the case of our channel, hence the suggestion leaves the capacity the same. □

Problem 1.17   (a) *Given a pair of discrete random variables X, Y, define the joint and conditional entropies $h(X,Y)$ and $h(X|Y)$.*
(b) *Prove that $h(X,Y) \ge h(X|Y)$ and explain when equality holds.*
(c) *Let $0 < \delta < 1$, and prove that*

$$h(X|Y) \ge (\log(\delta^{-1}))\mathbb{P}(q(X,Y) \le \delta),$$

where $q(x,y) = \mathbb{P}(X = x|Y = y)$. For which δ and for which X, Y does equality hold here?

Solution (a) The conditional entropy is given by

$$h(X|Y) = -\mathbb{E}\log q(x,y) = -\sum_{x,y} \mathbb{P}(X=x, Y=y)\log q(x,y)$$

where

$$q(x,y) = \mathbb{P}(X=x|Y=y).$$

The joint entropy is given by

$$h(X,Y) = -\sum_{x,y} \mathbb{P}(X=x, Y=y)\log \mathbb{P}(X=x, Y=y).$$

(b) From the definition,

$$h(X,Y) = h(X|Y) - \sum_{y} \mathbb{P}(Y=y)\log \mathbb{P}(Y=y) \geq h(X|Y).$$

The equality in (b) is achieved iff $h(Y) = 0$, i.e. Y is constant a.s.

(c) By Chebyshev's inequality,

$$\mathbb{P}(q(X,Y) \leq \delta) = \mathbb{P}(-\log q(X,Y) \geq \log 1/\delta)$$
$$\leq \frac{1}{\log 1/\delta} \mathbb{E}[-\log q(X,Y)] = \frac{1}{\log 1/\delta} h(X|Y).$$

Here equality holds iff

$$\mathbb{P}(q(X,Y) = \delta) = 1.$$

This requires that (i) $\delta = 1/m$ where m is a positive integer and (ii) for all $y \in$ support Y, there exists a set A_y of cardinality m such that

$$\mathbb{P}(X=x|Y=y) = \frac{1}{m}, \quad \text{for } x \in A_y.$$

□

Problem 1.18 *A text is produced by a Bernoulli source with alphabet $1, 2, \ldots, m$ and probabilities $p_1, p_2, \ldots, p_m$. It is desired to send this text reliably through a memoryless binary symmetric channel (MBSC) with the row error-probability p^*. Explain what is meant by the capacity C of the channel, and show that*

$$C = 1 - h(p^*, 1 - p^*).$$

Explain why reliable transmission is possible if

$$h(p_1, p_2, \ldots, p_m) + h(p^*, 1 - p^*) < 1$$

and is impossible if

$$h(p_1, p_2, \ldots, p_m) + h(p^*, 1 - p^*) > 1,$$

where $h(p_1, p_2, \ldots, p_m) = -\sum_{i=1}^{m} p_i \log_2 p_i.$

Solution The asymptotic equipartition property for a Bernoulli source states that the number of distinct strings (words) of length n emitted by the source is 'typically' $2^{nH+o(n)}$, and they have 'nearly equal' probabilities $2^{-nH+o(n)}$:

$$\lim_{n\to\infty} \mathbb{P}\left(2^{-n(H+\varepsilon)} \le \mathbb{P}_n(\mathbf{U}^{(n)}) \le 2^{-n(H-\varepsilon)}\right) = 1.$$

Here, $H = h(p_1, \ldots, p_n)$.

Denote

$$T_n(= T_n(\varepsilon)) = \left\{\mathbf{u}^{(n)} : 2^{-n(H+\varepsilon)} \le \mathbb{P}_n(\mathbf{u}^{(n)}) \le 2^{-n(H-\varepsilon)}\right\}$$

and observe that

$$\lim_{n\to\infty} \frac{1}{n} \log \sharp T_n = H, \quad \text{i.e. } \limsup_{n\to\infty} \frac{1}{n} \log \sharp T_n < H + \varepsilon.$$

By the definition of the channel capacity, the words $\mathbf{u}^{(n)} \in T_n(\varepsilon)$ may be encoded by binary codewords of length $\overline{R}^{-1}(H+\varepsilon)$ and sent reliably through a memoryless symmetric channel with matrix

$$\begin{pmatrix} 1-p^* & p^* \\ p^* & 1-p^* \end{pmatrix}$$

for any $\overline{R} < C$ where

$$C = \sup_{p_X} I(X:Y) = \sup_{p_X}[h(Y) - h(Y|X)].$$

The supremum here is taken over all distributions

$$p_X = (p_X(0), p_X(1))$$

of the input binary symbol X; the conditional distribution of the output symbol Y is given by

$$\mathbb{P}(Y = y | X = x) = \begin{cases} 1-p^*, & y = x, \\ p^*, & y \ne x. \end{cases}$$

We see that

$$h(Y|X) = -p_X(0)\left[-(1-p^*)\log(1-p^*) - p^*\log p^*\right]$$

$$+p_X(1)\left[-p^*\log p^* - (1-p^*)\log(1-p^*)\right] = h(p^*, 1-p^*),$$

independently of p_X. Hence,

$$C = \sup_{p_X} h(Y) - h(p^*, 1 - p^*) = 1 - h(p^*, 1 - p^*),$$

because $h(Y)$ is achieved for

$$p_Y(0) = p_Y(1) = 1/2 \text{ (occurring when } p_X(0) = p_X(1) = 1/2).$$

Therefore, if

$$H < C \qquad \Leftrightarrow \qquad h(p_1, \dots, p_n) + h(p^*, 1 - p^*) < 1,$$

then $\overline{R}^{-1}(H + \varepsilon)$ can be made < 1, for $\varepsilon > 0$ small enough and $\overline{R} < C$ close to C. This means that there exists a sequence of codes f_n of length n such that the error-probability, while using encoding f_n and the ML decoder, is

$$\leq \mathbb{P}\left(\mathbf{u}^{(n)} \notin T_n\right)$$
$$+ \mathbb{P}\left(\mathbf{u}^{(n)} \in T_n; \text{ an error while using } f_n(\mathbf{u}^{(n)}) \text{ and the ML decoder}\right)$$
$$\to 0, \quad \text{as } n \to \infty,$$

since both probabilities go to 0.

On the other hand,

$$H > C \qquad \Leftrightarrow \qquad h(p_1, \dots, p_n) + h(p^*, 1 - p^*) > 1,$$

then $\overline{R}^{-1}H > 1$ for all $\overline{R} < C$, and we cannot encode words $\mathbf{u}^{(n)} \in T_n$ by codewords of length n so that the error probability tends to 0. Hence, no reliable transmission is possible. □

Problem 1.19 *A Markov source with an alphabet of m characters has a transition matrix P_m whose elements p_{jk} are specified by*

$$p_{11} = p_{mm} = 2/3, \quad p_{jj} = 1/3 \quad (1 < j < m),$$
$$p_{jj+1} = 1/3 \quad (1 \leq j < m), \quad p_{jj-1} = 1/3 \quad (1 < j \leq m).$$

All other elements are zero. Determine the information rate of the source.

Denote the transition matrix thus specified by P_m. Consider a source in an alphabet of $m + n$ characters whose transition matrix is $\begin{pmatrix} P_m & 0 \\ 0 & P_n \end{pmatrix}$, where the zeros indicate zero matrices of appropriate size. The initial character is supposed uniformly distributed over the alphabet. What is the information rate of the source?

Solution The transition matrix

$$P_m = \begin{pmatrix} 2/3 & 1/3 & 0 & 0 & \cdots & 0 \\ 1/3 & 1/3 & 1/3 & 0 & \cdots & 0 \\ 0 & 1/3 & 1/3 & 1/3 & \cdots & 0 \\ \vdots & \vdots & \vdots & \vdots & \ddots & \vdots \\ 0 & 0 & 0 & 0 & \cdots & 2/3 \end{pmatrix}$$

is Hermitian and so has the equilibrium distribution $\pi = (\pi_i)$ with $\pi_i = 1/m$, $1 \le i \le m$ (equidistribution). The information rate equals

$$\begin{aligned} H_m &= -\sum_{j,k} \pi_j p_{jk} \log p_{jk} \\ &= -\frac{1}{m} \left[2 \left(\frac{2}{3} \log \frac{2}{3} + \frac{1}{3} \log \frac{1}{3} \right) + 3(m-2) \frac{1}{3} \log \frac{1}{3} \right] \\ &= \log 3 - \frac{4}{3m}. \end{aligned}$$

The source with transition matrix $\begin{pmatrix} P_m & 0 \\ 0 & P_n \end{pmatrix}$ is non-ergodic, and its information rate is the maximum of the two rates

$$\max \left[H_m, H_n \right] = H_{m \vee n}.$$

$\square$

Problem 1.20   *Consider a source in a finite alphabet. Define $J_n = n^{-1} h(\mathbf{U}^{(n)})$ and $K_n = h(U_{n+1} | \mathbf{U}^{(n)})$ for $n = 1, 2, \ldots$. Here U_n is the nth symbol in the sequence and $\mathbf{U}^{(n)}$ is the string constituted by the first n symbols, $h(\mathbf{U}^{(n)})$ is the entropy and $h(U_{n+1} | \mathbf{U}^{(n)})$ the conditional entropy. Show that, if the source is stationary, then J_n and K_n are non-increasing and have a common limit.*

Suppose the source is Markov and not necessarily stationary. Show that the mutual information between U_1 and U_2 is not smaller than that between U_1 and U_3.

Solution For the second part, the Markov property implies that

$$\mathbb{P}(U_1 = u_1 | U_2 = u_2, U_3 = u_3) = \mathbb{P}(U_1 = u_1 | U_2 = u_2).$$

Hence,

$$\begin{aligned} I(U_1 : (U_2, U_3)) &= \mathbb{E} \left[-\log \frac{\mathbb{P}(U_1 = u_1 | U_2 = u_2, U_3 = u_3)}{\mathbb{P}(U_1 = u_1)} \right] \\ &= \mathbb{E} \left[-\log \frac{\mathbb{P}(U_1 = u_1 | U_2 = u_2)}{\mathbb{P}(U_1 = u_1)} \right] = I(U_1 : U_2). \end{aligned}$$

Since

$$I(U_1 : (U_2, U_3)) \ge I(U_1 : U_3),$$

the result follows.

$\square$

Problem 1.21   *Construct a Huffman code for a set of 5 messages with probabilities as indicated below*

Message	1	2	3	4	5
Probability	0.1	0.15	0.2	0.26	0.29

Solution

Message	1	2	3	4	5
Probability	0.1	0.15	0.2	0.026	0.029
Codeword	101	100	11	01	00

The expected codeword-length equals 2.4.   ☐

Problem 1.22   *State the first coding theorem (FCT), which evaluates the information rate for a source with suitable long-run properties. Give an interpretation of the FCT as an asymptotic equipartition property. What is the information rate for a Bernoulli source?*

Consider a Bernoulli source that emits symbols 0, 1 *with probabilities* $1 - p$ *and p respectively, where $0 < p < 1$. Let $\eta(p) = -p \log p - (1 - p) \log(1 - p)$ and let $\varepsilon > 0$ be fixed. Let $\mathbf{U}^{(n)}$ be the string consisting of the first n symbols emitted by the source. Prove that there is a set S_n of possible values of $\mathbf{U}^{(n)}$ such that*

$$\mathbb{P}\left(\mathbf{U}^{(n)} \in S_n\right) \geq 1 - \left(\log \frac{p}{1-p}\right)^2 \frac{p(1-p)}{n\varepsilon^2},$$

and so that for each $\mathbf{u}^{(n)} \in S_n$ the probability that $\mathbb{P}\left(\mathbf{U}^{(n)} = \mathbf{u}^{(n)}\right)$ lies between $2^{-n(h+\varepsilon)}$ and $2^{-n(h-\varepsilon)}$.

Solution For the Bernoulli source

$$-\frac{1}{n} \log P_n(\mathbf{U}^{(n)}) = -\frac{1}{n} \sum_{1 \leq j \leq n} \log P(U_j) \to \eta(p),$$

in the sense that for all $\varepsilon > 0$, by Chebyshev,

$$\mathbb{P}\left(\left|-\frac{1}{n} \log P_n(\mathbf{U}^{(n)}) - h\right| > \varepsilon\right)$$

$$\leq \frac{1}{\varepsilon^2 n^2} \operatorname{Var}\left(\sum_{1 \leq j \leq n} \log P(U_j)\right)$$

$$= \frac{1}{\varepsilon^2 n} \operatorname{Var}\left[\log P(U_1)\right]. \tag{1.6.23}$$

Here

$$P(U_j) = \begin{cases} 1-p, & \text{if } U_j = 0, \\ p, & \text{if } U_j = 1, \end{cases} \quad P_n(\mathbf{U}^{(n)}) = \prod_{1 \le j \le n} P(U_j),$$

and

$$\text{Var}\left(\sum_{1 \le j \le n} \log P(U_j)\right) = \sum_{1 \le j \le n} \text{Var}\left(\log P(U_j)\right)$$

where

$$\text{Var}\left(\log P(U_j)\right) = \mathbb{E}\left[\log P(U_j)\right]^2 - \left[\mathbb{E}\log P(U_j)\right]^2$$
$$= p(\log p)^2 + (1-p)(\log(1-p))^2 - (p\log p + (1-p)\log(1-p))^2$$
$$= p(1-p)\left(\log \frac{p}{1-p}\right)^2.$$

Hence, the bound (1.6.23) yields

$$\mathbb{P}\left(2^{-n(h+\varepsilon)} \le P_n(\mathbf{U}^{(n)}) \le 2^{-n(h-\varepsilon)}\right)$$
$$\ge 1 - \frac{1}{n\varepsilon^2} p(1-p)\left(\log \frac{p}{1-p}\right)^2.$$

It now suffices to set

$$S_n = \{\mathbf{u}^{(n)} = u_1 \ldots u_n : \ 2^{-n(h+\varepsilon)} \le \mathbb{P}(\mathbf{U}^{(n)} = \mathbf{u}^{(n)}) \le 2^{-n(h-\varepsilon)}\},$$

and the result follows. □

Problem 1.23　　The alphabet $\{1,2,\ldots,m\}$ is to be encoded by codewords with letters taken from an alphabet of $q < m$ letters. State Kraft's inequality for the word-lengths $s_1, \ldots, s_m$ of a decipherable code. Suppose that a source emits letters from the alphabet $\{1,2,\ldots,m\}$, each letter occurring with known probability $p_i > 0$. Let S be the random codeword-length resulting from the letter-by-letter encoding of the source output. It is desired to find a decipherable code that minimises the expected value of q^S. Establish the lower bound $\mathbb{E}(q^S) \ge \left(\sum_{1 \le i \le m} \sqrt{p_i}\right)^2$, and characterise when equality occurs.

Prove also that an optimal code for the above criterion must satisfy $\mathbb{E}(q^S) <$
$$q\left(\sum_{1 \le i \le m} \sqrt{p_i}\right)^2.$$
Hint: Use the Cauchy–Schwarz inequality: for all positive x_i, y_i,

$$\sum_{1 \le i \le m} x_i y_i \le \left(\sum_{1 \le i \le m} x_i^2\right)^{1/2} \left(\sum_{1 \le i \le m} y_i^2\right)^{1/2},$$

with equality iff $x_i = c y_i$ for all i.

Solution By Cauchy–Schwarz,

$$\sum_{1\le i\le m} p_i^{1/2} = \sum_{1\le i\le m} p_i^{1/2} q^{s_i/2} q^{-s_i/2}$$

$$\le \left(\sum_{1\le i\le m} p_i q^{s_i}\right)^{1/2} \left(\sum_{1\le i\le m} q^{-s_i}\right)^{1/2} \le \left(\sum_{1\le i\le m} p_i q^{s_i}\right)^{1/2},$$

since, by Kraft, $\sum_{1\le i\le m} q^{-s_i} \le 1$. Hence,

$$\mathbb{E}q^S = \sum_{1\le i\le m} p_i q^{s_i} \ge \left(\sum_{1\le i\le m} p_i^{1/2}\right)^2.$$

Now take the probabilities p_i to be

$$p_i = (cq^{-x_i})^2, \quad x_i > 0,$$

where $\sum_{1\le i\le m} q^{-x_i} = 1$ (so, $\sum_{1\le i\le m} p_i^{1/2} = c$). Take s_i to be the smallest integer $\ge x_i$. Then $\sum_{1\le i\le m} q^{-s_i} \le 1$ and, again by Kraft, there exists a decipherable coding with the codeword-length s_i. For this code, $q^{s_i-1} < q^{x_i} = c/p_i^{1/2}$, and hence

$$\mathbb{E}q^S = \sum_{1\le i\le m} p_i q^{s_i} = q \sum_{1\le i\le m} p_i q^{s_i-1}$$

$$< q \sum_{1\le i\le m} p_i q^{x_i} = qc \sum_{1\le i\le m} p_i^{1/2} = q\left(\sum_{1\le i\le m} p_i^{1/2}\right)^2.$$

$\square$

Problem 1.24 A Bernoulli source of information of rate H is fed character-by-character into a transmission line which may be live or dead. If the line is live when a character is transmitted then that character is received faithfully; if the line is dead then the receiver learnt only that it is indeed dead. In shifting between its two states the line follows a Markov chain (DTMC) with constant transition probabilities, independent of the text being transmitted.

Show that the information rate of the source constituted by the received signal is $H_L + \pi_L H_S$ where H_S is the signal, H_L is the information rate of the DTMC governing the functioning of the line and π_L is the equilibrium probability that the line is alive.

Solution The rate of a Bernoulli source emitting letter $j = 1, 2, \dots$ with probability p_j is $H = -\sum_j p_j \log p_j$. The state of the line is a DTMC with a 2×2 transition matrix

$$\begin{array}{c} \text{dead} \\ \text{live} \end{array} \begin{pmatrix} 1-\alpha & \alpha \\ \beta & 1-\beta \end{pmatrix}$$

and the equilibrium probabilities

$$1 - \pi_{\mathrm{L}}(\text{dead}) = \frac{\beta}{\alpha+\beta}, \quad \pi_{\mathrm{L}}(\text{live}) = \frac{\alpha}{\alpha+\beta}$$

(assuming that $\alpha + \beta > 0$). The received signal sequence follows a DTMC with states 0 (dead), $1, 2, \ldots$ and transition probabilities

$$\begin{array}{ll} q_{00} = 1 - \alpha, & q_{0j} = \alpha p_j, \\ q_{j0} = \beta, & q_{jk} = (1-\beta) p_k \end{array} \quad j, k \geq 1.$$

This chain has a unique equilibrium distribution

$$\pi_{\mathrm{RS}}(0) = \frac{\beta}{\alpha+\beta}, \quad \pi_{\mathrm{RS}}(j) = \frac{\alpha}{\alpha+\beta} p_j, \ j \geq 1.$$

Then the information rate of the received signal equals

$$\begin{aligned} H_{\mathrm{RS}} &= - \sum_{j,k\geq 0} \pi_{\mathrm{RS}}(j) q_{jk} \log q_{jk} \\ &= -\frac{\beta}{\alpha+\beta} \left[(1-\alpha)\log(1-\alpha) + \sum_{j\geq 1} \alpha p_j \log(\alpha p_j) \right] \\ &\quad -\frac{\alpha}{\alpha+\beta} \left(\sum_{j\geq 1} p_j \left[\beta \log \beta + (1-\beta) \sum_{k\geq 1} p_k \log \big((1-\beta)p_k\big) \right] \right) \\ &= H_{\mathrm{L}} + \frac{\alpha}{\alpha+\beta} H_{\mathrm{S}}. \end{aligned}$$

Here H_{L} is the entropy rate of the line state DTMC:

$$\begin{aligned} H_{\mathrm{L}} = &-\frac{\beta}{\alpha+\beta} \left[(1-\alpha)\log(1-\alpha) + \alpha \log \alpha \right] \\ &-\frac{\alpha}{\alpha+\beta} \left[(1-\beta)\log(1-\beta) + \beta \log \beta \right], \end{aligned}$$

and $\pi = \alpha/(\alpha+\beta)$. $\qquad\qquad\qquad\qquad\qquad\qquad\qquad\qquad\qquad\qquad$ $\square$

Problem 1.25 *Consider a Bernoulli source in which the individual character can take value i with probability p_i $(i = 1, \ldots, m)$. Let n_i be the number of times the character value i appears in the sequence $\mathbf{u}^{(n)} = u_1 u_2 \cdots u_n$ of given length n. Let A_n be the smallest set of sequences $\mathbf{u}^{(n)}$ which has total probability at least $1 - \varepsilon$. Show that each sequence in A_n satisfies the inequality*

$$-\sum n_i \log p_i \leq nh + (nk/\varepsilon)^{1/2},$$

where k is a constant independent of n or ε. State (without proof) the analogous assertion for a Markov source.

Solution For a Bernoulli source with letters $1,\ldots,m$, the probability of a given string $\mathbf{u}^{(n)} = u_1\, u_2\, \ldots\, u_n$ is

$$\mathbb{P}(\mathbf{U}^{(n)} = \mathbf{u}^{(n)}) = \prod_{1 \le i \le m} p_i^{n_i}.$$

Set A_n consists of strings of maximal probabilities (selected in the decreasing order), i.e. of maximal value of $\log \mathbb{P}(\mathbf{U}^{(n)} = \mathbf{u}^{(n)}) = \sum_{1 \le i \le m} n_i \log p_i$. Hence,

$$A_n = \left\{ \mathbf{u}^{(n)} : -\sum_i n_i \log p_i \le c \right\},$$

for some (real) c, to be determined. To determine c, we use that

$$\mathbb{P}(A_n) \ge 1 - \varepsilon.$$

Hence, c is the value for which

$$\mathbb{P}\left(\mathbf{u}^{(n)} : -\sum_{1 \le i \le m} n_i \log p_i \ge c \right) < \varepsilon.$$

Now, for the random string $\mathbf{U}^{(n)} = U_1\, \ldots\, U_n$, let N_i is the number of appearances of value i. Then

$$-\sum_{1 \le i \le m} N_i \log p_i = \sum_{1 \le j \le n} \theta_j, \quad \text{where } \theta_j = -\log p_i \text{ when } U_j = i.$$

Since entries U_j are IID, so are random variables θ_j. Next,

$$\mathbb{E}\theta_j = -\sum_{1 \le i \le m} p_i \log p_i := h$$

and

$$\text{Var}\,\theta_j = \mathbb{E}(\theta_j)^2 - (\mathbb{E}\theta_j)^2 = \sum_{1 \le i \le m} p_i (\log p_i)^2 - \left(\sum_{1 \le i \le m} p_i \log p_i \right)^2 := v.$$

Then

$$\mathbb{E}\left[\sum_{1 \le j \le n} \theta_j \right] = nh \quad \text{and Var}\left[\sum_{1 \le j \le n} \theta_j \right] = nv.$$

Recall that $h = H$ is the information rate of the source.

By Chebyshev's inequality, for all $b > 0$,

$$\mathbb{P}\left(\left|-\sum_{1\leq i\leq m} N_i \log p_i - nh\right| > b\right) \leq \frac{nv}{b^2},$$

and with $b = \sqrt{nk/\varepsilon}$, we obtain

$$\mathbb{P}\left(\left|-\sum_{1\leq i\leq m} N_i \log p_i - nh\right| > \sqrt{\frac{nk}{\varepsilon}}\right) \leq \varepsilon.$$

Therefore, for all $\mathbf{u}^{(n)} \in A_n$,

$$-\sum_{1\leq i\leq m} n_i \log p_i \leq nh + \sqrt{\frac{nk}{\varepsilon}} := c.$$

For an irreducible and aperiodic Markov source the assertion is similar, with

$$H = -\sum_{1\leq i,j\leq m} \pi_i p_{ij} \log p_{ij},$$

and $v \geq 0$ a constant given by $v = \limsup_{n\to\infty} \frac{1}{n} \mathrm{Var}\left[\sum_{1\leq j\leq n} \theta_j\right].$ $\square$

Problem 1.26   *Demonstrate that an efficient and decipherable noiseless coding procedure leads to an entropy as a measure of attainable performance.*

Words of length s_i $(i = 1,\ldots,n)$ in an alphabet $\mathbb{F}_a = \{0,1,\ldots,a-1\}$ are to be chosen to minimise expected word-length $\sum_{i=1}^{n} p_i s_i$ subject not only to decipherability but also to the condition that $\sum_{i=1}^{n} q_i s_i$ should not exceed a prescribed bound, where q_i is a feasible alternative to the postulated probability distribution $\{p_i\}$ of characters in the original alphabet. Determine bounds on the minimal value of $\sum_{i=1}^{n} p_i s_i$.

Solution If we disregard the condition that $s_1,\ldots,s_n$ are positive integers, the minimisation problem becomes

$$\begin{aligned} \text{minimise} \quad &\sum_i s_i p_i \\ \text{subject to} \quad &s_i \geq 0 \text{ and } \sum_i a^{-s_i} \leq 1 \text{ (Kraft).} \end{aligned} \qquad (1.6.24)$$

This can be solved by the Lagrange method, with the Lagrangian

$$\mathscr{L}(s_1,\ldots,s_n,\lambda) = \sum_{1\leq i\leq n} s_i p_i - \lambda\left(1 - \sum_{1\leq i\leq n} a^{-s_i}\right).$$

The solution of the relaxed problem is unique and given by

$$s_i = -\log_a p_i, \quad 1 \le i \le n. \tag{1.6.25}$$

The relaxed optimal value v_{rel},

$$v_{\text{rel}} = -\sum_{1 \le i \le n} p_i \log_a p_i := h,$$

provides a lower bound for the optimal expected word-length $\sum_i s_i^* p_i$:

$$h \le \sum_i s_i^* p_i.$$

Now consider the additional constraint

$$\sum_{1 \le i \le n} q_i s_i \le b. \tag{1.6.26}$$

The relaxed problem (1.6.24) complemented with (1.6.26) again can be solved by the Lagrange method. Here, if

$$-\sum_i q_i \log_a p_i \le b$$

then adding the new constraint does not affect the minimiser (1.6.24), i.e. the optimal positive $s_1, \ldots, s_n$ are again given by (1.6.25), and the optimal value is h. Otherwise, i.e. when $-\sum_i q_i \log_a p_i > b$, the new minimiser $\tilde{s}_1, \ldots, \tilde{s}_n$ is still unique (since the problem is still strong Lagrangian) and fulfils both constraints

$$\sum_i a^{-\tilde{s}_i} = 1, \quad \sum_i q_i \tilde{s}_i = b.$$

In both cases, the optimal value $\tilde{v}_{\text{rel}}$ for the new relaxed problem satisfies $h \le \tilde{v}_{\text{rel}}$.
Finally, the solution $\tilde{s}_1^*, \ldots, \tilde{s}_n^*$ to the integer-valued word-length problem

$$\begin{aligned}
\text{minimise} \quad & \sum_i s_i p_i \\
\text{subject to} \quad & s_i \ge 1 \text{ integer} \quad \text{and} \quad \sum_i a^{-s_i} \le 1, \ \sum_i q_i s_i \le b
\end{aligned} \tag{1.6.27}$$

will satisfy

$$h \le \tilde{v}^{\text{rel}} \le \sum_i \tilde{s}_i^* p_i, \quad \sum_i \tilde{s}_i^* q_i \le b.$$

$\square$

Problem 1.27 *Suppose a discrete Markov source $\{X_t\}$ has transition probability*

$$p_{jk} = \mathbb{P}(X_{t+1} = k | X_t = j)$$

with equilibrium distribution (π_j). Suppose the letter can be obliterated by noise (in which case one observes only the event 'erasure') with probability $\beta = 1 - \alpha$, independent of current or previous letter values or previous noise. Show that the noise-corrupted source has information rate

$$-\alpha \log \alpha - \beta \log \beta - \alpha^2 \sum_j \sum_k \sum_{s \geq 1} \pi_j \beta^{s-1} p_{jk}^{(s)} \log p_{jk}^{(s)},$$

where $p_{jk}^{(s)}$ is the s-step transition probability of the original DTMC.

Solution Denote the corrupted source sequence $\{\widetilde{X}_t\}$, with $\widetilde{X}_t = *$ (a splodge) every time there was an erasure. Correspondingly, a string $\widetilde{x}_1^n$ from the corrupted source is produced from a string x_1^n of the original Markov source by replacing the obliterated digits with splodges. The probability $p_n(\widetilde{x}) = \mathbb{P}\left(\widetilde{X}_1^n = \widetilde{x}_1^n\right)$ of such a string is represented as

$$\sum_{x_1^n \text{ consistent with } \widetilde{x}_1^n} \mathbb{P}(X_1^n = x_1^n) \mathbb{P}(\widetilde{X}_1^n | X_1^n = x_1^n) \qquad (1.6.28)$$

and is calculated as the product where the initial factor is

$$\lambda_{x_1} \alpha \quad \text{or} \quad \sum_y \lambda_y p_{yx_s}^{(s)} \beta^{s-1} \alpha, \text{ where } 1 < s \leq n, \text{ or } 1,$$

depending on where the initial non-obliterated digit occurred in $\widetilde{x}_1^n$ (if at all). The subsequent factors contributing to (1.6.28) have a similar structure:

$$p_{x_{t-1} x_t} \beta \quad \text{or} \quad p_{x_{t-s} x_t}^{(s)} \beta^{s-1} \alpha \quad \text{or} \quad 1.$$

Consequently, the information $-\log p_n(\widetilde{x}_1^n)$ carried by string $\widetilde{x}_1^n$ is calculated as

$$\begin{aligned}
&-\log \mathbb{P}(X_{s_1} = x_{s_1}) - (s_1 - 1) \log \beta - \log \alpha \\
&-\log p_{x_{s_1} x_{s_2}}^{s_2 - s_1)} - (s_2 - s_1 - 1) \log \beta - \log \alpha - \cdots \\
&-\log p_{x_{s_{N-1}} x_{s_N}}^{(s_N - s_{N-1})} - (s_N - s_{N-1} - 1) \log \beta - \log \alpha
\end{aligned}$$

where $1 \leq s_1 < \cdots < s_N \leq n$ are the consecutive times of appearance of non-obliterated symbols in $\widetilde{x}_1^n$.

Now take $-\dfrac{1}{n} \log p_n\left(\widetilde{X}_1^n\right)$, the information rate provided by the random string $\widetilde{X}_1^n$. Ignoring the initial bit, we can write

$$-\frac{1}{n} \log p_n\left(\widetilde{X}_1^n\right) = -\frac{N(\beta)}{n} \log \beta - \frac{N(\alpha)}{n} \log \alpha - \sum_i \frac{M(i,j;s)}{n} \log p_{ij}^{(s)}.$$

Here

$N(\alpha)$ = number of non-obliterated digits in $\widetilde{X}_1^n$,

$N(\beta)$ = number of obliterated digits in $\widetilde{X}_1^n$,

$M(i,j;s)$ = number of series of digits $i * \cdots * j$ in $\widetilde{X}_1^n$ of length $s+1$

As $n \to \infty$, we have the convergence of the limiting frequencies (the law of large numbers applies):

$$\frac{N(\alpha)}{n} \to \alpha, \quad \frac{N(\beta)}{n} \to \beta, \quad \frac{M(i,j;s)}{n} \to \alpha \beta^{s-1} \pi_i p_{ij}^{(s)} \alpha.$$

This yields

$$-\frac{1}{n} \log p_n\left(\widetilde{X}_1^n\right)$$
$$\to -\alpha \log \alpha - \beta \log \beta - \alpha^2 \sum_{i,j} \pi_i \sum_{s \geq 1} \beta^{s-1} p_{ij}^{(s)} \log p_{ij}^{(s)},$$

as required. [The convergence holds almost surely (a.s.) and in probability.] According to the SCT, the limiting value gives the information rate of the corrupted source.   □

Problem 1.28   *A binary source emits digits 0 or 1 according to the rule*

$$\mathbb{P}(X_t = k | X_{t-1} = j, X_{t-2} = i) = q_r,$$

where k, j, i and r take values 0 or 1, $r = k - j - i \bmod 2$, and $q_0 + q_1 = 1$. Determine the information rate of this source.

Also derive the information rate of a Bernoulli source emitting digits 0 and 1 with probabilities q_0 and q_1. Explain the relationship between these two results.

Solution Re-write the conditional probabilities in a detailed form:

$$\mathbb{P}(X_t = 0 | X_{t-1} = j, X_{t-2} = i) = \begin{cases} q_0, & i = j, \\ q_1, & i \neq j, \end{cases}$$

$$\mathbb{P}(X_t = 1 | X_{t-1} = j, X_{t-2} = i) = \begin{cases} q_1, & i = j, \\ q_0, & i \neq j. \end{cases}$$

The source is a second-order Markov chain on $\{0, 1\}$, i.e. a DTMC with four states $\{00, 01, 10, 11\}$. The 4×4 transition matrix is

$$\begin{matrix} 00 \\ 01 \\ 10 \\ 11 \end{matrix} \begin{pmatrix} q_0 & q_1 & 0 & 0 \\ 0 & 0 & q_1 & q_0 \\ q_1 & q_0 & 0 & 0 \\ 0 & 0 & q_0 & q_1 \end{pmatrix}$$

The equilibrium probabilities are uniform:

$$\pi_{00} = \pi_{01} = \pi_{10} = \pi_{11} = \frac{1}{4}.$$

The information rate is calculated in a standard way:

$$H = - \sum_{\alpha,\beta=0,1} \pi_{\alpha\beta} \sum \sum_{\gamma=0,1} \pi_{\alpha\beta} P_{\alpha\beta,\beta\gamma} \log P_{\alpha\beta,\beta\gamma}$$

and equals

$$\frac{1}{4} \sum_{\alpha,\beta=0,1} h(q_0,q_1) = -q_0 \log q_0 - q_1 \log q_1.$$

□

Problem 1.29   *An input to a discrete memoryless channel has three letters 1, 2 and 3. The letter j is received as $(j-1)$ with probability p, as $(j+1)$ with probability p and as j with probability $1-2p$, the letters from the output alphabet ranging from 0 to 4. Determine the form of the optimal input distribution, for general p, as explicitly as possible. Compute the channel capacity in the three cases $p = 0$, $p = 1/3$ and $p = 1/2$.*

Solution The channel matrix is 3×5:

$$\begin{matrix} 1 \\ 2 \\ 3 \end{matrix} \begin{pmatrix} p & (1-2p) & p & 0 & 0 \\ 0 & p & (1-2p) & p & 0 \\ 0 & 0 & p & (1-2p) & p \end{pmatrix}.$$

The rows are permutations of each other, so the capacity equals

$$C = \max_{P_X} [h(Y) - h(Y|X)]$$
$$= (\max_{P_X} h(Y)) + [2p \log p + (1-2p) \log(1-2p)],$$

with the maximisation over the input-letter distribution P_X applied only to $h(Y)$, the entropy of the output-symbol.
   Next,

$$h(Y) = - \sum_{y=0,1,2,3,4} P_Y(y) \log P_Y(y),$$

where

$$\begin{aligned} P_Y(0) &= P_X(1)p, \\ P_Y(1) &= P_X(1)(1-2p) + P_X(2)p, \\ P_Y(2) &= P_X(1)p + P_X(2)(1-2p) + P_X(3)p, \\ P_Y(3) &= P_X(3)(1-2p) + P_X(2)p, \\ P_Y(4) &= P_X(3)p. \end{aligned} \right\} \qquad (1.6.29)$$

The symmetry in (1.6.29) suggests that $h(Y)$ is maximised when $P_X(0) = P_X(2) = q$ and $P_X(1) = 1 - 2q$. So:

$$\max h(Y) = \max \left[-2qp\log(qp) - 2\left[q(1 - 2p) + (1 - 2q)p\right]\right.$$
$$\times \log\left[q(1 - 2p) + (1 - 2q)p\right]$$
$$\left. -\left[2qp + (1 - 2q)(1 - 2p)\right]\log\left[2qp + (1 - 2q)(1 - 2p)\right]\right].$$

To find the maximum, differentiate and solve:

$$\frac{d}{dq}h(Y) = -2p\log(qp) - 2p - 2(1 - 4p)\log\left[q(1 - 2p) + (1 - 2q)p\right]$$
$$-2(1 - 4p) - (2p - 2)\log\left[2qp + (1 - 2q)(1 - 2p)\right] - (2p - 2)$$
$$= 4p - 2p\log(qp) - 2(1 - 4p)\log\left[q(1 - 2p) + (1 - 2q)p\right]$$
$$-2(1 - 4p) - 2(p - 1)\log\left[2qp + (1 - 2q)(1 - 2p)\right] = 0.$$

For $p = 0$ we have a perfect error-free channel, of capacity $\log 3$ which is achieved when $P_X(1) = P_X(2) = P_X(3) = 1/3$ (i.e. $q = 1/3$), and $P_Y(1) = P_Y(2) = P_Y(3) = 1/3$, $P_Y(0) = P_Y(4) = 0$.

For $p = 1/3$, the output probabilities are

$$p_Y(0) = p_Y(4) = q/3, \quad p_Y(1) = (1 - q)/3, \quad p_Y(2) = 1/3,$$

and $h(Y)$ simplifies to

$$h(Y) = -2\frac{q}{3}\log\frac{q}{3} - 2\frac{1 - q}{3}\log\frac{1 - q}{3} - \frac{1}{3}\log\frac{1}{3}.$$

The derivative $dh(Y)/dq = 0$ becomes

$$-\frac{2}{3}\log\frac{q}{3} - \frac{2}{3} + \frac{2}{3}\log\frac{1 - q}{3} + \frac{2}{3}$$

and vanishes when $q = 1/2$, i.e.

$$P_X(1) = P_X(3) = 1/2, \quad P_X(2) = 0,$$
$$P_Y(0) = P_Y(1) = P_Y(3) = P_Y(4) = 1/6, \quad P_Y(2) = 1/3.$$

Next, the conditional entropy

$$h(Y|X) = \log 3.$$

For the capacity this yields

$$C = -\frac{2}{3}\log\frac{1}{6} - \frac{1}{3}\log\frac{1}{3} - \log 3 = \frac{2}{3}.$$

Finally, for $p = 1/2$, we have $h(Y|X) = 1$ and

$$P_Y(0) = P_Y(4) = \frac{q}{2}, \quad P_Y(1) = P_Y(3) = \frac{1 - 2q}{2}, \quad P_Y(2) = q.$$

The output entropy is

$$h(Y) = -q\log\frac{q}{2} - \frac{1-2q}{2}\log\frac{1-2q}{2} - q\log q$$

$$= q - 2q\log q - \frac{1-2q}{2}\log\frac{1-2q}{2}$$

and is maximised when $q = 1/6$, i.e.

$$P_X(1) = P_X(2) = \frac{1}{6}, \ P_X(2) = \frac{2}{3},$$

$$P_Y(0) = P_Y(4) = \frac{1}{12}, \ P_Y(1) = P_Y(3) = \frac{1}{3}, \ P_Y(2) = \frac{1}{6}.$$

The capacity in this case equals

$$C = \log 3 - \frac{1}{2}.$$

$\square$

Problem 1.30   *A memoryless discrete-time channel produces outputs Y from non-negative integer-valued inputs X by*

$$Y = \varepsilon X,$$

where ε is independent of X, $\mathbb{P}(\varepsilon = 1) = p$, $\mathbb{P}(\varepsilon = 0) = 1 - p$, and inputs are restricted by the condition that $\mathbb{E}X \le 1$.

By considering input distributions $\{a_i, \ i = 0,1,\ldots\}$ of the form $a_i = cq^i$, $i = 1,2,\ldots$, or otherwise, derive the optimal input distribution and determine an expression for the capacity of the channel.

Solution The channel matrix is

$$\begin{pmatrix} 1 & 0 & 0 & \cdots & 0 \\ 1-p & p & 0 & \cdots & 0 \\ 1-p & 0 & p & \cdots & 0 \\ \vdots & \vdots & \vdots & \ddots & \vdots \end{pmatrix}.$$

For the input distribution with $q_i = \mathbb{P}(X = i)$, we have that

$$\mathbb{P}(Y = 0) = q_0 + (1-p)(1-q_0) = 1 - p + pq_0,$$
$$\mathbb{P}(Y = i) = pq_i, \ i \ge 1,$$

whence

$$h(Y) = -(1 - p + pq_0)\log(1 - p + pq_0) - \sum_{i\ge1} pq_i \log(pq_i).$$

With the conditional entropy being

$$h(Y|X) = -(1-q_0)\big[(1-p)\log(1-p) + p\log p\big]$$

the mutual entropy equals

$$I(Y:X) = -(1-p+pq_0)\log(1-p+pq_0)$$
$$- \sum_{i\geq 1} pq_i\log(pq_i) + (1-q_0)[(1-p)\log(1-p) + p\log p].$$

We have to maximise $I(Y:X)$ in $q_0, q_1, \ldots$, subject to $q_i \geq 0$, $\sum_i q_i = 1$, $\sum_i iq_i \leq 1$. First, we fix q_0 and maximise the sum $-\sum_{i\geq 1} pq_i\log(pq_i)$ in q_i with $i \geq 1$. By Gibbs, for all non-negative $a_1, a_2, \ldots$ with $\sum_{i\geq 1} a_i = 1 - q_0$,

$$-\sum_{i\geq 1} q_i\log q_i \leq -\sum_{i\geq 1} q_i\log a_i, \quad \text{with equality iff } q_i \equiv a_i.$$

For $a_i = cd^i$ with $\sum_i ia_i = 1$, the RHS becomes

$$-(1-q_0)\log c - \big(\log d\big)\sum_{i\geq 1} ia_i = -(1-q_0)\log c - \log d.$$

From $\sum_i icd^i = 1$, $cd/(1-d) = 1 - a_0$ and $d = a_0$, $c = (1-a_0)^2/a_0$.
  Next, we maximise, in $a_0 \in [0,1]$, the function

$$f(a_0) = -(1-p+pa_0)\log(1-p+pa_0) - p(1-a_0)\log\frac{(1-a_0)^2}{a_0}$$
$$- \log a_0 + (1-a_0)[(1-p)\log(1-p) + p\log p].$$

Requiring that

$$f'(a_0) = 0 \tag{1.6.30a}$$

and

$$f''(a_0) = \frac{-p^2}{q+pa_0} - \frac{2p}{1-a_0} - \frac{p}{a_0} \leq 0, \tag{1.6.30b}$$

one can solve equation (1.6.30a) numerically. Denote its root where (1.6.30b) holds by a_0^-. Then we obtain the following answer for the optimal input distribution:

$$a_i = \begin{cases} a_0^-, & i = 0, \\ (1-a_0^-)^2(a_0^-)^{i-1}, & i \geq 1. \end{cases}$$

with the capacity $C = f(a_0^-)$. □

Problem 1.31   *The representation known as binary-coded decimal encodes 0 as 0000, 1 as 0001 and so on up to 9, coded as 1001, with other 4-digit binary strings being discarded. Show that by encoding in blocks, one can get arbitrarily near the lower bound on word-length per decimal digit.*
Hint: Assume all integers to be equally probable.

Solution The code in question is obviously decipherable (and even prefix-free, as is any decipherable code with a fixed codeword-length). The standard block-coding procedure treats a string of n letters from the original source (U_n) operating with an alphabet $\mathscr{A}$ as a letter from $\mathscr{A}^n$. Given joint probabilities $p_n(u_1^{(n)}) = \mathbb{P}(U_1 = i_1,\ldots,U_n = i_n)$, of the blocks in a typical message, we look at the binary entropy

$$h^{(n)} = - \sum_{i_1,\ldots,i_n} \mathbb{P}(U_1 = i_1,\ldots,U_n = i_n)\log\mathbb{P}(U_1 = i_1,\ldots,U_n = i_n).$$

Denote by $S^{(n)}$ the random codeword-length while encoding in blocks. The minimal expected word-length per source letter is $e_n := \min \dfrac{1}{n}\mathbb{E}S^{(n)}$. By Shannon's NC theorem,

$$\frac{h^{(n)}}{n\log q} \le e_n \le \frac{h^{(n)}}{n\log q} + \frac{1}{n},$$

where q is the size of the original alphabet $\mathscr{A}$. We see that, for large n, $e_n \sim \dfrac{h^{(n)}}{n\log q}$. In the question, $q = 10$ and

$$h^{(n)} = hn, \quad \text{where } h = \log 10 \text{ (equiprobability)}.$$

Hence, the minimal expected word-length e_n can be made arbitrarily close to 1. □

Problem 1.32   Let $\{U_t\}$ be a discrete-time process with values u_t and let $\mathbb{P}(\mathbf{u}^{(n)})$ be the probability that a string $\mathbf{u}^{(n)} = u_1 \ldots u_n$ is produced. Show that if $-\log\mathbb{P}(\mathbf{U}^{(n)})/n$ converges in probability to a constant γ then γ is the information rate of the process.

Write down the formula for the information rate of an m-state DTMC and find the rate when the transition matrix has elements p_{jk} where

$$p_{jj} = p, \quad p_{jj+1} = 1 - p \ (j = 1,\ldots,m-1), \quad p_{m1} = 1 - p.$$

Relate this to the information rate of a two-state source with transition probabilities p and $1 - p$.

Solution The information rate of an m-state stationary DTMC with transition matrix $P = (p_{ij})$ and an equilibrium (invariant) distribution $\pi = (\pi_i)$ equals

$$h = - \sum_{i,j} \pi_i p_{ij} \log p_{ij}.$$

If matrix P is irreducible (i.e. has a unique communicating class) then this statement holds for the chain with any initial distribution λ (in this case the equilibrium distribution is unique).

In the example, the transition matrix is

$$
\begin{pmatrix}
p & 1-p & 0 & \cdots & 0 \\
0 & p & 1-p & \cdots & 0 \\
\vdots & \vdots & \vdots & \ddots & \vdots \\
1-p & 0 & 0\ldots & & p
\end{pmatrix}.
$$

The rows are permutations of each other, and each of them has entropy

$$
-p\log p - (1-p)\log(1-p).
$$

The equilibrium distribution is $\pi = (1/m, \ldots, 1/m)$:

$$
\sum_{1\le i\le m} \frac{1}{m}p_{ij} = \frac{1}{m}(p+1-p) = \frac{1}{m},
$$

and it is unique, as the chain has a unique communicating class. Therefore, the information rate equals

$$
h = \frac{1}{m}\sum_{1\le i\le m}\left[-p\log p - (1-p)\log(1-p)\right] = -p\log p - (1-p)\log(1-p).
$$

For $m=2$ we obtain precisely the matrix $\begin{pmatrix} p & 1-p \\ 1-p & p \end{pmatrix}$, so – with the equilibrium distribution $\pi = (1/2, 1/2)$ – the information rate is again $h = \eta(p)$. $\qquad\square$

Problem 1.33 *Define a symmetric channel and find its capacity.*
A native American warrior sends smoke signals. The signal is coded in puffs of smoke of different lengths: short, medium and long. One puff is sent per unit time. Assume a puff is observed correctly with probability p, and with probability $1-p$ (a) a short signal appears to be medium to the recipient, (b) a medium puff appears to be long, and (c) a long puff appears to be short. What is the maximum rate at which the warrior can transmit reliably, assuming the recipient knows the encoding system he uses?

It would be more reasonable to assume that a short puff may disperse completely rather than appear medium. In what way would this affect your derivation of a formula for channel capacity?

Solution Suppose we use an input alphabet $\mathscr{I}$, of m letters, to feed a memoryless channel that produces symbols from an output alphabet $\mathscr{J}$ of size n (including illegibles). The channel is described by its $m \times n$ matrix where entry p_{ij} gives the probability that letter $i \in \mathscr{I}$ is transformed to symbol $j \in \mathscr{J}$. The rows of the channel matrix form stochastic n-vectors (probability distributions over $\mathscr{J}$):

$$
\begin{pmatrix}
p_{11} & \cdots & p_{1j} & \cdots & p_{1n} \\
\vdots & \ddots & \vdots & \ddots & \vdots \\
p_{i1} & \cdots & p_{ij} & \cdots & p_{in} \\
\vdots & \ddots & \vdots & \ddots & \vdots \\
p_{m1} & \cdots & p_{mj} & \cdots & p_{mn}
\end{pmatrix}.
$$

The channel is called symmetric if its rows are permutations of each other (or, more generally, have the same entropy $E = h(p_{i1}, \ldots, p_{in})$, for all $i \in \mathscr{I}$). The channel is said to be double-symmetric if in addition its columns are permutations of each other (or, more generally, have the same column sum $\Sigma = \sum\limits_{1 \leq i \leq m} p_{ij}$, for all $j \in \mathscr{J}$).

For a memoryless channel, the capacity (the supremum of reliable transmission rates) is given by

$$
C = \max_{P_X} I(X : Y).
$$

Here, the maximum is taken over $P_X = (P_X(i), i \in \mathscr{I})$, the input-letter probability distribution, and $I(X : Y)$ is the mutual entropy between the input and output random letters X and Y tied through the channel matrix:

$$
I(X : Y) = h(Y) - h(Y|X) = h(X) - h(X|Y).
$$

For the symmetric channel, the conditional entropy

$$
h(Y|X) = -\sum_{i,j} P_X(i) p_{ij} \log p_{ij} \equiv h,
$$

regardless of the input probabilities $p_X(i)$. Hence,

$$
C = \left[\max_{P_X} h(Y) \right] - h(Y|X),
$$

and the maximisation needs only to be performed for the output symbol entropy

$$
h(Y) = -\sum_j P_Y(j) \log P_Y(j), \quad \text{where } P_Y(j) = \sum_i P_X(i) p_{ij}.
$$

For a double-symmetric channel, the latter problem becomes straightforward: $h(Y)$ is maximised by the uniform input equidistribution $P_X^{eq}(i) = 1/m$, as in this case P_Y is also uniform:

$$
P_Y(j) = \frac{1}{m} \sum_i p_{ij} = \frac{1}{m} \text{ as it doesn't depend on } j \in \mathscr{J}.
$$

Thus, for the double-symmetric channel:

$$
C = \log n - h(Y|X).
$$

In the example, the channel matrix is 3×3,

$$
\begin{array}{l}
1 \sim \text{short} \\
2 \sim \text{medium} \\
3 \sim \text{long}
\end{array}
\quad
\begin{pmatrix}
p & 1-p & 0 \\
0 & p & 1-p \\
1-p & 0 & p
\end{pmatrix},
$$

and double-symmetric. This yields

$$
C = \log 3 + p \log p + (1-p) \log(1-p).
$$

In the modified example, the matrix becomes 3×4:

$$
\begin{pmatrix}
p & 0 & 0 & 1-p \\
0 & p & 1-p & 0 \\
1-p & 0 & p & 0
\end{pmatrix} ;
$$

column 4 corresponds to a 'no-signal' output state (a 'splodge'). The maximisation problem loses its symmetry:

$$
\max \quad \left[-\sum_{j=1,2,3,4} \left(\sum_{i=1,2,3} P_X(i) p_{ij} \right) \log \left(\sum_{i=1,2,3} P_X(i) p_{ij} \right) \right.
$$

$$
\left. - \sum_{i=1,2,3} P_X(i) \sum_{j=1,2,3,4} p_{ij} \log p_{ij} \right]
$$

subject to $\quad P_X(1), P_X(2), P_X(3) \geq 0, \text{ and } \sum_{i=1,2,3} P_X(i) = 1,$

and requires a full-scale analysis. $\qquad\qquad\qquad\qquad\qquad\qquad\qquad\quad\Box$

Problem 1.34 *The entropy power inequality (EPI, see (1.5.10)) states: for* $\mathbf{X}$ *and* $\mathbf{Y}$ *independent d-dimensional random vectors,*

$$
2^{2h(\mathbf{X+Y})/d} \geq 2^{2h(\mathbf{X})/d} + 2^{2h(\mathbf{Y})/d}, \tag{1.6.31}
$$

with equality iff $\mathbf{X}$ *and* $\mathbf{Y}$ *are Gaussian with proportional covariance matrices.*

Let X be a real-valued random variable with a PDF f_X and finite differential entropy $h(X)$, and let function $g: \mathbb{R} \to \mathbb{R}$ have strictly positive derivative g' everywhere. Prove that the random variable $g(X)$ has differential entropy satisfying

$$
h(g(X)) = h(X) + \mathbb{E} \log_2 g'(X),
$$

assuming that $\mathbb{E} \log_2 g'(X)$ is finite.

Let Y_1 and Y_2 be independent, strictly positive random variables with densities. Show that the differential entropy of the product $Y_1 Y_2$ satisfies

$$
2^{2h(Y_1 Y_2)} \geq \alpha_1 2^{2h(Y_1)} + \alpha_2 2^{2h(Y_2)},
$$

where $\log_2(\alpha_1) = 2\mathbb{E} \log_2 Y_2$ and $\log_2(\alpha_2) = 2\mathbb{E} \log_2 Y_1$.

Solution The CDF of the random variable $g(X)$ satisfies

$$F_{g(X)}(y) = \mathbb{P}(g(X) \le y) = \mathbb{P}(X \le g^{-1}(y)) = F_X(g^{-1}(y)),$$

i.e. the PDF $f_{g(X)}(y) = \dfrac{\mathrm{d}F_{g(X)}(y)}{\mathrm{d}y}$ takes the form

$$f_{g(X)}(y) = f_X\left(g^{-1}(y)\right)\left(g^{-1}(y)\right)' = \frac{f_X\left(g^{-1}(y)\right)}{g'\left(g^{-1}(y)\right)}.$$

Then

$$
\begin{aligned}
h(g(X)) &= -\int f_{g(X)}(y)\log_2 f_{g(X)}(y)\mathrm{d}y \\
&= \int \frac{f_X\left(g^{-1}(y)\right)}{g'\left(g^{-1}(y)\right)}\log_2 \frac{f_X\left(g^{-1}(y)\right)}{g'\left(g^{-1}(y)\right)}\mathrm{d}y \\
&= -\int \frac{f_X(x)}{g'(x)}\left[\log_2 f_X(x) - \log_2 g'(x)\right]g'(x)\mathrm{d}x \\
&= h(X) + \mathbb{E}\left[\log_2 g'(X)\right].
\end{aligned}
\tag{1.6.32}
$$

When $g(t) = e^t$ then

$$\log_2 g'(t) = \log_2 e^t = t\log_2 e.$$

So, $Y_i = e^{X_i} = g(X_i)$ and (1.6.32) implies

$$h(e^{X_i}) = h(g(X_i)) = h(X_i) + \mathbb{E}X_i\log_2 e, \quad i = 1,2,3,$$

with $X_3 = X_1 + X_2$. Then

$$h(Y_1Y_2) = h(e^{X_1+X_2}) = h(X_1 + X_2) + \left(\mathbb{E}X_1 + \mathbb{E}X_2\right)\log_2 e.$$

Hence, in the entropy-power inequality,

$$
\begin{aligned}
2^{2h(Y_1Y_2)} &= 2^{2h(X_1+X_2)+2(\mathbb{E}X_1+\mathbb{E}X_2)\log_2 e} \\
&\ge \left(2^{2h(X_1)} + 2^{2h(X_2)}\right)2^{2(\mathbb{E}X_1+\mathbb{E}X_2)\log_2 e} \\
&= 2^{2\mathbb{E}X_2\log_2 e}\left(2^{2[h(X_1)+\mathbb{E}X_1\log_2 e]}\right) \\
&\quad + 2^{2\mathbb{E}X_1\log_2 e}\left(2^{2[h(X_2)+\mathbb{E}X_2\log_2 e]}\right) \\
&= \alpha_1 2^{2h(Y_1)} + \alpha_2 2^{2h(Y_2)}.
\end{aligned}
$$

Here $\alpha_1 = 2^{2\mathbb{E}X_2\log_2 e}$, i.e.

$$\log_2 \alpha_1 = 2\mathbb{E}X_2\log_2 e = 2\mathbb{E}\ln Y_2\log_2 e = 2\mathbb{E}\log_2 Y_2,$$

and similarly, $\log_2 \alpha_1 = 2\mathbb{E}\log_2 Y_1$. $\qquad\square$

Problem 1.35   *In this problem we work with the following functions defined for* $0 < a < b$:

$$G(a,b) = \sqrt{ab}, \ \ L(a,b) = \frac{b-a}{\log(b/a)}, \ \ I(a,b) = \frac{1}{e}(b^b/a^a)^{1/(b-a)}.$$

Check that

$$0 < a < G(a,b) < L(a,b) < I(a,b) < A(a,b) = \frac{a+b}{2} < b. \tag{1.6.33}$$

Next, for $0 < a < b$ *define*

$$\Lambda(a,b) = L(a,b)I(a,b)/G^2(a,b).$$

Let $\mathbf{p} = (p_i)$ *and* $\mathbf{q} = (q_i)$ *be the probability distributions of random variables* X *and* Y:

$$\mathbb{P}(X = i) = p_i > 0, \mathbb{P}(Y = i) = q_i, i = 1,\ldots,r, \sum p_i = \sum q_i = 1.$$

Let $m = \min[q_i/p_i], M = \max[q_i/p_i], \mu = \min[p_i], \nu = \max[p_i]$. *Prove the following bounds for the entropy* $h(X)$ *and Kullback–Leibler divergence* $D(\mathbf{p}\|\mathbf{q})$ *(cf. PSE II, p. 419):*

$$0 \le \log r - h(X) \le \log \Lambda(\mu, \nu). \tag{1.6.34}$$
$$0 \le D(\mathbf{p}\|\mathbf{q}) \le \log \Lambda(m, M). \tag{1.6.35}$$

Solution The inequality (1.6.33) is straightforward and left as an exercise. For $a \le x_i \le b$, set $\mathscr{A}(\mathbf{p},\mathbf{x}) = \sum p_i x_i$, $\mathscr{G}(\mathbf{p},\mathbf{x}) = \prod x_i^{p_i}$. The following general inequality holds:

$$1 \le \frac{\mathscr{A}(\mathbf{p},\mathbf{x})}{\mathscr{G}(\mathbf{p},\mathbf{x})} \le \Lambda(a,b). \tag{1.6.36}$$

It implies that

$$0 \le \log \left(\sum p_i x_i \right) - \sum p_i \log x_i \le \log \Lambda(a,b).$$

Selecting $x_i = q_i/p_i$ we immediately obtain (1.6.35). Taking $\mathbf{q}$ to be uniform, we obtain (1.6.34) from (1.6.35) since

$$\Lambda \left(\frac{1}{r\nu}, \frac{1}{r\mu} \right) = \Lambda \left(\frac{1}{\nu}, \frac{1}{\mu} \right) = \Lambda(\mu, \nu).$$

Next, we sketch the proof of (1.6.36); see details in [144], [50]. Let f be a convex function, $p, q \ge 0, p + q = 1$. Then for $x_i \in [a,b]$, we have

$$0 \le \sum p_i f(x_i) - f \left(\sum p_i x_i \right) \le \max_p [pf(a) + qf(b) - f(pa + qb)]. \tag{1.6.37}$$

Applying (1.6.37) for a convex function $f(x) = -\log x$ we obtain after some calculations that the maximum in (1.6.37) is achieved at $p_0 = (b - L(a,b))/(b - a)$, with $p_0 a + (1 - p_0)b = L(a,b)$, and

$$0 \le \log \frac{\mathscr{A}(\mathbf{p},\mathbf{x})}{\mathscr{G}(\mathbf{q},\mathbf{x})} \le \log\left(\frac{b-a}{\log(b/a)}\right) - \log(ab) + \frac{\log(b^b/a^a)}{b-a} - 1$$

which is equivalent to (1.6.36). Finally, we establish (1.6.37). Write $x_i = \lambda_i a + (1 - \lambda_i)b$ for some $\lambda_i \in [0,1]$. Then by convexity

$$0 \le \sum p_i f(x_i) - f\left(\sum p_i x_i\right)$$
$$\le \sum p_i(\lambda_i f(a) + (1 - \lambda_i)f(b)) - f\left(a\sum p_i \lambda_i + b \sum p_i(1 - \lambda_i)\right).$$

Denoting $\sum p_i \lambda_i = p$ and $1 - \sum p_i \lambda_i = q$ and maximising over p we obtain (1.6.37). □

Problem 1.36   Let f *be a strictly positive probability density function (PDF) on the line* $\mathbb{R}$, *define the Kullback–Leibler divergence* $D(g\|f)$ *and prove that* $D(g\|f) \ge 0$.

Next, assume that $\int e^x f(x)\mathrm{d}x < \infty$ *and* $\int |x|e^x f(x)\mathrm{d}x < \infty$. *Prove that the minimum of the expression*

$$-\int xg(x)\mathrm{d}x + D(g\|f) \tag{1.6.38}$$

over the PDFs g *with* $\int |x|g(x)\mathrm{d}x < \infty$ *is attained at the unique PDF* $g^* \propto e^x f(x)$ *and calculate this minimum.*

Solution   The Kullback–Leibler divergence $D(g\|f)$ is defined by

$$D(g\|f) = \int g(x)\ln\frac{g(x)}{f(x)}\mathrm{d}x, \quad \text{if} \quad \int g(x)\left|\ln\frac{g(x)}{f(x)}\right|\mathrm{d}x < \infty$$

and

$$D(g\|f) = \infty, \quad \text{if} \quad \int g(x)\left|\ln\frac{g(x)}{f(x)}\right|\mathrm{d}x = \infty.$$

The bound $D(g\|f) \ge 0$ is the Gibbs inequality.

Now take the PDF $g^*(x) = e^x f(x)/Z$ where $Z = \int e^z f(z)\mathrm{d}z$. Set $W = \int xe^x f(x)\mathrm{d}x$; then $W/Z = \int xg^*(x)\mathrm{d}x$. Further, write:

$$\begin{aligned} D(g^*\|f) &= \frac{1}{Z}\int e^x f(x)\ln\frac{e^x}{Z}\mathrm{d}x \\ &= \frac{1}{Z}\int e^x f(x)(x - \ln Z)\mathrm{d}x = \frac{1}{Z}(W - Z\ln Z) = \frac{W}{Z} - \ln Z \end{aligned}$$

and obtain that

$$- \int xg^*(x)dx + D(g^*||f) = -\ln Z.$$

This is the claimed minimum in the last part of the question.

Indeed, for any PDF g such that $\int |x|g(x)dx < \infty$, set $q(x) = g(x)/f(x)$ and write

$$\begin{aligned} D(g||g^*) &= \int g(x) \ln \frac{g(x)}{g^*(x)}dx = \int q(x) \ln \left[q(x)e^{-x}Z\right]f(x)dx \\ &= -\int xf(x)q(x)dx + \int f(x)q(x) \ln q(x)dx + \ln Z \\ &= -\int xg(x)dx + D(g||f) + \ln Z, \end{aligned}$$

implying that

$$-\int xg(x)dx + D(g||f) = -\int xg^*(x)dx + D(g^*||f) + D(g||g^*).$$

Since $D(g||g^*) > 0$ unless $g = g^*$, the claim follows. $\square$

Remark 1.6.1   The property of minimisation of (1.6.38) is far reaching and important in a number of disciplines, including statistical physics, ergodic theory and financial mathematics. We refer the reader to the paper [109] for further details.

2

Introduction to Coding Theory

2.1 Hamming spaces. Geometry of codes. Basic bounds on the code size

For presentational purposes, it is advisable to concentrate at the first reading of this section on the binary case where the symbols sent through a channel are 0 and 1.

As we saw earlier, in the case of an MBSC with the row error-probability $p \in (0, 1/2)$, the ML decoder looks for a codeword $\mathbf{x}_*^{(N)}$ that has the maximum number of digits coinciding with the received binary word $\mathbf{y}^{(N)}$. In fact, if $\mathbf{y}^{(N)}$ is received, the ML decoder compares the probabilities

$$\mathbf{P}\left(\mathbf{y}^{(N)} | \mathbf{x}^{(N)}\right) = p^{\delta(\mathbf{x}^{(N)}, \mathbf{y}^{(N)})} (1-p)^{N - \delta(\mathbf{x}^{(N)}, \mathbf{y}^{(N)})}$$

$$= (1-p)^N \left(\frac{p}{1-p}\right)^{\delta(\mathbf{x}^{(N)}, \mathbf{y}^{(N)})}$$

for different binary codewords $\mathbf{x}^{(N)}$. Here

$$\delta(\mathbf{x}^{(N)}, \mathbf{y}^{(N)}) = \text{ the number of digits } i \text{ with } x_i \neq y_i \qquad (2.1.1a)$$

is the so-called *Hamming distance* between words $\mathbf{x}^{(N)} = x_1 \ldots x_N$ and $\mathbf{y}^{(N)} = y_1 \ldots y_N$. Since the first factor $(1-p)^N$ does not depend on $\mathbf{x}^{(N)}$, the decoder seeks to maximise the second factor, that is to minimise $\delta(\mathbf{x}^{(N)}, \mathbf{y}^{(N)})$ (as $0 < p/(1-p) < 1$ for $p \in (0, 1/2)$).

The definition (2.1.1a) of Hamming distance can be extended to q-ary strings. The space of q-ary words $\mathcal{H}_{N,q} = \{0, 1, \ldots, q-1\}^{\times N}$ (the Nth Cartesian power of set $J_q = \{0, 1, \ldots, q-1\}$) with distance (2.1.1a) is called the q-ary *Hamming space* of length N. It contains q^N elements. In the binary case, $\mathcal{H}_{N,2} = \{0, 1\}^{\times N}$.

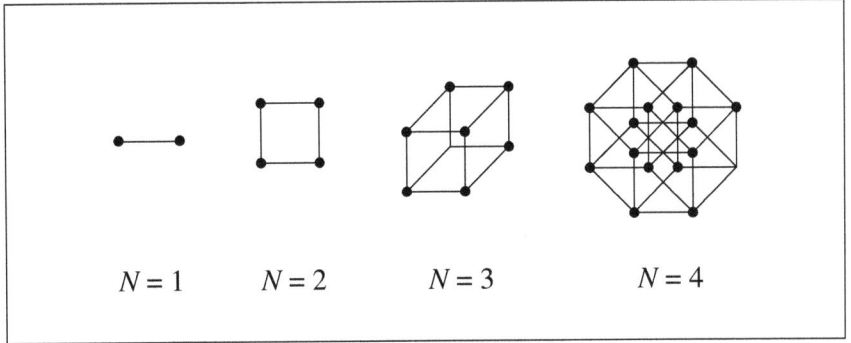

$$N = 1 \qquad N = 2 \qquad N = 3 \qquad N = 4$$

Figure 2.1

An important part is played by the distance $\delta(\mathbf{x}^{(N)}, \mathbf{0}^{(N)})$ between words $\mathbf{x}^{(N)} = x_1 \ldots x_N$ and $\mathbf{0}^{(N)} = 0 \ldots 0$; it is called the *weight* of word $\mathbf{x}^{(N)}$ and denoted by $w\left(\mathbf{x}^{(N)}\right)$:

$$w\left(\mathbf{x}^{(N)}\right) = \text{the number of digits } i \text{ with } x_i \neq 0. \qquad (2.1.1b)$$

Lemma 2.1.1 *The quantity $\delta(\mathbf{x}^{(N)}, \mathbf{y}^{(N)})$ defines a distance on $\mathscr{H}_{N,q}$. That is:*

(i) $0 \leq \delta(\mathbf{x}^{(N)}, \mathbf{y}^{(N)}) \leq N$ *and* $\delta(\mathbf{x}^{(N)}, \mathbf{y}^{(N)}) = 0$ *iff* $\mathbf{x}^{(N)} = \mathbf{y}^{(N)}$.
(ii) $\delta(\mathbf{x}^{(N)}, \mathbf{y}^{(N)}) = \delta(\mathbf{y}^{(N)}, \mathbf{x}^{(N)})$.
(iii) $\delta(\mathbf{x}^{(N)}, \mathbf{z}^{(N)}) \leq \delta(\mathbf{x}^{(N)}, \mathbf{y}^{(N)}) + \delta(\mathbf{y}^{(N)}, \mathbf{z}^{(N)})$ *(the triangle inequality).*

Proof The proof of (i) and (ii) is obvious. To check (iii), observe that any digit i with $z_i \neq x_i$ has either $y_i \neq x_i$ and then counted in $\delta(\mathbf{x}^{(N)}, \mathbf{y}^{(N)})$ or $z_i \neq y_i$ and then counted in $\delta(\mathbf{y}^{(N)}, \mathbf{z}^{(N)})$. $\square$

Geometrically, the binary Hamming space $\mathscr{H}_{N,2}$ may be identified with the collection of the vertices of a unit cube in N dimensions. The Hamming distance equals the lowest number of edges we have to pass from one vertex to another. It is a good practice to plot pictures for relatively low values of N: see Figure 2.1.

An important role is played below by geometric and algebraic properties of the Hamming space. Namely, as in any metric space, we can consider a ball of a given radius R around a given word $\mathbf{x}^{(N)}$:

$$\mathscr{B}_{N,q}(\mathbf{x}^{(N)}, R) = \{\mathbf{y}^{(N)} \in \mathscr{H}_{N,q} : \delta(\mathbf{x}^{(N)}, \mathbf{y}^{(N)}) \leq R\}. \qquad (2.1.2)$$

An important (and hard) problem is to calculate the maximal number of disjoint balls of a given radius which can be packed in a given Hamming space.

Observe that words admit an operation of addition mod q:

$$\mathbf{x}^{(N)} + \mathbf{y}^{(N)} = (x_1 + y_1) \bmod q \, \ldots \, (x_N + y_N) \bmod q. \qquad (2.1.3a)$$

This makes the Hamming space $\mathscr{H}_{N,q}$ a commutative group, with the zero code-word $\mathbf{0}^{(N)} = 0\ldots0$ playing the role of the zero of the group. (Words also may be multiplied which generates a powerful apparatus; see below.)

For $q = 2$, we have a two-point code alphabet $\{0,1\}$ that is actually a two-point *field*, $\mathbb{F}_2$, with the following arithmetic: $0 + 0 = 1 + 1 = 0 \cdot 1 = 1 \cdot 0 = 0$, $0 + 1 = 1 + 0 = 1 \cdot 1 = 1$. (Recall, a field is a set equipped with two commutative operations: addition and multiplication, satisfying standard axioms of associativity and distributivity.) Thus, each point in the binary Hamming space $\mathscr{H}_{N,2}$ is opposite to itself: $\mathbf{x}^{(N)} + \mathbf{x}'^{(N)} = \mathbf{0}^{(N)}$ iff $\mathbf{x}^{(N)} = \mathbf{x}'^{(N)}$. In fact, $\mathscr{H}_{N,2}$ is a linear space over the coefficient field $\mathbb{F}_2$, with $1 \cdot \mathbf{x}^{(N)} = \mathbf{x}^{(N)}$, $0 \cdot \mathbf{x}^{(N)} = \mathbf{0}^{(N)}$.

Henceforth, all additions of q-ary words are understood digit-wise and mod q.

Lemma 2.1.2 *The Hamming distance on $\mathscr{H}_{N,q}$ is invariant under group transla-tions:*

$$\delta(\mathbf{x}^{(N)} + \mathbf{z}^{(N)}, \mathbf{y}^{(N)} + \mathbf{z}^{(N)}) = \delta(\mathbf{x}^{(N)}, \mathbf{y}^{(N)}). \qquad (2.1.3b)$$

Proof   For all $i = 1,\ldots,N$ and $x_i, y_i, z_i \in \{0,1,\ldots,q-1\}$, the digits $x_i + z_i$ mod q and $y_i + z_i$ mod q are in the same relation ($=$ or $\neq$) as digits x_i and y_i. $\square$

A *code* is identified with a set of codewords $\mathscr{X}_N \subset \mathscr{H}_{N,q}$; this means that we dis-regard any particular allocation of codewords (which fits the assumption that the source messages are equidistributed). An assumption is that the code is **known** to both the sender and the receiver. Shannon's coding theorems guarantee that, under certain conditions, there exist asymptotically good codes attaining the limits im-posed by the information rate of a source and the capacity of a channel. Moreover, Shannon's SCT shows that almost all codes are asymptotically good. However, in a practical situation, these facts are of a limited use: one wants to have a good code in an explicit form. Besides, it is desirable to have a code that leads to fast encoding and decoding and maximises the rate of the information transmission.

So, assume that the source emits binary strings $\mathbf{u}^{(n)} = u_1 \ldots u_n$, $u_i = 0, 1$. To obtain the overall error-probability vanishing as $n \to \infty$, we have to encode words $\mathbf{u}^{(n)}$ by longer codewords $\mathbf{x}^{(N)} \in \mathscr{H}_{N,2}$ where $N \sim R^{-1}n$ and $0 < R < 1$. Word $\mathbf{x}^{(N)}$ is then sent to the channel and is transformed into another word, $\mathbf{y}^{(N)} \in \mathscr{H}_{N,2}$. It is convenient to represent the error occurred by the difference of the two words: $\mathbf{e}^{(N)} = \mathbf{y}^{(N)} - \mathbf{x}^{(N)} = \mathbf{x}^{(N)} + \mathbf{y}^{(N)}$, or equivalently, write $\mathbf{y}^{(N)} = \mathbf{x}^{(N)} + \mathbf{e}^{(N)}$, in the sense of (2.1.3a). Thus, the more digits 1 the error word $\mathbf{e}^{(N)}$ has, the more sym-bols are distorted by the channel. The ML decoder then produces a 'guessed' code-word $\mathbf{x}_\star^{(N)}$ that may or may not coincide with $\mathbf{x}^{(N)}$, and then reconstructs a string $\mathbf{u}_\star^{(n)}$. In the case of a one-to-one encoding rule, the last procedure is (theoretically) straightforward: we simply invert the map $\mathbf{u}^{(n)} \to \mathbf{x}^{(N)}$. Intuitively, a code is 'good' if it allows the receiver to 'correct' the error string $\mathbf{e}^{(N)}$, at least when word $\mathbf{e}^{(N)}$ does not contain 'too many' non-zero digits.

Going back to an MBSC with the row probability of the error $p < 1/2$: the ML decoder selects a codeword $\mathbf{x}_\star^{(N)}$ that leads to a word $\mathbf{e}^{(N)}$ with a minimal number of the unit digits. In geometric terms:

$$\mathbf{x}_\star^{(N)} \in \mathscr{X}_N \text{ is the codeword closest to } \mathbf{y}^{(N)} \tag{2.1.4}$$
$$\text{in the Hamming distance } \delta.$$

The same rule can be applied in the q-ary case: we look for the codeword closest to the received string. A drawback of this rule is that if several codewords have the same minimal distance from a received word we are 'stuck'. In this case we either choose one of these codewords arbitrarily (possibly randomly or in connection with the message's content; this is related to the so-called list decoding), or, when a high quality of transmission is required, refuse to decode the received word and demand a re-transmission.

Definition 2.1.3 We call N the *length* of a binary code $\mathscr{X}_N$, $M := \sharp \mathscr{X}_N$ the *size* and $\rho := \dfrac{\log_2 M}{N}$ the *information rate*. A code $\mathscr{X}_N$ is said to be *D-error detecting* if making up to D changes in any codeword does not produce another codeword, and *E-error correcting* if making up to E changes in any codeword $\mathbf{x}^{(N)}$ produces a word which is still (strictly) closer to $\mathbf{x}^{(N)}$ than to any other codeword (that is, $\mathbf{x}^{(N)}$ is correctly guessed from a distorted word under the rule (2.1.4)). A code has minimal distance (or briefly distance) d if

$$d = \min \left[\delta(\mathbf{x}^{(N)}, \mathbf{x}'^{(N)}) : \mathbf{x}^{(N)}, \mathbf{x}'^{(N)} \in \mathscr{X}_N, \ \mathbf{x}^{(N)} \neq \mathbf{x}'^{(N)} \right]. \tag{2.1.5}$$

The minimal distance and the information rate of a code $\mathscr{X}_N$ will be sometimes denoted by $d(\mathscr{X}_N)$ and $\rho(\mathscr{X}_N)$, respectively.

This definition can be repeated almost *verbatim* for the general case of a q-ary code $\mathscr{X}_N \subset \mathscr{H}_{N,q}$, with information rate $\rho = \dfrac{\log_q M}{N}$. Namely, a code $\mathscr{X}_N$ is called *E-error correcting* if, for all $r = 1, \ldots, E$, $\mathbf{x}^{(N)} \in \mathscr{X}_N$ and $\mathbf{y}^{(N)} \in \mathscr{H}_{N,q}$ with $\delta(\mathbf{x}^{(N)}, \mathbf{y}^{(N)}) = r$, the distance $\delta(\mathbf{y}^{(N)}, \mathbf{x}'^{(N)}) > r$ for all $\mathbf{x}'^{(N)} \in \mathscr{X}_N$ such that $\mathbf{x}'^{(N)} \neq \mathbf{x}^{(N)}$. In words, it means that making up to E errors in a codeword produces a word that is still closer to it than to any other codeword. Geometrically, this property means that the balls of radius E about the codewords do not intersect:

$$\mathscr{B}_{N,q}(\mathbf{x}^{(N)}, E) \cap \mathscr{B}_{N,q}(\mathbf{x}'^{(N)}, E) = \emptyset \quad \text{for all distinct } \mathbf{x}^{(N)}, \mathbf{x}'^{(N)} \in \mathscr{X}_N.$$

Next, a code $\mathscr{X}_N$ is called *D-error detecting* if the ball of radius D about a codeword does not contain another codeword. Equivalently, the intersection $\mathscr{B}_{N,q}(\mathbf{x}^{(N)}, D) \cap \mathscr{X}_N$ is reduced to a single point $\mathbf{x}^{(N)}$.

A code of length N, size M and minimal distance d is called an $[N, M, d]$ code. Speaking of an $[N, M]$ or $[N, d]$ code, we mean any code of length N and size M or minimal distance d.

To make sure we understand this definition, let us prove the aforementioned equivalence in the definition of an E-error correcting code. First, assume that the balls of radius E are disjoint. Then, making up to E changes in a codeword produces a word that is still in the corresponding ball, and hence is further apart from any other codeword. Conversely, let our code have the property that changing up to E digits in a codeword does not produce a word which lies at the same distance from or closer to another codeword. Then any word obtained by changing precisely E digits in a codeword cannot fall in any ball of radius E but in the one about the original codeword. If we make fewer changes we again do not fall in any other ball, for if we do, then moving towards the second centre will produce, sooner or later, a word that is at distance E from the original codeword and at distance $< E$ from the second one, which is impossible.   □

For a D-error detecting code, the distance $d \geq D + 1$. Furthermore, a code of distance d detects $d - 1$ errors, and it corrects $\lfloor (d-1)/2 \rfloor$ errors.

Remark 2.1.4   Formally, Definition 2.1.3 means that a code detects *at least D* and corrects *at least E* errors, and some authors make a point of this fact, specifying D and E as maximal values with the above properties. We followed an original tradition where the detection and correction abilities of codes are defined in terms of inequalities rather than equalities, although in a number of forthcoming constructions and examples the claim that a code detects D and/or corrects E errors means that D and/or E and no more. See, for instance, Definition 2.1.7.

Definition 2.1.5   In Section 2.3 we systematically study so-called *linear* codes. The linear structure is established in space $\mathcal{H}_{N,q}$ when the alphabet size q is of the form p^s where p is a prime and s a positive integer; in this case the alphabet $\{0, 1, \ldots, q-1\}$ can be made a field, $\mathbb{F}_q$, by introducing two suitable operations: addition and multiplication. See Section 3.1. When $s = 1$, i.e. q is a prime number, then both operations can be understood as standard ones, modulo q. When $\mathbb{F}_q$ is a field with addition $+$ and multiplication $\cdot$, set $\mathcal{H}_{N,q} = \mathbb{F}_q^{\times N}$ becomes a linear space over $\mathbb{F}_q$, with component-wise addition and multiplication by 'scalars' generated by the corresponding operations in $\mathbb{F}_q$. Namely, for $\mathbf{x}^{(N)} = x_1 \ldots x_N$, $\mathbf{y}^{(N)} = y_1 \ldots y_N$ and $\gamma \in \mathbb{F}_q$,

$$\mathbf{x}^{(N)} + \mathbf{y}^{(N)} = (x_1 + y_1) \ldots (x_N + y_N), \ \gamma \cdot \mathbf{x}^{(N)} = (\gamma \cdot x_1) \ldots (\gamma \cdot x_N). \quad (2.1.6a)$$

With $q = p^s$, a q-ary $[N, M, d]$ code $\mathscr{X}_N$ is called *linear* if it is a linear subspace of $\mathcal{H}_{N,q}$. That is, $\mathscr{X}_N$ has the property that if $\mathbf{x}^{(N)}, \mathbf{y}^{(N)} \in \mathscr{X}_N$ then $\mathbf{x}^{(N)} + \mathbf{y}^{(N)} \in \mathscr{X}_N$

and $\gamma \cdot \mathbf{x}^{(N)} \in \mathscr{X}_N$ for all $\gamma \in \mathbb{F}_q$. For a linear code $\mathscr{X}$, the size M is given by $M = q^k$ where k may take values $1, \ldots, N$ and gives the *dimension* of the code, i.e. the maximal number of linearly independent codewords. Accordingly, one writes $k = \dim \mathscr{X}$. As in the usual geometry, if $k = \dim \mathscr{X}$ then in $\mathscr{X}$ there exists a *basis* of size k, i.e. a linearly independent collection of codewords $\mathbf{x}^{(1)}, \ldots, \mathbf{x}^{(k)}$ such that any codeword $\mathbf{x} \in \mathscr{X}$ can be (uniquely) written as a linear combination $\sum_{1 \le j \le k} a_j \mathbf{x}^{(j)}$, where $a_j \in \mathbb{F}_q$. [In fact, if $k = \dim \mathscr{X}$ then any linearly independent collection of k codewords is a basis in $\mathscr{X}$.] In the linear case, we speak of $[N, k, d]$ or $[N, k]$ codes.

As follows from the definition, a linear $[N, k, d]$ code $\mathscr{X}_N$ always contains the zero string $\mathbf{0}^{(N)} = 0 \ldots 0$. Furthermore, owing to property (2.1.3b), the minimal distance $d(\mathscr{X}_N)$ in a linear code $\mathscr{X}$ equals the minimal weight $w\left(\mathbf{x}^{(N)}\right)$ of a non-0 codeword $\mathbf{x}^{(N)} \in \mathscr{X}_N$. See (2.1.1b).

Finally, we define the so-called wedge-product of codewords $\mathbf{x}$ and $\mathbf{y}$ as a word $\mathbf{w} = \mathbf{x} \wedge \mathbf{y}$ with components

$$w_i = \min[x_i, y_i], \quad i = 1, \ldots, N. \tag{2.1.6b}$$

A number of properties of linear codes can be mentioned already in this section, although some details of proofs will be postponed.

A simple example of a linear code is a *repetition* code $\mathscr{R}_N \subset \mathscr{H}_{N,q}$, of the form

$$\mathscr{R}_N = \left\{ \mathbf{x}^{(N)} = x \ldots x : x = 0, 1, \ldots, q-1 \right\}$$

detects $N - 1$ and corrects $\left\lfloor \dfrac{N-1}{2} \right\rfloor$ errors. A linear *parity-check* code

$$\mathscr{P}_N = \left\{ \mathbf{x}^{(N)} = x_1 \ldots x_N : x_1 + \cdots + x_N = 0 \right\}$$

detects a single error only, but does not correct it.

Observe that the 'volume' of the ball in the Hamming space $\mathscr{H}_{N,q}$ centred at $\mathbf{z}^{(N)}$ is

$$v_{N,q}(R) = \sharp \mathscr{B}_{N,q}(\mathbf{z}^{(N)}, R) = \sum_{0 \le k \le R} \binom{N}{k} (q-1)^k; \tag{2.1.7}$$

it does not depend on the choice of the centre $\mathbf{z}^{(N)} \in \mathscr{H}_{N,q}$.

It is interesting to consider large values of N (theoretically, $N \to \infty$), and analyse parameters of the code $\mathscr{X}_N$ such as the information rate $\rho(\mathscr{X}) = \dfrac{\log \sharp \mathscr{X}}{N}$ and the distance per digit $\bar{d}(\mathscr{X}) = \dfrac{d(\mathscr{X})}{N}$. Our aim is to focus on 'good' codes, with many codewords (to increase the information rate) and large distances (to increase the detecting and correcting abilities). From this point of view, it is important to understand basic bounds for codes.

Upper bounds are usually written for $M_q^*(N,d)$, the largest size of a q-ary code of length N and distance d. We begin with elementary facts: $M_q^*(N,1) = q^N$, $M_q^*(N,N) = q$, $M_q^*(N,d) \le qM_q^*(N-1,d)$ and – in the binary case – $M_2^*(N,2s) = M_2^*(N-1,2s-1)$ (easy exercises).

Indeed, the number of the codewords cannot be too high if we want to keep good an error-detecting and error-correcting ability. There are various bounds for parameters of codes; the simplest bound was discovered by Hamming in the late 1940s.

Theorem 2.1.6   (The Hamming bound)

(i) *If a q-ary code $\mathcal{X}_N$ corrects E errors then its size $M = \sharp\,\mathcal{X}_N$ obeys*

$$M \le q^N/v_{N,q}(E). \qquad (2.1.8a)$$

For a linear $[N,k]$ code this can be written in the form

$$N - k \ge \log_q\left(v_{N,q}(E)\right).$$

(ii) *Accordingly, with $E = \lfloor(d-1)/2\rfloor$,*

$$M_q^*(N,d) \le q^N/v_{N,q}(E). \qquad (2.1.8b)$$

Proof   (i) The E-balls about the codewords $\mathbf{x}^{(N)} \in \mathcal{X}_N$ must be disjoint. Hence, the total number of points covered equals the product $v_{N,q}(E)M$ which should not exceed q^N, the cardinality of the Hamming space $\mathcal{H}_{N,q}$.

(ii) Likewise, if $\mathcal{X}_N$ is an $[N,M,d]$ code then, as was noted above, for $E = \lfloor(d-1)/2\rfloor$, the balls $\mathcal{B}_{N,q}(\mathbf{x}^{(N)},E)$, $\mathbf{x}^{(N)} \in \mathcal{X}_N$, do not intersect. The volume $\sharp\,\mathcal{B}_{N,q}(\mathbf{x}^{(N)},E)$ is given by

$$v_{N,q}(E) = \sum_{0 \le k \le E} \binom{N}{k}(q-1)^k,$$

and the union of balls

$$\bigcup_{\mathbf{x}^{(N)} \in \mathcal{X}_N} \mathcal{B}_{N,q}(\mathbf{x}^{(N)},E)$$

must lie in $\mathcal{H}_{N,q}$, again with cardinality $\sharp\,\mathcal{H}_{N,q} = q^N$.   □

We see that the problem of finding good codes becomes a *geometric* problem, because a 'good' code $\mathcal{X}_N$ correcting E errors must give a 'close-packing' of the Hamming space by balls of radius E. A code $\mathcal{X}_N$ that gives a 'true' close-packing partition has an additional advantage: the code not only corrects errors, but never leads to a refusal of decoding. More precisely:

Definition 2.1.7 An E-error correcting code $\mathscr{X}_N$ of size $\sharp \mathscr{X}_N = M$ is called *perfect* when the equality is achieved in the Hamming bound:

$$M = q^N / v_{N,q}(E).$$

If a code $\mathscr{X}_N$ is perfect, every word $\mathbf{y}^{(N)} \in \mathscr{H}_{N,q}$ belongs to a (unique) ball $\mathscr{B}_E(\mathbf{x}^{(N)})$. That is, we are always able to decode $\mathbf{y}^{(N)}$ by a codeword: this leads to the correct answer if the number of errors is $\leq E$, and to a wrong answer if it is $> E$. But we never get 'stuck' in the case of decoding.

The problem of finding perfect binary codes was solved about 20 years ago. These codes exist only for

(a) $E = 1$: here $N = 2^l - 1$, $M = 2^{2^l - 1 - l}$, and these codes correspond to the so-called Hamming codes;
(b) $E = 3$: here $N = 23$, $M = 2^{12}$; they correspond to the so called (binary) Golay code.

Both the Hamming and Golay codes are discussed below. The Golay code is used (together with some modifications) in the US space programme: already in the 1970s the quality of photographs encoded by this code and transmitted from Mars and Venus was so excellent that it did not require any improving procedure. In the former Soviet Union space vessels (and early American ones) other codes were also used (and we also discuss them later): they generally produced lower-quality photographs, and further manipulations were required, based on statistics of the pictorial images.

If we consider non-binary codes then there exists one more perfect code, for three symbols (also named after Golay).

We will now describe a number of straightforward constructions producing new codes from existing ones.

Example 2.1.8 Constructions of new codes include:

(i) **Extension**: You add a digit x_{N+1} to each codeword $\mathbf{x}^{(N)} = x_1 \dots x_N$ from code $\mathscr{X}_N$, following an agreed rule. Viz., the so-called *parity-check extension* requires that $x_{N+1} + \sum_{1 \leq j \leq N} x_j = 0$ in the alphabet field $\mathbb{F}_q$. Clearly, the extended code, $\mathscr{X}_{N+1}^+$, has the same size as the original code $\mathscr{X}_N$, and the distance $d(\mathscr{X}_{N+1}^+)$ is equal to either $d(\mathscr{X}_N)$ or $d(\mathscr{X}_N) + 1$.

(ii) **Truncation**: Remove a digit from the codewords $\mathbf{x} \in \mathscr{X}\,(= \mathscr{X}_N)$. The resulting code, $\mathscr{X}_{N-1}^-$, has length $N-1$ and, if the distance $d(\mathscr{X}_N) \geq 2$, the same size as $\mathscr{X}_N$, while $d(\mathscr{X}_{N-1}^-) \geq d(\mathscr{X}_N) - 1$, provided that $d(\mathscr{X}_N) \geq 2$.

(iii) **Purge**: Simply delete some codewords $\mathbf{x} \in \mathscr{X}_N$. For example, in the binary case removing all codewords with an odd number of non-zero digits from a linear code leads to a linear subcode; in this case if the distance of the original code was odd then the purged code will have a strictly larger distance.

(iv) **Expansion**: Opposite to purging. Say, let us add the complement of each codeword to a binary code $\mathscr{X}_N$, i.e. the N-word where the 1s are replaced by the 0s and vice versa. Denoting the expanded code by $\overline{\mathscr{X}}_N$ one can check that $d(\overline{\mathscr{X}}_N) = \min[d(\mathscr{X}_N), N - \overline{d}(\mathscr{X}_N)]$ where

$$\overline{d}(\mathscr{X}_N) = \max[\delta(\mathbf{x}^{(N)}, \mathbf{x}'^{(N)}) : \mathbf{x}^{(N)}, \mathbf{x}'^{(N)} \in \mathscr{X}_N].$$

(v) **Shortening**: Take all codewords $\mathbf{x}^{(N)} \in \mathscr{X}_N$ with the ith digit 0, say, and delete this digit (shortening on $x_i = 0$). In this way the original binary linear $[N, M, d]$ code $\mathscr{X}_N$ is reduced to a binary linear code $\mathscr{X}_{N-1}^{\mathrm{sh},0}(i)$ of length $N-1$, whose size can be $M/2$ or M and distance $\geq d$ or, in a trivial case, 0.

(vi) **Repetition**: Repeat each codeword $\mathbf{x}(= \mathbf{x}^{(N)}) \in \mathscr{X}_N$ a fixed number of times, say m, producing a concatenated (Nm)-word $\mathbf{xx}\ldots\mathbf{x}$. The result is a code $\mathscr{X}_{Nm}^{\mathrm{re}}$, of length Nm and distance $d(\mathscr{X}_{Nm}^{\mathrm{re}}) = md(\mathscr{X}_N)$.

(vii) **Direct sum**: Given two codes $\mathscr{X}_N$ and $\mathscr{X}_{N'}$, form a code $\mathscr{X} + \mathscr{X}' = \{\mathbf{xx}' : \mathbf{x} \in \mathscr{X}, \mathbf{x}' \in \mathscr{X}'\}$. Both the repetition and direct-sum constructions are not very effective and neither is particularly popular in coding (though we will return to these constructions in examples and problems). A more effective construction is

(viii) **The bar-product** $(\mathbf{x}|\mathbf{x}+\mathbf{x}')$: For the $[N, M, d]$ and $[N, M', d']$ codes $\mathscr{X}_N$ and $\mathscr{X}_N'$ define a code $\mathscr{X}_N | \mathscr{X}_N'$ of length $2N$ as the collection

$$\left\{ \mathbf{x}(\mathbf{x}+\mathbf{x}') : \mathbf{x}(= \mathbf{x}^{(N)}) \in \mathscr{X}_N, \mathbf{x}'(= \mathbf{x}'^{(N)}) \in \mathscr{X}_N' \right\}.$$

That is, each codeword in $\mathscr{X}|\mathscr{X}'$ is a concatenation of the codeword from $\mathscr{X}_N$ and its sum with a codeword from $\mathscr{X}_N'$ (formally, neither of $\mathscr{X}_N$, $\mathscr{X}_N'$ in this construction is supposed to be linear). The resulting code is denoted by $\mathscr{X}_N | \mathscr{X}_N'$; it has size

$$\sharp \left(\mathscr{X}_N | \mathscr{X}_N' \right) = \left(\sharp\, \mathscr{X}_N \right) \left(\sharp\, \mathscr{X}_N' \right).$$

A useful exercise is to check that the distance

$$d \left(\mathscr{X}_N | \mathscr{X}_N' \right) = \min \left[2d(\mathscr{X}_N), d(\mathscr{X}'_N) \right].$$

(ix) **The dual code**. The concept of duality is based on the inner *dot-product* in space $\mathcal{H}_{N,q}$ (with $q = p^s$): for $\mathbf{x} = x_1 \ldots x_N$ and $\mathbf{y} = y_1 \ldots y_N$,

$$\left\langle \mathbf{x}^{(N)} \cdot \mathbf{y}^{(N)} \right\rangle = x_1 \cdot y_1 + \cdots + x_N \cdot y_N$$

which yields a value from field $\mathbb{F}_q$. For a linear $[N, k]$ code $\mathcal{X}_N$ its *dual*, $\mathcal{X}_N^{\perp}$, is a linear $[N, N-k]$ code defined by

$$\mathcal{X}_N^{\perp} = \left\{ \mathbf{y}^{(N)} \in \mathcal{H}_{N,q} : \left\langle \mathbf{x}^{(N)} \cdot \mathbf{y}^{(N)} \right\rangle = 0 \quad \text{for all } \mathbf{x} \in \mathcal{X}_N \right\}. \qquad (2.1.9)$$

Clearly, $(\mathcal{X}_N^{\perp})^{\perp} = \mathcal{X}_N$. Also dim $\mathcal{X}_N + \dim \mathcal{X}_N^{\perp} = N$. A code is called *self-dual* if $\mathcal{X}_N = \mathcal{X}_N^{\perp}$.

Worked Example 2.1.9

(a) *Prove that if the distance d of an $[N, M, d]$ code $\mathcal{X}_N$ is an odd number then the code may be extended to an $[N+1, M]$ code $\mathcal{X}^+$ with distance $d+1$.*

(b) *Show an E-error correcting code $\mathcal{X}_N$ can be extended to a code $\mathcal{X}^+$ that detects $2E+1$ errors.*

(c) *Show that the distance of a perfect binary code is an odd number.*

Solution (a) By adding the digit x_{N+1} to the codewords $\mathbf{x} = x_1 \ldots x_N$ of an $[N, M]$ code $\mathcal{X}_N$ so that $x_{N+1} = \sum\limits_{1 \leq j \leq N} x_j$, we obtain an $[N+1, M]$ code $\mathcal{X}^+$. If the distance d of $\mathcal{X}_N$ was odd, the distance of $\mathcal{X}^+$ is $d+1$. In fact, if a pair of codewords, $\mathbf{x}, \mathbf{x}' \in \mathcal{X}$, had $\delta(\mathbf{x}, \mathbf{x}') > d$, then the extended codewords, $\mathbf{x}_+$ and $\mathbf{x}'_+$, have $\delta(\mathbf{x}_+, \mathbf{x}'_+) \geq \delta(\mathbf{x}, \mathbf{x}') > d$. Otherwise, i.e. if $\delta(\mathbf{x}, \mathbf{x}') = d$, the distance increases: $\delta(\mathbf{x}_+, \mathbf{x}'_+) = d+1$.

(b) The distance d of an E-error correcting code is strictly greater than $2E$. Hence, the above extension gives a code with distance strictly greater than $2E+1$.

(c) For a perfect E-error correcting code the distance is at most $2E+1$ and hence equals $2E+1$. $\qquad \square$

Worked Example 2.1.10   *Show that there is no perfect 2-error correcting code of length 90 and size 2^{78} over $\mathbb{F}_2$.*

Solution We might be interested in the existence of a perfect 2-error correcting binary code of length $N = 90$ and size $M = 2^{78}$ because

$$v_{90,2}(2) = 1 + 90 + \frac{90 \cdot 89}{2} = 4096 = 2^{12}$$

and

$$M \times v_{90,2}(2) = 2^{78} \cdot 2^{12} = 2^{90} = 2^N.$$

However, such a code does not exist. Assume that it exists, and, the zero word $\mathbf{0} = 0\ldots0$ is a codeword. The code must have $d = 5$. Consider the 88 words with three non-zero digits, with 1 in the first two places:

$$1110\ldots00, \quad 1101\ldots00, \quad \ldots, \quad 110\ldots01. \tag{2.1.10}$$

Each of these words should be at distance ≤ 2 from a unique codeword. Say, the codeword for $1110\ldots00$ must contain 5 non-zero digits. Assume that it is

$$111110\ldots00.$$

This codeword is at distance 2 from two other subsequent words,

$$11010\ldots00 \quad \text{and} \quad 11001\ldots00.$$

Continuing with this construction, we see that any word from list (2.1.10) is 'attracted' to a codeword with 5 non-zero digits, along with two other words from (2.1.10). But 88 is not divisible by 3. □

Let us continue with bounds on codes.

Theorem 2.1.11   (The Gilbert–Varshamov (GV) bound) *For any $q \geq 2$ and $d \geq 2$, there exists a q-ary $[N,M,d]$ code $\mathscr{X}_N$ such that*

$$M = \sharp \mathscr{X}_N \geq q^N / v_{N,q}(d-1) . \tag{2.1.11}$$

Proof   Consider a code of maximal size among the codes of minimal distance d and length N. Then any word $\mathbf{y}^{(N)} \in \mathscr{H}_{N,q}$ must be distant $\leq d-1$ from some codeword: otherwise we can add $\mathbf{y}^{(N)}$ to the code without changing the minimal distance. Hence, the balls of radius $d-1$ about the codewords cover the whole Hamming space $\mathscr{H}_{N,q}$. That is, for the code of maximal size, $\mathscr{X}_N^{\max}$,

$$\left(\sharp \mathscr{X}_N^{\max}\right) v_{N,q}(d-1) \geq q^N.$$

□

As was listed before, there are ways of producing one code from another (or from a collection of codes). Let us apply truncation and drop the last digit x_N in each codeword $\mathbf{x}^{(N)}$ from an original code $\mathscr{X}_N$. If code $\mathscr{X}_N$ had the minimal distance $d > 1$ then the new code, $\mathscr{X}_{N-1}^-$, has the minimal distance $\geq d-1$ and the same size as $\mathscr{X}_N$. The truncation procedure leads to the following bound.

Theorem 2.1.12   (The Singleton bound) *Any q-ary code $\mathscr{X}_N$ with minimal distance d has*

$$M = \sharp \mathscr{X}_N \leq M_q^*(N,d) \leq q^{N-d+1}. \tag{2.1.12}$$

Proof As before, perform a truncation on an $[N,M,d]$ code $\mathscr{X}_N$: drop the last digit from each codeword $\mathbf{x} \in \mathscr{X}_N$. The new code is $[N,M,d^-]$ where $d^- \geq d-1$. Repeating this procedure $d-1$ times gives an $(N-d+1)$ code of the same size M and distance ≥ 1. This code must fit in Hamming space $\mathscr{H}_{N-d+1,q}$ with $\sharp \mathscr{H}_{N-d+1,q} = q^{N-d+1}$; hence the result. $\qquad\square$

As with the Hamming bound, the case of equality in the Singleton bound attracted a special interest:

Definition 2.1.13 A q-ary linear $[N,k,d]$ code is called *maximum distance separating* (MDS) if it gives equality in the Singleton bound:

$$d = N-k+1. \qquad (2.1.13)$$

We will see below that, similarly to perfect codes, the family of the MDS codes is rather 'thin'.

Corollary 2.1.14 *If $M_q^*(N,d)$ is the maximal size of a code $\mathscr{X}_N$ with minimal distance d then*

$$\frac{q^N}{v_{N,q}(d-1)} \leq M_q^*(N,d) \leq \min\left[\frac{q^N}{v_{N,q}(\lfloor (d-1)/2 \rfloor)}, q^{N-d+1}\right]. \qquad (2.1.14)$$

From now on we will omit indices N and (N) whenever it does not lead to confusion. The upper bound in (2.1.14) becomes too rough when $d \sim N/2$. Say, in the case of binary $[N,M,d]$-code with $N = 10$ and $d = 5$, expression (2.1.14) gives the upper bound $M_2^*(10,5) \leq 18$, whereas in fact there is no code with $M \geq 13$, but there exists a code with $M = 12$. The codewords of the latter are as follows:

0000000000, 1111100000, 1001011010, 0100110110,

1100001101, 0011010101, 0010011011, 1110010011,

1001100111, 1010111100, 0111001110, 0101111001.

The lower bound gives in this case the value 2 (as $2^{10}/v_{10,2}(4) = 2.6585$) and is also far from being satisfactory. (Some better bounds will be obtained below.)

Theorem 2.1.15 (The Plotkin bound) *For a binary code $\mathscr{X}$ of length N and distance d with $N < 2d$, the size M obeys*

$$M = \sharp \mathscr{X} \leq 2\left\lfloor \frac{d}{2d-N} \right\rfloor. \qquad (2.1.15)$$

Proof   The minimal distance cannot exceed the average distance, i.e.

$$M(M-1)d \leq \sum_{\mathbf{x} \in \mathscr{X}} \sum_{\mathbf{x}' \in \mathscr{X}} \delta(\mathbf{x}, \mathbf{x}').$$

On the other hand, write code $\mathscr{X}$ as an $M \times N$ matrix with rows as codewords. Suppose that column i of the matrix contains s_i zeros and $M - s_i$ ones. Then

$$\sum_{\mathbf{x} \in \mathscr{X}} \sum_{\mathbf{x}' \in \mathscr{X}} \delta(\mathbf{x}, \mathbf{x}') \leq 2 \sum_{1 \leq i \leq N} s_i(M - s_i). \qquad (2.1.16)$$

If M is even, the RHS of (2.1.16) is maximised when $s_i = M/2$ which yields

$$M(M-1)d \leq \frac{1}{2}NM^2, \ \ \text{or} \ \ M \leq \frac{2d}{2d-N}.$$

As M is even, this implies

$$M \leq 2 \left\lfloor \frac{d}{2d-N} \right\rfloor.$$

If M is odd, the RHS of (2.1.16) is $\leq N(M^2-1)/2$ which yields

$$M \leq \frac{N}{2d-N} = \frac{2d}{2d-N} - 1.$$

This implies in turn that

$$M \leq \left\lfloor \frac{2d}{2d-N} \right\rfloor - 1 \leq 2 \left\lfloor \frac{d}{2d-N} \right\rfloor,$$

because, for all $x > 0$, $\lfloor 2x \rfloor \leq 2\lfloor x \rfloor + 1$. $\qquad \qquad \square$

Theorem 2.1.16   *Let $M_2^*(N,d)$ be the maximal size of a binary $[N,d]$ code. Then, for any N and d,*

$$M_2^*(N, 2d-1) = M_2^*(N+1, 2d), \qquad (2.1.17)$$

and

$$2M_2^*(N-1, d) = M_2^*(N, d). \qquad (2.1.18)$$

Proof   To prove (2.1.17) let $\mathscr{X}$ be a code of length N, distance $2d-1$ and size $M_2^*(N, 2d-1)$. Take its parity-check extension $\mathscr{X}^+$. That is, add digit x_{N+1} to every codeword $\mathbf{x} = x_1 \dots x_N$ so that $\sum_{i=1}^{N+1} x_i = 0$. Then $\mathscr{X}^+$ is a code of length $N+1$, the same size $M_2^*(N, 2d-1)$ and distance $2d$. Therefore,

$$M_2^*(N, 2d-1) \leq M_2^*(N+1, 2d).$$

Similarly, deleting the last digit leads to the inverse:

$$M_2^*(N, 2d-1) \geq M_2^*(N+1, 2d).$$

Turning to the proof of (2.1.18), given an $[N,d]$ code, divide the codewords into two classes: those ending with 0 and those ending with 1. One class must contain at least half of the codewords. Hence the result.   □

Corollary 2.1.17   *If d is even and such that $2d > N$,*

$$M_2^*(N,d) \leq 2 \left\lfloor \frac{d}{2d-N} \right\rfloor \qquad (2.1.19)$$

and

$$M_2^*(2d,d) \leq 4d. \qquad (2.1.20)$$

If d is odd and $2d+1 > N$ then

$$M_2^*(N,d) \leq 2 \left\lfloor \frac{d+1}{2d+1-N} \right\rfloor \qquad (2.1.21)$$

and

$$M_2^*(2d+1,d) \leq 4d+4. \qquad (2.1.22)$$

Proof Inequality (2.1.19) follows from (2.1.17), and (2.1.20) follows from (2.1.18) and (2.1.19): if $d = 2d'$ then

$$M_2^*(4d',2d') = 2M_2^*(4d'-1,2d') \leq 8d' = 4d.$$

Furthermore, (2.1.21) follows from (2.1.17):

$$M_2^*(N,d) = M_2^*(N+1,d+1) \leq 2 \left\lfloor \frac{d+1}{2d+1-N} \right\rfloor.$$

Finally, (2.1.22) follows from (2.1.17) and (2.1.20).   □

Worked Example 2.1.18   *Prove the Plotkin bound for a q-ary code:*

$$M_q^*(N,d) \leq \left\lfloor d \Big/ \left(d - N\frac{q-1}{q} \right) \right\rfloor, \text{ if } d > N\frac{q-1}{q}. \qquad (2.1.23)$$

Solution Given a q-ary $[N,M,d]$ code $\mathscr{X}_N$, observe that the minimal distance d is bounded by the average distance

$$d \leq \frac{1}{M(M-1)} S, \text{ where } S = \sum_{\mathbf{x} \in \mathscr{X}} \sum_{\mathbf{x}' \in \mathscr{X}} \delta(\mathbf{x},\mathbf{x}').$$

As before, let k_{ij} denote the number of letters $j \in \{0,\dots,q-1\}$ in the ith position in all codewords from $\mathscr{X}$, $i = 1,\dots,N$. Then, clearly, $\sum_{0 \leq j \leq q-1} k_{ij} = M$ and the contribution of the ith position into S is

$$\sum_{0 \leq j \leq q-1} k_{ij}(M-k_{ij}) = M^2 - \sum_{0 \leq j \leq q-1} k_{ij}^2 \leq M^2 - \frac{M^2}{q}$$

as the quadratic function $(u_1, \ldots, u_q) \mapsto \sum\limits_{1 \le j \le q} u_j^2$ achieves its minimum on the set
$\{\mathbf{u} = u_1 \ldots u_q : u_j \ge 0, \sum u_j = M\}$ at $u_1 = \cdots = u_q = M/q$. Summing over all N digits, we obtain with $\theta = (q-1)/q$

$$M(M-1)d \le \theta M^2 N,$$

which yields the bound $M \le d(d - \theta N)^{-1}$. The proof is completed as in the binary case. $\qquad\square$

There exists a substantial theory related to the equality in the Plotkin bound (Hadamard codes) but it will not be discussed in this book. We would also like to point out the fact that all bounds established so far (Hamming, Singleton, GV and Plotkin) hold for codes that are not necessarily linear. As far as the GV bound is concerned, one can prove that it can be achieved by linear codes: see Theorem 2.3.26.

Worked Example 2.1.19   *Prove that a 2-error correcting binary code of length 10 can have at most 12 codewords.*

Solution The distance of the code must be ≥ 5. Suppose that it contains M codewords and extend it to an $[11, M]$ code of distance 6. The Plotkin bound works as follows. List all codewords of the extended code as rows of an $M \times 11$ matrix. If column i in this matrix contains s_i zeros and $M - s_i$ ones then

$$6(M-1)M \le \sum_{\mathbf{x} \in \mathscr{X}^+} \sum_{\mathbf{x}' \in \mathscr{X}^+} \delta(\mathbf{x}, \mathbf{x}') \le 2 \sum_{i=1}^{11} s_i(M - s_i).$$

The RHS is $\le (1/2) \cdot 11 M^2$ if M is even and $\le (1/2) \cdot 11 (M^2 - 1)$ if M is odd. Hence, $M \le 12$. $\qquad\square$

Worked Example 2.1.20   (Asymptotics of the size of a binary ball) *Let $q = 2$ and $\tau \in (0, 1/2)$. Then, with $\eta(\tau) = -\tau \log_2 \tau - (1 - \tau) \log_2(1 - \tau)$ (cf. (1.2.2a)),*

$$\lim_{N \to \infty} \frac{1}{N} \log v_{N,2}(\lfloor \tau N \rfloor) = \lim_{N \to \infty} \frac{1}{N} \log v_{N,2}(\lceil \tau N \rceil) = \eta(\tau). \qquad (2.1.24)$$

Solution Observe that with $R = \lceil \tau N \rceil$ the last term in the sum

$$v_{N,2}(R) = \sum_{i=0}^{R} \binom{N}{i}, \quad R = \lceil \tau N \rceil,$$

is the largest. Indeed, the ratio of two successive terms is

$$\binom{N}{i+1} \Big/ \binom{N}{i} = \frac{N-i}{i+1}$$

which remains ≥ 1 for $0 \leq i \leq R$. Hence,

$$\binom{N}{R} \leq v_{N,2}(R) \leq (R+1)\binom{N}{R}.$$

Now use Stirling's formula: $N! \sim N^{N+1/2}e^{-N}\sqrt{2\pi}$. Then

$$\log\binom{N}{R} = -(N-R)\log\frac{N-R}{N} - R\log\frac{R}{N} + O(\log N) \qquad (2.1.25)$$

and

$$-\left(1-\frac{R}{N}\right)\log\left(1-\frac{R}{N}\right) - \frac{R}{N}\log\frac{R}{N} + \frac{O(\log N)}{N}$$
$$\leq \frac{\log v_{N,2}(R)}{N} \leq \frac{1}{N}\log(R+1) + \text{the LHS}.$$

The limit $R/N \to \tau$ yields the result. The case where $R = \lfloor \tau N \rfloor$ is considered in a similar manner. $\qquad \square$

Worked Example 2.1.20 is useful in the study of the asymptotics of

$$\alpha(N,\tau) = \frac{1}{N}\log M_2^*(N,\lceil \tau N \rceil), \qquad (2.1.26)$$

the information rate for the maximum size of a code correcting $\sim \tau N$ and detecting $\sim 2\tau N$ errors (i.e. a linear portion of the total number of digits N). Set

$$\underline{a}(\tau) := \liminf_{N\to\infty}\alpha(N,\tau) \leq \limsup_{N\to\infty}\alpha(N,\tau) =: \bar{a}(\tau) \qquad (2.1.27)$$

For these limits we have

Theorem 2.1.21 *With $\eta(\tau) = -\tau\log_2\tau - (1-\tau)\log_2(1-\tau)$, the following asymptotic bounds hold for binary codes:*

$$\bar{a}(\tau) \leq 1 - \eta(\tau/2), \ 0 \leq \tau \leq 1/2 \ \text{(Hamming)}, \qquad (2.1.28)$$
$$\bar{a}(\tau) \leq 1 - \tau, \ 0 \leq \tau \leq 1/2 \ \text{(Singleton)}, \qquad (2.1.29)$$
$$\underline{a}(\tau) \geq 1 - \eta(\tau), \ 0 \leq \tau \leq 1/2 \ \text{(GV)}, \qquad (2.1.30)$$
$$\bar{a}(\tau) = 0, \ 1/2 \leq \tau \leq 1 \ \text{(Plotkin)}. \qquad (2.1.31)$$

By using more elaborate bounds (also due to Plotkin), we'll show in Problem 2.10 that

$$\bar{a}(\tau) \leq 1 - 2\tau, \qquad 0 \leq \tau \leq 1/2. \qquad (2.1.32)$$

The proof of Theorem 2.1.21 is based on a direct inspection of the above-mentioned bounds; for Hamming and GV bounds it is carried in Worked Example 2.1.22 later.

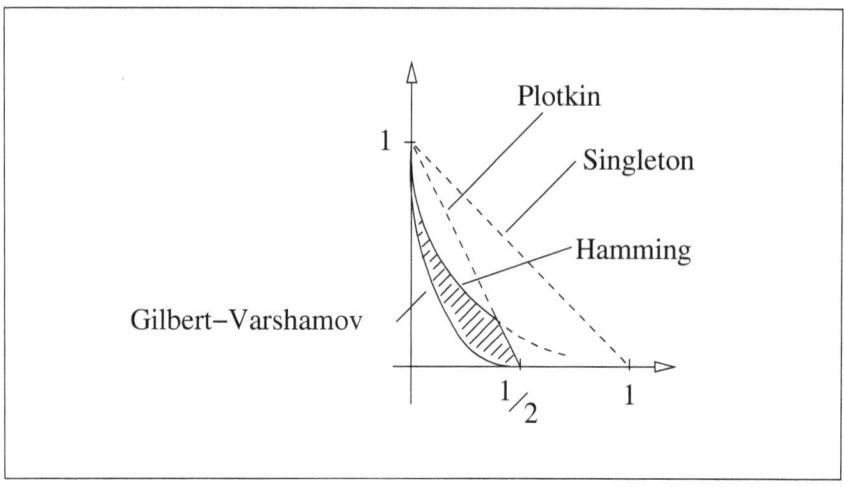

Figure 2.2

Figure 2.2 shows the behaviour of the bounds established. 'Good' sequences of codes are those for which the pair $(\tau, \alpha(N, \lceil \tau N \rceil))$ is asymptotically confined to the domain between the curves indicating the asymptotic bounds. In particular, a 'good' code should 'lie' above the curve emerging from the GV bound. Constructing such sequences is a difficult problem: the first examples achieving the asymptotic GV bound appeared in 1973 (the Goppa codes, based on ideas from algebraic geometry). All families of codes discussed in this book produce values below the GV curve (in fact, they yield $\alpha(\tau) = 0$), although these codes demonstrate quite impressive properties for particular values of N, M and d.

As to the upper bounds, the Hamming and Plotkin compete against each other, while the Singleton bound turns out to be asymptotically insignificant (although it is quite important for specific values of N, M and d). There are about a dozen various other upper bounds, some of which will be discussed in this and subsequent sections of the book.

The Gilbert–Varshamov bound itself is not necessarily optimal. Until 1982 there was no better lower bound known (and in the case of binary coding there is still no better lower bound known). However, if the alphabet used contains $q \geq 49$ symbols where $q = p^{2m}$ and $p \geq 7$ is a prime number, there exists a construction, again based on algebraic geometry, which produces a different lower bound and gives examples of (linear) codes that asymptotically exceed, as $N \to \infty$, the GV curve [159]. Moreover, the TVZ construction carries a polynomial complexity. Subsequently, two more lower bounds were proposed: (a) Elkies' bound, for $q = p^{2m} + 1$; and (b) Xing's bound, for $q = p^{m}$ [43, 175]. See N. Elkies, 'Excellent codes from modular curves', Manipulating with different coding constructions, the GV bound can also be improved for other alphabets.

Worked Example 2.1.22   *Prove bounds (2.1.28) and (2.1.30) (that is, those parts of Theorem 2.1.21 related to the asymptotical Hamming and GV bounds).*

Solution Picking up the Hamming and the GV parts in (2.1.14), we have

$$2^N/v_{N,2}(d-1) \leq M_2^*(N,d) \leq 2^N/v_{N,2}(\lfloor (d-1)/2 \rfloor). \tag{2.1.33}$$

The lower bound for the Hamming volume is trivial:

$$v_{N,2}(\lfloor (d-1)/2 \rfloor) \geq \binom{N}{\lfloor (d-1)/2 \rfloor}.$$

For the upper bound, observe that with $d/N \leq \tau < 1/2$,

$$v_{N,2}(d-1) \leq \sum_{0 \leq i \leq d-1} \left(\frac{d-1}{N-d+1} \right)^{d-1-i} \binom{N}{i}$$

$$\leq \sum_{0 \leq i \leq d-1} \left(\frac{\tau}{1-\tau} \right)^{d-1-i} \binom{N}{d-1} \leq \frac{1-\tau}{1-2\tau} \binom{N}{d-1}.$$

Then, for the information rate $(\log M_2^*(N,d))/N$,

$$1 - \frac{1}{N} \log \left[\frac{1-\tau}{1-2\tau} \binom{N}{d-1} \right]$$
$$\leq \frac{1}{N} \log M_2^*(N,d) \leq 1 - \frac{1}{N} \log \binom{N}{\lfloor (d-1)/2 \rfloor}.$$

By Stirling's formula, as $N \to \infty$ the logs in the previous inequalities obey

$$\frac{1}{N} \log \binom{N}{\lfloor (d-1)/2 \rfloor} \to \eta(\tau/2), \quad \frac{1}{N} \log \binom{N}{d-1} \to \eta(\tau).$$

The bounds (2.1.28) and (2.1.30) then readily follow.   □

Consider now the case of a general q-ary alphabet.

Example 2.1.23   Set $\theta := (q-1)/q$. By modifying the argument in Worked Example 2.1.22, prove that for any $q \geq 2$ and $\tau \in (0, \theta)$, the volume of the q-ary Hamming ball has the following logarithmic asymptote:

$$\lim_{N \to \infty} \frac{1}{N} \log_q v_{N,q}(\lfloor \tau N \rfloor) = \lim_{N \to \infty} \frac{1}{N} \log_q v_{N,q}(\lceil \tau N \rceil) = \eta^{(q)}(\tau) + \tau \kappa \tag{2.1.34}$$

where

$$\eta^{(q)}(\tau) := -\tau \log_q \tau - (1-\tau) \log_q (1-\tau), \ \kappa := \log_q (q-1). \tag{2.1.35}$$

Next, similarly to (2.1.26), introduce

$$\alpha^{(q)}(N,\tau) = \frac{1}{N} \log M_q^*(N, \lceil \tau N \rceil) \tag{2.1.36}$$

and the limits

$$\underline{a}^{(q)}(\tau) := \liminf_{N \to \infty} \alpha^{(q)}(N,\tau) \le \limsup_{N \to \infty} \alpha^{(q)}(N,\tau) =: \overline{a}^{(q)}(\tau). \tag{2.1.37}$$

Theorem 2.1.24   *For all* $0 < \tau < \theta$,

$$\overline{a}^{(q)}(\tau) \le 1 - \eta^{(q)}(\tau/2) - \kappa\tau/2 \quad \text{(Hamming)}, \tag{2.1.38}$$

$$\overline{a}^{(q)}(\tau) \le 1 - \tau \quad \text{(Singleton)}, \tag{2.1.39}$$

$$\underline{a}^{(q)}(\tau) \ge 1 - \eta^{(q)}(\tau) - \kappa\tau \quad \text{(GV)}, \tag{2.1.40}$$

$$\overline{a}^{(q)}(\tau) \le \max[1 - \tau/\theta, 0] \quad \text{(Plotkin)}. \tag{2.1.41}$$

Of course, the minimum of the right-hand sides of (2.1.38), (2.1.39) and (2.1.41) provides the better of the three upper bounds. We omit the proof of Theorem 2.1.24, leaving it as an exercise that is a repetition of the argument from Worked Example 2.1.22.

Example 2.1.25   Prove bounds (2.1.38) and (2.1.40), by modifying the solution to Worked Example 2.1.22.

2.2 A geometric proof of Shannon's second coding theorem. Advanced bounds on the code size

In this section we give alternative proofs of both parts of Shannon's second coding theorem (or Shannon's noisy coding theorem, SCT/NCT; cf. Theorems 1.4.14 and 1.4.15) by using the geometry of the Hamming space. We then apply the techniques that are developed in the course of the proof for obtaining some 'advanced' bounds on codes. The advanced bounds strengthen the Hamming bound established in Theorem 2.1.6 and its asymptotic counterparts in Theorems 2.1.21 and 2.1.24.

The direct part of the SCT/NCT is given in Theorem 2.2.1 below, in a somewhat modified form compared with Theorems 1.4.14 and 1.4.15. For simplicity, we only consider here memoryless binary symmetric channels (MBSC), working in space $\mathcal{H}_{N,2} = \{0,1\}^N$ (the subscript 2 will be omitted for brevity). As we learned in Section 1.4, the direct part of the SCT states that for any transmission rate $R < C$ there exist

(i) a sequence of codes $f_n : \mathscr{U}_n \to \mathscr{H}_N$, encoding a total of $\sharp\,\mathscr{U}_n = 2^n$ messages; and

(ii) a sequence of decoding rules $\widehat{f}_N : \mathscr{H}_N \to \mathscr{U}_n$, such that $n \sim NR$ and the probability of erroneous decoding vanishes as $n \to \infty$.

Here C is given by (1.4.11) and (1.4.27). For convenience, we reproduce the expression for C again:

$$C = 1 - \eta(p), \quad \text{where} \quad \eta(p) = -p \log p - (1-p) \log(1-p) \qquad (2.2.1)$$

and the channel matrix is

$$\Pi = \begin{pmatrix} 1-p & p \\ p & 1-p \end{pmatrix}. \qquad (2.2.2)$$

That is, we assume that the channel transmits a letter correctly with probability $1 - p$ and reverses with probability p, independently for different letters.

In Theorem 2.2.1, it is asserted that there exists a sequence of one-to-one coding maps f_n, for which the task of decoding is reduced to guessing the codewords $f_n(\mathbf{u}) \in \mathscr{H}_N$. In other words, the theorem guarantees that for all $R < C$ there exists a sequence of subsets $\mathscr{X}_N \subset \mathscr{H}_N$ with $\sharp\,\mathscr{X}_N \sim 2^{NR}$ for which the probability of incorrect guessing tends to 0, and the exact nature of the coding map f_n is not important. Nevertheless, it is convenient to keep the map f_n firmly in sight, as the existence will follow from a probabilistic construction (random coding) where sample coding maps are not necessarily one-to-one. Also, the decoding rule is *geometric*: upon receiving a word $\mathbf{a}^{(N)} \in \mathscr{H}_N$, we look for the nearest codeword $f_n(\mathbf{u}) \in \mathscr{X}_N$. Consequently, an error is declared every time such a codeword is not unique or is a result of multiple encodings or simply yields a wrong message. As we saw earlier, the geometric decoding rule corresponds with the ML decoder when the probability $p \in (0, 1/2)$. Such a decoder enables us to use geometric arguments constituting the core of the proof.

Again as in Section 1.4, the new proof of the direct part of the SCT/NCT only guarantees the existence of 'good' codes (and even their 'proliferation') but gives no clue on how to construct such codes [apart from running again a random coding scheme and picking its 'typical' realisation].

In the statement of the SCT/NCT given below, we deal with the maximum error-probability (2.2.4) rather than the averaged one over possible messages. However, a large part of the proof is still based on a direct analysis of the error-probabilities averaged over the codewords.

Theorem 2.2.1 (The SCT/NCT, the direct part) *Consider an MBSC with channel matrix Π as in (2.2.2), with $0 \le p < 1/2$, and let C be as in (2.2.1). Then for any*

$R \in (0,C)$ *there exists a sequence of one-to-one encoding maps* $f_n : \mathcal{U}_n \to \mathcal{H}_N$
such that

(i)

$$n = \lfloor NR \rfloor, \text{ and } \sharp \mathcal{U}_n = 2^n; \qquad (2.2.3)$$

(ii) *as* $n \to \infty$, *the maximum error-probability under the geometric decoding rule
vanishes:*

$$e^{\max}(f_n) = \max \left[\mathbf{P}_{ch} \Big(\text{error under geometric decoding} \right.$$
$$\left. | \, f_n(\mathbf{u}) \text{ sent} \Big) : \mathbf{u} \in \mathcal{U}_n \right] \to 0. \qquad (2.2.4)$$

Here $\mathbf{P}_{ch}\big(\, \cdot \, |f_n(\mathbf{u})\text{sent}\big)$ *stands for the probability distribution of the received
word in* $\mathcal{H}_N$ *generated by the channel, conditional on codeword* $f_n(\mathbf{u})$ *being
sent, with* $\mathbf{u} \in \mathcal{U}_n$ *being the original message emitted by the source.*

As an illustration of this result, consider the following.

Example 2.2.2   We wish to send a message $\mathbf{u} \in \mathscr{A}^n$, where the size of alphabet
$\mathscr{A}$ equals K, through an MBSC with channel matrix $\begin{pmatrix} 0.85 & 0.15 \\ 0.15 & 0.85 \end{pmatrix}$. What rate of
transmission can be achieved with an arbitrarily small probability of error?
   Here the value $C = 1 - \eta(0.15) = 0.577291$. Hence, by Theorem 2.2.1 any rate
of transmission < 0.577291 can be achieved for n large enough, with an arbitrarily
small probability of error. For example, if we want a rate of transmission $0.5 < R <$
0.577291 and $e_{\max} < 0.015$ then there exist codes $f_n \colon \mathscr{A}^n \to \{0,1\}^{\lceil n/R \rceil}$ achieving
this goal provided that n is sufficiently large: $n > n_0$.
   Suppose that we know such a code f_n. How do we encode the message m? First
divide m into blocks of length L where

$$L = \left\lceil \frac{0.577291N}{\log K} \right\rceil \text{ so that } |\mathscr{A}^L| = K^L \leq 2^{0.577291N}.$$

Then we can embed the blocks from $\mathscr{A}^L$ in the alphabet $\mathscr{A}^n$ and so encode the
blocks. The transmission rate is $\log |\mathscr{A}^L| / \lceil n/R \rceil \sim 0.577291$. As was mentioned,
the SCT tells us that there are such codes but gives no idea of how to find (or
construct) them, which is difficult to do.

   Before we embark on the proof of Theorem 2.2.1, we would like to explore
connections between the geometry of Hamming space $\mathcal{H}_N$ and the randomness
generated by the channel. As in Section 1.4, we use the symbol $\mathbf{P}\big(\, \cdot \, |f_n(\mathbf{u})\big)$ as a
shorthand for $\mathbf{P}_{ch}\big(\, \cdot \, |f_n(\mathbf{u}) \text{ sent}\big)$. The expectation and variance under this distribu-
tion will be denoted by $\mathbf{E}(\cdot \, |f_n(\mathbf{u}))$ and $\mathrm{Var}(\cdot \, |f_n(\mathbf{u}))$.

Observe that, under distribution $\mathbf{P}(\,\cdot\,|f_n(\mathbf{u}))$, the number of distorted digits in the (random) received word $\mathbf{Y}^{(N)}$ can be written as

$$\sum_{j=1}^{N} \mathbf{1}(\text{digit } j \text{ in } \mathbf{Y}^{(N)} \neq \text{digit } j \text{ in } f_n(\mathbf{u})).$$

This is a random variable which has a binomial distribution $\text{Bin}(N,p)$, with the mean value

$$\mathbf{E}\left[\sum_{j=1}^{N} \mathbf{1}(\text{digit } j \text{ in } \mathbf{Y}^{(N)} \neq \text{digit } j \text{ in } f_n(\mathbf{u})) \middle| f_n(\mathbf{u})\right]$$
$$= \sum_{j=1}^{N} \mathbf{E}\left[\mathbf{1}(\text{digit } j \text{ in } \mathbf{Y}^{(N)} \neq \text{digit } j \text{ in } f_n(\mathbf{u}))|f_n(\mathbf{u})\right] = Np,$$

and the variance

$$\text{Var}\left[\sum_{j=1}^{N} \mathbf{1}(\text{digit } j \text{ in } \mathbf{Y}^{(N)} \neq \text{digit } j \text{ in } f_n(\mathbf{u})) \middle| f_n(\mathbf{u})\right]$$
$$= \sum_{j=1}^{N} \text{Var}\left[\mathbf{1}(\text{digit } j \text{ in } \mathbf{Y}^{(N)} \neq \text{digit } j \text{ in } f_n(\mathbf{u}))|f_n(\mathbf{u})\right] = Np(1-p).$$

Then, by Chebyshev's inequality, for all given $\varepsilon \in (0, 1-p)$ and positive integer $N > 1/\varepsilon$, the probability that at least $\lfloor N(p+\varepsilon) \rfloor$ digits have been distorted given that the codeword $f_n(\mathbf{u})$ has been sent, is

$$\leq \mathbf{P}\left(\geq N(p+\varepsilon) - 1 \text{ distorted } |f_n(\mathbf{u})\right) \leq \frac{p(1-p)}{N(\varepsilon - 1/N)^2}. \qquad (2.2.5)$$

Proof of Theorem 2.2.1. Throughout the proof, we follow the set-up from (2.2.3). Subscripts n and N will be often omitted; viz., we set

$$2^n = M.$$

We will assume the ML/geometric decoder without any further mention. Similarly to Section 1.4, we identify the set of source messages $\mathcal{U}_n$ with Hamming space $\mathcal{H}_n$. As proposed by Shannon, we use again a random encoding. More precisely, a message $\mathbf{u} \in \mathcal{H}_n$ is mapped to a random codeword $F_n(\mathbf{u}) \in \mathcal{H}_N$, with IID digits taking values 0 and 1 with probability $1/2$ and independently of each other. In addition, we make codewords $F_n(\mathbf{u})$ independent for different messages $\mathbf{u} \in \mathcal{H}_n$; labelling the strings from $\mathcal{H}_n$ by $\mathbf{u}(1), \dots, \mathbf{u}(M)$ (in no particular order) we obtain a family of IID random strings $F_n(\mathbf{u}(1)), \dots, F_n(\mathbf{u}(M))$ from $\mathcal{H}_N$. Finally, we make the codewords independent of the channel. Again, in analogy with Section 1.4, we can think of the random code under consideration as a random megastring/codebook from $\mathcal{H}_{NM} = \{0,1\}^{NM}$ with IID digits 0, 1 of equal probability. Every given sample $f(= f_n)$ of this random codebook (i.e. any given megastring from $\mathcal{H}_{NM}$) specifies

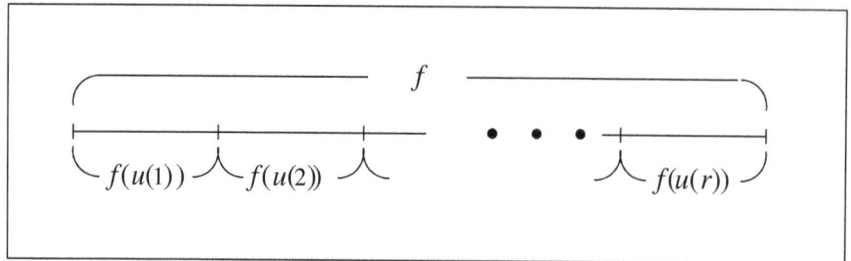

Figure 2.3

a deterministic encoding $f(\mathbf{u}(1)),\ldots,f(\mathbf{u}(M))$ of messages $\mathbf{u}(1),\ldots,\mathbf{u}(M)$, i.e. a code f; see Figure 2.3.

As in Section 1.4, we denote by $\mathscr{P}_n$ the probability distribution of the random code, with

$$\mathscr{P}_n(F_n = f) = \frac{1}{2^{NM}}, \quad \text{for all sample megastrings } f, \tag{2.2.6}$$

and by $\mathscr{E}_n$ the expectation relative to $\mathscr{P}_n$.

The plan of the rest of the proof is as follows. First, we will prove (by repeating in part arguments from Section 1.4) that, for the transmission rate $R \in (0,C)$, the expected average probability for the above random coding goes to zero as $n \to \infty$:

$$\lim_{n\to\infty} \mathscr{E}_n\left[e^{\text{ave}}(F_n)\right] = 0. \tag{2.2.7}$$

Here $e^{\text{ave}}(F_n)$, which is shorthand for $e^{\text{ave}}(F_n(\mathbf{u}(1)),\ldots,F_n(\mathbf{u}(M)))$, is a random variable taking values in $(0,1)$ and representing the aforementioned average error-probability for the random coding in question. More precisely, as in Section 1.4, for all given sample collections of codewords $f(\mathbf{u}(1)),\ldots,f(\mathbf{u}(M)) \in \mathscr{H}_N$ (i.e. for all given megastrings f from $\mathscr{H}_{NM}$), we define

$$e^{\text{ave}}(f_n) = \frac{1}{M} \sum_{1 \le i \le M} \mathbf{P}\left(\text{error while using codebook } f| f(\mathbf{u}(i))\right). \tag{2.2.8}$$

Then the expected average error-probability is given by

$$\mathscr{E}_n\left[e^{\text{ave}}(F_n)\right] = \frac{1}{2^{NM}} \sum_{f(\mathbf{u}(1)),\ldots,f(\mathbf{u}(M))\in\mathscr{H}_N} e^{\text{ave}}(f). \tag{2.2.9}$$

Relation (2.2.9) implies (again in a manner similar to Section 1.4) that there exists a sequence of deterministic codes f_n such that the average error-probability $e^{\text{ave}}(f_n) = e^{\text{ave}}(f_n(\mathbf{u}(1)),\ldots,f_n(\mathbf{u}(2^n)))$ obeys

$$\lim_{n\to\infty} e^{\text{ave}}(f_n) = 0. \tag{2.2.10}$$

Finally, we will deduce (2.2.4) from (2.2.10): see Lemma 2.2.6.

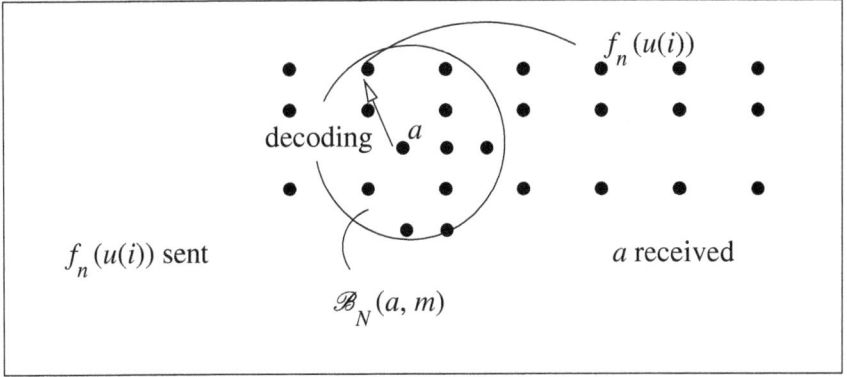

Figure 2.4

Remark 2.2.3 As the codewords $f(\mathbf{u}(1)),\dots,f(\mathbf{u}(M))$ are thought to come from a sample of the random codebook, we must allow that they may coincide $(f(\mathbf{u}(i)) = f(\mathbf{u}(j))$ for $i \neq j)$, in which case, by default, the ML decoder is reporting an error. This must be included when we consider probabilities in the RHS of (2.2.8). Therefore, for $i = 1,\dots,M$ we define

$$\mathbf{P}\Big(\text{error while using codebook } f \mid f(\mathbf{u}(i))\Big)$$

$$= \begin{cases} 1, & \text{if } f(\mathbf{u}(i)) = f(\mathbf{u}(i')) \text{ for some } i' \neq i, \\ \mathbf{P}\Big(\delta\big(\mathbf{Y}^{(N)}, f(\mathbf{u}(j))\big) \leq \delta\big(\mathbf{Y}^{(N)}, f(\mathbf{u}(i))\big) \\ \qquad \text{for some } j \neq i \mid f(\mathbf{u}(i))\Big), \\ \text{if } f(\mathbf{u}(i)) \neq f(\mathbf{u}(i')) \text{ for all } i' \neq i. \end{cases} \qquad (2.2.11)$$

Let us now go through the detailed argument. The first step is

Lemma 2.2.4 *Consider the channel matrix Π (cf. (2.2.2)) with $0 \leq p < 1/2$. Suppose that the transmission rate $R < C = 1 - \eta(p)$. Let N be $> 1/\varepsilon$. Then for any $\varepsilon \in (0, 1/2 - p)$, the expected average error-probability $\mathscr{E}_n\big[e^{\text{ave}}(F_n)\big]$ defined in (2.2.8), (2.2.9) obeys*

$$\mathscr{E}_n\big[e^{\text{ave}}(F_n)\big] \leq \frac{p(1-p)}{N(\varepsilon - 1/N)^2} + \frac{M-1}{2^N} v_N\big(\lceil N(p+\varepsilon)\rceil\big), \qquad (2.2.12)$$

where $v_N(b)$ stands for the number of points in the ball of radius b in the binary Hamming space $\mathscr{H}_N$.

Proof Set $m(= m_N(p,\varepsilon)) := \lceil N(p+\varepsilon)\rceil$. The ML decoder definitely returns the codeword $f_n(\mathbf{u}(i))$ sent through the channel when $f_n(\mathbf{u}(i))$ is the only codeword in the Hamming ball $\mathscr{B}_N(\mathbf{y},m)$ around the received word $\mathbf{y}(=\mathbf{y}^{(N)}) \in \mathscr{H}_N$ (see Figure 2.4). In any other situation (when $f_n(\mathbf{u}(i)) \notin \mathscr{B}_N(\mathbf{y},m)$ or $f_n(\mathbf{u}(k)) \in \mathscr{B}_N(\mathbf{y},m)$ for some $k \neq i$) there is a possibility of error.

Hence,

$$
\begin{aligned}
\mathbf{P}&\Big(\text{error while using codebook } f \,|\, f_n(\mathbf{u}(i))\Big) \\
&\leq \sum_{\mathbf{y} \in \mathscr{H}_N} \mathbf{P}\big(\mathbf{y} | f_n(\mathbf{u}(i))\big) \mathbf{1}\Big(f_n(\mathbf{u}(i)) \notin \mathscr{B}_N(\mathbf{y},m)\Big) \\
&\quad + \sum_{\mathbf{z} \in \mathscr{H}_N} \mathbf{P}\big(\mathbf{z} | f_n(\mathbf{u}(i))\big) \sum_{k \neq i} \mathbf{1}\Big(f_n(\mathbf{u}(k)) \in \mathscr{B}_N(\mathbf{z},m)\Big).
\end{aligned}
\tag{2.2.13}
$$

The first sum in the RHS is simple to estimate:

$$
\begin{aligned}
&\sum_{\mathbf{y} \in \mathscr{H}_N} \mathbf{P}\big(\mathbf{y} | f_n(\mathbf{u}(i))\big)\, \mathbf{1}\big(f_n(\mathbf{u}(i) \notin \mathscr{B}_N(\mathbf{y},m)\big) \\
&= \sum_{\mathbf{y} \in \mathscr{H}_N} \mathbf{P}\big(\mathbf{y} | f_n(\mathbf{u}(i))\big) \mathbf{1}\big(\text{distance } \delta\,(\mathbf{y}, f_n(\mathbf{u}(i))) \geq m\big) \\
&= \mathbf{P}\Big(\geq m \text{ digits distorted} | f(\mathbf{u}(i))\Big) \leq \frac{p(1-p)}{N(\varepsilon - 1/N)^2},
\end{aligned}
\tag{2.2.14}
$$

by virtue of (2.2.5). Observe that since the RHS in (2.2.14) does not depend on the choice of the sample code f, the bound (2.2.14) will hold when we take first the average $\dfrac{1}{M} \sum_{1 \leq i \leq M}$ and then expectation $\mathscr{E}_n$.

The second sum in the RHS of (2.2.13) is more tricky: it requires averaging and taking expectation. Here we have

$$
\begin{aligned}
\mathscr{E}_n&\left[\sum_{1 \leq i \leq M} \sum_{\mathbf{z} \in \mathscr{H}_N} \mathbf{P}\big(\mathbf{z} | F_n(\mathbf{u}(i))\big) \sum_{k \neq i} \mathbf{1}\big(F_n(\mathbf{u}(k)) \in \mathscr{B}_N(\mathbf{z},m)\big) \right] \\
&= \sum_{1 \leq i \leq M} \sum_{k \neq i} \sum_{\mathbf{z} \in \mathscr{H}_N} \mathscr{E}_n\left[\mathbf{P}\big(\mathbf{z} | F_n(\mathbf{u}(i))\big) \mathbf{1}\big(F_n(\mathbf{u}(k)) \in \mathscr{B}_N(\mathbf{z},m)\big) \right] \\
&= \sum_{1 \leq i \leq M} \sum_{k \neq i} \sum_{\mathbf{z} \in \mathscr{H}_N} \mathscr{E}_n\left[\mathbf{P}\big(\mathbf{z} | F_n(\mathbf{u}(i))\big) \right] \\
&\qquad\qquad \times \mathscr{E}_n\left[\mathbf{1}\big(F_n(\mathbf{u}(k)) \in \mathscr{B}_N(\mathbf{z},m)\big) \right],
\end{aligned}
\tag{2.2.15}
$$

since random codewords $F_n(\mathbf{u}(1)), \ldots, F_n(\mathbf{u}(M))$ are independent. Next, as each of these codewords is uniformly distributed over $\mathscr{H}_N$, the expectations $\mathscr{E}_n\big[\mathbf{P}(\mathbf{z} | F_n(\mathbf{u}(i)))\big]$ and $\mathscr{E}_n\big[\mathbf{1}(F_n(\mathbf{u}(k)) \in \mathscr{B}_N(\mathbf{z},m))\big]$ can be calculated as

$$
\mathscr{E}_n\big[\mathbf{P}(\mathbf{z} | F_n(\mathbf{u}(i)))\big] = \frac{1}{2^N} \sum_{\mathbf{x} \in \mathscr{H}_N} \mathbf{P}(\mathbf{z} | \mathbf{x})
\tag{2.2.16a}
$$

and

$$
\mathscr{E}_n\big[\mathbf{1}(F_n(\mathbf{u}(k)) \in \mathscr{B}_N(\mathbf{z},m))\big] = \frac{\nu_N(m)}{2^N}.
\tag{2.2.16b}
$$

Further, summing over $\mathbf{z}$ yields

$$\sum_{\mathbf{z}\in\mathcal{H}_N}\sum_{\mathbf{x}\in\mathcal{H}_N}\mathbf{P}(\mathbf{z}|\mathbf{x}) = \sum_{\mathbf{x}\in\mathcal{H}_N}\sum_{\mathbf{z}\in\mathcal{H}_N}\mathbf{P}(\mathbf{z}|\mathbf{x}) = 2^N. \tag{2.2.17}$$

Finally, after summation over $k \neq i$ we obtain

$$\text{the RHS of } (2.2.15) = \frac{1}{M}\sum_{1\leq i\leq M}\sum_{k\neq i}\frac{v_N(m)}{2^N}$$

$$= \frac{v_N(m)M(M-1)}{2^N M} = \frac{(M-1)v_N(m)}{2^N}. \tag{2.2.18}$$

Collecting (2.2.12)–(2.2.18) we have that $\mathscr{E}_N\left[e^{\mathrm{ave}}(F_n)\right]$ does not exceed the RHS of (2.2.12). $\qquad\square$

At the next stage we estimate the volume $v_N(m)$ in terms of entropy $h(p+\varepsilon)$ where, recall, $m = \lceil N(p+\varepsilon)\rceil$. The argument here is close to that from Section 1.4 and based on the following result.

Lemma 2.2.5   *Suppose that $0 < p < 1/2$, $\varepsilon > 0$ and positive integer N satisfy $p+\varepsilon+1/N < 1/2$. Then the following bound holds true:*

$$v_N(\lceil N(p+\varepsilon)\rceil) \leq 2^{N\eta(p+\varepsilon)}. \tag{2.2.19}$$

The proof of Lemma 2.2.5 will be given later, after Worked Example 2.2.7. For the moment we proceed with the proof of Theorem 2.2.1. Recall, we want to establish (2.2.7). In fact, if $p < 1/2$ and $R < C = 1 - \eta(p)$ then we set $\zeta = C - R > 0$ and take $\varepsilon > 0$ so small that (i) $p+\varepsilon < 1/2$ and (ii) $R+\zeta/2 < 1 - \eta(p+\varepsilon)$. Then we take N so large that (iii) $N > 2/\varepsilon$. With this choice of ε and N, we have

$$\varepsilon - \frac{1}{N} > \frac{\varepsilon}{2} \quad\text{and}\quad R - 1 + \eta(p+\varepsilon) < -\frac{\zeta}{2}. \tag{2.2.20}$$

Then, starting with (2.2.12), we can write

$$\mathscr{E}_N\left[e(F_n)\right] \leq \frac{4p(1-p)}{N\varepsilon^2} + \frac{2^{NR}}{2^N}2^{N\eta(p+\varepsilon)}$$

$$< \frac{4}{N\varepsilon^2}p(1-p) + 2^{-N\zeta/2}. \tag{2.2.21}$$

This implies (2.2.7) and hence the existence of a sequence of codes $f_n\colon \mathcal{H}_n \to \mathcal{H}_N$ obeying (2.2.10).

To finish the proof of Theorem 2.2.1, we deduce (2.2.4) from (2.2.7), in the form of Lemma 2.2.6:

Lemma 2.2.6   *Consider a binary channel (not necessarily memoryless), and let $C > 0$ be a given constant. With $0 < R < C$ and $n = \lfloor NR\rfloor$, define quantities $e_{\mathrm{max}}(f_n)$ and $e^{\mathrm{ave}}(\tilde{f}_n)$ as in (2.2.4), (2.2.8) and (2.2.11), for codes $f_n\colon \mathcal{H}_n \to \mathcal{H}_N$ and $\tilde{f}_n\colon \mathcal{H}_n \to \mathcal{H}_N$. Then the following statements are equivalent:*

(i) *For all $R \in (0,C)$, there exist codes f_n with $\lim\limits_{n\to\infty} e_{\max}(f_n) = 0$.*

(ii) *For all $R \in (0,C)$, there exist codes $\tilde{f}_n$ such that $\lim\limits_{n\to\infty} e^{\mathrm{ave}}(\tilde{f}_n) = 0$.*

Proof of Lemma 2.2.6. It is clear that assertion (i) implies (ii). To deduce (i) from (ii), take $R < C$ and set for N big enough

$$R' = R + \frac{1}{N} < C, \quad n' = \lfloor NR' \rfloor, \quad M' = 2^{n'}. \tag{2.2.22}$$

We know that there exists a sequence $\tilde{f}_n$ of codes $\mathcal{H}_{n'} \to \mathcal{H}_N$ with $e^{\mathrm{ave}}(\tilde{f}_n) \to 0$. Recall that

$$e^{\mathrm{ave}}(\tilde{f}_n) = \frac{1}{M'} \sum_{1 \le i \le M'} \mathbf{P}\left(\text{error while using } \tilde{f}_n \middle| \tilde{f}_n(\mathbf{u}(i))\right). \tag{2.2.23}$$

Here and below, $M' = 2^{\lfloor NR' \rfloor}$ and $\tilde{f}_n(\mathbf{u}(1)), \ldots, \tilde{f}_n(\mathbf{u}(M'))$ are the codewords for source messages $\mathbf{u}(1), \ldots, \mathbf{u}(M') \in \mathcal{H}_{n'}$.

Instead of $\mathbf{P}\left(\text{error while using } \tilde{f}_n \middle| \tilde{f}_n(\mathbf{u}(i))\right)$, we write $\mathbf{P}\left(f_n\text{-error} \middle| \tilde{f}_n(\mathbf{u}(i))\right)$, for brevity. Now, at least half of summands $\mathbf{P}\left(\tilde{f}_n\text{-error} \middle| \tilde{f}_n(\mathbf{u}(i))\right)$ in the RHS of (2.2.23) must be $< 2e^{\mathrm{ave}}(\tilde{f}_n)$. Observe that, in view of (2.2.22),

$$M'/2 \ge 2^{\lfloor NR \rfloor - 1}. \tag{2.2.24}$$

Hence we have at our disposal at least $2^{\lfloor NR \rfloor - 1}$ codewords $f(\mathbf{u}(i))$ with

$$\mathbf{P}\left(\text{error} \middle| \tilde{f}_n(\mathbf{u}(i))\right) < 2e^{\mathrm{ave}}(\tilde{f}_n).$$

List these codewords as a new binary code, of length N and information rate $\left(\log M'/2\right)/N$. Denoting this new code by f_n, we have

$$e^{\max}(f_n) \le 2e^{\mathrm{ave}}(\tilde{f}_n).$$

Hence, $e^{\max}(f_n) \to 0$ as $n \to \infty$ whereas $\left(\log M'/2\right)/N \to R$. This gives statement (i) and completes the proof of Lemma 2.2.6. □

Therefore, the proof of Theorem 2.2.1 is now complete (provided that we prove Lemma 2.2.5).

Worked Example 2.2.7 (cf. Worked Example 2.1.20.) *Prove that for positive integers N and m, with $m < N/2$ and $\beta = m/N$,*

$$2^{N\eta(\beta)}/(N+1) < v_N(m) < 2^{N\eta(\beta)}. \tag{2.2.25}$$

Solution Write

$$v_N(m) = \sharp\{\text{points at distance} \le m \text{ from } \mathbf{0} \text{ in } \mathcal{H}_N\} = \sum_{0 \le k \le m} \binom{N}{k}.$$

With $\beta = m/N < 1/2$, we have that $\beta/(1-\beta) < 1$, and so

$$\left(\frac{\beta}{1-\beta}\right)^m < \left(\frac{\beta}{1-\beta}\right)^k, \quad \text{for } 0 \le k < m.$$

Then, for $0 \le k < m$, the product

$$\beta^k(1-\beta)^{N-k} = \left(\frac{\beta}{1-\beta}\right)^k (1-\beta)^N$$

$$> \left(\frac{\beta}{1-\beta}\right)^m (1-\beta)^N = \beta^m(1-\beta)^{N-m}.$$

Hence,

$$1 = \sum_{0 \le k \le N} \binom{N}{k} \beta^k(1-\beta)^{N-k} > \sum_{0 \le k \le m} \binom{N}{k} \beta^k(1-\beta)^{N-k}$$

$$> \beta^m(1-\beta)^{N-m} \sum_{0 \le k \le m} \binom{N}{k} = v_N(m)\beta^m(1-\beta)^{N-m}$$

$$= v_N(m)2^{N[(m/N)\log\beta + (1-m/N)\log(1-\beta)]},$$

implying that $v_N(m) < 2^{N\eta(\beta)}$. To obtain the left-hand bound in (2.2.25), write

$$v_N(m) > \binom{N}{m};$$

then we aim to check that the RHS is $\ge 2^{N\eta(\beta)}/(N+1)$. Consider a binomial random variable $Y \sim \text{Bin}(N, \beta)$ with

$$p_k = \mathbb{P}(Y = k) = \binom{N}{k} \beta^k(1-\beta)^{N-k}, \quad k = 0, \dots, N.$$

It suffices to prove that p_k achieves its maximal value when $k = m$, since then

$$p_m = \binom{N}{m} \beta^m(1-\beta)^{N-m} \ge \frac{1}{N+1}, \quad \text{with } \beta^m(1-\beta)^{N-m} = 2^{-N\eta(\beta)}.$$

To this end, suppose first that $k \le m$ and write

$$\frac{p_k}{p_m} = \frac{m!(N-m)!(N-m)^{m-k}}{k!(N-k)!m^{m-k}}$$

$$= \frac{(k+1)\cdots m}{m^{m-k}} \cdot \frac{(N-m)^{m-k}}{(N-m+1)\cdots(N-k)}.$$

Here, the RHS is ≤ 1, as it is the product of $2(m-k)$ factors each of which is ≤ 1. Similarly, if $k \geq m$, we arrive at the product

$$\frac{m^{k-m}}{(m+1)\cdots k} \cdot \frac{(N-k+1)\cdots(N-m)}{(N-m)^{k-m}}$$

which is again ≤ 1 as the product of $2(k-m)$ factors ≤ 1. Thus, the ratio $p_k/p_m \leq 1$, and the desired bound follows.   □

We are now in position to prove Lemma 2.2.5.

Proof of Lemma 2.2.5 First, $p + \varepsilon < 1/2$ implies that $m = \lfloor N(p+\varepsilon) \rfloor < N/2$ and

$$\beta := \frac{m}{N} = \frac{\lfloor N(p+\varepsilon) \rfloor}{N} < p + \varepsilon,$$

which, in turn, implies that $\eta(\beta) < \eta(p+\varepsilon)$ as $x \mapsto \eta(x)$ is a strictly increasing function for x from the interval $(0, 1/2)$. This yields the assertion of Lemma 2.2.5.   □

The geometric proof of the direct part of SCT/NCT clarifies the meaning of the concept of capacity (of an MBSC at least). Physically speaking, in the expressions (1.4.11), (1.4.27) and (2.2.1) for capacity $C = \eta(p)$ of an MBSC, the positive term 1 points at the rate at which a random code produces an 'empty' volume between codewords whereas the negative term $-\eta(p)$ indicates the rate at which the codewords progressively fill this space. We continue with a working example of an essay type:

Worked Example 2.2.8   *Quoting general theorems on the evaluation of the channel capacity, deduce an expression for the capacity of a memoryless binary symmetric channel. Evaluate, in particular, the capacities of (i) a symmetric memoryless channel and (ii) a perfect channel with an input alphabet $\{0,1\}$ whose inputs are subject to the restriction that 0 should never occur in succession.*

Solution The channel capacity is defined as a supremum of transmission rates R for which the received message can be decoded correctly, with probability approaching 1 as the length of the message increases to infinity. A popular class is formed by memoryless channels where, for a given input word $\mathbf{x}^{(N)} = x_1 \ldots x_N$, the probability

$$\mathbf{P}^{(N)}\left(\mathbf{y}^{(N)} \text{ received}|\mathbf{x}^{(N)} \text{ sent}\right) = \prod_{1 \leq i \leq N} P(y_i|x_i).$$

In other words, the noise acts on each symbol x_i of the input string $\mathbf{x}$ independently, and $P(y|x)$ is the probability of having an output symbol y given that the input symbol is x.

Symbol x runs over $\mathscr{A}_{\text{in}}$, an input alphabet of a given size q, and y belongs to $\mathscr{A}_{\text{out}}$, an output alphabet of size r. Then probabilities $P(y|x)$ form a $q \times r$ stochastic matrix (the channel matrix). A memoryless channel is called symmetric if the rows of this matrix are permutations of each other, i.e. contain the same collection of probabilities, say $p_1, \ldots, p_r$. A memoryless symmetric channel is said to be double-symmetric if the columns of the channel matrix are also permutations of each other. If $m = n = 2$ (typically, $\mathscr{A}_{\text{in}} = \mathscr{A}_{\text{out}} = \{0, 1\}$) a memoryless channel is called binary. For a memoryless binary symmetric channel, the channel matrix entries $P(y|x)$ are $P(0|0) = P(1|1) = 1 - p$, $P(1|0) = P(0|1) = p$, $p \in (0,1)$ being the flipping probability and $1 - p$ the probability of flawless transmission of a single binary symbol.

A channel is characterised by its capacity: the value $C \geq 0$ such that:

(a) for all $R < C$, R is a reliable transmission rate; and
(b) for all $R > C$, R is an unreliable transmission rate.

Here R is called a reliable transmission rate if there exists a sequence of codes $f_n : \mathscr{H}_n \to \mathscr{H}_N$ and decoding rules $\widehat{f}_N : \mathscr{H}_N \to \mathscr{H}_n$ such that $n \sim NR$ and the (suitably defined) probability of error

$$e(f_n, \widehat{f}_N) \to 0, \quad \text{as } N \to \infty.$$

In other words,

$$C = \lim_{N \to \infty} \frac{1}{N} \log M_N$$

where M_N is the maximal number of codewords $\mathbf{x} \in \mathscr{H}_N$ for which the probability of erroneous decoding tends to 0.

The SCT asserts that, for a memoryless channel,

$$C = \max_{p_X} I(X : Y)$$

where $I(X : Y)$ is the mutual information between a (random) input symbol X and the corresponding output symbol Y, and the maximum is over all possible probability distributions p_X of X.

Now in the case of a memoryless symmetric channel (MSC), the above maximisation procedure applies to the output symbols only:

$$C = \left(\max_{p_X} h(Y) \right) + \sum_{1 \leq i \leq r} p_i \log p_i;$$

the sum $-\sum_i p_i \log p_i$ being the entropy of the row of channel matrix $(P(y|x))$. For a double-symmetric channel, the expression for C simplifies further:

$$C = \log M - h(p_1, \ldots, p_r)$$

as $h(Y)$ is achieved at equidistribution p_X, with $p_X(x) \equiv 1/q$ (and $p_Y(y) \equiv 1/r$). In the case of an MBSC we have

$$C = 1 - \eta(p).$$

This completes the solution to part (i).

Next, the channel in part (ii) is not memoryless. Still, the general definitions are applicable, together with some arguments developed so far. Let $n(j,t)$ denote the number of allowed strings of length t ending with letter j, $j = 0, 1$. Then

$$n(0,t) = n(1,t-1),$$
$$n(1,t) = n(0,t-1) + n(1,t-1),$$

whence

$$n(1,t) = n(1,t-1) + n(1,t-2).$$

Write it as a recursion

$$\begin{pmatrix} n(1,t) \\ n(1,t-1) \end{pmatrix} = A \begin{pmatrix} n(1,t-1) \\ n(1,t-2) \end{pmatrix},$$

with the recursion matrix

$$A = \begin{pmatrix} 1 & 1 \\ 1 & 0 \end{pmatrix}.$$

The general solution is

$$n(1,t) = c_1 \lambda_1^t + c_2 \lambda_2^t,$$

where λ_1, λ_2 are the eigenvalues of A, i.e. the roots of the characteristic equation

$$\det (A - \lambda \mathbf{I}) = (1 - \lambda)(-\lambda) - 1 = \lambda^2 - \lambda - 1 = 0.$$

So, $\lambda = \left(1 \pm \sqrt{5}\right)/2$, and

$$\frac{1}{t} \log n(1,t) = \log \left(\frac{\sqrt{5}+1}{2} \right).$$

The capacity of the channel is given by

$$\begin{aligned}
C &= \lim_{t \to \infty} \frac{1}{t} \log \sharp \text{ of allowed input strings of length } t \\
&= \lim_{t \to \infty} \frac{1}{t} \log \left[n(1,t) + n(0,t) \right] = \log \left(\frac{\sqrt{5}+1}{2} \right).
\end{aligned}$$

□

Remark 2.2.9   We can modify the last question, by considering an MBC with the channel matrix $\Pi = \begin{pmatrix} 1-p & p \\ p & 1-p \end{pmatrix}$ whose input is under a restriction that 0 should never occur in succession. Such a channel may be treated as a composition of two consecutive channels (cf. Worked Example 1.4.29(a)), which yields the following answer for the capacity:

$$C = \min \left[\log \left(\frac{\sqrt{5}+1}{2} \right), 1 - \eta(p) \right].$$

Next, we present the strong converse part of Shannon's SCT for an MBSC (cf. Theorem 1.4.14); again we are going to prove it by using geometry of Hamming's spaces. The term 'strong' indicates that for every transmission rate $R > C$, the channel capacity, the maximum probability of error actually gets arbitrarily close to 1. Again for simplicity, we prove the assertion for an MBSC.

Theorem 2.2.10   (The SCT/NCT, the strong converse part) *Let C be the capacity of an MBSC with the channel matrix* $\begin{pmatrix} 1-p & p \\ p & 1-p \end{pmatrix}$, *where $0 < p < 1/2$, and take $R > C$. Then, with $n = \lfloor NR \rfloor$, for all codes $f_n \colon \mathcal{H}_n \to \mathcal{H}_N$ and decoding rules $\widehat{f}_N \colon \mathcal{H}_N \to \mathcal{H}_n$, the maximum error-probability*

$$\varepsilon^{\max}(f_n, \widehat{f}_N) := \max \left[\mathbf{P}\left(\text{error under } \widehat{f}_N | f_n(\mathbf{u}) \right) : \mathbf{u} \in \mathcal{H}_n \right] \qquad (2.2.26a)$$

obeys

$$\limsup_{N \to \infty} \varepsilon^{\max}(f_n, \widehat{f}_N) = 1. \qquad (2.2.26b)$$

Proof   As in Section 1.4, we can assume that codes f_n are one-to-one and obey $\widehat{f}_N(f_n(\mathbf{u})) = \mathbf{u}$, for all $\mathbf{u} \in \mathcal{H}_n$ (otherwise, the chances of erroneous decoding will be even larger). Assume the opposite of (2.2.26b):

$$\varepsilon^{\max}(f_n, \widehat{f}_N) \leq c \text{ for some } c < 1 \text{ and all } N \text{ large enough.} \qquad (2.2.27)$$

Our aim is to deduce from (2.2.27) that $R \leq C$. As before, set $n = \lfloor NR \rfloor$ and let $f_n(\mathbf{u}(i))$ be the codeword for string $\mathbf{u}(i) \in \mathcal{H}_n$, $i = 1, \ldots, 2^n$. Let $\mathscr{D}_i \subset \mathcal{H}_N$ be the set of binary strings where $\widehat{f}_N$ returns $f_n(\mathbf{u}(i))$: $\widehat{f}_N(\mathbf{a}) = f_n(\mathbf{u}(i))$ if and only if $\mathbf{a} \in \mathscr{D}_i$. Then $\mathscr{D}_i \ni f(\mathbf{u}(i))$, sets $\mathscr{D}_i$ are pairwise disjoint, and if the union $\cup_i \mathscr{D}_i \neq \mathcal{H}_N$ then on the complement $\mathcal{H}_N \setminus \cup_i \mathscr{D}_i$ the channel declares an error. Set $s_i = \sharp \mathscr{D}_i$, the size of set $\mathscr{D}_i$.

Our first step is to 'improve' the decoding rule, by making it 'closer' to the ML rule. In other words, we want to replace each $\mathscr{D}_i$ with a new set, $\mathscr{C}_i \in \mathcal{H}_N$, of the same cardinality $\sharp \mathscr{C}_i = s_i$, but of a more 'rounded' shape (i.e. closer to a Hamming ball $B(f(\mathbf{u}(i)), b_i))$. That is, we look for pairwise disjoint sets $\mathscr{C}_i$, of cardinalities $\sharp\mathscr{C}_i = s_i$, satisfying

$$\mathscr{B}_N(f(\mathbf{u}(i)), b_i) \subseteq \mathscr{C}_i \subset \mathscr{B}_N(f(\mathbf{u}(i)), b_i + 1), \ \ 1 \leq i \leq 2^n, \tag{2.2.28}$$

for some values of radius $b_i \geq 0$, to be specified later. We can think that $\mathscr{C}_i$ is obtained from $\mathscr{D}_i$ by applying a number of 'disjoint swaps' where we remove a string $\mathbf{a}$ and add another string, $\mathbf{b}$, with the Hamming distance

$$\delta\left(\mathbf{b}, f_n(\mathbf{u}(i))\right) \leq \delta\left(\mathbf{a}, f_n(\mathbf{u}(i))\right). \tag{2.2.29}$$

Denote the new decoding rule by $\widehat{g}_N$. As the flipping probability $p < 1/2$, the relation (2.2.29) implies that

$$\mathbf{P}(\widehat{f}_N \text{ returns } f_n(\mathbf{u}(i)) | f_n(\mathbf{u}(i))) = \mathbf{P}(\mathscr{D}_i | f_n(\mathbf{u}(i)))$$
$$\leq \mathbf{P}(\mathscr{C}_i | f_n(\mathbf{u}(i))) = \mathbf{P}(\widehat{g}_N \text{ returns } f_n(\mathbf{u}(i)) | f_n(\mathbf{u}(i))),$$

which in turn is equivalent to

$$\mathbf{P}(\text{error when using } \widehat{g}_N | f_n(\mathbf{u}(i))) \leq \mathbf{P}(\text{error when using } \widehat{f}_N | f_n(\mathbf{u}(i))). \tag{2.2.30}$$

Then, clearly,

$$\varepsilon^{\max}(f_n, \widehat{g}_N) \leq \varepsilon^{\max}(f_n, \widehat{f}_N) \leq c. \tag{2.2.31}$$

Next, suppose that there exists $p' < p$ such that, for any N large enough,

$$b_i + 1 \leq \lceil Np' \rceil \text{ for some } 1 \leq i \leq 2^n. \tag{2.2.32}$$

Then, by virtue of (2.2.28) and (2.2.31), with $\mathscr{C}_i^c$ standing for the complement $\mathscr{H}_N \setminus \mathscr{C}_i$,

$$\mathbf{P}(\text{at least } Np' \text{ digits distorted} | f_n(\mathbf{u}(i)))$$
$$\leq \mathbf{P}(\text{at least } b_i + 1 \text{ digits distorted} | f_n(\mathbf{u}(i)))$$
$$\leq \mathbf{P}(\mathscr{C}_i^c | f_n(\mathbf{u}(i))) \leq \varepsilon^{\max}(f_n, \widehat{g}_N) \leq c.$$

This would lead to a contradiction, since, by the law of large numbers, as $N \to \infty$, the probability

$$\mathbf{P}(\text{at least } Np' \text{ digits distorted} | \mathbf{x} \text{ sent}) \to 1$$

uniformly in the choice of the input word $\mathbf{x} \in \mathscr{H}_N$. (In fact, this probability does not depend on $\mathbf{x} \in \mathscr{H}_N$.)

Thus, we cannot have $p' \in (0, p)$ such that, for N large enough, (2.2.32) holds true. That is, the opposite is true: for any given $p' \in (0, p)$, we can find an arbitrarily large N such that

$$b_i > Np', \quad \text{for all } i = 1, \ldots, 2^n. \tag{2.2.33}$$

(As we claim (2.2.33) for all $p' \in (0, p)$, it does not matter if in the LHS of (2.2.33) we put b_i or $b_i + 1$.)

At this stage we again use the explicit expression for the volume of the Hamming ball:

$$s_i = \sharp \mathcal{D}_i = \sharp \mathcal{C}_i \geq v_N(b_i) = \sum_{0 \leq j \leq b_i} \binom{N}{j} \geq \binom{N}{b_i}$$

$$\geq \binom{N}{\lfloor Np' \rfloor}, \quad \text{provided that } b_i > Np'. \tag{2.2.34}$$

A useful bound has been provided in Worked Example 2.2.7 (see (2.2.25)):

$$v_N(R) \geq \frac{1}{N+1} 2^{N\eta\left(\frac{R}{N}\right)}. \tag{2.2.35}$$

We are now in a position to finish the proof of Theorem 2.2.10. In view of (2.2.35), we have that, for all $p' \in (0, p)$, we can find an arbitrarily large N such that

$$s_i \geq 2^{N(\eta(p') - \varepsilon_N)}, \quad \text{for all } 1 \leq i \leq 2^n,$$

with $\lim_{N \to \infty} \varepsilon_N = 0$. As the original sets $\mathcal{D}_1, \ldots, \mathcal{D}_{2^n}$ are disjoint, we have that

$$s_1 + \cdots + s_{2^n} \leq 2^N, \quad \text{implying that } 2^{N(\eta(p') - \varepsilon_N)} \times 2^{\lfloor NR \rfloor} \leq 2^N,$$

or

$$\eta(p') - \varepsilon_N + \frac{\lfloor NR \rfloor}{N} \leq 1, \quad \text{implying that } R \leq 1 - \eta(p') + \varepsilon_N + \frac{1}{N}.$$

As $N \to \infty$, the RHS tends to $1 - \eta(p)$. So, given any $p' \in (0, p)$, $R \leq 1 - \eta(p')$. This is true for all $p' < p$, hence $R \leq 1 - \eta(p) = C$. This completes the proof of Theorem 2.2.10. $\square$

We have seen that the analysis of intersections of a given set $\mathcal{X}$ in a Hamming space $\mathcal{H}_N$ (and more generally, in $\mathcal{H}_{N,q}$) with various balls $\mathcal{B}_N(\mathbf{y}, s)$ reveals a lot about the set $\mathcal{X}$ itself. In the remaining part of this section such an approach will be used for producing some advanced bounds on q-ary codes: the Elias bound and the Johnson bound. These bounds are among the best-known general bounds for codes, and they are competing.

The Elias bound is proved in a fashion similar to Plotkin's: cf. Theorem 2.1.15 and Worked Example 2.1.18. We count codewords from a q-ary $[N, M, d]$ code $\mathcal{X}$ in balls $\mathcal{B}_{N,q}(\mathbf{y}, s)$ of radius s about words $\mathbf{y} \in \mathcal{H}_{N,q}$. More precisely, we count pairs $(\mathbf{x}, \mathcal{B}_{N,q}(\mathbf{y}, s))$ where $\mathbf{x} \in \mathcal{X} \cap \mathcal{B}_{N,q}(\mathbf{y}, s)$. If ball $\mathcal{B}_{N,q}(\mathbf{y}, s)$ contains $K_\mathbf{y}$ codewords then

$$\sum_{\mathbf{y} \in \mathcal{H}_N} K_\mathbf{y} = M v_{N,q}(s) \tag{2.2.36}$$

as each word $\mathbf{x}$ falls in $v_{N,q}(s)$ of the balls $\mathscr{B}_{N,q}(\mathbf{y},s)$.

Lemma 2.2.11   If $\mathscr{X}$ is a q-ary $[N,M]$ code then for all $s = 1,\ldots,N$ there exists a ball $\mathscr{B}_{N,q}(\mathbf{y},s)$ about an N-word $\mathbf{y} \in \mathscr{H}_{N,q}$ with the number $K_{\mathbf{y}} = \sharp\left(\mathscr{X} \cap \mathscr{B}_{N,q}(\mathbf{y},s)\right)$ of codewords in $\mathscr{B}_{N,q}(\mathbf{y},s)$ obeying

$$K_{\mathbf{y}} \geq M v_{N,q}(s)/q^N. \tag{2.2.37}$$

Proof   Divide both sides of (2.2.36) by q^N. Then $\dfrac{1}{q^N}\sum_{\mathbf{y}} K_{\mathbf{y}}$ gives the average number of codewords in ball $\mathscr{B}_{N,q}(\mathbf{y},s)$. But there must be a ball containing at least as many as the average number of codewords.   □

A ball $\mathscr{B}_{N,q}(\mathbf{y},s)$ with property (2.2.37) is called critical (for code $\mathscr{X}$).

Theorem 2.2.12   (The Elias bound) Set $\theta = (q-1)/q$. Then for all integers $s \geq 1$ such that $s < \theta N$ and $s^2 - 2\theta Ns + \theta Nd > 0$, the maximum size $M_q^*(N,d)$ of a q-ary code of length N and distance d satisfies

$$M_q^*(N,d) \leq \frac{\theta Nd}{s^2 - 2\theta Ns + \theta Nd} \cdot \frac{q^N}{v_{N,q}(s)}. \tag{2.2.38}$$

Proof   Fix a critical ball $\mathscr{B}_{N,q}(\mathbf{y},s)$ and consider code $\mathscr{X}'$ obtained by subtracting word $\mathbf{y}$ from the codewords of $\mathscr{X}$: $\mathscr{X}' = \{\mathbf{x} - \mathbf{y} : \mathbf{x} \in \mathscr{X}\}$. Then $\mathscr{X}'$ is again an $[N,M,d]$ code. So, we can assume that $\mathbf{y} = \mathbf{0}$ and $\mathscr{B}_{N,q}(\mathbf{0},s)$ is a critical ball.

Then take $\mathscr{X}_1 = \mathscr{X} \cap \mathscr{B}_{N,q}(\mathbf{0},s) = \{\mathbf{x} \in \mathscr{X} : w(\mathbf{x}) \leq s\}$. The code $\mathscr{X}_1$ is $[N,K,e]$ where $e \geq d$ and $K\ (= K_0) \geq M v_{N,q}(s)/q^N$. As in the proof of the Plotkin bound, consider the sum of the distances between the codewords in $\mathscr{X}_1$:

$$S_1 = \sum_{\mathbf{x} \in \mathscr{X}_1} \sum_{\mathbf{x}' \in \mathscr{X}_1} \delta(\mathbf{x},\mathbf{x}').$$

Again, we have that $S_1 \geq K(K-1)e$. On the other hand, if k_{ij} is the number of letters $j \in J_q = \{0,\ldots,q-1\}$ in the ith position in all codewords $\mathbf{x} \in \mathscr{X}_1$ then

$$S_1 = \sum_{1 \leq i \leq N} \sum_{0 \leq j \leq q-1} k_{ij}(K - k_{ij}).$$

Note that the sum $\sum_{0 \leq j \leq q-1} k_{ij} = K$. Besides, as $w(\mathbf{x}) \leq s$, the number of 0s in every word $\mathbf{x} \in \mathscr{X}_1$ is $\geq N-s$. Then the total number of 0s in all codewords equals $\sum_{1 \leq i \leq N} k_{i0} \geq K(N-s)$. Now write

$$S = NK^2 - \sum_{1 \leq i \leq N} \left(k_{i0}^2 + \sum_{1 \leq j \leq q-1} k_{ij}^2 \right),$$

and use the Cauchy–Schwarz inequality to estimate

$$\sum_{1 \le j \le q-1} k_{ij}^2 \ge \frac{1}{q-1} \left(\sum_{1 \le j \le q-1} k_{ij} \right)^2 = \frac{1}{q-1}(K - k_{i0})^2.$$

Then

$$S \le NK^2 - \sum_{1 \le i \le N} \left[k_{i0}^2 + \frac{1}{q-1}(K - k_{i0})^2 \right]$$

$$= NK^2 - \frac{1}{q-1} \sum_{1 \le i \le N} \left[(q-1)k_{i0}^2 + K^2 - 2Kk_{i0} + k_{i0}^2 \right]$$

$$= NK^2 - \frac{1}{q-1} \sum_{1 \le i \le N} (qk_{i0}^2 + K^2 - 2Kk_{i0})$$

$$= NK^2 - \frac{N}{q-1}K^2 - \frac{q}{q-1} \sum_{1 \le i \le N} k_{i0}^2 + \frac{2}{q-1}K \sum_{1 \le i \le N} k_{i0}$$

$$= \frac{q-2}{q-1}NK^2 - \frac{q}{q-1} \sum_{1 \le i \le N} k_{i0}^2 + \frac{2}{q-1}KL,$$

where $L = \sum_{1 \le i \le N} k_{i0}$. Use Cauchy–Schwarz once again:

$$\sum_{1 \le i \le N} k_{i0}^2 \ge \frac{1}{N} \left(\sum_{1 \le i \le N} k_{i0} \right)^2 = \frac{1}{N}L^2.$$

Then

$$S \le \frac{q-2}{q-1}NK^2 - \frac{q}{q-1}\frac{1}{N}L^2 + \frac{2}{q-1}KL$$

$$= \frac{1}{q-1} \left[(q-2)NK^2 - \frac{q}{N}L^2 + 2KL \right].$$

The maximum of the quadratic expression in the square brackets occurs at $L = NK/q$. Recall that $L \ge K(N - s)$. So, choosing $K(N - s) \ge NK/q$, i.e. $s \le N(q - 1)/q$, we can estimate

$$S \le \frac{1}{q-1} \left[(q-2)NK^2 - \frac{q}{N}K^2(N-s)^2 + 2K^2(N-s) \right]$$

$$= \frac{1}{q-1}K^2 s \left[2(q-1) - \frac{qs}{N} \right].$$

This yields the inequality $K(K - 1)e \le \frac{1}{q-1}K^2 s \left[2(q-1) - \frac{qs}{N} \right]$ which can be solved for K:

$$K \le \frac{\theta N e}{s^2 - 2\theta Ns + \theta Ne},$$

provided that $s < N\theta$ and $s^2 - 2\theta Ns + \theta Ne > 0$. Finally, recall that $\mathscr{X}^{(1)}$ arose from an $[N, M, d]$ code $\mathscr{X}$, with $K \geq Mv(s)/q^N$ and $e \geq d$. As a result, we obtain that

$$\frac{Mv_{N,q}(s)}{q^N} \leq \frac{\theta Nd}{s^2 - 2\theta Ns + \theta Nd}.$$

This leads to the Elias bound (2.2.38). $\qquad\qquad\qquad\qquad\qquad\qquad\qquad\square$

The ideas used in the proof of the Elias bound (and earlier in the proof of the Plotkin bounds) are also helpful in obtaining bounds for $W_2^*(N, d, \ell)$, the maximal size of a binary (non-linear) code $\mathscr{X} \in \mathscr{H}_{N,2}$ of length N, distance $d(\mathscr{X}) \geq d$ and with the property that the weight $w(\mathbf{x}) \equiv \ell$, $\mathbf{x} \in \mathscr{X}$. First, three obvious statements:

(i) $\quad W_2^*(N, 2k, k) = \left\lfloor \dfrac{N}{k} \right\rfloor$,

(ii) $\quad W_2^*(N, 2k, \ell) = W_2^*(N, 2k, N - \ell)$,

(iii) $\quad W_2^*(N, 2k-1, \ell) = W_2^*(N, 2k, \ell), \quad \ell/2 \leq k \leq \ell$.

[The reader is advised to prove these as an exercise.]

Worked Example 2.2.13 *Prove that for all positive integers $N \geq 1$, $k \leq \dfrac{N}{2}$ and*

$$\ell < \frac{N}{2} - \sqrt{\frac{N^2}{4} - kN},$$

$$W_2^*(N, 2k, \ell) \leq \left\lfloor \frac{kN}{\ell^2 - \ell N + kN} \right\rfloor. \qquad\qquad (2.2.39)$$

Solution Take an $[N, M, 2k]$ code $\mathscr{X}$ such that $w(\mathbf{x}) \equiv \ell$, $\mathbf{x} \in \mathscr{X}$. As before, let k_{i1} be the number of 1s in position i in all codewords. Consider the sum of the dot-products $D = \sum\limits_{\mathbf{x},\mathbf{x}' \in \mathscr{X}} \mathbf{1}\left(\mathbf{x} \neq \mathbf{x}'\right)\langle \mathbf{x} \cdot \mathbf{x}' \rangle$. We have

$$\langle \mathbf{x} \cdot \mathbf{x}' \rangle = w(\mathbf{x} \wedge \mathbf{x}') = \frac{1}{2}\left(w(\mathbf{x}) + w(\mathbf{x}') - \delta\left(\mathbf{x}, \mathbf{x}'\right)\right)$$
$$\leq \frac{1}{2}(2\ell - 2k) = \ell - k$$

and hence

$$D \leq (\ell - k)M(M - 1).$$

On the other hand, the contribution to D from position i equals $k_{i1}(k_{i1} - 1)$, i.e.

$$D = \sum_{1 \leq i \leq N} k_{i1}(k_{i1} - 1) = \sum_{1 \leq i \leq N} (k_{i1}^2 - k_{i1}) = \sum_{1 \leq i \leq N} k_{i1}^2 - \ell M.$$

Again, the last sum is minimised at $k_{i1} = \ell M/N$, i.e.

$$\frac{\ell^2 M^2}{N} - \ell M \leq D \leq (\ell - k)M(M-1).$$

This immediately leads to (2.2.39). □

Another useful bound is given now.

Worked Example 2.2.14   *Prove that for all positive integers $N \geq 1$, $k \leq \dfrac{N}{2}$ and $2k \leq \ell \leq 4k$,*

$$W_2^*(N, 2k, \ell) \leq \left\lfloor \frac{N}{\ell} W_2^*(N-1, 2k, \ell-1) \right\rfloor. \qquad (2.2.40)$$

Solution Again take an $[N, M, 2k]$ code $\mathscr{X}$ such that $w(\mathbf{x}) \equiv \ell$ for all $\mathbf{x} \in \mathscr{X}$. Consider the shortening code $\mathscr{X}$ on $x_1 = 1$ (cf. Example 2.1.8(v)): it gives a code of length $(N-1)$, distance $\geq 2k$ and constant weight $(\ell-1)$. Hence, the size of the cross-section is $\leq W_2^*(N-1, 2k, \ell-1)$. Therefore, the number of 1s at position 1 in the codewords of $\mathscr{X}$ does not exceed $W_2^*(N-1, 2k, \ell-1)$. Repeating this argument, we obtain that the total number of 1s in all positions is $\leq NW_2^*(N-1, 2k, \ell-1)$. But this number equals ℓM, i.e. $\ell M \leq NW_2^*(N-1, 2k, \ell-1)$. The bound (2.2.40) then follows. □

Corollary 2.2.15   *For all positive integers $N \geq 1$, $k \leq \dfrac{N}{2}$ and $2k \leq \ell \leq 4k - 2$,*

$$W_2^*(N, 2k-1, \ell) = W_2^*(N, 2k, \ell)$$
$$< \left\lfloor \frac{N}{\ell} \left\lfloor \frac{N-1}{\ell-1} \right\rfloor \cdots \left\lfloor \frac{N-\ell+k}{k} \right\rfloor \cdots \right\rfloor \right\rfloor. \qquad (2.2.41)$$

The remaining part of Section 2.2 focuses on the Johnson bound. This bound aims at improving the binary Hamming bound (cf. (2.1.8b) with $q-2$):

$$M_2^*(N, 2E+1) \leq 2^N/v_N(E) \quad \text{or} \quad v_N(E)M_2^*(N, 2E+1) \leq 2^N. \qquad (2.2.42)$$

Namely, the Johnson bound asserts that

$$M_2^*(N, 2E+1) \leq 2^N/v_N^*(E) \quad \text{or} \quad v_N^*(E)M_2^*(N, 2E+1) \leq 2^N, \qquad (2.2.43)$$

where

$$v_N^*(E) = v_N(E) + \frac{1}{\lceil N/(E+1) \rceil} \left[\binom{N}{E+1} \right.$$
$$\left. - W_2^*(N, 2E+1, 2E+1) \binom{2E+1}{E} \right]. \qquad (2.2.44)$$

Recall that $v_N(E) = \sum_{0 \leq s \leq E} \binom{N}{s}$ stands for the volume of the binary Hamming ball of radius E. We begin our derivation of bound (2.2.43) with the following result.

Lemma 2.2.16 *If $\mathbf{x}, \mathbf{y}$ are binary words, with $\delta(\mathbf{x}, \mathbf{y}) = 2\ell + 1$, then there exists $\binom{2\ell+1}{\ell}$ binary words $\mathbf{z}$ such that $\delta(\mathbf{x}, \mathbf{z}) = \ell + 1$ and $\delta(\mathbf{y}, \mathbf{z}) = \ell$.*

Proof Left as an exercise. □

Consider the set $\mathscr{T}(= \mathscr{T}_{N,E+1})$ of all binary N-words at distance exactly $E+1$ from the codewords from $\mathscr{X}$:

$$\mathscr{T} = \Big\{ \mathbf{z} \in \mathscr{H}_N : \delta(\mathbf{z}, \mathbf{x}) = E+1 \text{ for some } \mathbf{x} \in \mathscr{X}$$

$$\text{and } \delta(\mathbf{z}, \mathbf{y}) \geq E+1 \text{ for all } \mathbf{y} \in \mathscr{X} \Big\}. \tag{2.2.45}$$

Then we can write that

$$M v_N(E) + \sharp \mathscr{T} \leq 2^N, \tag{2.2.46}$$

as none of the words $\mathbf{z} \in \mathscr{T}$ falls in any of the balls of radius E about the codewords $\mathbf{y} \in \mathscr{X}$. The bound (2.2.43) will follow when we solve the next worked example.

Worked Example 2.2.17 *Prove that the cardinality $\sharp \mathscr{T}$ is greater than or equal to the second term from the RHS of (2.2.44):*

$$\frac{M}{\lfloor N/(E+1) \rfloor} \left[\binom{N}{E+1} - W_2^*(N, 2E+1, 2E+1) \binom{2E+1}{E} \right]. \tag{2.2.47}$$

Solution We want to find a lower bound on $\sharp \mathscr{T}$. Consider the set $\mathscr{W}(= \mathscr{W}_{N,E+1})$ of 'matched' pairs of N-words defined by

$$\mathscr{W} = \Big\{ (\mathbf{x}, \mathbf{z}) : \mathbf{x} \in \mathscr{X}, \mathbf{z} \in \mathscr{T}_{E+1}, \delta(\mathbf{x}, \mathbf{z}) = E+1 \Big\}$$

$$= \Big\{ (\mathbf{x}, \mathbf{z}) : \mathbf{x} \in \mathscr{X}, \mathbf{z} \in \mathscr{H}_N : \delta(\mathbf{x}, \mathbf{z}) = E+1, \tag{2.2.48}$$

$$\text{and } \delta(\mathbf{y}, \mathbf{z}) \geq E+1 \quad \text{for all } \mathbf{y} \in \mathscr{X} \Big\}.$$

Given $\mathbf{x} \in \mathscr{X}$, the $\mathbf{x}$-section $\mathscr{W}^{\mathbf{x}}$ is defined as

$$\mathscr{W}^{\mathbf{x}} = \{ \mathbf{z} \in \mathscr{H}_N : (\mathbf{x}, \mathbf{z}) \in \mathscr{W} \}$$
$$= \{ \mathbf{z} : \delta(\mathbf{x}, \mathbf{z}) = E+1, \delta(\mathbf{y}, \mathbf{z}) \geq E+1 \quad \text{for all } \mathbf{y} \in \mathscr{X} \}. \tag{2.2.49}$$

Observe that if $\delta(\mathbf{x}, \mathbf{z}) = E+1$ then $\delta(\mathbf{y}, \mathbf{z}) \geq E$ for all $\mathbf{y} \in \mathscr{X} \setminus \{\mathbf{x}\}$, as otherwise $\delta(\mathbf{x}, \mathbf{y}) < 2E+1$. Hence:

$$\mathscr{W}^{\mathbf{x}} = \{ \mathbf{z} : \delta(\mathbf{x}, \mathbf{z}) = E+1, \delta(\mathbf{y}, \mathbf{z}) \neq E \quad \text{for all } \mathbf{y} \in \mathscr{X} \}. \tag{2.2.50}$$

We see that, to evaluate $\sharp \mathcal{W}^{\mathbf{x}}$, we must detract, from the number of binary N-words lying at distance $E+1$ from $\mathbf{x}$, i.e. from $\binom{N}{E+1}$, the number of those lying also at distance E from some other codeword $\mathbf{y} \in \mathcal{X}$. But if $\delta(\mathbf{x},\mathbf{z}) = E+1$ and $\delta(\mathbf{y},\mathbf{z}) = E$ then $\delta(\mathbf{x},\mathbf{y}) = 2E+1$. Also, no two distinct codewords can have distance E from a single N-word $\mathbf{z}$. Hence, by the previous remark,

$$\sharp \mathcal{W}^{\mathbf{x}} = \binom{N}{E+1} - \binom{2E+1}{E} \times \sharp \{\mathbf{y} \in \mathcal{X} : \delta(\mathbf{x},\mathbf{y}) = 2E+1\}.$$

Moreover, if we subtract $\mathbf{x}$ from every $\mathbf{y} \in \mathcal{X}$ with $\delta(\mathbf{x},\mathbf{y}) = 2E+1$, the result is a code of length N whose codewords $\mathbf{z}$ have weight $w(\mathbf{z}) \equiv 2E+1$. Hence, there are at most $W^*(N, 2E+1, 2E+1)$ codewords $\mathbf{y} \in \mathcal{X}$ with $\delta(\mathbf{x},\mathbf{y}) = 2E+1$. Consequently,

$$\sharp \mathcal{W}^{\mathbf{x}} \geq \binom{N}{E+1} - W^*(N, 2E+1, 2E+1)\binom{2E+1}{E} \tag{2.2.51}$$

and

$$\sharp \mathcal{W} \geq M \times \text{ the RHS of (2.2.51).} \tag{2.2.52}$$

Now fix $\mathbf{v} \in \mathcal{T}$ and consider the $\mathbf{v}$-section

$$\mathcal{W}^{\mathbf{v}} = \{\mathbf{y} \in \mathcal{X} : (\mathbf{y},\mathbf{v}) \in \mathcal{W}\} = \{\mathbf{y} \in \mathcal{X} : \delta(\mathbf{y},\mathbf{v}) = E+1\}. \tag{2.2.53}$$

If $\mathbf{y},\mathbf{z} \in \mathcal{W}^{\mathbf{v}}$ then $\delta(\mathbf{y},\mathbf{u}) = \delta(\mathbf{z},\mathbf{u}) = E+1$. Thus,

$$w(\mathbf{y} - \mathbf{u}) = w(\mathbf{z} - \mathbf{u}) = E+1$$

and

$$\begin{aligned}
2E+1 \leq \delta(\mathbf{y},\mathbf{z}) &= \delta(\mathbf{y}-\mathbf{v}, \mathbf{z}-\mathbf{v}) \\
&= w(\mathbf{y}-\mathbf{v}) + w(\mathbf{z}-\mathbf{v}) - 2w((\mathbf{y}-\mathbf{v}) \wedge (\mathbf{z}-\mathbf{v})) \\
&= 2E+2 - 2w((\mathbf{y}-\mathbf{v}) \wedge (\mathbf{z}-\mathbf{v})).
\end{aligned}$$

This implies that

$$w((\mathbf{y}-\mathbf{v}) \wedge (\mathbf{z}-\mathbf{v})) = 0 \text{ and } \delta(\mathbf{y},\mathbf{z}) = 2E+2.$$

So, $\mathbf{y} - \mathbf{v}$ and $\mathbf{z} - \mathbf{v}$ have no digit 1 in common. Hence, there exist at most $\left\lceil \dfrac{N}{E+1} \right\rceil$ words of the form $\mathbf{y} - \mathbf{v}$ where $\mathbf{y} \in \mathcal{W}^{\mathbf{v}}$, i.e. at most $\left\lceil \dfrac{N}{E+1} \right\rceil$ words in $\mathcal{W}^{\mathbf{v}}$. Therefore,

$$\sharp \mathcal{W} \leq \left\lceil \frac{N}{E+1} \right\rceil \sharp \mathcal{T}. \tag{2.2.54}$$

Collecting (2.2.51), (2.2.52) and (2.2.54) yields inequality (2.2.47). $\qquad\square$

Corollary 2.2.18   *In view of Corollary 2.2.15 the following bound holds true:*

$$M^*(N, 2E+1) \le 2^N \left[v_N(E) \right.$$

$$\left. - \frac{1}{\lfloor N/(E+1) \rfloor} \binom{N}{E} \left(\frac{N-E}{E+1} - \left\lfloor \frac{N-E}{E+1} \right\rfloor \right) \right]^{-1}. \qquad (2.2.55)$$

Example 2.2.19   Let $N = 13$ and $E = 2$, i.e. $d = 5$. Inequality (2.2.41) implies

$$W^*(13, 5, 5) \le \left\lfloor \frac{13}{5} \left\lfloor \frac{12}{4} \left\lfloor \frac{11}{3} \right\rfloor \right\rfloor \right\rfloor = 23, \text{ and the Johnson bound in (2.2.43) yields}$$

$$M^*(13, 5) \le \left\lfloor \frac{2^{13}}{1 + 13 + 78 + (286 - 10 \times 23)/4} \right\rfloor = 77.$$

This bound is much better than Hamming's which gives $M^*(13, 5) \le 89$. In fact, it is known that $M^*(13, 5) = 64$. Compare Section 3.4.

2.3 Linear codes: basic constructions

In this section we explore further the class of linear codes. To start with, we consider binary codes, with digits 0 and 1. Accordingly, $\mathcal{H}_N$ will denote the binary Hamming space of length N; words $\mathbf{x}^{(N)} = x_1 \dots x_N$ from $\mathcal{H}_N$ will be also called (row) vectors. All operations over binary digits are performed in the binary arithmetic (that is, mod 2). When it does not lead to a confusion, we will omit subscripts N and superscripts (N). Let us repeat the definition of a linear code (cf. Definition 2.1.5).

Definition 2.3.1   A binary code $\mathcal{X} \subseteq \mathcal{H}_N$ is called *linear* if, together with a pair of vectors, $\mathbf{x} = x_1 \dots x_N$ and $\mathbf{x}' = x_1' \dots x_N'$, code $\mathcal{X}$ contains the sum $\mathbf{x} + \mathbf{x}'$, with digits $x_i + x_i'$. In other words, a linear code is a *linear subspace* in $\mathcal{H}_N$, over field $\mathbb{F}_2 = \{0, 1\}$. Consequently, a linear code always contains a zero row-vector $\mathbf{0} = 0 \dots 0$. A *basis* of a linear code $\mathcal{X}$ is a maximal linearly independent set of words from $\mathcal{X}$; the linear code is *generated* by its basis in the sense that every vector $\mathbf{x} \in \mathcal{X}$ is (uniquely) represented as a sum of (some) vectors from the basis. All bases of a given linear code $\mathcal{X}$ contain the same number of vectors; the number of vectors in the basis is called the *dimension* or the *rank* of $\mathcal{X}$. A linear code of length N and rank k is also called an $[N, k]$ code, or an $[N, k, d]$ code if its distance is d.

Practically all codes used in modern practice are linear. They are popular because they are easy to work with. For example, to identify a linear code it is enough to fix its basis, which yields a substantial economy as the subsequent material shows.

Lemma 2.3.2   *Any binary linear code of rank k contains 2^k vectors, i.e. has size $M = 2^k$.*

Proof A basis of the code contains k linearly independent vectors. The code is generated by the basis; hence it consists of the sums of basic vectors. There are precisely 2^k sums (the number of subsets of $\{1,\dots,k\}$ indicating the summands), and they all give different vectors. □

Consequently, a binary linear code $\mathscr{X}$ of rank k may be used for encoding *all* possible source strings of length k; the information rate of a binary linear $[N,k]$ code is k/N. Thus, indicating $k \le N$ linearly independent words $\mathbf{x} \in \mathscr{H}_N$ identifies a (unique) linear code $\mathscr{X} \subset \mathscr{H}_N$ of rank k. In other words, a linear binary code of rank k is characterised by a $k \times N$ matrix of 0s and 1s with linearly independent rows:

$$
G = \begin{pmatrix}
g_{11} & \cdots & \cdots & \cdots & g_{1N} \\
g_{21} & \cdots & \cdots & \cdots & g_{2N} \\
& \vdots & & & \vdots \\
g_{k1} & \cdots & \cdots & \cdots & g_{kN}
\end{pmatrix}
$$

Namely, we take the rows $\mathbf{g}(i) = g_{i1}\dots g_{iN}$, $1 \le i \le k$, as the basic vectors of a linear code.

Definition 2.3.3 A matrix G is called a *generating matrix* of a linear code. It is clear that the generating matrix is not unique.

Equivalently, a linear $[N,k]$ code $\mathscr{X}$ may be described as the kernel of a certain $(N-k) \times N$ matrix H, again with the entries 0 and 1: $\mathscr{X} = \ker H$ where

$$
H = \begin{pmatrix}
h_{11} & h_{12} & \cdots & \cdots & h_{1N} \\
h_{21} & h_{22} & \cdots & \cdots & h_{2N} \\
\vdots & \vdots & \ddots & \ddots & \vdots \\
h_{(N-k)1} & h_{(N-k)2} & \cdots & \cdots & h_{(N-k)N}
\end{pmatrix}
$$

and

$$
\ker H = \left\{ \mathbf{x} = x_1 \dots x_N : \mathbf{x} H^{\mathrm{T}} = \mathbf{0}^{(N-k)} \right\}. \tag{2.3.1}
$$

It is plain that the rows $\mathbf{h}(j)$, $1 \le j \le N-k$, of matrix H are vectors *orthogonal* to $\mathscr{X}$, in the sense of the inner *dot-product*:

$$
\langle \mathbf{x} \cdot \mathbf{h}(j) \rangle = 0, \quad \text{for all } \mathbf{x} \in \mathscr{X} \text{ and } 1 \le j \le N-k.
$$

Here, for $\mathbf{x}, \mathbf{y} \in \mathscr{H}_N$,

$$
\langle \mathbf{x} \cdot \mathbf{y} \rangle = \langle \mathbf{y} \cdot \mathbf{x} \rangle = \sum_{i=1}^{N} x_i y_i, \quad \text{where } \mathbf{x} = x_1 \dots x_N, \ \mathbf{y} = y_1 \dots y_N; \tag{2.3.2}
$$

cf. Example 2.1.8(ix).

The inner product (2.3.2) possesses all properties of the Euclidean scalar product in $\mathbb{R}^N$, but one: it is not *positive definite* (and therefore does not define a norm). That is, there are non-zero vectors $\mathbf{x} \in \mathcal{H}_N$ with $\langle \mathbf{x} \cdot \mathbf{x} \rangle = 0$. Luckily, we do not need the positive definiteness.

However, the key *rank–nullity* property holds true for the dot-product: if $\mathscr{L}$ is a linear subspace in $\mathcal{H}_N$ of rank k then its orthogonal complement $\mathscr{L}^\perp$ (i.e. the collection of vectors $\mathbf{z} \in \mathcal{H}_N$ such that $\langle \mathbf{x} \cdot \mathbf{z} \rangle = 0$ for all $\mathbf{x} \in \mathscr{L}$) is a linear subspace of rank $N - k$. Thus, the $(N - k)$ rows of H can be considered as a basis in $\mathscr{X}^\perp$, the orthogonal complement to $\mathscr{X}$.

The matrix H (or sometimes its transpose H^{T}) with the property $\mathscr{X} = \ker \mathscr{H}$ or $\langle \mathbf{x} \cdot \mathbf{h}(j) \rangle \equiv 0$ (cf. (2.3.1)) is called a *parity-check* (or, simply, check) matrix of code $\mathscr{X}$. In many cases, the description of a code by a check matrix is more convenient than by a generating one.

The parity-check matrix is again not unique as the basis in $\mathscr{X}^\perp$ can be chosen non-uniquely. In addition, in some situations where a family of codes is considered, of varying length N, it is more natural to identify a check matrix where the number of rows can be greater than $N - k$ (but some of these rows will be linearly dependent); such examples appear in Chapter 3. However, for the time being we will think of H as an $(N - k) \times N$ matrix with linearly independent rows.

Worked Example 2.3.4 Let $\mathscr{X}$ be a binary linear $[N, k, d]$ code of information rate $\rho = k/N$. Let G and H be, respectively, the generating and parity-check matrices of $\mathscr{X}$. In this example we refer to constructions introduced in Example 2.1.8.

(a) The parity-check extension of $\mathscr{X}$ is a binary code $\mathscr{X}^+$ of length $N + 1$ obtained by adding, to each codeword $\mathbf{x} \in \mathscr{X}$, the symbol $x_{N+1} = \sum_{1 \le i \le N} x_i$ so that the sum $\sum_{1 \le i \le N+1} x_i$ is zero. Prove that $\mathscr{X}^+$ is a linear code and find its rank and minimal distance. How are the information rates and generating and parity-check matrices of $\mathscr{X}$ and $\mathscr{X}^+$ related?

(b) The truncation $\mathscr{X}^-$ of $\mathscr{X}$ is defined as a linear code of length $N - 1$ obtained by omitting the last symbol of each codeword $\mathbf{x} \in \mathscr{X}$. Suppose that code $\mathscr{X}$ has distance $d \ge 2$. Prove that $\mathscr{X}^-$ is linear and find its rank and generating and parity-check matrices. Show that the minimal distance of $\mathscr{X}^-$ is at least $d - 1$.

(c) The m-repetition of $\mathscr{X}$ is a code $\mathscr{X}^{\mathrm{re}}(m)$ of length Nm obtained by repeating each codeword $\mathbf{x} \in \mathscr{X}$ a total of m times. Prove that $\mathscr{X}^{\mathrm{re}}(m)$ is a linear code and find its rank and minimal distance. How are the information rates and generating and parity-check matrices of $\mathscr{X}^{\mathrm{re}}(m)$ related to ρ, G and H?

Solution (a) The generating and parity-check matrices are

$$
G^+ = \begin{pmatrix} & & | & \sum_{1 \le i \le N} g_{1i} \\ & G & | & \vdots \\ & & | & \vdots \\ & & | & \sum_{1 \le i \le N} g_{ki} \end{pmatrix}, H^+ = \begin{pmatrix} & & & | & 1 \\ & & & | & 1 \\ & H & & | & \cdot \\ & & & | & \cdot \\ & & & | & 1 \\ - & - & - & | & -- \\ 0 & \dots & 0 & | & 1 \end{pmatrix}.
$$

The rank of $\mathcal{X}^+$ equals the rank of $\mathcal{X} = k$. If the minimal distance of $\mathcal{X}$ was even it is not changed; if odd it increases by 1. The information rate $\rho^+ = (N-1)\rho/N$.

(b) The generating matrix

$$
G^- = \begin{pmatrix} g_{11} & \cdots & g_{1N-1} \\ & \vdots & \\ & \vdots & \\ & \vdots & \\ g_{k1} & \cdots & g_{kN-1} \end{pmatrix}.
$$

The parity-check matrix H of $\mathcal{X}$, after suitable column operations, may be written as

$$
H = \begin{pmatrix} & & | & \cdot \\ & & | & \cdot \\ & H^- & | & \cdot \\ & & | & \cdot \\ & & | & \cdot \\ - & - & - & - & | & -- \\ 0 & \dots & 0 & | & \star \end{pmatrix}.
$$

The parity-check matrix of $\mathcal{X}^-$ is then identified with H^-. The rank is unchanged; the distance may decrease maximum by 1. The information rate $\rho^- = N\rho/(N-1)$.

(c) The generating and parity-check matrices are

$$
G^{\mathrm{re}}(m) = (G \ \dots \ G) \ (m \text{ times}),
$$

and

$$
H^{\mathrm{re}}(m) = \begin{pmatrix} H & 0 & 0 & \dots & 0 \\ I & I & 0 & \dots & 0 \\ I & 0 & I & \dots & 0 \\ \vdots & \vdots & \vdots & \ddots & \vdots \\ I & 0 & 0 & \dots & I \end{pmatrix}.
$$

Here, **I** is a unit $N \times N$ matrix and the zeros mean the zero matrices (of size $(N - k) \times N$ and $N \times N$, accordingly). The number of the unit matrices in the first column equals $m - 1$. (This is not a unique form of $H^{\mathrm{re}}(m)$.) The size of $H^{\mathrm{re}}(m)$ is $(Nm - k) \times Nm$.

The rank is unchanged, the minimal distance in $\mathscr{X}^{\mathrm{re}}(m)$ is md and the information rate ρ/m.   □

Worked Example 2.3.5   *A dual code of a linear binary $[N,k]$ code $\mathscr{X}$ is defined as the set $\mathscr{X}^{\perp}$ of the words $\mathbf{y} = y_1 \dots y_N$ such that the dot-product*

$$\langle \mathbf{y} \cdot \mathbf{x} \rangle = \sum_{1 \leq i \leq N} y_i \cdot x_i = 0 \quad \text{for every } \mathbf{x} = x_1 \dots x_N \text{ from } \mathscr{X}.$$

Compare Example 2.1.8(ix). Prove that an $(N-k) \times N$ matrix H is a parity-check matrix of code $\mathscr{X}$ iff H is a generating matrix for the dual code. Hence, derive that G and H are generating and parity-check matrices, respectively, for a linear code iff:

 (i) *the rows of G are linearly independent;*
 (ii) *the columns of H are linearly independent;*
 (iii) *the number of rows of G plus the number of rows of H equals the number of columns of G which equals the number of columns of H;*
 (iv) *$GH^{\mathrm{T}} = 0$.*

Solution The rows $\mathbf{h}(j)$, $j = 1,\dots,N-k$, of the matrix H obey $\langle \mathbf{x} \cdot \mathbf{h}(j) \rangle \equiv 0$, $\mathbf{x} \in \mathscr{X}$. Furthermore, if a vector $\mathbf{y}$ obeys $\langle \mathbf{x} \cdot \mathbf{y} \rangle \equiv 0$, $\mathbf{x} \in \mathscr{X}$, then $\mathbf{y}$ is a linear combination of the $\mathbf{y}(j)$. Hence, H is a generating matrix of $\mathscr{X}^{\perp}$. On the other hand, any generating matrix of $\mathscr{X}^{\perp}$ is a parity-check matrix for $\mathscr{X}$.

Therefore, for any pair G, H representing generating and parity-check matrices of a linear code, (i), (ii) and (iv) hold by definition, and (iii) comes from the rank–nullity formula

$$N = \dim(\mathrm{Row} - \mathrm{Range}\ G) + \dim(\mathrm{Row} - \mathrm{Range}\ H)$$

that follows from (iv) and the maximality of G and H.

On the other hand, any pair G, H of matrices obeying (i)–(iv) possesses the maximality property (by virtue of (i)–(iii)) and the orthogonality property (iv). Thus, they are generating and parity-check matrices for $\mathscr{X} = \mathrm{Row} - \mathrm{Range}\ G$.   □

Worked Example 2.3.6   *What is the number of codewords in a linear binary $[N,k]$ code? What is the number of different bases in it? Calculate the last number for $k = 4$. List all bases for $k = 2$ and $k = 3$.*

Show that the subset of a linear binary code consisting of all words of even weight is a linear code. Prove that, for d even, if there exists a linear $[N,k,d]$ code then there exists a linear $[N,k,d]$ code with codewords of even weight.

Solution The size is 2^k and the number of different bases $\dfrac{1}{k!}\displaystyle\prod_{i=0}^{k-1}\left(2^k-2^i\right)$. Indeed, if the l first basis vectors are selected, all their 2^l linear combinations should be excluded on the next step. This gives 840 for $k=4$, and 28 for $k=3$.

Finally, for d even, we can truncate the original code and then use the parity-check extension. □

Example 2.3.7 The binary Hamming $[7,4]$ code is determined by a 3×7 parity-check matrix. The columns of the check matrix are all non-zero words of length 3. Using lexicographical order of these words we obtain

$$H_{\text{lex}}^{\text{Ham}}=\begin{pmatrix}1&0&1&0&1&0&1\\0&1&1&0&0&1&1\\0&0&0&1&1&1&1\end{pmatrix}.$$

The corresponding generating matrix may be written as

$$G_{\text{lex}}^{\text{Ham}}=\begin{pmatrix}0&0&1&1&0&0&1\\0&1&0&0&1&0&1\\0&0&1&0&1&1&0\\1&1&1&0&0&0&0\end{pmatrix}. \tag{2.3.3}$$

In many cases it is convenient to write the check matrix of a linear $[N,k]$ code in a *canonical* (or *standard*) form:

$$H_{\text{can}}=\left(\,I_{N-k}\ H'\,\right). \tag{2.3.4a}$$

In the case of the Hamming $[7,4]$ code it gives

$$H_{\text{can}}^{\text{Ham}}=\begin{pmatrix}1&0&0&1&1&0&1\\0&1&0&1&0&1&1\\0&0&1&0&1&1&1\end{pmatrix},$$

with a generating matrix also in a canonical form:

$$G_{\text{can}}=\left(\,G'\ I_k\,\right); \tag{2.3.4b}$$

namely,

$$G_{\text{can}}^{\text{Ham}}=\begin{pmatrix}1&1&0&1&0&0&0\\1&0&1&0&1&0&0\\1&1&1&0&0&1&0\\1&1&1&0&0&0&1\end{pmatrix}.$$

Formally, G_{lex} and G_{can} determine *different* codes. However, these codes are *equivalent*:

Definition 2.3.8   Two codes are called *equivalent* if they differ only in permutation of digits. For linear codes, equivalence means that their generating matrices can be transformed into each other by permutation of columns and by row-operations including addition of columns multiplied by scalars. It is plain that equivalent codes have the same parameters (length, rank, distance).

In what follows, unless otherwise stated, we do *not* distinguish between equivalent linear codes.

Remark 2.3.9   An advantage of writing G in a canonical form is that a source string $\mathbf{u}^{(k)} \in \mathscr{H}_k$ is encoded as an N-vector $\mathbf{u}^{(k)} G^{\text{can}}$; according to (2.3.4b), the last k digits in $\mathbf{u}^{(k)} G^{\text{can}}$ form word $\mathbf{u}^{(k)}$ (they are called information digits), whereas the first $N-k$ are used for the parity-check (and called parity-check digits). Pictorially, the parity-check digits carry the redundancy that allows the decoder to detect and correct errors.

> *Like following life thro' creatures you dissect*
> *You lose it at the moment you detect.*
> Alexander Pope (1668–1744), English poet

Definition 2.3.10   The *weight* $w(\mathbf{x})$ of a binary word $\mathbf{x} = x_1 \ldots x_N$ is the number of the non-zero digits in $\mathbf{x}$:

$$w(\mathbf{x}) = \sharp \{i : 1 \leq i \leq N, \; x_i \neq 0\}. \tag{2.3.5}$$

Theorem 2.3.11

(i) *The distance of a linear binary code equals the minimal weight of its non-zero codewords.*

(ii) *The distance of a linear binary code equals the minimal number of linearly dependent columns in the check matrix.*

Proof   (i) As the code $\mathscr{X}$ is linear, the sum $\mathbf{x} + \mathbf{y} \in \mathscr{X}$ for each pair of codewords $\mathbf{x}, \mathbf{y} \in \mathscr{X}$. Owing to the shift invariance of the Hamming distance (see Lemma 2.1.1), $\delta(\mathbf{x}, \mathbf{y}) = \delta(\mathbf{0}, \mathbf{x} + \mathbf{y}) = w(\mathbf{x} + \mathbf{y})$ for any pair of codewords. Hence, the minimal distance of $\mathscr{X}$ equals the minimal distance between $\mathbf{0}$ and the rest of the code, i.e. the minimal weight of a non-zero codeword from $\mathscr{X}$.

(ii) Let $\mathscr{X}$ be a linear code with a parity-check matrix H and minimal distance d. Then there exists a codeword $\mathbf{x} \in \mathscr{X}$ with exactly d non-zero digits. Since $\mathbf{x} H^{\mathsf{T}} = 0$, we conclude that there are d columns of H which are linearly dependent (they correspond to non-zero digits in $\mathbf{x}$). On the other hand, if there exist $(d-1)$ columns of H which are linearly dependent then their sum is zero. But that means that there exists a word $\mathbf{y}$, of weight $w(\mathbf{y}) = d-1$, such that $\mathbf{y}H^{\mathsf{T}} = 0$. Then $\mathbf{y}$ must belong to $\mathcal{X}$ which is impossible, since $\min[w(\mathbf{x}) : \mathbf{x} \in \mathcal{X}, \mathbf{x} \neq \mathbf{0}] = d$. $\qquad\square$

Theorem 2.3.12 *The Hamming* $[7,4]$ *code has minimal distance 3, i.e. it detects 2 errors and corrects 1. Moreover, it is a perfect code correcting a single error.*

Proof For any pair of columns the parity-check matrix H^{lex} contains their sum to obtain a linearly dependent triple (viz. look at columns 1, 6, 7). No two columns are linearly dependent because they are distinct ($\mathbf{x}+\mathbf{y} = \mathbf{0}$ means that $\mathbf{x} = \mathbf{y}$). Also, the volume $v_7(1)$ equals $1+7 = 2^3$, and the code is perfect as its size is 2^4 and $2^4 \times 2^3 = 2^7$. $\qquad\square$

The construction of the Hamming $[7,4]$ code admits a straightforward generalisation to any length $N = 2^l - 1$; namely, consider a $(2^l - 1) \times l$ matrix H^{Ham} with columns representing all possible non-zero binary vectors of length l:

$$
H^{\text{Ham}} = \begin{pmatrix} 1 & 0 & \cdots & 0 & 1 & \cdots & 1 \\ 0 & 1 & \cdots & 0 & 1 & \cdots & 1 \\ 0 & 0 & \cdots & 0 & 0 & \cdots & 1 \\ \vdots & \vdots & \ddots & \vdots & \vdots & \ddots & 1 \\ 0 & 0 & \cdots & 1 & 0 & \cdots & 1 \end{pmatrix}. \tag{2.3.6}
$$

The rows of H^{Ham} are linearly independent, and hence H^{Ham} may be considered as a check matrix of a linear code of length $N = 2^l - 1$ and rank $N - l = 2^l - 1 - l$. Any two columns of H^{Ham} are linearly independent but there exist linearly dependent triples of columns, e.g. $\mathbf{x}, \mathbf{y}$ and $\mathbf{x}+\mathbf{y}$. Hence, the code $\mathcal{X}^{\text{Ham}}$ with the check matrix H^{Ham} has a minimal distance 3, i.e. it detects 2 errors and corrects 1.

This code is called the Hamming $[2^l - 1, 2^l - 1 - l]$ code. It is a perfect one-error correcting code: the volume of the 1-ball $v_{2^l-1}(1)$ equals $1 + 2^l - 1 = 2^l$, and size $\times$ volume $= 2^{2^l-1-l} \times 2^l = 2^{2^l-1} = 2^N$. The information rate is $\dfrac{2^l - l - 1}{2^l - 1} \to 1$ as $l \to \infty$. This proves

Theorem 2.3.13 *The above construction defines a family of* $[2^l - 1, 2^l - 1 - l, 3]$ *linear binary codes* $\mathcal{X}^{\text{Ham}}_{2^l-1}$, $l = 1, 2, \ldots$, *which are perfect one-error correcting codes.*

Example 2.3.14   Suppose that the probability of error in any digit is $p \ll 1$, independently of what occurred to other digits. Then the probability of an error in transmitting a non-encoded $(4N)$-digit message is

$$1 - (1-p)^{4N} \simeq 4Np.$$

But if we use the $[7,4]$ code, we need to transmit $7N$ digits. An erroneous transmission requires at least two wrong digits, which occurs with probability

$$\approx 1 - \left(1 - \binom{7}{2}p^2\right)^N \simeq 21Np^2 \ll 4Np.$$

We see that the extra effort of using 3 check digits in the Hamming code is justified.

A standard decoding procedure for linear codes is based on the concepts of coset and syndrome. Recall that the ML rule decodes a vector $\mathbf{y} = y_1 \dots y_N$ by the closest codeword $\mathbf{x} \in \mathcal{X}$.

Definition 2.3.15   Let $\mathcal{X}$ be a binary linear code of length N and $\mathbf{w} = w_1 \dots w_N$ be a word from $\mathcal{H}_N$. A *coset* of $\mathcal{X}$ determined by $\mathbf{y}$ is the collection of binary vectors of the form $\mathbf{w} + \mathbf{x}$ where $\mathbf{x} \in \mathcal{X}$. We denote it by $\mathbf{w} + \mathcal{X}$.

An easy (and useful) exercise in linear algebra and counting is

Example 2.3.16   Let $\mathcal{X}$ be a linear code and $\mathbf{w}, \mathbf{v}$ be words of length N. Then:

(1) If $\mathbf{w}$ is in the coset $\mathbf{v} + \mathcal{X}$, then $\mathbf{v}$ is in the coset $\mathbf{w} + \mathcal{X}$; in other words, each word in a coset determines this coset.
(2) $\mathbf{w} \in \mathbf{w} + \mathcal{X}$.
(3) $\mathbf{w}$ and $\mathbf{v}$ are in the same coset iff $\mathbf{w} + \mathbf{v} \in \mathcal{X}$.
(4) Every word of length N belongs to one and only one coset. That is, the cosets form a partition of the whole Hamming space $\mathcal{H}_N$.
(5) All cosets contain the same number of words which equals $\sharp \mathcal{X}$. If the rank of $\mathcal{X}$ is k then there are 2^{N-k} different cosets, each containing 2^k words. The code $\mathcal{X}$ is itself a coset of any of the codewords.
(6) The coset determined by $\mathbf{w} + \mathbf{v}$ coincides with the set of elements of the form $\mathbf{x} + \mathbf{y}$, where $\mathbf{x} \in \mathbf{w} + \mathcal{X}, \mathbf{y} \in \mathcal{X} + \mathbf{v}$.

Now the decoding rule for a linear code: you know the code $\mathcal{X}$ beforehand, hence you can calculate all cosets. Upon receiving a word $\mathbf{y}$, you find its coset $\mathbf{y} + \mathcal{X}$ and find a word $\mathbf{w} \in \mathbf{y} + \mathcal{X}$ of least weight. Such a word is called a *leader* of the coset $\mathbf{y} + \mathcal{X}$. A leader may not be unique: in that case you have to make a choice among the list of leaders (list decoding) or refuse to decode and demand a re-transmission. Suppose you have chosen a leader $\mathbf{w}$. You then decode $\mathbf{y}$ by the word

$$\mathbf{x}_* = \mathbf{y} + \mathbf{w}. \tag{2.3.7}$$

Worked Example 2.3.17   Show that word $\mathbf{x}_*$ is always a codeword that min-imises the distance between $\mathbf{y}$ and the words from $\mathcal{X}$.

Solution As $\mathbf{y}$ and $\mathbf{w}$ are in the same coset, $\mathbf{y} + \mathbf{w} \in \mathcal{X}$ (see Example 2.3.16(3)). All other words from $\mathcal{X}$ are obtained as the sums $\mathbf{y} + \mathbf{v}$ where $\mathbf{v}$ runs over coset $\mathbf{y} + \mathcal{X}$. Hence, for any $\mathbf{x} \in \mathcal{X}$,

$$\delta(\mathbf{y},\mathbf{x}) = w(\mathbf{y}+\mathbf{x}) \geq \min_{\mathbf{v} \in \mathbf{y}+\mathcal{X}} w(\mathbf{v}) = w(\mathbf{w}) = d(\mathbf{y},\mathbf{x}_*).$$

□

The parity-check matrix provides a convenient description of the cosets $\mathbf{y} + \mathcal{X}$.

Theorem 2.3.18   Cosets $\mathbf{w} + \mathcal{X}$ are in one-to-one correspondence with vectors of the form $\mathbf{y}H^\mathsf{T}$: two vectors, $\mathbf{y}$ and $\mathbf{y}'$ are in the same coset iff $\mathbf{y}H^\mathsf{T} = \mathbf{y}'H^\mathsf{T}$. In other words, cosets are identified with the rank (or range) space of the parity-check matrix.

Proof The vectors $\mathbf{y}$ and $\mathbf{y}'$ are in the same coset iff $\mathbf{y} + \mathbf{y}' \in \mathcal{X}$, i.e.

$$(\mathbf{y}+\mathbf{y}')H^\mathsf{T} = \mathbf{y}H^\mathsf{T} + \mathbf{y}'H^\mathsf{T} = 0, \ \text{i.e.} \ \mathbf{y}H^\mathsf{T} = \mathbf{y}'H^\mathsf{T}.$$

□

In practice, the decoding rule is implemented as follows. Vectors of the form $\mathbf{y}H^\mathsf{T}$ are called *syndromes*: for a linear (N,k) code there are 2^{N-k} syndromes. They are all listed in the syndrome 'table', and for each syndrome a leader of the corre-sponding coset is calculated. Upon receiving a word $\mathbf{y}$, you calculate the syndrome $\mathbf{y}H^\mathsf{T}$ and find, in the syndrome table, the corresponding leader $\mathbf{w}$. Then follow (2.3.7): decode $\mathbf{y}$ by $\mathbf{x}_* = \mathbf{y} + \mathbf{w}$.

The procedure described is called *syndrome decoding*; although it is relatively simple, one has to write a rather long table of the leaders. Moreover, it is desirable to make the whole procedure of decoding algorithmically independent on a con-crete choice of the code, i.e. of its generating matrix. This goal is achieved in the case of the Hamming codes:

Theorem 2.3.19   For the Hamming code, for each syndrome the leader $\mathbf{w}$ is unique and contains not more than one non-zero digit. More precisely, if the syn-drome $\mathbf{y}\left(H^{\mathrm{Ham}}\right)^\mathsf{T} = \mathbf{s}$ gives column i of the check matrix H^{Ham} then the leader of the corresponding coset has the only non-zero digit i.

Proof The leader minimises the distance between the received word and the code. The Hamming code is perfect 1-error correcting. Hence, every word is either a codeword or within distance 1 of a unique codeword. Hence, the leader is unique and contains at most one non-zero digit. If the syndrome $\mathbf{y}H^{\mathrm{T}} = s$ occupies position i among the columns of the parity-check matrix then, for word $\mathbf{e}_i = 0 \ldots 1\,0 \ldots 0$ with the only non-zero digit i,

$$(\mathbf{y}+\mathbf{e}_i)H^{\mathrm{T}} = s+s = \mathbf{0}.$$

That is, $(\mathbf{y}+\mathbf{e}_i) \in \mathcal{X}$ and $\mathbf{e}_i \in \mathbf{y}+\mathcal{X}$. Obviously, $\mathbf{e}_i$ is the leader.   □

The duals $\mathcal{X}^{\mathrm{Ham}\perp}$ of binary Hamming codes form a particular class, called *simplex codes*. If $\mathcal{X}^{\mathrm{Ham}}$ is $[2^\ell - 1, 2^\ell - 1 - \ell]$, its dual $(\mathcal{X}^{\mathrm{Ham}})^{\perp}$ is $[2^\ell - 1, \ell]$, and the original parity-check matrix H^{Ham} serves as a generating matrix for $\mathcal{X}^{\mathrm{Ham}\perp}$.

Worked Example 2.3.20 *Prove that each non-zero codeword in a binary simplex code $\mathcal{X}^{\mathrm{Ham}\perp}$ has weight $2^{\ell-1}$ and the distance between any two codewords equals $2^{\ell-1}$. Hence justify the term 'simplex'.*

Solution If $\mathcal{X} = \mathcal{X}^{\mathrm{Ham}}$ is the binary Hamming $[2^l - 1, 2^l - l - 1]$ code then the dual $\mathcal{X}^\perp$ is $[2^l - 1, l]$, and its $l \times (2^l - 1)$ generating matrix is H^{Ham}. The weight of any row of H^{Ham} equals 2^{l-1} (and so $d(\mathcal{X}^\perp) = 2^{l-1}$). Indeed, the weight of row j of H^{Ham} equals the number of non-zero vectors of length l with 1 at position j. This gives 2^{l-1} as the weight, as half of all 2^l vectors from $\mathcal{H}_l$ have 1 at any given position.

Consider now a general codeword from $\mathcal{X}^{\mathrm{Ham}\perp}$. It is represented by the sum of rows $j_1,\ldots,j_s$ of H^{Ham} where $s \leq l$ and $1 \leq j_1 < \cdots < j_s \leq l$. This word again has weight 2^{l-1}; this gives the number of non-zero words $\mathbf{v} = v_1\ldots v_l \in \mathcal{H}_{l,2}$ such that the sum $v_{j_1} + \cdots + v_{j_s} = 1$. Moreover, 2^{l-1} gives the weight of half of all vectors in $\mathcal{H}_{l,2}$. Indeed, we require that $v_{j_1} + \cdots + v_{j_s} = 1$, which results in 2^{s-1} possibilities for the s involved digits. Next, we impose no restriction on the remaining $l - s$ digits which gives 2^{l-s} possibilities. Then $2^{s-1} \times 2^{l-s} = 2^{l-1}$, as required. So, $w(\mathbf{x}) = 2^{l-1}$ for all non-zero $\mathbf{x} \in \mathcal{X}^\perp$. Finally, for any $\mathbf{x},\mathbf{x}' \in \mathcal{X}, \mathbf{x} \neq \mathbf{x}'$, the distance $\delta(\mathbf{x},\mathbf{x}') = \delta(\mathbf{0},\mathbf{x}+\mathbf{x}') = w(\mathbf{x}+\mathbf{x}')$ which is always equal to 2^{l-1}. So, the codewords $\mathbf{x} \in \mathcal{X}^\perp$ form a geometric pattern of a 'simplex' with 2^l 'vertices'.   □

Next, we briefly summarise basic facts about linear codes over a finite-field alphabet $\mathbb{F}_q = \{0, 1, \ldots, q-1\}$ of size $q = p^s$. We now switch to the notation $\mathbb{F}_q^{\times N}$ for the Hamming space $\mathcal{H}_{N,q}$.

Definition 2.3.21 A q-ary code $\mathcal{X} \subseteq \mathbb{F}^{\times N}$ is called *linear* if, together with a pair of vectors, $\mathbf{x} = x_1\ldots x_N$ and $\mathbf{x}' = x'_1\ldots x'_N$, $\mathcal{X}$ contains the linear combinations $\gamma\cdot\mathbf{x}+\gamma'\cdot\mathbf{x}'$, with digits $\gamma\cdot x_i + \gamma'\cdot x'_i$, for all coefficients $\gamma, \gamma' \in \mathbb{F}_q$. That is, $\mathcal{X}$ is a *linear subspace* in $\mathbb{F}^{\times N}$. Consequently, as in the binary case, a linear code always contains the vector $\mathbf{0} = 0\ldots 0$. A *basis* of a linear code is again defined as a maximal linearly independent set of its words; the linear code is *generated* by its basis in the sense that every codevector is (uniquely) represented as a linear combination of the basis codevectors. The number of vectors in the basis is called, as before, the *dimension* or the *rank* of the code; because all bases of a given linear code contain the same number of vectors, this object is correctly defined. As in the binary case, the linear code of length N and rank k is referred to as an $[N,k]$ code, or an $[N,k,d]$ code when its distance equals d.

As in the binary case, the minimal distance of a linear code $\mathscr{X}$ equals the minimal non-zero *weight*:

$$d(\mathscr{X}) = \min \left[w(\mathbf{x}) : \mathbf{x} \in \mathscr{X}, \mathbf{x} \neq \mathbf{0} \right],$$

where

$$w(\mathbf{x}) = \sharp \{ j : 1 \leq j \leq N, \, x_j \neq 0 \text{ in } \mathbb{F}_q \},$$
$$\mathbf{x} = x_1 \ldots x_N \in \mathbb{F}_q^{\times N}. \tag{2.3.8}$$

A linear code $\mathscr{X}$ is defined by a *generating* matrix G or a *parity-check* matrix H. The generating matrix of a linear $[N,k]$ code is a $k \times N$ matrix G, with entries from $\mathbb{F}_q$, whose rows $\mathbf{g}(i) = g_{i1} \ldots g_{iN}$, $1 \leq i \leq k$, form a basis of $\mathscr{X}$. A *parity-check matrix* is an $(N-k) \times N$ matrix H, with entries from $\mathbb{F}_q$, whose rows $\mathbf{h}(j) = h_{j1} \ldots h_{jN}$, $1 \leq j \leq N-k$, are linearly independent and dot-orthogonal to $\mathscr{X}$: for all $j = 1, \ldots, N-k$ and codeword $\mathbf{x} = x_1 \ldots x_N$ from $\mathscr{X}$,

$$\langle \mathbf{x} \cdot \mathbf{h}(j) \rangle = \sum_{1 \leq l \leq N} x_l h_{jl} = 0.$$

In other words, all q^k codewords of $\mathscr{X}$ are obtained as linear combinations of rows of G. That is, subspace $\mathscr{X}$ can be viewed as a result of acting on Hamming space $\mathbb{F}_q^{\times k}$ (of length k) by matrix G: symbolically, $\mathscr{X} = \mathbb{F}_q^{\times k} G$. This shows how code $\mathscr{X}$ can be used for encoding q^k 'messages' of length k (and justifies the term information rate for $\rho(\mathscr{X}) = k/N$). On the other hand, $\mathscr{X}$ is determined as the kernel (the null-space) of H^{T}: $\mathscr{X} H^{\mathsf{T}} = 0$. A useful exercise is to check that for the dual code, $\mathscr{X}^{\perp}$, the situation is opposite: H is a generating matrix and G the parity-check. Compare with Worked Example 2.3.5.

Of course, both the generating and parity-check matrices of a given code are not unique, e.g. we can permute rows $\mathbf{g}(j)$ of G or perform row operations, replacing a row by a linear combination of rows in which the original row enters with a non-zero coefficient. Permuting columns of G gives a different but equivalent code, whose basic geometric parameters are identical to those of $\mathscr{X}$.

Lemma 2.3.22 *For any $[N,k]$ code, there exists an equivalent code whose generating matrix G has a 'canonical' form: $G = \left(\, G' \, \mathrm{I}_k \, \right)$ where I_k is the identity $k \times k$ matrix and G' is an $k \times (N-k)$ matrix. Similarly, the parity-check matrix H may have a standard form which is $\left(\, \mathrm{I}_{N-k} \, H' \, \right)$.*

We now discuss the decoding procedure for a general linear code $\mathscr{X}$ of rank k. As was noted before, it may be used for encoding source messages (strings) $\mathbf{u} = u_1 \ldots u_k$ of length k. The source encoding $\mathbf{u} \in \mathbb{F}_q^k \mapsto \mathscr{X}$ becomes particularly simple when the generating and parity-check matrices are used in the canonical (or standard) form.

Theorem 2.3.23 *For any linear code $\mathscr{X}$ there exists an equivalent code $\mathscr{X}'$ with the generating matrix G^{can} and the check matrix H^{can} in standard form (2.3.4a), (2.3.4b) and $G' = -(H')^{\mathrm{T}}$.*

Proof Assume that code $\mathscr{X}$ is non-trivial (i.e. not reduced to the zero word $\mathbf{0}$). Write a basis for $\mathscr{X}$ and take the corresponding generating matrix G. By performing row-operations (where a pair of rows i and j are exchanged or row i is replaced by row i plus row j) we can change the basis, but do *not* change the code. Our matrix G contains a non-zero column, say l_1: perform row operations to make g_{1l_1} the only non-zero entry in this column. By permuting digits (columns), place column l_1 at position $N - k$. Drop row 1 and column $N - k$ (i.e. the old column l_1) and perform a similar procedure with the rest, ending up with the only non-zero entry g_{2l_2} in a column l_2. Place column l_2 at position $N - k + 1$. Continue until an upper triangular $k \times k$ submatrix emerges. Further operations may be reduced to this matrix only. If this matrix is a unit matrix, stop. If not, pick the first column with more than one non-zero entry. Add the corresponding rows from the bottom to 'kill' redundant non-zero entries. Repeat until a unit submatrix emerges. Now a generating matrix is in a standard form, and new code is equivalent to the original one.

To complete the proof, observe that matrices G^{can} and H^{can} figuring in (2.3.4a), (2.3.4b) with $G' = -(H')^{\mathrm{T}}$, have k independent rows and $N - k$ independent columns, correspondingly. Besides, the $k \times (N - k)$ matrix $G^{\mathrm{can}} (H^{\mathrm{can}})^{\mathrm{T}}$ vanishes. In fact,

$$(G^{\mathrm{can}} (H^{\mathrm{can}})^{\mathrm{T}})_{ij} = \langle \text{ row } i \text{ of } G' \cdot \text{column } j \text{ of } (H')^{\mathrm{T}} \rangle = g'_{ij} - g'_{ij} = 0.$$

Hence, H^{can} is a check matrix for G^{can}. □

Returning to source encoding, select the generating matrix in the canonical form G^{can}. Then, given a string $\mathbf{u} = u_1 \ldots u_k$, we set $\mathbf{x} = \sum_{i=1}^{k} u_i \mathbf{g}^{\mathrm{can}}(i)$, where $\mathbf{g}^{\mathrm{can}}(i)$ represents row i of G^{can}. The last k digits in $\mathbf{x}$ give string $\mathbf{u}$; they are called the information digits. The first $N - k$ digits are used to ensure that $\mathbf{x} \in \mathscr{X}$; they are called the parity-check digits.

The standard form is convenient because in the above representation $\mathscr{X} = \mathbb{F}^{\times k} G$, the initial $(N - k)$ string of each codeword is used for encoding (enabling the detection and correction of errors), and the final k string yields the message from $F_q^{\times k}$. As in the binary case, the parity-check matrix H satisfies Theorem 2.3.11. In particular, *the minimal distance of a code equals the minimal number of linearly dependent columns in its parity-check matrix H.*

Definition 2.3.24   Given an $[N,k]$ linear q-ary code $\mathcal{X}$ with parity-check matrix H, the *syndrome* of an N vector $\mathbf{y} \in F_q^{\times N}$ is the k vector $\mathbf{y}H^\mathsf{T} \in F_q^{\times k}$, and the syndrome subspace is the image $F_q^{\times N} H^\mathsf{T}$. A *coset* of $\mathcal{X}$ by vector $\mathbf{w} \in F_q^{\times N}$ is denoted by $\mathbf{w} + \mathcal{X}$ and formed by words of the form $\mathbf{w} + \mathbf{x}$ where $\mathbf{x} \in \mathcal{X}$. All cosets have the same number of elements equal to q^k and partition the whole Hamming space $F_q^{\times N}$ into q^{N-k} disjoint subsets; code $\mathcal{X}$ is one of them. The cosets are in one-to-one correspondence with syndromes $\mathbf{y}H^\mathsf{T}$. The *syndrome decoding* procedure is carried as in the binary case: a received vector $\mathbf{y}$ is decoded by $\mathbf{x}_* = \mathbf{y} + \mathbf{w}$ where $\mathbf{w}$ is the leader of coset $\mathbf{y} + \mathcal{X}$ (i.e. the word from $\mathbf{y} + \mathcal{X}$ with minimum weight).

All drawbacks we had in the case of binary syndrome decoding persist in the general q-ary case, too (and in fact are more pronounced): the coset tables are bulky, the leader of a coset may be not unique. However, for q-ary Hamming codes the syndrome decoding procedure works well, as we will see in Section 2.4.

In the case of linear codes, some of the bounds can be improved (or rather new bounds can be produced).

Worked Example 2.3.25   Let $\mathcal{X}$ be a binary linear $[N,k,d]$ code.

(a) *Fix a codeword $\mathbf{x} \in \mathcal{X}$ with exactly d non-zero digits. Prove that truncating $\mathcal{X}$ on the non-zero digits of $\mathbf{x}$ produces a code $\mathcal{X}'_{N-d}$ of length $N-d$, rank $k-1$ and distance d' for some $d' \geq \lceil d/2 \rceil$.*
(b) *Deduce the Griesmer bound improving the Singleton bound (2.1.12):*

$$N \geq d + \sum_{1 \leq \ell \leq k-1} \left\lceil \frac{d}{2^\ell} \right\rceil. \tag{2.3.9}$$

Solution (a) Without loss of generality, assume that the non-zero digits in $\mathbf{x}$ are $x_1 = \cdots = x_d = 1$. Truncating on digits $1,\ldots, d$ will produce the code $\mathcal{X}'_{N-d}$ with the rank reduced by 1. Indeed, suppose that a linear combination of $k-1$ vectors vanishes on positions $d+1,\ldots,N$. Then on the positions $1,\ldots,d$ all the values equal either 0s or 1s because d is the minimal distance. But the first case is impossible, unless the vectors are linearly dependent. The second case also leads to contradiction by adding the string $\mathbf{x}$ and obtaining k linearly dependent vectors in the code $\mathcal{X}$. Next, suppose that $\mathcal{X}'$ has distance $d' < \left\lceil \dfrac{d}{2} \right\rceil$ and take $\mathbf{y}' \in \mathcal{X}'$ with

$$w(\mathbf{y}') = \sum_{j=d+1}^{N} y'_j = d'.$$

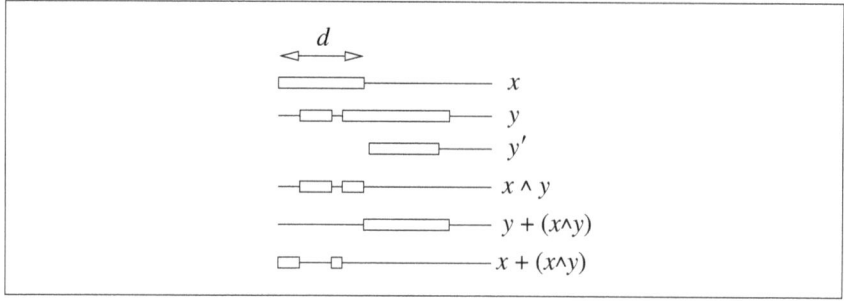

Figure 2.5

Let $\mathbf{y} \in \mathcal{X}$ be an inverse image of $\mathbf{y}'$ under truncation. Referring to (2.1.6b), we write the following property of the binary wedge-product:

$$w(\mathbf{y}) = w(\mathbf{x} \wedge \mathbf{y}) + w(\mathbf{y} + (\mathbf{x} \wedge \mathbf{y})) \geq d.$$

Consequently, we must have that $w(\mathbf{x} \wedge \mathbf{y}) > d - \lceil d/2 \rceil$. See Figure 2.5.
   Then

$$w(\mathbf{x}) = w(\mathbf{x} \wedge \mathbf{y}) + w(\mathbf{x} + (\mathbf{x} \wedge \mathbf{y})) = d$$

implies that $w(\mathbf{x} + (\mathbf{x} \wedge \mathbf{y})) < \lceil d/2 \rceil$. But this is a contradiction, because

$$w(\mathbf{x} + \mathbf{y}) = w(\mathbf{x} + (\mathbf{x} \wedge \mathbf{y})) + w(\mathbf{y} + (\mathbf{x} \wedge \mathbf{y})) < d.$$

We conclude that $d' \geq \lceil d/2 \rceil$.
(b) Iterating the argument in (a) yields

$$N \geq d + d_1 + \cdots + d_{k-1},$$

where $d_l \geq \left\lceil \dfrac{d_{l-1}}{2} \right\rceil$. With $\left\lceil \dfrac{\lceil d/2 \rceil}{2} \right\rceil \geq \left\lceil \dfrac{d}{4} \right\rceil$, we obtain that

$$N \geq d + \sum_{1 \leq \ell \leq k-1} \left\lceil \frac{d}{2^\ell} \right\rceil.$$

$\square$

   Concluding this section, we provide a specification of the GV bound for linear codes.

Theorem 2.3.26   (Gilbert bound) *If $q = p^s$ is a prime power then for all integers N and d such that $2 \leq d \leq N/2$, there exists a q-ary linear $[N, k, d]$ code with minimum distance $\geq d$ provided that*

$$q^k \geq q^N / v_{N,q}(d-1). \tag{2.3.10}$$

Proof Let $\mathcal{X}$ be a linear code of maximal rank with distance at least d of maximal size. If inequality (2.3.10) is violated the union of all Hamming spheres of radius $d-1$ centred on codewords cannot cover the whole Hamming space. So, there must be a point $\mathbf{y}$ that is not in any Hamming sphere around a codeword. Then for any codeword $\mathbf{x}$ and any scalar $b \in \mathbb{F}_q$ the vectors $\mathbf{y}$ and $\mathbf{y} + b \cdot \mathbf{x}$ are in the same coset by $\mathcal{X}$. Also $\mathbf{y} + b \cdot \mathbf{x}$ cannot be in any Hamming sphere of radius $d-1$. The same is true for $\mathbf{x} + b \cdot \mathbf{y}$ because if it were, then $\mathbf{y}$ would be in a Hamming sphere around another codeword. Here we use the fact that $\mathbb{F}_q$ is a field. Then the vector subspace spanned by $\mathcal{X}$ and $\mathbf{y}$ is a linear code larger than $\mathcal{X}$ and with a minimal distance at least d. That is a contradiction, which completes the proof. $\qquad\square$

For example, let $q=2$ and $N=10$. Then $2^5 < v_{10,2}(2) = 56 < 2^6$. Upon taking $d=3$, the Gilbert bound guarantees the existence of a binary $[10,5]$ code with $d \geq 3$.

2.4 The Hamming, Golay and Reed–Muller codes

In this section we systematically study codes with a general finite alphabet $\mathbb{F}_q$ of q elements which is assumed to be a field. Let us repeat that q has to be of the form p^s where p is a prime and s a natural number; the operations of addition $(+)$ and multiplication $(\cdot)$ must also be specified. (As was said above, if $q=p$ is prime, we can think that $\mathbb{F}_q = \{0,1,\ldots,q-1\}$ and addition and multiplication in $\mathbb{F}_q$ are standard, mod q.) See Section 3.1. Correspondingly, the Hamming space $\mathcal{H}_{N,q}$ of length N with digits from $\mathbb{F}_q$ is identified, as before, with the Cartesian power $\mathbb{F}_q^{\times N}$ and inherits the component-wise addition and multiplication by scalars.

Definition 2.4.1 Given positive integers q, $\ell \geq 2$, set $N = \dfrac{q^\ell - 1}{q - 1}$, $k = N - \ell$, and construct the q-ary $[N,k,3]$ *Hamming code* $\mathcal{X}_{N,q}^{\mathrm{Ham}}$ with alphabet $\mathbb{F}_q$ as follows. (a) Pick any non-zero q-ary ℓ-word $\mathbf{h}^{(1)} \in \mathcal{H}_{\ell,q}$. (b) Pick any non-zero q-ary ℓ-word $\mathbf{h}^{(2)} \in \mathcal{H}_{\ell,q}$ that is not a scalar multiple of $\mathbf{h}^{(1)}$. (c) Continue: if $\mathbf{h}^{(1)},\ldots,\mathbf{h}^{(s)}$ is a collection of q-ary ℓ-words selected so far, pick any non-zero vector $\mathbf{h}^{(s+1)} \in \mathcal{H}_{\ell,q}$ which is not a scalar multiple of $\mathbf{h}^{(1)},\ldots,\mathbf{h}^{(s)}$, $1 \leq s \leq N-1$. (d) This process ends up with a selection of N vectors $\mathbf{h}^{(1)},\ldots,\mathbf{h}^{(N)}$; form an $\ell \times N$ matrix H^{Ham} with the columns $\mathbf{h}^{(1)\mathrm{T}},\ldots,\mathbf{h}^{(N)\mathrm{T}}$. Code $\mathcal{X}_{N,q}^{\mathrm{Ham}} \subset \mathbb{F}_q^{\times N}$ is defined by the parity-check matrix H^{Ham}. [In fact, we deal with the whole family of equivalent codes here, modulo choices of words $\mathbf{h}^{(j)}$, $1 \leq j \leq N$.]

For brevity, we will now write $\mathcal{X}^{\mathrm{H}}$ and H^{H} (or even simply H when possible) instead of $\mathcal{X}_{N,q}^{\mathrm{Ham}}$ and H^{Ham}. In the binary case (with $q=2$), matrix H^{H} is composed by all non-zero binary column-vectors of length ℓ. For general q we have to exclude columns that are multiples of each other. To this end, we can choose as columns all non-zero ℓ-words that have 1 in their top-most non-0 component. Such columns are linearly independent, and their total equals $\dfrac{q^\ell - 1}{q - 1}$. Next, as in the binary case, one can arrange words with digits from $\mathbb{F}_q$ in the lexicographic order. By construction, any two columns of H^{H} are linearly independent, but there exist triples of linearly dependent columns. Hence, $d(\mathcal{X}^{\mathrm{H}}) = 3$, and $\mathcal{X}^{\mathrm{H}}$ detects two errors and corrects one. Furthermore, $\mathcal{X}^{\mathrm{H}}$ is a perfect code correcting a single error, as

$$M(1 + (q-1)N) = q^k \left(1 + (q-1)\frac{q^\ell - 1}{q - 1}\right) = q^{k+\ell} = q^N.$$

As in the binary case, the general Hamming codes admit an efficient (and elegant) decoding procedure. Suppose a parity-check matrix $H (= H^{\mathrm{H}})$ has been constructed as above. Upon receiving a word $\mathbf{y} \in \mathbb{F}_q^{\times N}$ we calculate the syndrome $\mathbf{y}H^{\mathrm{T}} \in \mathbb{F}_q^{\times \ell}$. If $\mathbf{y}H^{\mathrm{T}} = 0$ then $\mathbf{y}$ is a codeword. Otherwise, the column-vector $H\mathbf{y}^{\mathrm{T}}$ is a scalar multiple of a column $\mathbf{h}^{(j)}$ of H: $H\mathbf{y}^{\mathrm{T}} = a \cdot \mathbf{h}^{(j)}$, for some $j = 1, \ldots, N$ and $a \in \mathbb{F}_q \setminus \{0\}$. In other words, $\mathbf{y}H^{\mathrm{T}} = a \cdot \mathbf{e}(j)H^{\mathrm{T}}$ where word $\mathbf{e}(j) = 0 \ldots 1 \ldots 0 \in \mathcal{H}_{N,q}$ (with the jth digit 1, all others 0). Then we decode $\mathbf{y}$ by $\mathbf{x}_* = \mathbf{y} - a \cdot \mathbf{e}(j)$, i.e. simply change digit y_j in $\mathbf{y}$ to $y_j - a$.

Summarising, we have the following

Theorem 2.4.2   *The q-ary Hamming codes form a family of*

$$\left[\frac{q^\ell - 1}{q - 1}, \frac{q^\ell - 1}{q - 1} - \ell, 3\right] \text{ perfect codes } \mathcal{X}_N^{\mathrm{H}}, \text{ for } N = \frac{q^\ell - 1}{q - 1}, \ell = 1, 2, \ldots,$$

correcting one error, with a decoding rule that changes the digit y_j to $y_j - a$ in a received word $\mathbf{y} = y_1 \ldots y_N \in \mathbb{F}_q^{\times N}$, where $1 \leq j \leq N$, and $a \in \mathbb{F}_q \setminus \{0\}$ are determined from the condition that $H\mathbf{y}^{\mathrm{T}} = a \cdot \mathbf{h}(j)$, the a-multiple of column j of the parity-check matrix H.

Hamming codes were discovered by R. Hamming and M. Golay in the late 1940s. At that time Hamming, an electrical engineer turned computer scientist during the Jurassic computers era, was working at Los Alamos ("as an intellectual janitor" to local nuclear physicists, in his own words). This discovery shaped the theory of codes for more than two decades: people worked hard to extend properties of Hamming codes to wider classes of codes (with variable success). Most of the topics on codes discussed in this book are related, in one way or another, to Hamming codes. Richard Hamming was not only an outstanding scientist but also an illustrious personality; his writings (and accounts of his life) are entertaining and thought-provoking.

Until the late 1950s, the Hamming codes were a unique family of codes exist-ing in dimensions $N \to \infty$, with 'regular' properties. It was then discovered that these codes have a deep *algebraic* background. The development of the algebraic methods based on these observations is still a dominant theme in modern coding theory.

Another important example is the four *Golay codes* (two binary and two ternary). Marcel Golay (1902–1989) was a Swiss electrical engineer who lived and worked in the USA for a long time. He had an extraordinary ability to 'see' the discrete geometry of the Hamming spaces and 'guess' the construction of various codes without bothering about proofs.

The binary Golay code $\mathscr{X}_{24}^{\mathrm{Gol}}$ is a $[24,12]$ code with the generating matrix $G = (\mathbf{I}_{12}|G')$ where $\mathbf{I}_{12}$ is a 12×12 identity matrix, and $G'\left(= G'_{(2)}\right)$ has the following form:

$$G' = \begin{pmatrix} 0 & 1 & 1 & 1 & 1 & 1 & 1 & 1 & 1 & 1 & 1 & 1 \\ 1 & 1 & 1 & 0 & 1 & 1 & 1 & 0 & 0 & 0 & 1 & 0 \\ 1 & 1 & 0 & 1 & 1 & 1 & 0 & 0 & 0 & 1 & 0 & 1 \\ 1 & 0 & 1 & 1 & 1 & 0 & 0 & 0 & 1 & 0 & 1 & 1 \\ 1 & 1 & 1 & 1 & 0 & 0 & 0 & 1 & 0 & 1 & 1 & 0 \\ 1 & 1 & 1 & 0 & 0 & 0 & 1 & 0 & 1 & 1 & 0 & 1 \\ 1 & 1 & 0 & 0 & 0 & 1 & 0 & 1 & 1 & 0 & 1 & 1 \\ 1 & 0 & 0 & 0 & 1 & 0 & 1 & 1 & 0 & 1 & 1 & 1 \\ 1 & 0 & 0 & 1 & 0 & 1 & 1 & 0 & 1 & 1 & 1 & 0 \\ 1 & 0 & 1 & 0 & 1 & 1 & 0 & 1 & 1 & 1 & 0 & 0 \\ 1 & 1 & 0 & 1 & 1 & 0 & 1 & 1 & 1 & 0 & 0 & 0 \\ 1 & 0 & 1 & 1 & 0 & 1 & 1 & 1 & 0 & 0 & 0 & 1 \end{pmatrix}. \qquad (2.4.1)$$

The rule of forming matrix G' is *ad hoc* (and this is how it was determined by M. Golay in 1949). There will be further *ad hoc* arguments in the analysis of Golay codes.

Remark 2.4.3 Interestingly, there is a systematic way of constructing all code-words of $\mathscr{X}_{24}^{\mathrm{Gol}}$ (or its equivalent) by fitting together two versions of Hamming $[7,4]$ code $\mathscr{X}_7^{\mathrm{H}}$. First, observe that reversing the order of all the digits of a Hamming code $\mathscr{X}_7^{\mathrm{H}}$ yields an equivalent code which we denote by $\mathscr{X}_7^{\mathrm{K}}$. Then add a parity-check to both $\mathscr{X}_7^{\mathrm{H}}$ and $\mathscr{X}_7^{\mathrm{K}}$, producing codes $\mathscr{X}_8^{\mathrm{H},+}$ and $\mathscr{X}_8^{\mathrm{K},+}$. Finally, select two different words $\mathbf{a},\mathbf{b} \in \mathscr{X}_8^{\mathrm{H},+}$ and a word $\mathbf{x} \in \mathscr{X}_8^{\mathrm{K},+}$. Then all 2^{12} codewords of $\mathscr{X}_{24}^{\mathrm{Gol}}$ of length 24 could be written as concatenation $(\mathbf{a}+\mathbf{x})(\mathbf{b}+\mathbf{x})(\mathbf{a}+\mathbf{b}+\mathbf{x})$. This can be checked by inspection of generating matrices.

Lemma 2.4.4 *The binary Golay code $\mathscr{X}_{24}^{\mathrm{Gol}}$ is self-dual, with $\mathscr{X}_{24}^{\mathrm{Gol}\perp} = \mathscr{X}_{24}^{\mathrm{Gol}}$. The code $\mathscr{X}_{24}^{\mathrm{Gol}}$ is also generated by the matrix $\widetilde{G} = (G'|\mathbf{I}_{12})$.*

Proof A direct calculation shows that any two rows of matrix G are dot-orthogonal. Thus $\mathscr{X}_{24}^{\text{Gol}} \subset \mathscr{X}_{24}^{\text{Gol}\perp}$. But the dimensions of $\mathscr{X}_{24}^{\text{Gol}}$ and $\mathscr{X}_{24}^{\text{Gol}\perp}$ coincide. Hence, $\mathscr{X}_{24}^{\text{Gol}} = \mathscr{X}_{24}^{\text{Gol}\perp}$. The last assertion of the lemma now follows from the property $(G')^{\text{T}} = G'$. □

Worked Example 2.4.5   *Show that the distance* $d(\mathscr{X}_{24}^{\text{Gol}}) = 8$.

Solution First, we check that for all $\mathbf{x} \in \mathscr{X}_{24}^{\text{Gol}}$ the weight $w(\mathbf{x})$ is divisible by 4 . This is true for every row of $G = (\mathrm{I}_{12}|G')$: the number of 1s is either 12 or 8. Next, for all binary N-words $\mathbf{x}, \mathbf{x}'$,

$$w(\mathbf{x}+\mathbf{x}') = w(\mathbf{x}) + w(\mathbf{x}') - 2w(\mathbf{x} \wedge \mathbf{x}')$$

where $(\mathbf{x} \wedge \mathbf{y})$ is the wedge-product, with digits $(\mathbf{x} \wedge \mathbf{y})_j = \min(\mathbf{x}_j, \mathbf{y}_j)$, $1 \leq j \leq N$ (cf. (2.1.6b)). But for any pair $\mathbf{g}(j)$, $\mathbf{g}(j')$ of the rows of G, $w(\mathbf{g}(j) \wedge \mathbf{g}(j')) = 0$ mod 2. So, 4 divides $w(\mathbf{x})$ for all $\mathbf{x} \in \mathscr{X}_{24}^{\text{Gol}}$.

On the other hand, $\mathscr{X}_{24}^{\text{Gol}}$ does not have codewords of weight 4. To prove this, compare two generating matrices, $(\mathrm{I}_{12}|G')$ and $((G')^{\text{T}}|\mathrm{I}_{12})$. If $\mathbf{x} \in \mathscr{X}_{24}^{\text{Gol}}$ has $w(\mathbf{x}) = 4$, write $\mathbf{x}$ as a concatenation $\mathbf{x}_L\mathbf{x}_R$. Any non-trivial sum of rows of $(\mathrm{I}_{12}|G')$ has the weight of the L-half of at least 1, so $w(\mathbf{x}_L) \geq 1$. Similarly, $w(\mathbf{x}_R) \geq 1$. But if $w(\mathbf{x}_L) = 1$ then $\mathbf{x}$ must be one of the rows of $(\mathrm{I}_{12}|G)$, none of which has weight $w(\mathbf{x}_R) = 3$. Hence, $w(\mathbf{x}_L) \geq 2$. Similarly, $w(\mathbf{x}_R) \geq 2$. But then the only possibility is that $w(\mathbf{x}_L) = w(\mathbf{x}_R) = 2$ which is impossible by a direct check. Thus, $w(\mathbf{x}) \geq 8$. But $(\mathrm{I}_{12}|G')$ has rows of weight 8. So, $d(\mathscr{X}_{24}^{\text{Gol}}) = 8$. □

When we truncate $\mathscr{X}_{24}^{\text{Gol}}$ at any digit, we get $\mathscr{X}_{23}^{\text{Gol}}$, a $[23,12,7]$ code. This code is perfect 3 error correcting. We recover $\mathscr{X}_{24}^{\text{Gol}}$ from $\mathscr{X}_{23}^{\text{Gol}}$ by adding a parity-check.

The Hamming $[2^\ell-1, 2^\ell-1-\ell, 3]$ and the Golay $[23,12,7]$ are the only possible perfect binary linear codes.

The ternary Golay code $\mathscr{X}_{12,3}^{\text{Gol}}$ of length 12 has the generating matrix $\left(\mathrm{I}_6|G'_{(3)}\right)$ where

$$G'_{(3)} = \begin{pmatrix} 0 & 1 & 1 & 1 & 1 & 1 \\ 1 & 0 & 1 & 2 & 2 & 1 \\ 1 & 1 & 0 & 1 & 2 & 2 \\ 1 & 2 & 1 & 0 & 1 & 2 \\ 1 & 2 & 2 & 1 & 0 & 1 \\ 1 & 1 & 2 & 2 & 1 & 0 \end{pmatrix}, \text{ with } (G'_{(3)})^{\text{T}} = G'_{(3)}. \qquad (2.4.2)$$

The ternary Golay code $\mathscr{X}_{11,3}^{\text{Gol}}$ is a truncation of $\mathscr{X}_{12,3}^{\text{Gol}}$ at the last digit.

Theorem 2.4.6 *The ternary Golay code $\mathscr{X}_{12,3}^{(Gol)\perp} = \mathscr{X}_{12,3}^{(Gol)}$ is $[12,6,6]$. The code $\mathscr{X}_{11,3}^{(Gol)}$ is $[11,6,5]$, hence perfect.*

Proof The code $[11,6,5]$ is perfect since $v_{11,3}(2) = 1 + 11 \times 2 + \dfrac{11 \times 10}{2} \times 2^2 = 3^5$. The rest of the assertions of the theorem are left as an exercise. $\qquad\square$

The Hamming $\left[\dfrac{3^\ell - 1}{2}, 3^\ell - 1 - \ell, 3\right]$ and the Golay $[11,6,5]$ codes are the only possible perfect ternary linear codes. Moreover, the Hamming and Golay are the only perfect linear codes, occurring in any alphabet $\mathbb{F}_q$ where $q = p^s$ is a prime power. Hence, these codes are the only possible perfect linear codes. And even non-linear perfect codes do not bring anything essentially new: they all have the same parameters (length, size and distance) as the Hamming and Golay codes. The Golay codes were used in the 1980s in the American Voyager spacecraft program, to transmit close-up photographs of Jupiter and Saturn.

The next popular examples are the Reed–Muller codes. For $N = 2^m$ consider binary Hamming spaces $\mathscr{H}_{m,2}$ and $\mathscr{H}_{N,2}$. Let $\mathbf{M}(= \mathbf{M}_m)$ be an $m \times N$ matrix where the columns are the binary representations of the integers $j = 0, 1, \ldots, N-1$, with the least significant bit in the first place:

$$j = j_1 \cdot 2^0 + j_2 \cdot 2^1 + \cdots + j_m 2^{m-1}. \tag{2.4.3}$$

So,

$$
\mathbf{M} = \begin{pmatrix}
0 & 1 & 0 & \cdots & 1 \\
0 & 0 & 1 & \cdots & 1 \\
\vdots & \vdots & \vdots & \ddots & \vdots \\
0 & 0 & 0 & \cdots & 1 \\
0 & 0 & 0 & \cdots & 1
\end{pmatrix}
\begin{matrix}
\mathbf{v}^{(1)} \\
\mathbf{v}^{(2)} \\
\vdots \\
\mathbf{v}^{(m-1)} \\
\mathbf{v}^{(m)}
\end{matrix}. \tag{2.4.4}
$$

(with column labels $0\ \ 1\ \ 2\ \ \cdots\ \ 2^m-1$)

The columns of $\mathbf{M}$ list all vectors from $\mathscr{H}_{m,2}$ and the rows are vectors from $\mathscr{H}_{N,2}$ denoted by $\mathbf{v}^{(1)}, \ldots, \mathbf{v}^{(m)}$. In particular, $\mathbf{v}^{(m)}$ has the first 2^{m-1} entries 0, the last 2^{m-1} entries 1. To pass from $\mathbf{M}_m$ to $\mathbf{M}_{m-1}$, one drops the last row and takes one of the two identical halves of the remaining $(m-1) \times N$ matrix. Conversely, to pass from $\mathbf{M}_{m-1}$ to $\mathbf{M}_m$, one concatenates two copies of $\mathbf{M}_{m-1}$ and adds row $\mathbf{v}^{(m)}$:

$$\mathbf{M}_m = \begin{pmatrix} \mathbf{M}_{m-1} & \mathbf{M}_{m-1} \\ 0 \ldots 0 & 1 \ldots 1 \end{pmatrix}. \tag{2.4.5}$$

Consider the columns $\mathbf{w}^{(1)}, \ldots, \mathbf{w}^{(m)}$ of $\mathbf{M}_m$ corresponding to numbers $1, 2, 4, \ldots, 2^{m-1}$. They form the standard basis in $\mathcal{H}_{m,2}$:

$$
\begin{matrix}
1 & 0 & \ldots & 0 \\
0 & 1 & \ldots & 0 \\
\vdots & \vdots & \ddots & \vdots \\
0 & 0 & \ldots & 1
\end{matrix}
\ .
$$

Then the column at position $j = \sum\limits_{1 \le i \le m} j_i 2^{i-1}$ is $\sum\limits_{1 \le i \le m} j_i \mathbf{w}^{(i)}$.

The vector $\mathbf{v}^{(i)}$, $i = 1, \ldots, m$, can be interpreted as the indicator function of the set $\mathcal{A}_i \subset \mathcal{H}_{m,2}$ where the ith digit is 1:

$$
\mathcal{A}_i = \{ \mathbf{j} \in \mathcal{H}_{m,2} : j_i = 1 \}. \tag{2.4.6}
$$

In terms of the wedge-product (cf. (2.1.6b)) $\mathbf{v}^{(i_1)} \wedge \mathbf{v}^{(i_2)} \wedge \cdots \wedge \mathbf{v}^{(i_k)}$ is the indicator function of the intersection $\mathcal{A}_{i(1)} \cap \cdots \cap \mathcal{A}_{i(k)}$. If all $i_1, \ldots, i_k$ are distinct, the cardinality $\sharp \left(\cap_{1 \le j \le k} \mathcal{A}_{i(j)} \right) = 2^{m-k}$. In other words, we have the following.

Lemma 2.4.7   *The weight* $w \left(\wedge_{1 \le j \le k} \mathbf{v}^{(i_j)} \right) = 2^{m-k}$.

An important fact is

Theorem 2.4.8   *The vectors* $\mathbf{v}^{(0)} = 11 \ldots 1$ *and* $\wedge_{1 \le j \le k} \mathbf{v}^{(i_j)}$, $1 \le i_1 < \cdots < i_k \le m$, $k = 1, \ldots, m$, *form a basis in* $\mathcal{H}_{N,2}$.

Proof   It suffices to check that the standard basis N-words $\mathbf{e}(j) = 0 \ldots 1 \ldots 0$ (1 in position j, 0 elsewhere) can be written as linear combinations of the above vectors. But

$$
\mathbf{e}(j) = \wedge_{1 \le i \le m} (\mathbf{v}^{(i)} + (1 + v_j^{(i)}) \mathbf{v}^{(0)}), \quad 0 \le j \le N - 1. \tag{2.4.7}
$$

[All factors in position j are equal to 1 and at least one factor in any position $l \ne j$ is equal to 0.] $\square$

Example 2.4.9   For $m = 4, N = 16$,

$$\mathbf{v}^{(0)} = 1111111111111111$$
$$\mathbf{v}^{(1)} = 0101010101010101$$
$$\mathbf{v}^{(2)} = 0011001100110011$$
$$\mathbf{v}^{(3)} = 0000111100001111$$
$$\mathbf{v}^{(4)} = 0000000011111111$$
$$\mathbf{v}^{(1)} \wedge \mathbf{v}^{(2)} = 0001000100010001$$
$$\mathbf{v}^{(1)} \wedge \mathbf{v}^{(3)} = 0000010100000101$$
$$\mathbf{v}^{(1)} \wedge \mathbf{v}^{(4)} = 0000000001010101$$
$$\mathbf{v}^{(2)} \wedge \mathbf{v}^{(3)} = 0000001100000011$$
$$\mathbf{v}^{(2)} \wedge \mathbf{v}^{(4)} = 0000000000110011$$
$$\mathbf{v}^{(3)} \wedge \mathbf{v}^{(4)} = 0000000000001111$$
$$\mathbf{v}^{(1)} \wedge \mathbf{v}^{(2)} \wedge \mathbf{v}^{(3)} = 0000000100000001$$
$$\mathbf{v}^{(1)} \wedge \mathbf{v}^{(2)} \wedge \mathbf{v}^{(4)} = 0000000000010001$$
$$\mathbf{v}^{(1)} \wedge \mathbf{v}^{(3)} \wedge \mathbf{v}^{(4)} = 0000000000000101$$
$$\mathbf{v}^{(2)} \wedge \mathbf{v}^{(3)} \wedge \mathbf{v}^{(4)} = 0000000000000011$$
$$\mathbf{v}^{(1)} \wedge \mathbf{v}^{(2)} \wedge \mathbf{v}^{(3)} \wedge \mathbf{v}^{(4)} = 0000000000000001$$

Definition 2.4.10   Given $0 \leq r \leq m$, the *Reed–Muller* (RM) code $\mathscr{X}^{RM}(r,m)$ of order r is a binary code of length $N = 2^m$ spanned by all wedge-products $\wedge_{1 \leq j \leq k} \mathbf{v}^{(i_j)}$ and $\mathbf{v}^{(0)}$ where $1 \leq k \leq r$ and $1 \leq i_1 < \cdots < i_k \leq m$. The rank of $\mathscr{X}^{RM}(r,m)$ equals $1 + \binom{m}{1} + \cdots + \binom{m}{r}$.

So, $\mathscr{X}^{RM}(0,m) \subset \mathscr{X}^{RM}(1,m) \subset \cdots \subset \mathscr{X}^{RM}(m-1,m) \subset \mathscr{X}^{RM}(m,m)$. Here $\mathscr{X}^{RM}(m,m) = \mathscr{H}_{N,2}$, the whole Hamming space, and $\mathscr{X}^{RM}(0,m) = \{00\ldots00, 11\ldots1\}$, the repetition code. Next, $\mathscr{X}^{RM}(m-1,m)$ consists of all words $\mathbf{x} \in \mathscr{H}_{N,2}$ of even weight (shortly: even words). In fact, any basis vector is even, by Lemma 2.4.7. Further, if $\mathbf{x}, \mathbf{x}'$ are even then

$$w(\mathbf{x} + \mathbf{x}') = w(\mathbf{x}) + w(\mathbf{x}') - 2w(\mathbf{x} \wedge \mathbf{x}')$$

is again even. Hence, all codewords $\mathbf{x} \in \mathscr{X}^{RM}(m-1,m)$ are even. Finally, $\dim \mathscr{X}^{RM}(m-1,m) = N - 1$ coincides with the dimension of the subspace of even words. This proves the claim. As $\mathscr{X}^{RM}(r,m) \subset \mathscr{X}^{RM}(m-1,m)$, $r \leq m-1$, any RM code consists of even words.

The dual code is $\mathscr{X}^{RM}(r,m)^{\perp} = \mathscr{X}^{RM}(m-r-1,m)$. Indeed, if $\mathbf{a} \in \mathscr{X}^{RM}(r,m)$, $\mathbf{b} \in \mathscr{X}^{RM}(m-r-1,m)$ then the wedge-product $\mathbf{a} \wedge \mathbf{b}$ is an even word, and hence the dot-product $\langle \mathbf{a} \cdot \mathbf{b} \rangle = 0$. But

$$\dim(\mathscr{X}^{RM}(r,m)) + \dim(\mathscr{X}^{RM}(m-r-1,m)) = N,$$

hence the claim. As a corollary, code $\mathscr{X}^{RM}(m-2,m)$ is the parity-check extension of the Hamming code.

By definition, codewords $\mathbf{x} \in \mathscr{X}^{RM}(r,m)$ are associated with $\wedge$-polynomials in idempotent 'variables' $\mathbf{v}^{(1)}, \ldots, \mathbf{v}^{(m)}$, with coefficients $0, 1$, of degrees $\leq r$ (here, the degree of a polynomial is counted by taking the maximal number of variables $\mathbf{v}^{(1)}, \ldots, \mathbf{v}^{(m)}$ in the summand monomials). The 0-degree monomial in such a polynomial is proportional to $\mathbf{v}^{(0)}$.

Write this correspondence as

$$\mathbf{x} \in \mathscr{X}^{RM}(r,m) \leftrightarrow p_{\mathbf{x}}(\mathbf{v}^{(1)}, \ldots, \mathbf{v}^{(m)}), \deg p_{\mathbf{x}} \leq r. \tag{2.4.8}$$

Each such polynomial can be written in the form

$$p_x(\mathbf{v}^{(1)}, \ldots, \mathbf{v}^{(m)}) = \mathbf{v}^{(m)} \wedge q(\mathbf{v}^{(1)}, \ldots, \mathbf{v}^{(m-1)}) + l(\mathbf{v}^{(1)}, \ldots, \mathbf{v}^{(m-1)}),$$

with $\deg q \leq r-1, \deg l \leq r$. The word $\mathbf{v}^{(m)} \wedge q(\mathbf{v}^{(1)}, \ldots, \mathbf{v}^{(m-1)})$ has zeros on the first 2^{m-1} positions.

By the same token, as above,

$$\begin{aligned} q(\mathbf{v}^{(1)}, \ldots, \mathbf{v}^{(m-1)}) &\leftrightarrow \mathbf{b} \in \mathscr{X}^{RM}(r-1,m-1), \\ l(\mathbf{v}^{(1)}, \ldots, \mathbf{v}^{(m-1)}) &\leftrightarrow \mathbf{a} \in \mathscr{X}^{RM}(r,m-1). \end{aligned} \tag{2.4.9}$$

Furthermore, 2^m-word $\mathbf{x}$ can be written as the sum of concatenated 2^{m-1}-words:

$$\mathbf{x} = (\mathbf{a}|\mathbf{a}) + (\mathbf{0}|\mathbf{b}) = (\mathbf{a}|\mathbf{a}+\mathbf{b}). \tag{2.4.10}$$

This means that the Reed–Muller codes are related via the bar-product construction (cf. Example 2.1.8(viii)):

$$\mathscr{X}^{RM}(r,m) = \mathscr{X}^{RM}(r,m-1) | \mathscr{X}^{RM}(r-1,m-1). \tag{2.4.11}$$

Therefore, inductively,

$$d(\mathscr{X}^{RM}(r,m)) = 2^{m-r}. \tag{2.4.12}$$

In fact, for $m = r = 0$, $d(\mathscr{X}^{RM}(0,0)) = 2^m$ and for all m, $d(\mathscr{X}^{RM}(m,m)) = 1 = 2^0$. Assume $d(\mathscr{X}^{RM}(r-1,\tilde{m})) = 2^{\tilde{m}-r+1}$ for all $\tilde{m} \geq r-1$, and $d(\mathscr{X}^{RM}(r-1,m-1)) = 2^{m-r}$. Then (cf. (2.4.14) below)

$$\begin{aligned} d(\mathscr{X}^{RM}(r,m)) &= \min[2d(\mathscr{X}^{RM}(r,m-1)), d(\mathscr{X}^{RM}(r-1,m-1))] \\ &= \min[2 \cdot 2^{m-1-r}, 2^{m-1-r+1}] = 2^{m-r}. \end{aligned} \tag{2.4.13}$$

Summarising,

Theorem 2.4.11   *The RM code $\mathscr{X}^{RM}(r,m), 0 \leq r \leq m$, is a binary code of length $N = 2^m$, rank $k = \sum\limits_{0 \leq l \leq r} \binom{N}{l}$ and distance $d = 2^{m-r}$. Furthermore,*

(1) $\mathscr{X}^{RM}(0,m) = \{0\ldots0, 1\ldots1\} \subset \mathscr{X}^{RM}(1,m) \subset \cdots \subset \mathscr{X}^{RM}(m-1,m) \subset \mathscr{X}^{RM}(m,m) = \mathscr{H}_{N,2}$; $\mathscr{X}^{RM}(m-1,m)$ *is the set of all even N-words and* $\mathscr{X}^{RM}(m-2,m)$ *the parity-check extension of the Hamming $[2^m - 1, 2^m - 1 - m]$ code.*
(2) $\mathscr{X}^{RM}(r,m) = \mathscr{X}^{RM}(r,m-1)|\mathscr{X}^{RM}(r-1,m-1), 1 \leq r \leq m-1.$
(3) $\mathscr{X}^{RM}(r,m)^{\perp} = \mathscr{X}^{RM}(m-r-1,m), 0 \leq r \leq m-1.$

Worked Example 2.4.12   *Define the bar-product $\mathscr{X}_1|\mathscr{X}_2$ of binary linear codes $\mathscr{X}_1$ and $\mathscr{X}_2$, where $\mathscr{X}_2$ is a subcode of $\mathscr{X}_1$. Relate the rank and minimum distance of $\mathscr{X}_1|\mathscr{X}_2$ to those of $\mathscr{X}_1$ and $\mathscr{X}_2$. Show that if $\mathscr{X}^{\perp}$ denotes the dual code of $\mathscr{X}$, then*

$$(\mathscr{X}_1|\mathscr{X}_2)^{\perp} = \mathscr{X}_2^{\perp}|\mathscr{X}_1^{\perp}.$$

Using the bar-product construction, or otherwise, define the Reed–Muller code $\mathscr{X}^{RM}(r,m)$ for $0 \leq r \leq m$. Show that if $0 \leq r \leq m-1$, then the dual of $\mathscr{X}^{RM}(r,m)$ is again a Reed–Muller code.

Solution The bar-product $\mathscr{X}_1|\mathscr{X}_2$ of two linear codes $\mathscr{X}_1 \subseteq \mathscr{X}_2 \subseteq \mathbb{F}_2^N$ is defined as

$$\mathscr{X}_1|\mathscr{X}_2 = \{(\mathbf{x}|\mathbf{x}+\mathbf{y}) : \mathbf{x} \in \mathscr{X}_1, \mathbf{y} \in \mathscr{X}_2\};$$

it is a linear code of length $2N$. If $\mathscr{X}_1$ has basis $\mathbf{x}_1, \ldots, \mathbf{x}_k$ and $\mathscr{X}_2$ basis $\mathbf{y}_1, \ldots, \mathbf{y}_l$ then $\mathscr{X}_1|\mathscr{X}_2$ has basis

$$(\mathbf{x}_1|\mathbf{x}_1), \ldots, (\mathbf{x}_k|\mathbf{x}_k), (\mathbf{0}, \mathbf{y}_1), \ldots, (\mathbf{0}|\mathbf{y}_l),$$

and the rank of $\mathscr{X}_1|\mathscr{X}_2$ equals the sum of ranks of $\mathscr{X}_1$ and $\mathscr{X}_2$.

Next, we are going to check that the minimum distance

$$d(\mathscr{X}_1|\mathscr{X}_2) = \min\left[2d(\mathscr{X}_1), d(\mathscr{X}_2)\right]. \tag{2.4.14}$$

Indeed, let $\mathbf{0} \neq (\mathbf{x}|\mathbf{x}+\mathbf{y}) \in \mathscr{X}_1|\mathscr{X}_2$. If $\mathbf{y} \neq \mathbf{0}$ then the weight $w(\mathbf{x}|\mathbf{x}+\mathbf{y}) \geq w(\mathbf{y}) \geq d(\mathscr{X}_2)$. If $\mathbf{y} = \mathbf{0}$ then $w(\mathbf{x}|\mathbf{x}+\mathbf{y}) = 2w(\mathbf{x}) \geq 2d(\mathscr{X}_1)$. This implies that

$$d(\mathscr{X}_1|\mathscr{X}_2) \geq \min\left[2d(\mathscr{X}_1), d(\mathscr{X}_2)\right]. \tag{2.4.15}$$

On the other hand, if $\mathbf{x} \in \mathscr{X}_1$ has $w(\mathbf{x}) = d(\mathscr{X}_1)$ then $d(\mathscr{X}_1|\mathscr{X}_2) \leq w(\mathbf{x}|\mathbf{x}) = 2d(\mathscr{X}_1)$. Finally, if $\mathbf{y} \in \mathscr{X}_2$ has $w(\mathbf{y}) = d(\mathscr{X}_2)$ then $d(\mathscr{X}_1|\mathscr{X}_2) \leq w(\mathbf{0}|\mathbf{y}) = d(\mathscr{X}_2)$. We conclude that

$$d(\mathscr{X}_1|\mathscr{X}_2) \leq \min\left[2d(\mathscr{X}_1), d(\mathscr{X}_2)\right], \tag{2.4.16}$$

proving (2.4.14).

Now, we will check that

$$\left(\mathcal{X}_2^\perp | \mathcal{X}_1^\perp\right) \subseteq (\mathcal{X}_1 | \mathcal{X}_2)^\perp.$$

Indeed, let $(\mathbf{u}|\mathbf{u}+\mathbf{v}) \in \mathcal{X}_2^\perp | \mathcal{X}_1^\perp$ and $(\mathbf{x}|\mathbf{x}+\mathbf{y}) \in (\mathcal{X}_1 | \mathcal{X}_2)$. The dot-product

$$\langle (\mathbf{u}|\mathbf{u}+\mathbf{v}) \cdot (\mathbf{x}|\mathbf{x}+\mathbf{y}) \rangle = \mathbf{u} \cdot \mathbf{x} + (\mathbf{u}+\mathbf{v}) \cdot (\mathbf{x}+\mathbf{y})$$
$$= \mathbf{u} \cdot \mathbf{y} + \mathbf{v} \cdot (\mathbf{x}+\mathbf{y}) = 0,$$

since $\mathbf{u} \in \mathcal{X}_2^\perp, \mathbf{y} \in \mathcal{X}_2, \mathbf{v} \in \mathcal{X}_1^\perp$ and $(\mathbf{x}+\mathbf{y}) \in \mathcal{X}_1$. In addition, we know that

$$\text{rank}\left(\mathcal{X}_2^\perp | \mathcal{X}_1^\perp\right) = N - \text{rank}(\mathcal{X}_2) + N - \text{rank}(\mathcal{X}_1)$$
$$= 2N - \text{rank}(\mathcal{X}_1 | \mathcal{X}_2) = \text{rank}(\mathcal{X}_1 | \mathcal{X}_2)^\perp.$$

This implies that in fact

$$\left(\mathcal{X}_2^\perp | \mathcal{X}_1^\perp\right) = (\mathcal{X}_1 | \mathcal{X}_2)^\perp. \tag{2.4.17}$$

Turning to the RM codes, they are determined as follows:

$\mathcal{X}^{\text{RM}}(0,m) =$ the repetition binary code of length $N = 2^m$,

$\mathcal{X}^{\text{RM}}(m,m) =$ the whole space $\mathcal{H}_{N,2}$ of length $N = 2^m$,

$\mathcal{X}^{\text{RM}}(r,m)$ for $0 < r < m$ is defined recursively by
$$\mathcal{X}^{\text{RM}}(r,m) = \mathcal{X}^{\text{RM}}(r,m-1) | \mathcal{X}(r-1,m-1).$$

By construction, $\mathcal{X}^{\text{RM}}(r,m)$ has rank $\sum_{j=0}^{r} \binom{m}{j}$ and the minimum distance 2^{m-r}.
In particular, $\mathcal{X}^{\text{RM}}(m-1,m)$ is the parity-check code and hence dual of $\mathcal{X}(0,m)$.
We will show that in general, for $0 \le r \le m-1$,

$$\mathcal{X}^{\text{RM}}(r,m)^\perp = \mathcal{X}^{\text{RM}}(m-r-1,m).$$

It is done by induction in $m \ge 3$. By the above, we can assume that
$\mathcal{X}^{\text{RM}}(r,m-1)^\perp = \mathcal{X}^{\text{RM}}(m-r-2,m-1)$ holds for $0 \le r < m-1$. Then for
$0 \le r < m$:

$$\mathcal{X}^{\text{RM}}(r,m)^\perp = \left(\mathcal{X}^{\text{RM}}(r,m-1) | \mathcal{X}^{\text{RM}}(r-1,m-1)\right)^\perp$$
$$= \mathcal{X}^{\text{RM}}(r-1,m-1)^\perp | \mathcal{X}^{\text{RM}}(r,m-1)^\perp$$
$$= \mathcal{X}^{\text{RM}}(m-r-1,m-1) | \mathcal{X}^{\text{RM}}(m-r-2,m-1)$$
$$= \mathcal{X}^{\text{RM}}(m-r-1,m).$$

$\square$

Encoding and decoding of RM codes is based on the following observation. By virtue of (2.4.5), the product $\mathbf{v}^{(i_1)} \wedge \cdots \wedge \mathbf{v}^{(i_k)}$ occurs in the expansion for $\mathbf{e}(j) \in \mathcal{H}_{m,2}$ iff $v_j^{(i)} = 0$ for all $i \notin \{i_1, \ldots, i_k\}$.

Definition 2.4.13   For $1 \le i_1 < \cdots < i_k \le m$, define

$$C(i_1, \ldots, i_k) := \text{the set of all integers } j = \sum_{1 \le i \le m} j_i 2^{i-1} \tag{2.4.18}$$
$$\text{with } j_i = 0 \text{ for } i \notin \{i_1, \ldots, i_k\}.$$

For an empty set $(k = 0)$, $C(\emptyset) = \{1, \ldots, 2^m - 1\}$. Furthermore, set

$$C(i_1, \ldots, i_k) + t = \{j + t : j \in C(i_1, \ldots, i_k)\}. \tag{2.4.19}$$

Then, again in view of (2.4.5), for all $\mathbf{y} = y_0 \ldots y_{N-1} \in \mathcal{H}_{N,2}$,

$$\mathbf{y} = \sum_{0 \le k \le m} \sum_{1 \le i_1 < \cdots < i_k \le m} \left(\sum_{j \in C(i_1, \ldots, i_k)} y_j \right) \mathbf{v}^{(i_1)} \wedge \cdots \wedge \mathbf{v}^{(i_k)} \tag{2.4.20}$$

(for $k = 0$, take $\mathbf{v}^{(0)}$).

For encoding a sequence $a = a_0 \ldots a_{k-1}$ of information symbols from $\mathcal{H}_{k,2}$, with $k = 1 + \binom{m}{1} + \cdots + \binom{m}{r}$, with $\mathcal{X}_{r,m}^{\mathrm{RM}}$, rewrite it as $(a_{i_1, \ldots, i_\ell})$; here $i_1, \ldots, i_\ell$ are the successive positions of the 1s. Then construct a codeword as $\mathbf{x} = (x_0, \ldots, x_{N-1}) \in \mathcal{X}_{r,m}^{\mathrm{RM}}$ where

$$\mathbf{x} = \sum_{0 \le l \le r} \sum_{1 \le i_1 < \cdots < i_l \le m} a_{i_1, \ldots, i_l} \mathbf{v}^{(i_1)} \wedge \cdots \wedge \mathbf{v}^{(i_l)}. \tag{2.4.21}$$

We see that the 'information space' $\mathcal{H}_{k,2}$ is embedded into $\mathcal{H}_{N,2}$, by identifying entries $a_j \sim a_{i_1, \ldots, i_l}$ where $j = j_0 2^0 + j_1 2^1 + \cdots + j_{m-1} 2^{m-1}$ and $i_1, \ldots, i_l$ are the successive positions of the 1s among $j_1, \ldots, j_m$, $1 \le l \le r$. With such an identification we obtain:

Lemma 2.4.14   *For all $0 \le l \le m$ and $1 \le i_1 < \cdots < i_l \le m$,*

$$\sum_{j \in C(i_1, \ldots, i_l)} x_j = a_{i_1, \ldots, i_l}, \quad \text{if } l \le r, \tag{2.4.22}$$
$$= 0, \quad \text{if } l > r.$$

Proof   The result follows from (2.4.20).   □

Lemma 2.4.15   *For all $1 \le i_1 < \cdots < i_r \le m$ and for any $1 \le t \le m$ such that $t \notin \{i_1, \ldots, i_r\}$,*

$$a_{i_1, \ldots, i_r} = \sum_{j \in C(i_1, \ldots, i_r) + 2^{t-1}} x_j. \tag{2.4.23}$$

Proof The proof follows from the fact that $C(i_1,\ldots,i_r,t)$ is the disjoint union $C(i_1,\ldots,i_r)\cup(C(i_1,\ldots,i_r)+2^{t-1})$ and the equation $\sum\limits_{j\in C(i_1,\ldots,i_r,t)} x_j=0$ (cf. (2.4.19)).

$\square$

Moreover:

Theorem 2.4.16   *For any information symbol* $a_{i_1,\ldots,i_r}$ *corresponding to* $\mathbf{v}^{(i_1,\ldots,i_r)}$, *we can split the set* $\{0,\ldots,N-1\}$ *into* 2^{m-r} *disjoint subsets* S, *each containing* 2^r *elements, such that, for all such* S, $a_{i_1,\ldots,i_r}=\sum\limits_{j\in S}x_j$.

Proof The list of sets S begins with $C(i_1,\ldots,i_r)$ and continues with $(m-r)$ disjoint sets $C(i_1,\ldots,i_r)+2^{t-1}$ where $1\le t\le m$, $t\notin\{i_1,\ldots,i_r\}$. Next, we take any pair $1\le t_1<t_2\le m$ such that $\{t_1,t_2\}\cap\{i_1,\ldots,i_r\}=\emptyset$. Then $C(i_1,\ldots,i_r,t_1,t_2)$ contains disjoint sets $C(i_1,\ldots,i_r)$, $C(i_1,\ldots,i_r)+2^{t_1-1}$ and $C(i_1,\ldots,i_r)+2^{t_2-1}$, and for each of them, $a_{i_1,\ldots,i_r}=\sum\limits_{j\in C(i_1,\ldots,i_r)+2^{t_k-1}} x_j,k=1,2$. Then the same is true for the remaining sets

$$C(i_1,\ldots,i_r)+2^{t_1-1}+2^{t_2-1}=C(i_1,\ldots,i_r,t_1,t_2)\setminus$$
$$\left[C(i_1,\ldots,i_r)\cup(C(i_1,\ldots,i_r)+2^{t_1-1})\cup(C(i_1,\ldots,i_r)+2^{t_2-1})\right];\quad (2.4.24)$$

there are $\binom{m-r}{2}$ of them and they are still disjoint with each other and the previous ones. The sets (2.4.24) form a further bunch of sets S.

And so on: a general form of set S is

$$C(i_1,\ldots,i_r)+2^{t_1-1}+\cdots+2^{t_s-1}$$

which is the same as the set-theoretic difference

$$C(i_1,\ldots,i_r,t_1,\ldots,t_s)$$
$$\setminus\left[\bigcup_{\{t_1',\ldots,t_{s'}'\}\subset\{t_1,\ldots t_s\}}\left(C(i_1,\ldots,i_r)+2^{t_1'-1}+\cdots+2^{t_{s'}'-1}\right)\right].\quad (2.4.25)$$

Here each such set is labelled by a collection $\{t_1,\ldots,t_s\}$ where $0\le s\le m-r$, $t_1<\cdots<t_s$ and $\{t_1,\ldots,t_s\}\cap\{i_1,\ldots,i_r\}=\emptyset$. [The union $\bigcup_{\{t_1',\ldots,t_{s'}'\}\subset\{t_1,\ldots t_s\}}$ in (2.4.25) is over all ('strict') subsets $\{t_1',\ldots,t_{s'}'\}$ of $\{t_1,\ldots,t_s\}$, with $t_1'<\cdots<t_{s'}'$ and $s'=0,\ldots,s-1$ ($s'=0$ gives the empty subset).] The total number of sets $C(i_1,\ldots,i_r)$ equals 2^{m-r} and each of them has 2^r elements by construction. $\square$

Theorem 2.4.16 provides a rationale for the so-called *majority* decoding for the Reed–Muller codes. Namely, upon receiving a word $\mathbf{y}=(y_0,\ldots,y_{N-1})$, produced from a codeword $\mathbf{x}^\wedge\in\mathscr{X}_{r,m}^{\mathrm{RM}}$, we take any $1\le i_1<\cdots<i_r\le m$ and consider the sums $\sum\limits_{j\in C}y_j$ along the 2^{m-r} above sets S. If $\mathbf{y}\in\mathscr{X}_{r,m}^{\mathrm{RM}}$, all these sums coincide

and give $a_{i_1,\ldots,i_r}$. If the number of errors in $\mathbf{y}$ (i.e. the Hamming distance $\delta(\mathbf{x}^\wedge,\mathbf{y})$) $< 2^{m-r-1} = d\left(\mathscr{X}_{r,m}^{\mathrm{RM}}\right)/2$, the majority of sums will still give a correct $a_{i_1,\ldots,i_r}$ (the worst case is where each set S contains no or a single error). By varying $\{i_1,\ldots,i_r\}$, we will determine a codeword $\mathbf{x}^{(1)} \in \mathscr{X}_{r,m}^{\mathrm{RM}}$ containing only monomials of degree r. Note that $\mathbf{x}^\wedge - \mathbf{x}^{(1)}$ will be a codeword in $\mathscr{X}_{r-1,m}^{\mathrm{RM}}$.

Then $\mathbf{y}$ can be 'reduced' to $\mathbf{y} - \mathbf{x}^{(1)}$. Compared with $\mathbf{x}^\wedge - \mathbf{x}^{(1)}$, the reduced word $\mathbf{y} - \mathbf{x}^{(1)}$ will have $\delta(\mathbf{x}^\wedge - \mathbf{x}^{(1)}, \mathbf{y} - \mathbf{x}^{(1)}) = \delta(\mathbf{x}^\wedge,\mathbf{y})$ errors, which is $< 2^{m-r} = d(\mathscr{X}_{r-1,m}^{\mathrm{RM}})/2$. We can repeat the above procedure and obtain the correct $a_{i_1,\ldots,i_{r-1}}$ for any $1 \le i_1 < \cdots < i_{r-1} \le m$, etc. At the end, we recover the whole sequence of information symbols $a_{i_1,\ldots,i_r}$.

Therefore, any word $\mathbf{y} \in \mathscr{H}_{N,2}$ with distance $\delta\left(\mathbf{y}, \mathscr{X}_{r,m}^{\mathrm{RM}}\right) < d\left(\mathscr{X}_{r,m}^{\mathrm{RM}}\right)/2$ is uniquely decoded.

> ... *correct, insert, refine,*
> *enlarge, diminish, interline.*
> Jonathan Swift (1667–1745), Anglo–Irish writer

Reed–Muller codes were discovered at the beginning of the 1950s by David Muller (1924–2008); Irwin Reed (1923–2012) proposed the above decoding procedure. In the early 1970s, the RM codes were used to transmit pictures from space (as far as the Moon) by the spacecrafts. The quality of transmission was then considered as exceptionally good. However, later on, NASA engineers decided in favour of the Golay codes while photographing Jupiter and Saturn.

Worked Example 2.4.17 *A maximum distance separable (MDS) code was defined earlier as a q-ary linear $[N,k,d]$ code with $d = N - k + 1$ (equality in the Singleton bound; see Definition 2.1.13).*

(a) *Prove that $\mathscr{X}$ is MDS iff*

    (i) *any $N - k$ columns of its parity-check matrix H are linearly independent, and*

    (ii) *there exist $N - k + 1$ columns of H that are linearly dependent.*

(b) *Prove that the dual of an MDS code is MDS and deduce that $\mathscr{X}$ is MDS iff any k columns of its generating matrix G are linearly independent and k is the maximal such number.*

(c) *Hence prove that when G is written in the standard form $(\mathrm{I}_k|G')$ then $\mathscr{X}$ is MDS iff any square sub-matrix of G' is non-singular.*

(d) *Finally, check that $[N,k,d]$ code $\mathscr{X}$ is MDS iff for any d positions $1 \le i_1 < \cdots < i_d \le N$, there exists a codeword of weight d with non-zero digits at digits $i_1,\ldots,i_d$.*

Solution (a) An MDS $[N,k,d]$ code has $d = N - k + 1$. If a linear code $\mathscr{X}$ has $d(\mathscr{X}) = d$ then any $(d-1)$ columns of its parity-check matrix H are linearly independent, and $(d-1)$ is the maximal number with this property, and vice versa. So, any $(N-k)$ columns are linearly independent and $(N-k)$ is the maximal such number, and vice versa. Equivalently, any $(N-k) \times (N-k)$ submatrix of H is invertible.

(b) Let $\mathscr{X}$ be $[N,k,d]$ MDS code with a parity-check matrix H. Then H is a generating matrix for $\mathscr{X}^{\perp}$. Any $(N-k) \times (N-k)$ submatrix of H is invertible. Then any non-trivial combination of rows of $H^{\top}$ has $\leq N - k - 1$ zero entries, i.e. weight $\geq k + 1$; the minimal weight is equal to $k + 1$. So, $d(\mathscr{X}^{\perp}) = k + 1 = N - (N - k) + 1$. As $\mathscr{X}^{\perp}$ is $[N, N - k]$ code, it is MDS.

Then, clearly, $[N,k]$ code $\mathscr{X}$ is MDS iff k is the maximal number l such that any l columns of its generating matrix G are linearly independent. Equivalently, $\mathscr{X}$ is systematic on any k positions.

(c) Again, let $\mathscr{X}$ be $[N,k,d]$ MDS code, and write $G = (\mathbf{I}_k | G')$. Take a $(u \times u)$ submatrix $\widetilde{G}_u$ of G'. By using row and column permutations, we may assume that $\widetilde{G}_u$ occupies the top left corner in G'. Then consider the last $(k - u)$ columns of $\mathbf{I}_k$ and u columns of G' containing $\widetilde{G}_u$; the corresponding $k \times k$ matrix is non-singular and forms a $k \times k$ submatrix G_k,

$$G_k = \begin{pmatrix} 0 & \widetilde{G}_u \\ \mathbf{I}_{k-u} & * \end{pmatrix},$$

with

$$\det G_k = \pm \det \widetilde{G}_u \det \mathbf{I}_{k-u} = \pm \det \widetilde{G}_u \neq 0, \text{ by (b)}.$$

So, $\widetilde{G}_u$ is non-singular. The proof of the inverse statement is similar.

(d) Finally, choose $d = N - k + 1$ digits, say $i_1, \ldots, i_d$. Consider i_1 together with the remaining digits $j_1, \ldots, j_{k-1}$. Then $i_1, j_1, \ldots, j_{k-1}$ are information symbols. So, there exists a codeword $\mathbf{x}$ with digit i_1 non-zero and digits $j_1, \ldots, j_{k-1}$ zero. Then $\mathbf{x}$ must have digits $i_1, \ldots, i_d$ non-zero.

The converse: consider an $(N - d + 1) \times N$ matrix

$$\widetilde{G} = [\mathbf{I}_{N-d+1} | E_{(N-d+1)\times(d-1)}]$$

where $\mathbf{I}_{N-d+1}$ is a unit matrix and E is an $(N - d + 1) \times (d - 1)$ matrix with all entries 1 (the unit of $\mathbb{F}_2$). The rows of $\widetilde{G}$ are linearly independent and have weight d, and for any row there exists a codeword $\mathbf{x}^{(i)} \in \mathscr{X}$ with non-zero digits at the same positions (and, possibly, elsewhere). Then k, the rank of the code, is $\geq N - d + 1$. Thus, $k = N - d + 1$. $\square$

Worked Example 2.4.18   The MDS codes $[N,N,1]$, $[N,1,N]$ and $[N,N-1,2]$ always exist and are called trivial. Any $[N,k]$ MDS code with $2 \le k \le N-2$ is called non-trivial. Show that there is no non-trivial MDS code over $\mathbb{F}_q$ with $q \le k \le N-q$. In particular, there is no non-trivial binary MDS code (which causes a discernible lack of enthusiasm about binary MDS codes).

Solution Indeed, the $[N,N,1]$, $[N,N-1,2]$ and $[N,1,N]$ codes are MDS. Take $q \le k \le N-q$ and assume $\mathcal{X}$ is a q-ary MDS. Take its generating matrix G in the standard form $(\mathrm{I}_k | G')$ where G' is $k \times (N-k)$, $N-k \ge q$.

If some entries in a column of G' are zero then this column is a linear combination of $k-1$ columns of I_{k-1}. This is impossible by (b) in the previous example; hence G' has no 0 entry. Next, assume that the first row of G' is $1\dots1$: otherwise we can perform scalar multiplication of columns maintaining codes' equivalence.

Now take the second row of G': it is of length $N-k \ge q$ and has no 0 entry. Then these must be repeated entries. That is,

$$G = \left(\mathrm{I}_k \left|
\begin{array}{ccccccc}
1 & \dots & 1 & \dots & 1 & \dots & 1 \\
\dots & \dots & a & \dots & a & \dots & \dots \\
& & \dots & & \dots & &
\end{array}
\right. \right), a \ne 0.$$

Then take the codeword

$$x = \text{ row } 1 - a^{-1}(\text{row } 2);$$

it has $w(\mathbf{x}) \le N-k-2+2 = N-k$ and $\mathcal{X}$ cannot be MDS.

By using the dual code, obtain that there exists no non-trivial q-ary MDS code with $k \ge q$. Hence, non-trivial MDS code can only have

$$N-q+1 \le k \text{ or } k \le q-1.$$

That is, there exists no non-trivial binary MDS code, but there exists a non-trivial $[3,2,2]$ ternary MDS code.   □

Remark 2.4.19   It is interesting to find, given k and q, the largest value of N for which there exists a q-ary MDS $[N,k]$ code. We demonstrated that N must be $\le k+q-1$, but computational evidence suggests this value is $q+1$.

2.5 Cyclic codes and polynomial algebra. Introduction to BCH codes

A useful class of linear codes is formed by the so-called *cyclic* codes (in particular, the Hamming, Golay and Reed–Muller codes are cyclic). Cyclic codes were proposed by Eugene Prange in 1957; their importance was immediately recognised, and they generated a large literature. But more importantly, the idea of cyclic codes,

together with some other sharp observations made at the end of the 1950s, particularly the invention of BCH codes, opened a connection from the theory of linear
codes (which was then at its initial stage) to algebra, particularly to the theory of finite fields. This created algebraic coding theory, a thriving direction in the modern
theory of linear codes.

We begin with binary cyclic codes. The coding and decoding procedures for
binary cyclic codes of length N are based on the related algebra of polynomials
with binary coefficients:

$$a(X) = a_0 + a_1X + \cdots + a_{N-1}X^{N-1}, \text{ where } a_k \in \mathbb{F}_2 \text{ for } k = 0,\ldots,N-1. \quad (2.5.1)$$

Such polynomials can be added and multiplied in the usual fashion, except that
$X^k + X^k = 0$. This defines a binary *polynomial algebra* $\mathbb{F}_2[X]$; the operations over
binary polynomials refer to this algebra. The degree $\deg a(X)$ of polynomial $a(X)$
equals the maximal label of its non-zero coefficient. The degree of the zero polynomial is set to be 0. Thus, the representation (2.5.1) covers polynomials of degree
$< N$.

Theorem 2.5.1   (a) $(1+X)^{2^l} = 1 + X^{2^l}$ *(A freshman's dream).*

(b) *(The division algorithm) Let $f(X)$ and $h(X)$ be two binary polynomials with
$h(X) \neq 0$. Then there exist unique polynomials $g(X)$ and $r(X)$ such that*

$$f(X) = g(X)h(X) + r(X) \quad with \quad \deg r(X) < \deg h(X). \quad (2.5.2)$$

The polynomial $g(X)$ is called the ratio, or quotient, and $r(X)$ the remainder.

Proof   (a) The statement follows from the binomial decomposition where all
intermediate terms vanish.

(b) If $\deg h(X) > \deg f(X)$ we simply set

$$f(X) = 0 \cdot h(X) + f(X).$$

If $\deg h(X) \le \deg f(X)$, we can perform the 'standard' procedure of long division, with the rules of the binary addition and multiplication.   □

Example 2.5.2   For binary polynomials:

(a) $(1 + X + X^3 + X^4)(X + X^2 + X^3) = X + X^7.$
(b) $1 + X^N = (1+X)(1 + X + \cdots + X^{N-1}).$
(c) The quotient $(X + X^2 + X^6 + X^7 + X^8) / (1 + X + X^2 + X^4) = X^3 + X^4$; the
     remainder equals $X + X^2 + X^3$.

Definition 2.5.3   Two polynomials, $f_1(X)$ and $f_2(X)$, are called *equivalent* mod $h(X)$, or $f_1(X) \equiv f_2(X)$ mod $h(X)$, if their remainders, after division by $h(X)$, coincide. That is,

$$f_i(X) = g_i(X)h(X) + r(X), \quad i = 1, 2,$$

and $\deg r(X) < \deg h(X)$.

Theorem 2.5.4   *Addition and multiplication of polynomials respect the equivalence. That is, if*

$$f_1(X) \equiv f_2(X) \bmod h(X) \quad and \quad p_1(X) \equiv p_2(X) \bmod h(X), \tag{2.5.3}$$

then

$$\begin{cases} f_1(X) + p_1(X) \equiv f_2(X) + p_2(X) \bmod h(X), \\ f_1(X)p_1(X) \equiv f_2(X)p_2(X) \bmod h(X). \end{cases} \tag{2.5.4}$$

Proof   We have, for $i = 1, 2$,

$$f_i(X) = g_i(X)h(X) + r(X), \quad p_i(X) = q_i(X)h(X) + s(X),$$

with

$$\deg r(X), \deg s(X) < \deg h(X).$$

Hence

$$f_i(X) + p_i(X) = (g_i(X) + q_i(X))h(X) + (r(X) + s(X))$$

with

$$\deg(r(X) + s(X)) \leq \max[r(X), s(X)] < \deg h(X).$$

Thus

$$f_1(X) + p_1(X) \equiv f_2(X) + p_2(X) \bmod h(X).$$

Furthermore, for $i = 1, 2$, the product $f_i(X)p_i(X)$ is represented as

$$\Big(g_i(X)q_i(X)h(X) + r(X)q_i(X) + s(X)g_i(X) \Big) h(X) + r(X)s(X).$$

Hence, the remainder for both polynomials $f_1(X)p_1(X)$ and $f_2(X)p_2(X)$ may come only from $r(X)s(X)$. Thus it is the same for both of them.   □

Note that every linear binary code $\mathcal{X}_N$ corresponds to a set of polynomials, with coefficients $0, 1$, of degree $N - 1$ which is closed under addition mod 2:

$$\begin{aligned} a(X) &= a_0 + a_1X + \cdots + a_{N-1}X^{N-1} \leftrightarrow \mathbf{a}^{(N)} = a_0 \ldots a_{N-1}, \\ b(X) &= b_0 + b_1X + \cdots + b_{N-1}X^{N-1} \leftrightarrow \mathbf{b}^{(N)} = b_0 \ldots b_{N-1}, \\ a(X) + b(X) &\leftrightarrow \mathbf{a}^{(N)} + \mathbf{b}^{(N)} = (a_0 + b_0) \ldots (a_{N-1} + b_{N-1}). \end{aligned} \tag{2.5.5}$$

[The numeration of the digits in a word of length N using $0, \ldots, N-1$ instead of $1, \ldots, N$ is more convenient.]

We systematically write $a(X) \in \mathscr{X}$ when the word $\mathbf{a}^{(N)} = a_0 \ldots a_{N-1}$, representing polynomial $a(X)$, belongs to code $\mathscr{X}$.

Definition 2.5.5 Given a binary word $\mathbf{a} = a_0 a_1 \ldots a_{N-1}$, we define the *cyclic shift* $\pi \mathbf{a}$ as a word $a_{N-1} a_0 \ldots a_{N-2}$. A linear binary code $\mathscr{X}$ is called *cyclic* if the cyclic shift of each codeword is again a codeword.

A 'straightforward' way to form a cyclic code is as follows: take a word a, then its subsequent cyclic shifts $\pi \mathbf{a}$, $\pi^2 \mathbf{a}$, etc., and finally all sums of the vectors obtained. Such a construction allows one to build a code from a single word, and eventually all the properties of the code may be inferred from the properties of word $\mathbf{a}$. It turns out that *every* cyclic code may be obtained in such a way: the corresponding word is called a generator of a cyclic code.

Lemma 2.5.6 *A binary linear code $\mathscr{X}$ is cyclic iff, for any vector $\mathbf{u}$ from a basis of $\mathscr{X}$, $\pi \mathbf{u} \in \mathscr{X}$.*

Proof Each codeword in $\mathscr{X}$ is a sum of vectors of the basis, but $\pi(\mathbf{u} + \mathbf{v}) = \pi \mathbf{u} + \pi \mathbf{v}$; hence the result. □

A useful property of a cyclic shift is established below:

Lemma 2.5.7 *If the word $\mathbf{a}$ corresponds to a polynomial $a(X)$ then the word $\pi \mathbf{a}$ corresponds to $Xa(X) \bmod (1 + X^N)$.*

Proof The relations

$$Xa(X) = a_0 X + a_1 X^2 + \cdots + a_{N-2} X^{N-1} + a_{N-1} X^N$$
$$= a_{N-1} + a_0 X + a_1 X^2 + \cdots + a_{N-2} X^{N-1} \bmod (1 + X^N)$$

mean that the polynomial

$$a_{N-1} + a_0 X + \cdots + a_{N-2} X^{N-1}$$

corresponding to $\pi \mathbf{a}$ equals $Xa(X) \bmod (1 + X^N)$. □

A similar argument implies that the word $\pi^2 \mathbf{a}$ corresponds to $X^2 a(X) \bmod (1 + X^N)$, etc. More generally, we have the following.

Example 2.5.8 The inverse cyclic shift $\pi^{-1}: a_0 \ldots a_{N-2} a_{N-1} \in \{0,1\}^N \mapsto a_1 a_2 \ldots a_{N-1} a_0$ acts on polynomials $a(X)$ of degree at least $N-1$ by

$$\pi^{-1} a(X) = \frac{1}{X} \left[a(X) + a_0 \right] + a_0 X^{N-1}.$$

Theorem 2.5.9   *A binary cyclic code contains, with each pair of polynomials $a(X)$ and $b(X)$, the sum $a(X)+b(X)$ and any polynomial $v(X)a(X)$ mod $(1+X^N)$.*

Proof   By linearity the sum $a(X)+b(X) \in \mathcal{X}$. If $v(X) = v_0 + v_1 X + \cdots v_{N-1} X^{N-1}$ then each polynomial $X^k a(X)$ mod $(1+X^N)$ corresponds to $\pi^k a$ and hence belongs to $\mathcal{X}$. As

$$v(X)a(X) \text{ mod } (1+X^N) = \sum_{i=0}^{N-1} v_i X^i a(X) \text{ mod } (1+X^N),$$

the LHS belongs to $\mathcal{X}$.   □

In other words, the binary polynomials of degree at most $N-1$ with the $\star$-multiplication defined by

$$a \star b(X) = a(X)b(X) \text{ mod } (1+X^N), \tag{2.5.6}$$

and the usual $\mathbb{F}_2[X]$-addition, form a commutative *ring*, denoted by $\mathbb{F}_2[X]/(1+X^N)$. The binary cyclic codes are precisely the *ideals* of this ring.

Theorem 2.5.10   *Let $g(X) = \sum\limits_{i=0}^{N-k} g_i X^i$ be a non-zero polynomial of minimum degree in a binary cyclic code $\mathcal{X}$. Then:*

(i) *$g(X)$ is a unique polynomial of minimal degree;*
(ii) *the code $\mathcal{X}$ has rank k;*
(iii) *the codewords corresponding to $g(X), Xg(X), \ldots, X^{k-1}g(X)$, form a basis in $\mathcal{X}$; they are cyclic shifts of word $\mathbf{g} = g_0 \ldots g_{N-k}0 \ldots 0$;*
(iv) *$a(X) \in \mathcal{X}$ iff $a(X) = v(X)g(X)$ for some polynomial $v(x)$ of degree $< k$ (that is, $g(X)$ is a divisor of every polynomial from $\mathcal{X}$).*

Proof   (i) Suppose $c(X) = \sum\limits_{i=0}^{N-k} c_i X^i$ is another polynomial of minimal degree $N-k$ in $\mathcal{X}$. Then $g_{N-k} = c_{N-k} = 1$, and hence $\deg(c(X)+g(X)) < N-k$. But as $N-k$ is the minimal degree, $c(X)+g(X)$ should equal zero. This happens iff $g(X) = c(X)$. Hence, $g(X)$ is unique.

(ii) follows from (iii).

(iii) Assume that property (iv) holds. Then each polynomial $a(X) \in \mathcal{X}$ has the form

$$g(X)v(X) = \sum_{i=1}^{r} v_i X^i g(X), \quad r < k.$$

Hence, each polynomial $a(X) \in \mathcal{X}$ is a linear combination of polynomials $g(X), Xg(X), \ldots, X^{k-1}g(X)$ (all of which belong to $\mathcal{X}$). On the other hand, polynomials $g(X), Xg(X), \ldots, X^{k-1}g(X)$ have distinct degrees and hence are linearly independent. Therefore words $\mathbf{g}, \pi\mathbf{g}, \ldots, \pi^{k-1}\mathbf{g}$, corresponding to $g(X), Xg(X), \ldots, X^{k-1}g(X)$, form a basis in $\mathcal{X}$.

(iv) We know that each polynomial $a(X) \in \mathcal{X}$ has degree $> \deg g(X)$. By the division algorithm,

$$a(X) = v(X)g(X) + r(X).$$

Here, we must have

$$\deg v(X) < k \quad \text{and} \quad \deg r(X) < \deg g(X) = N - k.$$

But then $v(X)g(X)$ belongs to $\mathcal{X}$ owing to Theorem 2.5.9 (as $v(X)g(X)$ has degree $\leq N - 1$, it coincides with $v(X)g(X) \bmod (1 + X^N)$). Hence,

$$r(X) = a(X) + v(X)g(X) \in \mathcal{X}$$

by linearity. As $g(X)$ is a unique polynomial from $\mathcal{X}$ of minimum degree, $r(X) = 0$. □

Corollary 2.5.11   *Every binary cyclic code is obtained from the codeword corresponding to a polynomial of minimum degree, by cyclic shifts and linear combinations.*

Definition 2.5.12   A polynomial $g(X)$ of minimal degree in $\mathcal{X}$ is called a *minimal degree generator* of a (cyclic) binary code $\mathcal{X}$, or briefly a *generator* of $\mathcal{X}$.

Remark 2.5.13   There may be other polynomials that generate $\mathcal{X}$ in the sense of Corollary 2.5.11. But the minimum degree polynomial is unique.

Theorem 2.5.14   *A polynomial $g(X)$ of degree $\leq N - 1$ is the generator of a binary cyclic code of length N iff $g(X)$ divides $1 + X^N$. That is,*

$$1 + X^N = h(X)g(X) \qquad (2.5.7)$$

for some polynomial $h(X)$ (of degree $N - \deg g(X)$).

Proof   (The only if part.) By the division algorithm,

$$1 + X^N = h(X)g(X) + r(X), \quad \text{where } \deg r(X) < \deg g(X).$$

That is,

$$r(X) = h(X)g(X) + 1 + X^N, \quad \text{i.e. } r(X) = h(X)g(X) \bmod (1 + X^N).$$

By Theorem 2.5.10, $r(X)$ belongs to the cyclic code $\mathscr{X}$ generated by $g(X)$. But $g(X)$ must be the unique polynomial of minimum degree in $\mathscr{X}$. Hence, $r(X) = 0$ and $1 + X^N = h(X)g(X)$.

(The if part.) Suppose that $1 + X^N = h(X)g(X)$, $\deg h(X) = N - \deg g(X)$. Consider the set $\{a(X): a(X) = u(X)g(X) \bmod (1 + X^N)\}$, i.e. the *principal ideal* in the $\star$-multiplication polynomial ring corresponding to $g(X)$. This set forms a linear code; it contains $g(X), Xg(X), \ldots, X^{k-1}g(X)$ where $k = \deg h(X)$. It suffices to prove that $X^k g(X)$ also belongs to the set. But $X^k g(X) = 1 + X^N + \sum_{j=0}^{k-1} h_j X^j g(X)$, that is, $X^k g(X)$ is equivalent to a linear combination of $g(X), Xg(X), \ldots, X^{k-1}g(X)$. $\qquad\square$

Corollary 2.5.15 *All cyclic binary codes of length N are in a one-to-one correspondence with the divisors of polynomial $1 + X^N$.*

Hence, the cyclic codes are described through the factorisation of the polynomial $1 + X^N$. More precisely, we are interested in decomposing $1 + X^N$ into irreducible factors; combining these factors into products yields all possible cyclic codes of length N.

Definition 2.5.16 A polynomial $a(X) = a_0 + a_1 X + \cdots + a_{N-1} X^{N-1}$ is called *irreducible* if $a(X)$ cannot be written as a product of two polynomials, $b(X)$ and $b'(X)$, with $\min[\deg b(X), \deg b'(X)] \geq 1$.

The importance (and convenience) of irreducible polynomials for describing cyclic codes is obvious: every generator polynomial of a cyclic code of length N is a product of irreducible factors of $(1 + X^N)$.

Example 2.5.17 (a) The polynomial $1 + X^N$ has two 'standard' divisors:

$$1 + X^N = (1 + X)(1 + X + \cdots + X^{N-1}).$$

The first factor $1 + X$ generates the binary parity-check code $\mathscr{P}_N = \left\{ \mathbf{x} = x_0 \cdots x_{N-1} : \sum_i x_i = 0 \right\}$, whereas polynomial $1 + X + \cdots + X^{N-1}$ (it may be reducible) generates the repetition code $\mathscr{R}_N = \{ 00\ldots0, \ 11\ldots1 \}$.

(b) Select the generating and check matrices of the Hamming [7,4] code in the lexicographic form. If we re-order the digits $x_4 x_7 x_5 x_3 x_2 x_6 x_1$ (which leads to an equivalent code) then the rows of the generating matrix become subsequent cyclic shifts of each other:

$$G_{\text{cycl}}^{\text{H}} = \begin{pmatrix} 1101000 \\ 0110100 \\ 0011010 \\ 0001101 \end{pmatrix}$$

and the cyclic shift of the last row is again in the code:

$$\pi(0\,0\,0\,1\,1\,0\,1) = (1\,0\,0\,0\,1\,1\,0)$$
$$= (1\,1\,0\,1\,0\,0\,0) + (0\,1\,1\,0\,1\,0\,0) + (0\,0\,1\,1\,0\,1\,0).$$

By Lemma 2.5.6, the code is cyclic. By Theorem 2.5.10(iii), the generating polynomial $g(X)$ corresponds to the framed part in matrix $G^{\mathrm{H}}_{\mathrm{cycl}}$:

$$\underline{1101} \sim g(X) = 1 + X + X^3 = \text{the generator.}$$

But a similar argument can be used to show that an equivalent cyclic code is obtained from the word $1011 \sim 1 + X^2 + X^3$. There is no contradiction: it was not claimed that the polynomial ideal of a cyclic code is the principal ideal of a unique element.

If we choose a different order of the columns in the parity-check matrix, the code will be equivalent to the original code; that is, the code with the generator polynomial $1 + X^2 + X^3$ is again a Hamming $[7,4]$ code.

In Problem 2.3 we will check that the Golay $[23,7]$ code is generated by the polynomial $g(X) = 1 + X + X^5 + X^6 + X^7 + X^9 + X^{11}$.

Worked Example 2.5.18   *Using the factorisation*

$$X^7 + 1 = (X+1)(X^3 + X + 1)(X^3 + X^2 + 1) \tag{2.5.8}$$

in $\mathbb{F}_2[X]$, *find all cyclic binary codes of length 7. Identify those which are Hamming codes and their duals.*

Solution See the table below.

code $\mathcal{X}$	generator for $\mathcal{X}$	generator for $\mathcal{X}^{\perp}$
$\{0,1\}^7$	1	$1 + X^7$
parity-check	$1 + X$	$\sum_{0 \leq i \leq 6} X^i$
Hamming	$1 + X + X^3$	$1 + X^2 + X^3 + X^4$
Hamming	$1 + X^2 + X^3$	$1 + X + X^2 + X^4$
dual Hamming	$1 + X^2 + X^3 + X^4$	$1 + X + X^3$
dual Hamming	$1 + X + X^2 + X^4$	$1 + X^2 + X^3$
repetition	$\sum_{0 \leq i \leq 6} X^i$	$1 + X$
zero	$1 + X^7$	1

It is easy to check that all factors in (2.5.8) are irreducible. Any irreducible factor could be included or not included in decomposition of the generator polynomial. This argument proves that there exist exactly 8 binary codes in $\mathscr{H}_{7,2}$ as demonstrated in the table. □

Example 2.5.19 (a) Polynomials of the first degree, $1+X$ and X, are irreducible (but X does not appear in the decomposition for $1+X^N$). There is one irreducible binary polynomial of degree 2: $1+X+X^2$, two of degree 3: $1+X+X^3$ and $1+X^2+X^3$, and three of degree 4:

$$1+X+X^4, \quad 1+X^3+X^4 \quad \text{and} \quad 1+X+X^2+X^3+X^4, \tag{2.5.9}$$

each of which appears in the decomposition of $1+X^N$ for various values of N (see below). A further distinction is that polynomials $1+X+X^3$ and $1+X^2+X^3$ are 'primitive' whereas $1+X+X^2+X^3+X^4$ is not; see Example 2.5.34 below and Sections 3.1–3.3. On the other hand, polynomials

$$1+X^8, \quad 1+X^4+X^6+X^7+X^8 \quad \text{and} \quad 1+X^2+X^6+X^8 \tag{2.5.10}$$

are reducible. The polynomial $1+X^N$ is always reducible:

$$1+X^N = (1+X)(1+X+\cdots+X^{N-1}).$$

(b) Generally, the factorisation of polynomial $1+X^N$ into the irreducible factors is not easy to achieve. Among the first 13 odd values of N, the list of polynomials $1+X^N$ which admit only the trivial decomposition into two irreducible factors is as follows:

$$1+X, \quad 1+X^3, \quad 1+X^5, \quad 1+X^{11}, \quad 1+X^{13}.$$

Further, the polynomial $1+X^{19}$ admits only a trivial decomposition $(1+X)(1+X+\cdots+X^{18})$, while others have the following factors (the common factor $(1+X)$ is omitted):

$$
\begin{aligned}
&1+X^7 : (1+X+X^3)(1+X^2+X^3),\\
&1+X^9 : (1+X+X^2)(1+X^3+X^6),\\
&1+X^{15} : (1+X+X^2)(1+X+X^4)\\
&\qquad \times (1+X^3+X^4)(1+X+X^2+X^3+X^4),\\
&1+X^{17} : (1+X^3+X^4+X^5+X^8)\\
&\qquad \times (1+X+X^2+X^4+X^6+X^7+X^8),\\
&1+X^{21} : (1+X+X^2)(1+X+X^3)(1+X^2+X^3)\\
&\qquad \times (1+X+X^2+X^4+X^6)(1+X^2+X^4+X^5+X^6),\\
&1+X^{23} : (1+X+X^5+X^6+X^7+X^9+X^{11})\\
&\qquad \times (1+X^2+X^4+X^5+X^6+X^{10}+X^{11}),
\end{aligned}
$$

and

$$1 + X^{25} : (1 + X + X^2 + X^3 + X^4)(1 + X^5 + X^{10} + X^{15} + X^{20}).$$

For N even, $1 + X^N$ can have multiple roots (see Example 2.5.35(c)).

Example 2.5.20   Irreducible polynomials of degree 2 and 3 over the field $\mathbb{F}_3$ (that is, from $\mathbb{F}_3[X]$) are as follows. There exist three irreducible polynomials of degree 2 over $\mathbb{F}_3$: $X^2 + 1$, $X^2 + X + 2$ and $X^2 + 2X + 2$. There exist eight irreducible polynomials of degree 3 over $\mathbb{F}_3$: $X^3 + 2X + 2$, $X^3 + X^2 + 2$, $X^3 + X^2 + X + 2$, $X^3 + 2X^2 + 2X + 2$, $X^3 + 2X + 1$, $X^3 + X^2 + 2X + 1$, $X^3 + 2X^2 + 1$ and $X^3 + 2X^2 + X + 1$.

Cyclic codes admit encoding and decoding procedures in terms of the polynomials. It is convenient to have a generating matrix of a cyclic code $\mathcal{X}$ in a form similar to G_{cycl} for the Hamming $[7,4]$ code (see above). That is, we want to find the basis in $\mathcal{X}$ which gives the following picture in the corresponding generating matrix:

$$G_{\text{cycl}} = \begin{pmatrix} \rule{1cm}{0.3em} & & \\ & \rule{1cm}{0.3em} & \mathbf{0} \\ & \rule{1cm}{0.3em} & \\ \mathbf{0} & & \ddots \\ & & \rule{1cm}{0.3em} \end{pmatrix} \tag{2.5.11}$$

Such a basis is provided by Theorem 2.5.10(iii): take the generator polynomial $g(X)$ and its multiples:

$$g(X), Xg(X), \ldots, X^{k-1}g(X), \deg g(X) = N - k.$$

Symbolically,

$$G_{\text{cycl}} = \begin{pmatrix} g(X) \\ Xg(X) \\ \vdots \\ X^{k-1}g(X) \end{pmatrix}. \tag{2.5.12}$$

The code has rank k and may be used for encoding words of length k as follows. Given a word $a = a_0 \ldots a_{k-1}$, we form the polynomial $a(X) = \sum_{0 \le i < k} a_i X^i$ and take the product $a(X)g(X)$. It belongs to $\mathcal{X}$ by Theorem 2.5.9, and hence defines a codeword. So all we have to do is to store polynomial $g(X)$: the encoding will correspond to polynomial multiplication. If encoding is given by multiplication, decoding must be related to division. Recall that under the geometric decoder, we decode the received word by the closest codeword in the Hamming distance. Such a codeword is related to a leader of the corresponding coset: we have seen that the cosets are in a one-to-one correspondence with the syndrome words of the form

$\mathbf{y}H^{\mathrm{T}}$. In the case of a cyclic code, the syndromes are calculated straightforwardly. Recall that, if $g(X)$ is a generator polynomial of a cyclic code $\mathscr{X}$ and $\deg g(X) = N - k$, then the rank of $\mathscr{X}$ equals k, and there must be 2^{N-k} distinct cosets (see Theorem 2.5.10(v)).

Theorem 2.5.21   *The cosets* $\mathbf{y} + \mathscr{X}$ *are in a one-to-one correspondence with the remainders* $\mathbf{y}(X) = u(X) \bmod g(X)$. *In other words, two words* $\mathbf{y}, \mathbf{y}'$ *belong to the same coset iff, in the division algorithm representation,*

$$y(X) = a(X)g(X) + u(X), \ y'(X) = a'(X)g(X) + u'(X), \text{ and } u(X) = u'(X).$$

Proof $\mathbf{y}$ and $\mathbf{y}'$ belong to the same coset iff $\mathbf{y} + \mathbf{y}' \in \mathscr{X}$. This is equivalent to $u(X) + u'(X) = 0$, i.e. $u(X) = u'(X)$ by Theorem 2.5.14. $\square$

Hence the cosets are labelled by the polynomials $u(X)$ of $\deg u(X) < \deg g(X) = N - k$: there are exactly 2^{N-k} such polynomials. To determine the coset $\mathbf{y} + \mathscr{X}$ it is enough to compute the remainder $u(X) = y(X) \bmod g(X)$. Unfortunately, there is still a task to find a leader in each case: there is no simple algorithm for finding leaders, for a general cyclic code. However, there are known particular classes of cyclic codes which admit a relatively simple decoding: the first such class was discovered in 1959 and is formed by BCH codes (see Section 2.6).

As was observed, a cyclic code may be generated not only by its polynomial of minimum degree: for some purposes other polynomials with this property may be useful. However, they all are divisors of $1 + X^N$:

Theorem 2.5.22   *Let* $\mathscr{X}$ *be a binary cyclic code of length* N. *Then any polynomial* $\widetilde{g}(X)$ *such that* $\mathscr{X}$ *is the principal ideal of* $\widetilde{g}(X)$ *is a divisor of* $1 + X^N$.

Proof   An exercise from algebra. $\square$

We see that the cyclic codes are naturally labelled by their generator polynomials.

Definition 2.5.23   Let $\mathscr{X}$ be the cyclic binary code of length N generated by $g(X)$. The *check polynomial* $h(X)$ of $\mathscr{X}$ is defined as the ratio $(1 + X^N)/g(X)$. That is, $h(X)$ is a unique polynomial for which $h(X)g(X) = 1 + X^N$.

We will use the standard notation $\gcd(f(X), g(X))$ for the greatest common divisor of polynomials $f(X)$ and $g(X)$ and $\mathrm{lcm}(f(X), g(X))$ for their least common multiple. Denote by $\mathscr{X}_1 + \mathscr{X}_2$ the direct sum of two linear codes $\mathscr{X}_1, \mathscr{X}_2 \subset \mathscr{H}_{N,2}$. That is, $\mathscr{X}_1 + \mathscr{X}_2$ consists of the linear combinations $\alpha_1 \mathbf{a}^{(1)} + \alpha_2 \mathbf{a}^{(2)}$ where $\alpha_1, \alpha_2 = 0, 1$ and $\mathbf{a}^{(i)} \in \mathscr{X}_i$, $i = 1, 2$. Compare Example 2.1.8(vii).

Worked Example 2.5.24   Let $\mathscr{X}_1$ and $\mathscr{X}_2$ be two binary cyclic codes of length N, with generators $g_1(X)$ and $g_2(X)$. Prove that:

(a) $\mathscr{X}_1 \subset \mathscr{X}_2$ iff $g_2(X)$ divides $g_1(X)$;

(b) the intersection $\mathscr{X}_1 \cap \mathscr{X}_2$ yields a cyclic code generated by $\mathrm{lcm}\left[g_1(X), g_2(X)\right]$;

(c) the direct sum $\mathscr{X}_1 + \mathscr{X}_2$ is a cyclic code with the generator $\gcd\left[g_1(X), g_2(X)\right]$.

Solution   (a) We know that $a(X) \in \mathscr{X}_i$ iff, in the ring $\mathbb{F}_2[X]/(1+X^N)$, polynomial $a(X) = f_i \star g_i(X)$, $i = 1, 2$. Suppose $g_2(X)$ divides $g_1(X)$ and write $g_1(X) = r(X)g_2(X)$. Then every polynomial $a(X)$ of the form $f_1 \star g_1(X)$ is of the form $f_1 \star r \star g_2(X)$. That is, if $a(X) \in \mathscr{X}_1$ then $a(X) \in \mathscr{X}_2$, so $\mathscr{X}_1 \subset \mathscr{X}_2$.

Conversely, suppose that $\mathscr{X}_1 \subset \mathscr{X}_2$. Let d_i be the degree of $g_i(X)$, $1 \le d_i < N$, $i = 1, 2$, and write

$$g_1(X) = f(X)g_2(X) + r(X), \quad \text{where } \deg r(X) < d_2.$$

We have that every polynomial $\star$-divisible by $g_1(X)$ in $\mathbb{F}_2[X]/(1+X^N)$ is also $\star$-divisible by $g_2(X)$. In particular, the basis polynomials $X^i g_1(X)$, $0 \le i \le N - d_1 - 1$, are $\star$-divisible by $g_2(X)$, i.e. have the form

$$X^i g_1(X) = h^{(i)}(X)g_2(X) + \alpha_i(X^N - 1) \quad \text{where } \alpha_i = 0 \text{ or } 1.$$

If, for some i, the coefficient $\alpha_i = 0$ then we compare two identities,

$$X^i g_1(X) = X^i f(X)g_2(X) + X^i r(X) \quad \text{and} \quad X^i g_1(X) = h^{(i)}(X)g_2(X),$$

and conclude that $X^i r(X) = 0$. This implies that $r(X) = 0$ and hence $g_2(X)$ divides $g_1(X)$.

The remaining case is that all coefficients $\alpha_i \equiv 1$. Then we compare

$$X g_1(X) = X h^{(0)}(X)g_2(X) + X + X^{N+1}$$

and

$$X g_1(X) = h^{(1)}(X)g_2(X) + 1 + X^N$$

and see that this case is impossible.

(b) This part becomes straightforward: the intersection $\mathscr{X}_1 \cap \mathscr{X}_2$ is a subcode of both $\mathscr{X}_1$ and $\mathscr{X}_2$. It is obviously a cyclic code; hence, by part (a), its generator $g(X)$ is divisible by both $g_1(X)$ and $g_2(X)$. Then it is divisible by the $\mathrm{lcm}(g_1(X), g_2(X))$. We must exclude the case where $g(X)$ produces a non-trivial ratio after this division. But the $\mathrm{lcm}(g_1(X), g_2(X))$ is itself a generator of a cyclic code (of the same original length) contained in both $\mathscr{X}_1$ and $\mathscr{X}_2$. So, in the case $g(X) \neq$

$\text{lcm}(g_1(X), g_2(X))$, the code generated by $\text{lcm}(g_1(X), g_2(X))$ must be strictly larger than $\mathscr{X}_1 \cap \mathscr{X}_2$. This contradicts the definition of $\mathscr{X}_1 \cap \mathscr{X}_2$.

(c) Similarly, $\mathscr{X}_1 + \mathscr{X}_2$ is the minimal linear code containing both $\mathscr{X}_1$ and $\mathscr{X}_2$. Hence, its generator divides both $g_1(X)$ and $g_2(X)$, i.e. is their common divisor. And if it is not equal to the $\gcd(g_1(X), g_2(X))$ then it contradicts the above minimality property. □

Worked Example 2.5.25   Let $\mathscr{X}$ be a binary cyclic code of length N with the generator $g(X)$ and the check polynomial $h(X)$. Prove that $a(X) \in \mathscr{X}$ iff the polynomial $(1 + X^N)$ divides $a(X)h(X)$, i.e. $a \star h(X) = 0$ in $\mathbb{F}_2[X]/(1 + X^N)$.

Solution   If $a(X) \in \mathscr{X}$ then $a(X) = f(X)g(X)$ for some polynomial $f(X) \in \mathbb{F}_2[X]/(1 + X^N)$. Then

$$a(X)h(X) = f(X)g(X)h(X) = f(X)(1 + X^N)$$

which equals 0 in $\mathbb{F}_2[X]/(1 + X^N)$. Conversely, let $a(X) \in \mathbb{F}_2[X]/(1 + X^N)$ and assume that $a(X)h(X) = 0 \bmod (1 + X^N)$. Write $a(X) = f(X)g(X) + r(X)$ where $\deg r(X) < \deg g(X)$. Then

$$a(X)h(X) = f(X)(1 + X^N) + r(X)h(X) = r(X)h(X) \bmod (1 + X^N).$$

Hence, $r(X)h(X) = 0 \bmod (1 + X^N)$ which is only possible when $r(X) = 0$ (since $\deg r(X)h(X) < N$). Thus, $a(X) = f(X)g(X)$ and $a(X) \in \mathscr{X}$. □

Worked Example 2.5.26   Prove that the dual of a cyclic code is again cyclic and find its generating matrix.

Solution If $\mathbf{y} \in \mathscr{X}^{\perp}$, the dual code, then the dot-product $\langle \pi \mathbf{x} \cdot \mathbf{y} \rangle = 0$ for all $\mathbf{x} \in \mathscr{X}$. But $\langle \pi \mathbf{x} \cdot \mathbf{y} \rangle = \langle \mathbf{x} \cdot \pi \mathbf{y} \rangle$, i.e. $\pi \mathbf{y} \in \mathscr{X}^{\perp}$, which means that $\mathscr{X}^{\perp}$ is cyclic.

Let $g(X) = g_0 + g_1 X + \cdots + g_{N-k} X^{N-k}$ be the generating polynomial for $\mathscr{X}$, where $N - k = d$ is the degree of $g(X)$ and k gives the rank of $\mathscr{X}$. We know that the generating matrix G of $\mathscr{X}$ may be written as

$$G \sim \begin{pmatrix} g(X) \\ Xg(X) \\ \cdot \\ \cdot \\ \cdot \\ X^{k-1}g(X) \end{pmatrix} \sim \begin{pmatrix} \rule{1cm}{0.4pt} & & \\ & \rule{1cm}{0.4pt} & \mathbf{0} \\ & & \rule{1cm}{0.4pt} \\ \mathbf{0} & & \ddots \\ & & \rule{1cm}{0.4pt} \end{pmatrix}. \qquad (2.5.13)$$

Take $h(X) = (1+X^N)/g(X)$ and write $h(X) = \sum\limits_{j=0}^{k} h_j X^j$ and $\mathbf{h} = h_0 \ldots h_{N-1}$. Then

$$\sum_{j=0}^{i} g_j h_{i-j} \begin{cases} = 1, & i = 0, N, \\ = 0, & 1 \leq i < N. \end{cases}$$

Indeed, for $i = 0, N$, we have $h_0 g_0 = 1$ and $h_k g_{N-k} = 1$. For $1 \leq i < N$ we obtain that the dot-product

$$\langle \pi^j \mathbf{g} \cdot \pi^{j'} \mathbf{h}^\perp \rangle = 0 \ \text{ for } j = 0, 1, \ldots, N-k-1, \ j' = 0, \ldots, k-1,$$

where $\mathbf{h}^\perp = h_k h_{k-1} \ldots h_0$. It is then easy to see that $\mathbf{h}^\perp$ gives rise to the generator $h^\perp(X)$ of $\mathscr{X}^\perp$. $\qquad\square$

An alternative solution is based on Worked Example 2.5.25. We know that $a(X) \in \mathscr{X}$ iff $a \star h(X) = 0$. Let k be the degree of $g(X)$ then the degree of $h(X)$ equals $N - k$. The degree $\deg[a(X)h(X)]$ is $< 2N - k$, so the coefficients of X^{N-k}, $X^{N-k+1}, \ldots, X^{N-1}$ in $a(X)h(X)$ all vanish. That is:

$$\begin{aligned}
a_0 h_{N-k} + a_1 h_{N-k-1} + \cdots + a_{N-k} h_0 &= 0, \\
a_1 h_{N-k} + a_2 h_{N-k-1} + \cdots + a_{N-k+1} h_0 &= 0, \\
&\vdots \\
a_{k-1} h_{N-k} + a_k h_{N-k-1} + \cdots + a_{N-1} h_0 &= 0.
\end{aligned}$$

In other words, $\mathbf{a}H^{\mathsf{T}} = 0$ where $\mathbf{a} = a_0, \ldots, a_{N-1}$ is the word of the binary coefficients for $a(X)$ and H is an $(N-k) \times N$ matrix

$$H \sim \begin{pmatrix} h^\perp(X) \\ Xh^\perp(X) \\ \vdots \\ X^{N-k-1}h^\perp(X) \end{pmatrix} \sim \begin{pmatrix} \rule[0.5ex]{1.5em}{0.4pt} \quad\ \ & & \\ & \rule[0.5ex]{1.5em}{0.4pt} \ \ & \mathbf{0} \\ & & \rule[0.5ex]{1.5em}{0.4pt} \quad \\ \mathbf{0} & & \ddots \\ & & \rule[0.5ex]{1.5em}{0.4pt} \end{pmatrix} \qquad (2.5.14)$$

and $h^\perp(X) = X^{N-k}h(X^{-1})$, with the coefficient string $\mathbf{h}^\perp = h_k h_{k-1} \ldots h_0$.

We conclude that matrix H generates a code $\mathscr{X}' \subseteq \mathscr{X}^\perp$. But since $h_{N-k} = 1$, the rank of $\mathscr{X}'$ equals $N - k$. Hence, $\mathscr{X}' = \mathscr{X}^\perp$.

It remains to check that polynomial $h^\perp(X)$ divides $1 + X^N$. To this end, we deduce from $g(X)h(X) = 1 + X^N$ that $h(X^{-1})g(X^{-1}) = X^{-N} + 1$. Hence $h^\perp(X)X^k g(X^{-1}) = 1 + X^N$, and as $X^k g(X^{-1})$ equals the polynomial $g_k + g_{k-1}X + \cdots + g_0 X^k$, the required fact follows. $\qquad\square$

Worked Example 2.5.27   *Let $\mathscr{X}$ be a binary cyclic code of length N with generator $g(X)$.*

(a) *Show that the set of codewords* $\mathbf{a} \in \mathcal{X}$ *of even weight is a cyclic code and find its generator.*

(b) *Show that* $\mathcal{X}$ *contains a codeword of odd weight iff* $g(1) \neq 0$ *or, equivalently, word* $\mathbf{1} \in \mathcal{X}$.

Solution (a) If code $\mathcal{X}$ is even (i.e. contains only words of even weight) then every polynomial $a(X) \in \mathcal{X}$ has $a(1) = \sum_{0 \leq i < N-1} a_i = 0$. Hence, $a(X)$ contains a factor $(X+1)$. Therefore, the generator $g(X)$ has a factor $(X+1)$. The converse is also true: if $(X+1)$ divides $g(X)$, or, equivalently, $g(1) = 0$, then every codeword $\underline{a} \in \mathcal{X}$ is of even weight.

Now assume that $\mathcal{X}$ contains a word with an odd weight, i.e. $g(1) = 1$; that is, $(1+X)$ does not divide $g(X)$. Let $\mathcal{X}^{\text{ev}}$ be the subcode in $\mathcal{X}$ formed by the even codewords. A cyclic shift does not change the weight, so $\mathcal{X}^{\text{ev}}$ is a cyclic code. For the corresponding polynomials $a(X)$ we have, as before, that $(1+X)$ divides $a(X)$. Thus, the generator $g^{\text{ev}}(X)$ of $\mathcal{X}^{\text{ev}}$ is divisible by $(1+X)$, hence $g^{\text{ev}}(X) = g(X)(X+1)$.

(b) It remains to show that $g(1) = 1$ iff the word $\underline{1} \in \mathcal{X}$. The corresponding polynomial is $1 + \cdots + X^{N-1}$, the complementary factor to $(1+X)$ in the decomposition $1 + X^N = (1+X)(1+\cdots+X^{N-1})$. So, if $g(1) = 1$, i.e. $g(X)$ does not contain the factor $(1+X)$, then $g(X)$ must be a divisor of $1 + \cdots + X^{N-1}$. This implies that $\underline{1} \in \mathcal{X}$. The inverse statement is established in a similar manner. $\square$

Worked Example 2.5.28 *Let* $\mathcal{X}$ *be a binary cyclic code of length N with generator $g(X)$ and check polynomial $h(X)$.*

(a) *Prove that* $\mathcal{X}$ *is self-orthogonal iff $h^{\perp}(X)$ divides $g(X)$ and self-dual iff $h^{\perp}(X) = g(X)$ where $h^{\perp}(X) = h_k + h_{k-1}X + \cdots + h_0 X^k$ ^{1} and $h(X) = h_0 + \cdots + h_{k-1}X^{k-1} + h_k X^k$ is the check polynomial, with $g(X)h(X) = 1 + X^N$.*

(b) *Let r be a divisor of N: $r|N$. A binary code $\mathcal{X}$ is called r-degenerate if every codeword $\mathbf{a} \in \mathcal{X}$ is a concatenation $\mathbf{c} \ldots \mathbf{c}$ where $\mathbf{c}$ is a string of length r. Prove that $\mathcal{X}$ is r-degenerate iff $h(X)$ divides $(1 + X^r)$.*

Solution (a) Self-orthogonality means that $\mathcal{X} \subseteq \mathcal{X}^{\perp}$, i.e. $\langle \mathbf{a} \cdot \mathbf{b} \rangle = 0$ for all $\mathbf{a}, \mathbf{b} \in \mathcal{X}$. From Worked Example 2.5.26 we know that $h^{\perp}(X)$ gives the generator polynomial of $\mathcal{X}^{\perp}$. Then, by virtue of Worked Example 2.5.26, $\mathcal{X} \subseteq \mathcal{X}^{\perp}$ iff $h^{\perp}(X)$ divides $g(X)$.

Self-duality means that $\mathcal{X} = \mathcal{X}^{\perp}$, that is $h^{\perp}(X) = g(X)$.

(b) For $N = rs$, we have the decomposition

$$1 + X^N = (1 + X^r)(1 + X^r + \cdots + X^{r(s-1)}).$$

Now assume cyclic code $\mathcal{X}$ of length N with generator $g(X)$ is r-degenerate. Then the word $\mathbf{g}$ is of the form $1\widetilde{\mathbf{c}}1\widetilde{\mathbf{c}}\ldots1\widetilde{\mathbf{c}}$ for some string $\widetilde{\mathbf{c}}$ of length $r-1$ (with $\mathbf{c} = 1\widetilde{\mathbf{c}}$). Let $\widetilde{c}(X)$ be the polynomial corresponding to $\widetilde{\mathbf{c}}$ (of degree $\leq r-2$). Then $g(X)$ is given by

$$
\begin{aligned}
1 + X\widetilde{c}(X) + X^r + X^{r+1}\widetilde{c}(X) + \cdots + X^{r(s-1)} + X^{r(s-1)+1}\widetilde{c}(X) \\
= (1 + X^r + \cdots + X^{r(s-1)})[1 + X\widetilde{c}(X)].
\end{aligned}
$$

For the check polynomial $h(X)$ we obtain

$$
\begin{aligned}
h(X) &= (1 + X^N) \Big/ \left((1 + X^r + \cdots + X^{r(s-1)})\,[1 + X\widetilde{c}(X)] \right) \\
&= (1 + X^r) \big/ [1 + X\widetilde{c}(X)],
\end{aligned}
$$

i.e. $h(X)$ is a divisor of $(1 + X^r)$.

Conversely, let $h(X)|(1 + X^r)$, with $h(X)g(X) = 1 + X^r$ where $g(X) = \sum_{0 \leq j \leq r-1} c_j X^j$, with $c_0 = 1$. Take $\mathbf{c} = c_0 \ldots c_{r-1}$; repeating the above argument in the reverse order, we conclude that the word $\mathbf{g}$ is the concatenation $\mathbf{c} \ldots \mathbf{c}$. Then the cyclic shift $\pi\mathbf{g}$ is the concatenation $\mathbf{c}^{(1)} \ldots \mathbf{c}^{(1)}$ where $\mathbf{c}^{(1)} = c_{r-1}c_0 \ldots c_{r-2}$ ($= \pi\mathbf{c}$, the cyclic shift of $\mathbf{c}$ in $\{0,1\}^r$). Similarly, for subsequent cyclic shift iterations $\pi^2\mathbf{g}, \ldots$. Hence, the basis vectors in $\mathcal{X}$ are r-degenerate, and so is the whole of $\mathcal{X}$. $\square$

In the 'standard' arithmetic, a (real or complex) polynomial $p(X)$ of a given degree d is conveniently identified through its roots (or zeros) $\alpha_1, \ldots, \alpha_d$ (in general, complex), by means of the monomial decomposition: $p(X) = p_d \prod_{1 \leq i \leq d}(X - \alpha_i)$. In the binary arithmetic (and, more generally, the q-ary arithmetic), the roots of polynomials are still an extremely useful concept. In our situation, the roots help to construct the generator polynomial $g(X) = \sum_{0 \leq i \leq d} g_i X^i$ of a binary cyclic code with important predicted properties. Assume for the moment that the roots $\alpha_1, \ldots, \alpha_d$ of $g(X)$ are a well-defined object, and the representation

$$
g(X) = \prod_{1 \leq i \leq d}(X - \alpha_i)
$$

has a consistent meaning (which is provided within the framework of finite fields). Even without knowing the formal theory, we are able to make a couple of helpful observations.

The first observation is that the α_i are Nth roots of unity, as they should be among the zeros of polynomial $1 + X^N$. Hence, they could be multiplied and inverted, i.e. would form an Abelian multiplicative group of size N, perhaps cyclic. Second, in the binary arithmetic, if α is a zero of $g(X)$ then so is α^2, as $g(X)^2 = g(X^2)$. Then α^2 is also a zero, as well as α^4, and so on. We conclude that the sequence $\alpha, \alpha^2, \ldots$ begins cycling: $\alpha^{2^d} = \alpha$ (or $\alpha^{2^d-1} = 1$) where d is the degree of $g(X)$. That is, all

*N*th roots of unity split into disjoint classes, of the form $\mathscr{C} = \left\{ \alpha, \alpha^2, \ldots, \alpha^{2^{c-1}} \right\}$, of size c where $c = c(\mathscr{C})$ is a positive integer (with $2^c - 1$ dividing N). The notation $\mathscr{C}(\alpha)$ is instructive, with $c = c(\alpha)$. The members of the same class are said to be *conjugate* to each other. If we want a generating polynomial with root α then all conjugate roots of unity $\alpha' \in \mathscr{C}(\alpha)$ will also be among the roots of $g(X)$.

Thus, to form a generator $g(X)$ we have to 'borrow' roots from classes $\mathscr{C}$ and enlist, with each borrowed root of unity, all members of their classes. Then, since any polynomial $a(X)$ from the cyclic code generated by $g(X)$ is a multiple of $g(X)$ (see Theorem 2.5.10(iv)), the roots of $g(X)$ will be among the roots of $a(X)$. Conversely, if $a(X)$ has roots α_i of $g(X)$ among its roots then $a(X)$ is in the code. We see that cyclic codes are conveniently described in terms of roots of unity.

Example 2.5.29 (The Hamming [7,4] code) Recall that the parity-check matrix H for the binary Hamming [7,4] code $\mathscr{X}^{\mathrm{H}}$ is 3×7; it enlists as its columns all non-zero binary words of length 3: different orderings of these rows define equivalent codes. Later in this section we explain that the sequence of non-zero binary words of any given length $2^\ell - 1$ written in some particular order (or orders) can be interpreted as a sequence of powers of a single element ω: $\omega^0, \omega, \omega^2, \ldots, \omega^{2^\ell - 2}$. The multiplication rule generating these powers is of a special type (multiplication of polynomials modulo a particular irreducible polynomial of degree ℓ). To stress this fact, we use in this section the notation $*$ for this multiplication rule, writing ω^{*i} in place of ω^i. Anyway, for $l = 3$, one appropriate order of the binary non-zero 3-words (out of the two possible orders) is

$$H = \begin{pmatrix} 0 & 0 & 1 & 0 & 1 & 1 & 1 \\ 0 & 1 & 0 & 1 & 1 & 1 & 0 \\ 1 & 0 & 0 & 1 & 0 & 1 & 1 \end{pmatrix} \sim (\omega^{*0} \; \omega \; \omega^{*2} \; \omega^{*3} \; \omega^{*4} \; \omega^{*5} \; \omega^{*6}).$$

Then, with this interpretation, the equation $\mathbf{a}H^{\mathrm{T}} = 0$, determining that the word $\mathbf{a} = a_0 \ldots a_6$ (or its polynomial $a(X) = \sum\limits_{0 \leq i < 7} a_i X^i$) lies in $\mathscr{X}^{\mathrm{H}}$, can be rewritten as

$$\sum_{0 \leq i < 7} a_i \omega^{*i} = 0, \quad \text{or} \quad a(*\omega) = 0.$$

In other words, $a(X) \in \mathscr{X}^{\mathrm{H}}$ iff ω is a root of $a(X)$ under the multiplication rule $*$ (which in this case is multiplication of binary polynomials of degree ≤ 2 modulo the polynomial $1 + X + X^3$).

The last statement can be rephrased in this way: the Hamming [7,4] code is equivalent to the cyclic code with the generator $g(X)$ that has ω among its roots; in this case the generator $g(X) = 1 + X + X^3$, with $g(*\omega) = \omega^{*0} + \omega + \omega^{*3} = 0$. The alternative ordering of the rows of H^{H} is related in the same fashion to the polynomial $1 + X^2 + X^3$.

We see that the Hamming $[7,4]$ code is defined by a single root ω, provided that we establish proper terms of operation with its powers. For that reason we can call ω the defining root (or defining zero) for this code. There are reasons to call element ω 'primitive'; cf. Sections 3.1–3.3.

Worked Example 2.5.30 *A code $\mathscr{X}$ is called reversible if $\mathbf{a} = a_0 a_1 \ldots a_{N-1} \in \mathscr{X}$ implies that $\mathbf{a}^{\leftarrow} = a_{N-1} \ldots a_1 a_0 \in \mathscr{X}$. Prove that a cyclic code with generator $g(X)$ is reversible iff $g(\alpha) = 0$ implies $g(\alpha^{-1}) = 0$.*

Solution For the generator polynomial $g(X) = \sum\limits_{0 \le i \le d} g_i X^i$, with $\deg g(X) = d < N$ and $g_0 = g_d = 1$, the reversed polynomial is $g^{\mathrm{rev}}(X) = X^{N-1} g(X^{-1})$, so if the cyclic code $\mathscr{X}$ is reversible and α is a root of $g(X)$ then α is also a root of $g^{\mathrm{rev}}(X)$. This is possible only when $g(\alpha^{-1}) = 0$.

Conversely, let $g(X)$ satisfy the property that $g(\alpha) = 0$ implies $g(\alpha^{-1}) = 0$. The above formula holds for all polynomial $a(X)$ of degree $< N$: $a^{\mathrm{rev}}(X) = X^{N-1} a(X^{-1})$. If $a(X) \in \mathscr{X}$ then $a(\alpha) = a(\alpha^{-1}) = 0$ for all root α of $g(X)$. Then $a^{\mathrm{rev}}(\alpha) = a^{\mathrm{rev}}(\alpha^{-1}) = 0$ for all roots α of $g(X)$. Thus, $a^{\mathrm{rev}}(X)$ is a multiple of $g(X)$, and $a^{\mathrm{rev}}(X) \in \mathscr{X}$. □

The natural framework for studying roots of polynomials is provided by the theory of finite fields or Galois theory (we have seen already how polynomial fields can be used). In the rest of this section we give an initial (and brief) introduction into some aspects of Galois theory to understand better some examples of codes introduced so far. In Chapter 3 we will dive deeper into the Galois theory to gain enough knowledge in order to proceed further with code constructions.

Remark 2.5.31 A field is a commutative ring where each non-zero element has an inverse. In other words, a ring is a field if the multiplication generates a group. In fact, a multiplication group of non-zero elements of a field is cyclic.

Theorem 2.5.32 *Let $g(X) \in \mathbb{F}_2[X]$ be an irreducible binary polynomial of degree d. Then multiplication mod $g(X)$ makes the set of the binary polynomials of degree $\le d - 1$ (i.e. the space $\mathbb{F}_2^{\times d}$) a field with 2^d elements. Conversely, if multiplication mod $g(X)$ leads to a field then $g(X)$ is irreducible.*

Proof The only non-trivial property to check is the existence of the inverse element. Take a non-zero polynomial $f(X)$, with $\deg f(X) \le d - 1$, and consider all polynomials of the form $f(X)h(X)$ (the usual multiplication) where $h(X)$ runs over the whole set of the polynomials of degree $\le d - 1$. These products must be distinct mod $g(X)$. Indeed, if

$$f(X)h_1(X) = f(X)h_2(X) \bmod g(X),$$

then, for some polynomial $v(X)$ of degree $\leq d-2$,

$$f(X)(h_1(X) - h_2(X)) = v(X)g(X). \qquad (2.5.15)$$

This implies that either $g(X)|f(X)$ or $g(X)|h_1(X) - h_2(X)$. We conclude that if polynomial $g(X)$ is irreducible, (2.5.15) is impossible, unless $h_1(X) = h_2(X)$ and $v(X) = 0$. For one and only one polynomial $h(X)$, we have

$$f(X)h(X) = 1 \bmod g(X);$$

$h(X)$ represents the inverse for $f(X)$ in multiplication mod $g(X)$. We write $h(X) = f(X)^{*-1}$.

On the other hand, if $g(X)$ is reducible, then $g(X) = b(X)b'(X)$ where both $b(X)$ and $b'(X)$ are non-zero and have degree $< d$. That is, $b(X)b'(X) = 0 \bmod g(X)$. If the multiplication mod q led to a field both $b(X)$ and $b'(X)$ would have inverses, $b(X)^{-*1}$ and $b'(X)^{-*1}$. But then

$$b(X)^{-*1} * b(X) * b'(X) = b'(X) = 0,$$

and similarly $b(X) = 0$.   □

A field obtained via the above construction is called a *polynomial* field and is often denoted by $\mathbb{F}_2[X]/\langle g(X) \rangle$. It contains 2^d elements where $d = \deg g(X)$ (representing polynomials of degree $< d$). We will call $g(X)$ the *core polynomial* of the field. For the rest of this section we denote the multiplication in a given polynomial field by $*$. The zero polynomial and the unit polynomial are denoted correspondingly, by $\mathbf{0}$ and $\mathbf{1}$: they are obviously the zero and the unity of the polynomial field. A key role is played by the following result.

Theorem 2.5.33   (a) *The multiplicative group of non-zero elements in polynomial field $\mathbb{F}_2[X]/\langle g(X) \rangle$ is isomorphic to the cyclic group $\mathbb{Z}_{2^d-1}$ of size $2^d - 1$.*
(b) *The polynomial fields obtained by picking different irreducible polynomials of degree d are all isomorphic.*

Proof   We will only prove here assertion (a); assertion (b) will be established in Section 3.1. Take any element from the field, $a(X) \in \mathbb{F}_2[X]/\langle g(X) \rangle$, and observe that

$$a^{*i}(X) := \underbrace{a * \ldots * a}_{i \text{ times}}(X)$$

(the multiplication in the field) takes at most $2^d - 1$ values (the number of elements in the field less one, as the zero $\mathbf{0}$ is excluded). Hence there exists a positive integer r such that $a^{*r}(X) = \mathbf{1}$; the smallest value of r is called the *order* of $a(X)$.

Choose a polynomial $a(X) \in \mathbb{F}_2[X]/\langle g(X)\rangle$ with the largest order r. Then we claim that the order of any other element $b(X)$ divides r. Indeed, let s be the order of $b(X)$. Pick a prime factor p of s and write

$$s = p^{c'} l', \text{ and } r = p^c l,$$

with integers $c', c \geq 0$ and $l, l' \geq 1$, where l and l' are not divisible by p. We want to show that $c \geq c'$. Indeed, element $a^{*p^c}(X)$ has order l, $b^{*l'}(X)$ has order $p^{c'}$ and the product $a^{*p^b} * b^{*l'}(X)$ has order $l p^{c'}$. Hence, $c' \leq c$ or else r would not be maximal. This is true for any prime p, hence s divides r.

Thus, with r being the maximal order, every element $b(X)$ in the field obeys $b^{*r}(X) = \mathbf{1}$. By using the pigeon-hole principle, we conclude that $r = 2^d - 1$, the number of non-zero elements of the field. Hence, with $a(X)$ being an element of order r, the powers $\mathbf{1}, a(X), \ldots, a^{*(2^d-1)}(X)$ exhaust the multiplicative groups of the field. $\square$

In the wake of Theorem 2.5.33, we can use the notation $\mathbb{F}_{2^d}$ for any polynomial field $\mathbb{F}_2[X]/\langle g(X)\rangle$ where $g(X)$ is an irreducible binary polynomial of degree d. Further, the multiplicative group of non-zero elements in $\mathbb{F}_{2^d}$ is denoted by $\mathbb{F}_{2^d}^*$; it is cyclic ($\simeq \mathbb{Z}_{2^d-1}$, according to Theorem 2.5.33). Any generator of group $\mathbb{F}_{2^d}^*$ (whose $*$-powers exhaust $\mathbb{F}_{2^d}^*$) is called a *primitive element* of field $\mathbb{F}_{2^d}$.

Example 2.5.34   We can see the importance of writing down the full list of irreducible polynomials. There are six irreducible binary polynomials of degree 5 (each of which is primitive):

$$\begin{aligned} 1 + X^2 + X^5, \ 1 + X^3 + X^5, \ 1 + X + X^2 + X^3 + X^5, \\ 1 + X + X^2 + X^4 + X^5, \ 1 + X + X^3 + X^4 + X^5, \\ 1 + X^2 + X^3 + X^4 + X^5 \end{aligned} \qquad (2.5.16)$$

and nine of degree 6 (of which six are primitive):

$$\begin{aligned} 1 + X + X^6, \ 1 + X + X^3 + X^4 + X^6, \ 1 + X^5 + X^6, \\ 1 + X + X^2 + X^5 + X^6, \ 1 + X^2 + X^3 + X^5 + X^6, \\ 1 + X + X^4 + X^5 + X^6, \\ 1 + X + X^2 + X^4 + X^6, \ 1 + X^2 + X^4 + X^5 + X^6, \\ 1 + X^3 + X^6. \end{aligned} \qquad (2.5.17)$$

The number of irreducible polynomials grows significantly with the degree: there are 18 of degree 7, 30 of degree 8, and so on. However, there exist and are available quite extensive tables of irreducible polynomials over various finite fields.

X^{*i}	$1+X+X^3$ polynomial	word	X^{*i}	$1+X^2+X^3$ polynomial	word
$--$	**0**	000	$--$	**0**	000
X^{*0}	**1**	100	X^{*0}	**1**	100
X	X	010	X	X	010
X^{*2}	X^2	001	X^{*2}	X^2	001
X^{*3}	$1+X$	110	X^{*3}	$1+X^2$	101
X^{*4}	$X+X^2$	011	X^{*4}	$1+X+X^2$	111
X^{*5}	$1+X+X^2$	111	X^{*5}	$1+X$	110
X^{*6}	$1+X^2$	101	X^{*6}	$X+X^2$	011

Figure 2.6

Example 2.5.35   (a) The field $\mathbb{F}_2[X]/\langle 1+X+X^2\rangle$ has four elements: **0**, **1**, X, $1+X$, with the multiplication table:

$$X*X = 1+X, \qquad \text{as } X^2 = 1+X \bmod (1+X+X^2),$$
$$X*(1+X) = X+X*X = \mathbf{1},$$
$$(1+X)*(1+X) = 1+X+X+X*X = 1+1+X = X.$$

Since $X^{*3} = (1+X)*X = \mathbf{1}$, the group is isomorphic to $\mathbb{Z}_3$. An alternative notation for this field is $\mathbb{F}_4$.

(b) The fields $\mathbb{F}_2[X]/\langle 1+X+X^3\rangle$ and $\mathbb{F}_2[X]/\langle 1+X^2+X^3\rangle$ contain eight elements each, representing all polynomials of degree ≤ 2. Every such polynomial $a_0 + a_1X + a_2X^2$ is identified via the string of its coefficients $a_0a_1a_2$ (a binary word). The field tables are found by looking at the subsequent powers X^{*i}: see Figure 2.6.

In both cases the multiplicative group of non-zero elements is $\mathbb{Z}_7$. The two fields are obviously isomorphic, as they share the common multiplicative cyclic group formalism. The common notation for these fields is $\mathbb{F}_8$. Note that the two field tables coincide for the powers X^{*i} with $0 \leq i < 3$; in fact, this is a general pattern: see Sections 3.1–3.3.

Moreover, the element $X = X^{*1} \in \mathbb{F}_2[X]/\langle 1+X+X^3\rangle$ can be identified as a root of the core polynomial $1+X+X^3$ and element $X = X^{*1} \in \mathbb{F}_2[X]/\langle 1+X^2+X^3\rangle$ as a root of $1+X^2+X^3$, as these polynomials yield zeros in their respective fields. The remaining two roots are X^{*2} and X^{*4} (again calculated in their respective fields).

Applying this example to the Hamming $[7,4]$ code (cf. Example 2.5.29), the field $\mathbb{F}_2[X]/\langle 1+X+X^3\rangle$ leads to the roots of the generator $1+X+X^3$, and the field $\mathbb{F}_2[X]/\langle 1+X^2+X^3\rangle$ to those of $1+X^2+X^3$. That is, the Hamming $[7,4]$ code is equivalent to the cyclic code of length 7 with the defining root $\omega = X$ in either of the two isomorphic fields $\mathbb{F}_2[X]/\langle 1+X+X^3\rangle$ or $\mathbb{F}_2[X]/\langle 1+X^2+X^3\rangle$. With some ambiguity (which will be removed in Section 3.1) we may say that this code is defined by its root ω which is a primitive element of $\mathbb{F}_8$.

powers X^{*i}	polynomials	coefficient strings
$--$	**0**	0000
X^{*0}	**1**	1000
X	X	0100
X^{*2}	X^2	0010
X^{*3}	X^3	0001
X^{*4}	$1+X$	1100
X^{*5}	$X+X^2$	0110
X^{*6}	X^2+X^3	0011
X^{*7}	$1+X+X^3$	1101
X^{*8}	$1+X^2$	1010
X^{*9}	$X+X^3$	0101
X^{*10}	$1+X+X^2$	1110
X^{*11}	$X+X^2+X^3$	0111
X^{*12}	$1+X+X^2+X^3$	1111
X^{*13}	$1+X^2+X^3$	1011
X^{*14}	$1+X^3$	1001

Figure 2.7

(c) The field $\mathbb{F}_2[X]/\langle 1+X+X^4 \rangle$ contains 16 elements. The field table is given in Figure 2.7. In this case, the multiplicative group is $\mathbb{Z}_{15}$, and the field can be denoted by $\mathbb{F}_{16}$. As above, element $X \in \mathbb{F}_2[X]/\langle 1+X+X^4 \rangle$ yields a root of polynomial $1+X+X^4$; other roots are X^{*2}, X^{*4} and X^{*8}.

This example can be used to identify the Hamming $[15,11]$ code as (an equivalent to) the cyclic code with generator $g(X) = 1+X+X^4$. We can now say that the Hamming $[15,11]$ code is (modulo equivalence) the cyclic code of length 15 with the defining root $\omega(=X)$ in field $\mathbb{F}_2[X]/\langle 1+X+X^4 \rangle$. As X is a generator of the multiplicative group of the field, we again could say that the defining root ω is a primitive element in $\mathbb{F}_{16}$. $\square$

In general, take the field $\mathbb{F}_2[X]/\langle g(X) \rangle$, where $g(X) = \sum\limits_{0 \le i \le d} g_i X^i$ is an irreducible binary polynomial of degree m. Then the elements $X, X^{*2}, X^{*4}, \ldots, X^{*2^{d-1}}$ will satisfy the equation

$$\sum_{0 \le i \le d} g_i \left(X^{*s} \right)^{*i} = \mathbf{0}, \; s = 1, 2, \ldots, 2^{d-1}.$$

In other words, $X, X^{*2}, \ldots, X^{2^{d-1}}$ are precisely the zeros, in field $\mathbb{F}_2[X]/\langle g(X) \rangle$, of the irreducible core polynomial q.

Another feature emerging from Example 2.5.35 is that in all parts (a)–(c), element X represented the root of the core polynomial $g(X)$. However, this is not true in general: it only happens when $g(X)$ is a 'primitive' binary polynomial; for the detailed discussion of this property see Sections 3.1–3.3. For a primitive core polynomial $g(X)$ we have, in addition, that the powers X^i for $i < d = \deg g(X)$ coincide with X^{*i}, while further powers X^{*i}, $m \le i \le 2^d - 1$, are relatively easy to calculate. With this in mind, we can pass to a general binary Hamming code.

Example 2.5.36   Let $\mathscr{X}^{\mathrm{H}}$ be the binary Hamming $[2^\ell - 1, 2^\ell - 1 - \ell]$ code. We know that its parity-check matrix H features all non-zero column-vectors of length ℓ. These vectors, written in a particular order, list the consecutive powers ω^{*i}, $i = 0, 1, \ldots, 2^\ell - 2$, in the field $\mathbb{F}_2[X]/\langle g(X)\rangle$ where $\omega = X$ and $g(X) = g_0 + g_1 X + \cdots + g_\ell X^{\ell-1} + X^\ell$ is a primitive polynomial of degree ℓ. Thus,

$$H = \begin{pmatrix} 1 & 0 & \cdots & 0 & g_0 & \cdots \\ 0 & 1 & \cdots & 0 & g_1 & \cdots \\ \cdots & \cdots & \cdots & \cdots & \cdots & \cdots \\ 0 & 0 & \cdots & 0 & g_{\ell-1} & \cdots \\ 0 & 0 & \cdots & 1 & 0 & \cdots \end{pmatrix}, \tag{2.5.18}$$

or $H \sim \left(\mathbf{1}\; \omega \cdots \omega^{*(\ell-1)}\; \omega^{*\ell} \cdots \omega^{*(2^\ell-2)}\right)$.

Hence, as before, the equation $\mathbf{a}H^{\mathrm{T}} = \mathbf{0}$ for the codeword is equivalent to the equation $a(*\omega) = \mathbf{0}$ for the corresponding polynomial. So, we can say that $a(X) \in \mathscr{X}^{\mathrm{H}}$ iff ω is among the roots of $a(X)$.

On the other hand, by construction, ω is a root of $g(X)$: $g(*\omega) = 0$. Thus, we identify the Hamming $[2^\ell - 1, 2^\ell - 1 - \ell]$ code as equivalent to the cyclic code of length $2^\ell - 1$ with the generator polynomial $g(X)$, with the defining root ω. The role of ω can be played by any conjugate element, from $\left\{\omega, \omega^{*2}, \ldots, \omega^{*2^{\ell-1}}\right\}$.

The above idea leads to an immediate (and far-reaching) generalisation. Take $N = 2^\ell - 1$ and let ω be a primitive element of field $\mathbb{F}_{2^\ell}^* \simeq \mathbb{F}_2[X]/\langle g(X)\rangle$ where $g(X)$ is a primitive polynomial. (In all the examples and problems from this chapter, this requirement is fulfilled.) Consider a defining set of roots, to start with, of the form $\omega, \omega^2, \omega^3$, but more generally, $\omega, \omega^2, \ldots, \omega^{(\delta-1)}$. (Using parameter δ which is an integer > 3 is a tradition here.) Consider the cyclic code with these roots: what can we say about it? With the length $N = 2^\ell - 1$, we can guess that it will yield a subcode of the Hamming $[2^\ell - 1, 2^\ell - 1 - \ell]$ code, and it may correct more than a single error. This is the gist of the so-called (binary) BCH code construction (Bose–Choudhury, Hocquenguem, 1959).

In this section we restrict ourselves to a brief introduction to the BCH codes; in greater detail and generality these codes are discussed in Section 3.2. For $N = 2^\ell - 1$ field $\mathbb{F}_{2^\ell} \simeq \mathbb{F}_2[X]/\langle g(X)\rangle$ has the property that its non-zero elements are the Nth roots of unity (i.e. the zeros of the polynomial $1 + X^N$). In other words, polynomial $1 + X^N$ factorises into the product of linear factors $\prod_{1 \le j \le N} (X - \omega_j)$ where all ω_j list the whole of $\mathbb{F}_{2^\ell}^*$. (In the terminology of Section 3.1, $\mathbb{F}_{2^\ell}$ is the splitting field for $1 + X^N$ over $\mathbb{F}_2$.) As before, we use the notation $\omega := X$ for the generator of the multiplicative cyclic group $\mathbb{F}_{2^\ell}^*$. (In fact, it could be any generator of this group.)

Because $\omega^N = 1$ and the power N is minimal with this property, the element ω is often called a primitive Nth root of unity. Consequently, the powers ω^k for $0 \le k < N$ yield distinct elements of the field. This fact is used below when we conclude that the product

$$\prod_{1 \le i < j \le \delta-1} \left(\omega^{k_j} - \omega^{k_i} \right) \neq 0$$

for every collection of powers $\omega^{k_1}, \ldots, \omega^{k_{\delta-1}}$. (Such a collection extracts a $(\delta - 1) \times (\delta - 1)$ submatrix from the $(\delta - 1) \times N$ parity-check matrix in the proof of Theorem 2.5.39.)

Definition 2.5.37   Given $N = 2^\ell - 1$ and $\delta = 3, \ldots N$, define a *narrow-sense binary BCH* code $\mathscr{X}_{N,\delta}^{\mathrm{BCH}}$ of length N and with designed distance δ as the cyclic code formed by binary polynomials $a(X)$ of degree $< N$ such that

$$a(\omega) = a(\omega^2) = \cdots = a(\omega^{(\delta-1)}) = 0. \qquad (2.5.19)$$

In other words, $\mathscr{X}_{N,\delta}^{\mathrm{BCH}}$ is the cyclic code of length N whose generator $g(X)$ is the minimal binary polynomial with roots including $\omega, \omega^2, \ldots, \omega^{(\delta-1)}$:

$$g(X) = \mathrm{lcm} \left\{ (X - \omega), \ldots, (X - \omega^{(\delta-1)}) \right\}$$
$$= \mathrm{lcm} \left\{ M_\omega(X), \ldots, M_{\omega^{(\delta-1)}}(X) \right\}. \qquad (2.5.20)$$

Here lcm stands for the least common multiple and $M_\alpha(X)$ denotes the minimal binary polynomial with root α. For brevity, we will use in this chapter the term binary BCH codes. (A more general class of BCH codes will be introduced in Section 3.2.)

Example 2.5.38   For $N = 7$, the appropriate polynomial field is $\mathbb{F}_2[X]/\langle 1 + X + X^3 \rangle$ or $\mathbb{F}_2[X]/\langle 1 + X^2 + X^3 \rangle$, i.e. one of two realisations of field $\mathbb{F}_8$. Since 7 is a prime number, any non-zero polynomial from the field has the multiplicative order 7, i.e. is a generator of the multiplicative group in $\mathbb{F}_2[X]/\langle 1 + X^2 + X^3 \rangle$. In fact, we have the decomposition of polynomial $1 + X^7$ into irreducible factors:

$$1 + X^7 = (1 + X)(1 + X + X^3)(1 + X^2 + X^3).$$

Further, if we choose polynomial field $\mathbb{F}_2[X]/\langle 1 + X + X^3 \rangle$ then $\omega = X$ satisfies

$$\omega^3 = 1 + \omega, \quad (\omega^2)^3 = 1 + \omega^2, \quad (\omega^4)^3 = 1 + \omega^4,$$

i.e. the conjugates ω, ω^2 and ω^4 are the roots of the core polynomial $1 + X + X^3$:

$$1 + X + X^3 = (X - \omega)(X - \omega^2)(X - \omega^4).$$

Next, ω^3, ω^6 and $\omega^{12} = \omega^5$ are the roots of $1 + X^2 + X^3$:

$$1 + X^2 + X^3 = (X - \omega^3)(X - \omega^5)(X - \omega^6).$$

Hence, the binary BCH code of length 7 with designed distance 3 is formed by binary polynomials $a(X)$ of degree ≤ 6 such that

$$a(\omega) = a(\omega^2) = 0, \text{ that is, } a(X) \text{ is a multiple of } 1 + X + X^3.$$

This code is equivalent to the Hamming $[4, 7]$ code; in particular its 'true' distance equals 3.

Next, the binary BCH code of length 7 with designed distance 4 is formed by binary polynomials $a(X)$ of degree ≤ 6 such that

$$a(\omega) = a(\omega^2) = a(\omega^3) = 0, \text{ that is, } a(X) \text{ is a multiple of}$$
$$(1 + X + X^3)(1 + X^2 + X^3) = 1 + X + X^2 + X^3 + X^4 + X^5 + X^6.$$

This is simply the repetition code $\mathscr{R}_7$.

The staple of the theory of the BCH codes is

Theorem 2.5.39   (The BCH bound) *The minimal distance of a binary BCH code with designed distance δ is $\geq \delta$.*

The proof of Theorem 2.5.39 (sometimes referred to as the BCH theorem) is based on the following result.

Lemma 2.5.40   *Consider the $m \times m$ Vandermonde determinant with entries from a commutative ring:*

$$\det \begin{pmatrix} \alpha_1 & \alpha_2 & \cdots & \alpha_m \\ \alpha_1^2 & \alpha_2^2 & \cdots & \alpha_m^2 \\ \vdots & \vdots & \ddots & \vdots \\ \alpha_1^m & \alpha_2^m & \cdots & \alpha_m^m \end{pmatrix} = \det \begin{pmatrix} \alpha_1 & \alpha_1^2 & \cdots & \alpha_1^m \\ \alpha_2 & \alpha_2^2 & \cdots & \alpha_2^m \\ \vdots & \vdots & \ddots & \vdots \\ \alpha_m & \alpha_m^2 & \cdots & \alpha_m^m \end{pmatrix}. \tag{2.5.21}$$

The value of this determinant is

$$\prod_{1 \leq l \leq m} \alpha_l \times \prod_{1 \leq i < j \leq m} (\alpha_i - \alpha_j). \tag{2.5.22}$$

Proof of Lemma 2.5.40 Both determinants in (2.5.21) are polynomial expressions in $\alpha_1, \ldots, \alpha_m$. If $\alpha = \alpha_j$ for $i < j$ then the determinant has repeated rows (columns), and hence vanishes (as in the standard arithmetic). Hence, the determinant divides the product $\prod\limits_{1 \le i < j \le m} (\alpha_i - \alpha_j)$. Next, we compare the powers of α_i in (2.5.21) and (2.5.22): this immediately leads to the assertion of Lemma 2.5.40. ☐

Proof of Theorem 2.5.39 Let the polynomial $a(X) \in \mathscr{X}$. Then $a(*\omega^{*j}) = 0$ for all $j = 1, \ldots, \delta - 1$. That is,

$$
\begin{pmatrix}
1 & \omega & \omega^{*2} & \cdots & \omega^{*(N-1)} \\
1 & \omega^{*2} & \omega^{*4} & \cdots & \omega^{*2(N-1)} \\
\vdots & & \vdots & \ddots & \vdots \\
1 & \omega^{*(\delta-1)} & \omega^{*2(\delta-1)} & \cdots & \omega^{*(N-1)(\delta-1)}
\end{pmatrix}
\begin{pmatrix}
a_0 \\
a_1 \\
\vdots \\
a_{N-1}
\end{pmatrix}
= \mathbf{0}.
$$

Due to Lemma 2.5.40, any $(\delta - 1)$ columns of this $((\delta - 1) \times N)$ matrix are linearly independent. Hence, there must be at least δ non-zero coefficients in $a(X)$. Thus, the distance of $\mathscr{X}$ is $\ge \delta$. ☐

Example 2.5.41 (Here a mistake in [18], p. 106, is corrected.) Consider a BCH code with $N = 15$ and $\delta = 5$. Use the following decomposition into irreducible polynomials:

$$X^{15} - 1 = (X+1)(X^2 + X + 1)(X^4 + X + 1)(X^4 + X^3 + 1)$$
$$\times (X^4 + X^3 + X^2 + X + 1).$$

The generator of the code is

$$g(x) = (X^4 + X + 1)(X^4 + X^3 + X^2 + X + 1) = X^8 + X^7 + X^6 + X^4 + 1.$$

Indeed, $g(\omega^3) = g(\omega^9) = 0$. The set of zeros of $X^4 + X^3 + X^2 + X + 1$ is $(\omega^3, \omega^9, \omega^{12}, \omega^9)$. The set of zeros of $X^4 + X + 1$ is $(\omega, \omega^2, \omega^4, \omega^8)$. The set of zeros of $X^4 + X^3 + 1$ is $(\omega^7, \omega^{14}, \omega^{13}, \omega^{11})$. The set of zeros of $X^2 + X + 1$ is (ω^5, ω^{10}).

(b) Let $N = 31$ and ω be a primitive element of $\mathbb{F}_{32}$. The minimal polynomial with root ω is

$$M_\omega(X) = (X - \omega)(X - \omega^2)(X - \omega^4)(X - \omega^8)(X - \omega^{16}).$$

We find also the minimal polynomial for ω^5:

$$M_{\omega^5}(X) = (X - \omega^5)(X - \omega^{10})(X - \omega^{20})(X - \omega^9)(X - \omega^{18}).$$

By definition, the generator of the BCH code of length 31 with a designed distance $\delta = 8$ is $g(X) = \mathrm{lcm}(M_\omega(X), M_{\omega^3}(X), M_{\omega^5}(X), M_{\omega^7}(X))$. In fact, the minimal distance of the BCH code (which is, obviously, at least 9) is in fact at least 11. This follows from Theorem 2.5.39 because all the powers $\omega, \omega^2, \ldots, \omega^{10}$ are listed among the roots of $g(X)$.

There exists a decoding procedure for a BCH code which is simple to implement: it generalises the Hamming code decoding procedure. In view of Theorem 2.5.39, the BCH code with designed distance δ corrects at least $t = \left\lfloor \dfrac{\delta - 1}{2} \right\rfloor$ errors. Suppose a codeword $\mathbf{c} = c_0 \ldots c_{N-1}$ has been sent and corrupted to $\mathbf{r} = \mathbf{c} + \mathbf{e}$ where $\mathbf{e} = e_0 \ldots e_{N-1}$. Assume that $\mathbf{e}$ has at most t non-zero entries. Introduce the corresponding polynomials $c(X)$, $r(X)$ and $e(X)$, all of degrees $< N$. For $c(X)$ we have that $c(\omega) = c(\omega^2) = \cdots = c(\omega^{(\delta-1)}) = 0$. Then, clearly,

$$r(\omega) = e(\omega), \quad r(\omega^2) = e(\omega^2), \ldots, r\left(\omega^{(\delta-1)}\right) = e\left(\omega^{(\delta-1)}\right). \qquad (2.5.23)$$

So, we calculate $r(\omega^i)$ for $i = 1, \ldots, \delta - 1$. If these are all 0, $r(X) \in \mathcal{X}$ (no error or at least $t+1$ errors). Otherwise, let $E = \{i : e_i = 1\}$ indicate the erroneous digits and assume that $0 < \sharp E \leq t$. Introduce the *error locator polynomial*

$$\sigma(X) = \prod_{i \in E}(1 - \omega^i X), \qquad (2.5.24)$$

with binary coefficients, of degree $\sharp E$ and with the lowest coefficient 1. If we know $\sigma(X)$, we can find which powers ω^{-i} are its roots and hence find the erroneous digits $i \in E$. We then simply change these digits and correct the errors.

In order to calculate $\sigma(X)$, consider the formal power series

$$\zeta(X) = \sum_{j \geq 1} e\left(\omega^j\right) X^j.$$

(Observe that, as $\omega^N = 1$, the coefficients of this power series recur.) For the initial $(\delta - 1)$ coefficients, we have equalities, by virtue of (2.5.23):

$$e\left(\omega^j\right) = r\left(\omega^j\right), \quad j = 1, \ldots, \delta - 1;$$

these are the only ones needed for our purpose, and they are calculated in terms of the received word $\mathbf{r}$.

Now set

$$\omega(X) = \sum_{i \in E} \omega^i X \prod_{j \in E:\, j \neq i} (1 - \omega^j X). \qquad (2.5.25)$$

Next, rewrite the above formal series as

$$\zeta(X) = \sum_{j \geq 1} \sum_{i \in E} \omega^{ij} X^j = \sum_{i \in E} \sum_{j \geq 1} \omega^{ij} X^j = \sum_{i \in E} \frac{\omega^i X}{1 - \omega^i X} = \frac{\omega(X)}{\sigma(X)}. \qquad (2.5.26)$$

Observe that both polynomials $\omega(X)$ and $\sigma(X)$ are of degree $\sharp E \leq t$.

Now, the equation $\zeta(X)\sigma(X) = \omega(X)$ from (2.5.26) can be written in terms of the coefficients, with the help of the fact that

$$e\left(\omega^j\right) = r\left(\omega^j\right), \quad j = 1, \ldots, 2t;$$

namely,

$$\begin{aligned}\left(\sigma_0 + \sigma_1 X + \cdots + \sigma_t X^t\right) \\ \times \left(r(\omega)X + + \cdots + r\left(\omega^{2t}\right)X^{2t} + e\left(\omega^{(2t+1)}\right)X^{2t+1} + \cdots\right) \\ = \omega_0 + \omega_1 X + \cdots + \omega_t X^t.\end{aligned} \qquad (2.5.27)$$

We are interested in the coefficients of X^k for $t < k \le 2t$: these satisfy

$$\sum_{0 \le j \le t} \sigma_j r\left(\omega^{(k-j)}\right) = 0, \qquad (2.5.28)$$

which does not involve any of the terms $e\left(\omega^l\right)$. We obtain the following equations:

$$\begin{pmatrix} r\left(\omega^{(t+1)}\right) & r\left(\omega^t\right) & \cdots & r(\omega) \\ r\left(\omega^{(t+2)}\right) & r\left(\omega^{(t+1)}\right) & \cdots & r(\omega^2) \\ \vdots & \vdots & \ddots & \vdots \\ r\left(\omega^{(2t)}\right) & r\left(\omega^{(2t-1)}\right) & \cdots & r(\omega^t) \end{pmatrix} \begin{pmatrix} \sigma_0 \\ \sigma_1 \\ \vdots \\ \sigma_t \end{pmatrix} = \mathbf{0}.$$

The above matrix is $t \times (t+1)$, so it always has a non-zero vector in the kernel; this vector identifies the error locator polynomial $\sigma(X)$. We see that the above routine (called the *Berlekamp–Massey* decoding algorithm) enables us to specify the set E and hence correct $\le t$ errors.

Unfortunately, the BCH codes are asymptotically 'bad': for any sequence of BCH codes of length $N \to \infty$, either k/N or $d/N \to 0$. In other words, they lie at the bottom of Figure 2.2. To obtain codes that meet the Gilbert–Varshamov (GV) bound, one needs more powerful methods, based on algebraic geometry. Such codes were constructed in the early 1970s (the Goppa and Justesen codes). It remains a problem to construct codes that lie above the Gilbert–Varshamov curve. As was mentioned just before on page 160, a new class of codes was invented in 1982 by Tsfasman, Vlǎdut and Zink; these codes lie *above* the GV curve when the number of symbols in the code alphabet is large. However, for binary codes, the problem is still waiting for solution.

Worked Example 2.5.42   *Compute the rank and minimum distance of the cyclic code with generator polynomial $g(X) = X^3 + X + 1$ and parity-check polynomial $h(X) = X^4 + X^2 + X + 1$. Now let ω be a root of $g(X)$ in the field $\mathbb{F}_8$. We receive the word $r(X) = X^5 + X^3 + X \pmod{X^7 - 1}$. Verify that $r(\omega) = \omega^4$, and hence decode $r(X)$ using minimum-distance decoding.*

Solution   A cyclic code $\mathscr{X}$ of length N has generator polynomial $g(X) \in \mathbb{F}_2[X]$ and parity-check polynomial $h(X) \in \mathbb{F}_2[X]$ with $g(X)h(X) = 1 + X^N$. Recall that if $g(X)$ has degree k, i.e. $g(X) = a_0 + a_1 X + \cdots + a_k X^k$ where $a_k \ne 0$, then $g(X)$,

$Xg(X), \ldots, X^{N-k-1} g(X)$ form a basis for $\mathcal{X}$. In particular, the rank of $\mathcal{X}$ equals $N - k$. In this example, $N = 7$, $k = 3$ and rank$(\mathcal{X}) = 4$.

If $h(X) = b_0 + b_1 X + \cdots + b_{N-k} X^{N-k}$ then the parity-check matrix H for $\mathcal{X}$ has the form

$$\begin{pmatrix} b_{N-k} & b_{N-k-1} & \cdots & b_1 & b_0 & 0 & \cdots & 0 & 0 \\ 0 & b_{N-k} & b_{N-k-1} & \cdots & b_1 & b_0 & \cdots & 0 & 0 \\ & & & & & & & & \\ 0 & \ddots & \ddots & \ddots & \ddots & & \ddots & \ddots & \\ 0 & 0 & \cdots & 0 & b_{N-k} & b_{N-k-1} & \cdots & b_1 & b_0 \end{pmatrix}.$$

$$\underbrace{}_{N}$$

The codewords of $\mathcal{X}$ are linear dependence relations between the columns of H. In the example,

$$H = \begin{pmatrix} 1 & 0 & 1 & 1 & 1 & 0 & 0 \\ 0 & 1 & 0 & 1 & 1 & 1 & 0 \\ 0 & 0 & 1 & 0 & 1 & 1 & 1 \end{pmatrix}$$

and we have the following implications:

$$\begin{array}{ll} \text{no zero column} & \Rightarrow \quad \text{no codewords of weight 1,} \\ \text{no repeated column} & \Rightarrow \quad \text{no codewords of weight 2.} \end{array}$$

The minimum distance $d(\mathcal{X})$ of a linear code $\mathcal{X}$ is the minimum non-zero weight of a codeword. In the example, $d(\mathcal{X}) = 3$. [In fact, $\mathcal{X}$ is equivalent to the Hamming $[7,4]$ code.]

Since $g(X) \in \mathbb{F}_2[X]$ is irreducible, the code $\mathcal{X} \in \mathbb{F}_2[X] / \langle X^7 - 1 \rangle$ is the cyclic code defined by ω. The multiplicative cyclic group $\mathbb{Z}_7^\times$ of non-zero elements of field $\mathbb{F}_8$ is

$$\begin{aligned} & \omega^0 = 1, \; \omega, \; \omega^2, \; \omega^3 = \omega + 1, \; \omega^4 = \omega^2 + \omega, \\ & \omega^5 = \omega^3 + \omega^2 = \omega^2 + \omega + 1, \; \omega^6 = \omega^3 + \omega^2 + \omega = \omega^2 + 1, \\ & \omega^7 = \omega^3 + \omega = 1. \end{aligned}$$

Next, the value $r(\omega)$ is

$$\begin{aligned} r(\omega) &= \omega + \omega^3 + \omega^5 \\ &= \omega + (\omega + 1) + (\omega^2 + \omega + 1) \\ &= \omega^2 + \omega = \omega^4, \end{aligned}$$

as required. Let $c(X) = r(X) + X^4 \bmod (X^7 - 1)$. Then $c(\omega) = 0$, i.e. $c(X)$ is a codeword. Since $d(\mathcal{X}) = 3$ the code is 1-error correcting. We just found a codeword $c(X)$ at distance 1 from $r(X)$. Then $r(X) = X + X^3 + X^5$ should be decoded by

$$c(X) = X + X^3 + X^4 + X^5 \bmod (X^7 - 1)$$

under minimum-distance decoding. □

We conclude this section with two useful statements.

Worked Example 2.5.43   *(The Euclid algorithm for polynomials) The Euclid algorithm is a method for computing the greatest common divisor of two polynomials, $f(X)$ and $g(X)$, over the same finite field $\mathbb{F}$. Assume that $\deg g(X) \le \deg f(X)$ and set $f(X) = r_{-1}(X)$, $g(X) = r_0(X)$. Then*

(1) divide $r_{-1}(X)$ by $r_0(X)$:

$$r_{-1}(X) = q_1(X)r_0(X) + r_1(X) \text{ where } \deg r_1(X) < \deg r_0(X),$$

(2) divide $r_0(X)$ by $r_1(X)$:

$$r_0(X) = q_2(X)r_1(X) + r_2(X) \text{ where } \deg r_1(X) < \deg r_1(X),$$

$$\vdots$$

(k) divide $r_{k-1}(X)$ by $r_{k-1}(X)$:

$$r_{k-2}(X) = q_k(X)r_{k-1}(X) + r_k(X) \text{ where } \deg r_k(X) < \deg R_{k-1}(X),$$

$$\vdots$$

The algorithm continues until the remainder is 0:

(s) divide $r_{s-2}(X)$ by $r_{s-1}(X)$:

$$r_{s-2}(X) = q_s(X)r_{s-1}(X).$$

Then

$$\gcd\bigl(f(X), g(X)\bigr) = r_{s-1}(X). \tag{2.5.29}$$

At each stage, the equation for the current remainder $r_k(X)$ involves two previous remainders. Hence, all remainders, including $\gcd(f(X), g(X))$, can be written in terms of $f(X)$ and $g(X)$. In fact,

Lemma 2.5.44   *The remainders $r_k(X)$ in the Euclid algorithm satisfy*

$$r_k(X) = a_k(X)f(X) + b_k(X)g(X), \ k \le -1,$$

where

$$a_{-1}(X) = b_{-1}(X) = 0,$$
$$a_0(X) = 0, b_0(X) = 1,$$
$$a_k(X) = -q_k(X)a_{k-1}(X) + a_{k-2}(X), k \geq 1,$$
$$b_k(X) = -q_k(X)b_{k-1}(X) + b_{k-2}(X), k \geq 1.$$

In particular, there exist polynomials $a(X), b(X)$ such that

$$\gcd(f(X), g(X)) = a(X)f(X) + b(X)g(X).$$

Furthermore:

(1) $\deg a_k(X) = \sum\limits_{2 \leq i \leq k} \deg q_i(X)$, $\deg b_k(X) = \sum\limits_{1 \leq i \leq k} \deg q_k(X)$.

(2) $\deg r_k(X) = \deg f(X) - \sum\limits_{1 \leq i \leq k+1} \deg q_k(X)$.

(3) $\deg b_k(X) = \deg f(X) - \deg r_{k-1}(X)$.

(4) $a_k(X)b_{k+1}(X) - a_{k+}(X)b_k(X) = (-1)^{k+1}$.

(5) $a_k(X)$ and $b_k(X)$ are co-prime.

(6) $r_k(X)b_{k+1}(X) - r_{k+1}(X)b_k(X) = (-1)^{k+1}f(X)$.

(7) $r_{k+1}(X)a_k(X) - r_k(X)a_{k+1}(X) = (-1)^{k+1}g(X)$.

Proof   The proof is left as an exercise.   ☐

2.6 Additional problems for Chapter 2

Problem 2.1   *A check polynomial $h(X)$ of a binary cyclic code $\mathcal{X}$ of length N is defined by the condition $a(X) \in \mathcal{X}$ if and only if $a(X)h(X) = 0 \bmod (1 + X^N)$. How is the check polynomial related to the generator of $\mathcal{X}$? Given $h(X)$, construct the parity-check matrix and interpret the cosets $\mathcal{X} + \mathbf{y}$ of $\mathcal{X}$.*

Describe all cyclic codes of length 16 and 15. Find the generators and the check polynomials of the repetition and parity-check codes. Find the generator and the check polynomial of Hamming code of length 7.

Solution All cyclic codes of length 16 are divisors of $1 + X^{16} = (1 + X)^{16}$, i.e. are generated by $g(X) = (1 + X)^k$ where $k = 0, 1, \ldots, 16$. Here $k = 0$ gives the whole $\{0, 1\}^{16}$, $k = 1$ the parity-check code, $k = 15$ the repetition code $\{00\ldots0, 11\ldots1\}$ and $k = 16$ the zero code. For $N = 15$, the decomposition into irreducible polynomials looks as follows:

$$1 + X^{15} = (1 + X)(1 + X + X^2)(1 + X + X^4)(1 + X^3 + X^4)$$
$$\times (1 + X + X^2 + X^3 + X^4).$$

Any product of the listed irreducible polynomials generates a cyclic code.

244 — *Introduction to Coding Theory*

In general, $1+X^N = (1+X)(1+X+\cdots+X^{N-1})$; $g(X)=1+X$ generates the parity-check code and $g(X)=1+X+\cdots+X^{N-1}$ the repetition code. In the case of a Hamming $[7,4]$ code, the generator is $g(X)=1+X+X^3$, by inspection.

The check polynomial $h(X)$ equals the ratio $(1+X^N)/g(X)$. In fact, for all $a(X)\in\mathscr{X}$, $a(X)h(X)=v(X)g(X)h(X)=v(X)(1+X^N)=0 \bmod (1+X^N)$. Conversely, if $a(X)h(X)=v(X)(1+X^N)$ then $a(X)$ must be of the form $v(X)g(X)$, by the uniqueness of the irreducible decomposition.

The cosets $\mathbf{y}+\mathscr{X}$ are in a one-to-one correspondence with the remainders $y(X)=u(X)\bmod g(X)$. In other words, two words $\mathbf{y}^{(1)},\mathbf{y}^{(2)}$ belong to the same coset iff, in the division algorithm representation,

$$y^{(i)}(X)=v^i(X)g(X)+u^{(i)}(X),\quad i=1,2,\ \text{where}\ u^{(1)}(X)=u^{(2)}(X).$$

In fact, $\mathbf{y}^{(1)}$ and $\mathbf{y}^{(2)}$ belong to the same coset iff $\mathbf{y}^{(1)}+\mathbf{y}^{(2)}\in\mathscr{X}$. This is equivalent to $u^{(1)}(X)+u^{(2)}(X)=0$, i.e. $u^{(1)}(X)=u^{(2)}(X)$.

If we write $h(X)=\sum_{j=0}^k h_j X^j$, then the dot-product

$$\sum_{j=0}^i g_j h_{i-j}=\begin{cases}1,& i=0,N,\\ 0,& 1\le i<N.\end{cases}$$

So, $\langle g(X)\cdot h^{\perp}(X)\rangle=0$ where $h^{\perp}(X)=h_k+h_{k-1}X+\cdots+h_0X^k$. Therefore, the parity-check matrix H for $\mathscr{X}$ is formed by rows that are cyclic shifts of $\mathbf{h}=h_k\,h_{k-1}\cdots h_0 0\cdots 0$. The check polynomials for the repetition and parity-check codes then are $1+X$ and $1+X+\cdots+X^{N-1}$, and they are dual of each other. The check polynomial for the Hamming $[7,4]$ code equals $1+X+X^2+X^4$, by inspection. □

Problem 2.2 (a) *Prove the Hamming and Gilbert–Varshamov bounds on the size of a binary $[N,d]$ code in terms of $v_N(d)$, the volume of an N-dimensional Hamming ball of radius d.*

Suppose that the minimum distance is $\lfloor\lambda N\rfloor$ for some fixed $\lambda\in(0,1/4)$. Let $\alpha(N,\lfloor\lambda N\rfloor)$ be the largest information rate of any binary code correcting $\lfloor\lambda N\rfloor$ errors. Show that

$$1-\eta(\lambda)\le\liminf_{N\to\infty}\alpha(N,\lfloor\lambda N\rfloor)\le\limsup_{N\to\infty}\alpha(N,\lfloor\lambda N\rfloor)\le1-\eta(\lambda/2).\quad(2.6.1)$$

(b) *Fix $R\in(0,1)$ and suppose we want to send one of a collection U_N of messages of length N, where the size $\sharp U_N=2^{NR}$. The message is transmitted through an*

MBSC with error-probability $p < 1/2$, so that we expect about pN errors. According to the asymptotic bound of part (a), for which values of p can we correct pN errors, for large N?

Solution (a) A code $\mathscr{X} \subset \mathbb{F}_2^N$ is said to be E-error correcting if $B(\mathbf{x}, E) \cap B(\mathbf{y}, E) = \emptyset$ for all $\mathbf{x}, \mathbf{y} \in \mathscr{X}$ with $\mathbf{x} \neq \mathbf{y}$. The Hamming bound for a code of size M, distance d, correcting $E = \lfloor \dfrac{d-1}{2} \rfloor$ errors is as follows. The balls of radius E about the codewords are disjoint: their total volume equals $M \times v_N(E)$. But their union lies inside $\mathbb{F}_2^N$, so $M \leq 2^N/v_N(E)$.

On the other hand, take an E-correcting code $\mathscr{X}^*$ of maximum size $\sharp \mathscr{X}$. Then there will be no word

$$\mathbf{y} \in \mathbb{F}_2^N \setminus \cup_{\mathbf{x} \in \mathscr{X}^*} B(\mathbf{x}, 2E+1)$$

or we could add such a word to $\mathscr{X}^*$, increasing the size but preserving the error-correcting property. Since every word $\mathbf{y} \in \mathbb{F}_2^N$ is less than $d-1$ from a codeword, we can add $\mathbf{y}$ to the code. Hence, balls of radius $d-1$ cover the whole of $\mathbb{F}_2^N$, i.e. $M \times v_N(d-1) \geq 2^N$, or

$$M \geq 2^N/v_N(d-1) \text{ (the Varshamov–Gilbert bound).}$$

Combining these bounds yields, for $\alpha(N, E) = (\log \sharp \mathscr{X})/N$:

$$1 - \frac{\log v_N(2E+1)}{N} \leq \alpha(N, E) \leq 1 - \frac{\log v_N(E)}{N}.$$

Observe that for any $s < \kappa N$ with $0 < \kappa < 1/2$

$$\binom{N}{s-1} = \frac{s}{N-s+1} \binom{N}{s} < \frac{\kappa}{1-\kappa} \binom{N}{s}.$$

Consequently,

$$\binom{N}{E} \leq v_N(E) \leq \binom{N}{E} \sum_{j=0}^{E} \left(\frac{\kappa}{1-\kappa} \right)^j.$$

Now, by the Stirling formula as $N, E \to \infty$ and $E/N \to \lambda \in (0, 1/4)$

$$\frac{1}{N} \log \binom{N}{E} \to \eta(\lambda/2).$$

So, we proved that $\lim_{N \to \infty} \frac{1}{N} \log v_N([\lambda N]) = \eta(\lambda)$, and

$$1 - \eta(\lambda) \leq \liminf_{N \to \infty} \frac{1}{N} \log M \leq \limsup_{N \to \infty} \frac{1}{N} \log M \leq 1 - \eta(\lambda/2).$$

(b) We can correct pN errors if the minimum distance d satisfies $\left\lfloor \dfrac{d-1}{2} \right\rfloor \geq pN$, i.e. $\lambda/2 \geq p$. Using the asymptotic Hamming bound we obtain $R \leq 1 - \eta(\lambda/2) \leq 1 - \eta(p)$. So, the reliable transmission is possible if $p \leq \eta^{-1}(1-R)$,

The Shannon SCT states:

$$\text{capacity } C \text{ of a memoryless channel} = \sup_{px} I(X:Y).$$

Here $I(X:Y) = h(Y) - h(Y|X)$ is the mutual entropy between the single-letter random input and output of the channel, maximised over all distributions of the input letter X. For an MBSC with the error-probability p, the conditional entropy $h(Y|X)$ equals $\eta(p)$. Then

$$C = \sup_{px} h(Y) - \eta(p).$$

But $h(Y)$ attains its maximum 1, by using the equidistributed input X (then Y is also equidistributed). Hence, for the MBSC, $C = 1 - \eta(p)$. So, a reliable transmission is possible via MBSC with $R \leq 1 - \eta(p)$, i.e. $p \leq \eta^{-1}(1-R)$. These two arguments lead to the same answer. $\qquad\square$

Problem 2.3   *Prove that the binary code of length 23 generated by the polynomial $g(X) = 1 + X + X^5 + X^6 + X^7 + X^9 + X^{11}$ has minimum distance 7, and is perfect. Hint: Observe that by the BCH bound (see Theorem 2.5.39) if a generator polynomial of a cyclic code has roots $\{\omega, \omega^2, \ldots, \omega^{\delta-1}\}$ then the code has distance $\geq \delta$, and check that $X^{23} + 1 \equiv (X+1)g(X)g^{\mathrm{rev}}(X) \bmod 2$, where $g^{\mathrm{rev}}(X) = X^{11}g(1/X)$ is the reversal of $g(X)$.*

Solution   First, show that the code is BCH, of designed distance 5. Recall that if ω is a root of a polynomial $p(X) \in \mathbb{F}_2[X]$ then so is ω^2. Thus, if ω is a root of $g(X) = 1 + X + X^5 + X^6 + X^7 + X^9 + X^{11}$ then so are ω^2, ω^4, ω^8, ω^{16}, ω^9, ω^{18}, ω^{13}, ω^3, ω^6, ω^{12}. This yields the design sequence $\{\omega, \omega^2, \omega^3, \omega^4\}$. By the BCH theorem, the code $\mathscr{X} = \langle g(X) \rangle$ has distance ≥ 5.

Next, the parity-check extension, $\mathscr{X}^+$, is self-orthogonal. To check this, we need only to show that any two rows of the generating matrix of $\mathscr{X}^+$ are orthogonal. These are represented by

$$(X^i g(X)|1) \quad \text{and} \quad (X^j g(X)|1)$$

and their dot-product is

$$1 + (X^i g(X))(X^j g(X)) = 1 + \sum_r g_{i+r} g_{j+r} = 1 + \sum_r g_{i+r} g^{\text{rev}}_{11-j-r}$$
$$= 1 + \text{coefficient of } X^{11+i-j} \text{ in } \underbrace{g(X) \times g^{\text{rev}}(X)}_{\substack{\| \\ 1 + \cdots + X^{22}}}$$

$$= 1 + 1 = 0.$$

So,

$$\text{any two words in } \mathscr{X}^+ \text{ are dot-orthogonal.} \tag{2.6.2}$$

This implies that all words in $\mathscr{X}^+$ have weight divisible by 4. Indeed, by in-spection, all rows $(X^i g(X)|1)$ of the generating matrix of $\mathscr{X}^+$ have weight 8. Then, by induction on the number of rows involved in the sum, if $\mathbf{c} \in \mathscr{X}^+$ and $\mathbf{g}^{(i)} \sim (X^i g(X)|1)$ is a row of the generating matrix of $\mathscr{X}^+$ then

$$w(\mathbf{g}^{(i)} + \mathbf{c}) = w(\mathbf{g}^{(i)}) + w(\mathbf{c}) - 2w(\mathbf{g}^{(i)} \wedge \mathbf{c}),$$

where $(\mathbf{g}^{(i)} \wedge \mathbf{c})_l = \min\ [(g^{(i)})_l, c_l]$, $l = 1, \ldots, 24$. We know that $8|w(\mathbf{g}^{(i)})$ and by the induction hypothesis, $4|w(\mathbf{c})$. Next, $w(\mathbf{g}^{(i)} \wedge \mathbf{c})$ is even, so $2w(\mathbf{g}^{(i)} \wedge \mathbf{c})$ is divis-ible by 4. Then the LHS, $w(\mathbf{g}^{(i)} + \mathbf{c})$, is divisible by 4. Therefore, the distance of $\mathscr{X}^+$ is 8, as it is ≥ 5 and is divisible by 4. (Clearly, it cannot be bigger than 8 as then it would be 12.) Then the distance of the original code, $\mathscr{X}$, equals 7.

Finally, the code $\mathscr{X}$ is perfect 3-error correcting, since the volume of the 3-ball in $\mathbb{F}_2^{23}$ equals

$$\binom{23}{0} + \binom{23}{1} + \binom{23}{2} + \binom{23}{3} = 1 + 23 + 253 + 1771 = 2048 = 2^{11},$$

and $2^{12} \times 2^{11} = 2^{23}$. Here, obviously, 12 represents the rank and 23 the length. $\square$

Problem 2.4   Show that the Hamming code is cyclic with check polynomial $X^4 + X^2 + X + 1$. What is its generator polynomial? Does Hamming's original code contain a subcode equivalent to its dual? Let the decomposition into irreducible monic polynomials $M_j(X)$ be

$$X^N + 1 = \prod_{j=1}^{l} M_j(X)^{k_j}. \tag{2.6.3}$$

Prove that the number of cyclic code of length N is $\prod_{j=1}^{l}(k_j + 1)$.

Solution In $\mathbb{F}_2^7$ we have

$$X^7 - 1 = (X^3 + X + 1)(X^4 + X^2 + X + 1).$$

The cyclic code with generator $g(X) = X^3 + X + 1$ has check polynomial $h(X) = X^4 + X^2 + X + 1$. The parity-check matrix of the code is

$$\begin{pmatrix} 1 & 0 & 1 & 1 & 1 & 0 & 0 \\ 0 & 1 & 0 & 1 & 1 & 1 & 0 \\ 0 & 0 & 1 & 0 & 1 & 1 & 1 \end{pmatrix}. \tag{2.6.4}$$

The columns of this matrix are the non-zero elements of $\mathbb{F}_2^3$. So, this is equivalent to Hamming's original $[7,4]$ code.

The dual of Hamming's $[7,4]$ code has the generator polynomial $X^4 + X^3 + X^2 + 1$ (the reverse of $h(X)$). Since $X^4 + X^3 + X^2 + 1 = (X+1)g(X)$, it is a subcode of Hamming's $[7,4]$ code.

Finally, any irreducible polynomial $M_j(X)$ could be included in a generator of a cyclic code in any power $0, \ldots, k_j$. So, the number of possibilities to construct this generator equals $\prod_{j=1}^{l}(k_j + 1)$. $\qquad\qquad\square$

Problem 2.5 *Describe the construction of a Reed–Muller code. Establish its information rate and its distance.*

Solution The space $\mathbb{F}_2^m$ has $N = 2^m$ points. If $A \subseteq \mathbb{F}_2^m$, let $\mathbf{1}_A$ be the indicator function of A. Consider the collection of hyperplanes

$$\Pi_j = \{\mathbf{p} \in \mathbb{F}_2^m : p_j = 0\}.$$

Set $h^j = \mathbf{1}_{\Pi_j}$, $j = 1, \ldots, m$, and $h^0 = \mathbf{1}_{\mathbb{F}_2^m} \equiv 1$. Define sets of functions $\mathbb{F}_2^m \to \mathbb{F}_2$:

$$\begin{aligned}
\mathscr{A}_0 &= \{h^0\}, \\
\mathscr{A}_1 &= \{h^j ; j = 1, 2, \ldots, m\}, \\
\mathscr{A}_2 &= \{h^i \cdot h^j ; i, j = 1, 2, \ldots, m, i < j\}, \\
&\quad\vdots \\
\mathscr{A}_{k+1} &= \{a \cdot h^j ; a \in \mathscr{A}_k, \ j = 1, 2, \ldots, m, \ h^j \nmid a\}, \\
&\quad\vdots \\
\mathscr{A}_m &= \{h^1 \cdots h^m\}.
\end{aligned}$$

The union of these sets has cardinality $N = 2^m$ (there are 2^m functions altogether). Therefore, functions from $\cup_{i=0}^{m}\mathscr{A}_i$ can be taken as a basis in $\mathbb{F}_2^N$.

Then the Reed–Muller code $\mathrm{RM}(r, m) = \mathscr{X}_{r,m}^{\mathrm{RM}}$ of length $N = 2^m$ is defined as the span of $\cup_{i=0}^{r}\mathscr{A}_i$ and has rank $\sum_{i=0}^{r}\binom{m}{i}$. Its information rate is

$$\frac{1}{2^m}\sum_{i=0}^{r}\binom{m}{i}.$$

Next, if $\mathbf{a} \in \text{RM}(r,m)$ then

$$\mathbf{a} = (\mathbf{y},\mathbf{y})h^j + (\mathbf{x},\mathbf{x}) = (\mathbf{x},\mathbf{x}+\mathbf{y}),$$

for some $\mathbf{x} \in \text{RM}(m-1,r)$ and $\mathbf{y} \in \text{RM}(m-1,r-1)$. Thus, $\text{RM}(m,r)$ coincides with the bar-product $(R(m-1,r)|R(m-1,r-1))$. By the bar-product bound,

$$d\big[\text{RM}(m,k)\big] \geq \min\big(2d\big[\text{RM}(m-1,k)\big], d\big[\text{RM}(m-1,k-1)\big]\big),$$

which, by induction, yields

$$d\big[\text{RM}(r,m)\big] \geq 2^{m-r}.$$

On the other hand, the vector $h^1 \cdot h^2 \cdot \,\cdots\, \cdot h^m$ is at distance 2^{m-r} from $\text{RM}(m,r)$. Hence,

$$d\big[\text{RM}(r,m)\big] = 2^{m-r}.$$

$\square$

Problem 2.6 (a) *Define a parity-check code of length N over the field $\mathbb{F}_2$. Show that a code is linear iff it is a parity-check code. Define the original Hamming code in terms of parity-checks and then find a generating matrix for it.*
(b) *Let $\mathscr{X}$ be a cyclic code. Define the dual code*

$$\mathscr{X}^{\perp} = \{\mathbf{y} = y_1 \ldots y_N : \sum_{i=1}^{N} x_i y_i = 0 \text{ for all } \mathbf{x} = x_1 \ldots x_N \in \mathscr{X}\}.$$

Prove that $\mathscr{X}^{\perp}$ is cyclic and establish how the generators of $\mathscr{X}$ and $\mathscr{X}^{\perp}$ are related to each other. Show that the repetition and parity-check codes are cyclic, and determine their generators.

Solution (a) The parity-check code $\mathscr{X}^{\text{PC}}$ of a (not necessarily linear) code $\mathscr{X}$ is the collection of vectors $\mathbf{y} = y_1 \ldots y_N \in \mathbb{F}_2^N$ such that the dot-product

$$\mathbf{y} \cdot \mathbf{x} = \sum_{i=1}^{N} x_i y_i = 0 \text{ (in } \mathbb{F}_2), \quad \text{for all } \mathbf{x} = x_1 \ldots x_N \in \mathscr{X}.$$

From the definition it is clear that $\mathscr{X}^{\text{PC}}$ is also the parity-check code for $\overline{\mathscr{X}}$, the linear code spanned by $\mathscr{X}$: $\mathscr{X}^{\text{PC}} = \overline{\mathscr{X}}^{\text{PC}}$. Indeed, if $\mathbf{y} \cdot \mathbf{x} = 0$ and $\mathbf{y} \cdot \mathbf{x}' = 0$ then $\mathbf{y} \cdot (\mathbf{x}+\mathbf{x}') = 0$. Hence, the parity-check code $\mathscr{X}^{\text{PC}}$ is always linear, and it forms a subspace dot-orthogonal to $\mathscr{X}$. Thus, a given code $\mathscr{X}$ is linear iff it is a parity-check code. A pair of linear codes $\mathscr{X}$ and $\mathscr{X}^{\text{PC}}$ form a dual pair: $\mathscr{X}^{\text{PC}}$ is the dual of $\mathscr{X}$ and vice versa. The generating matrix H for $\mathscr{X}^{\text{PC}}$ serves as a parity-check matrix for $\mathscr{X}$ and vice versa.

The Hamming code of length $N = 2^l - 1$ is the one whose check matrix is $l \times N$ and lists all non-zero columns from $\mathbb{F}_2^l$ (in some agreed order). So, the Hamming $[7,4]$ code corresponds to $l = 3$; its parity-checks are

$$x_1 + x_3 + x_5 + x_7 = 0,$$
$$x_2 + x_3 + x_6 + x_7 = 0,$$
$$x_4 + x_5 + x_6 + x_7 = 0,$$

and the generating matrix equals

$$\begin{pmatrix} 1 & 1 & 0 & 1 & 0 & 0 & 0 \\ 0 & 1 & 1 & 0 & 1 & 0 & 0 \\ 0 & 0 & 1 & 1 & 0 & 1 & 0 \\ 0 & 0 & 0 & 1 & 1 & 0 & 1 \end{pmatrix}.$$

(b) The generator of dual code $g^\perp(X) = X^{N-1} g(X^{-1})$. The repetition code has $g(X) = 1 + X + \cdots + X^{N-1}$ and the rank 1. The parity-check code has $g(X) = 1 + X$ and the rank $N - 1$. $\qquad\square$

Problem 2.7   (a) *How does coding theory apply when the error rate $p > 1/2$?*
(b) *Give an example of a code which is not a linear code.*
(c) *Give an example of a linear code which is not a cyclic code.*
(d) *Define the binary Hamming code and its dual. Prove that the Hamming code is perfect. Explain why the Hamming code cannot always correct two errors.*
(e) *Prove that in the dual code:*

(i) *The weight of any non-zero codeword equals $2^{\ell-1}$.*
(ii) *The distance between any pair of words equals $2^{\ell-1}$.*

Solution (a) If $p > 1/2$, we reverse the output to get $p' = 1 - p$.
  (b) The code $\mathscr{X} \subset \mathbb{F}_2^2$ with $\mathscr{X} = \{11\}$ is not linear as $00 \notin \mathscr{X}$.
  (c) The code $\mathscr{X} \subset \mathbb{F}_2^2$ with $\mathscr{X} = \{00, 10\}$ is linear, but not cyclic, as $01 \notin \mathscr{X}$.
  (d) The original Hamming $[7,4]$ code has distance 3 and is perfect one-error correcting. Thus, making two errors in a codeword will always lead outside the ball of radius 1 about the codeword, i.e. to a ball of radius 1 about a different codeword (at distance 1 of the nearest, at distance 2 from the initial word). Thus, one detects two errors but never corrects them.
  (e) The dual of a Hamming $[2^\ell - 1, 2^\ell - \ell - 1, 3]$ code is linear, of length $N = 2^\ell - 1$ and rank ℓ, and its generating matrix is $\ell \times (2^\ell - 1)$, with columns listing all non-zero vectors of length ℓ (the parity-check matrix of the original code). The rows of this matrix are linearly independent; moreover, any row $i = 1, \ldots, \ell$ has $2^{\ell-1}$ digits 1. This is because each such digit comes from a column, i.e. a non-zero vector of length ℓ, with 1 in position i; there are exactly $2^{\ell-1}$ such vectors. Also any pair of

columns of this matrix are linearly independent, but there are triples of columns
that are linearly dependent (a pair of columns complemented by their sum).

Every non-zero dual codeword $\mathbf{x}$ is a sum of rows of the above generating matrix.
Suppose these summands are rows $i_1, \ldots, i_s$ where $1 \le i_1 < \cdots < i_s \le \ell$. Then, as
above, the number of digits 1 in the sum equals the number of columns of this
matrix for which the sum of digits $i_1, \ldots, i_s$ is 1. We have no restriction on the
remaining $\ell - s$ digits, so for them there are $2^{\ell-s}$ possibilities. For digits $i_1, \ldots, i_s$
we have 2^{s-1} possibilities (a half of the total of 2^s). Thus, again $2^{\ell-s} \times 2^{s-1} = 2^{\ell-1}$.

We proved that the weight of every non-zero dual codeword equals $2^{\ell-1}$. That is,
the distance from the zero vector to any dual codeword is $2^{\ell-1}$. Because the dual
code is linear, the distance between any pair of distinct dual codewords $\mathbf{x}, \mathbf{x}'$ equals
$2^{\ell-1}$:

$$\delta(\mathbf{x}, \mathbf{x}') = \delta(\mathbf{0}, \mathbf{x}' - \mathbf{x}) = w(\mathbf{x} - \mathbf{x}') = 2^{\ell-1}.$$

Let $J \subset \{1, \ldots, \ell\}$ be the set of contributing rows:

$$\mathbf{x} = \sum_{i \in J} g^{(i)}.$$

Then $\delta(\mathbf{0}, \mathbf{x}) = \sharp$ of non-zero digits in $\mathbf{x}$ is calculated as

$$2^{\ell-|J|} \qquad \times \qquad \left(\sharp \text{ of subsets } K \subseteq J \text{ with } |K| \text{ odd}\right)$$

$$\uparrow \qquad\qquad\qquad\qquad \uparrow$$

$\sharp$ of ways to place $\sharp$ of ways to get $\sum_{l \in J} x_i = 1 \mod 2$

0s and 1s outside J with $x_i = 0$ or 1

which yields $2^{\ell-|J|} \, 2^{|J|-1} = 2^{\ell-1}$. In other words, to get a contribution from a digit
$x_j = \sum_{i \in J} g^{(i)}_j = 1$, we must fix (i) a configuration of 0s and 1s over $\{1, \ldots, \ell\} \setminus J$ (as it
is a part of the description of a non-zero vector of length N), and (ii) a configuration
of 0s and 1s over J, with an odd number of 1s.

To check that $d\left(\mathscr{X}^{\mathrm{H}^{\perp}}\right) = 2^{\ell-1}$, it suffices to establish that the distance between
the zero word and any other word $\mathbf{x} \in \mathscr{X}^{\mathrm{H}^{\perp}}$ equals $2^{\ell-1}$.   □

Problem 2.8   (a) *What is a necessary and sufficient condition for a polynomial*
$g(X)$ *to be the generator of a cyclic code of length N? What is the BCH code?*
Show that the BCH code associated with $\{\omega, \omega^2\}$, where ω is a root of $X^3 + X + 1$
in an appropriate field, is Hamming's original code.
(b) *Define and evaluate the Vandermonde determinant. Define the BCH code and*
obtain a good estimate for its minimum distance.

Solution (a) The necessary and sufficient condition for $g(X)$ being the generator of a cyclic code of length N is $g(X)|(X^N - 1)$. The generator $g(X)$ may be irreducible or not; in the latter case it is represented as a product $g(X) = M_1(X) \cdots M_k(X)$ of its irreducible factors, with $k \leq d = \deg g$. Let s be the minimal number such that $N|2^s - 1$. Then $g(X)$ is factorised into the product of first-degree monomials in a field $\mathbb{K} = \mathbb{F}_{2^s} \supseteq \mathbb{F}_2$: $g(X) = \prod\limits_{i=1}^{d}(X - \omega_j)$ with $\omega_1, \ldots, \omega_d \in \mathbb{K}$. [Usually one refers to the minimal field – the splitting field for g, but this is not necessary.] Each element ω_i is a root of $g(X)$ and also a root of at least one of its irreducible factors $M_1(X), \ldots, M_k(X)$. [More precisely, each $M_i(X)$ is a sub-product of the above first-degree monomials.]

We want to select a defining set D of roots among $\omega_1, \ldots, \omega_d \in \mathbb{K}$: it is a collection comprising at least one root ω_{j_i} for each factor $M_i(X)$. One is naturally tempted to take a minimal defining set where each irreducible factor is represented by one root, but this set may not be easy to describe exactly. Obviously, the cardinality $|D|$ of defining set D is between k and d. The roots forming D are all from field $\mathbb{K}$ but in fact there may be some from its subfield, $\mathbb{K}' \subset \mathbb{K}$ containing all ω_{j_i}. [Of course, $\mathbb{F}_2 \subset \mathbb{K}'$.] We then can identify the cyclic code $\mathscr{X}$ generated by $g(X)$ with the set of polynomials

$$\{f(X) \in \mathbb{F}_2[X]/\langle X^N - 1 \rangle : f(\omega) = 0 \quad \text{for all } \omega \in D\}.$$

It is said that $\mathscr{X}$ is a cyclic code with defining set of roots (or zeros) D.

(b) A binary BCH code of length N (for N odd) and designed distance δ is a cyclic code with defining set $\{\omega, \omega^2, \ldots, \omega^{\delta-1}\}$ where $\delta \leq N$ and ω is a primitive Nth root of unity, with $\omega^N = 1$. It is helpful to note that if ω is a root of a polynomial $p(X)$ then so are $\omega^2, \omega^4, \ldots, \omega^{2^{s-1}}$. By considering a defining set of the form $\{\omega, \omega^2, \ldots, \omega^{\delta-1}\}$ we 'fill the gaps' in the above diadic sequence and produce an ideal of polynomials whose properties can be analytically studied.

The simplest example is where $N = 7$ and $D = \{\omega, \omega^2\}$ where ω is a root of $X^3 + X + 1$. Here, $\omega^7 = (\omega^3)^2 \omega = (\omega + 1)^2 \omega = \omega^3 + \omega = 1$, so ω is a 7th root of unity. [We used the fact that the characteristic is 2.] In fact, it is a primitive root. Also, as was said, ω^2 is a root of $X^3 + X + 1$: $(\omega^2)^3 + \omega^2 + 1 = (\omega^3 + \omega + 1)^2 = 0$, and so is ω^4. Then the cyclic code with defining set $\{\omega, \omega^2\}$ has generator $X^3 + X + 1$ since all roots of this polynomial are engaged. We know that it coincides with the Hamming $[7,4]$ code.

The Vandermonde determinant is

$$\Delta = \det \begin{pmatrix} 1 & 1 & 1 & \ldots & 1 \\ x_1 & x_2 & x_3 & \ldots & x_n \\ \ldots & \ldots & \ldots & \ldots & \ldots \\ x_1^{n-1} & x_2^{n-1} & x_3^{n-1} & \ldots & x_n^{n-1} \end{pmatrix}.$$

Observe that if $x_i = x_j (i \neq j)$ the determinant vanishes (two rows are the same). Thus $x_i - x_j$ is a factor of Δ,

$$\Delta = P(\mathbf{x}) \prod_{i<j}(x_i - x_j),$$

with P a polynomial in $x_1, \ldots, x_n$. Now consider terms in expansion Δ in the sum of terms of form $a \prod_i x_i^{m(i)}$ with $\sum m(i) = 0 + 1 + \cdots + (n-1) = n(n-1)/2$. But $\prod_{i<j}(x_i - x_j)$ is a sum of terms $a \prod_i x_i^{m(i)}$ with $\sum m(i) = n(n-1)/2$, so $P(\mathbf{x}) = const.$ Considering $x_2 x_3^2 \ldots x_n^{n-1}$ we have $const = 1$, so

$$\Delta = \prod_{i<j}(x_i - x_j). \tag{2.6.5}$$

Suppose N is odd and K is a field containing $\mathbb{F}_2$ in which $X^N - 1$ factorises into linear factors. [This field can be selected as $\mathbb{F}_{2^s}$ where $N|2^s - 1$.] A cyclic code consisting of words $\mathbf{c} = c_0 c_1 \ldots c_{N-1}$ with $\sum_{j=0}^{N-1} c_j \omega^{rj} = 0$ for all $r = 1, 2, \ldots, \delta - 1$ where ω is a primitive Nth root of unity is called a BCH code of design distance $\delta < N$. Next, $\mathcal{X}^{BCH}$ is a vector space over $\mathbb{F}_2$ and $\mathbf{c} \in \mathcal{X}^{BCH}$ iff

$$\mathbf{c}H^{\mathrm{T}} = 0 \tag{2.6.6}$$

where

$$H = \begin{pmatrix} 1 & \omega & \omega^2 & \cdots & \omega^{N-1} \\ 1 & \omega^2 & \omega^4 & \cdots & \omega^{2N-2} \\ 1 & \omega^3 & \omega^6 & \cdots & \omega^{3N-3} \\ \vdots & \vdots & \vdots & \ddots & \vdots \\ 1 & \omega^{\delta-1} & \omega^{2\delta-2} & \cdots & \omega^{(N-1)(\delta-1)} \end{pmatrix}. \tag{2.6.7}$$

Now rank $H = \delta$. Indeed, by (2.6.5) for any $\delta \times \delta$ minor $\widetilde{H}$

$$\det \widetilde{H} = \prod_{i<j}(\omega^i - \omega^j) \neq 0.$$

Thus (2.6.6) tells us that

$$\mathbf{c} \in \mathcal{X}, \mathbf{c} \neq \mathbf{0} \Rightarrow \sum |c_j| \geq \delta.$$

So, the minimum distance in $\mathcal{X}^{BCH}$ is not smaller than δ. $\qquad\square$

Problem 2.9 A subset $\mathcal{X}$ of the Hamming space $\{0,1\}^N$ of cardinality $\sharp \mathcal{X} = M$ and with the minimal Hamming distance $d = \min[\delta(\mathbf{x}, \mathbf{x}') : \mathbf{x}, \mathbf{x}' \in \mathcal{X}, \mathbf{x} \neq \mathbf{x}']$ is called an $[N, M, d]$ code (not necessarily linear). An $[N, M, d]$ code is called maximal if it is not contained in any $[N, M+1, d]$ code. Prove that an $[N, M, d]$ code is maximal if and only if for any $\mathbf{y} \in \{0,1\}^N$ there exists $\mathbf{x} \in \mathcal{X}$ such that

$\delta(\mathbf{x}, \mathbf{y}) < d$. Conclude that if d or more changes are made in a codeword then the new word is closer to some other codeword than to the original one.

Suppose that a maximal $[N, M, d]$ code is used for transmitting information via a binary memoryless channel with the error-probability p, and the receiver uses the maximum likelihood decoder. Prove that the probability of erroneous decoding, π_{err}^{ML}, obeys the bounds

$$1 - b(N, d-1) \leqslant \pi_{err}^{ML} \leqslant 1 - b(N, \lfloor (d-1)/2 \rfloor),$$

where $b(N, m)$ is a partial binomial sum

$$b(N, m) = \sum_{0 \leq k \leq m} \binom{N}{k} p^k (1-p)^{N-k}.$$

Solution If a code is maximal then adding one more word will reduce the distance. Hence, for all $\mathbf{y}$ there exists $\mathbf{x} \in \mathcal{X}$ such that $\delta(\mathbf{x}, \mathbf{y}) < d$. Conversely, if this property holds then the code cannot be enlarged without reducing d. Then making d or more changes in a codeword gives a word that is closer to a different codeword. This will certainly not give the correct guess under the ML decoder as it chooses the closest codeword.

Therefore,

$$\pi_{err}^{ML} \geq \sum_{d \leq k \leq N} \binom{N}{k} p^k (1-p)^{N-k} = 1 - b(N, d-1).$$

On the other hand, the code corrects $\lfloor (d-1)/2 \rfloor$ errors. Hence,

$$\pi_{err}^{ML} \leq 1 - b(N, \lfloor d - 1/2 \rfloor).$$

$\square$

Problem 2.10   The Plotkin bound for an $[N, M, d]$ binary code states that $M \leq \dfrac{d}{d - N/2}$ if $d > N/2$. Let $M_2^*(N, d)$ be the maximum size of a code of length N and distance d, and let

$$\alpha(\lambda) = \lim_{N \to \infty} \frac{1}{N} \log_2 M_2^*(N, \lfloor \lambda N \rfloor).$$

Deduce from the Plotkin bound that $\alpha(\lambda) = 0$ for $\lambda \geq \frac{1}{2}$.

Assuming the above bound, show that if $d \leq N/2$, then

$$M \leq 2^{N-(2d-1)} \frac{d}{d - (2d-1)/2} = 2d \, 2^{N-(2d-1)}.$$

Deduce the asymptotic Plotkin bound: $\alpha(\lambda) \leq 1 - 2\lambda, 0 \leq \lambda < \frac{1}{2}$.

Solution If $d > N/2$ apply the Plotkin bound and conclude that $\alpha(\lambda) = 0$. If $d \leq N/2$ consider the partition of a code $\mathscr{X}$ of length N and distance $d \leq N/2$ according to the last $N - (2d - 1)$ digits, i.e. divide $\mathscr{X}$ into disjoint subsets, with fixed $N - (2d - 1)$ last digits. One of these subsets, $\mathscr{X}'$, must have size M' such that $M'2^{N-(2d-1)} \geq M$.

Hence, $\mathscr{X}'$ is a code of length $N' = 2d - 1$ and distance $d' = d$, with $d' > N'/2$. Applying Plotkin's bound to $\mathscr{X}'$ gives

$$M' \leq \frac{d'}{d' - N/2} = \frac{d}{d - (2d - 1)/2} = 2d.$$

Therefore,

$$M \leq 2^{N-(2d-1)}2d.$$

Taking $d = \lfloor \lambda N \rfloor$ with $N \to \infty$ yields $\alpha(\lambda) \leq 1 - 2\lambda, 0 \leq \lambda \leq 1/2$.   □

Problem 2.11   *State and prove the Hamming, Singleton and Gilbert–Varshamov bounds. Give (a) examples of codes for which the Hamming bound is attained, (b) examples of codes for which the Singleton bound is attained.*

Solution The Hamming bound states that the size M of an E-error correcting code $\mathscr{X}$ of length N,

$$M \leq \frac{2^N}{v_N(E)},$$

where $v_N(E) = \sum\limits_{0 \leq i \leq E} \binom{N}{i}$ is the volume of an E-ball in the Hamming space $\{0, 1\}^N$. It follows from the fact that the E-balls about the codewords $\mathbf{x} \in \mathscr{X}$ must be disjoint:

$$M \times v_N(E) \quad = \quad \sharp \text{ of points covered by } M \ E\text{-balls}$$
$$\leq 2^N = \sharp \text{ of points in } \{0, 1\}^N.$$

The Singleton bound is that the size M of a code $\mathscr{X}$ of length N and distance d obeys

$$M \leq 2^{N-d+1}.$$

It follows by observing that truncating $\mathscr{X}$ (i.e. omitting a digit from the codewords $\mathbf{x} \in \mathscr{X}$) $d - 1$ times still does not merge codewords (i.e. preserves M) while the resulting code fits in $\{0, 1\}^{N-d+1}$.

The Gilbert–Varshamov bound is that the maximal size $M^* = M_2^*(N, d)$ of a binary $[N, d]$ code satisfies

$$M^* \geq \frac{2^N}{v_N(d - 1)}.$$

This bound follows from the observation that any word $\mathbf{y} \in \{0,1\}^N$ must be within distance $\leq d-1$ from a maximum-size code $\mathscr{X}^*$. So,

$$M^* \times v_N(d-1) \geq \sharp \text{ of points within distance } d-1 = 2^N.$$

Codes attaining the Hamming bound are called perfect codes, e.g. the Hamming $[2^\ell - 1, 2^\ell - 1 - \ell, 3]$ codes. Here, $E = 1$, $v_N(1) = 1 + 2^\ell - 1 = 2^\ell$ and $M = 2^{2^\ell - \ell - 1}$. Apart from these codes, there is only one example of a (binary) perfect code: the Golay $[23, 12, 7]$ code.

Codes attaining the Singleton bound are called maximum distance separable (MDS): their check matrices have any $N-M$ rows linearly independent. Examples of such codes are (i) the whole $\{0,1\}^N$, (ii) the repetition code $\{0\ldots0, 1\ldots1\}$ and the collection of all words $\mathbf{x} \in \{0,1\}^N$ of even weight. In fact, these are all examples of binary MDS codes. More interesting examples are provided by Reed–Solomon codes that are non-binary; see Section 3.2. Binary codes attaining the Gilbert–Varshamov bound for general N and d have not been constructed so far (though they have been constructed for non-binary alphabets). $\qquad\square$

Problem 2.12   (a) *Explain the existence and importance of error correcting codes to a computer engineer using Hamming's original code as your example.*

(b) *How many codewords in a Hamming code are of weight 1? 2? 3? 4? 5?*

Solution (a) Consider the linear map $\mathbb{F}_2^7 \to \mathbb{F}_2^3$ given by the matrix H of the form (2.6.4). The Hamming code $\mathscr{X}$ is the kernel $\ker H$, i.e. the collection of words $\mathbf{x} = x_1x_2x_3x_4x_5x_6x_7 \in \{0,1\}^7$ such that $\mathbf{x}H^\mathrm{T} = \mathbf{0}$. Here, we can choose four digits, say x_4, x_5, x_6, x_7, arbitrarily from $\{0,1\}$; then x_1, x_2, x_3 will be determined:

$$x_1 = x_4 + x_5 + x_7,$$
$$x_2 = x_4 + x_6 + x_7,$$
$$x_3 = x_5 + x_6 + x_7.$$

It means that code $\mathscr{X}$ can be used for encoding 16 binary 'messages' of length 4. If $\mathbf{y} = y_1y_2y_3y_4y_5y_6y_7$ differs from a codeword $\mathbf{x} \in \mathscr{X}$ in one place, say $\mathbf{y} = \mathbf{x} + \mathbf{e}_k$ then the equation $\mathbf{y}H^\mathrm{T} = \mathbf{e}_kH^\mathrm{T}$ gives the binary decomposition of number k, which leads to decoding $\mathbf{x}$. Consequently, code $\mathscr{X}$ allows a single error to be corrected.

Suppose that the probability of error in any digit is $p \ll 1$, independently of what occurred to other digits. Then the probability of an error in transmitting a non-encoded $(4N)$-digit message is

$$1 - (1-p)^{4N} \simeq 4Np.$$

But using the Hamming code we need to transmit $7N$ digits. An erroneous transmission requires at least two wrong digits, which occurs with probability

$$\approx 1 - \left(1 - \binom{7}{2} p^2\right)^N \simeq 21Np^2 << 4Np.$$

So, the extra effort of using 3 check digits in the Hamming code is justified.

(b) A Hamming code $\mathscr{X}_{H,\ell}$ of length $N = 2^\ell - 1$ $(\ell \geq 3)$ consists of binary words $\mathbf{x} = x_1 \ldots x_N$ such that $\mathbf{x} H^T = 0$ where H is an $\ell \times N$ matrix whose columns $h^{(1)}, \ldots, h^{(N)}$ are all non-zero binary vectors of length l. Hence, the number of codewords of weight $w(\mathbf{x}) = \sum_{j=1}^{N} x_j = s$ equals the number of (non-ordered) collections of s binary, non-zero, pair-wise distinct ℓ-vectors of total sum 0. In fact, if $\mathbf{x} H^T = 0$ and $w(\mathbf{x}) = s$ and $x_{j_1} = x_{j_2} = \cdots = x_{j_s} = 1$ then the sum of row-vectors $h^{(j_1)} + \cdots + h^{(j_s)} = \mathbf{0}$.

Thus, one codeword has weight 0, no codeword has weight 1 or 2, $N(N-1)/3!$ codewords have weight 3 (i.e. 7 and 35 words of weights 3 for $l = 3$ and $l = 4$). Further we have $[N(N-1)(N-2) - N(N-1)]/4! = N(N-1)(N-3)/4!$ words of weight 4 (i.e. 7 and 105 words of weights 4 for $\ell = 3$ and $\ell = 4$). Finally, we have $N(N-1)(N-3)(N-7)/5!$ words weight 5 (i.e. 0 and 168 words of weight 5 for $\ell = 3$ and $\ell = 4$). Each time when we add a factor, we should avoid ℓ-vectors equal to a linear combination of previously selected vectors. In Problem 3.9 we will compute the enumerator polynomial for $N = 15$:

$$1 + 35X^3 + 105X^4 + 168X^5 + 280X^6 + 435X^7 + 435X^8$$

$$+280X^9 + 168X^{10} + 105X^{11} + 35X^{12} + X^{15}.$$

□

Problem 2.13   (a) *The dot-product of vectors* $\mathbf{x}, \mathbf{y}$ *from a binary Hamming space* $\mathscr{H}_N$ *is defined as* $\mathbf{x} \cdot \mathbf{y} = \sum_{i=1}^{N} x_i y_i$ *(mod 2), and* $\mathbf{x}$ *and* $\mathbf{y}$ *are said to be orthogonal if* $\mathbf{x} \cdot \mathbf{y} = 0$. *What does it mean to say that* $\mathscr{X} \subseteq \mathscr{H}_N$ *is a linear* $[N,k]$ *code with generating matrix G and parity-check matrix H? Show that*

$$\mathscr{X}^\perp = \{\mathbf{x} \in \mathscr{H}_N : \mathbf{x} \cdot \mathbf{y} = 0 \quad \text{for all } \mathbf{y} \in \mathscr{X}\}$$

is a linear $[N, N-k]$ *code and find its generator and parity-check matrices.*

(b) *A linear code* $\mathscr{X}$ *is called self-orthogonal if* $\mathscr{X} \subseteq \mathscr{X}^\perp$. *Prove that* $\mathscr{X}$ *is self-orthogonal if the rows of G are self and pairwise orthogonal. A linear code is called self-dual if* $\mathscr{X} = \mathscr{X}^\perp$. *Prove that a self-dual code has to be an* $[N, N/2]$ *code (and hence N must be even). Conversely, prove that a self-orthogonal* $[N, N/2]$ *code, for N even, is self-dual. Give an example of such a code for any even N and prove that a self-dual code always contains the word* $1 \ldots 1$.

(c) *Consider now a Hamming* $[2^\ell - 1, 2^\ell - \ell - 1]$ *code* $\mathscr{X}_{H,\ell}$. *Describe the generating matrix of* $\mathscr{X}_{H,\ell}^\perp$. *Prove that the distance between any two codewords in* $\mathscr{X}_{H,\ell}^\perp$ *equals* $2^{\ell-1}$.

Solution By definition, $\mathscr{X}^\perp$ is preserved under the linear operations; hence $\mathscr{X}^\perp$ is a linear code. From algebraic considerations, dim $\mathscr{X}^\perp = N - k$. The generating matrix $G^\perp$ of $\mathscr{X}^\perp$ coincides with H, and the parity-check matrix $H^\perp$ with G.

If $\mathscr{X} \subseteq \mathscr{X}^\perp$ then the rows $g^{(1)}, \ldots, g^{(k)}$ of G are self- and pairwise orthogonal. The converse is also true. From the previous observation, if $\mathscr{X}$ is self-dual then $k = N - k$, i.e. $k = N/2$, and N should be even. Similarly, if $\mathscr{X}$ is self-orthogonal and $k = N/2$ then $\mathscr{X}$ is self-dual.

Let $\mathbf{1} = 1 \ldots 1$. If $\mathscr{X} = \mathscr{X}^\perp$ then $\mathbf{1} \cdot g^{(i)} = g^{(i)} \cdot g^{(i)} = 0$. So, $\mathbf{1} \in \mathscr{X}^\perp$ and hence $\mathbf{1} \in \mathscr{X}$. An example is a code with the generating matrix

$$
G = \left.
\begin{pmatrix}
1 & 1 & 1 & \cdots & 1 & 1 & \cdots & 1 \\
1 & 1 & 0 & \cdots & 0 & 1 & \cdots & 0 \\
1 & 0 & 1 & \cdots & 0 & 1 & \cdots & 0 \\
\vdots & \vdots & \vdots & \ddots & \vdots & \vdots & \ddots & \vdots \\
1 & 0 & 0 & \cdots & 1 & 1 & \cdots & 1
\end{pmatrix}
\right\} N/2.
$$
$$\leftarrow \quad N/2 \quad \rightarrow \quad \leftarrow N/2 \rightarrow$$

The dual $\mathscr{X}_{H^\perp}$ of a Hamming code $\mathscr{X}_H$ is called a simplex code. By the above, it has length $2^\ell - 1$ and rank ℓ, and its generating matrix $G_{H^\perp}$ is $\ell \times (2^\ell - 1)$, with columns listing all non-zero vectors of length ℓ. To check that dist $\mathscr{X}_H^\perp = 2^{\ell-1}$, it suffices to establish that the weight of non-zero word $\mathbf{x} \in \mathscr{X}_H^\perp$ equals $2^{\ell-1}$. But a non-zero word $\mathbf{x} \in \mathscr{X}_H^\perp$ is a non-zero linear combination of rows of $G_H^\perp$. Let $J \subset \{1, \ldots, \ell\}$ be the set of contributing rows:

$$\mathbf{x} = \sum_{i \in J} \mathbf{g}^{(i)}.$$

Clearly, $w(\mathbf{g}^{(i)}) = 2^{\ell-1}$ as exactly half of all 2^ℓ vectors have 1 on any given position. The proof is finished by induction on $\sharp J$.

A simple and elegant way is to use the MacWilliams identity (cf. Lemma 3.4.4) which immediately gives

$$W_{\mathscr{X}^\perp}(s) = 1 + (2^\ell - 1)s^{2^{\ell-1}}. \tag{2.6.8}$$

It is instructive to present this derivation. We will establish in Problem 3.9 the formula for a weight enumeration polynomial of Hamming code. Then substituting this expression into the MacWilliams identity one gets

$$
\begin{aligned}
W_{\mathcal{X}^\perp}(s) &= \frac{1}{2^{N-\ell}} \left[\frac{1}{N+1} \left(1 + \frac{1-s}{1+s}\right)^N \right. \\
&\qquad \left. + \frac{N}{N+1} \left(1 + \frac{1-s}{1+s}\right)^{(N-1)/2} \left(1 - \frac{1-s}{1+s}\right)^{(N+1)/2} \right] (1+s)^N \\
&= 2^\ell \left(\frac{1}{2^\ell} + \frac{2^\ell - 1}{2^\ell} s^{2^{\ell-1}} \right)
\end{aligned}
$$

which is equivalent to (2.6.8). ☐

Problem 2.14   *Describe briefly the decoding procedure for the Hamming $[2^\ell - 1,$ $2^\ell - 1 - \ell]$ code.*

The codewords of the Hamming $[7,4]$ code, with the lexicographical parity-check matrix H of the form (2.3.4a), are used for encoding 16 symbols, the first 15 letters of the alphabet and the space character $*$. The encoding rule is

A	0011001	E	0111100	I	1010101	M	1111111
B	0100101	F	0001111	J	1100110	N	1000011
C	0010110	G	1101001	K	0101010	O	0000000
D	1110000	H	0110001	L	1001100	$*$	1011010

You have received a 105-digit message

> 1000110 0000000 0110001 1000011 1000011 1110101
> 0111100 0011010 0100101 0111100 1011000 1101001
> 0000000 0010000 1010000

where some words are corrupted. Decode the received message.

Solution   The Hamming $[2^\ell - 1, 2^\ell - 1 - \ell]$ code, $\ell = 2, 3, \ldots$, is obtained as a collection of binary 'strings' $\mathbf{x} = x_1 \ldots x_N$ of length $N = 2^\ell - 1$ such that $\mathbf{x} H^{\mathrm{T}} = \mathbf{0}$. Matrix H is $(\ell \times 2^\ell - 1)$, with $2^\ell - 1$ non-zero binary strings as columns; that is,

$$
H = \begin{pmatrix}
1 & 0 & \cdots & 0 & 1 & \cdots & 1 \\
0 & 1 & \cdots & 0 & 0 & \cdots & 1 \\
\cdots & \cdots & \cdots & \cdots & \cdots & \cdots & \cdots \\
0 & 0 & \cdots & 1 & 1 & \cdots & 1
\end{pmatrix}.
$$

Here the columns are meant to be lexicographically ordered. Different matrices obtained from the above by permuting the rows define different, but equivalent, codes: they are all named Hamming codes.

To perform decoding, we have to fix a matrix H (the check matrix) and let it be known to both the sender and the receiver. Upon receiving a word (string) $\mathbf{y} = y_1 \ldots y_N$ we form a syndrome vector $\mathbf{y} H^{\mathrm{T}}$. If $\mathbf{y} H^{\mathrm{T}} = \mathbf{0}$, we decode $\mathbf{y}$ by itself. (We have no means to determine if the original codeword was corrupted by the channel or not.)

If $\mathbf{y}H^{\mathrm{T}} \neq 0$ then $\mathbf{y}H^{\mathrm{T}}$ coincides with a column of H. Suppose $\mathbf{y}H^{\mathrm{T}}$ gives column j of H; then we decode $\mathbf{y}$ by

$$\mathbf{x}^* = \mathbf{y} + \mathbf{e}_j \text{ where } \mathbf{e}_j = 0\dots1\dots0 \text{ (1 in digit } j\text{).}$$

In other words, we change digit j in $\mathbf{y}$ and decide that it was the word sent through the channel. This works well when errors in the channel are rare.

If $\ell = 3$ a Hamming $[7,4]$ code contains $2^4 = 16$ codewords. These codewords are fixed when H is fixed: in the example they are used for encoding 15 letters from A to O and the space character $*$. Upon receiving a message we divide it into words of length 7: in the example there are 15 words altogether. Performing the decoding procedure leads to

JOHNNIE∗BE∗GOOD

□

Problem 2.15   *A (binary) Hamming code of length* $N = 2^\ell - 1$, *where* $\ell \geq 2$, *is defined as a linear binary code with a parity-check matrix* H *whose columns consist of all non-zero binary vectors of length* ℓ. *Find the rank of such a code (i.e. the dimension of the corresponding linear subspace) and the number of the codewords. Find the minimum distance for the code and prove that it is single-error correcting. Prove that the code is perfect (i.e. the union of the one-balls around the codewords covers the space of all words).*

Give a parity-check matrix and a generating matrix for a Hamming code with $\ell = 3$. *What is the information rate of this code? Why is the case* $\ell = 2$ *not interesting?*

Solution   The parity-check matrix H for the Hamming code is $\ell \times 2^\ell - 1$ and formed by all non-zero columns of length ℓ; in particular, it includes all l columns of weight 1. The latter are linearly independent; hence the l columns of H are linearly independent. Since $\mathcal{X}_{\mathrm{Ham}} = \ker H$, we have $\dim \mathcal{X} = 2^\ell - 1 - \ell = \mathrm{rank}\,\mathcal{X}$. The number of codewords then equals $2^{2^\ell - \ell - 1}$.

Since all columns of H are distinct, any pair of columns are linearly independent. So, the minimal distance of $\mathcal{X}$ is > 2. But $\mathcal{X}$ contains three columns that are linearly dependent, e.g.

$$1\,0\,0\dots0^{\mathrm{T}}, \quad 0\,1\,0\dots0^{\mathrm{T}}, \quad \text{and} \quad 1\,1\,0\dots0^{\mathrm{T}}.$$

Hence, the minimal distance equals 3. Therefore, if a single error occurs, i.e. the received word is at distance 1 from a codeword, then this codeword is uniquely determined. Hence, the Hamming code is single-error correcting.

To prove that it is perfect, we must check that

$$\text{the } \# \text{ of codewords} \times \text{the volume of a one-ball}$$
$$= \text{the total } \# \text{ of words.}$$

In fact, denoting $2^l - 1 = N$, we have

$$\text{the } \# \text{ of codewords} = 2^{2^l - 1 - l} = 2^{N-l},$$

$$\text{the volume of a one-ball} = \binom{N}{0} + \binom{N}{1} = 1 + N,$$

$$\text{the total } \# \text{ of words} = 2^N,$$

and

$$(1+N)2^{N-\ell} = 2^{\ell}2^{N-\ell} = 2^N.$$

The information rate of the code equals

$$\text{rank} \, / \, \text{length} = \frac{2^\ell - \ell - 1}{2^\ell - 1}.$$

The code with $\ell = 3$ has the 3×7 parity-check matrix of the form (2.6.4); any permutation of rows leads to an equivalent code. The generating matrix is 4×7:

$$\begin{pmatrix} 1 & 0 & 0 & 0 & 1 & 1 & 1 \\ 0 & 1 & 0 & 0 & 0 & 1 & 1 \\ 0 & 0 & 1 & 0 & 1 & 0 & 1 \\ 0 & 0 & 0 & 1 & 0 & 1 & 1 \end{pmatrix}$$

and the information rate $4/7$. The Hamming code with $\ell = 2$ is trivial: it contains a single non-zero codeword 1 1 1. □

Problem 2.16   Define a BCH code of length N over the field $\mathbb{F}_q$ with designed distance δ. Show that the minimum weight of such a code is at least δ.

Consider a BCH code of length 31 over the field $\mathbb{F}_2$ with designed distance 8. Show that the minimum distance is at least 11.

Solution A BCH code of length N over the field $\mathbb{F}_q$ is defined as a cyclic code $\mathcal{X}$ whose minimum degree generator polynomial $g(X) \in \mathbb{F}_q[X]$, with $g(X)|(X^N - 1)$ (and hence $\deg g(X) \leq N$), contains among its roots the subsequent powers ω, $\omega^2, \ldots, \omega^{\delta-1}$ where $\omega \in \mathbb{F}_{q^s}$ is a primitive Nth root of unity. (This root ω lies in an extension field $\mathbb{F}_{q^s}$ – the splitting field for $X^N - 1$ over $\mathbb{F}_q$, i.e. $N|q^s - 1$.) Then δ is called the designed distance for $\mathcal{X}$; the actual distance (which may be difficult to calculate in a general situation) is $\geq \delta$.

If we consider the binary BCH code $\mathcal{X}$ of length 31, ω should be a primitive root of unity of degree 31, with $\omega^{31} = 1$ (the root ω lies in an extension field $\mathbb{F}_{32}$).

We know that in the binary arithmetic, if a polynomial $f(X) \in \mathbb{F}_2[X]$, of order s, has a root ω, it has roots $\omega^2, \omega^4, \ldots, \omega^{2^{s-1}}$, i.e.

$$\left(X - \omega^{2^r}\right) \Big| f(X), \quad r = 0, \ldots, s-1.$$

Thus, given that the generator $g(X)$ of $\mathcal{X}$ has roots $\omega, \omega^2, \omega^3, \omega^4, \omega^5, \omega^6, \omega^7$, it will also have roots

$$\omega^8 = (\omega^4)^2, \quad \omega^9 = (\omega^5)^8, \text{ and } \omega^{10} = (\omega^5)^2.$$

That is, the defining set can be extended to

$$\omega, \omega^2, \omega^3, \omega^4, \omega^5, \omega^6, \omega^7, \omega^8, \omega^9, \omega^{10}$$

(all these elements are distinct, as ω is a primitive 31st root of unity). In fact, code $\mathcal{X}$ has designed distance ≥ 11. Hence, the minimum distance in $\mathcal{X}$ is ≥ 11. ☐

Problem 2.17   *Let $\mathcal{X}$ be a linear $[N, k, d]$ code over the binary field $\mathbb{F}_2$, and G be a generating matrix of $\mathcal{X}$, with k rows and N columns, such that exactly d of the first row's entries are 1. Let G_1 be the matrix, of $k-1$ rows and $N-d$ columns, formed by deleting the first row of G and those columns of G with a non-zero entry in the first row. Show that $\mathcal{X}_1$, the linear code generated by G_1, has minimum distance $d' \geq \lceil d/2 \rceil$. Here, for a real number x, $\lceil x \rceil$ is the integer satisfying $x \leq \lceil x \rceil < x+1$.*

Show also that $\mathcal{X}_1$ has rank $k-1$. Deduce that

$$N \geq \sum_{0 \leq i \leq k-1} \lceil d/2^i \rceil .$$

Solution Let $\mathbf{x}$ be the codeword in $\mathcal{X}$ represented by the first row of G and pick a pair of other rows, say $\mathbf{y}$ and $\mathbf{z}$. After the first deleting they become $\mathbf{y}'$ and $\mathbf{z}'$, correspondingly. Both weights $w(\mathbf{y}')$ and $w(\mathbf{z}')$ must be $\geq \lceil d/2 \rceil$: otherwise at least one of the original words $\mathbf{y}$ and $\mathbf{z}$, say $\mathbf{y}$, would have had minimum $\lceil d/2 \rceil$ digits 1 among deleted d digits (as $w(\mathbf{y}) \geq d$ by condition). But then

$$w(\mathbf{x}+\mathbf{y}) = w(\mathbf{y}') + d - \lceil d/2 \rceil < d$$

which contradicts the condition that the distance of $\mathcal{X}$ is d.

We want to check that the weight $w(\mathbf{y}' + \mathbf{z}') \geq \lceil d/2 \rceil$. Assume the opposite:

$$w(\mathbf{y}' + \mathbf{z}') = m < \lceil d/2 \rceil .$$

Then $m' = w(\mathbf{y}^0 + \mathbf{z}^0)$ must be $\geq d - m \geq \lceil d/2 \rceil$ where $\mathbf{y}^0$ is the deleted part of $\mathbf{y}$, of length d, and $\mathbf{z}^0$ is the deleted part of $\mathbf{z}$, also of length d. In fact, as before, if $m' < d - m$ then $w(\mathbf{y} + \mathbf{z}) < d$ which is impossible. But if $m' \geq d - m$ then

$$w(\mathbf{x}+\mathbf{y}+\mathbf{z}) = d - m' + m < d,$$

again impossible. Hence, the sum of any two rows of G_1 has weight $\geq \lceil d/2 \rceil$.

This argument can be repeated for the sum of any number of rows of G_1 (not exceeding $k-1$). In fact, in the case of such a sum $\mathbf{x}+\mathbf{y}+\cdots+\mathbf{z}$, we can pass to new matrices, $\widetilde{G}$ and $\widetilde{G}_1$, with this sum among the rows. We conclude that $\mathscr{X}_1$ has minimum distance $d' \geq \lceil d/2 \rceil$. The rank of $\mathscr{X}_1$ is $k-1$, for any $k-1$ rows of G_1 are linearly independent. (The above sum cannot be $\mathbf{0}$.)

Now, the process of deletion can be applied to $\mathscr{X}_1$ (you delete d' columns in G_1 yielding digits 1 in a row of G_1 with exactly d' digits 1). And so on, until you exhaust the initial rank k by diminishing it by 1. This leads to the required bound

$$N \geq d + \lceil d/2 \rceil + \lceil d/2^2 \rceil + \cdots + \lceil d/2^{k-1} \rceil.$$

$\square$

Problem 2.18   Define a *cyclic linear code* $\mathscr{X}$ and show that it has a codeword of minimal length which is unique, under normalisation to be stated. The polynomial $g(X)$ whose coefficients are the symbols of this codeword is the (minimum degree) generator polynomial of this code: prove that all words of the code are related to $g(X)$ in a particular way.

Show further that $g(X)$ can be the generator polynomial of a cyclic code with words of length N iff it satisfies a certain condition, to be stated.

There are at least three ways of determining the parity-check matrix of the code from a knowledge of the generator polynomial. Explain one of them.

Solution Let $\mathscr{X}$ be the cyclic code of length N with generator polynomial $g(X) = \sum_{0 \leq i \leq d} g_i X^i$ of degree d. Without loss of generality, assume the code is non-trivial, with $1 < d < N-1$. Let $\mathbf{g}$ denote the corresponding codeword $g_0 \ldots g_d 0 \ldots 0$ (there are $d+1$ coefficients g_i completed with $N-d-1$ zeros). Then:

(a) $g(X)|(X^N-1)$, i.e. $g(X)h(X) = X^N - 1$ for some polynomial $h(X) = \sum_{0 \leq i \leq k} h_i X^i$

     of degree $k = N-d$;

(b) a string $\mathbf{a} = a_0 \ldots a_{N-1} \in \mathscr{X}$ iff the polynomial $a(X) = \sum_{0 \leq i \leq N-1} a_i X^i$ has the

     form $a(X) = f(X)g(X) \bmod (X^N - 1)$;

(c) a string $\mathbf{g}$ and its cyclic shifts $\pi\mathbf{g}, \ldots, \pi^{k-1}\mathbf{g}$ (corresponding to polynomials $g(X), Xg(X), \ldots, X^{k-1}g(X)$) form a basis in $\mathscr{X}$.

By virtue of (a), $g_0 = h_0 = 1$ and the sum $\sum_{0 \leq i \leq l} g_i h_{l-i}$ representing the lth coefficient of $g(X)h(X)$ is equal to 0, for all $l = 1, \ldots, N-1$. By virtue of (c), the rank of $\mathscr{X}$ equals k.

One way to specify the parity-check matrix is to take the ratio $(X^N - 1)/g(X) = h(X) = h_0 + h_1 X + \cdots + h_k X^k$. Then form the $N \times (N - k)$ matrix

$$
H = \begin{pmatrix}
h_k & h_{k-1} & \cdots & 0 & \cdots & 0 & 0 \\
0 & h_k & h_{k-1} & \cdots & h_1 & \cdots & 0 \\
\cdots & \cdots & \cdots & \cdots & \cdots & \cdots & \cdots \\
0 & 0 & \cdots & h_k & \cdots & h_1 & h_0
\end{pmatrix}. \tag{2.6.9}
$$

The rows of H are the cyclic shifts $(\pi^j \mathbf{h}^\downarrow), 0 \leq j \leq d - 1 = N - k - 1$, of the string $\mathbf{h}^\downarrow = h_k \ldots h_0 0 \ldots 0$.

We claim that for all $\mathbf{a} \in \mathscr{X}$, $\mathbf{a}H^{\mathrm{T}} = 0$. In fact, it suffices to check that for the basis words $\pi^j \mathbf{g}$, $\pi^j \mathbf{g}H^{\mathrm{T}} = 0$, $j = 0, \ldots, k - 1$. That is, the dot-product

$$
\pi^{j_1} \mathbf{g} \cdot \pi^{j_2} \mathbf{h}^\downarrow = 0, \ \ 0 \leq j_1 < k, \ 0 \leq j_2 < N - k - 1. \tag{2.6.10}
$$

But for $j_1 = k - 1$ and $j_2 = 0$, we have

$$
\pi^{k-1} \mathbf{g} \cdot \mathbf{h}^\downarrow = g_0 h_k + g_1 h_{k-1} = 0
$$

since it gives the first coefficient (at monomial X) of the product $g(X)h(X) = X^N - 1$. Similarly, for $j_1 = k - 2$ and $j_2 = 0$, $\pi^{k-2} \mathbf{g} \cdot \mathbf{h}^\downarrow$ gives the second coefficient of $g(X)h(X)$ (at monomial X^2) and is again equal to 0. And so on: for $j_1 = j_2 = 0$, $\mathbf{g} \cdot \mathbf{h}^\downarrow = 0$ as the kth-degree coefficient in $g(X)h(X)$.

Continuing, $\mathbf{g} \cdot \pi \mathbf{h}^\downarrow$ equals the $(k+1)$st-degree coefficient in $g(X)h(X)$, $\mathbf{g} \cdot \pi^2 \mathbf{h}^\downarrow$ the $(k+2)$nd, and so on; $\mathbf{g} \cdot \pi^{N-k-1} \mathbf{h}^\downarrow = g_{d-1} h_k + g_d h_{k-1}$ the $(N-1)$st. As before, they all vanish.

The same holds true when we simultaneously shift both words cyclically (when possible) which leads to (2.6.10).

Conversely, suppose that $\mathbf{a}H^{\mathrm{T}} = 0$ for some word $\mathbf{a} = a_0 \ldots a_{N-1}$. Write the corresponding polynomial $a(X)$ as $a(X) = f(X)g(X) + r(X)$ where the ratio $f(X) = \sum\limits_{0 \leq i \leq k-1} f_i X^i$ and $r(X)$ is the remainder. Then either $r(X) = 0$ or $1 \leq \deg r(X) = d' < d$ (and $r_{d'} = 1$ and $r_l = 0$ for $d' < l \leq n - 1$). Then set $\mathbf{r} = r_0 \ldots r_{d'}$.

Assume that $r(X) \neq 0$. By the above argument,

(i) $\mathbf{a}H^{\mathrm{T}} = \mathbf{r}H^{\mathrm{T}}$ and hence $\mathbf{r}H^{\mathrm{T}} = 0$,

(ii) the entries of vector $\mathbf{r}H^{\mathrm{T}}$ coincide with the coefficients of the product $r(X)h(X)$, beginning with $r_0 h_k + \cdots + r_{d'} h_{k-d'}$ and ending with $r_{d'} h_k$. So, these coefficients must be 0. But the equality $r_{d'} h_k = 0$ is impossible since $r_{d'} = h_k = 1$. Hence, $r(X) = 0$ and $a(X) = f(X)g(X)$, i.e. $\mathbf{a} \in \mathscr{X}$. We conclude that H is the parity-check matrix for $\mathscr{X}$.

Equivalently, H is the matrix formed by the words corresponding to polynomials $X^i h^\downarrow(X)$ where

$$
h^\downarrow(X) = \sum_{0 \leq i \leq k} h_i X^{k-i}.
$$

Alternatively, let $h(X)$ be the check polynomial for the cyclic code $\mathscr{X}$ length N with a generator polynomial $g(X)$ so that $g(X)h(X) = X^N - 1$. Then:

(a) $\mathscr{X} = \{f(X): f(X)h(X) = 0 \bmod (X^N - e)\}$;

(b) if $h(X) = h_0 + h_1 X + \cdots + h_{N-r} X^{N-r}$ then the parity-check matrix H of $\mathscr{X}$ has the form (2.6.9);

(c) the dual code $\mathscr{X}^\perp$ is a cyclic code of dim $\mathscr{X}^\perp = r$, and $\mathscr{X}^\perp = \langle h^\perp(X) \rangle$, where $h^\perp(X) = h_0^{-1} X^{N-r} h(X^{-1}) = h_0^{-1}(h_0 X^{N-r} + h_1 X^{N-r-1} + \cdots + h_{N-r})$. □

Problem 2.19   Consider the parity-check matrix H of a Hamming $[2^\ell - 1, 2^\ell - \ell - 1]$ binary code. Form the parity-check matrix H^* of a $[2^\ell, 2^\ell - \ell - 1]$ code by augmenting H with a column of zeros and then with a row of ones. The dual of the resulting code is called a first-order Reed–Muller code. Show that a first-order Reed–Muller code can correct errors of up to $2^{\ell-2} - 1$ bits per codeword.

For the photographs of Mars taken by the Mariner spacecraft such code with $\ell = 5$ was used in 1972. What was the code rate? Why is this likely to have been much less than the capacity of the channel?

Solution The code in question is $[2^\ell, \ell + 1, 2^{\ell-1}]$; with $\ell = 5$, the information rate equals $6/32 \approx 1/5$. Let us check that all codewords except $\mathbf{0}$ and $\mathbf{1}$ have weight $2^{\ell-1}$. For $\ell \geq 1$ the code $\mathscr{R}(\ell)$ is defined by recursion

$$\mathscr{R}(\ell+1) = \{\mathbf{xx} | \mathbf{x} \in \mathscr{R}(\ell)\} \vee \{\mathbf{x}, \mathbf{x}+\mathbf{1} | \mathbf{x} \in \mathscr{R}(\ell)\}.$$

So, the length of codewords in $\mathscr{R}(\ell+1)$ is obviously $2^{\ell+1}$. As $\{\mathbf{xx} | \mathbf{x} \in \mathscr{R}(\ell)\}$ and $\{\mathbf{x}, \mathbf{x}+\mathbf{1} | \mathbf{x} \in \mathscr{R}(\ell)\}$ are disjoint, the number of codewords is doubled, i.e. $\sharp \mathscr{R}(\ell+1) = 2^{\ell+2}$. Finally, assuming that all codewords in $\mathscr{R}(\ell)$ except $\mathbf{0}$ and $\mathbf{1}$ have weight $2^{\ell-1}$, consider a codeword $\mathbf{y} \in \mathscr{R}(l+1)$. If $\mathbf{y} - \mathbf{xx}$ is different from $\mathbf{0}$ or $\mathbf{1}$, then $\mathbf{x} \neq \mathbf{0}$ or $\mathbf{1}$, and so $w(\mathbf{y}) = 2w(\mathbf{x}) = 2 \times 2^{\ell-1} = 2^\ell$.

If $\mathbf{y} = \mathbf{x}, \mathbf{x}+\mathbf{1}$ we must consider some cases. If $\mathbf{x} = \mathbf{0}$ then $\mathbf{y} = \mathbf{01}$, which has weight 2^l. If $\mathbf{x} = \mathbf{1}$ then $\mathbf{y} = \mathbf{10}$, which also has weight 2^l. Finally, if $\mathbf{x} \neq \mathbf{0}$ or $\mathbf{1}$ then $w(\mathbf{x}+\mathbf{1}) = 2^\ell - 2^{\ell-1} = 2^{\ell-1}$ and $w(\mathbf{y}) = 2 \times 2^{\ell-1} = 2^\ell$. It is clear now that codewords $\mathbf{xx}$ and $\mathbf{x}, \mathbf{x}+\mathbf{1}$ with $w(\mathbf{x}) = 2^{\ell-1}$ are orthogonal to rows of parity-check matrix H^*.

Up to 7 bits may be in error, thus the probability of a transmission error p_e (for a binary symmetric memoryless channel with the error-probability p) obeys

$$p_e \leq 1 - \sum_{0 \leq i \leq 7} \binom{32}{i} p^i (1-p)^{32-i},$$

which is small when p is small. (As an estimate of an acceptable p, we can take the solution to $1 + p\log p + (1-p)\log(1-p) = 26/32$.) If the block length is fixed (and rather small), with a low value of p we can't get near the capacity.

Indeed, for $\ell = 5$, the code is $[32, 6, 16]$, detecting 15 and correcting 7 errors. That is, the code can correct a fraction $> 1/5$ of the total of 32 digits. Its information rate is $6/32$ and if the capacity of the (memoryless) channel is $C = 1 - \eta(p)$ (where p stands for the symbol-probability of error), we need the bound $C > 6/32$; that is, $\eta(p) + 6/32 < 1$, for a reliable transmission. This yields $|p - 1/2| > |p^* - 1/2|$ where $p^* \in (0, 1)$ solves $26/32 = \eta(p^*)$. Definitely $0 \le p < 1/5$ and $4/5 < p \le 1$ would do. In reality the error-probability was much less.   □

Problem 2.20   *Prove that any binary $[5, M, 3]$ code must have $M \le 4$. Verify that there exists, up to equivalence, exactly one $[5, 4, 3]$ code.*

Solution By the Plotkin bound, if d is odd and $d > \frac{1}{2}(N - 1)$ then

$$M_2^*(N, d) \le 2 \lfloor \frac{d + 1}{2d + 1 - N} \rfloor.$$

In fact,

$$M_2^*(5, 3) \le 2 \lfloor \frac{4}{6 + 1 - 5} \rfloor = 2 \cdot 2 = 4.$$

All $[5, 4, 3]$ codes are equivalent to $00000, 00111, 11001, 11110$.   □

Problem 2.21   *Let $\mathscr{X}$ be a binary $[N, k, d]$ linear code with generating matrix G. Verify that we may assume that the first row of G is $1 \ldots 1 0 \ldots 0$ with d ones. Write:*

$$G = \begin{pmatrix} 1 \ldots 1 & 0 \ldots 0 \\ G_1 & G_2 \end{pmatrix}.$$

Show that if d_2 is the distance of the code with generating matrix G_2 then $d_2 \ge d/2$.

Solution Let $\mathscr{X}$ be $[N, k, d]$. We can always form a generating matrix G of X where the first row is a codeword x with $w(x) = d$; by permuting columns of G we can have the first row in the form $\underbrace{1 \ldots 1_d}\underbrace{0 \ldots 0_{N-d}}$. So, up to equivalence,

$$G = \begin{pmatrix} 1 \ldots 1 & 0 \ldots 0 \\ G_1 & G_2 \end{pmatrix}.$$

Suppose $d(G_2) < d/2$ then, without loss of generality, we may assume that there exists a row of $(G_1 G_2)$ where the number of ones among digits $d + 1, \ldots, N$ is $< d/2$. Then the number of ones among digits $1, \ldots d$ in this row is $> d/2$, as its total weight is $\ge d$. Then adding this row and $1 \ldots 1 0 \ldots 0$ gives a codeword with weight $< d$. So, $d(G_2) \ge d/2$.   □

Problem 2.22   (Gilbert–Varshamov bound) *Prove that there exists a p-ary linear* $[N,k,d]$ *code if* $p^k < 2^N/v_{N-1}(d-2)$. *Thus, if* p^k *is the largest power of p satisfying this inequality, we have* $M_p^*(N,d) \geq p^k$.

Solution We construct a parity-check matrix by selecting N columns of length $N-k$ with the requirement that no $d-1$ columns are linearly dependent. The first column may be any non-zero string in $\mathbb{Z}_p^{N-k}$. On the step $i \geq 2$ we must choose a column which is not a linear combination of any $d-2$ (or fewer) of previously selected columns. The number of such linear combinations (with non-zero coefficients) is

$$S_i = \sum_{j=1}^{d-2} \binom{i-1}{j} (p-1)^j.$$

So, the parity-check matrix may be constructed iff $S_N + 1 < p^{N-k}$. Finally, observe that $S_N + 1 = v_{N-1}(d-2)$. Say, there exists $[5, 2^k, 3]$ code if $2^k < 32/5$, so $k = 2$ and $M_2^*(5,3) \geq 4$, which is, in fact, sharp.   □

Problem 2.23   *An element* $b \in \mathbb{F}_q^*$ *is called primitive if its order (i.e. the minimal k such that $b^k = 1 \bmod q$) is $q-1$. It is not difficult to find a primitive element of the multiplicative group $\mathbb{F}_q^*$ explicitly. Consider the prime factorisation*

$$q - 1 = \prod_{j=1}^{s} p_j^{v_j}.$$

For any $j = 1, \ldots, s$ *select* $a_j \in \mathbb{F}_q$ *such that* $a_j^{(q-1)/p_j} \neq e$. *Set* $b_j = a_j^{(q-1)/p_j^{v_j}}$ *and check that* $b = \prod_{j=1}^{s} b_j$ *has the order* $q-1$.

Solution Indeed, the order of b_j is $p_j^{v_j}$. Next, if $b^n = 1$ for some n then $n = 0$ $\bmod p_j^{v_j}$ because $b^{n \prod_{i \neq j} p_i^{v_i}} = 1$ implies $b_j^{n \prod_{i \neq j} p_i^{v_i}} = 1$, i.e. $n \prod_{i \neq j} p_i^{v_i} = 0 \bmod p_j^{v_j} = 0$. Because p_j are distinct primes, it follows that $n = 0 \bmod p_j^{v_j}$ for any j. Hence, $n = \prod_{j=1}^{s} p_j^{v_j}$.   □

Problem 2.24   *The minimal polynomial with a primitive root is called a primitive polynomial. Check that among irreducible binary polynomials of degree 4 (see (2.5.9)), $1 + X + X^4$ and $1 + X^3 + X^4$ are primitive and $1 + X + X^2 + X^3 + X^4$ is not. Check that all six irreducible binary polynomials of degree 5 (see (2.5.15)) are primitive; in practice, one prefers to work with $1 + X^2 + X^5$ as the calculations modulo this polynomial are slightly shorter. Check that among the nine irreducible polynomials of degree 6 in (2.5.16), there are six primitive: they are listed in the upper three lines. Prove that a primitive polynomial exists for every given degree.*

Solution For the solution to the last part, see Section 3.1.   □

Problem 2.25   *A cyclic code $\mathscr{X}$ of length N with the generator polynomial $g(X)$ of degree $d = N - k$ can be described in terms of the roots of $g(X)$, i.e. the elements $\alpha_1, \ldots \alpha_{N-k}$ such that $g(\alpha_j) = 0$. These elements are called zeros of code $\mathscr{X}$ and belong to a Galois field $\mathbb{F}_{2^d}$. As $g(X)|(1+X^N)$, they are also among roots of $1+X^N$. That is, $\alpha_j^N = 1$, $1 \le j \le N - k$, i.e. the α_j are Nth roots of unity. The remaining k roots of unity $\alpha_1', \ldots, \alpha_k'$ are called non-zeros of $\mathscr{X}$. A polynomial $a(X) \in \mathscr{X}$ iff, in Galois field $\mathbb{F}_{2^d}$, $a(\alpha_j) = 0$, $1 \le j \le N - k$.*
(a) Show that if $\mathscr{X}^\perp$ is the dual code then the zeros of $\mathscr{X}^\perp$ are $\alpha_1'^{-1}, \ldots, \alpha_k'^{-1}$, i.e. the inverses of the non-zeros of $\mathscr{X}$.
(b) A cyclic code $\mathscr{X}$ with generator $g(X)$ is called reversible if, for all $\mathbf{x} = x_0 \ldots x_{N-1} \in \mathscr{X}$, the word $x_{N-1} \ldots x_0 \in \mathscr{X}$. Show that $\mathscr{X}$ is reversible iff $g(\alpha) = 0$ implies that $g(\alpha^{-1}) = 0$.
(c) Prove that a q-ary cyclic code $\mathscr{X}$ of length N with $(q, N) = 1$ is invariant under the permutation of digits such that $\pi_q(i) = qi \bmod N$ (i.e. $x \to x^q$). If $s = \mathrm{ord}_N(q)$ then the two permutations $i \to i+1$ and $\pi_q(i)$ generate a subgroup of order Ns in the group $\mathrm{Aut}(\mathscr{X})$ of the code automorphisms.

Solution Indeed, since $a(x^q) = a(x)^q$ is proportional to the same generator polynomial it belongs to the same cyclic code as $a(x)$.   □

Problem 2.26   *Prove that there are 129 non-equivalent cyclic binary codes of length 128 (including the trivial codes, $\{0 \ldots 0\}$ and $\{0,1\}^{128}$). Find all cyclic binary codes of length 7.*

Solution The equivalence classes of the cyclic codes of length 2^k are in a one-to-one correspondence with the divisors of $1 + X^{2^k}$; the number of those equals $2^k + 1$. Furthermore, there are eight codes listed by their generators which are divisors of $X^7 - 1$ as

$$X^7 - 1 = (1+X)(1+X+X^3)(1+X^2+X^3).$$

                    □

3

Further Topics from Coding Theory

3.1 A primer on finite fields

In this section we present a summary of the theory of finite fields, limiting our scope by material needed in the subsequent sections and following standard texts (see [92], [93], [131]). A *finite field* is a (finite) set $\mathbb{F}$ possessing two distinct elements, 0 (zero) and e (unity), and equipped with two commutative group operations of addition and multiplication (where $0 \cdot b = 0$ for all $b \in \mathbb{F}$) related by a standard distributivity rule.

A vector space over a field $\mathbb{F}$ is a (finite) set $\mathbb{V}$, equipped with a commutative group operation of addition, and an operation of scalar multiplication by elements of $\mathbb{F}$, again obeying standard distributivity rules. The dimension $\dim \mathbb{V}$ of $\mathbb{V}$ is the minimal number d such that any collection of distinct elements $v_1, \ldots, v_{d+1} \in \mathbb{V}$ is linearly dependent, i.e. one can find elements $k_1, \ldots, k_{d+1} \in \mathbb{F}$, not all equal to 0, such that $k_1 v_1 + \cdots + k_{d+1} v_{d+1} = 0$. Then there exists a collection of elements $b_1, \ldots, b_d \in \mathbb{V}$, called a basis, such that every $v \in \mathbb{V}$ can be written as a linear combination $a_1 b_1 + \cdots + a_d b_d$ where $a_1, \ldots, a_d$ are elements of $\mathbb{F}$ (uniquely) determined by v. Unless the opposite is stated, we consider fields up to an isomorphism.

An important parameter of a field is its *characteristic*, i.e. the minimal integer number $p \geq 1$ such that $pe = e + \cdots + e \, (p \text{ times}) = 0$. Such a number, denoted by $\mathrm{char}(\mathbb{F})$, exists by a standard pigeon-hole principle. Furthermore, the characteristic is a prime number: if $p = q_1 q_2$ then $pe = (q_1 q_2)e = (q_1 e)(q_2 e) = 0$ which implies that $q_1 e = 0$ or $q_2 e = 0$ leading to a contradiction.

Example 3.1.1 Let p be a prime number. An additive cyclic group $\mathbb{Z}_p = \{0, 1, \ldots, p-1\}$, with a generator 1, becomes a field with the multiplication $(qe)(q'e) = (qq')e$. The characteristic of this field equals p.

Let $\mathbb{K}$ and $\mathbb{F}$ be fields. If $\mathbb{F} \subseteq \mathbb{K}$ we say that $\mathbb{K}$ is an extension of $\mathbb{F}$. Then $\mathbb{K}$ is also a vector space over $\mathbb{F}$ whose dimension is denoted by $[\mathbb{K} : \mathbb{F}]$.

Lemma 3.1.2 *Let $\mathbb{K}$ be an extension of $\mathbb{F}$, and $d = [\mathbb{K} : \mathbb{F}]$. Then $\sharp \mathbb{K} = (\sharp \mathbb{F})^d$.*

Proof   Let $b_1, \ldots, b_d$ be a basis for $\mathbb{K}$ over $\mathbb{F}$, with a unique representation $k = \sum\limits_{1 \le j \le d} a_j b_j$ for all $k \in \mathbb{K}$. Then for all j, we have $\sharp \mathbb{F}$ possibilities for a_j. So, altogether there exists precisely $(\sharp \mathbb{F})^d$ ways to write all combinations. □

Lemma 3.1.3   *If* $\mathrm{char}(\mathbb{F}) = p$ *then* $\sharp \mathbb{F} = p^d$, *for some integer* $d \ge 1$.

Proof   Consider elements $0, e, 2e, \ldots, (p-1)e$. They form $\mathbb{Z}_p$, i.e. $\mathbb{Z}_p \subseteq \mathbb{F}$. Then $\sharp \mathbb{F} = p^d$ by Lemma 3.1.2. □

Corollary 3.1.4   *The number of elements in a finite field* $\mathbb{F}$ *must be* $q = p^s$ *where* $p = \mathrm{char}(\mathbb{F})$ *and* $s \ge 1$ *is a natural number.*

From now on, unless otherwise stated, p stands for a prime and $q = p^s$ for a prime power.

Lemma 3.1.5   (A freshman's dream) *If* $\mathrm{char}(\mathbb{F}) = p$ *then for all* $a, b \in \mathbb{F}$ *and integers* $n \ge 1$,

$$(a \pm b)^{p^n} = a^{p^n} + (\pm b)^{p^n}. \tag{3.1.1}$$

Proof   Use induction in n: for $n = 1$,

$$(a \pm b)^p = \sum_{0 \le k \le p} \binom{p}{k} a^k (\pm b)^{p-k}.$$

For $1 \le k \le p-1$, the value $\binom{p}{k}$ is a multiple of p and the corresponding term vanishes. Therefore, $(a \pm b)^p = a^p + (\pm b)^p$. The inductive step is completed by the same argument, with a and $\pm b$ replaced by $a^{p^{n-1}}$ and $(\pm b)^{p^{n-1}}$. □

Lemma 3.1.6   *The multiplicative group* $\mathbb{F}^*$ *of non-zero elements of a field* $\mathbb{F}$ *of size* q *is isomorphic to the cyclic group* $\mathbb{Z}_{q-1}$.

Proof   Observe that for any divisor $d | (q-1)$, group $\mathbb{F}^*$ contains exactly $\phi(d)$ elements of multiplicative order d where ϕ is Euler's totient phi-function. (Recall that $\phi(d) = \sharp\{k : k < d, \gcd(k, d) = 1\}$.) We'll see that all elements of order d have the form $a^{\frac{q-1}{d} r}$ where a is a primitive element, $r \le d$ and r, d are co-prime. In fact, $q - 1 = \sum\limits_{d: d | (q-1)} \phi(d)$, and $\mathbb{F}^*$ will have at least one element of order $q - 1$ which implies that $\mathbb{F}^*$ is cyclic, of order $q - 1$. □

Let $a \in \mathbb{F}^*$ be an element of order d where $d | (q-1)$. Take the cyclic subgroup $\{e, a, \ldots, a^{d-1}\}$. Every element of this subgroup has multiplicative order dividing d, i.e. is a root of the polynomial $X^d - e$ (a dth root of unity). But $X^d - e$ has $\le d$ distinct roots in $\mathbb{F}$ (because $\mathbb{F}$ is a field). So, $\{e, a, \ldots, a^{d-1}\}$ is the set of all roots of $X^d - e$ in $\mathbb{F}$. In particular, each element from $\mathbb{F}$ of order d belongs to

$\{e, a, \ldots, a^{d-1}\}$. Observe that the cyclic group $\mathbb{Z}_d$ has exactly $\phi(d)$ elements of order d. So, the whole $\mathbb{F}^*$ has exactly $\phi(d)$ elements of order d; in other words, if $\psi(d)$ is the number of elements in $\mathbb{F}$ of order d then either $\psi(d) = 0$ or $\psi(d) = \phi(d)$ and

$$q - 1 = \sum_{d:d(n)} \psi(d) \le \sum_{d:d|n} \phi(d) = q - 1,$$

which implies that for all $d|n$,

$$\psi(d) = \phi(d).$$

Definition 3.1.7   A (multiplicative) generator of $\mathbb{F}^*$ (i.e. an element of multiplicative order $q - 1$) is called a *primitive* element of field $\mathbb{F}$. Although such an element is non-unique, we will usually single out one such element and denote it by ω; of course a power ω^r where r is coprime with $(q - 1)$ will also give a primitive element.

If $a \in \mathbb{F}^*$ with $\sharp \mathbb{F}^* = q - 1$ then $a^{q-1} = e$ (the order of every element divides the order of the group). Hence, $a^q = a$, i.e. a is a root of the polynomial $X^q - X$ in $\mathbb{F}$. But $X^q - X$ can have only $\le q$ roots (including zero 0), so $\mathbb{F}$ gives the set of all roots of $X^q - X$.

Definition 3.1.8   Given fields $\mathbb{K}$ and $\mathbb{F}$, with $\mathbb{F} \subseteq \mathbb{K}$, field $\mathbb{K}$ is called *the splitting field* for a polynomial $g(X)$ with coefficients from $\mathbb{F}$ if (a) $\mathbb{K}$ contains all roots of $g(X)$, (b) there is no field $\tilde{\mathbb{K}}$ with $\mathbb{F} \subset \tilde{\mathbb{K}} \subset \mathbb{K}$ satisfying (a). We will write $\mathrm{Spl}(g(X))$ for the splitting field for $g(X)$.

Thus, if $\sharp \mathbb{F} = q$ then $\mathbb{F}$ contains all roots of polynomial $X^q - X$ and is the splitting field for this polynomial.

Lemma 3.1.9   *Any two splitting fields $\mathbb{K}$, $\mathbb{K}'$ for the same polynomial $g(X)$ with coefficients from $\mathbb{F}$ coincide.*

Proof   In fact, take the intersection $\mathbb{K} \cap \mathbb{K}'$: it contains $\mathbb{F}$ and is a subfield of both $\mathbb{K}$ and $\mathbb{K}'$. It must then coincide with each of $\mathbb{K}$, $\mathbb{K}'$.   □

Corollary 3.1.10   *For any prime p and natural $s \ge 1$, there exists at most one field with p^s elements.*

Proof   Each such field is splitting for polynomial $X^q - X$ with coefficients from $\mathbb{Z}_p$ and $q = p^s$. So any two such fields coincide.   □

On the other hand, we will prove later the following.

Theorem 3.1.11   *For any non-constant polynomial with coefficients from $\mathbb{F}$, there exists a splitting field.*

Corollary 3.1.12 *For any prime p and natural $s \geq 1$, there exists precisely one field with p^s elements.*

Proof of Corollary 3.1.12 Take again the polynomial $X^q - X$ with coefficients from $\mathbb{Z}_p$ and $q = p^s$. By Theorem 3.1.11, there exists the splitting field $\mathrm{Spl}(X^q - X)$ where $X^q - X = X(X^{q-1} - e)$ is factorised into linear polynomials. So, $\mathrm{Spl}(X^q - X)$ contains the roots of $X^q - X$ and has characteristic p (as it contains $\mathbb{Z}_p$).

However, the roots of $(X^q - X)$ form a subfield: if $a^q = a$ and $b^q = b$ then $(a \pm b)^q = a^q + (\pm b^q)$ (Lemma 3.1.5) which coincides with $a \pm b$. Also, $(ab^{-1})^q = a^q (b^q)^{-1} = ab^{-1}$. This field cannot be strictly contained in $\mathrm{Spl}(X^q - X)$ thus it coincides with $\mathrm{Spl}(X^q - X)$.

It remains to check that all roots of $(X^q - X)$ are distinct: then the cardinality $\sharp \, \mathrm{Spl}(X^q - X)$ will be equal to q. In fact, if $X^q - X$ had a multiple root then it would have had a common factor with its 'derivative' $\partial_X (X^q - X) = qX^{q-1} - e$. However, $qX^{q-1} = 0$ in $\mathrm{Spl}(X^q - X)$ and thus cannot have such factors. $\qquad \square$

Summarising, we have the two characterisation theorems for finite fields.

Theorem 3.1.13 *All finite fields have size p^s where p is prime and $s \geq 1$ integer. For all such p, s, there exists a unique field of this size.*

The field of size $q = p^s$ will be denoted by $\mathbb{F}_q$ (a popular alternative notation is $\mathrm{GF}(q)$ (a Galois field)). In the case of the simplest fields $\mathbb{F}_p = \{0, 1, \ldots, p-1\}$ (for p is prime) we use symbol 1 instead of e for the unit.

Theorem 3.1.14 *All finite fields can be arranged into sequences ('towers'). For a prime p and positive integers $s_1, s_2, \ldots$,*

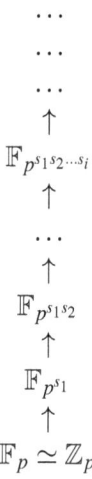

$$
\begin{array}{c}
\cdots \\
\cdots \\
\cdots \\
\uparrow \\
\mathbb{F}_{p^{s_1 s_2 \cdots s_i}} \\
\uparrow \\
\cdots \\
\uparrow \\
\mathbb{F}_{p^{s_1 s_2}} \\
\uparrow \\
\mathbb{F}_{p^{s_1}} \\
\uparrow \\
\mathbb{F}_p \simeq \mathbb{Z}_p
\end{array}
$$

Here each arrow is a uniquely defined injective homomorphism.

Example 3.1.15 In Section 2.5 we worked with polynomial fields $\mathbb{F}_2[X]/\langle q(X)\rangle$ where $q(X)$ is an irreducible binary polynomial; see Theorems 2.5.32 and 2.5.33. Continuing Example 2.5.35(c), consider the field $\mathbb{F}_{16}$ realised as $\mathbb{F}_2[X]/\langle 1+X^3+X^4\rangle$. The field structure is as follows:

power of X mod $1+X^3+X^4$	polynomial	vector (string)	
$--$	0	0000	
X^0	1	1000	
X	X	0100	
X^2	X^2	0010	
X^3	X^3	0001	
X^4	$1+X^3$	1001	
X^5	$1+X+X^3$	1101	(3.1.2)
X^6	$1+X+X^2+X^3$	1111	
X^7	$1+X+X^2$	1110	
X^8	$X+X^2+X^3$	0111	
X^9	$1+X^2$	1010	
X^{10}	$X+X^3$	0101	
X^{11}	$1+X^2+X^3$	1011	
X^{12}	$1+X$	1100	
X^{13}	$X+X^2$	0110	
X^{14}	X^2+X^3	0011	

Finally, if we choose to specify the table for $\mathbb{F}_2[X]/\langle 1+X+X^2+X^3+X^4\rangle$, the calculations will be considerably longer (and organised differently). The point is that in this case monomial X will not be a primitive element, since $X^5 = 1$ mod $(1+X+X^2+X^3+X^4)$. Instead, a generator of the multiplicative group will be a sum of monomials, viz. $1+X$.

Worked Example 3.1.16 (a) *How many elements are in the smallest extension of $\mathbb{F}_5$ which contains all roots of polynomials X^2+X+1 and X^3+X+1?*

(b) *Determine the number of subfields of $\mathbb{F}_{1024}$, $\mathbb{F}_{729}$. Find all primitive elements of $\mathbb{F}_7$, $\mathbb{F}_9$, $\mathbb{F}_{16}$. Compute $(\omega^{10}+\omega^5)(\omega^4+\omega^2)$ where ω is a primitive element of $\mathbb{F}_{16}$.*

Solution (a) Clearly, 5^6.

(b) $\mathbb{F}_{1024} = \mathbb{F}_{2^{10}}$ has 4 subfields: $\mathbb{F}_2, \mathbb{F}_4, \mathbb{F}_{32}$ and $\mathbb{F}_{1024}$. $\mathbb{F}_{729} = \mathbb{F}_{3^6}$ has 4 subfields: $\mathbb{F}_3, \mathbb{F}_9, \mathbb{F}_{27}$ and $\mathbb{F}_{729}$. $\mathbb{F}_7$ has 2 primitive elements: ω, ω^5 (with $(\omega^5)^5 = \omega$). $\mathbb{F}_9$ has

4 primitive elements, of the form $\omega, \omega^3, \omega^5, \omega^7$. $\mathbb{F}_{16}$ has 8 primitive elements: ω, $\omega^2, \omega^4, \omega^7, \omega^8, \omega^{11}, \omega^{13}, \omega^{14}$.

By using the table for $\mathbb{F}_2[X]/\langle 1+X+X^4 \rangle$ (see Example 2.5.35(c)), we can find that

$$(\omega^{10}+\omega^5)(\omega^4+\omega^2) = \omega^{14}+\omega^9+\omega^{12}+\omega^7$$
$$= 1001+0101+1111+1101 = 1110$$
$$= \omega^{10}.$$

However, by taking $\omega' = \omega^7$, the RHS becomes

$$\omega^8 + \omega^3 + \omega^9 + \omega^4 = 1010 + 0001 + 0101 + 1100 = 0010 = \omega^2 = (\omega')^{11}.$$

$\square$

From now on we will focus on polynomial representations of finite fields. Generalising concepts introduced in Section 2.5, consider

Definition 3.1.17   The set of all polynomials with coefficients from $\mathbb{F}_q$ is a commutative ring denoted by $\mathbb{F}_q[X]$. A *quotient* ring $\mathbb{F}_q[X]/\langle g(X) \rangle$ is where the operation is modulo a fixed polynomial $g(X) \in \mathbb{F}_q[X]$.

Definition 3.1.18   A polynomial $g(X) \in \mathbb{F}_q[X]$ is called *irreducible* (over $\mathbb{F}_q$) if it admits no representation

$$g(X) = g_1(X)g_2(X)$$

with $g_1(X), g_2(X) \in \mathbb{F}_q[X]$.

A generalisation of Theorem 2.5.32 is presented in Theorem 3.1.19 below.

Theorem 3.1.19   Let $g(X) \in \mathbb{F}_q[X]$ have degree $\deg g(X) = d$. Then $\mathbb{F}_q[X]/\langle g(X) \rangle$ is a field $\mathbb{F}_{q^d}$ iff $g(X)$ is irreducible.

Proof   Let $g(X)$ be an irreducible polynomial over $\mathbb{F}_q$. To show that $\mathbb{F}_q[X]/\langle g(X) \rangle$ is a field we should check that each non-zero element $f(X) \in \mathbb{F}_q[X]/\langle g(X) \rangle$ has an inverse. Consider the set $\mathbb{F}(f)$ of polynomials of the form $f(X)h(X) \bmod g(X)$ where $h(X) \in \mathbb{F}_q[X]/\langle g(X) \rangle$ (the principal ideal generated by $f(X)$). If $\mathbb{F}(f)$ contains the unity $e \in \mathbb{F}_q$ (the constant polynomial equal to e) then the corresponding $h(X) = f(X)^{-1}$. If not, the map $h(X) \mapsto f(X)h(X) \bmod g(X)$, from $\mathbb{F}_q[X]/\langle g(X) \rangle$ to itself, is not a surjection. That is, $f(X)h_1(X) = f(X)h_2(X) \bmod g(X)$ for some distinct $h_1(X), h_2(X)$, i.e.

$$f(X)\big(h_1(X) - h_2(X)\big) = r(X)g(X).$$

Then either $g(X)|f(X)$ or $g(X)|(h_1(X) - h_2(X))$ as $g(X)$ is irreducible. So, either $f(X) = 0 \mod g(X)$ (a contradiction) or $h_1(X) = h_2(X) \mod g(X)$. Hence, $\mathbb{F}_q[X]/\langle g(X)\rangle$ is a field.

The inverse assertion is proved similarly: if $g(X)$ is reducible then $\mathbb{F}_q[X]/\langle g(X)\rangle$ contains non-zero $g_1(X)$, $g_2(X)$ with $g_1(X)g_2(X) = 0$. Then $\mathbb{F}_q[X]/\langle g(X)\rangle$ cannot be a field.

The dimension $\left[\mathbb{F}_q[X]/\langle g(X)\rangle : \mathbb{F}_q\right]$ is equal to d, the degree of $g(X)$, so $\mathbb{F}_q[X]/\langle g(X)\rangle = \mathbb{F}_{q^d}$. $\qquad\square$

Worked Example 3.1.20   *Prove that $g(X)$ has an inverse in the polynomial ring $\mathbb{F}_q[X]/\langle X^N - e\rangle$ iff $\gcd\left(g(X), X^N - e\right) = e$.*

Solution Consider the map $\mathbb{F}_q[X]/\langle X^N - e\rangle \to \mathbb{F}_q[X]/\langle X^N - e\rangle$ given by $h(X) \mapsto h(X)g(X) \mod (X^N - e)$. If it is a surjection then there exists $h(X)$ with $h(X)g(X) = e$ and $h(X) = g(X)^{-1}$. Suppose it is not. Then there exist $h^{(1)}(X) \neq h^{(2)}(X) \mod (X^N - e)$ such that $h^{(1)}(X)g(X) = h^{(2)}(X)g(X) \mod (X^N - e)$, i.e.

$$(h^{(1)}(X) - h^{(2)}(X))g(X) = s(X)(X^N - e).$$

As $(X^N - e) \nmid (h^{(1)}(X) - h^{(2)}(X))$, this means that $\gcd(g(X), X^N - e) \neq e$.

Conversely, if $\gcd(g(X), X^N - e) = d(X) \neq e$ then the equation $h(X)g(X) = e \mod (X^N - e)$ gives

$$h(X)g(X) = e + q(X)(X^N - e)$$

where $d(X)|$LHS and $d(X)|q(X)(X^N - e))$. Therefore, $d(X)|e$: a contradiction. Hence, $g(X)^{-1}$ does not exist. $\qquad\square$

Example 3.1.21   (Continuing Example 2.5.19) There are six irreducible binary polynomials of degree 5:

$$
\begin{gathered}
1 + X^2 + X^5, \ \ 1 + X^3 + X^5, \ \ 1 + X + X^2 + X^3 + X^5, \\
1 + X + X^2 + X^4 + X^5, \ \ 1 + X + X^3 + X^4 + X^5, \\
1 + X^2 + X^3 + X^4 + X^5.
\end{gathered}
\tag{3.1.3}
$$

Then there are nine irreducible polynomials of degree 6, and so on. Calculating irreducible polynomials of a large degree is a demanding task, although extensive tables of such polynomials are now available on the web.

We are now going to prove Theorem 3.1.11.

Proof of Theorem 3.1.11 The key fact is that any non-constant polynomial $g(X) \in \mathbb{F}_q[X]$ has a root in some extension of $\mathbb{F}_q$. Without loss of generality, assume that $g(X)$ is irreducible, with $\deg g(X) = d$. Take $\mathbb{F}_q[X]/\langle g(X)\rangle = \mathbb{F}_{q^d}$ as an extension field. In this field, $g(\alpha) = 0$ where α is polynomial $X \in \mathbb{F}_q[X]/\langle g(X)\rangle$, so $g(X)$

has a root. We can divide $g(X)$ by $X - \alpha$ in $\mathbb{F}_{q^d}$ and use the same construction to prove that $g_1(X) = g(X)/(X - \alpha)$ has a root in some extension of $\mathbb{F}_{q^t}, t < d$. Finally, we obtain a field containing all d roots of $g(X)$, i.e. construct the splitting field $\text{Spl}(g(X))$. $\qquad\square$

Definition 3.1.22 Given a field $\mathbb{F} \subset \mathbb{K}$ and an element $\gamma \in \mathbb{K}$, we denote by $\mathbb{F}(\gamma)$ the smallest field containing $\mathbb{F}$ and γ (obviously, $\mathbb{F} \subset \mathbb{F}(\gamma) \subset \mathbb{K}$). Similarly, $\mathbb{F}(\gamma_1, \ldots, \gamma_r)$ is the smallest field containing $\mathbb{F}$ and elements $\gamma_1, \ldots, \gamma_r \in \mathbb{K}$. For $\mathbb{F} = \mathbb{F}_q$ and $\alpha \in \mathbb{K}$, set

$$M_{\alpha,\mathbb{F}}(X) = (X - \alpha)(X - \alpha^q) \ldots \left(X - \alpha^{q^{d-1}}\right), \qquad (3.1.4)$$

where d is the smallest positive integer such that $\alpha^{q^d} = \alpha$ (such a d exists as will be proved in Lemma 3.1.24).

A *monic* polynomial is the one with the highest coefficient. The *minimal* polynomial for $\alpha \in \mathbb{K}$ over $\mathbb{F}$ is a unique monic polynomial $M_\alpha(X)$ $(= M_{\alpha,\mathbb{F}}(X)) \in \mathbb{F}[X]$ such that $M_\alpha(\alpha) = 0$ and $M_\alpha(X)|g(X)$ for each $g(X) \in \mathbb{F}[X]$ with $g(\alpha) = 0$. When ω is a primitive element of $\mathbb{K}$ (generating $\mathbb{K}^*$), $M_\omega(X)$ is called a *primitive polynomial* (over $\mathbb{F}$). The *order* of a polynomial $p(X) \in \mathbb{F}[X]$ is the smallest n such that $p(X)|(X^n - e)$.

Example 3.1.23 (Continuing Example 3.1.21.) In this example we deal with polynomials over $\mathbb{F}_2$. The irreducible polynomial $X^2 + X + 1$ is primitive and has order 3. The irreducible polynomials $X^3 + X + 1$ and $X^3 + X^2 + 1$ are primitive and of order 7. The polynomials $X^4 + X^3 + 1$ and $X^4 + X + 1$ are primitive and have order 15 whereas $X^4 + X^3 + X^2 + X + 1$ is not primitive and of order 5. (It is helpful to note that with $d = 4$, the order of $X^4 + X^3 + 1$ and $X^4 + X + 1$ equals $2^d - 1$; on the other hand, the order of element X in the field $\mathbb{F}_2[X]/\langle 1 + X + X^2 + X^3 + X^4 \rangle$ equals 5, but its order, say, in the field $\mathbb{F}_2[X]/\langle 1 + X + X^4 \rangle$ equals 15.) All six polynomials listed in (3.1.3) are primitive and have order 31 (i.e. appear in the decomposition of $X^{31} + 1$).

Lemma 3.1.24 Let $\mathbb{F}_q \subset \mathbb{F}_{q^d}$ and $\alpha \in \mathbb{F}_{q^d}$. Let $M_\alpha(X) \in \mathbb{F}[X]$ be the minimal polynomial for α, of degree $\deg M_\alpha(X) = d$. Then:

(a) $M_\alpha(X)$ is the only irreducible polynomial in $\mathbb{F}_q[X]$ with a root at α.

(b) $M_\alpha(X)$ is the only monic polynomial in $\mathbb{F}_q[X]$ of degree d with a root at α.

(c) $M_\alpha(X)$ has the form (3.1.4).

Proof Assertions (a), (b) follow from the definition. To prove (c), assume $\gamma \in \mathbb{K}$ is a root of a polynomial $f(X) = a_0 + a_1 X + \cdots + a_d X^d$ from $\mathbb{F}[X]$, i.e. $\sum_{0 \le i \le d} a_i \gamma^i = 0$.

As $a_i^q = a_i$ (which is true for all $a \in \mathbb{F}$) and by virtue of Lemma 3.1.5,

$$f(\gamma^q) = \sum_{0 \le i \le d} a_i \gamma^{qi} = \sum_{0 \le i \le d} (a_i \gamma^i)^q = \left(\sum_{0 \le i \le d} a_i \gamma^i \right)^q = 0,$$

so γ^q is a root. Similarly, $(\gamma^q)^q = \gamma^{q^2}$ is a root, and so on.

For $M_\alpha(X)$ it yields that α, α^q, $\alpha^{q^2}, \dots$ are roots. This will end when $\alpha^{q^s} = \alpha$ for the first time (which proves the existence of such an s). Finally, $s = d$ as all α, $\alpha^q, \dots, \alpha^{q^{d-1}}$ are distinct: if not then $\alpha^{q^i} = \alpha^{q^j}$ where, say, $i < j$. Taking q^{d-j} power of both sides, we get $\alpha^{q^{d+i-j}} = \alpha^{q^d} = \alpha$. So, α is a root of polynomial $P(X) = X^{q^{d+i-j}} - X$, and $\mathrm{Spl}(P(X)) = \mathbb{F}_{q^{d+i-j}}$. On the other hand, α is a root of an irreducible polynomial of degree d, and $\mathrm{Spl}(M_\alpha(X)) = \mathbb{F}_{q^d}$. Hence, $d | (d + i - j)$ or $d | (i - j)$, which is impossible. This means that all the roots $\alpha^{q^i}, i < d$, are distinct. $\qquad\square$

Theorem 3.1.25 *For any field $\mathbb{F}_q$ and integer $d \ge 1$, there exists an irreducible polynomial $f(X) \in \mathbb{F}_q[X]$ of degree d.*

Proof Take a primitive element $\omega \in \mathbb{F}_{q^d}$. Then $\mathbb{F}_q(\omega)$, the minimal extension of $\mathbb{F}_q$ containing ω, coincides with $\mathbb{F}_{q^d}$. The dimension $[\mathbb{F}_q(\omega) : \mathbb{F}_q]$ of vector space $\mathbb{F}_q(\omega)$ over $\mathbb{F}_q$ equals $[\mathbb{F}_{q^d} : \mathbb{F}_q] = d$. The minimal polynomial $M_\omega(X)$ for ω over $\mathbb{F}_q$ has distinct roots $\omega, \omega^q, \dots, \omega^{q^{d-1}}$ and therefore is of degree d. $\qquad\square$

Although proving irreducibility of a given polynomial is a problem with no general solution, the number of irreducible polynomials of a given degree can be evaluated by using an elegant (and not very complicated) method invoking the so-called Möbius function.

Definition 3.1.26 The Möbius function μ on the set $\mathbb{Z}_+$ is given by

$$\mu(1) = 1, \quad \mu(n) = 0 \text{ if } n \text{ is divisible by a square of a prime number,}$$

and

$$\mu(n) = (-1)^k \text{ if } n \text{ is a product of } k \text{ distinct prime numbers.}$$

Theorem 3.1.27 *The number $N_q(n)$ of irreducible polynomials of degree n in the polynomial ring $\mathbb{F}_q[X]$ is given by*

$$N_q(n) = \frac{1}{n} \sum_{d: \, d|n} \mu(d) q^{n/d}. \qquad (3.1.5)$$

For example, $N_q(20)$ equals

$$\frac{1}{20}\left(\mu(1)q^{20} + \mu(2)q^{10} + \mu(4)q^5 + \mu(5)q^4 + \mu(10)q^2 + \mu(20)q\right)$$
$$= \frac{1}{20}\left(q^{20} - q^{10} - q^4 + q^2\right).$$

Proof First, we establish the additive Möbius inversion formula. Let ψ and Ψ be two functions from $\mathbb{Z}_+$ to an Abelian group G with an additive group operation. Then the following equations are equivalent:

$$\Psi(n) = \sum_{d|n} \psi(d) \tag{3.1.6}$$

and

$$\psi(n) = \sum_{d|n} \mu(d)\Psi\left(\frac{n}{d}\right). \tag{3.1.7}$$

This equivalence follows when we observe that (a) the sum $\sum_{d|n} \mu(d)$ is equal to 0 if $n > 1$ and to 1 if $n = 1$, and (b) for all n,

$$\sum_{d:\, d|n} \mu(d)\Psi(n/d) = \sum_{d:\, d|n} \mu(d) \sum_{c:\, c|n/d} \psi(c)$$
$$= \sum_{c:\, c|n} \psi(c) \sum_{d:\, d|n/c} \mu(d) = \psi(n).$$

To check (a), let $p_1, \ldots, p_k$ be different prime factors in decomposition of n then

$$\sum_{d|n} \mu(d) = \mu(1) + \sum_{i=1}^{k} \mu(p_i) + \cdots + \mu(p_1 \ldots p_k)$$
$$= 1 + \binom{k}{1}(-1) + \binom{k}{2}(-1)^2 + \cdots + \binom{k}{k}(-1)^k = 0.$$

Applying (3.1.7) to $G = \mathbf{Z}$, the additive group of integer numbers, with $\psi(n) = nN_q(n)$ and $\Psi(n) = q^n$, gives (3.1.5).

Now, decompose the polynomial $X^{q^n} - X$ into the product of irreducible polynomials. Then (3.1.6) holds true as the degree q^n of $X^{q^n} - X$ coincides with the sum of degrees of all irreducible polynomials whose degrees divide n. Indeed, we simply write $X^{q^n} - X$ as the product of all irreducible polynomials and observe that an irreducible polynomial enters the decomposition iff its degree divides n (cf. Corollary 3.1.30). $\qquad\qquad\square$

Worked Example 3.1.28 *Find all irreducible polynomials of degree 2 and 3 over $\mathbb{F}_3$ and determine their orders.*

Solution Over $\mathbb{F}_3 = \{0,1,2\}$ there are three irreducible polynomials of degree 2: $X^2 + 1$, of order 4, with

$$(X^4 - 1)/(X^2 + 1) = X^2 - 1,$$

and $X^2 + X + 2$ and $X^2 + 2X + 2$, of order 8, with

$$(X^8 - 1)/(X^2 + X + 2)(X^2 + 2X + 2) = X^4 - 1.$$

Next, there exist $(3^3 - 3)/3 = 8$ irreducible polynomials over $\mathbb{F}_3$ of degree 3. Four of them have order 13 (hence, are not primitive):

$$X^3 + 2X + 2, X^3 + X^2 + 2, X^3 + X^2 + X + 2, X^3 + 2X^2 + 2X + 2.$$

The remaining four have order 26 (hence, are primitive):

$$X^3 + 2X + 1, X^3 + X^2 + 2X + 1, X^3 + 2X^2 + 1, X^3 + 2X^2 + X + 1.$$

Indeed, if $p(X)$ denotes the product of the first four polynomials then $(X^{13} - 1)/p(X) = X - 1$. On the other hand, if $r(X)$ stands for the product of the remaining four then $(X^{26} - 1)/r(X)$ equals

$$(X - 1)(X + 1)(X^3 + 2X + 2)(X^3 + X^2 + 2)$$
$$\times (X^3 + X^2 + X + 2)(X^3 + 2X^2 + 2X + 2).$$

$\square$

Theorem 3.1.29 *If $g(X) \in \mathbb{F}_q[X]$ is irreducible and of degree d and α is a root of $g(X)$ then the splitting field $\mathrm{Spl}(g(X))$ and the minimal extension $\mathbb{F}_q(\alpha)$ both coincide with $\mathbb{F}_{q^d}$.*

Proof We know that $g(X) = M_{\alpha,\mathbb{F}_q}(X) = \mathrm{irr}_{\alpha,\mathbb{F}_q}(X)$ (by Lemma 3.1.24, as $g(X)$ is irreducible). We then have that $\mathbb{F}_q \subset \mathbb{F}_q(\alpha) = \mathbb{F}_{q^d} \subseteq \mathrm{Spl}(g(X))$. It is left to check that any root γ of $g(X)$ lies in $\mathbb{F}_q(\alpha)$: this will imply that $\mathrm{Spl}(g(X)) \subseteq \mathbb{F}_q(\alpha)$. By Theorem 3.1.13, the unique Galois field with q^d elements $\mathbb{F}_q(\alpha) = \mathbb{F}_{q^d}$ is the splitting field $\mathrm{Spl}(X^{q^d} - X)$, i.e. contains all roots of $X^{q^d} - X$ (one of which is α). Then $g(X)|(X^{q^d} - X)$ as $g(X) = M_{\alpha,\mathbb{F}_q}(X)$. Therefore, all roots of $g(X)$ are roots of $X^{q^d} - X$ and hence lie in $\mathbb{F}_q(\alpha)$. $\square$

Corollary 3.1.30 *Suppose that $g(X) \in \mathbb{F}_q[X]$ is an irreducible polynomial of degree d. Then $g(X)|(X^{q^n} - X)$ iff $d|n$.*

Proof We have the splitting fields $\mathrm{Spl}(g(X)) = \mathbb{F}_{q^d}$ and $\mathrm{Spl}(X^{q^n} - X) = \mathbb{F}_{q^n}$. By Theorem 3.1.29, $\mathrm{Spl}(g(X)) \subseteq \mathrm{Spl}(X^{q^n} - X)$ iff $d|n$.

Now if $g(X)|(X^{q^n} - X)$, each root of $g(X)$ is a root of $(X^{q^n} - X)$. Then $\mathrm{Spl}(g(X)) \subseteq \mathrm{Spl}(X^{q^n} - X)$ and hence $d|n$.

Conversely, if $d|n$, i.e. $\mathrm{Spl}(g(X)) \subseteq \mathrm{Spl}\left(X^{q^n} - X\right)$, then each root of $g(X)$ lies in $\mathrm{Spl}\left(X^{q^n} - X\right)$. But $\mathrm{Spl}\left(X^{q^n} - X\right)$ is precisely the set of the roots of $\left(X^{q^n} - X\right)$, so each root of $g(X)$ is that of $\left(X^{q^n} - X\right)$. Then $g(X)|\left(X^{q^n} - X\right)$.   □

Theorem 3.1.31   *If $g(X) \in \mathbb{F}_q[X]$ is an irreducible polynomial of degree d and $\alpha \in \mathrm{Spl}(g(X)) = \mathbb{F}_{q^d}[X]$ is its root then all the roots of $g(X)$ are $\alpha, \alpha^q, \ldots, \alpha^{q^{d-1}}$. Furthermore, d is the smallest positive integer such that $\alpha^{q^d} = \alpha$.*

Proof   As in the proof of Lemma 3.1.24, $\alpha, \alpha^q, \ldots, \alpha^{q^{d-1}}$ are distinct roots. Thus all the roots are listed and d is the smallest positive integer with the above property.

□

Corollary 3.1.32   *All roots of an irreducible polynomial $g(X) \in \mathbb{F}_q[X]$ with $\deg g(X) = d$ have in $\mathrm{Spl}(g(X))$ the same multiplicative order dividing $q^d - 1$, and it gives the order of polynomial $g(X)$ (see Definition 3.1.22).*

The order of irreducible polynomial $g(X)$ will be denoted by $\mathrm{ord}(g(X))$.

Worked Example 3.1.33   (a) *Prove that for natural n, q such that $\mathrm{lcm}(n, q) = 1$ there exists a natural s such that $n|(q^s - 1)$.*

(b) *Prove that if a polynomial $g(X) \in \mathbb{F}_2[X]$ is irreducible then $g(X)|(X^n - 1)$ iff $\mathrm{ord}(g(X))|n$.*

Solution (a) Set $q^l - 1 = na_l + b_l$ where $b_l \leq n$ and $l = 1, 2, \ldots$. By the pigeon-hole principle, $b_{l_1} = b_{l_2}$ for some $l_1 < l_2$. Then $n|q^{l_1}(q^{l_2 - l_1} - 1)$. Owing to the condition $\mathrm{lcm}(n, q) = 1$, $n|(q^s - 1)$ with $s = l_2 - l_1$.

(b) For an irreducible $g(X)$, the order $\mathrm{ord}\, g(X)$ was introduced in Definition 3.1.22:

$$\mathrm{ord}(g(X)) = \min[n : g(X)|(X^n - 1)].$$

First, our goal is to check that if $m = \mathrm{ord}(g(X))$ then $m|n$ iff $g(X)|(X^n - 1)$. Indeed, suppose $m|n$: $n = mr$. Then $X^n - 1 = (X^m - 1)(1 + X^m + \cdots + X^{m(r-1)})$. As $g(X)|(X^m - 1)$, this implies $g(X)|(X^n - 1)$.

Conversely, if $g(X)|(X^n - 1)$ then the roots of $\alpha_1, \ldots, \alpha_d$ of $g(X)$ are among those of $X^n - 1$ in $\mathrm{Spl}(X^n - 1)$. So, $\alpha_j^m = \alpha_j^n = 1$ in $\mathrm{Spl}(X^n - 1)$, $1 \leq j \leq d$. Write $n = mb + a$ where $0 \leq a < m$. Then $\alpha_j^n = \alpha_j^{bm}\alpha_j^a = \alpha_j^a = 1$, i.e. each α_j is a root of $X^a - 1$. Hence, if $a > 0$ then $g(X)|(X^a - 1)$: a contradiction. So, $a = 0$ and $m|n$.   □

Calculating an irreducible polynomial $g(X) \in \mathbb{F}_q[X]$ with a given root $\alpha \in \mathbb{F}_{q^n}$, in particular, the minimal polynomial $M_{\alpha, \mathbb{F}_q}(X)$, is not easy. This is because the relation between q, n, α and $d = \deg M_\alpha(X)$ is complicated. However, if $\alpha = \omega$ is a primitive element of $\mathbb{F}_{q^n}$ then $d = n$ as $\omega^{q^n - 1} = e$, $\omega^{q^n} = \omega$ and n is the least positive integer with this property. In this case $M_\omega(X) = \prod_{b \in \mathbb{F}_{q^n}}(X - b)$.

For a general irreducible polynomial, the notion of conjugacy is helpful: see Definition 3.1.34 below. This concept was introduced (and used) informally in Section 2.5 for fields $\mathbb{F}_{2^s}$.

Definition 3.1.34 Elements $\alpha, \alpha' \in \mathbb{F}_{q^n}$ are called *conjugate* over $\mathbb{F}_q$ if $M_{\alpha,\mathbb{F}_q}(X) = M_{\alpha',\mathbb{F}_q}(X)$.

Summarising what was said above, we deduce the following assertion.

Theorem 3.1.35 *The conjugates of $\alpha \in \mathbb{F}_{q^n}$ over $\mathbb{F}_q$ are $\alpha, \alpha^q, \ldots, \alpha^{q^{d-1}} \in \mathbb{F}_{q^n}$ where d is as before. In particular,* $\prod\limits_{0 \le j \le d-1} (X - \alpha^{q^j})$ *has all its coefficients in $\mathbb{F}_q$ and is a unique irreducible polynomial from $\mathbb{F}_q[X]$ with a root at α. It is also a unique monic polynomial of minimum degree in $\mathbb{F}_q[X]$ with a root at α.*

Worked Example 3.1.36 *Continuing Worked Example 3.1.28, we identify $\mathbb{F}_{16}$ with $\mathbb{F}_2(\omega)$, the smallest field containing a root ω of a primitive polynomial of order 4. So, if we choose $1 + X + X^4$, ω will satisfy $\omega^4 = 1 + \omega$, and if we choose $1 + X^3 + X^4$, ω will satisfy $\omega^4 = 1 + \omega^3$. In both cases, the conjugates are ω, ω^2, ω^4 and ω^8.*

Correspondingly, the table in (3.1.2) will take the form

power of ω	$1+X+X^4$ vector (word)	$1+X^3+X^4$ vector (word)	
$--$	0000	0000	
0	1000	1000	
1	0100	0100	
2	0010	0010	
3	0001	0001	
4	1100	1001	
5	0110	1101	(3.1.8)
6	0011	1111	
7	1101	1110	
8	1010	0111	
9	0101	1010	
10	1110	0101	
11	0111	1011	
12	1111	1100	
13	1011	0110	
14	1001	0011	

Under the left table addition rule, the minimal polynomial $M_{\omega^i}(X)$ for the power ω^i is $1+X+X^4$ for $i = 1,2,4,8$ and $1+X^3+X^4$ for $i = 7,14,13,11$, while for $i = 3,6,12,9$ it is $1+X+X^2+X^3+X^4$ and for $i = 5,10$ it is $1+X+X^2$. Under the right table addition rule, we have to swap polynomials $1+X+X^4$ and $1+X^3+X^4$. Polynomials $1+X+X^4$ and $1+X^3+X^4$ are of order 15, polynomial $1+X+X^2+X^3+X^4$ is of order 5 and $1+X+X^2$ of order 3.

A short way to produce these answers is to find the expression for $(\omega^i)^4$ as a linear combination of 1, ω^i, $(\omega^i)^2$ and $(\omega^i)^3$. For example, from the left table we have for ω^7:

$$(\omega^7)^4 = \omega^{28} = \omega^3 + \omega^2 + 1,$$
$$(\omega^7)^3 = \omega^{21} = \omega^3 + \omega^2,$$

and readily see that $(\omega^7)^4 = 1 + (\omega^7)^3$, which yields $1+X^3+X^4$. For completeness, write down the unused expression for $(\omega^7)^2$:

$$(\omega^7)^2 = \omega^{14} = \omega^{12}\omega^2 = (1+\omega)^3\omega^2 = (1+\omega+\omega^2+\omega^3)\omega^2$$
$$= \omega^2 + \omega^3 + \omega^4 + \omega^5 = \omega^2 + \omega^3 + 1 + \omega + (1+\omega)\omega = 1 + \omega^3.$$

For $M_{\omega^5}(X)$ the 'standard' approach gives a shortcut:

$$M_{\omega^5}(X) = (X - \omega^5)(X - \omega^{10}) = X^2 + (\omega^5 + \omega^{10})X + \omega^{15} = X^2 + X + 1.$$

So, the full list of minimal polynomials for $\mathbb{F}_{16}$ is

$$M_{\omega^0}(X) = 1+X, \quad M_{\omega}(X) = 1+X+X^4,$$
$$M_{\omega^3}(X) = 1+X+X^2+X^3+X^4,$$
$$M_{\omega^5}(X) = 1+X+X^2, \quad M_{\omega^7}(X) = 1+X^3+X^4.$$

Example 3.1.37 For the field $\mathbb{F}_{32} \simeq \mathbb{F}_2[X]/\langle 1+X^2+X^5 \rangle$, the addition table is calculated below. The minimal polynomials are

(i) $1+X^2+X^5$ for conjugates $\{\omega, \omega^2, \omega^4, \omega^8, \omega^{16}\}$,

(ii) $1+X^2+X^3+X^4+X^5$ for $\{\omega^3, \omega^6, \omega^{12}, \omega^{24}, \omega^{17}\}$,

(iii) $1+X+X^2+X^4+X^5$ for $\{\omega^5, \omega^{10}, \omega^{20}, \omega^9, \omega^{18}\}$,

(iv) $1+X+X^2+X^3+X^5$ for $\{\omega^7, \omega^{14}, \omega^{28}, \omega^{25}, \omega^{19}\}$,

(v) $1+X+X^3+X^4+X^5$ for $\{\omega^{11}, \omega^{22}, \omega^{13}, \omega^{26}, \omega^{21}\}$,

(vi) $1+X^3+X^5$ for $\{\omega^{15}, \omega^{30}, \omega^{29}, \omega^{27}, \omega^{23}\}$.

All minimal polynomials have order 31.

power of ω	vector (word)	power of ω	vector (word)	
$--$	00000	15	11111	
0	10000	16	11011	
1	01000	17	11001	
2	00100	18	11000	
3	00010	19	01100	
4	00001	20	00110	
5	10100	21	00011	(3.1.9)
6	01010	22	10101	
7	00101	23	11110	
8	10110	24	01111	
9	01011	25	10011	
10	10001	26	11101	
11	11100	27	11010	
12	01110	28	01101	
13	00111	29	10010	
14	10111	30	01001	

Definition 3.1.38 An *automorphism* of $\mathbb{F}_{q^n}$ over $\mathbb{F}_q$ (in short, an $(\mathbb{F}_{q^n}, \mathbb{F}_q)$-automorphism) is a bijection $\sigma : \mathbb{F}_{q^n} \to \mathbb{F}_{q^n}$ with: (a) $\sigma(a+b) = \sigma(a) + \sigma(b)$; (b) $\sigma(ab) = \sigma(a)\sigma(b)$; (c) $\sigma(c) = c$, for all $a, b \in \mathbb{F}_{q^n}, c \in \mathbb{F}_q$.

Theorem 3.1.39 *The set of $(\mathbb{F}_{q^n}, \mathbb{F}_q)$-automorphisms is isomorphic to the cyclic group $\mathbb{Z}_n$ and generated by the Frobenius map $\sigma_q(a) = a^q, a \in \mathbb{F}_{q^n}$.*

Proof Let $\omega \in \mathbb{F}_{q^n}$ be a primitive element. Then $\omega^{q^n-1} = e$ and $M_\omega(X) \in \mathbb{F}_q[X]$ has roots $\omega, \omega^q, \omega^{q^2}, \ldots, \omega^{q^{n-1}}$. An $(\mathbb{F}_{q^n}; \mathbb{F}_q)$-automorphism τ fixes the coefficients of $M_\omega(X)$, thus it permutes the roots, and $\tau(\omega) = \omega^{q^j}$ for some $j, 0 \le j \le n-1$. But as ω is primitive, τ is completely determined by $\tau(\omega)$. Then as $\sigma_{q^j}(\omega) = \omega^{q^j} = \tau(\omega)$, we have that $\tau = \sigma_{q^j}$. $\square$

The rest of this section is devoted to a study of roots of unity, i.e. the roots of the polynomial $X^n - e$ over field $\mathbb{F}_q$ where $q = p^s$ and $p = $char $(\mathbb{F}_q)$. Without loss of generality, we suppose from now on that

$$\gcd(n, q) = 1, \text{ i.e. } n \text{ and } q \text{ are co-prime.} \qquad (3.1.10)$$

Indeed, if n and q are not co-prime, we can write $n = mp^k$. Then, by Lemma 3.1.5

$$X^n - e = X^{mp^k} - e = (X^m - e)^{p^k},$$

and our analysis is reduced to the polynomial $X^m - e$.

Definition 3.1.40   The roots of polynomial $(X^n - e) \in \mathbb{F}_q[X]$ in the splitting field $\mathrm{Spl}\,(X^n - e) = \mathbb{F}_{q^s}$ are called the *nth roots of unity* over $\mathbb{F}_q$ (or the $(n, \mathbb{F}_q)$-roots of unity). The set of all $(n, \mathbb{F}_q)$-roots of unity is denoted by $\mathbb{E}^{(n)}$. It turns out that the value s is the least integer $s \geq 1$ such that $q^s \equiv 1 \bmod n$ (cf. Theorem 3.1.44 below). This fact is reflected in denoting the value s by $\mathrm{ord}_n(q)$ and calling it the *order* of $q \bmod n$.

Under assumption (3.1.10), there is no multiple root (as the derivative ∂_X $(X^n - e) = nX^{n-1}$ does not have roots in $\mathrm{Spl}(X^n - e) = \mathbb{F}_{q^s}$). Thus, $\sharp\,\mathbb{E}^{(n)} = n$.

Theorem 3.1.41 $\mathbb{E}^{(n)}$ *is a cyclic subgroup of* $\mathbb{F}_{q^s}^*$.

Proof   Suppose $\alpha, \beta \in \mathbb{E}^{(n)}$. Then $(\alpha\beta^{-1})^n = \alpha^n(\beta^n)^{-1} = e$, i.e. $\alpha\beta^{-1} \in \mathbb{E}^{(n)}$. So, $\mathbb{E}^{(n)}$ is a subgroup of the cyclic group $\mathbb{F}_{q^s}^*$ and so is cyclic. $\square$

Definition 3.1.42   A generator of group $\mathbb{E}^{(n)}$ (i.e. an nth root of unity whose multiplicative order equals n) is called a *primitive* $(n, \mathbb{F}_q)$-*root of unity*; it will be denoted by β.

Corollary 3.1.43   *There are precisely* $\phi(n)$ *primitive* $(n, \mathbb{F}_q)$-*roots of unity. In particular, primitive* $(n, \mathbb{F}_q)$-*roots of unity exist for any n co-prime to q.*

This allows us to calculate s in the splitting field $\mathbb{F}_{q^s} = \mathrm{Spl}(X^n - e)$. If β is a primitive $(n, \mathbb{F}_q)$-root of unity then its multiplicative order equals n. As $\omega \neq 0$, we have that $\omega \in \mathbb{F}_{q^r}$ if $\beta^{q^r} = \beta$, i.e. $\beta^{q^r - 1} = e$. This happens iff $n | (q^r - 1)$. But s is the least r with $\mathbb{F}_{q^r} \ni \omega$.

Theorem 3.1.44 $\mathrm{Spl}(X^n - e) = \mathbb{F}_{q^s}$, *where* $s = \mathrm{ord}_n(q)$ *is the least integer* ≥ 1 *for which* $n | (q^s - 1)$, *i.e. the least integer* $s \geq 1$ *with* $q^s \equiv 1 \bmod n$.

It is instructive to stress similarities and differences between primitive elements and primitive $(n, \mathbb{F}_q)$-roots of unity in field $\mathbb{F}_{q^s}$ with $s = \mathrm{ord}_n(q)$. A primitive field element, ω, generates the multiplicative cyclic group $\mathbb{F}_{q^s}^*$: $\mathbb{F}_{q^s}^* = \{e, \omega, \ldots, \omega^{q^s - 2}\}$; its multiplicative order equals $q^s - 1$. A primitive root of unity, β, generates the multiplicative cyclic group $\mathbb{E}^{(n)}$: $\mathbb{E}^{(n)} = \{e, \beta, \ldots, \beta^{n-1}\}$; its multiplicative order equals n. [On the other hand, β generates $\mathbb{F}_{q^s}$ as a field element: $\mathbb{F}_{q^s} = \mathbb{F}_q(\beta) = \mathbb{F}_q(\mathbb{E}^{(n)})$.] This suggests that $\beta = \omega$ happens iff $n = q^s - 1$. In fact, let us ask under what condition a power ω^k is a primitive nth root of unity. As was established in Worked Example 3.1.33 this happens when $n | (q^s - 1)$, i.e. $q^s - 1 = nr$. In fact, if $k \geq 1$ is such that

$$\gcd(k, nr) = \gcd(k, q^s - 1) = r$$

then element ω^k is a primitive nth root of unity as its multiplicative order equals

$$\frac{q^s - 1}{\gcd(k, q^s - 1)} = \frac{nr}{r} = n.$$

This holds when $k = ru$ and u is co-prime with n. Conversely, if ω^k is a primitive root of unity then $\gcd(k, q^s - 1) = (q^s - 1)/n$. Hence we obtain the following.

Theorem 3.1.45 *Let $\mathbb{P}^{(n)}$ be the set of the primitive $(n, \mathbb{F}_q)$-roots of unity and $\mathbb{T}^{(n)}$ the set of primitive elements in $\mathbb{F}_{q^s} = \mathrm{Spl}(X^n - e)$. Then either (i) $\mathbb{P}^{(n)} \cap \mathbb{T}^{(n)} = \emptyset$ or (ii) $\mathbb{P}^{(n)} = \mathbb{T}^{(n)}$; case (ii) occurs iff $n = q^s - 1$.*

Now we can factorise polynomial $(X^n - e)$ over $\mathbb{F}_q$ by taking the product of the distinct minimal polynomials for the $(n, \mathbb{F}_q)$-roots of unity:

$$X^n - e = \mathrm{lcm}\left(M_\beta(X) : \beta \in \mathbb{E}^{(n)}\right). \tag{3.1.11}$$

If we begin with a primitive element $\omega \in \mathbb{F}_{q^s}$ where $s = \mathrm{ord}_n(q)$ then $\beta = \omega^{(q^s-1)/n}$ is a primitive root of unity and $\mathbb{E}^{(n)} = \{e, \beta, \ldots, \beta^{n-1}\}$.

This enables us to calculate the minimal polynomial $M_{\beta^i}(X)$. For all $i = 0, \ldots, n-1$, the conjugates of β^i are $\beta^i, \beta^{iq}, \ldots, \beta^{iq^{d-1}}$ where $d(= d(i))$ is the least positive integer for which $\beta^{iq^d} = \beta^i$, i.e. $\beta^{iq^d - i} = e$. This is equivalent to $n | (iq^d - i)$, i.e. $iq^d = i \bmod n$. Therefore,

$$M_i(X)\left(= M_{\beta^i}(X)\right) = \left(X - \beta^i\right)\left(X - \beta^{iq}\right) \cdots \left(X - \beta^{iq^{d-1}}\right). \tag{3.1.12}$$

Definition 3.1.46 The set of exponents $i, iq, \ldots, iq^{d-1}$ where $d(= d(i))$ is the minimal positive integer such that $iq^d = i \bmod n$ is called a *cyclotomic coset* (for i) and denoted by $C_i(= C_i(n, q))$ (alternatively, C_{ω^i} is defined as the set of non-zero field elements $\omega^i, \omega^{iq}, \ldots, \omega^{iq^{d-1}}$).

Worked Example 3.1.47 *Check that polynomials $X^2 + X + 2$ and $X^3 + 2X^2 + 1$ are primitive over $\mathbb{F}_3$ and compute the field tables for $\mathbb{F}_9$ and $\mathbb{F}_{27}$ generated by these polynomials.*

Solution The field $\mathbb{F}_9$ is isomorphic to $\mathbb{F}_3[X]/\langle X^2 + X + 2\rangle$. The multiplicative powers of $\omega \sim X$ are

$$\omega^2 \sim 2X + 1, \quad \omega^3 \sim 2X + 2, \quad \omega^4 \sim 2,$$
$$\omega^5 \sim 2X, \quad \omega^6 \sim X + 2, \quad \omega^7 \sim X + 1, \quad \omega^8 \sim 1.$$

The cyclotomic coset of ω is $\{\omega, \omega^3\}$ (as $\omega^9 = \omega$). Then the minimal polynomial

$$M_\omega(X) = (X - \omega)(X - \omega^3) = X^2 - (\omega + \omega^3)X + \omega^4$$
$$= X^2 - 2X + 2 = X^2 + X + 2.$$

Hence, $X^2 + X + 2$ is primitive.

Next, $\mathbb{F}_{27} \simeq \mathbb{F}_3[X]/\langle X^3 + 2X^2 + 1 \rangle$, and with $\omega \sim X$, we have

$$\omega^2 \sim X^2, \; \omega^3 \sim X^2 + 2, \; \omega^4 \sim X^2 + 2X + 2, \; \omega^5 \sim 2X + 2,$$
$$\omega^6 \sim 2X^2 + 2X, \; \omega^7 \sim X^2 + 1, \; \omega^8 \sim X^2 + X + 2,$$
$$\omega^9 \sim 2X^2 + 2X + 2, \; \omega^{10} \sim X^2 + 2X + 1, \; \omega^{11} \sim X + 2,$$
$$\omega^{12} \sim X^2 + 2X, \; \omega^{13} \sim 2, \; \omega^{14} \sim 2X, \; \omega^{15} \sim 2X^2, \; \omega^{16} \sim 2X^2 + 1,$$
$$\omega^{17} \sim 2X^2 + X + 1, \; \omega^{18} \sim X + 1, \; \omega^{19} \sim X^2 + X,$$
$$\omega^{20} \sim 2X^2 + 2, \; \omega^{21} \sim 2X^2 + 2X + 1, \; \omega^{22} \sim X^2 + X + 1,$$
$$\omega^{23} \sim 2X^2 + X + 2, \; \omega^{24} \sim 2X + 1, \; \omega^{25} \sim 2X^2 + X, \; \omega^{26} \sim 1.$$

The cyclotomic coset of ω in $\mathbb{F}_{27}$ is $\{\omega, \omega^3, \omega^9\}$. Consequently, the primitive polynomial

$$\begin{aligned} M_\omega(X) &= (X - \omega)(X - \omega^3)(X - \omega^9) \\ &= X^3 - (\omega + \omega^3 + \omega^9)X^2 + (\omega^4 + \omega^{10} + \omega^{12})X - \omega^{13} \\ &= X^3 + 2X^2 + 1 \end{aligned}$$

as required. $\qquad\qquad\qquad\qquad\qquad\qquad\qquad\qquad\qquad\qquad\qquad\qquad \square$

Worked Example 3.1.48 (a) *Consider the polynomial $X^{15} - 1$ over $\mathbb{F}_2$ (with $n = 15$, $q = 2$). Then $\omega = 2$, $s = \mathrm{ord}_{15}(2) = 4$ and $\mathrm{Spl}(X^{15} - 1) = \mathbb{F}_{2^4} = \mathbb{F}_{16}$.*

The polynomial $g(X) = 1 + X + X^4$ is primitive: any of its roots β are primitive in $\mathbb{F}_{16}$. So, the primitive $(15, \mathbb{F}_2)$-root of unity is

$$\beta = \omega^{(2^4 - 1)/15} = \omega.$$

Hence, the roots of $X^{15} - 1$ are $1, \beta, \ldots, \beta^{14}$. The minimal polynomials for them have been calculated in Worked Example 3.1.36. So, we have the factorisation

$$\begin{aligned} X^{15} - 1 = (1 + X)(1 + X + X^4)(1 + X + X^2 + X^3 + X^4) \\ \times (1 + X + X^2)(1 + X^3 + X^4). \end{aligned}$$

(b) *Knowing the cyclotomic cosets we can show that a particular factorisation of $X^n - e$ contains irreducible factors. Explicitly, take the polynomial $X^9 - 1$ over $\mathbb{F}_2$ (with $n = 9$, $q = 2$). There are three cyclotomic cosets:*

$$C_0 = \{0\}, C_1 = \{1, 2, 4, 8, 7, 5\}, C_3 = \{3, 6\};$$

the corresponding minimal polynomials are of degree $1, 6$ and 2, respectively:

$$1 + X, \quad 1 + X^3 + X^6 \text{ and } 1 + X + X^2.$$

This yields

$$X^9 - 1 = (1 + X)(1 + X + X^2)(1 + X^3 + X^6).$$

(c) *Let us check primitivity of the polynomial*

$$f(X) = 1 + X + X^6$$

over $\mathbb{F}_2$, *with* $n = 6$, $q = 2$. *Here,* $2^6 - 1 = 63 = 3^2 \cdot 7$. *As* $63/3 = 21$, $3^2 | \mathrm{ord}(f(X)) \Leftrightarrow X^{21} - 1 \neq 0 \bmod (1 + X + X^6)$. *But* $X^{21} = 1 + X + X^3 + X^4 + X^5 \neq 1 \bmod (1 + X + X^6)$, *so* $3^2 | \mathrm{ord}(f(X))$.

Next, as $63/7 = 9$, $7 | \mathrm{ord}(f(X)) \Leftrightarrow X^9 - 1 \neq 0 \bmod (1 + X + X^6)$. *But* $X^9 = 1 + X^3 + X^4 \neq 1 \bmod (1 + X + X^6)$, *so* $7 | \mathrm{ord}(f(X))$. *Therefore,* $\mathrm{ord}(f(X)) = 63$, *and* $f(X)$ *is primitive. Theorem 3.1.53 below shows that any irreducible polynomial of order 63 has degree 6 as* $2^6 = 1 \bmod 63$.

(d) *Now consider the polynomial*

$$g(X) = 1 + X + X^2 + X^4 + X^6,$$

again over $\mathbb{F}_2$ *(here $n = 6$ and $q = 2$, as before). Again* $3^2 | \mathrm{ord}(g(X)) \Leftrightarrow X^{21} \neq 1$ *mod* $(1 + X + X^2 + X^4 + X^6)$. *However, in* $\mathbb{F}_2$

$$X^{21} - 1 = (1+X)(1+X+X^2)(1+X+X^3)(1+X^2+X^3)$$
$$\times (1+X+X^2+X^4+X^6)(1+X^2+X^4+X^5+X^6).$$

Hence, $X^{21} - 1 = 0 \bmod (1 + X + X^2 + X^4 + X^6) = 1$, *and so* 3^2 *does not divide* $\mathrm{ord}(g(X))$.

Next, $3 | \mathrm{ord}(g(X)) \Leftrightarrow X^7 \neq 1 \bmod (1 + X + X^2 + X^4 + X^6)$. *As* $X^7 = X + X^2 + X^3 + X^5 \neq 1 \bmod (1 + X + X^2 + X^4 + X^6)$, *3 is a divisor for* $\mathrm{ord}(g(X))$.

Finally, $7 | \mathrm{ord}(g(X)) \Leftrightarrow X^9 \neq 1 \bmod (1 + X + X^2 + X^4 + X^6)$, *and as* $X^9 = 1 + X^2 + X^4 \neq 1 \bmod (1 + X + X^2 + X^4 + X^6)$, *7 divides* $\mathrm{ord}(g(X))$. *So,* $\mathrm{ord}(g(X)) = 21$.

Let us summarise results about minimal polynomials and roots of unity. We know from Theorem 3.1.25 that for all integers $d \geq 1$ and for all $q = p^d$, where p is prime and $s \geq 1$ integer, there exists a primitive polynomial of degree d, say $M_\omega(X)$, where ω is a primitive element in the field $\mathbb{F}_{q^d}$. On the other hand, for all irreducible polynomials $f(X) \in \mathbb{F}_q[X]$ of degree d, the roots of $f(X)$ lie in the field $\mathrm{Spl}(f(X)) = \mathbb{F}_{q^d}$ and have the same multiplicative order $\mathrm{ord}(f(X))$.

Theorem 3.1.49 *Let polynomial $f(X) \in \mathbb{F}_q[X]$ be irreducible, of degree d, and* $\mathrm{ord}(f(X)) = \ell$. *Then:*

(a) $\ell | (q^d - 1)$,
(b) $\sharp f(X) | (X^\ell - e)$,
(c) $\ell | n$ *iff* $f(X) | (X^n - e)$,
(d) ℓ *is the least positive integer such that* $f(X) | (X^\ell - e)$.

Proof (a) $\mathrm{Spl}(f(X)) = \mathbb{F}_{q^d}$, hence every root α of $f(X)$ is a root of $X^{q^d-1} - e$. So, it has $\mathrm{ord}(\alpha)|(q^d - 1)$.

(b) Each root α of $f(X)$ in $\mathrm{Spl}(f(X))$ has $\mathrm{ord}(\alpha) = \ell$ and hence is a root of $(X^\ell - e)$. So, $f(X)|(X^\ell - e)$.

(c) If $f(X)|(X^n - e)$ then each root of $f(X)$ is a root of $X^n - e$, i.e. $\mathrm{ord}(\alpha)|n$. So, $\ell|n$. Conversely, if $n = k\ell$ then $(X^\ell - e)|(X^{k\ell} - e)$ and $f(X)|(X^n - e)$ by (b).

(d) Follows from (c). □

Theorem 3.1.50 *If $f(X) \in \mathbb{F}_q[X]$ is an irreducible polynomial of degree d and order ℓ then $d = \mathrm{ord}_\ell(q)$.*

Proof If $\alpha \in \mathbb{F}_{q^d}$ has $f(\alpha) = 0$ then by Theorem 3.1.29, $\mathbb{F}_q(\alpha) = \mathbb{F}_{q^d} = \mathrm{Spl}(f(X))$. But α is also a primitive $(\ell, \mathbb{F}_q)$-root of unity, so $\mathbb{F}_q(\alpha) = \mathbb{F}_q(\mathbb{E}^{(\ell)}) = \mathrm{Spl}(X^\ell - e) = \mathbb{F}_{q^s}$ where $s = \mathrm{ord}_\ell(q)$. Hence, $d = \mathrm{ord}_\ell(q)$. □

Worked Example 3.1.51 *Use the Frobenius map $\sigma : a \mapsto a^q$ to prove that every element $a \in \mathbb{F}_{q^n}$ has a unique q^jth root, for $j = 1, \ldots, n-1$.*

Suppose that $q = p^s$ is odd. Show that exactly a half of the non-zero elements of $\mathbb{F}_q$ have square roots.

Solution The Frobenius map $\sigma : a \mapsto a^q$ is a bijection $\mathbb{F}_{q^n} \to \mathbb{F}_{q^n}$. So, for all $b \in \mathbb{F}_{q^n}$ there exists unique a with $a^q = b$ (the qth root). The jth power iteration $\sigma^j : a \mapsto a^{q^j}$ is also a bijection, so again for all $b \in \mathbb{F}_{q^n}$ there exists unique a with $a^{q^j} = b$. Observe that for all $c \in \mathbb{F}_q$, $c^{1/q^j} = c$.

Now take $\tau : a \mapsto a^2$, a multiplicative homomorphism $\mathbb{F}_q^* \to \mathbb{F}_q^*$. If q is odd then $\mathbb{F}_q^* \simeq \mathbb{Z}_{q-1}$ has an even number of elements $q - 1$. We want to show that if $\tau(a) = b$ then $\tau^{-1}(b)$ consists of two elements, a and $-a$. In fact, $\tau(-a) = b$. Also, if $\tau(a') = b$ then $\tau(a'a^{-1}) = e$.

So, we want to analyse $\tau^{-1}(e)$. Clearly, $\pm e \in \tau^{-1}(e)$. On the other hand, if ω is a primitive element then $\tau(\omega^{(q-1)/2}) = \omega^{q-1} = e$ and $\tau^{-1}(e)$ consists of $e = \omega^0$ and $\omega^{(q-1)/2}$. So, $\omega^{(q-1)/2} = -e$.

Now if $\tau(a'a^{-1}) = e$ then $a'a^{-1} = \pm e$ and $a' = \pm a$. Hence, τ sends precisely two elements, a and $-a$, into the same image, and its range $\tau(\mathbb{F}_q^*)$ is a half of $\mathbb{F}_q^*$. □

Theorem 3.1.52 *(cf. [92], Theorem 3.46.) Let polynomial $p(X) \in \mathbb{F}_q[X]$ be irreducible, of degree n. Set $m = \gcd(d, n)$. Then $m|n$ and $p(X)$ factorises over $\mathbb{F}_{q^d}$ into m irreducible polynomials of degree n/m each. Hence, $p(X)$ is irreducible over $\mathbb{F}_{q^d}$ iff $m = 1$.*

Theorem 3.1.53 *(cf. [92], Theorem 3.5.) Let $\gcd(d, q) = 1$. The number of monic irreducible polynomials of order ℓ and degree d equals $\phi(\ell)/d$ if $\ell \geq 2$, and the* degree $d = \mathrm{ord}_\ell(q)$, equals 2 if $\ell = d = 1$, equals 0 *in all other cases. In particular, the degree of an order ℓ irreducible polynomial always equals $\mathrm{ord}_\ell(q)$, i.e. the minimal s such that $q^s = 1 \bmod \ell$. Here $\phi(\ell)$ is the Euler totient function.*

The proofs of Theorems 3.1.52 and 3.1.53 are omitted (see [92]). We only make a short comment about Theorem 3.1.53. If $p(0) \neq 0$, the order of irreducible polynomial $p(X)$ of degree d coincides with the order of any of its roots in the multiplicative group $\mathbb{F}^*_{q^d}$. So, the order is ℓ iff $d = \mathrm{ord}_\ell(q)$ and $p(X)$ divides the so-called circular polynomial

$$Q_\ell(X) = \prod_{s:\gcd(s,\ell)=1} (X - \omega^s).$$

In fact, the circular polynomial could be decomposed into a product of irreducible polynomials, all of degree $d = \mathrm{ord}_\ell(q)$, and their number equals $\phi(\ell)/d$. (In the case $d = \ell = 1$ the polynomial $p(X) = X$ should be accounted for as well.)

Concluding this section, we give short summaries of the facts of the theory of finite fields discussed above.

Summary 1.55. A *field* is a ring such that its non-zero elements form a commutative group under multiplication. (i) Any finite field $\mathbb{F}$ has the number of elements $q = p^s$ where p is prime, and the characteristic $\mathrm{char}(\mathbb{F}) = p$. (ii) Any two finite fields with the same number of elements are isomorphic. Thus, for a given $q = p^s$, there exists, up to isomorphism, a unique field of cardinality q; such a field is denoted by $\mathbb{F}_q$ (it is often called a Galois field of size q). When q is prime, the field $\mathbb{F}_q$ is isomorphic to the additive cyclic group $\mathbb{Z}_p$ of p elements, equipped with multiplication mod p. (iii) The multiplicative group $\mathbb{F}^*_q$ of non-zero elements from $\mathbb{F}_q$ is isomorphic to the additive cyclic group $\mathbb{Z}_{q-1}$ of $q-1$ elements. (iv) Field $\mathbb{F}_q$ contains $\mathbb{F}_r$ as a subfield iff $r|q$; in this case $\mathbb{F}_q$ is isomorphic to a linear space over (i.e. with coefficients from) $\mathbb{F}_r$, of dimension $\log_p(q/r)$. So, each prime number p gives rise to an increasing sequence of finite fields $\mathbb{F}_{p^s}$, $s = 1, 2, \dots$ An element $\omega \in \mathbb{F}_q$ generating the multiplicative group $\mathbb{F}^*_q$ is called a *primitive* element of $\mathbb{F}_q$.

Summary 1.56. The polynomial ring over $\mathbb{F}_q$ is denoted by $\mathbb{F}_q[X]$; if the polynomials are considered mod $g(X)$, a fixed polynomial from $\mathbb{F}_q[X]$, the corresponding ring is denoted by $\mathbb{F}_q[X]/\langle g(X)\rangle$. (i) Ring $\mathbb{F}_q[X]/\langle g(X)\rangle$ is a field iff $g(X)$ is irreducible over $\mathbb{F}_q$ (i.e. does not admit a decomposition $g(X) = g_1(X)g_2(X)$ where $\deg(g_1(X)), \deg(g_2(X)) < \deg(g(X))$). (ii) For any q and a positive integer d there exists an irreducible polynomial $g(X)$ over $\mathbb{F}_q$ of degree d. (iii) If $g(X)$ is irreducible and $\deg g(X) = d$ then the cardinality of field $\mathbb{F}_q[X]/\langle g(X)\rangle$ is q^d, i.e. $\mathbb{F}_q[X]/\langle g(X)\rangle$ is isomorphic to $\mathbb{F}_{q^d}$ and belongs to the same series of fields as $\mathbb{F}_q$ (that is, $\mathrm{char}(\mathbb{F}_{q^d}) = \mathrm{char}(\mathbb{F}_q)$).

Summary 1.57. An extension of a field $\mathbb{F}_q$ by a finite family of elements $\alpha_1, \ldots, \alpha_u$ (contained in a larger field from the same series) is the smallest field containing $\mathbb{F}_q$ and α_i, $1 \leq i \leq u$. Such a field is denoted by $\mathbb{F}_q(\alpha_1, \ldots, \alpha_u)$. (i) For any monic polynomial $p(X) \in \mathbb{F}_q[X]$ there exists a larger field $\mathbb{F}_{q'}$ from the same series as $\mathbb{F}_q$ such that $p(X)$ factors over $\mathbb{F}_{q'}$:

$$p(X) = \prod_{1 \leq j \leq u} (X - \alpha_j), \quad u = \deg p(X), \ \alpha_1, \ldots, \alpha_u \in \mathbb{F}_{q'}. \qquad (3.1.13)$$

The smallest field $\mathbb{F}_{q'}$ with this property (i.e. field $\mathbb{F}_q(\alpha_1, \ldots, \alpha_u)$) is called a *splitting* field for $p(X)$; we also say that $p(X)$ splits over $\mathbb{F}_q(\alpha_1, \ldots, \alpha_u)$. The splitting field for $p(X)$ is denoted by $\mathrm{Spl}(p(X))$; an element $\alpha \in \mathrm{Spl}(p(X))$ takes part in decomposition (3.1.13) iff $p(\alpha) = 0$. Field $\mathrm{Spl}(p(X))$ is described as the set $\{g(\alpha_j)\}$ where $j = 1, \ldots, u$, and $g(X) \in \mathbb{F}_q[X]$ are polynomials of degree $< \deg(p(X))$. (ii) Field $\mathbb{F}_q$ is splitting for the polynomial $X^q - X$. (iii) If polynomial $p(X)$ of degree d is irreducible over $\mathbb{F}_q$ and α is a root of $p(X)$ in field $\mathrm{Spl}(p(X))$ then $\mathbb{F}_{q^d} \simeq \mathbb{F}_q[X]/\langle p(X) \rangle$ is isomorphic to $\mathbb{F}_q(\alpha)$ and all the roots of $p(X)$ in $\mathrm{Spl}(p(X))$ are given by the conjugate elements $\alpha, \alpha^q, \alpha^{q^2}, \ldots, \alpha^{q^{d-1}}$. Thus, d is the smallest positive integer for which $\alpha^{q^d} = \alpha$. (iv) Suppose that, for a given field $\mathbb{F}_q$, a monic polynomial $p(X) \in \mathbb{F}_q[X]$ and an element α from a larger field we have $p(\alpha) = 0$. Then there exists a unique minimal polynomial $M_\alpha(X)$ with the property that $M_\alpha(\alpha) = 0$ (i.e. such that any other polynomial $p(X)$ with $p(\alpha) = 0$ is divided by $M_\alpha(X)$). Polynomial $M_\alpha(X)$ is the unique irreducible polynomial over $\mathbb{F}_q$ vanishing at α. It is also the unique polynomial of the minimum degree vanishing at α. We call $M_\alpha(X)$ the minimal polynomial of α over $\mathbb{F}_q$. If ω is a primitive element of $\mathbb{F}_{q^d}$ then $M_\omega(X)$ is called a *primitive* polynomial for $\mathbb{F}_{q^d}$ over $\mathbb{F}_q$. We say that elements $\alpha, \beta \in \mathbb{F}_{q^d}$ are *conjugate* over $\mathbb{F}_q$ if they have the same minimal polynomial over $\mathbb{F}_q$. Then (v) the conjugates of $\alpha \in \mathbb{F}_{q^d}$ over $\mathbb{F}_q$ are $\alpha, \alpha^q, \ldots, \alpha^{q^{d-1}}$, where d is the smallest positive integer with $\alpha^{q^d} = \alpha$. When $\alpha = \omega^i$ where ω is a primitive element, the conjugacy class is associated with a cyclotomic coset $C_{\omega^i} = \{\omega^i, \omega^{iq}, \ldots, \omega^{iq^{d-1}}\}$.

Summary 1.58. Now assume that n and $q = p^s$ are co-prime and take polynomial $X^n - e$. The roots of $X^n - e$ in the splitting field $\mathrm{Spl}(X^n - e)$ are called nth roots of unity over $\mathbb{F}_q$. The set of all nth roots of unity is denoted by $\mathbb{E}_n$. (i) Set $\mathbb{E}_n$ is a cyclic subgroup of order n in the multiplicative group of field $\mathrm{Spl}(X^n - e)$. An nth root of unity generating $\mathbb{E}_n$ is called a primitive nth root of unity. (ii) If $\mathbb{F}_{q^s}$ is $\mathrm{Spl}(X^n - e)$ then s is the smallest positive integer with $n | (q^s - 1)$. (iii) Let Π_n be the set of primitive nth roots of unity over field $\mathbb{F}_q$ and Φ_n the set of primitive elements of the splitting field $\mathbb{F}_{q^s} = \mathrm{Spl}(X^s - e)$. Then either $\Pi_n \cap \Phi_n = \emptyset$ or $\Pi_n = \Phi_n$, the latter happening iff $n = q^s - 1$.

3.2 Reed–Solomon codes. The BCH codes revisited

From now on we consider finite fields $\mathbb{F}_q$ up to isomorphism, but from time to time refer to a specific field table (e.g. by specifying $\mathbb{F}_{p^s}$ as $\mathbb{F}_p[X]/\langle P(X) \rangle$ where $P(X) \in \mathbb{F}_p[X]$ is an irreducible polynomial of degree s).

In Definition 2.5.37 we introduced narrow-sense binary BCH codes. Our study will continue in this section with general q-ary BCH codes $\mathscr{X}^{\mathrm{BCH}}_{q,N,\delta,\omega,b}$, of length N, designed distance δ and zeros $\omega^b, \ldots, \omega^{b+\delta-1}$; see in Definition 3.2.7 below. Prior to that, we discuss an interesting special class of BCH codes formed by the Reed–Solomon (RS) codes; as we shall see, their analysis is facilitated by the fact that the RS codes are MDS (maximum distance separable).

Definition 3.2.1 Given $q \geq 3$, a q-ary Reed–Solomon code is defined as a cyclic code of length $N = q - 1$ with the generator

$$g(X) = (X - \omega^b)(X - \omega^{b+1}) \ldots (X - \omega^{b+\delta-2}), \qquad (3.2.1)$$

where δ and b are integers, $1 \leq \delta$, $b < q - 1$, and ω is a primitive element of $\mathbb{F}_q$ (or equivalently, a primitive Nth root of unity). Such a code is denoted by $\mathscr{X}^{\mathrm{RS}}$ $\left(= \mathscr{X}^{\mathrm{RS}}_{q,\delta,\omega,b} \right)$.

According to Definition 3.2.7, the RS code is identified as $\mathscr{X}^{\mathrm{BCH}}_{q,q-1,\delta,\omega,b}$, i.e. as a q-ary BCH code of length $q - 1$ and designed distance δ. There are no reasonable binary RS codes, as in this case the length $q - 1 = 1$. Observe that $q - 1$ gives the number of non-zero elements in the alphabet field $\mathbb{F}_q$. Moreover, for $N = q - 1$ we have

$$X^N - e = X^{q-1} - e = \prod_{\alpha \in \mathbb{F}_q^*} (X - \alpha)$$

(as the splitting field Spl $(X^q - X)$ is $\mathbb{F}_q$). Furthermore, owing to the fact that ω is a primitive $(q - 1, \mathbb{F}_q)$ root of unity (or, equivalently, a primitive element of $\mathbb{F}_q$), the minimal polynomial $M_i(X)$ is just $X - \omega^i$, for all $i = 0, \ldots, N - 1$.

An important property is that the RS codes are MDS. Indeed, the generator $g(X)$ of $\mathscr{X}^{\mathrm{RS}}_{q,\delta,\omega,b}$ has $\deg g(X) = \delta - 1$. Hence, the rank k is given by

$$k = \dim(\mathscr{X}^{\mathrm{RS}}_{q,\delta,\omega,b}) = N - \deg g(X) = N - \delta + 1. \qquad (3.2.2)$$

By the generalised BCH bound (see Theorem 3.2.9 below), the minimal distance

$$d\left(\mathscr{X}^{\mathrm{RS}}_{q,\delta,\omega,b}\right) \geq \delta = N - k + 1.$$

But the Singleton bound states that $d(\mathscr{X}^{\mathrm{RS}}) \leq N - k + 1$. Hence,

$$d(\mathscr{X}^{\mathrm{RS}}_{q,d,\omega,b}) = N - k + 1 = \delta. \qquad (3.2.3)$$

Thus the RS codes have the largest possible minimal distance among all q-ary codes of length $q-1$ and dimension $k = q - \delta$. Summarising, we obtain

Theorem 3.2.2	*The code $\mathscr{X}^{\mathrm{RS}}_{q,\delta,\omega,b}$ is MDS and has distance δ and rank $q - \delta$.*

The dual of a BCH code is not always BCH. However,

Theorem 3.2.3	*The dual of an RS code is an RS code.*

Proof	The proof is straightforward, as $(\mathscr{X}^{\mathrm{RS}}_{q,\delta,\omega,b})^{\perp} = \mathscr{X}^{\mathrm{RS}}_{q,q-\delta,\omega,b+\delta-1}.$	□

Theorem 3.2.4	*Let $\mathscr{X}^{\mathrm{RS}}$ be a $[N,k,\delta]$ RS code. Then its parity-check extension is a $[N+1,k,\delta+1]$ code, with distance one more than that of $\mathscr{X}^{\mathrm{RS}}$.*

Proof	Let $c(X) = c_0 + c_1 X + \cdots + c_{N-1} X^{N-1} \in \mathscr{X}^{\mathrm{RS}}$, with weight $w(c(X)) = \delta$. Its extension is $\widehat{c}(X) = c(X) + c_N X^N$, with $c_N = - \sum_{0 \le i \le N-1} c_i = -c(e)$. We want to show that $c(e) \ne 0$ and hence $w(\widehat{c}(X)) = \delta + 1$.

To simplify notation assume that $b = 1$ and let $g(X) = (X - \omega)(X - \omega^2) \ldots (X - \omega^{\delta-1})$ be the generator of $\mathscr{X}^{\mathrm{RS}}$. Write $c(X) = g(X)p(X)$ for some $p(X)$, yielding that $c(e) = p(e)g(e)$. Clearly, $g(e) \ne 0$, as $\omega^i \ne e$ for all $i = 1, \ldots, \delta - 1$. If $p(e) = 0$, the polynomial $g_1(X) = (X - e)g(X)$ divides $c(X)$. Then $c(X) \in \langle g_1(X) \rangle$ where $g_1(X) = (X - e)(X - \omega) \ldots (X - \omega^{\delta-1})$. That is, $\langle g_1(X) \rangle$ is BCH, with the designed distance $\ge \delta + 1$. But this contradicts the choice of $c(X)$.	□

RS codes admit specific (and elegant) encoding and decoding procedures. Let $\mathscr{X}^{\mathrm{RS}}$ be an $[N,k,\delta]$ RS code, with $N = q - 1$. For a message string $a_0 \ldots a_{k-1}$ set $a(X) = \sum_{0 \le i \le k-1} a_i X^i$ and encode $a(X)$ as $c(X) = \sum_{0 \le j \le N-1} a(\omega^j)X^j$. To show that $c(X) \in \mathscr{X}^{\mathrm{RS}}$, we have to check that $c(\omega) = \cdots = c(\omega^{\delta-1}) = 0$. Think of $a(X)$ as a polynomial $\sum_{0 \le i \le N-1} a_i X^i$ with $a_i = 0$ for $i \ge k$, and use

Lemma 3.2.5	*Let $a(X) = a_0 + a_1 X + \cdots + a_{N-1} X^{N-1} \in \mathbb{F}_q[X]$ and ω be a primitive $(N, \mathbb{F}_q)$ root of unity over $\mathbb{F}_q$, $N = q - 1$. Then*

$$a_i = \frac{1}{N} \sum_{0 \le j \le N-1} a(\omega^j)\omega^{-ij}. \tag{3.2.4}$$

We postpone the proof till after Lemma 3.2.12.
Indeed, by Lemma 3.2.5

$$a_i = \frac{1}{N} \sum_{0 \le j \le N-1} a(\omega^j)\omega^{-ij} = \frac{1}{N}c(\omega^{-i}) = \frac{1}{N}c(\omega^{N-i}),$$

so $c(\omega^j) = N a_{N-j}$. For $0 \leq j \leq \delta - 1 = N - k$, $c(\omega^j) = N a_{N-j} = 0$. Therefore, $c(X) \in \mathscr{X}^{RS}$. In addition, the original message is easy to recover from $c(X)$: $a_i = \frac{1}{N} c(\omega^{N-i})$.

To decode the received word $u(X) = c(X) + e(X)$, write

$$u_i = c_i + e_i = e_i + a(\omega^i), \quad 0 \leq i \leq N - 1.$$

Then obtain

$$u_0 = e_0 + a_0 + a_1 + \cdots + a_{k-1},$$
$$u_1 = e_1 + a_0 + a_1 \omega + \cdots + a_{k-1} \omega^{k-1},$$
$$u_2 = e_2 + a_0 + a_1 \omega^2 + \cdots + a_{k-1} \omega^{2(k-1)},$$
$$\vdots$$
$$u_{N-1} = e_{N-1} + a_0 + a_1 \omega^{N-1} + \cdots + a_{k-1} \omega^{(N-1)(k-1)}.$$

If there are no errors, i.e. $e_0 = \cdots = e_{N-1} = 0$, any k of these equations can be solved in the k unknowns $a_0, \ldots, a_{k-1}$, as the corresponding matrix is Vandermonde. In fact, any subsystem of k equations can be solved for any error vector (it is a different matter if the solution will give the correct string $a_0, \ldots, a_{k-1}$ or not).

Now suppose that t errors have occurred, $t < N - k$. Call the equations with $e_i = 0$ good and $e_i \neq 0$ bad, then we have t bad and $N - t$ good ones. If we solve all subsystems of k equations then the $\binom{N-t}{k}$ subsystems consisting of k good equations will give the correct values of the a_is. Moreover, a given incorrect solution cannot satisfy any set of k good equations; it can satisfy at most $k - 1$ correct equations. In addition, it can satisfy at most t incorrect equations. So, it is a solution to $\leq t + k - 1$ equations, i.e. can be obtained $\leq \binom{t+k-1}{k}$ times from subsystems of k equations. Hence, if

$$\binom{N-t}{k} > \binom{t+k-1}{k},$$

the majority solution from among $\binom{N}{k}$ solutions gives the true values of the a_is. The last inequality holds iff $N - t > t + k - 1$, i.e. $\delta = N - k + 1 > 2t$. Therefore we get:

Theorem 3.2.6 *For a $[N, k, \delta]$ RS code $\mathscr{X}^{RS}$, the majority logic decoding corrects up to $t < \delta/2$ errors, at the cost of having to solve $\binom{N}{k}$ systems of equations of size $k \times k$.*

Reed–Solomon codes were discovered in 1960 by Irving S. Reed and Gustave Solomon, both working at that time in the Lincoln Laboratory of MIT. When their joint article was published, an efficient decoding algorithm for these codes was not known. Such an algorithm solution for the latter was found in 1969 by El-wyn Berlekamp and James Massey, and is known since as the Berlekamp–Massey decoding algorithm (cf. [20]); see Section 3.3. Later on, other algorithms were proposed: continued fraction algorithm and Euclidean algorithm (see [112]).

Reed–Solomon codes played an important role in transmitting digital pictures from American spacecraft throughout the 1970s and 1980s, often in combination with other code constructions. These codes still figure prominently in modern space missions although the advent of *turbo-codes* provides a much wider choice of coding and decoding procedures.

Reed–Solomon codes are also a key component in compact disc and digital game production. The encoding and decoding schemes employed here are capable of correcting bursts of up to 4000 errors (which makes about 2.5mm on the disc surface).

Definition 3.2.7   A *BCH code* $\mathscr{X}^{\text{BCH}}_{q,N,\delta,\omega,b}$ with parameters q, N, ω, δ and b is the q-ary cyclic code $\mathscr{X}_N = \langle g(X) \rangle$ with length N, designed distance δ, such that its generating polynomial is

$$g(X) = \text{lcm}\big(M_{\omega^b}(X), M_{\omega^{b+1}}(X), \ldots, M_{\omega^{b+\delta-2}}(X)\big), \tag{3.2.5}$$

i.e.

$$\mathscr{X}^{\text{BCH}}_{q,N,\delta,\omega,b} = \big\{ f(X) \in \mathbb{F}_q[X] \bmod (X^N - 1) : \\ f(\omega^{b+i}) = 0, \ 0 \le i \le \delta - 2 \big\}.$$

If $b = 1$, this is a *narrow sense* BCH code. If ω is a primitive Nth root of unity, i.e. a primitive root of the polynomial $X^N - 1$, the BCH code is called *primitive*. (Recall that under condition $\gcd(q,N) = 1$ these roots form a commutative multiplicative group which is cyclic, of order N, and ω is a generator of this group.)

Lemma 3.2.8   The BCH code $\mathscr{X}^{\text{BCH}}_{q,N,\delta,\omega,b}$ has minimum distance $\ge \delta$.

Proof   Without loss of generality consider a narrow sense code. Set the parity-check $(\delta - 1) \times N$ matrix

$$H = \begin{pmatrix} 1 & \omega & \omega^2 & \cdots & \omega^{N-1} \\ 1 & \omega^2 & \omega^4 & \cdots & \omega^{2(N-1)} \\ \vdots & \vdots & \ddots & & \vdots \\ 1 & \omega^{\delta-1} & \omega^{2(\delta-1)} & \cdots & \omega^{(\delta-1)(N-1)} \end{pmatrix}.$$

The codewords of $\mathscr{X}$ are linear dependence relations between the columns of H. Then Lemma 2.5.40 implies that any $\delta - 1$ columns of H are linearly independent. In fact, select columns with top (row 1) entries $\omega^{k_1}, \ldots, \omega^{k_{\delta-1}}$ where $0 \le k_1 < \cdots < k_{\delta-1} \le N - 1$. They form a square $(\delta - 1) \times (\delta - 1)$ matrix

$$
D = \begin{pmatrix}
\omega^{k_1} \cdot 1 & \omega^{k_2} \cdot 1 & \cdots & \omega^{k_{\delta-1}} \cdot 1 \\
\omega^{k_1} \cdot \omega^{k_1} & \omega^{k_2} \cdot \omega^{k_2} & \cdots & \omega^{k_{\delta-1}} \cdot \omega^{k_{\delta-1}} \\
\vdots & \vdots & \ddots & \vdots \\
\omega^{k_1} \cdot \omega^{k_1(\delta-2)} & \omega^{k_2} \cdot \omega^{k_2(\delta-2)} & \cdots & \omega^{k_{\delta-1}} \cdot \omega^{k_{\delta-1}(\delta-2)}
\end{pmatrix}
$$

that differs from the Vandermonde matrix by factors ω^{k_s} in front of the sth column. Then the determinant of D is the product

$$
\det D = \left(\prod_{s=1}^{\delta-1} \omega^{k_s} \right)
\begin{vmatrix}
1 & 1 & \cdots & 1 \\
\omega^{k_1} & \omega^{k_2} & \cdots & \omega^{k_{\delta-1}} \\
\vdots & \vdots & \ddots & \vdots \\
\omega^{k_1(\delta-2)} & \alpha^{k_2(\delta-2)} & \cdots & \omega^{k_{\delta-1}(\delta-2)}
\end{vmatrix}
$$

$$
= \left(\prod_{s=1}^{\delta-1} \omega^{k_s} \right) \times \left(\prod_{i>j} \left(\omega^{k_i} - \omega^{k_j} \right) \right) \neq 0,
$$

and any $\delta - 1$ columns of H are indeed linearly independent. In turn, this means that any non-zero codeword in $\mathscr{X}$ has weight at least δ. Thus, $\mathscr{X}$ has minimum distance $\geq \delta$. $\qquad\square$

Theorem 3.2.9 (A generalisation of the BCH bound) *Let ω be a primitive Nth root of unity and $b \geq 1$, $r \geq 1$ and $\delta > 2$ integers, with $\gcd(r, N) = 1$. Consider a cyclic code $\mathscr{X} = \langle g(X) \rangle$ of length N where $g(X)$ is a monic polynomial of smallest degree with $g(\omega^b) = g(\omega^{b+r}) = \cdots = g(\omega^{b+(\delta-2)r}) = 0$. Prove that $\mathscr{X}$ has $d(\mathscr{X}) \geq \delta$.*

Proof As $\gcd(r, N) = 1$, ω^r is a primitive root of unity. So, we can repeat the proof given above, with b replaced by bru where ru is found from $ru + Nv = 1$. An alternative solution: the matrix $N \times (\delta - 1)$

$$
\begin{pmatrix}
1 & 1 & \cdots & 1 \\
\omega^b & \omega^{b+r} & \cdots & \omega^{b+(\delta-2)r} \\
\omega^{2b} & \omega^{2(b+r)} & \cdots & \omega^{2(b+(\delta-2)r)} \\
\vdots & \vdots & \ddots & \vdots \\
\omega^{(N-1)b} & \omega^{(N-1)b+r} & \cdots & \omega^{(N-1)(b+(\delta-2)r)}
\end{pmatrix}
$$

checks the code $X = \langle g(X) \rangle$. Take any of its $(\delta - 1) \times (\delta - 1)$ submatrices, say, with rows $i_1 < i_2 < \cdots < i_{\delta-1}$. Denote it by $D = (D_{jk})$. Then

$$
\det D = \prod_{1 \leq l \leq \delta-1} \omega^{(i_l-1)b} \det(\omega^{r(i_j-1)(\delta-2)})
$$

$$
= \prod_{1 \leq l \leq \delta-1} \omega^{(i_l-1)b} \det (\text{Vandermonde}) \neq 0,
$$

because $\gcd(r, N) = 1$. So, $d(X) \geq \delta$.

Worked Example 3.2.10   *Let ω be a primitive n-root of unity in an extension field of $\mathbb{F}_q$ and $a(X) = \sum\limits_{0 \le i \le n-1} a_i X^i$ be a polynomial of degree at most $n-1$. The Mattson–Solomon polynomial is defined by*

$$a_{MS}(X) = \sum_{j=1}^{n} a(\omega^j) X^{n-j}. \qquad (3.2.6)$$

Let $q = 2$ and $a(X) \in \mathbb{F}_2[X]/\langle X^n - 1 \rangle$. Prove that the Mattson–Solomon polynomial $a_{MS}(X)$ is idempotent, i.e. $a_{MS}(X)^2 = a_{MS}(X)$ in $\mathbb{F}_2[X]/\langle X^n - 1 \rangle$.

Solution Let $a(X) = \sum\limits_{0 \le i \le n-1} a_i X^i$, then $na_i = a_{MS}(\omega^i), 0 \le i \le n-1$, by Lemma 3.2.5. In $\mathbb{F}_2$, $(na_i)^2 = na_i$, so $a_{MS}(\omega^i)^2 = a_{MS}(\omega^i)$. For polynomials, write $b^{(2)}(X)$ for the square in $\mathbb{F}_2[X]$ and $b(X)^2$ for the square in $\mathbb{F}_2[X]/\langle X^n - 1 \rangle$:

$$b^{(2)}(X) = c(X)(X^n - 1) + b(X)^2.$$

Then

$$a_{MS}(X) \restriction_{X=\omega^i} = (a_{MS}(X) \restriction_{X=\omega^i})^2 = a_{MS}^{(2)}(X) \restriction_{X=\omega^i}$$
$$= a_{MS}(X)^2 \restriction_{X=\omega^i},$$

i.e. polynomials $a_{MS}(X)$ and $a_{MS}(X)^2$ agree at $\omega^0 = e, \omega, \ldots, \omega^{n-1}$. Write this in the matrix form, with $a_{MS}(X) = a_{0,MS} + a_{1,MS}X + \cdots + a_{n-1,MS}X^{n-1}$, $a_{MS}(X)^2 = a'_{0,MS}X + \cdots + a'_{n-1,MS}X^{n-1}$:

$$\left(a_{MS} - a_{MS}^{(2)'}\right) \begin{pmatrix} e & e & \cdots & e \\ e & \omega & \cdots & \omega^{n-1} \\ \vdots & \vdots & \ddots & \vdots \\ e & \omega^{n-1} & \cdots & \omega^{(n-1)^2} \end{pmatrix} = 0.$$

As the matrix is Vandermonde, its determinant is

$$\prod_{0 \le i < j \le n-1} (\omega^j - \omega^i) \ne 0,$$

and $a_{MS} = a_{MS}^{(2)'}$. So, $a_{MS}(X) = a_{MS}(X)^2$. $\square$

Definition 3.2.11   Let $v = v_0 v_1 \ldots v_{N-1}$ be a vector over $\mathbb{F}_q$, and let ω be a primitive $(N, \mathbb{F}_q)$ root of unity over $\mathbb{F}_q$. The *Fourier transform* of the vector v is the vector $V = V_0 V_1 \ldots V_{N-1}$ with components given by

$$V_j = \sum_{i=0}^{N-1} \omega^{ij} v_i, \quad j = 0, \ldots, N-1. \qquad (3.2.7)$$

Lemma 3.2.12 (The inversion formula) *The vector v is recovered from its Fourier transform V by the formula*

$$v_i = \frac{1}{N} \sum_{j=0}^{N-1} \omega^{-ij} V_j. \tag{3.2.8}$$

Proof In any field $X^N - 1 = (X-1)(X^{N-1} + \cdots + X + 1)$. As the order of ω is N, for any r, ω^r is a zero of LHS. Hence for all $r \neq 0 \bmod N$, ω^r is a zero of the last term, i.e.

$$\sum_{j=0}^{N-1} \omega^{rj} = 0 \bmod N.$$

On the other hand, for $r = 0$

$$\sum_{j=0}^{N-1} \omega^{rj} = N \bmod p$$

which is not zero if N is not a multiple of the field characteristic p. But $q - 1 = p^s - 1$ is a multiple of N, so N is not a multiple of p. Hence, $N \neq 0 \bmod p$. Finally, change the order of summation to obtain that

$$\frac{1}{N} \sum_{j=0}^{N-1} \omega^{-ij} V_j = \frac{1}{N} \sum_{k=0}^{N-1} v_k \sum_{j=0}^{N-1} \omega^{(k-i)j} = v_i.$$

$\square$

Proof of Lemma 3.2.5 Let $a(X) = a_0 + a_1 X + \cdots + a_{N-1} X^{N-1} \in \mathbb{F}_q[X]$ and ω be a primitive $(N, \mathbb{F}_q)$ root of unity over $\mathbb{F}_q$. Then write

$$N^{-1} \sum_{0 \leq j \leq N-1} a(\omega^j) \omega^{-ij} = N^{-1} \sum_{0 \leq j \leq N-1} \sum_{0 \leq k \leq N-1} a_k \omega^{jk} \omega^{-ij}$$

$$= N^{-1} \sum_{0 \leq k \leq N-1} a_k \sum_{0 \leq j \leq N-1} \omega^{j(k-i)} = N^{-1} \sum_{0 \leq k \leq N-1} a_k N \delta_{ki} = a_i.$$

Here we used the fact that, for $1 \leq \ell \leq N - 1$, $\omega^\ell \neq 1$, and

$$\sum_{0 \leq j \leq N-1} \omega^{j\ell} = \sum_{0 \leq j \leq N-1} (\omega^\ell)^j = (e - (\omega^\ell)^N)(e - \omega^\ell)^{-1} = 0.$$

Hence

$$a_i = \frac{1}{N} \sum_{0 \leq j \leq N-1} a(\omega^j) \omega^{-ij}. \tag{3.2.9}$$

$\square$

Worked Example 3.2.13   *Give an alternative proof of the BCH bound: Let ω be a primitive $(N, \mathbb{F}_q)$ root of unity and $b \geq 1$ and $\delta \geq 2$ integers. Let $\mathscr{X}_N = \langle g(X) \rangle$ be a cyclic code where $g(X) \in \mathbb{F}_q[X]/\langle X^N - e \rangle$ is a monic polynomial of smallest degree having $\omega^b, \omega^{b+1}, \ldots, \omega^{b+\delta-2}$ among its roots. Then $\mathscr{X}_N$ has minimum distance at least δ.*

Solution   Let $a(X) = \sum\limits_{0 \leq j \leq N-1} a_j X^j \in \mathscr{X}_N$ satisfy condition $g(X)|a(X)$ and $a(\omega^i) = 0$ for $i = b, \ldots, b+\delta-2$. Consider the Mattson–Solomon polynomial $c_{MS}(X)$ for $a(X)$:

$$
\begin{aligned}
c_{MS}(X) &= \sum_{0 \leq i \leq N-1} a(\omega^{-i}) X^i = \sum_{0 \leq i \leq N-1} a(\omega^{N-i}) X^i \\
&= \sum_{1 \leq i \leq N} a(\omega^i) X^{N-i} \\
&= \sum_{1 \leq j \leq b-1} a(\omega^j) X^{N-j} + 0 + \cdots + 0 \text{ (from } \omega^b, \ldots, \omega^{b+\delta-2}) \\
&\quad + a(\omega^{b+\delta-1}) X^{N-b-\delta+1} + \cdots + a(\omega^N).
\end{aligned}
\tag{3.2.10}
$$

Multiply by X^{b-1} and group:

$$
\begin{aligned}
X^{b-1} c_{MS}(X) &= a(\omega) X^{N+b-2} + \cdots + a(\omega^{b-1}) X^N \\
&\quad + a(\omega^{b+\delta-1}) X^{N-\delta} + \cdots + a(\omega^N) X^{b-1} \\
&= X^N \left[a(\omega) X^{b-2} + \cdots + a(\omega^{b-1}) \right] \\
&\quad + \left[a(\omega^{b+\delta-1}) X^{N-\delta} + \cdots + a(\omega^N) X^{b-1} \right] \\
&= X^N p_1(X) + q(X) \\
&= (X^N - e) p_1(X) + p_1(X) + q(X).
\end{aligned}
$$

We see that $c_{MS}(\omega^i) = 0$ iff $p_1(\omega^i) + q(\omega^i) = 0$. But $p_1(X) + q(X)$ is a polynomial of degree $\leq N - \delta$ so it has at most $N - \delta$ roots. Thus, $c_{MS}(X)$ has at most $N - \delta$ roots of the form ω^i.

Therefore, the inversion formula (3.2.8) implies that the weight $w(a(X))$ (i.e. the weight of the coefficient string $a_0 \ldots a_{N-1}$) obeys

$$
w(a(X)) \geq N - \text{ the number of roots of } c_{MS}(X) \text{ of the form } \omega^i. \tag{3.2.11}
$$

That is,

$$
w(a(X)) \geq N - (N - \delta) = \delta.
$$

□

We finish this section with a brief discussion of the *Guruswami–Sudan decoding algorithm*, for list decoding of Reed–Solomon codes. First, we have to provide an alternative description of the Reed–Solomon codes (as Reed and Solomon have done it in their joint paper). For brevity, we take the value $b = 1$ (but will be able to extend the definition to values of $N > q - 1$).

Given $N \leq q$, let $S = \{x_1, \ldots, x_N\} \subset \mathbb{F}_q$ be a set of N distinct points in $\mathbb{F}_q$ (a supporting set). Let Ev denote the evaluation map

$$\text{Ev} : f \in \mathbb{F}_q[X] \mapsto \text{Ev}(f) = (f(x_1), \ldots, f(x_N)) \in \mathbb{F}_q^N \qquad (3.2.12)$$

and take

$$L = \{f \in \mathbb{F}_q[X] : \deg f < k\}. \qquad (3.2.13)$$

Then the q-ary Reed–Solomon code of length N and dimension k can be defined as

$$\mathscr{X} = \text{Ev}(L); \qquad (3.2.14)$$

it has the minimum distance $d = d(\mathscr{X}) = N - k + 1$ and corrects up to $\left\lfloor \dfrac{d-1}{2} \right\rfloor$ er-

rors. The encoding of a source message $\mathbf{u} = u_0 \ldots u_{k-1} \in \mathbb{F}_q^k$ consists in calculating the values of the polynomial $f(X) = u_0 + u_1 X + \cdots + u_k X^{k-1}$ at points $x_i \in S$.

Definition 3.2.1 (where $\mathscr{X}$ was defined as the set of polynomials $c(X) = \sum_{0 \leq l < q-1} c_l X^l \in \mathbb{F}_q[X]$ with $c(\omega) = c(\omega^2) = \cdots = c(\omega^{\delta-1}) = 0$) emerges when $N = q - 1$, $k = N - \delta + 1 = q - \delta$, the supporting set $S = \{e, \omega, \ldots, \omega^{N-1}\}$ and the coefficients $c_0, c_1, \ldots, c_{N-1}$ are related to the polynomial $f(X)$ by

$$c_i = f(\omega^i), \;\; 0 \leq i \leq N - 1.$$

This determines uniquely the coefficients f_l in the representation $f(X) = \sum_{0 \leq l < N} f_l X^l$, via the discrete inverse Fourier transform relation

$$N f_l = c(\omega^{N-l}), \;\; \text{or} \;\; N f_{N-l-1} = c(\omega^{l+1}), \; l = 0, \ldots, N - 1,$$

guaranteeing, in particular, that $f_k = \cdots = f_{N-1} = 0$.

Given $f \in \mathbb{F}_q[X]$ and $\mathbf{y} = y_1 \ldots y_N \in \mathbb{F}_q^N$, set

$$\text{dist}(f, \mathbf{y}) = \sum_{1 \leq i \leq N} \mathbf{1}(f(x_i) \neq y_i).$$

Now assume $\mathbf{y} = y_1 \ldots y_N$ is a received word and set $t = \left\lfloor \dfrac{d-1}{2} \right\rfloor$. The above-mentioned 'conventional' decoding algorithms (the Berlekamp–Massey algorithm, the continued fractions algorithm and the Euclidean algorithm) follow the same principle: the algorithm either finds a unique f such that $\text{dist}(f, \mathbf{y}) \leq t$ or reports that such f does not exist. On the other hand, given $s > t$, list decoding attempts to find all f with $\text{dist}(f, \mathbf{y}) \leq s$; the hope is that if we are lucky, the codeword with this property will be unique, and we will be able to correct s errors, exceeding the 'conventional' limit of t errors.

This idea goes back to Shannon's bounded distance decoding: upon receiving a word $\mathbf{y}$, you inspect the Hamming balls around $\mathbf{y}$ until you encounter a closest codeword (or a collection of closest codewords) to $\mathbf{y}$. Of course, we want two things: that (i) when we take s 'moderately' larger than t, the chance of finding two or more codewords within distance s is small, and (ii) the algorithm has a reasonable computational complexity.

Example 3.2.14   The $[32, 8]$ RS code over $\mathbb{F}_{32}$ has $d = 25$ and $t = 12$. If we take $s = 13$, the Hamming ball about the received word $\mathbf{y}$ may contain two codewords. However, assuming that all error vectors $\mathbf{e}$ of weight 13 are equally likely, the probability of this event is 2.08437×10^{-12}.

The Guruswami–Sudan list decoding algorithm (see [59]) performs the task of finding the codewords within distance s for $t \leq s \leq t_{GS}$ in a polynomial time. Here

$$t_{GS} = n - 1 - \left\lfloor \sqrt{(k-1)n} \right\rfloor,$$

and t_{GS} can considerably exceed t.

In the above example, $t_{GS} = 17$. Asymptotically, for RS codes of rate R, the conventional decoding algorithms will correct a fraction $(1 - R)/2$ of errors, while the GS algorithm can correct up to $1 - \sqrt{R}$. The expected number of codewords in a ball of radius $s \leq t_{GS}$ (under the assumption of error-vector equidistribution) can also be assessed.

The Guruswami–Sudan algorithm works not only for the RS codes. In the original GS paper, the algorithm was shown to perform well for several classes of codes; later on it was extended to cover the RM codes as well (see [7]).

3.3 Cyclic codes revisited. Decoding the BHC codes

Let us begin afresh. As before, we assume that $\gcd(N, q) = 1$ (so if $q = 2$, N is odd), and write words $\mathbf{x} \in \mathcal{H}_{N,q}$ as $x_0 \ldots x_{N-1}$. Remind that a linear code $\mathcal{X} \subseteq \mathcal{H}_N$ is called *cyclic* if, for all $\mathbf{x} = x_0 \ldots x_{N-1} \in \mathcal{X}$, the cyclic shift $\pi \mathbf{x} = x_{N-1} x_0 \ldots x_{N-2} \in \mathcal{X}$. With each word $c = c_0 \ldots c_{N-1}$ we associate a polynomial $c(X) \in \mathbb{F}_q[X]$:

$$c(X) = c_0 + c_1 X + \cdots + c_{N-1} X^{N-1}.$$

The map $c \leftrightarrow c(X)$ is an isomorphism between $\mathcal{X}$ and a linear subspace of $\mathbb{F}_q[X]$. Writing $c(X) \in \mathcal{X}$ simply means that the coefficient string $c_0 \ldots c_{N-1} \in \mathcal{X}$.

Lemma 3.3.1   *The code $\mathcal{X}$ is cyclic iff its image under the above isomorphism is an ideal in the quotient ring $\mathbb{F}_q[X]/\langle X^N - e \rangle$.*

Proof   Cyclic shift corresponds to multiplying a polynomial $c(X)$ by X. Hence, multiplication by any polynomial preserves $\mathcal{X}$. ∎

It is fruitful to think of $\mathcal{X}$ as the ideal in $\mathbb{F}_q[X]/\langle X^N - e \rangle$ and consider all poly-nomials mod $(X^N - e)$. Moreover, $\mathbb{F}_q[X]/\langle X^N - e \rangle$ is a *principal ideal ring*: each its ideal is of the form

$$\langle g(X) \rangle = \{ f(X) : \ f(X) = g(X)h(X), \ h(X) \in \mathbb{F}_q[X]/\langle X^N - e \rangle \} \qquad (3.3.1)$$

where $g(X)$ is a fixed polynomial.

Theorem 3.3.2   *If the code $\mathcal{X} \subseteq \mathcal{H}_{N,q}$ is cyclic then there exists a unique monic polynomial $g(X) \in \mathcal{X}$ such that:*

(i) $\mathcal{X} = \langle g(X) \rangle$;
(ii) $g(X)$ *has the minimum degree among all polynomials $f(X) \in \mathcal{X}$. Further-more,*

(a) $g(X)|(X^N - e)$,
(b) *if* $\deg g(X) = d$ *then* $\dim \mathcal{X} = N - d$,
(c) $\mathcal{X} = \{ f(X) : f(X) = g(X)h(X), \ h(X) \in \mathbb{F}_q[X], \ \deg h(X) < N - d \}$,
(d) *if* $g(X) = g_0 + g_1 X + g_2 X^2 + \cdots + g_d X^d$, *with* $g_d = e$, *then* $g_0 \neq 0$ *and*

$$G = \begin{pmatrix} g_0 & g_1 & g_2 & \cdots & g_d & 0 & 0 & \cdots & 0 \\ 0 & g_0 & g_1 & \cdots & g_{d-1} & g_d & 0 & \cdots & 0 \\ & & & \cdots & & & & \cdots & \\ 0 & 0 & 0 & \cdots & g_0 & g_1 & & \cdots & g_d \end{pmatrix}$$

is a generating matrix for $\mathcal{X}$, with row i being the cyclic shift of row $i - 1, i = 2, \ldots, N - d$.

Conversely, for any polynomial $g(X)|(X^N - e)$, the set $\langle g(X) \rangle = \{ f(X) : f(X) = g(X)h(X), \ h(X) \in \mathbb{F}_q[X]/\langle X^N - e \rangle \}$ is an ideal in $\mathbb{F}_q[X]/\langle X^N - e \rangle$, i.e. a cyclic code $\mathcal{X}$, and the above properties (b)–(d) hold.

Proof   Take $g(X) \in \mathbb{F}_2[X]$ a non-zero polynomial of the least degree in $\mathcal{X}$. Take $p(X) \in \mathcal{X}$ and write

$$p(X) = q(X)g(X) + r(X), \ \text{with} \deg r(X) < \deg g(X).$$

Then $r(X) \bmod (X^N - 1)$ belongs to $\mathcal{X}$. This contradicts the choice of $g(X)$ unless $r(X) = 0$. Therefore, $g(X)|p(x)$ which proves (i). Taking $p(X) = X^N - 1$ proves (ii). Finally, if $g(X)$ and $\tilde{g}(X)$ both satisfy (i) and (ii) then $g(X)|\tilde{g}(X)$ and $\tilde{g}(X)|g(X)$, implying $g(X) = \tilde{g}(X)$. $\qquad\square$

Corollary 3.3.3   *The cyclic codes of length N are in a one-to-one correspondence with factors of $X^N - e$. In other words, the map*

$$\{ \text{cyclic codes of length } N \} \ \rightarrow \ \{ \text{divisors of } X^N - 1 \},$$
$$\mathcal{X} \ \mapsto \ g(X),$$

is a bijection.

With the identification

$$\mathbb{F}_2[X]/\langle (X^N - 1) \rangle = \{ f \in \mathbb{F}_2[X] : \deg(f) < N \} = \mathbb{F}_2^N$$

the cyclic codes become ideals in the polynomial ring $\mathbb{F}_2[X]/\langle (X^N - 1) \rangle$. They are in a one-to-one correspondence with the ideals in $\mathbb{F}_2[X]$ containing polynomial $X^N - 1$. Because $\mathbb{F}_2[X]$ is a Euclidean domain, all ideals in $\mathbb{F}_2[X]$ are principal, i.e. of the form $\{ f(X)g(X) : f(X) \in \mathbb{F}_2[X] \}$. In fact, all ideals in $\mathbb{F}_2[X]/\langle (X^N - 1) \rangle$ are also principal ideals.

Definition 3.3.4   The polynomial $g(X)$ is called the *minimal degree generator* (or simply the *generator*) of the cyclic code $\mathcal{X}$. The ratio $h(X) = (X^N - e)/g(X)$, of degree $N - \deg g(X)$, is called the *check polynomial* for the cyclic code $\mathcal{X} = \langle g(X) \rangle$.

Example 3.3.5 $X - e$ generates the parity-check code $\{ \mathbf{x} : \sum_i x_i = 0 \}$ and $e + X + \cdots + X^{N-1}$ the repetition code $\{ a \ldots a, \ a \in \mathbb{F}_q \}$; $X \equiv e$ generates $\mathcal{X} = \mathcal{H}$.

Worked Example 3.3.6   (a) *A cyclic code $\mathcal{X} = \langle g(X) \rangle$ of length N is called reversible if $c_0 \ldots c_{N-1} \in \mathcal{X}$ implies $c_{N-1} \ldots c_0 \in \mathcal{X}$. Prove that $\mathcal{X}$ is reversible iff $g(\alpha) = 0$ implies $g(\alpha^{-1}) = 0$.*

(b) *A cyclic code is called degenerate if, for some $r|N$, each codeword $\mathbf{c} \in \mathcal{X}$ is a concatenation $c'c' \cdots c'$ of N/r copies of some string c' of length r. Prove that $\mathcal{X}$ is degenerate iff its check polynomial $h(X)|(X^r - 1)$.*

[Hint: Prove that the generating polynomial $g(X) = a(X)(1 + X^r + X^{2r} + \cdots + X^{N-r})$.]

Solution   (a) If the code $\mathcal{X} = \langle g(X) \rangle$ is reversible and $\mathbf{g} = g_0 \ldots g_{N-k} 0 \ldots 0$ then $X^{N-1} g(X^{-1}) \sim 0 \ldots 0 g_{N-k} \ldots g_0 \in \mathcal{X}$, i.e. $X^{N-1} g(X^{-1}) = g(X)q(X)$. Thus, if $g(\alpha) = 0$ then $\alpha^{N-1} g(\alpha^{-1}) = 0$, i.e. $g(\alpha^{-1}) = 0$.

Conversely, $g(\alpha) = 0$ implies $g(\alpha^{-1}) = 0$. Suppose that $c(X) \in \mathcal{X}$ then $g(X)|c(X)$. Moreover, $X^{N-1} c(X^{-1})$ has all zeros of $g(X)$ among its roots, and so belongs to $\mathcal{X}$. But $X^{N-1} c(X^{-1}) \sim c_{N-1} \ldots c_0$, so $\mathcal{X}$ is reversible.

(b) The condition $\mathbf{g} = a' \ldots a'$ means $g(X) = a(X)(e + X^r + X^{2r} + \cdots + X^{N-r})$. On the other hand,

$$X^N - e = (X^r - e)(X^{N-r} + \cdots + X^r + e) = h(X)g(X).$$

Thus, if $\mathcal{X} = \langle g(X) \rangle$ is degenerate then $X^r - e = h(X)a(X)$, i.e. $h(X)|(X^r - e)$.

Conversely, if $h(X)|(X^r - e)$ then $X^r - e = a(X)h(X)$ and

$$X^N - e = (X^r - e)(X^{N-r} + \cdots + X^r + e)$$
$$= h(X)a(X)(X^{N-r} + \cdots + X^r + e).$$

Then

$$g(X) = a(X)(X^{N-r} + \cdots + X^r + e),$$

i.e. $\mathbf{g} = a' \ldots a'$. Furthermore, any $c(X) \in \mathcal{X}$ is of the form $c(X) = q(X)g(X)$ where $\deg q(X) \le N - \deg g(X)$. Write

$$c(X) = q(X)g(X) = a(X)q(X)(X^{N-r} + \cdots + X^r + e);$$

we conclude that $\deg a(X)q(X) < r$ (after multiplying by X^{N-r} the degree cannot exceed $N - 1$). Then $\mathbf{c} = c' \ldots c'$ is the concatenation $c' \ldots c'$ where $c' \sim a(X)q(X)$. $\qquad\square$

Worked Example 3.3.7 *Show that Hamming's $[7,4]$ code is a cyclic code with check polynomial $X^4 + X^2 + X + 1$. What is its generator polynomial? Does Hamming's original code contain a subcode equivalent to its dual?*

Solution In $\mathbb{F}_2^7$ we have

$$X^7 - 1 = (X^3 + X + 1)(X^4 + X^2 + X + 1).$$

The cyclic code with generator $g(X) = X^3 + X + 1$ has check polynomial $h(X) = X^4 + X^2 + X + 1$. The parity-check matrix of the code is

$$\begin{pmatrix} 1 & 0 & 1 & 1 & 1 & 0 & 0 \\ 0 & 1 & 0 & 1 & 1 & 1 & 0 \\ 0 & 0 & 1 & 0 & 1 & 1 & 1 \end{pmatrix}.$$

The columns of this matrix are the non-zero elements of $\mathbb{F}_2^3$. So, it is equivalent to Hamming's $[7,4]$ code.

The dual of Hamming's $[7,4]$ code has generator polynomial $X^4 + X^3 + X^2 + 1$ (the reverse of $h(X)$). Since $X^4 + X^3 + X^2 + 1 = (X + 1)g(X)$, it is a subcode of Hamming's $[7,4]$ code. $\qquad\square$

Worked Example 3.3.8 *Let ω be a primitive Nth root of unity. Let $\mathcal{X} = \langle g(X) \rangle$ be a cyclic code of length N. Show that the dimension $\dim(\mathcal{X})$ equals the number of powers ω^j such that $g(\omega^j) \ne 0$.*

Solution Denote $\mathbb{E}^{(N)} = \{\omega, \omega^2, \ldots, \omega^N = e\}$, $\dim\langle g(X) \rangle = N - d$, $d = \deg g(X)$. But $g(X) = \prod_{1 \le j \le d} (X - \omega^{i_j})$ where $\omega^{i_1}, \ldots, \omega^{i_d}$ are the zeros of $\langle g(X) \rangle$. Hence, the remaining $N - d$ roots of unity ω^l satisfy the condition $g(\omega^l) \ne 0$. $\qquad\square$

It is important to note that the generator polynomial of a cyclic code $\mathscr{X} = \langle g(X) \rangle$ is not unique. In particular, there exists a unique polynomial $i(X) \in \mathscr{X}$ such that $i(X)^2 = i(X)$ and $\mathscr{X} = \langle i(X) \rangle$ (an idempotent generator).

Theorem 3.3.9 *If $\mathscr{X}_1 = \langle g_1(X) \rangle$ and $\mathscr{X}_2 = \langle g_2(X) \rangle$ are cyclic codes with generators $g_1(X)$ and $g_2(X)$ then*

(a) $\mathscr{X}_1 \subset \mathscr{X}_2$ iff $g_2(X) | g_1(X)$,
(b) $\mathscr{X}_1 \cap \mathscr{X}_2 = \langle \mathrm{lcm}(g_1(X), g_2(X)) \rangle$,
(c) $\mathscr{X}_1 | \mathscr{X}_2 = \langle \mathrm{gcd}(g_1(X), g_2(X)) \rangle$.

Theorem 3.3.10 *Let $h(X)$ be the check polynomial for $\mathscr{X}$. Then*

(a) $\mathscr{X} = \{ f(X) : f(X)h(X) = 0 \bmod (X^N - e) \}$,
(b) *if $h(X) = h_0 + h_1 X + \cdots + h_{N-r} X^{N-r}$ then the parity-check matrix H of $\mathscr{X}$ is*

$$
H = \begin{pmatrix}
h_{N-r} & h_{N-r-1} & \cdots & h_1 & h_0 & 0 & 0 & \cdots & 0 \\
0 & h_{N-r} & \cdots & \cdots & h_1 & h_0 & 0 & \cdots & 0 \\
\cdots & \cdots & \cdots & \cdots & \cdots & \cdots & \cdots & \cdots & \cdots \\
0 & 0 & \cdots & h_{N-r} & h_{N-r-1} & \cdots & \cdots & \cdots & h_0
\end{pmatrix},
$$

(c) *the dual code $\mathscr{X}^\perp$ is a cyclic code of dim $\mathscr{X}^\perp = r$, and $\mathscr{X}^\perp = \langle g^\perp(X) \rangle$, where*
$$g^\perp(X) = h_0^{-1} X^{N-r} h(X^{-1}) = h_0^{-1}(h_0 X^{N-r} + h_1 X^{N-r-1} + \cdots + h_{N-r}).$$

The generator $g(X)$ of a cyclic code is specified, in terms of factorisation of $X^N - e$, as a 'sub-product',

$$X^N - e = \mathrm{lcm}(M_\omega(X) : \omega \in \mathbb{E}^{(N)}), \tag{3.3.2}$$

of some minimal polynomials $M_\omega(X)$. A convenient way is to characterise a cyclic code via roots of $g(X)$. If ω is a root of $M_\omega(X)$ in an extension field $\mathbb{F}_q(\omega)$ then $M_\omega(X)$ is the minimal polynomial for ω over $\mathbb{F}_q$. For any polynomial $f(X) \in \mathbb{F}_q[X]$ we have $f(\omega) = 0$ iff $f(X) = a(X)M_\omega(X)$, and if in addition $f(X) \in \mathbb{F}_q[X]/\langle X^N - e \rangle$ then $f(\omega) = 0$ iff $f(X) \in \langle M_\omega(X) \rangle$. Hence we get

Theorem 3.3.11 *Let $g(X) = q_1(X) \ldots q_t(X)$ be a product of irreducible factors of $X^N - e$, and $\omega_1, \ldots, \omega_u$ be the roots of $g(X)$ in $\mathrm{Spl}(X^N - e)$ over $\mathbb{F}_q$. Then*

$$\langle g(X) \rangle = \{ f(X) \in \mathbb{F}_q[X]/\langle X^N - e \rangle : f(\omega_1) = \cdots = f(\omega_u) = 0 \}. \tag{3.3.3}$$

Furthermore, it is enough to pick up a single root of each irreducible factor: if ω_j' is any root of $M_\omega(X)$, $1 \le j \le t$, then

$$\langle g(X) \rangle = \{ f(X) \in \mathbb{F}_q[X]/\langle X^N - e \rangle : f(\omega_1') = \cdots = f(\omega_t') = 0 \}. \tag{3.3.4}$$

Conversely, if $\omega_1, \ldots, \omega_u$ is a set of roots of $X^N - e$ then the code $\{f(X) \in \mathbb{F}_q[X]/\langle X^N - e \rangle : f(\omega_1) = \cdots = f(\omega_u) = 0\}$ has a generator which is the lcm of the minimal polynomials for $\omega_1, \ldots, \omega_u$.

Definition 3.3.12 The roots of generator $g(X)$ are called the *zeros* of the cyclic code $\langle g(X) \rangle$. Other roots of unity are often called non-zeros of the code.

Let $\{\omega_1, \ldots, \omega_u\}$ be a set of roots of $X^N - e$ lying in an extension field $\mathbb{F}_{q^l}$. Recall that l is the minimal integer such that $N | q^l - 1$. If $f(X) = \sum f_i X^i$ is a polynomial in $\mathbb{F}_q[X]/\langle X^N - e \rangle$ then $f(\omega_j) = 0$ iff $\sum_{0 \le i \le u} f_i \omega_j^i = 0$. Representing $\mathbb{F}_{q^l}$ as a vector space over $\mathbb{F}_q$ of dimension l, we associate ω_j^i with a (column) vector $\overrightarrow{\omega}_j^i$ of length l over $\mathbb{F}_q$, writing the last equality as $\sum_i f_i \overrightarrow{\omega}_j^i = \overrightarrow{\sum_i f_i \omega_j^i} = 0$. So, the $(ul) \times N$ matrix

$$
\tilde{H}^T = \begin{pmatrix}
\overrightarrow{\omega}_1^0 & \overrightarrow{\omega}_1^1 & \cdots & \overrightarrow{\omega}_1^{N-1} \\
\overrightarrow{\omega}_2^0 & \overrightarrow{\omega}_2^1 & \cdots & \overrightarrow{\omega}_2^{N-1} \\
\vdots & \vdots & \ddots & \vdots \\
\overrightarrow{\omega}_u^0 & \overrightarrow{\omega}_u^1 & \cdots & \overrightarrow{\omega}_\mathbf{u}^{N-1}
\end{pmatrix}
\tag{3.3.5}
$$

can be considered as a parity-check matrix for the code with zeros $\omega_1, \ldots, \omega_u$ (with the proviso that its rows may not be linearly independent).

Theorem 3.3.13 *For $q = 2$, the Hamming $[2^l - 1, 2^l - l - 1, 3]$ code is equivalent to a cyclic code $\langle M_\omega(X) \rangle = \prod_{0 \le i \le l-1}(X - \omega^{2^i})$ where ω is a primitive element in $\mathbb{F}_{2^l}$.*

Proof Let ω be a primitive $(N, \mathbb{F}_2)$ root of unity where $N = 2^l - 1$. The splitting field Spl $(X^N - e)$ is $\mathbb{F}_{2^l}$ (as $\mathrm{ord}_N(2) = l$). So, ω is a primitive element in $\mathbb{F}_{2^l}$. Take $M_\omega(X) = (X - \omega)(X - \omega^2) \cdots (X - \omega^{2^{l-1}})$, of degree l. The powers $\omega^0 = e, \omega, \ldots, \omega^{N-1}$ form $\mathbb{F}_{2^l}^*$, the list of the non-zero elements and the columns of the $l \times N$ matrix

$$
H = (\overrightarrow{\omega}^0, \overrightarrow{\omega}, \ldots, \overrightarrow{\omega}^{N-1})
\tag{3.3.6}
$$

consist of all non-zero binary vectors of length l. Hence, the Hamming $[2^l - 1, 2^l - l - 1, 3]$ code is (equivalent to) the cyclic code $\langle M_\omega(X) \rangle$ whose zeros consist of a primitive $(2^l - 1; \mathbb{F}_2)$ root of unity ω and (necessarily) all the other roots of the minimal polynomial for ω. $\square$

Theorem 3.3.14 *If $\gcd(l, q - 1) = 1$ then the q-ary Hamming $\left[\dfrac{q^l - 1}{q - 1}, \dfrac{q^l - 1}{q - 1} - l, 3\right]$ code is equivalent to the cyclic code.*

Proof Write $\mathrm{Spl}(X^N - e) = \mathbb{F}_{q^l}$ where $l = \mathrm{ord}_N(q)$, $N = \frac{q^l-1}{q-1}$. To justify the selection of l observe that $\frac{q^l-1}{N} = q - 1$ and l is the least positive integer with this property as $\frac{q^l-1}{q-1} > q^{l-1} - 1$.

Therefore, $\mathrm{Spl}(X^N - e) = \mathbb{F}_{q^l}$. Take a primitive $\beta \in \mathbb{F}_{q^l}$. Then $\omega = \beta^{(q^l-1)/N} = \beta^{q-1}$ is a primitive $(N, \mathbb{F}_q)$ root of unity. As before, take the minimal polynomial $M_\omega(X) = (X - \omega)(X - \omega^q) \cdots (X - \omega^{q^{l-1}})$ and consider the cyclic code $\langle M_\omega(X) \rangle$ with the zero ω (and necessarily $\omega^q, \ldots, \omega^{q^{l-1}}$). Consider again the $l \times N$ matrix (3.3.6). We want to check that any two distinct columns of H are linearly independent. If not, there exist $i < j$ such that ω^i and ω^j are scalar multiples of the element $\omega^{j-i} \in \mathbb{F}_q$. But then $(\omega^{j-i})^{q-1} = \omega^{(j-i)(q-1)} = e$ in $\mathbb{F}_q$; as ω is a primitive Nth root of unity, this holds iff $(j-i)(q-1) \equiv 0 \bmod N$. Write

$$N = \frac{q^l-1}{q-1} = 1 + \cdots + q^{l-1}.$$

As $(q-1) | (q^r - 1)$ for all $r \geq 1$, we have $q^r = (q-1)v_r + 1$ for some natural v_r. Summing over $0 \leq r \leq s-1$ yields

$$N = (q-1) \sum_r v_r + l. \tag{3.3.7}$$

As $\gcd(q-1, l) = 1$ we have $\gcd(q-1, N) = 1$. But then the equality $(j-i)(q-1) = 0 \bmod N$ is impossible.

So, the code with the parity-check matrix H has length N, rank $k \geq N - l$ and distance $d \geq 3$. But the Hamming bound says that

$$q^k \leq q^N \left(\sum_{0 \leq m \leq E} \binom{N}{m} (q-1)^m \right)^{-1}, \quad E = \lfloor \frac{d-1}{2} \rfloor.$$

As the volume of the ball $v_{N,q}(E) \geq q^l$, this implies that in fact $k = N - l, E = 1$ and $d = 3$. So, this code is equivalent to a Hamming code. $\qquad \square$

Next, we look in more detail on BCH codes correcting several errors. Recall that if $\omega_1, \ldots, \omega_u \in \mathbb{E}_{(N,q)}$ are $(N, \mathbb{F}_q)$ roots of unity then

$$\mathscr{X}_N = \{ f(X) \in \mathbb{F}_q[X] / \langle X^N - e \rangle : f(\omega_1) = \cdots = f(\omega_u) = 0 \}$$

is a cyclic code $\langle g(X) \rangle$ where

$$g(X) = \mathrm{lcm} \left(M_{\omega_1, \mathbb{F}_q}(X), \ldots, M_{\omega_u, \mathbb{F}_q(X)}(X) \right) \tag{3.3.8}$$

is the product of distinct minimal polynomials for $\omega_1, \ldots, \omega_u$ over $\mathbb{F}_q$. In particular, if $q = 2$, $N = 2^l - 1$, and ω is a primitive element in $\mathbb{F}_{2^l}$ then the cyclic code with roots $\omega, \omega^2, \ldots, \omega^{2^{l-1}}$ (which is the same as with a single root ω) coincides with

$\langle M_{\omega}(X) \rangle$ and is equivalent to the Hamming code. We could try other possibilities for zeros of $\mathcal{X}$ to see if it leads to interesting examples. This is the way to discover the BCH codes [25], [70].

Recall the factorisation into minimal polynomials $M_i(X)(= M_{\omega^i, \mathbb{F}_q}(X))$,

$$X^N - 1 = \text{lcm}(M_i(X): i = 0, \dots, t), \qquad (3.3.9)$$

where ω is a primitive $(N, \mathbb{F}_q)$ root of unity. The roots of $M_i(X)$ are conjugate, i.e. have the form $\omega^i, \omega^{iq}, \dots, \omega^{iq^{d-1}}$ where $d(= d(i))$ is the least integer ≥ 1 such that $iq^d = i \bmod N$. The set $C_i = \{i, iq, \dots, iq^{d-1}\}$ is the ith cyclotomic coset of q mod N. So,

$$M_i(X) = \prod_{j \in C_i}(X - \omega^j). \qquad (3.3.10)$$

In Section 3.2, we obtained a cyclic code of minimal distance $\geq \delta$ by requiring that the generator $g(X)$ has $(\delta - 1)$ successive roots (with successive exponents). Compare Theorem 3.3.16 below.

Example 3.3.15 A binary Hamming code is a binary primitive narrow sense BCH of designed distance $\delta = 3$.

By Lemma 3.2.8, the distance $d(\mathcal{X}_{q,N,\delta}^{\text{BCH}}) \geq \delta$. As $\text{Spl}(X^N - e) = \mathbb{F}_{q^s}$ where $s = \text{ord}_N(q)$, we have that

$$\deg M_{\omega^{b+j}}(X) \leq s. \qquad (3.3.11)$$

Hence, the rank $(\mathcal{X}_{q,N,\delta}^{\text{BCH}}) = N - \deg(g(X)) \geq N - (\delta - 1)s$. So:

Theorem 3.3.16 *The q-ary BCH code $\mathcal{X}_{q,N,\delta}^{\text{BCH}}$ has distance $\geq \delta$ and rank $\geq N - (\delta - 1)\text{ord}_N(q)$.*

As before, we can form a parity-check matrix for $\mathcal{X}^{\text{BCH}}$ by writing $\omega^b, \omega^{b+1}, \dots, \omega^{b+\delta-2}$ and their powers as vectors from $\mathbb{F}_q^s$ where $s = \text{ord}_N(q)$. Set

$$\tilde{H}^{\mathsf{T}} = \begin{pmatrix} \overrightarrow{e} & \overrightarrow{e} & \dots & \overrightarrow{e} \\ \overrightarrow{\omega}^b & \overrightarrow{\omega}^{b+1} & \dots & \overrightarrow{\omega}^{b+\delta-2} \\ \overrightarrow{\omega}^{2b} & \overrightarrow{\omega}^{2(b+1)} & \dots & \overrightarrow{\omega}^{2(b+\delta-2)} \\ & & \dots & \\ \overrightarrow{\omega}^{(N-1)b} & \overrightarrow{\omega}^{(N-1)(b+1)} & \dots & \overrightarrow{\omega}^{(N-1)(b+\delta-2)} \end{pmatrix}. \qquad (3.3.12)$$

The 'proper' parity-check matrix H is obtained by removing redundant rows.

The binary BCH codes are simplest to deal with. Let $C_i = \{i, 2i, \dots, i2^{d-1}\}$ be the ith cyclotomic coset (with $d(= d(i))$ being the smallest non-zero integer such that $i \cdot 2^d = i \bmod N$). Then $u \in C_i$ iff $2u \bmod N \in C_i$. So, $M_i(X) = M_{2i}(X)$, and for all $s \geq 1$ the polynomials

$$g_{2s-1}(X) = g_{2s}(X) = \text{lcm}\{M_1(X), M_2(X), \dots, M_{2s}(X)\}.$$

We immediately deduce that $\mathscr{X}_{2,N,2s+1}^{\text{BCH}} = \mathscr{X}_{2,N,2s}^{\text{BCH}}$. So we can focus on the narrow sense BCH codes with odd designed distance $\delta = 2E + 1$, and obtain an improvement of Theorem 3.3.16:

Theorem 3.3.17   *The rank of a binary BCH code $\mathscr{X}_{2,N,2E+1}^{\text{BCH}}$ is $\geq N - E\,\text{ord}_N(2)$.*

The problem of determining exactly the minimum distance of a BCH code has been solved only partially (although a number of results exist in the literature). We present the following theorem without proof.

Theorem 3.3.18   *The minimum distance of a binary primitive narrow sense BCH code is an odd number.*

The previous results can be sharpened in a number of particular cases.

Worked Example 3.3.19   *Prove that $\log_2(N+1) > 1 + \log_2(E+1)!$ implies*

$$(N+1)^E < \sum_{0 \leq i \leq E+1} \binom{N}{i}. \qquad (3.3.13)$$

Solution For $i \leq E+1$ we obtain $i! \leq (E+1)! < (N+1)/2$. Hence, (3.3.13) follows from

$$(N+1)^{E+1} \leq 2 \sum_{0 \leq i \leq E+1} N(N-1)\ldots(N-i+1) = S(E). \qquad (3.3.14)$$

Inequality (3.3.14) holds for $E = 0$, and is checked by induction in E. Write the RHS of (3.3.14) as $S(E+1) = S(E) + N(N-1)\ldots(N-E)$. Then $S(E) > (N+1)^{E+1}$ by the induction hypothesis and it remains to check

$$N(N+1)^{E+1} < 2N(N-1)\ldots(N-E)(N-E-1), \text{ for } N+1 > 2(E+2)!. \quad (3.3.15)$$

Consider the polynomial $(y+1)^{E+1} - 2(y-1)\ldots(y-E)(y-E-1)$ and group together the monomials of degrees $E+1$ and E. Clearly, they are negative for $y > 2(E+2)!$. Continue this procedure, concluding that (3.3.13) holds.   □

Theorem 3.3.20   *Let $N = 2^s - 1$. If $2^{sE} < \sum_{0 \leq i \leq E+1} \binom{N}{i}$ then a primitive binary narrow sense BCH code $\mathscr{X}_{2,2^s-1,2E+1}^{\text{BCH}}$ has distance $2E + 1$.*

Proof   By Theorem 3.3.18, the distance is odd. So, $d(\mathscr{X}_{2,2^s-1,2E+1}^{\text{BCH}}) \neq 2E+2$. Suppose the distance is $\geq 2E+3$. Observe that the rank $\mathscr{X}_{2,2^s-1,2E+1}^{\text{BCH}} \geq N - sE$, and use the Hamming bound

$$2^{N-sE} \sum_{0 \leq i \leq E+1} \binom{N}{i} \leq 2^N, \text{ i.e. } 2^{sE} \geq \sum_{0 \leq i \leq E+1} \binom{N}{i}.$$

The contradiction implies $d(\mathscr{X}_{2,2^s-1,2E+1}^{\text{BCH}}) = 2E + 1$.   □

Corollary 3.3.21 *If* $N = 2^s - 1$ *and* $s > 1 + \log_2(E+1)!$ *then* $d(\mathcal{X}^{BCH}_{2,2^s-1,2E+1}) = 2E + 1$. *In particular, let* $N = 31$ *and* $s = 5$. *Then we easily verify that*

$$2^{5E} < \sum_{0 \le i \le E+1} \binom{31}{i}$$

for $E = 1, 2$ *and* 3. *This proves that the actual distance of* $\mathcal{X}^{BCH}_{2,31,d}$ *in fact equals* δ *for* $\delta = 3, 5, 7$.

Proof $s > 1 + \log_2(E+1)!$ *implies that* $2^{sE} < \sum_{0 \le i \le E+1} \binom{N}{i}$. $\qquad\square$

Theorem 3.3.22 *If* $\delta | N$, *the minimum distance of primitive binary narrow sense BCH code of designed distance* δ *equals* δ.

Proof Set $N = \delta m$, then

$$X^N - 1 = X^{\delta m} - 1 = (X^m - 1)(1 + X^m + \cdots + X^{(\delta-1)m}).$$

As $\omega^{jm} \ne 1$ for $j = 1, \ldots, \delta - 1$, none of $\omega, \omega^2, \ldots, \omega^{\delta-1}$ is a root of $X^m - 1$. So, they must be roots of $1 + X^m + \cdots + X^{(\delta-1)m}$. Then this polynomial gives a codeword, of weight δ. So, δ is the distance. $\qquad\square$

Two more results on the minimal distance of a BCH code are presented in Theorems 3.3.23 and 3.3.25. The full proofs are beyond the scope of this book and omitted.

Theorem 3.3.23 *Let* $N = q^s - 1$. *The minimal distance of a primitive q-ary narrow sense BCH code* $\mathcal{X}^{BCH}_{q,q^s-1,q^k-1,\omega,1}$ *of designed distance* $q^k - 1$ *equals* $q^k - 1$.

Theorem 3.3.24 *The minimal distance of a primitive q-ary narrow sense BCH code* $\mathcal{X}^{BCH} = \mathcal{X}^{BCH}_{q,q^s-1,\delta,\omega,1}$ *of designed distance* δ *is at most* $q\delta - 1$.

Proof Take k to be an integer ≥ 1 with $q^{k-1} \le \delta \le q^k - 1$. Set $\delta' = q^k - 1$ and consider $\mathcal{X}' (= \mathcal{X}^{BCH}_{q,q^s-1,\delta',\omega,1})$, the q-ary primitive narrow sense BCH code of the same length $N = q^s - 1$ and designed distance δ'. The roots of the generator of $\mathcal{X}$ are among those of $\mathcal{X}'$, so $\mathcal{X}' \subseteq \mathcal{X}$. But according to Theorem 3.3.22, $d(\mathcal{X}') = \delta'$ which is $\le \delta q - 1$. $\qquad\square$

The following result shows that BCH codes are not 'asymptotically good'. However, for small N (a few thousand or less), the BCH are among the best codes known.

Theorem 3.3.25   *There exists no infinite sequence of q-ary primitive BCH codes $\mathscr{X}_N^{\text{BCH}}$ of length N such that $d(\mathscr{X}_N)/N$ and $\text{rank}(\mathscr{X}_N)/N$ are bounded away from 0.*

Decoding BCH codes can be done by using the so-called *Berlekamp–Massey* algorithm. To begin with, consider a binary primitive narrow sense BCH code $\mathscr{X}^{BCH}$ $(= \mathscr{X}_{2,N,5}^{BCH})$ of length $N = 2^s - 1$ and designed distance 5. With $E = 2$ and $s \geq 4$, inequality $2^{sE} < \sum_{0 \leq i \leq E+1} \binom{N}{i}$ holds, and by Theorem 3.3.20, the distance $d(\mathscr{X}^{BCH})$ equals 5. Thus, the code is two-error correcting. Also, by Theorem 3.3.17, the rank of $\mathscr{X}^{BCH}$ is $\geq N - 2s$. [For $s = 4$, the rank is actually equal to $N - 2s = 15 - 8 = 7$.] So, $\mathscr{X}^{BCH}$ is $[2^s - 1, \geq 2^s - 1 - 2s, 5]$.

The defining zeros are ω, ω^2, ω^3, ω^4 where ω is a primitive Nth root of unity over $\mathbb{F}_2$ (which is also a primitive element ω of $\mathbb{F}_{2^s}$). We know that ω and ω^3 suffice as defining zeros: $\mathscr{X}^{BCH} = \{c(X) \in \mathbb{F}_2[X]/\langle X^N - 1\rangle: c(\omega) = c(\omega^3) = 0\}$. So, the parity-check matrix $\widetilde{H}$ in (3.3.12) can be taken in the form

$$\widetilde{H}^{\text{T}} = \begin{pmatrix} \overrightarrow{e} & \overrightarrow{\omega} & \overrightarrow{\omega}^2 & \cdots & \overrightarrow{\omega}^{N-1} \\ \overrightarrow{e} & \overrightarrow{\omega}^3 & \overrightarrow{\omega}^6 & \cdots & \overrightarrow{\omega}^{3(N-1)} \end{pmatrix}. \tag{3.3.16}$$

It is instructive to compare the situation with the binary Hamming $[2^l - 1, 2^l - 1 - l]$ code $\mathscr{X}^{(\text{H})}$. In the case of code $\mathscr{X}^{BCH}$, suppose again that a codeword $c(X) \in \mathscr{X}$ was sent and the received word $r(X)$ has ≤ 2 errors. Write $r(X) = c(X) + e(X)$ where the error polynomial $e(X)$ now has weight ≤ 2. There are three cases to consider: $e(X) = 0$, $e(X) = X^i$ or $e(X) = X^i + X^j$, $0 \leq i \neq j \leq N - 1$. If $r(\omega) = r_1$ and $r(\omega^3) = r_3$ then $e(\omega) = r_1$ and $e(\omega^3) = r_3$. In the case of no error $(e(X) = 0)$, $r_1 = r_3 = 0$, and vice versa. In the single-error case $(e(X) = X^i)$,

$$r_3 = e(\omega^3) = \omega^{3i} = (\omega^i)^3 = (e(\omega))^3 = r_1^3 \neq 0.$$

Conversely, if $r_3 = r_1^3 \neq 0$ then $e(\omega^3) = e(\omega)^3$. If $e(X) = X^i + X^j$ with $i \neq j$ then

$$\omega^{3i} + \omega^{3j} = (\omega^i + \omega^j)^3 = \omega^{3i} + \omega^{2i}\omega^j + \omega^i\omega^{2j} + \omega^{3j},$$

i.e. $\omega^{2i}\omega^j + \omega^i\omega^{2j} = 0$ or $\omega^i + \omega^j = 0$ which implies $i = j$, a contradiction. So, the single error occurs iff $r_3 = r_1^3 \neq 0$, and the wrong digit is i such that $r_1 = \omega^i$. So, in the single-error case we identify a column of $\widetilde{H}$, i.e. a pair $(\omega^i, \omega^{3i}) = (r_1, r_3)$ and change digit i in $r(X)$. This is completely similar to the decoding procedure for Hamming codes.

In the two-error case $(e(X) = X^i + X^j, i \neq j)$, in the spirit of the Hamming codes, we try to find a pair of columns (ω^i, ω^{3i}) and (ω^j, ω^{3j}) such that the sum $(\omega^i + \omega^j, \omega^{3i} + \omega^{3j}) = (r_1, r_3)$, i.e. solve the equation

$$r_1 = \omega^i + \omega^j, \quad r_3 = \omega^{3i} + \omega^{3j}.$$

Then find i, j such that $y_1 = \omega^i$, $y_2 = \omega^j$ (y_1, y_2 are called error locators). If such i, j (or equivalently, error locators y_1, y_2) are found, we know that errors occurred at positions i and j.

It is convenient to introduce an *error-locator* polynomial $\sigma(X)$ whose roots are y_1^{-1}, y_2^{-1}:

$$
\begin{aligned}
\sigma(X) &= (1 - y_1 X)(1 - y_2 X) = 1 - (y_1 + y_2)X + y_1 y_2 X^2 \\
&= 1 - r_1 X + (r_3 r_1^{-1} - r_1^2)X^2.
\end{aligned}
\tag{3.3.17}
$$

As $y_1 + y_2 = r_1$, we check that $y_1 y_2 = r_3 r_1^{-1} - r_1^2$. Indeed,

$$
r_3 = y_1^3 + y_2^3 = (y_1 + y_2)(y_1^2 + y_1 y_2 + y_2^2) = r_1(r_1^2 + y_1 y_2).
$$

If N is not large, the roots of $\sigma(X)$ can be found by trying all $2^s - 1$ non-zero elements of $\mathbb{F}_{2^s}^*$. (The standard formula for the roots of a quadratic polynomial does *not* apply over $\mathbb{F}_2$.) Thus, the following assertion arises:

Theorem 3.3.26   For $N = 2^l - 1$, consider a two-error correcting binary primitive narrow sense BCH code $\mathscr{X}$ (which equals $\mathscr{X}^{\mathrm{BCH}}$) of length N and designed distance 5, with the parity-check matrix produced from

$$
\widetilde{H}^{\mathsf{T}} = \begin{pmatrix} e & \omega & \omega^2 & \cdots & \omega^{N-1} \\ e & \omega^3 & \omega^6 & \cdots & \omega^{3(N-1)} \end{pmatrix},
$$

where ω is the primitive element of $\mathbb{F}_{2^s}$. [The rank of the code is $\geq N - 2l$ and for $l \geq 4$ the distance equals 5, i.e. $\mathscr{X}$ is $[2^l - 1, \geq 2^l - 1 - 2l, 5]$ and corrects two errors.] Assume that at most two errors occurred in a received word $r(X)$ and let $r(\omega) = r_1$, $r(\omega^3) = r_3$. Then:

(a) if $r_1 = 0$ then $r_3 = 0$ and no error occurred;
(b) if $r_3 = r_1^3 \neq 0$ then a single error occurred at position i where $r_1 = \omega^i$;
(c) if $r_1 \neq 0$ and $r_3 \neq r_1^3$ then two errors occurred: the error locator polynomial $\sigma(X) = 1 - r_1 X + (r_3 r_1^{-1} - r_1^2)X^2$ has two distinct roots $\omega^{N-1-i}, \omega^{N-1-j}$ and the errors occurred at positions i and j.

For a binary BCH code with a general designed distance δ ($\delta = 2t + 1$ is assumed odd), we follow the same idea: compute

$$
r_1 = e(\omega), r_3 = e(\omega^3), \ldots, r_{\delta-2} = e(\omega^{\delta-2})
$$

for the received word $r(X) = c(X) + e(X)$. Suppose that errors occurred at places $i_1, \ldots, i_t$. Then

$$
e(X) = \sum_{1 \leq j \leq t} X^{i_j}.
$$

As before, consider the system

$$\sum_{1 \le j \le t} \omega^{ij} = r_1, \quad \sum_{1 \le j \le t} \omega^{3ij} = r_3, \dots, \quad \sum_{1 \le j \le t} \omega^{(\delta-2)ij} = r_{\delta-2},$$

and introduce the error locators $y_j = \omega^{ij}$:

$$\sum_{1 \le j \le t} y_j = r_1, \quad \sum_{1 \le j \le t} y_j^3 = r_3, \dots, \quad \sum_{1 \le j \le t} y_j^{\delta-2} = r_{\delta-2}.$$

The error locator polynomial

$$\sigma(X) = \prod_{1 \le j \le t} (1 - y_j X)$$

has the roots y_j^{-1}. The coefficients σ_i in $\sigma(X) = \sum_{0 \le i \le t} \sigma_i X^i$ can be determined from the equations below

$$\begin{pmatrix} 1 & 0 & 0 & 0 & 0 & \cdots & 0 \\ r_2 & r_1 & 1 & 0 & 0 & \cdots & 0 \\ r_4 & r_3 & r_2 & r_1 & 1 & \cdots & 0 \\ \vdots & \vdots & \vdots & \vdots & \vdots & \ddots & \vdots \\ r_{2t-4} & r_{2t-5} & \vdots & \vdots & \vdots & \ddots & r_{t-3} \\ r_{2t-2} & r_{2t-3} & \cdots & \cdots & \cdots & \cdots & r_{t-1} \end{pmatrix} \begin{pmatrix} \sigma_1 \\ \sigma_2 \\ \sigma_3 \\ \vdots \\ \sigma_{2t-3} \\ \sigma_{2t-1} \end{pmatrix}$$

$$= \begin{pmatrix} r_1 \\ r_3 \\ r_5 \\ \vdots \\ r_{2t-3} \\ r_{2t-1} \end{pmatrix}$$

This requires computing r_k only for k odd as

$$r_{2j} = e(\omega^{2j}) = e(\omega^j)^2 = r_j^2.$$

Once the σ_i are found, the roots y_j^{-1} can be determined by trial and error.

Example 3.3.27   Consider $\mathscr{X}^{BCH}_{2,16,\omega,5}$ where ω is a primitive element of $\mathbb{F}^*_{16}$. We know that the primitive polynomial is $M_1(X) = X^4 + X + 1$ and $M_3(X) = X^4 + X^3 + X^2 + X + 1$. Hence, the generator of the code

$$g(X) = M_1(X)M_3(X) = X^8 + X^7 + X^6 + X^4 + 1.$$

Let us introduce two errors in the codeword $c = 10001011100000000$ at the 4th and 12th positions by taking $a(X) = X^{12} + X^8 + X^7 + X^6 + 1$. Then

$$r_1 = a(\omega) = \omega^{12} + \omega^8 + \omega^7 + \omega^6 + 1 = \omega^6,$$
$$r_3 = a(\omega^3) = \omega^{36} + \omega^{24} + \omega^{21} + \omega^{18} + 1 = \omega^9 + \omega^3 + 1 = \omega^4.$$

Since $r_3 \neq r_1^3$, consider the location polynomial

$$\sigma(X) = 1 + \omega^6 X + (\omega^{13} + \omega^{12})X^2.$$

The roots of $l(X)$ are ω^3 and ω^{11} by the direct check. Hence we discover the errors at the 4th and 12th positions.

3.4 The MacWilliams identity and the linear programming bound

The MacWilliams identity for linear codes deals with the so-called weight-enumerator polynomials $W_{\mathscr{X}}(z)$ and $W_{\mathscr{X}^{\perp}}(z)$ where $\mathscr{X}$ and $\mathscr{X}^{\perp}$ are a pair of dual codes of a given length N. The polynomials $W_{\mathscr{X}}(z)$ and $W_{\mathscr{X}^{\perp}}(z)$ are defined by

$$W_{\mathscr{X}}(z) = \sum_{0 \leq k \leq N} A_k z^k \text{ and } W_{\mathscr{X}^{\perp}}(z) = \sum_{0 \leq k \leq N} A_k^{\perp} z^k \qquad (3.4.1)$$

where $A_k (= A_k(\mathscr{X}))$ equals the number of codewords of weight k in $\mathscr{X}$, and $A_k^{\perp}$ $(= A_k(\mathscr{X}^{\perp}))$ the number in $\mathscr{X}^{\perp}$. The identity for q-ary codes reads

$$W_{\mathscr{X}^{\perp}}(z) = \frac{1}{\sharp \mathscr{X}}[1 + (q-1)z]^N W_{\mathscr{X}}\left(\frac{1-z}{1+(q-1)z}\right), \ z \in \mathbb{C}, \qquad (3.4.2)$$

and takes a particularly elegant form in the binary case ($q = 2$):

$$W_{\mathscr{X}^{\perp}}(z) = \frac{1}{\sharp \mathscr{X}}(1+z)^n W_{\mathscr{X}}\left(\frac{1-z}{1+z}\right). \qquad (3.4.3)$$

A short derivation of the abstract MacWilliams identity is rather algebraic. It may be skipped at the first reading as only its specification for linear codes will be used later on.

Definition 3.4.1 Let $(G, +)$ be a group. A homomorphism $\chi : G$ to the multiplicative group of complex numbers $\mathbb{S}' = \{z \in \mathscr{C} : |z| = 1\}$ is called a (one-dimensional) *character* of G. Since χ is a homomorphism

$$\chi(g_1 + g_2) = \chi(g_1)\chi(g_2), \chi(0) = 1. \qquad (3.4.4)$$

We say χ is *trivial* (or *principal*) if $\chi(\cdot) \equiv 1$.

More generally, a *linear representation D* of a group G over a field $\mathbb{F}$ (not necessarily finite) is defined as a homomorphism

$$D : G \to \mathrm{GL}(V) : g \to D(g) \tag{3.4.5}$$

from G into the group $\mathrm{GL}(V)$ of invertible linear mappings of a finite-dimensional space V over $\mathbb{F}$. The vector space V is called the *representation space* and its dimension $\dim(V)$ is called the *dimension of representation*.

Let D be a representation of a group G. Then the map

$$\chi^D : G \to \mathbb{F} : g \to \sum d_{ii}(g) = \mathrm{trace}\big(D(g)\big), \tag{3.4.6}$$

which takes $g \in G$ to $\chi^D(g)$, the trace of $D(g) = (d_{ij}(g))$, is called the *character* of D. Representations and characters over the field $\mathbb{C}$ of complex numbers are called *ordinary*. In the situation where the underlying field $\mathbb{F}$ is finite, they are called *modular*.

In our case $G = \mathbb{F}_q$ with additive group operation. Fix a primitive qth root of unity $\omega = e^{2\pi i/q} \in \mathbb{S}'$ and for any $j \in \mathbb{F}_q$ define a one-dimensional representation of the group $\mathbb{F}_q$ as follows:

$$\chi^{(j)} : \mathbb{F}_q \to \mathbb{S}' : u \to \omega^{ju}.$$

The character $\chi^{(j)}$ is non-trivial for $j \neq 0$. In fact, all characters of $\mathbb{F}_q$ can be described in this way, but we omit the proof of this assertion.

Next, we define a character of the group $G' = \mathbb{F}_q^N$. Fix a non-trivial one-dimensional ordinary character $\chi : \mathbb{F}_q \to \mathbb{S}'$ and a non-zero element $\mathbf{v} \in \mathbb{F}_q^N$ and define a character of the additive group $G' = \mathbb{F}_q^N$ as follows:

$$\chi_{(\mathbf{v})} : \mathbb{F}_q^N \to \mathbb{S}' : \mathbf{u} \to \chi(\mathbf{v} \cdot \mathbf{u}), \tag{3.4.7}$$

where $\mathbf{v} \cdot \mathbf{u}$, as before, is the dot-product.

Lemma 3.4.2 *Let χ be a non-trivial (i.e. $\chi \not\equiv 1$) character of a finite group G. Then*

$$\sum_{g \in G} \chi(g) = 0. \tag{3.4.8}$$

If χ is trivial then $\sum\limits_{g \in G} \chi(g) = \sharp G$.

Proof Since χ is non-trivial, there exists an element $h \in G$ such that $\chi(h) \neq 1$. From

$$\chi(h) \sum_{g \in G} \chi(g) = \sum_{g \in G} \chi(hg) = \sum_{g \in G} \chi(g),$$

we obtain that $(\chi(h) - 1) \sum\limits_{g \in G} \chi(g) = 0$. Therefore, $\sum\limits_{g \in G} \chi(g) = 0$. □

In the case $G = \mathbb{F}_q^N$, $\sum\limits_{x \in \mathbb{F}_q^N} \chi(x) = q^N$ for a trivial.

Definition 3.4.3   The *discrete Fourier transform* (in short, DFT) of a function f on $\mathbb{F}_q^N$ is defined by

$$\hat{f} = \sum_{v \in \mathbb{F}_q^N} f(v) \chi_{(v)}. \tag{3.4.9}$$

Sometimes, the weight enumerator polynomial of code $\mathscr{X}$ is defined as a function of two formal variables x, y:

$$W_{\mathscr{X}}(x, y) = \sum_{v \in \mathscr{X}} x^{w(v)} y^{N-w(v)} \tag{3.4.10}$$

(if one sets $x = z, y = 1$, (3.4.10) coincides with (3.4.1)). So, we want to apply the DFT to the function (no harm to say that $x, y \in \mathbb{S}'$)

$$g : \mathbb{F}_q^N \to \mathscr{C}[x, y] : v \to x^{w(v)} y^{N-w(v)}. \tag{3.4.11}$$

Lemma 3.4.4   (The abstract MacWilliams identity) *For* $v \in \mathbb{F}_q^N$ *let*

$$g : \mathbb{F}_q^N \to \mathscr{C}[x, y] : v \to x^{w(v)} y^{N-w(v)}. \tag{3.4.12}$$

Then

$$\hat{g}(u) = (y - x)^{w(u)} (y + (q-1)x)^{N-w(u)}. \tag{3.4.13}$$

Proof   Let χ denote a non-trivial ordinary character of the additive group $G = \mathbb{F}_q$. Given $\alpha \in \mathbb{F}_q$, set $|\alpha| = 0$ if $\alpha = 0$ and $|\alpha| = 1$ otherwise. Then for all $u \in \mathbb{F}_q^N$ we compute

$$\hat{g}(u) = \sum_{v \in \mathbb{F}_q^N} \chi(\langle v, u \rangle) g(v)$$

$$= \sum_{v \in \mathbb{F}_q^N} \chi(\langle v, u \rangle) x^{w(v)} y^{N-w(v)}$$

$$= \sum_{v_0 \in \mathbb{F}_q} \cdots \sum_{v_{N-1} \in \mathbb{F}_q} \chi\Big(\sum_{i=0}^{N-1} v_i u_i \Big) x^{|v_0| + \cdots + |v_{N-1}|} y^{(1-|v_0|) + \cdots + (1-|v_{N-1}|)}$$

$$= \sum_{v_0 \in \mathbb{F}_q} \cdots \sum_{v_{N-1} \in \mathbb{F}_q} \prod_{i=0}^{N-1} \chi(v_i u_i) x^{|v_i|} y^{1-|v_i|}$$

$$= \prod_{i=0}^{N-1} \sum_{g \in G} \chi(g u_i) x^{|g|} y^{1-|g|}.$$

If $u_i = 0$ then $\chi(g u_i) = \chi(0) = 1$ and so

$$\sum_{g \in G} x^{|g|} y^{1-|g|} = y + (q-1)x.$$

If $u_i \neq 0$ then

$$\sum_{g \in G} \chi(gu_i) x^{|g|} y^{1-|g|} = y + \sum_{g \in G \setminus 0} \chi(gu_i) x = y - \chi(0) x = y - x.$$

□

Lemma 3.4.5   (MacWilliams identity for linear codes) *If $\mathscr{X}$ is a linear $[N,k]$ code over $\mathbb{F}_q$ then*

$$\sum_{\mathbf{x} \in \mathscr{X}} \widehat{f}(\mathbf{x}) = q^k \sum_{\mathbf{y} \in \mathscr{X}^\perp} f(\mathbf{y}). \qquad (3.4.14)$$

Proof   Consider the following sum:

$$\sum_{\mathbf{x} \in \mathscr{X}} \widehat{f}(x) = \sum_{\mathbf{x} \in \mathscr{X}} \sum_{\mathbf{v} \in \mathbb{F}_q^N} \chi_{(\mathbf{v})}(\mathbf{x}) f(\mathbf{v})$$

$$= \sum_{\mathbf{v} \in \mathbb{F}_q^N} \sum_{\mathbf{x} \in \mathscr{X}} \chi\Big(\langle \mathbf{v}, \mathbf{x} \rangle\Big) f(\mathbf{v})$$

$$= \sum_{\mathbf{v} \in \mathscr{X}^\perp} \sum_{\mathbf{x} \in \mathscr{X}} \chi\Big(\langle \mathbf{v}, \mathbf{x} \rangle\Big) f(\mathbf{v})$$

$$+ \sum_{\mathbf{v} \in \mathbb{F}_q^N \setminus \mathscr{X}^\perp} \sum_{\mathbf{x} \in \mathscr{X}} \chi\Big(\langle \mathbf{v}, \mathbf{x} \rangle\Big) f(\mathbf{v}).$$

In the first sum we have $\chi\Big(\langle \mathbf{v}, \mathbf{x} \rangle\Big) = \chi(0) = 1$ for all $\mathbf{v} \in \mathscr{X}^\perp$ and all $\mathbf{x} \in \mathscr{X}$. In the second sum we study the linear form

$$\mathscr{X} \to \mathbb{F}_q : \mathbf{x} \to \langle \mathbf{v}, \mathbf{x} \rangle.$$

Since $\mathbf{v} \in \mathbb{F}_q^N \setminus \mathscr{X}^\perp$, this linear form is surjective, whence its kernel has dimension $k - 1$, i.e. for any $g \in \mathbb{F}_q$ there exist q^{k-1} vectors $\mathbf{x} \in \mathscr{X}$ such that $\langle \mathbf{v}, \mathbf{x} \rangle = g$. This implies

$$\sum_{\mathbf{x} \in \mathscr{X}} \widehat{f}(\mathbf{x}) = q^k \sum_{\mathbf{y} \in \mathscr{X}^\perp} f(\mathbf{y}) + q^{k-1} \sum_{\mathbf{v} \in \mathbb{F}_q^n \setminus \mathscr{X}^\perp} f(\mathbf{v}) \sum_{g \in G} \chi(g)$$

$$= q^k \sum_{\mathbf{y} \in \mathscr{X}^\perp} f(\mathbf{y})$$

as the second term vanishes by Lemma 3.4.2.   □

Lemma 3.4.6   *The weight enumerator of an $[N,k]$ code $\mathscr{X}$ over $\mathbb{F}_q$ is related to the weight enumerator of its dual as follows:*

$$W_{\mathscr{X}^\perp}(x,y) = q^{-k} W_{\mathscr{X}}(y - x, y + (q-1)x). \qquad (3.4.15)$$

Proof By Lemma 3.4.5 with $g(\mathbf{v}) = x^{w(\mathbf{v})}y^{N-w(\mathbf{v})}$

$$W_{\mathscr{X}^{\perp}}(x,y) = \sum_{\mathbf{v}\in\mathscr{X}^{\perp}} g(\mathbf{v}) = q^{-k}\sum_{\mathbf{v}\in\mathscr{X}} \widehat{g}(\mathbf{v})$$

$$= q^{-k}W_{\mathscr{X}}(y-x,y+(q-1)x).$$

Substituting $x = z, y = 1$ we obtain (3.4.3). $\qquad\square$

Example 3.4.7 (i) For all codes $\mathscr{X}$, $W_{\mathscr{X}}(0) = A_0 = 1$ and $W_{\mathscr{X}}(1) = \sharp\,\mathscr{X}$. When $\mathscr{X} = \mathbb{F}_q^{\times N}$, $W_{\mathscr{X}}(z) = [1+z(q-1)]^N$.

(ii) For a binary repetition code $\mathscr{X} = \{0000, 1111\}$, $W_{\mathscr{X}}(x,y) = x^4 + y^4$. Hence,

$$W_{\mathscr{X}^{\perp}}(x,y) = \frac{1}{2}\left((y-x)^4 + (y+x)^4\right) = y^4 + 6x^2y^2 + x^4.$$

(iii) Let $\mathscr{X}$ be the Hamming $[7,4]$ code. The dual code $\mathscr{X}^{\perp}$ has 8 codewords; all except $\mathbf{0}$ are of weight 4. Hence, $W_{\mathscr{X}^{\perp}}(x,y) = x^7 + 7x^4y^3$, and, by the MacWilliams identity,

$$W_{\mathscr{X}} = \frac{1}{2^3}W_{\mathscr{X}^{\perp}}(x-y,x+y) = \frac{1}{2^3}\left((x-y)^7 + 7(x-y)^4(x+y)^3\right)$$

$$= x^7 + 7x^4y^3 + 7x^3y^4 + y^4.$$

Hence, $\mathscr{X}$ has 7 words of weight 3 and 4 each. Together with the $\mathbf{0}$ and $\mathbf{1}$ words, this accounts for all 16 words of the Hamming $[7,4]$ code.

Another way to derive the identity (3.4.1) is to use an abstract result related to group algebras and character transforms for Hamming spaces $\mathbb{F}_q^{\times N}$ (which are linear spaces over field $\mathbb{F}_q$ of dimension N). For brevity, the subscript q and superscript (N) will be often omitted.

Definition 3.4.8 The (complex) *group algebra* $\mathbb{C}\mathbb{F}^{\times N}$ for space $\mathbb{F}^{\times N}$ is defined as the linear space of complex functions $G: \mathbf{x}\in\mathbb{F}^{\times N}\mapsto G(\mathbf{x})\in\mathbb{C}$ equipped by a complex involution (conjugation) and multiplication. Thus, we have four operations for functions $G(\mathbf{x})$; addition and scalar (complex) multiplication are standard (point-wise), with $(G+G')(\mathbf{x}) = G(\mathbf{x}) + G'(\mathbf{x})$ and $(aG)(\mathbf{x}) = aG(\mathbf{x})$, $G, G' \in \mathbb{C}\mathbb{F}^{\times N}$, $a \in \mathbb{C}$, $\mathbf{x}\in\mathbb{F}^{\times N}$. The involution is just the (point-wise) complex conjugation: $G^*(\mathbf{x}) = G(\mathbf{x})^*$; it is an idempotent operation, with $G^{**} = G$. However, the multiplication (denoted by $\star$) is a convolution:

$$(G\star G')(\mathbf{x}) = \sum_{\mathbf{y}\in\mathbb{F}^{\times N}} G(\mathbf{y})G'(\mathbf{x}-\mathbf{y}), \quad \mathbf{x}\in\mathbb{F}^{\times N}. \qquad (3.4.16)$$

This makes $\mathbb{C}\mathbb{F}^{\times N}$ a commutative ring and at the same time a (complex) linear space, of dimension $\dim \mathbb{C}\mathbb{F}^{\times N} = q^N$, with involution. (A set that is a commutative ring and a linear space is called an algebra.) The natural basis in $\mathbb{CF}^{\times N}$ is formed by Dirac's (or Kronecker's) delta-functions $\delta^{\mathbf{y}}$, with $\delta^{\mathbf{y}}(\mathbf{x}) = \mathbf{1}(\mathbf{x} = \mathbf{y})$, $\mathbf{x}, \mathbf{y} \in \mathcal{H}$.

If $\mathcal{X} \subseteq \mathbb{F}^{\times N}$ is a linear code, we set $G_{\mathcal{X}}(\mathbf{x}) = \mathbf{1}(\mathbf{x} \in \mathcal{X})$.

The multiplication rule (3.4.16) requires an explanation. If we rewrite the RHS in a symmetric form $\sum\limits_{\mathbf{y}, \mathbf{y}' \in \mathbb{F}^{\times N}: \mathbf{y} + \mathbf{y}' = \mathbf{x}} G(\mathbf{y}) G(\mathbf{y}')$ (which makes the commutativity of the $\star$-multiplication obvious) then there will be an analogy with the multiplication of polynomials. In fact, if $A(t) = a_0 + a_1 t + \cdots + a_{l-1} t^{l-1}$ and $A'(t) = a_0' + a_1' t + \cdots + a_{l'-1}' t^{l'-1}$ are two polynomials, with coefficient strings $(a_0, \ldots, a_{l-1})$ and $(a_0', \ldots, a_{l'-1}')$, then the product $B(t) = A(t)A'(t)$ has a string of coefficients $(b_0, \ldots, b_{l-1+l'-1})$ where $b_k = \sum\limits_{m, m' \geq 0: m + m' = k} a_m a_{m'}'$.

From this point of view, rule (3.4.16) is behind some polynomial-type multiplication. Polynomials of degree $\leq n - 1$ form of course a (complex) linear space of dimension n. However, they do not form a group (or even a semi-group). To make a group, we should affiliate inverse monomials $1/t$, $1/t^2$, and so on, and either consider infinite series or make an agreement that $t^n = 1$ (i.e. treat t as an element of a cyclic group, not a 'free' variable). Similar constructions can be done for polynomials of several variables, but there we have a variety of possible agreements on relations between variables.

Returning to our group algebra $\mathbb{C}\mathcal{H}$, we make the following steps:
(i) Produce a 'multiplicative version' of the Hamming group $\mathcal{H}$. That is, take a collection of 'formal' variables $t^{(x)}$ labelled by elements $x \in \mathcal{H}$ and postulate the rule $t^{(\mathbf{x})} t^{(\mathbf{x}')} = t^{(\mathbf{x} + \mathbf{x}')}$ for all $\mathbf{x}, \mathbf{x}' \in \mathbb{C}\mathcal{H}$.
(ii) Then consider the set $\mathbb{T}\mathcal{H}$ of all (complex) linear combinations $G = \sum_{\mathbf{x} \in \mathcal{H}} \gamma_{\mathbf{x}} t^{(\mathbf{x})}$ and introduce (ii1) the addition $G + G' = \sum_{\mathbf{x} \in \mathcal{H}} (\gamma_{\mathbf{x}} + \gamma_{\mathbf{x}}') t^{(\mathbf{x})}$ and (ii2) the scalar multiplication $aG = \sum_{\mathbf{x} \in \mathcal{H}} (a\gamma_{\mathbf{x}}) t^{(\mathbf{x})}$, $G, G' \in \mathbb{T}\mathcal{H}$, $a \in \mathbb{C}$. We again obtain a linear space of dimension q^N, with the basis formed by 'basic' combinations $t^{(\mathbf{x})}$, $\mathbf{x} \in \mathcal{H}$. Obviously, $\mathbb{T}\mathcal{H}$ and $\mathbb{C}\mathcal{H}$ are isomorphic as linear spaces, with $G \Longleftrightarrow g$.
(iii) Now remove brackets in $t^{(\mathbf{x})}$ (but keep the rule $t^{\mathbf{x}} t^{\mathbf{x}'} = t^{\mathbf{x} + \mathbf{x}'}$) and write $\sum_{\mathbf{x} \in \mathcal{H}} \gamma_{\mathbf{x}} t^{\mathbf{x}}$ as $g(t)$ thinking that this is a function (in fact, a 'polynomial') of some 'variable' t obeying the above rule. Finally, consider the polynomial multiplication $g(t) g'(t)$ in $\mathbb{T}\mathcal{H}$. Then $\mathbb{T}\mathcal{H}$ and $\mathbb{C}\mathcal{H}$ become isomorphic not only as linear spaces but also as rings, i.e. as algebras.

The above construction is very powerful and can be used for any group, not just for $\mathcal{H}_N$. Its power will be manifested in the derivation of the MacWilliams identity.

So, we will think of $\mathbb{C}\mathcal{H}$ as a set of functions

$$g(t) = \sum_{\mathbf{x} \in \mathcal{H}_n} \gamma_{\mathbf{x}} t^{\mathbf{x}} \qquad (3.4.17)$$

of a formal variable t obeying an 'exponentiation rule': $t^{\mathbf{x}+\mathbf{x}'} = t^{\mathbf{x}}t^{\mathbf{x}'}$, with addition and multiplication of formal polynomials.

In agreement with (3.4.17), for a linear code $\mathcal{X} \subset \mathcal{H}_n$ we set

$$g_{\mathcal{X}}(t) = \sum_{\mathbf{x} \in \mathcal{X}} t^{\mathbf{x}}; \tag{3.4.18}$$

$g_{\mathcal{X}}(t)$ is often called the generating function of $\mathcal{X}$.

Definition 3.4.3 admits a straightforward generalisation for any non-principal character $\chi \colon \mathbb{F} \to \mathbb{S}$. Note the similarity with the Fourier transform (and other types of popular transforms (viz. the Hadamard transform in the group theory)).

Definition 3.4.9   The *character transform* $g \mapsto \hat{g}$ of the group algebra $\mathbb{C}\mathcal{H}_n$ is defined by

$$\hat{g}(t) = \sum_{\mathbf{x} \in \mathcal{H}_n} X_{\mathbf{x}}(g)t^{\mathbf{x}}, \tag{3.4.19a}$$

where $g \sim (\gamma_{\mathbf{x}}, \mathbf{x} \in \mathcal{H}_n)$ and

$$X_{\mathbf{x}}(g) = \sum_{\mathbf{y} \in \mathcal{H}_n} \gamma_{\mathbf{y}} \chi(\mathbf{x} \cdot \mathbf{y}) \tag{3.4.19b}$$

and $\mathbf{x} \cdot \mathbf{y}$ is the dot-product $\sum_{1 \le j \le n} x_j y_j$ in $\mathcal{H}_n$.

Now define the *weight enumerator* of a group algebra element $g \in \mathbb{C}\mathcal{H}$ as a polynomial $W_g(s)$ in a variable s (which may be thought of as a complex variable):

$$W_g(s) = \sum_{\mathbf{x} \in \mathcal{H}} \gamma_{\mathbf{x}} s^{w(\mathbf{x})} = \sum_{k=0}^{n} \left[\sum_{\mathbf{x} : w(\mathbf{x}) = k} \gamma_{\mathbf{x}} \right] s^k = \sum_{0 \le k \le n} A_k s^k, \ s \in \mathbb{C}. \tag{3.4.20}$$

Here

$$A_k = \sum_{\mathbf{x} \in \mathcal{H} : w(\mathbf{x}) = k} \gamma_{\mathbf{x}}. \tag{3.4.21}$$

For a linear code $\mathcal{X}$, with generating function $g_{\mathcal{X}}(t)$ (see 3.4.18)), A_k gives the number of codewords of weight k:

$$A_k = \#\{\mathbf{x} \in \mathcal{X} : w(\mathbf{x}) = k\}. \tag{3.4.22}$$

The weight enumerator $W_{\hat{g}}(s)$ of the character transform $\hat{g}$ of $g \sim (\gamma_{\mathbf{x}}, \mathbf{x} \in \mathcal{H})$ is given by

$$W_{\hat{g}}(s) = \sum_{\mathbf{x} \in \mathcal{H}} X_{\mathbf{x}}(g) s^{w(\mathbf{x})} = \sum_{0 \le k \le n} \left[\sum_{\mathbf{x} : w(\mathbf{x}) = k} X_{\mathbf{x}}(g) \right] s^k = \sum_{k} \hat{A}_k s^k, \tag{3.4.23}$$

where

$$\hat{A}_k = \sum_{\mathbf{x} \in \mathcal{H}: w(\mathbf{x})=k} X_{\mathbf{x}}(g). \tag{3.4.24}$$

The 'abstract' MacWilliams identity is established in the following result.

Theorem 3.4.10   *We have*

$$W_{\hat{g}}(s) = (1 + (q-1)s)^n W_g \left(\frac{1-s}{1+(q-1)s} \right). \tag{3.4.25}$$

Proof   Basically coincides with that of Lemma 3.4.4.   □

Rewrite (3.4.25) in terms of coefficients A_k and $\hat{A}_k$:

$$\sum_{k=0}^{n} \hat{A}_k s^k = \sum_{k=0}^{n} A_k (1-s)^k (1 + (q-1)s)^{n-k} \tag{3.4.26}$$

and expand:

$$(1-s)^k (1+(q-1)s)^{n-k} = \sum_{i=0}^{n} K_i(k) s^i. \tag{3.4.27}$$

Here $K_i(k)(= K_i(k,n,q))$ is a *Kravchuk polynomial*: for all $i,k = 0,1,\dots,n$,

$$K_i(k) = \sum_{j=0 \vee (i+k-n)}^{i \wedge k} \binom{k}{j} \binom{n-k}{i-j} (-1)^j (q-1)^{i-j}, \tag{3.4.28}$$

$$0 \vee (i+k-n) = \max[0, i+k-n], \quad i \wedge k = \min[i,k].$$

Then

$$\sum_{0 \le k \le n} \hat{A}_k s^k = \sum_{0 \le k \le n} A_k \sum_{0 \le i \le n} K_i(k) s^i = \sum_{0 \le i \le n} \sum_{0 \le k \le n} A_k K_i(k) s^i$$

$$= \sum_{0 \le k \le n} \sum_{0 \le i \le n} A_i K_k(i) s^k,$$

i.e.

$$\hat{A}_k = \sum_{0 \le i \le n} A_i K_k(i). \tag{3.4.29}$$

Lemma 3.4.11   *For any (linear) code $\mathscr{X} \subseteq \mathcal{H}_n$, with generating function $g_{\mathscr{X}} \sim 1(x \in \mathscr{X})$, the character transform coefficients are related by*

$$X_{\mathbf{u}}(g_{\mathscr{X}}) = \# \mathscr{X} \mathbf{1}(\mathbf{u} \in \mathscr{X}^{\perp}) \tag{3.4.30}$$

and the character transform

$$\hat{g}_{\mathscr{X}} = \# \mathscr{X} g_{\mathscr{X}^{\perp}}. \tag{3.4.31}$$

Here, $\mathscr{X}^{\perp}$ is the dual code.

Proof By Lemma 3.4.2

$$X_{\mathbf{u}}(g\mathcal{X}) = X_{\mathbf{u}}\left(\sum_{\mathbf{x}\in\mathcal{X}} t^{\mathbf{x}}\right) = \sum_{\mathbf{y}\in\mathcal{X}} \chi(\mathbf{y}\cdot\mathbf{u}) = \#\mathcal{X}\,\mathbf{1}(\mathbf{u}\in\mathcal{X}^{\perp}).$$

In fact, the character $\mathbf{y}\in\mathcal{X}\mapsto\chi(\mathbf{y}\cdot\mathbf{u})$ is principal iff $\mathbf{u}\in\mathcal{X}^{\perp}$. Consequently,

$$\hat{g}(t) = \sum_{\mathbf{x}\in\mathcal{H}} X_{\mathbf{x}}(g\mathcal{X})t^{\mathbf{x}} = \sum_{\mathbf{x}\in\mathcal{H}} \#\mathcal{X}\,\mathbf{1}(\mathbf{x}\in\mathcal{X}^{\perp})t^{\mathbf{x}}$$

$$= \#\mathcal{X}\sum_{\mathbf{x}\in\mathcal{X}^{\perp}} t^{\mathbf{x}} = \#\mathcal{X}\,g_{\mathcal{X}^{\perp}}(t).$$

$\square$

Hence,

$$W_{\hat{g}\mathcal{X}}(s) = \#\mathcal{X}\,W_{g_{\mathcal{X}^{\perp}}}(s), \tag{3.4.32}$$

and we obtain the MacWilliams identity for linear codes:

Theorem 3.4.12 Let $\mathcal{X}\subset\mathcal{H}_n$ be a linear code, $\mathcal{X}^{\perp}$ its dual, and

$$W_{\mathcal{X}}(s) = \sum_{k=0}^{n} A_k s^k, \quad W_{\mathcal{X}^{\perp}}(s) = \sum_{k=0}^{n} A_k^{\perp} s^k \tag{3.4.33}$$

the w-enumerators for $\mathcal{X}$ and $\mathcal{X}^{\perp}$, respectively, with $A_k = \#\{\mathbf{x}\in\mathcal{X}: w(\mathbf{x}) = k\}$ and $\hat{A}_k = \#\{\mathbf{x}\in\mathcal{X}^{\perp}: w(\mathbf{x}) = k\}$. Then

$$W_{\mathcal{X}^{\perp}}(s) = \frac{1}{\#\mathcal{X}}(1+(q-1)s)^n W_{\mathcal{X}}\left(\frac{1-s}{1+(q-1)s}\right), \quad s\in\mathbb{C}, \tag{3.4.34}$$

or, equivalently,

$$A_k^{\perp} = \frac{1}{\#\mathcal{X}}\sum_{0\le i\le n} A_i K_k(i), \tag{3.4.35}$$

where $K_k(i)$ are Kravchuk polynomials (see (3.4.28)).

For a binary code, i.e. $q=2$, (3.4.34) takes the form (3.4.3). Sometimes the weight enumerators are defined as

$$W_{\mathcal{X}^{\perp}}(s,r) = \sum_k A_k s^k r^{n-k}. \tag{3.4.36}$$

Then the MacWilliams identity (3.4.33) takes the form

$$W_{\mathcal{X}^{\perp}}(s,r) = \frac{1}{\#\mathcal{X}}W_{\mathcal{X}}(r-s,r+(q-1)s). \tag{3.4.37}$$

The MacWilliams identity is a powerful result providing a deep insight into the structure of a (linear) code, particularly when the code is self-dual.

The MacWilliams identity helps to establish an interesting bound on linear codes called the linear programming (LP) bound. First, we discuss some immediate consequences of this identity. If $\mathscr{X} \subset \mathscr{H}_{N,q}$ is a code of size M, set

$$B_k = \frac{1}{M} \sharp\{(\mathbf{x},\mathbf{y}) \; : \; \mathbf{x},\mathbf{y} \in \mathscr{X}, \delta(\mathbf{x},\mathbf{y}) = k\},$$
$$k = 0,1,\ldots,N$$

(each pair $\mathbf{x},\mathbf{y}$ is counted two times). The numbers $B_0, B_1, \ldots, B_N$ form the *distance distribution* of code $\mathscr{X}$. The expression

$$B_{\mathscr{X}}(s) = \sum_{0 \le k \le N} B_k s^k \tag{3.4.38}$$

is called the distance enumerator of $\mathscr{X}$. Clearly, the w- and d-distributions of a linear code coincide. Furthermore we have

Lemma 3.4.13   *The d-enumerator of an $[N,M]$ code $\mathscr{X}$ coincides with the w-enumerator of the group algebra element*

$$h_{\mathscr{X}}(s) := \frac{1}{M} \zeta_{\mathscr{X}}(s) \zeta_{\mathscr{X}}(s^{-1})) \tag{3.4.39}$$

where the generating function of $\mathscr{X}$ is

$$\zeta_{\mathscr{X}}(s) = \sum_{\mathbf{x} \in \mathscr{X}} s^{\mathbf{x}}. \tag{3.4.40}$$

Proof   Using the notation $(s^{-1})^{\mathbf{x}}$, write

$$h_{\mathscr{X}}(s) = \frac{1}{M} \sum_{\mathbf{x} \in \mathscr{X}} s^{\mathbf{x}} \sum_{\mathbf{y} \in \mathscr{X}} s^{-\mathbf{y}} = \frac{1}{M} \sum_{\mathbf{x},\mathbf{y} \in \mathscr{X}} s^{\mathbf{x}-\mathbf{y}}$$

and hence

$$W_{h_{\mathscr{X}}}(s) = \frac{1}{M} \sum_{0 \le k \le N} \sum_{\mathbf{x},\mathbf{y} \in \mathscr{X}:} \mathbf{1}\big(w(\mathbf{x}-\mathbf{y}) = k\big) s^k = \sum_{0 \le k \le N} B_k s^k$$
$$= B_{\mathscr{X}}(s).$$

$\square$

Now by the MacWilliams identity, for a given non-trivial character χ and the corresponding transform $\zeta \mapsto \widehat{\zeta}$, we obtain

Theorem 3.4.14   *For $h_{\mathscr{X}}(s)$ as above, if $\widehat{h}_{\mathscr{X}}(s)$ is the character transform and $W_{\widehat{h}_{\mathscr{X}}}(s)$ its w-enumerator, with*

$$W_{\widehat{h}_{\mathscr{X}}}(s) = \sum_{0 \le k \le N} \widehat{B}_k s^k = \sum_{0 \le k \le N} \left[\sum_{w(\mathbf{x})=k} \chi_{\mathbf{x}}(h_{\mathscr{X}}) \right] s^k,$$

then

$$\widehat{B}_k = \sum_{0 \leq i \leq N} B_i K_k(i),$$

where $K_k(i)$ are Kravchuk polynomials.

The following assertion is straightforward.

Lemma 3.4.15 *The following identity holds: $\chi_{\mathbf{x}}(\zeta_{\mathscr{X}}(s^{-1})) = \overline{\chi_{\mathbf{x}}(\zeta_{\mathscr{X}}(s))}$, where the bar denotes the complex conjugate.*

With the help of Lemma 3.4.15, we can write

$$\chi_{\mathbf{x}}(h_{\mathscr{X}}(t)) = \frac{1}{M}\chi_{\mathbf{x}}(\zeta_{\mathscr{X}}(s)\zeta_{\mathscr{X}}(s^{-1})) = \frac{1}{M}\chi_{\mathbf{x}}(\zeta_{\mathscr{X}}(s))\chi_{\mathbf{x}}(\zeta_{\mathscr{X}}(s^{-1}))$$

$$= \frac{1}{M}\chi_{\mathbf{x}}(\zeta_{\mathscr{X}}(s))\overline{\chi_{\mathbf{x}}(\zeta_{\mathscr{X}}(s))} = \frac{1}{M}|\chi_{\mathbf{x}}(\zeta_{\mathscr{X}}(s))|^2,$$

and so,

$$\widehat{B}_k = \sum_{\mathbf{x}:w(\mathbf{x})=k} \chi_{\mathbf{x}}(h_{\mathscr{X}}) = \frac{1}{M} \sum_{w(\mathbf{x})=k} |\chi_{\mathbf{x}}(\zeta_{\mathscr{X}})|^2 \geq 0.$$

Thus:

Theorem 3.4.16 *For all $[N,M]$ codes $\mathscr{X}$ and $k = 0,\ldots,N$,*

$$\sum_{0 \leq i \leq N} B_i K_k(i) \geq 0. \tag{3.4.41}$$

Now counting the number of pairs $(\mathbf{x},\mathbf{y}) \in \mathscr{X} \times \mathscr{X}$:

$$\sum_{0 \leq i \leq N} B_i = M^2$$

or

$$\sum_{0 \leq i \leq N} E_i = M, \text{ with } E_i = \frac{1}{M}B_i \tag{3.4.42}$$

(sometimes $E_0, E_1,\ldots,E_N$ are called the d-distribution of $\mathscr{X}$). Then, by (3.4.41)–(3.4.42),

$$\sum_{0 \leq i \leq N} E_i K_k(i) \geq 0.$$

In addition, by definition, $E_i \geq 0, 0 \leq i \leq N$, and $E_0 = 1$ and $E_i = 0, 1 \leq i < d$.

Proof Let ω be a primitive qth root of unity and $\mathbf{x} \in \mathbb{F}_q^N$ be a fixed word of weight i. Then

$$\sum_{\mathbf{y} \in \mathbb{F}_q^N : w(\mathbf{y})=k} \omega^{\langle \mathbf{x},\mathbf{y} \rangle} = K_k(i). \tag{3.4.43}$$

Indeed, we may assume that $\mathbf{x} = x_1 x_2 \dots x_i 0 \dots 0$ where the coordinates x_i are not 0. Let D be a set of words that have their non-zero coordinates in a given set of k positions. Suppose that exactly j positions $h_1, \dots, h_k$ belong to $[0, i]$ and $k - j$ positions belong to $[i + 1, N]$. For such, a set could be selected in $\binom{i}{j} \binom{N-i}{k-j}$ choices. Then

$$\sum_{\mathbf{y} \in D} \omega^{\langle \mathbf{x}, \mathbf{y} \rangle} = \sum_{y_{h_1} \in \mathbb{F}_q^*} \dots \sum_{y_{h_k} \in \mathbb{F}_q^*} \omega^{x_{h_1} y_{h_1} + \dots + x_{h_k} y_{h_k}}$$

$$= (q-1)^{k-j} \prod_{i=1}^{j} \sum_{y \in \mathbb{F}_q^*} \omega^{x_{h_i} y} = (-1)^j (q-1)^{k-j}.$$

Hence,

$$M \sum_{i=0}^{N} B_i K_k(i) = \sum_{i=0}^{N} \sum_{\mathbf{x}, \mathbf{y} \in \mathscr{X} : \delta(\mathbf{x}, \mathbf{y}) = i} \sum_{\underline{z} \in \mathbb{F}_q^N : w(\underline{z}) = k} \omega^{\langle \mathbf{x} - \mathbf{y}, \underline{z} \rangle}$$

$$= \sum_{\underline{z} \in \mathbb{F}_q^N : w(\underline{z}) = k} \left| \sum_{\mathbf{x} \in \mathscr{X}} \omega^{\langle \mathbf{x}, \underline{z} \rangle} \right|^2 \geq 0.$$

$\square$

This leads us to the so-called linear programming (LP) bound stated in Theorem 3.4.17 below.

Theorem 3.4.17 (The LP bound) *The following inequality holds:*

$$M_q^*(N, d) \leq \max \left[\sum_{0 \leq i \leq N} \tilde{E}_i : \tilde{E}_i \geq 0, \tilde{E}_0 = 1, \tilde{E}_i = 0 \text{ for } 1 \leq i < d \right.$$

$$\left. \text{and } \sum_{0 \leq i \leq N} \tilde{E}_i K_k(i) \geq 0 \text{ for } 0 \leq k \leq N \right]. \quad (3.4.44)$$

For $q = 2$, the LP bound will be slightly improved in Theorem 3.4.19. First, an auxiliary result whose proof is straightforward and left as an exercise.

Lemma 3.4.18

(a) *If there exists a binary $[N, M, d]$ code, with d even, then there exists a binary $[N, M, d]$ code where any codeword has even weight, and so all distances are even. So, if $q = 2$ and d is even, we may assume that $E_i = 0$ for all odd values of i.*

(b) *For $q = 2$,*

$$K_i(2k) = K_{N-i}(2k).$$

Hence, for d even, as we can assume that $E_{2i+1} = 0$, the constraint in (3.4.44) need only be considered for $k = 0, \ldots, [N/2]$.

(c) $K_0(i) = 1$ for all i, and thus the bound $\sum\limits_{0 \leq i \leq N} \tilde{E}_i K_0(i) \geq 0$ follows from $\tilde{E}_i \geq 0$.

Lemma 3.4.18 directly implies

Theorem 3.4.19 (The LP for $q = 2$) *If d is even then*

$$M_2^*(N,d) \leq \max \left[\sum_{0 \leq i \leq N} \tilde{E}_i : \tilde{E}_i \geq 0, \tilde{E}_0 = 1, \tilde{E}_i = 0 \text{ for } 1 < i < d, \right.$$

$$\tilde{E}_i = 0 \text{ for } i \text{ odd, and } \binom{N}{k} + \sum_{d \leq i \leq N} \tilde{E}_i K_k(i) \geq 0 \qquad (3.4.45)$$

$$\left. \text{for } k = 1, \ldots, \left\lfloor \frac{N}{2} \right\rfloor \right].$$

Since $M_2^*(N, 2t+1) = M_2^*(N+1, 2t+2)$, Theorem 3.4.19 provides a useful bound also when d is odd. We will explore the MacWilliams identity further on.

The LP bound represents a rather universal tool in the theory of codes. For instance, the Singleton, Hamming and Plotkin bounds can all be derived from the LP bound. However, we will not exploit this avenue in detail.

Worked Example 3.4.20 *For positive integers N and $d \leq N$, let*

$$f(x) = 1 + \sum_{j=1}^{N} f_j K_j(x)$$

be a polynomial such that $f_j \geq 0, 1 \leq j \leq N$ and $f(i) \leq 0$ for $d \leq i \leq N$. Prove that

$$M_q^*(N,d) \leq f(0). \qquad (3.4.46)$$

Derive the Singleton bound from (3.4.46).

Solution Let $M = M_q^*(N,d)$ and $\mathcal{X}$ be a q-ary $[N,M]$ code with the distance distribution $B_i(\mathcal{X})$, $i = 0, \ldots, N$. The condition $f(i) \leq 0$ for $d \leq i \leq N$ implies

$\sum_{j=d}^{N} B_j(\mathcal{X}) f(j) \le 0$. Using the LP bound (3.4.45) for $k = 0$ obtain $K_i(0) \ge$
$-\sum_{j=d}^{N} B_j(\mathcal{X}) K_i(j)$. Hence,

$$f(0) = 1 + \sum_{j=1}^{N} f_j K_j(0)$$

$$\ge 1 - \sum_{k=1}^{N} f_k \sum_{i=d}^{N} B_i(\mathcal{X}) K_k(i)$$

$$= 1 - \sum_{i=d}^{N} B_i(\mathcal{X}) \sum_{k=1}^{N} f_k K_k(i)$$

$$= 1 - \sum_{i=d}^{N} B_i(\mathcal{X})(f(i) - 1)$$

$$\ge 1 + \sum_{i=d}^{N} B_i(\mathcal{X})$$

$$= M = M_q^*(N, d).$$

To obtain the Singleton bound select

$$f(x) = q^{N-d+1} \prod_{j=d}^{N} \left(1 - \frac{x}{j}\right).$$

Then by the identity

$$\sum_{i=0}^{j} \binom{N-i}{N-j} K_i(k) = q^j \binom{N-k}{j}$$

with $j = d - 1$ we have that

$$f_k = \frac{1}{q^N} \sum_{i=0}^{N} f(i) K_i(k)$$

$$= \frac{1}{q^{d-1}} \sum_{i=0}^{d-1} \binom{N-i}{N-d+1} K_i(k) / \binom{N}{d-1}$$

$$= \binom{N-k}{d-1} / \binom{N}{d-1} \ge 0.$$

Here we use the identity

$$\sum_{k=0}^{j} \binom{N-k}{N-j} K_k(x) = q^j \binom{N-x}{j}. \tag{3.4.47}$$

Clearly, $f(i) = 0$ for $d \le i \le N$. Hence, $R_q(N, d) \le f(0) = q^{N-d+1}$. In a similar manner Hamming's and Plotkin's bounds may be derived as well, cf. [97]. $\qquad \square$

Worked Example 3.4.21   *Using the linear programming bound, prove that* $M_2^*(13,5) = M_2^*(14,6) \leq 64$. *Compare it with the Elias bound.* [*Hint:* $E_6 = 42$, $E_8 = 7$, $E_{10} = 14$, $E_{12} = E_{14} = 0$. *You may need a computer to get the solution.*]

Solution The LP bound for linear codes reads

$$M_2^*(N,d) = \max_{0 \leq i \leq N} \sum E_i$$

$$\text{subject to } E_i \geq 0, \ E_0 = 1, \ E_j = 0 \text{ for } 1 \leq j < d,$$
$$E_i = 0 \text{ for } j \text{ odd, and}$$
$$\binom{N}{k} + \sum_{\substack{d \leq i \leq N \\ i \text{ even}}} E_i K_k(i) \geq 0 \text{ for } k = 1, \ldots, \left\lfloor \frac{N}{2} \right\rfloor.$$

For $N = 14$, $d = 6$, the constraints are

$$E_0 = 1, \ E_1 = E_2 = E_3 = E_4 = E_5 = E_7 = E_9 = E_{11} = E_{13} = 0,$$
$$E_6, E_8, E_{10}, E_{12}, E_{14} \geq 0,$$
$$14 + 2E_6 - 2E_8 - 6E_{10} - 10E_{12} - 14E_{14} \geq 0,$$
$$91 - 5E_6 - 5E_8 + 11E_{10} + 43E_{12} + 91E_{14} \geq 0,$$
$$364 - 12E_6 + 12E_8 + 4E_{10} - 100E_{12} - 364E_{14} \geq 0,$$
$$1001 + 9E_6 + 9E_8 - 39E_{10} + 121E_{12} + 1001E_{14} \geq 0,$$
$$2002 + 30E_6 - 30E_8 + 38E_{10} - 22E_{12} - 2002E_{14} \geq 0,$$
$$3003 - 5E_6 - 5E_8 + 27E_{10} - 165E_{12} + 3003E_{14} \geq 0,$$
$$3432 - 40E_6 + 40E_8 - 72E_{10} + 264E_{12} - 3432E_{14} \geq 0,$$

and the maximiser of the objective function $S = E_6 + E_8 + E_{10} + E_{12} + E_{14}$ is

$$E_6 = 42, \ E_8 = 7, \ E_{10} - 14, \ E_{12} - E_{14} - 0,$$

with $S = 63$, $E_0 + S = 1 + 63 = 64$. So, the LP bound yields

$$M_2^*(13,5) = M_2^*(14,6) \leq 64.$$

Note that the bound is sharp as a $[13, 64, 5]$ binary code actually exists. Compare the LP bound with the Hamming bound:

$$M_2^*(13,5) \leq 2^{13}/(1 + 13 + 13 \cdot 6) = 2^{13}/92 = 2^{11}/23,$$

i.e.

$$M_2^*(13,5) \leq 91.$$

Next, the Singleton bound gives $k \leq 13 - 5 - 1 = 7$,

$$M_2^*(13,5) \leq 2^7 = 128.$$

It is also interesting to see what the Elias bound gives:

$$M_2^*(13,5) \leq \frac{65/2}{s^2 - 13s + 65/2} \; \frac{2^{13}}{1 + 13 + \cdots + \binom{13}{5}}$$

for all $s < 13$ such that $s^2 - 13s + 65/2 > 0$.

Substituting $s = 2$ yields $s^2 - 13s + 65/2 = 4 - 26 + 65/2 = 21/2 > 0$ and

$$M_2^*(13,5) \leq \frac{65}{21} 2^{13} / (1 + 13 + 13 \cdot 6) = 2.33277 \times 10^6:$$

not good enough. Next, $s = 3$ yields $s^2 - 13s + 65/2 = 9 - 39 + 65/2 = 5/2 > 0$ and

$$M_2^*(13,5) \leq \frac{65}{5} 2^{13} / (1 + 13 + 13 \cdot 6 + 13 \cdot 2 \cdot 5) = 13 \times \frac{2^{12}}{111} \geq \frac{13}{66} 2^{11}:$$

not as good as Hamming's. Finally, observe that $4^2 - 13 \times 4 + 65/2 < 0$, and the procedure stops. $\qquad\qquad\square$

3.5 Asymptotically good codes

In this section we briefly discuss some families of codes where the number of corrected errors gives a non-zero fraction of the codeword length. For more details, see [54], [71], [131].

Definition 3.5.1　A sequence of $[N_i, k_i, d_i]$ codes with $N_i \to \infty$ is said to be *asymptotically good* if k_i/N_i and d_i/N_i are both bounded away from 0.

Theorem 3.3.25 showed that there is no asymptotically good sequence of primitive BCH codes (in fact, there is no asymptotically good sequence of BCH codes of any form). Theoretically, an elegant way to produce an asymptotically good family are the so-called Justensen codes. As the first attempt to define a good code take $0 \neq \alpha \in \mathbb{F}_{2^m} \simeq \mathbb{F}_2^m$ and define the set

$$\mathcal{X}_\alpha = \{(\mathbf{a}, \alpha\mathbf{a}) : \mathbf{a} \in \mathbb{F}_2^m\}. \qquad (3.5.1)$$

Then $\mathcal{X}_\alpha$ is a $[2m, m]$ linear code and has information rate $1/2$. We can recover α from any non-zero codeword $(\mathbf{a}, \mathbf{b}) \in \mathcal{X}_\alpha$, as $\alpha = \mathbf{b}\mathbf{a}^{-1}$ (division in $\mathbb{F}_{2^m}$). Hence, if $\alpha \neq \alpha'$ then $\mathcal{X}_\alpha \cap \mathcal{X}_{\alpha'} = \{0\}$.

Now, given $\lambda = \lambda_m \in (0, 1/2]$, we want to find $\alpha = \alpha_m$ such that code $\mathcal{X}_\alpha$ has minimum weight $\geq 2m\lambda$. Since a non-zero binary $(2m)$-word can enter at most one of the $\mathcal{X}_\alpha$'s, we can find such α if the number of the non-zero $(2m)$-words of weight $< 2m\lambda$ is $< 2^m - 1$, the number of distinct codes $\mathcal{X}_\alpha$. That is, we can manage if

$$\sum_{1 \leq i \leq 2m\lambda - 1} \binom{2m}{i} < 2^m - 1$$

or even better, $\sum_{1 \leq i \leq 2m\lambda} \binom{2m}{i} < 2^m - 1$. Now use the following:

Lemma 3.5.2 *For $0 \leq \lambda \leq 1/2$,*

$$\sum_{0 \leq k \leq \lfloor \lambda N \rfloor} \binom{N}{k} \leq 2^{N\eta(\lambda)}, \tag{3.5.2}$$

where $\eta(\lambda)$ is the binary entropy.

Proof Observe that (3.5.2) holds for $\lambda = 0$ (here both sides are equal to 1) and $\lambda = 1/2$ (where the RHS equals 2^N). So we will assume that $0 < \lambda < 1/2$. Consider a random variable ξ with the binomial distribution

$$\mathbb{P}(\xi = k) = \binom{N}{k}(1/2)^N, 0 \leq k \leq N.$$

Given $t \in \mathbb{R}_+$, use the following Chebyshev-type inequality:

$$\sum_{0 \leq k \leq \lambda N} \binom{N}{k}\left(\frac{1}{2}\right)^N = \mathbb{P}(\xi \leq \lambda N)$$

$$= \mathbb{P}\left(\exp(-t\xi) \geq e^{-\lambda Nt}\right)$$

$$\leq e^{-\lambda Nt}\mathbb{E}e^{-t\xi}$$

$$= e^{-\lambda Nt}\left(\frac{1}{2} + \frac{1}{2}e^{-t}\right)^N. \tag{3.5.3}$$

Minimise the RHS of (3.5.3) in $x = e^{-t}$ for $t > 0$, i.e. for $0 < x < 1$. This yields the minimiser $e^{-t} = \lambda/(1-\lambda)$ and the minimal value

$$\left(\frac{\lambda}{1-\lambda}\right)^{-\lambda N}\left(\frac{1}{2}\right)^N\left(1 + \frac{\lambda}{1-\lambda}\right)^N$$

$$= \lambda^{-\lambda N}\mu^{-\mu N}\left(\frac{1}{2}\right)^N = 2^{N\eta(\lambda)}\left(\frac{1}{2}\right)^N,$$

with $\mu = 1 - \lambda$. Hence, (3.5.2) implies

$$\sum_{0 \leq k \leq \lambda N} \binom{N}{k}\left(\frac{1}{2}\right)^N \leq 2^{N\eta(\lambda)}\left(\frac{1}{2}\right)^N.$$

□

Owing to Lemma 3.5.2, inequality (3.5.1) occurs when

$$2^{2m\eta(\lambda)} < 2^m - 1. \tag{3.5.4}$$

Now if, for example,

$$\lambda = \lambda_m = \eta^{-1}\left(1/2 - \frac{1}{\log m}\right)$$

(with $0 < \lambda < \frac{1}{2}$), bound (3.5.4) becomes $2^{m-2m/\log m} < 2^m - 1$ which is true for m large enough. And $\lambda_m \to h^{-1}(1/2) > 0$, as $m \to \infty$. Here and below, η^{-1} is the inverse function to $\lambda \in (0, 1/2] \mapsto \eta(\lambda)$. In the code (3.5.1) with a fixed α the information rate is $1/2$ but one cannot guarantee that $d/2m$ is bounded away from 0. Moreover, there is no effective way of finding a proper $\alpha = \alpha_m$. However, in 1972, Justensen [81] showed how to obtain a good sequence of codes cleverly using the concatenation of words from an RS code.

More precisely, consider a binary $(k_1 k_2)$-word $\underline{a}$ organised as k_1 separate k_2-words: $\underline{a} = a^{(0)} a^{(1)} \ldots a^{(k_1-1)}$. Pictorially,

$$a^{(i)} \in \mathbb{F}_{2^{k_2}}, \ 0 \le i \le k_1 - 1.$$

We fix an $[N_1, k_1, d_1]$ code $\mathscr{X}_1$ over $\mathbb{F}_{2^{k_2}}$ called an outer code: $\mathscr{X}_1 \subset \mathbb{F}_{2^{k_2}}^{N_1}$. Then string $\underline{a}$ is encoded into a codeword $\underline{c} = c_0 c_1 \ldots c_{N_1-1} \in \mathscr{X}_1$. Next, each $c_i \in \mathbb{F}_{2^{k_2}}$ is encoded by a codeword b_i from an $[N_2, k_2, d_2]$ code $\mathscr{X}_2$ over $\mathbb{F}_2$, called an inner code. The result is a string $\underline{b} = b^{(0)} \ldots b^{(N_1-1)} \in \mathbb{F}_q^{N_1 N_2}$ of length $N_1 N_2$:

$$b^{(i)} \in \mathbb{F}_{2^{N_2}}, \ 0 \le i \le N_1 - 1.$$

The encoding is represented by the diagram:

input: a $(k_1 k_2)$ string $\underline{a}$,   **output:** an $(N_1 N_2)$ codeword $\underline{b}$.

Observe that different symbols c_i can be encoded by means of different inner codes. Let the outer code $\mathscr{X}_1$ be a $[2^m - 1, k, d]$ RS code $\mathscr{X}^{\text{RS}}$ over $\mathbb{F}_{2^m}$. Write a binary $(k2^m)$-word $\underline{a}$ as a concatenation $a^{(0)} \ldots a^{(k-1)}$, with $a^{(i)} \in \mathbb{F}_{2^m}$. Encoding $\underline{a}$ using $\mathscr{X}^{\text{RS}}$ gives a codeword $\underline{c} = c_0 \ldots c_{N-1}$, with $N = 2^m - 1$ and $c_i \in \mathbb{F}_{2^m}$. Let β be a primitive element in $\mathbb{F}_{2^m}$. Then for all $j = 0, \ldots, N-1 = 2^m - 2$, consider the inner code

$$\mathscr{X}^{(j)} = \left\{ (c, \beta^j c) : c \in \mathbb{F}_{2^m} \right\}. \tag{3.5.5}$$

The resulting codeword (a 'super-codeword') is

$$\underline{b} = (c_0, c_0)(c_1, \beta c_1)(c_2, \beta^2 c_2) \ldots (c_{N-1}, \beta^{N-1} c_{N-1}). \tag{3.5.6}$$

Definition 3.5.3 The *Justensen code* $\mathcal{X}_{m,k}^{\mathrm{Ju}}$ is the collection of binary super-words $\underline{b}$ obtained as above, with the $[2^m - 1, k, d]$ RS code as an outer code $\mathcal{X}_1$, and $\mathcal{X}^{(j)}$ (see (3.5.6)) as the inner codes, where $0 \leq j \leq 2^m - 2$. Code $\mathcal{X}_{m,k}^{\mathrm{Ju}}$ has length $2m(2^m - 1)$, rank mk and hence rate $\dfrac{k}{2(2^m - 1)} < 1/2$.

A convenient parameter describing $\mathcal{X}_{m,k}^{\mathrm{Ju}}$ is $N = 2^m - 1$, the length of the outer RS code. We want to construct a sequence $\mathcal{X}_{m,k}^{\mathrm{Ju}}$ with $N \to \infty$, but $k/(2m(2^m - 1))$ and $d/(2m(2^m - 1))$ bounded away from 0. Fix $R_0 \in (0, 1/2)$ and choose a sequence of outer RS codes $\mathcal{X}_N^{\mathrm{RS}}$ of length N, with $N = 2^m - 1$ and $k = [2NR_0]$. Then the rate of $\mathcal{X}_{m,k}^{\mathrm{Ju}}$ is $k/(2N) \geq R_0$.

Now consider the minimum weight

$$w\left(\mathcal{X}_{m,k}^{\mathrm{Ju}}\right) = \min \left[w(\mathbf{x}) : \mathbf{x} \in \mathcal{X}_{m,k}^{\mathrm{Ju}}, \mathbf{x} \neq 0\right] \left(= d\left(\mathcal{X}_{m,k}^{\mathrm{Ju}}\right)\right). \tag{3.5.7}$$

For any fixed m, if the outer RS code $\mathcal{X}_N^{\mathrm{RS}}$, $N = 2^m - 1$, has minimum weight d then any super-codeword $\underline{b} = (c_0, c_0)(c_1, \beta c_1) \ldots (c_{N-1}, \beta^{N-1} c_{N-1}) \in \mathcal{X}_{m,k}^{\mathrm{Ju}}$ has $\geq d$ non-zero first components $c_0, \ldots, c_{N-1}$. Furthermore, any two inner codes among $\mathcal{X}^{(0)}, \mathcal{X}^{(1)}, \ldots, \mathcal{X}^{(N-1)}$ have only 0 in common. So, the corresponding d ordered pairs, being from different codes, must be distinct. That is, super-codeword $\underline{b}$ has $\geq d$ distinct non-zero binary $(2m)$-strings.

Next, the weight of super-codeword $\underline{b} \in \mathcal{X}_{m,k}^{\mathrm{Ju}}$ is at least the sum of the weights of the above d distinct non-zero binary $(2m)$-strings. So, we need to establish a lower bound on such a sum. Note that

$$d = N - k + 1 = N\left(1 - \frac{k-1}{N}\right) \geq N(1 - 2R_0).$$

Hence, a super-codeword $\underline{b} \in \mathcal{X}_{m,k}^{\mathrm{Ju}}$ has at least $N(1 - 2R_0)$ distinct non-zero binary $(2m)$-strings.

Lemma 3.5.4 *The sum of the weights of any $N(1 - 2R_0)$ distinct non-zero binary $(2m)$-strings is*

$$\geq 2mN(1 - 2R_0) \left(\eta^{-1}\left(\frac{1}{2}\right) - o(1)\right). \tag{3.5.8}$$

Proof By Lemma 3.5.2, for any $\lambda \in [0, 1/2]$, the number of non-zero binary $(2m)$-strings of weight $\leq 2m\lambda$ is

$$\leq \sum_{1 \leq i \leq 2m\lambda} \binom{2m}{i} \leq 2^{2m\eta(\lambda)}.$$

Discarding these lightweight strings, the total weight is

$$\geq 2m\lambda\left(N(1-2R_0) - 2^{2mn(\lambda)}\right) = 2mN\lambda(1-2R_0)\left(1 - \frac{2^{mn(\lambda)}}{N(1-2R_0)}\right).$$

Select $\lambda_m = \eta^{-1}\left(\frac{1}{2} - \frac{1}{\log(2m)}\right) \in (0, 1/2)$. Then $\lambda_m \to \eta^{-1}(1/2)$, as h^{-1} is continuous on $[0, 1/2]$. So,

$$\lambda_m = \eta^{-1}\left(\frac{1}{2} - \frac{1}{\log(2m)}\right) = \eta^{-1}\left(\frac{1}{2}\right) - o(1).$$

Since $N = 2^m - 1$, we have that as $m \to \infty$, $N \to \infty$, and

$$\frac{2^{mn(\lambda)}}{N(1-2R_0)} = \frac{1}{1-2R_0}\frac{2^{m-2m/\log(2m)}}{2^m-1}$$

$$= \frac{1}{1-2R_0}\frac{2^m}{2^m-1}\frac{1}{2^{2m/\log(2m)}} \to 0.$$

So the total weight of the $N(1-2R_0)$ distinct $(2m)$-strings is bounded below by

$$2mN(1-2R_0)\left(\eta^{-1}(1/2) - o(1)\right)(1-o(1)) = 2mN(1-2R_0)\left(\eta^{-1}(1/2) - o(1)\right).$$

Thus the result follows.   □

Lemma 3.5.4 demonstrates that $\mathcal{X}_{m,k}^{\mathrm{Ju}}$ has

$$w\left(\mathcal{X}_{m,k}^{\mathrm{Ju}}\right) \geq 2mN(1-2R_0)\left(\eta^{-1}\left(\frac{1}{2}\right) - o(1)\right). \tag{3.5.9}$$

Then

$$\frac{w\left(\mathcal{X}_{m,k}^{\mathrm{Ju}}\right)}{\text{length}\left(\mathcal{X}_{m,k}^{\mathrm{Ju}}\right)} \geq (1-2R_0)(\eta^{-1}(1/2) - o(1)) \to (1-2R_0)\eta^{-1}(1/2)$$

$$\approx c(1-2R_0) > 0.$$

So, the sequence $\mathcal{X}_{m,k}^{\mathrm{Ju}}$ with $k = [2NR_0]$, $N = 2^m - 1$ and a fixed $0 < R_0 < 1/2$, has information rate $\geq R_0 > 0$ and

$$\frac{w\left(\mathcal{X}_{m,k}^{\mathrm{Ju}}\right)}{\text{length}\left(\mathcal{X}_{m,k}^{\mathrm{Ju}}\right)} \to c(1-2R_0) > 0, \quad c = \eta^{-1}(1/2) > 0.3. \tag{3.5.10}$$

In the construction, $R_0 \in (0, 1/2)$. However, by truncating one can achieve any given rate $R_0 \in (0, 1)$; see [110].

The next family of codes to be discussed in this section is formed by *alternant* codes. Alternant codes are a generalisation of BCH (though in general not cyclic). Like Justesen codes, alternant codes also form an asymptotically good family.

Let M be a $(r \times n)$ matrix over field $\mathbb{F}_{q^m}$:

$$M = \begin{pmatrix} c_{11} & \cdots & c_{1n} \\ \vdots & \ddots & \vdots \\ c_{r1} & \cdots & c_{rn} \end{pmatrix}.$$

As before, each c_{ij} can be written as $\vec{c_{ij}} \in (\mathbb{F}_q)^m$, a column vector of length m over $\mathbb{F}_q$. That is, we can think of M as an $(mr \times n)$ matrix over $\mathbb{F}_q$ (denoted again by M).

Given elements $a_1, \ldots, a_n \in \mathbb{F}_{q^m}$, we have

$$M \begin{pmatrix} a_1 \\ \vdots \\ a_n \end{pmatrix} = \begin{pmatrix} c_{11} & \cdots & c_{1n} \\ \vdots & \ddots & \vdots \\ c_{r1} & \cdots & c_{rn} \end{pmatrix} \begin{pmatrix} a_1 \\ \vdots \\ a_n \end{pmatrix} = \begin{pmatrix} \sum\limits_{1 \le j \le n} a_j c_{ij} \\ \vdots \\ \sum\limits_{1 \le j \le n} a_j c_{rj} \end{pmatrix}.$$

Furthermore, if $b \in \mathbb{F}_q$ and $c, d \in \mathbb{F}_{q^m}$ then $b\vec{c} = \overrightarrow{bc}$ and $\vec{c} + \vec{d} = \overrightarrow{(c+d)}$. Thus, if $a_1, \ldots, a_n \in \mathbb{F}_q$, then

$$M \begin{pmatrix} a_1 \\ \vdots \\ a_n \end{pmatrix} = \begin{pmatrix} \vec{c_{11}} & \cdots & \vec{c_{1n}} \\ \vdots & \ddots & \vdots \\ \vec{c_{r1}} & \cdots & \vec{c_{rn}} \end{pmatrix} \begin{pmatrix} a_1 \\ \vdots \\ a_n \end{pmatrix} = \begin{pmatrix} \sum\limits_{1 \le j \le n} \overrightarrow{a_i c_{ij}} \\ \vdots \\ \sum\limits_{1 \le j \le n} \overrightarrow{a_i c_{rj}} \end{pmatrix}.$$

So, if the columns of M are linearly independent as r-vectors over $\mathbb{F}_{q^m}$, they are also linearly independent as (rm)-vectors over $\mathbb{F}_q$. That is, the columns of M are linearly independent over $\mathbb{F}_q$.

Recall that if ω is a primitive $(n, \mathbb{F}_{q^m})$ root of unity and $\delta \ge 2$ then the $n \times (m\delta)$ Vandermonde matrix over $\mathbb{F}_q$

$$H^{\mathrm{T}} = \begin{pmatrix} \vec{e} & \vec{e} & \cdots & \vec{e} \\ \vec{\omega} & \vec{\omega}^2 & \cdots & \vec{\omega}^{\delta-1} \\ \vec{\omega}^2 & \vec{\omega}^4 & \cdots & \vec{\omega}^{2(\delta-1)} \\ & & \cdots & \\ \vec{\omega}^{n-1} & \vec{\omega}^{2(n-1)} & \cdots & \vec{\omega}^{(\delta-1)(n-1)} \end{pmatrix}$$

checks a narrow-sense BCH code $\mathscr{X}_{q,n,\omega,\delta}^{\mathrm{BCH}}$ (a proper parity-check matrix emerges after column purging). Generalise it by taking an $n \times r$ matrix over $\mathbb{F}_{q^m}$

$$A = \begin{pmatrix} h_1 & h_1 \alpha_1 & \cdots & h_1 \alpha_1^{r-2} & h_1 \alpha_1^{r-1} \\ h_2 & h_2 \alpha_2 & \cdots & h_2 \alpha_2^{r-2} & h_2 \alpha_2^{r-1} \\ \vdots & \vdots & \ddots & \vdots & \vdots \\ h_n & h_n \alpha_n & \cdots & h_n \alpha_n^{r-2} & h_n \alpha_n^{r-1} \end{pmatrix}, \tag{3.5.11}$$

or its $n \times (mr)$ version over $\mathbb{F}_q$:

$$\overrightarrow{A} = \begin{pmatrix} \overrightarrow{h_1} & \overrightarrow{h_1 \alpha_1} & \cdots & \overrightarrow{h_1 \alpha_1}^{r-2} & \overrightarrow{h_1 \alpha_1}^{r-1} \\ \overrightarrow{h_2} & \overrightarrow{h_2 \alpha_2} & \cdots & \overrightarrow{h_2 \alpha_2}^{r-2} & \overrightarrow{h_2 \alpha_2}^{r-1} \\ \vdots & \vdots & \ddots & \vdots & \vdots \\ \overrightarrow{h_n} & \overrightarrow{h_n \alpha_n} & \cdots & \overrightarrow{h_n \alpha_n}^{r-2} & \overrightarrow{h_n \alpha_n}^{r-1} \end{pmatrix}. \tag{3.5.12}$$

Here $r < n$, $h_1, \ldots, h_n$ are non-zero elements and $\alpha_1, \ldots, \alpha_n$ are distinct elements from $\mathbb{F}_q$.

Note that any r rows of A in (3.5.11) form a square sub-matrix K that is similar to Vandermonde's. It has a non-zero determinant and hence any r rows of A are linearly independent over $\mathbb{F}_{q^m}$ and hence over $\mathbb{F}_q$. Also the columns of A in (3.5.11) are linearly independent over $\mathbb{F}_{q^m}$. However, columns of $\overrightarrow{A}$ in (3.5.12) can be linearly dependent and purging such columns may be required to produce a 'genuine' parity-check matrix H.

Definition 3.5.5 Let $\underline{\alpha} = (\alpha_1, \ldots, \alpha_n)$ and $\underline{h} = (h_1, \ldots, h_n)$ where $\alpha_1, \ldots, \alpha_n$ are distinct and $h_1, \ldots, h_n$ non-zero elements of $\mathbb{F}_{q^m}$. Given $r < n$, an alternant code $\mathscr{X}_{\underline{\alpha},\underline{h}}^{\mathrm{Alt}}$ is the kernel of the $n \times (rm)$ matrix A in (3.5.12).

Theorem 3.5.6 $\mathscr{X}_{\underline{\alpha},\underline{h}}^{\mathrm{Alt}}$ *has length* n, *rank* k *satisfying* $n - mr \leq k \leq n - r$ *and minimum distance* $d\left(\mathscr{X}_{\underline{\alpha},\underline{h}}^{\mathrm{Alt}}\right) \geq r + 1$.

We see that the alternant codes are indeed generalisations of BCH. The main outcome of the theory of alternant codes is the following Theorem 3.5.7 (not to be proven here).

Theorem 3.5.7 *There exist arbitrarily long alternant codes* $\mathscr{X}_{\underline{\alpha},\underline{h}}^{\mathrm{Alt}}$ *meeting the Gilbert–Varshamov bound.*

So, alternant codes are asymptotically good. More precisely, a sequence of asymptotically good alternant codes is formed by the so-called *Goppa codes*. See below.

The Goppa codes are particular examples of alternant codes. They were invented by a Russian coding theorist, Valery Goppa, in 1972 by following an elegant idea that has its origin in algebraic geometry. Here, we perform the construction by using methods developed in this section.

Let $G(X) \in \mathbb{F}_{q^m}[X]$ be a polynomial over $\mathbb{F}_{q^m}$ and consider $\mathbb{F}_{q^m}[X]/\langle G(X)\rangle$, the polynomial ring $\mathrm{mod}\, G(X)$ over $\mathbb{F}_{q^m}$. Then $\mathbb{F}_{q^m}[X]/\langle G(X)\rangle$ is a field iff $G(X)$ is irreducible. But if, for a given $\alpha \in \mathbb{F}_{q^m}$, $G(\alpha) \neq 0$, the linear polynomial $X - \alpha$ is invertible in $\mathbb{F}_{q^m}[X]/\langle G(X)\rangle$. In fact, write

$$G(X) = q(X)(X - \alpha) + G(\alpha), \tag{3.5.13}$$

with $q(X) \in \mathbb{F}_q[X]$, $\deg q(X) = \deg G(X) - 1$.
So, $q(X)(X - \alpha) = -G(\alpha) \bmod G(X)$ or

$$(-G(\alpha)^{-1}q(X))(X - \alpha) = e \bmod G(X)$$

and

$$(X - \alpha)^{-1} = (-G(\alpha)^{-1}q(X)) \bmod G(X). \tag{3.5.14a}$$

As $q(X) = (G(X) - G(\alpha))(X - \alpha)^{-1}$, we have that

$$(X - \alpha)^{-1} = -(G(X) - G(\alpha))(X - \alpha)^{-1}G(\alpha)^{-1} \bmod G(X). \tag{3.5.14b}$$

So we define $(X - \alpha)^{-1}$ as a polynomial in $\mathbb{F}_{q^m}[X]/\langle G(X) \rangle$ given by (3.5.14a).

Definition 3.5.8   Fix a polynomial $G(X) \in \mathbb{F}_q[X]$ and a set $\underline{\alpha} = \{\alpha_1, \ldots, \alpha_n\}$ of distinct elements of $\mathbb{F}_{q^m}$, $q^m \geq n > \deg G(X)$, where $G(\alpha_j) = 0$, $1 \leq j \leq n$. Given a word $\underline{b} = b_1 \ldots b_n$ with $b_i \in \mathbb{F}_q$, $1 \leq i \leq n$, set

$$R_b(X) = \sum_{1 \leq i \leq n} b_i(X - \alpha_i)^{-1} \in \mathbb{F}_{q^m}[X]/\langle G(X) \rangle. \tag{3.5.15}$$

The q-ary Goppa code $\mathscr{X}^{\text{Go}}$ $(= \mathscr{X}^{\text{Go}}_{\underline{\alpha},G})$ is defined as the set

$$\{b \in \mathbb{F}_q^n : R_b(X) = 0 \bmod G(X)\}. \tag{3.5.16}$$

Clearly, $\mathscr{X}^{\text{Go}}_{\underline{\alpha},G}$ is a linear code. The polynomial $G(X)$ is called the Goppa polynomial; if $G(X)$ is irreducible, we say that $\mathscr{X}^{\text{Go}}$ is irreducible.

So, $\underline{b} = b_1 \ldots b_n \in \mathscr{X}^{\text{Go}}$ iff in $\mathbb{F}_{q^m}[X]$

$$\sum_{1 \leq i \leq n} b_i(G(X) - G(\alpha_i))(X - \alpha_i)^{-1}G(\alpha_i)^{-1} = 0. \tag{3.5.17}$$

Write $G(X) = \sum_{0 \leq i \leq r} g_i X^i$ where $\deg G(X) = r$, $g_r = 1$ and $r < n$. Then in $\mathbb{F}_{q^m}[X]$

$$(G(X) - G(\alpha_i))(X - \alpha_i)^{-1}$$

$$= \sum_{0 \leq j \leq r} g_j(X^j - \alpha_i^j)(X - \alpha_i)^{-1}$$

$$= \sum_{0 \leq j \leq r} g_j \sum_{0 \leq u \leq j-1} X^u \alpha_i^{j-1-u}$$

and so

$$\sum_{1\le i\le n} b_i(G(X)-G(\alpha_i))(X-\alpha_i)^{-1}G(\alpha_i)^{-1}$$

$$= \sum_{1\le i\le n} b_i \sum_{0\le j\le r} g_j \sum_{0\le u\le j-1} \alpha_i^{j-1-u} X^u G(\alpha_j)^{-1}$$

$$= \sum_{0\le u\le r-1} X^u \sum_{1\le i\le n} b_i G(\alpha_i)^{-1} \sum_{u+1\le j\le r} g_j \alpha_i^{j-1-u}.$$

Hence, $\underline{b} \in \mathscr{X}^{\mathrm{Go}}$ iff in $\mathbb{F}_{q^m}$

$$\sum_{1\le i\le n} b_i G(\alpha_i)^{-1} \sum_{u+1\le j\le r} g_j \alpha_i^{j-1-u} = 0 \tag{3.5.18}$$

for all $u = 0,\dots,r-1$.

Equation (3.5.18) leads to the parity-check matrix for $\mathscr{X}^{\mathrm{Go}}$. First, we see that the matrix

$$\begin{pmatrix}
G(\alpha_1)^{-1} & G(\alpha_2)^{-1} & \cdots & G(\alpha_n)^{-1} \\
\alpha_1 G(\alpha_1)^{-1} & \alpha_2 G(\alpha_2)^{-1} & \cdots & \alpha_n G(\alpha_n)^{-1} \\
\alpha_1^2 G(\alpha_1)^{-1} & \alpha_2^2 G(\alpha_2)^{-1} & \cdots & \alpha_n^2 G(\alpha_n)^{-1} \\
\vdots & \vdots & \ddots & \vdots \\
\alpha_1^{r-1} G(\alpha_1)^{-1} & \alpha_2^{r-1} G(\alpha_2)^{-1} & \cdots & \alpha_n^{r-1} G(\alpha_n)^{-1}
\end{pmatrix}, \tag{3.5.19}$$

which is $(n \times r)$ over $\mathbb{F}_q^m$, provides a parity-check. As before, any r rows of matrix (3.5.19) are linearly independent over $\mathbb{F}_{q^m}$ and so are its columns. Then again we write (3.5.19) as an $n \times (mr)$ matrix over $\mathbb{F}_q$; after purging linearly dependent columns it will give the parity-check matrix H.

We see that $\mathscr{X}^{\mathrm{Go}}$ is an alternant code $\mathscr{X}_{\underline{\alpha},\underline{h}}^{\mathrm{Alt}}$ where $\underline{\alpha} = (\alpha_1,\dots,\alpha_n)$ and $\underline{h} = (G(\alpha_1)^{-1},\dots,G(\alpha_n)^{-1})$. So, Theorem 3.5.6 implies

Theorem 3.5.9   *The q-ary Goppa code $\mathscr{X} = \mathscr{X}_{\underline{\alpha},G}^{\mathrm{Go}}$, where $\underline{\alpha} = \{\alpha_1,\dots,\alpha_n\}$ and $\deg G(X) = r < n$, has length n, rank k satisfying $n - mr \le k \le n - r$ and minimum distance $d(\mathscr{X}) \ge r+1$.*

As before, the above bound on minimum distance can be improved in the binary case. Suppose that a binary word $\underline{b} = b_1\dots b_n \in \mathscr{X}$ where $\mathscr{X}$ is a Goppa code $\mathscr{X}_{\underline{\alpha},G}^{\mathrm{Go}}$, where $\underline{\alpha} \subset \mathbb{F}_{2^m}$ and $G(X) \in \mathbb{F}_2[X]$. Suppose $w(\underline{b}) = w$ and $b_{i_1} = \cdots = b_{i_w} = 1$. Take $f_b(X) = \prod_{1\le j\le w} (X - \alpha_{i_j})$ and write the derivative $\partial_X f_b(X)$ as

$$\partial_X f_b(X) = R_b(X) f_b(X) \tag{3.5.20}$$

where $R_b(X) = \sum_{1\le j\le w} (X - \alpha_{i_j})^{-1}$ (cf. (3.5.15)). As polynomials $f_b(X)$ and $R_b(X)$ have no common roots in any extension $\mathbb{F}_{2^K}$, they are co-prime. Then $\underline{b} \in \mathscr{X}^{\mathrm{Go}}$

iff $G(X)$ divides $R_b(X)$ which is the same as $G(X)$ divides $\partial_X f_b(X)$. For $q = 2$, $\partial_X f_b(X)$ has only even powers of X (as its monomials are of the form $\ell X^{\ell-1}$ times a product of some α_{i_j}'s: this vanishes when ℓ is even). In other words, $\partial_X f_b = h(X^2) = (h(X))^2$ for some polynomial $h(X)$. Hence if $g(X)$ is the polynomial of lowest degree which is a square and divisible by $G(X)$ then $G(X)$ divides $\partial_X f_b(X)$ iff $g(X)$ divides $\partial_X f_b(X)$. So,

$$\underline{b} \in \mathscr{X}^{\text{Go}} \Leftrightarrow g(X)|\partial_X f_b(X) \Leftrightarrow R_b(X) = 0 \bmod g(X). \tag{3.5.21}$$

Theorem 3.5.10 Let $\mathscr{X}$ be a binary Goppa code $\mathscr{X}_{\underline{\alpha},G}^{\text{Go}}$. If $g(X)$ is a polynomial of the lowest degree which is a square and divisible by $G(X)$ then $\mathscr{X} = \mathscr{X}_{\underline{\alpha},g}^{\text{Go}}$. Hence, $d(\mathscr{X}^{\text{Go}}) \geq \deg g(X) + 1$.

Corollary 3.5.11 Suppose that the Goppa polynomial $G(X) \in \mathbb{F}_2[X]$ has no multiple roots in any extension field. Then $\mathscr{X}_{\underline{\alpha},G}^{\text{Go}} = \mathscr{X}_{\underline{\alpha},G^2}^{\text{Go}}$, and the minimum distance $d(\mathscr{X}_{\underline{\alpha},G}^{\text{Go}})$ is $\geq 2 \deg G(X) + 1$. Therefore, $\mathscr{X}_{\underline{\alpha},G}^{\text{Go}}$ can correct $\geq \deg G(X)$ errors.

A binary Goppa code $\mathscr{X}_{\underline{\alpha},G}^{\text{Go}}$ where polynomial $G(X)$ has no multiple roots is called separable.

It is interesting to discuss a particular decoding procedure applicable for alternant codes and based on the Euclid algorithm; cf. Section 2.5.

The initial setup for decoding an alternant code $\mathscr{X}_{\underline{\alpha},h}^{\text{Alt}}$ over $\mathbb{F}_q$ is as follows. As in (3.5.12), we take the $n \times (mr)$ matrix $\overrightarrow{A} = \left(\overrightarrow{h_j \alpha_j^{i-1}}\right)$ over $\mathbb{F}_q$ obtained from the $n \times r$ matrix $A = \left(h_j \alpha_j^{i-1}\right)$ over $\mathbb{F}_{q^m}$ by replacing the entries with rows of length m. Then purge linearly dependent columns from $\overrightarrow{A}$. Recall that $h_1, \ldots, h_n$ are non-zero and $\alpha_1, \ldots, \alpha_n$ are distinct elements of $\mathbb{F}_{q^m}$. Suppose a word $u = c + e$ is received, where c is the right codeword and e an error vector. We assume that r is even and that $t \leq r/2$ errors have occurred, at digits $1 \leq i_1 < \cdots < i_t \leq n$. Let the i_jth entry of e be $e_{i_j} \neq 0$. It is convenient to identify the error locators with elements α_{i_j}: as $\alpha_i \neq \alpha_{i'}$ for $i \neq i'$ (the α_i are distinct), we will know the erroneous positions if we determine $\alpha_{i_1}, \ldots, \alpha_{i_t}$. Moreover, if we introduce the error locator polynomial

$$\ell(X) = \prod_{j=1}^{t}(1 - \alpha_{i_j}X) = \sum_{0 \leq i \leq t} \ell_i X^i, \tag{3.5.22}$$

with $\ell_0 = 1$ and the roots $\alpha_{i_j}^{-1}$, then it will be enough to find $\ell(X)$ (i.e. coefficients ℓ_i).

We have to calculate the syndrome (we will call it an A-syndrome) emerging by acting on matrix A:

$$uA = eA = 0 \ldots 0 e_{i_1} \ldots e_{i_t} 0 \ldots 0 A.$$

Suppose the A-syndrome is $s = s_0 \ldots s_{r-1}$, with $s(X) = \sum\limits_{0 \leq i \leq r-1} s_i X^i$. It is convenient to introduce the error evaluator polynomial $\varepsilon(X)$, by

$$\varepsilon(X) = \sum_{1 \leq k \leq t} h_{i_k} e_{i_k} \prod_{1 \leq j \leq t:\, j \neq k} (1 - \alpha_{i_j} X). \tag{3.5.23}$$

Lemma 3.5.12   For all $u = 1, \ldots, t$,

$$e_{i_j} = \frac{\varepsilon(\alpha_{i_j}^{-1})}{h_{i_j} \prod\limits_{1 \leq \tilde{j} \leq t,\, \tilde{j} \neq j} (1 - \alpha_{i_{\tilde{j}}} \alpha_{i_j}^{-1})}. \tag{3.5.24}$$

Proof   Straightforward.   □

The crucial fact is that $\ell(X)$, $\varepsilon(X)$ and $s(X)$ are related by

Lemma 3.5.13   *The following formula holds true:*

$$\varepsilon(X) = \ell(X)s(X) \bmod X^r. \tag{3.5.25}$$

Proof   Write the following sequence:

$$
\begin{aligned}
\varepsilon(X) - \ell(X)s(X) &= \sum_{1 \leq k \leq t} h_{i_k} e_{i_k} \prod_{1 \leq j \leq t: j \neq k} (1 - \alpha_{i_j} X) - \ell(X) \sum_{0 \leq l \leq r-1} s_l X^l \\
&= \sum_k h_{i_k} e_{i_k} \prod_{1 \leq j \leq t: j \neq k} (1 - \alpha_{i_j} X) - \ell(X) \sum_l \sum_{1 \leq k \leq t} h_{i_k} \alpha_{i_k}^l e_{i_k} X^l \\
&= \sum_k h_{i_k} e_{i_k} \prod_{j \neq k} (1 - \alpha_{i_j} X) - \ell(X) \sum_k h_{i_k} e_{i_k} \sum_l \alpha_{i_k}^l X^l \\
&= \sum_k h_{i_k} e_{i_k} \left(\prod_{j \neq k} (1 - \alpha_{i_j} X) - \ell(X) \sum_l \alpha_{i_k}^l X^l \right) \\
&= \sum_k h_{i_k} e_{i_k} \prod_{j \neq k} (1 - \alpha_{i_j} X)(1 - (1 - e_{i_k} X) \sum_l \alpha_{i_k}^l X^l) \\
&= \sum_k h_{i_k} e_{i_k} \prod_{j \neq k} (1 - \alpha_{i_j} X) \left(1 - (1 - \alpha_{i_k} X) \frac{1 - \alpha_{i_k}^r X^r}{1 - \alpha_{i_k} X} \right) \\
&= \sum_k h_{i_k} \alpha_{i_k} \prod_{j \neq k} (1 - \alpha_{i_j} X)\, \alpha_{i_k}^r X^r = 0 \bmod X^r.
\end{aligned}
$$

□

Lemma 3.5.13 shows the way of decoding alternant codes. We know that there exists a polynomial $q(X)$ such that

$$\varepsilon(X) = q(X)X^r + \ell(X)s(X). \tag{3.5.26}$$

We also have $\deg \varepsilon(X) \leq t - 1 < r/2$, $\deg \ell(X) = t \leq r/2$ and that $\varepsilon(X)$ and $\ell(X)$ are co-prime as they have no common roots in any extension. Suppose we apply the

Euclid algorithm to the known polynomials $f(X) = X^r$ and $g(X) = s(X)$ with the aim to find $\varepsilon(X)$ and $\ell(X)$. By Lemma 2.5.44, a typical step produces a remainder

$$r_k(X) = a_k(X)X^r + b_k(X)s(X). \qquad (3.5.27)$$

If we want $r_k(X)$ and $b_k(X)$ to give $\varepsilon(X)$ and $\ell(X)$, their degrees must match: at least we must have $\deg r_k(X) < r/2$ and $\deg b_k(X) \le r/2$. So, the algorithm is repeated until $\deg r_{k-1}(X) \ge r/2$ and $\deg r_k(X) < r/2$. Then, according to Lemma 2.5.44, statement (3), $\deg b_k(X) = \deg X^r - \deg r_{k-1}(X) \le r - r/2 = r/2$. This is possible as the algorithm can be iterated until $r_k(X) = \gcd(X^r, s(X))$. But then $r_k(X) | \varepsilon(X)$ and hence $\deg r_k(X) \le \deg \varepsilon(X) < r/2$. So we can assume $\deg r_k(X) \le r/2$, $\deg b_k(X) \le r/2$.

The relevant equations are

$$\varepsilon(X) = q(X)X^r + \ell(X)s(X),$$
$$\deg \varepsilon(X) < r/2, \quad \deg \ell(X) \le r/2,$$
$$\gcd(\varepsilon(X), \ell(X)) = 1,$$

and also

$$r_k(X) = a_k(X)X^r + b_k(X)s(X), \ \deg r_k(X) < r/2, \ \deg b_k(X) \le r/2.$$

We want to show that polynomials $r_k(X)$ and $b_k(X)$ are scalar multiples of $\varepsilon(X)$ and $\ell(X)$. Exclude $s(X)$ to get

$$b_k(X)\varepsilon(X) - r_k(X)\ell(X) = (b_k(X)q(X) - a_k(X)\ell(X))X^r.$$

As

$$\deg b_k(X)\varepsilon(X) = \deg b_k(X) + \deg \varepsilon(X) < r/2 + r/2 = r$$

and

$$\deg r_k(X)\ell(X) = \deg r_k(X) + \deg \ell(X) < r/2 + r/2 = r,$$

$\deg(b(X)\varepsilon(X) - r_k(X)\ell(X)) < r$. Hence, $b_k(X)\varepsilon(X) - r_k(X)\ell(X)$ must be 0, i.e.

$$\ell(X)r_k(X) = \varepsilon(X)b_k(X), \ b_k(X)q(X) = a_k(X)\ell(X).$$

So, $\ell(X) | \varepsilon(X)b_k(X)$ and $b_k(X) | a_k(X)\ell(X)$. But $\ell(X)$ and $\varepsilon(X)$ are co-primes as well as $a_k(X)$ and $b_k(X)$ (by statement (5) of Lemma 2.5.44). Therefore, $\ell(X) = \lambda b_k(X)$ and hence $\varepsilon(X) = \lambda r_k(X)$. As $l(0) = 1$, $\lambda = b_k(0)^{-1}$.

To summarise:

Theorem 3.5.14 (The decoding algorithm for alternant codes) *Suppose $\mathscr{X}_{\alpha,h}^{\mathrm{Alt}}$ is an alternant code, with even r, and that $t \le r/2$ errors occurred in a received word u. Then, upon receiving word u:*

(a) Find the A-syndrome $uA = s_0 \ldots s_{r-1}$, with the corresponding polynomial $s(X) = \sum_l s_l X^l$.

(b) Use the Euclid algorithm, beginning with $f(X) = X^r$, $g(X) = s(X)$, to obtain $r_k(X) = a_k(X)X^r + b_k(X)s(X)$ with $\deg r_{k-1}(X) \geq r/2$ and $\deg r_k(X) < r/2$.

(c) Set $\ell(X) = b_k(0)^{-1}b_k(X)$, $\varepsilon(X) = b_k(0)^{-1}r_k(X)$.

Then $\ell(X)$ is the error locator polynomial whose roots are the inverses of $\alpha_{i_1}, \ldots, y_t = \alpha_{i_t}$, and $i_1, \ldots, i_t$ are the error digits. The values e_{i_j} are given by

$$e_{i_j} = \frac{\varepsilon(\alpha_{i_j}^{-1})}{h_{i_j} \prod_{l \neq j}(1 - \alpha_{i_l}\alpha_{i_j}^{-1})}. \tag{3.5.28}$$

The ideas discussed in this section found a far-reaching development in *algebraic-geometric* codes. Algebraic geometry provided powerful tools in modern code design; see [98], [99], [158], [160].

3.6 Additional problems for Chapter 3

Problem 3.1   *Define Reed–Solomon codes and prove that they are maximum distance separable. Prove that the dual of a Reed–Solomon code is a Reed–Solomon code.*

Find the minimum distance of a Reed–Solomon code of length 15 and rank 11 and the generator polynomial $g_1(X)$ over $\mathbb{F}_{16}$ for this code. Use the provided $\mathbb{F}_{16}$ field table to write $g_1(X)$ in the form $\omega^{i_0} + \omega^{i_1}X + \omega^{i_2}X^2 + \cdots$, identifying each coefficient as a single power of a primitive element ω of $\mathbb{F}_{16}$.

Find the generator polynomial $g_2(X)$ and the minimum distance of a Reed–Solomon code of length 10 and rank 6. Use the provided $\mathbb{F}_{11}$ field table to write $g_2(X)$ in the form $a_0 + a_1X + a_2X^2 + \cdots$, where each coefficient is a number from $\{0, 1, \ldots, 10\}$.

Determine a two-error correcting Reed–Solomon code over $\mathbb{F}_{16}$ and find its length, rank and generator polynomial.

The field table for $\mathbb{F}_{11} = \{0, 1, 2, 3, 4, 5, 6, 7, 8, 9, 10\}$, with addition and multiplication mod 11:

i	0	1	2	3	4	5	6	7	8	9
ω^i	1	2	4	8	5	10	9	7	3	6

The field table for $\mathbb{F}_{16} = \mathbb{F}_2^4$:

i	0	1	2	3	4	5	6	7	8
ω^i	0001	0010	0100	1000	0011	0110	1100	1011	0101

i	9	10	11	12	13	14
ω^i	1010	0111	1110	1111	1101	1001

Solution A q-ary RS code $\mathcal{X}^{RS}$ of designed distance $\delta \le q - 1$ is defined as a cyclic code of length $N = q - 1$ over $\mathbb{F}_q$, with a generator polynomial

$$g(X) = (X - \omega^b)(X - \omega^{b+1}) \ldots (X - \omega^{b+\delta-2})$$

of $\deg(g(X)) = \delta - 1$. Here ω is a primitive $(q-1, \mathbb{F}_q)$ root of unity (i.e. a primitive element of $\mathbb{F}_q^*$) and $b = 0, 1, \ldots, q - 2$. The powers $\omega^b, \ldots, \omega^{b+\delta-2}$ are called the (defining) zeros and the remaining $N - \delta + 1$ powers of ω non-zeros of $\mathcal{X}^{RS}$.

The rank of $\mathcal{X}^{RS}$ equals $k = N - \delta + 1$. The distance is $\ge \delta = N - k + 1$, but by the Singleton bound should be $\le \delta = N - k + 1$. So, the distance equals $\delta = N - k + 1$, i.e. $\mathcal{X}^{RS}$ is maximum distance separable.

The dual $(\mathcal{X}^{RS})^{\perp}$ of $\mathcal{X}^{RS}$ is cyclic and its zeros are the inverses of the non-zeroes of $\mathcal{X}^{RS}$:

$$\omega^{q-1-j} = (\omega^j)^{-1}, j \ne b, \ldots, b + \delta - 2.$$

That is, they are

$$\omega^{q-b}, \omega^{q-b+1}, \ldots, \omega^{q-b+q-\delta-1}$$

and the generator polynomial $g^{\perp}(X)$ for $(\mathcal{X}^{RS})^{\perp}$ is

$$g^{\perp}(X) = (X - \omega^{b^{\perp}})(X - \omega^{b^{\perp}+1}) \ldots (X - \omega^{b^{\perp}+q-\delta-1})$$

where $b^{\perp} = q - b$. That is $(\mathcal{X}^{RS})^{\perp}$ is an RS code of designed distance $q - \delta + 1$.

In the example, length 15 means $q = 15 + 1 = 16$ and rank 11 yields distance $\delta = 15 - 11 + 1 = 5$. The generator $g_1(X)$ over $\mathbb{F}_{16} = \mathbb{F}_2^4$, for the code with $b = 1$, reads

$$\begin{aligned}
g_1(X) &= (X - \omega)(X - \omega^2)(X - \omega^3)(X - \omega^4) \\
&= X^4 - (\omega + \omega^2 + \omega^3 + \omega^4)X^3 \\
&\quad + (\omega^3 + \omega^4 + \omega^5 + \omega^5 + \omega^6 + \omega^7)X^2 \\
&\quad - (\omega^6 + \omega^7 + \omega^8 + \omega^9)X + \omega^{10} \\
&= X^4 + \omega^{13}X^3 + \omega^6 X^2 + \omega^3 X + \omega^{10}
\end{aligned}$$

where the calculation is accomplished by using the $\mathbb{F}_{16}$ field table.

Similarly, length 10 means $q = 11$ and rank 6 yields distance $\delta = 10 - 6 + 1 = 5$. The generator $g_2(X)$ is over $\mathbb{F}_{11}$ and, again for $b = 1$, reads

$$
\begin{aligned}
g_2(X) &= (X - \omega)(X - \omega^2)(X - \omega^3)(X - \omega^4) \\
&= X^4 - (\omega + \omega^2 + \omega^3 + \omega^4)X^3 \\
&\quad + (\omega^3 + \omega^4 + \omega^5 + \omega^5 + \omega^6 + \omega^7)X^2 \\
&\quad - (\omega^6 + \omega^7 + \omega^8 + \omega^9)X + \omega^{10} \\
&= X^4 + 3X^3 + 5X^2 + 8X + 1
\end{aligned}
$$

where the calculation is accomplished by using the $\mathbb{F}_{11}$-field table.

Finally, a two-error correcting RS code over $\mathbb{F}_{16}$ must have length $N = 15$ and distance $\delta = 5$, hence rank 11. So, it coincides with the above 16-ary $[15, 11]$ RS code. □

Problem 3.2   Let $\mathscr{X}$ be a binary linear $[N, k]$ code and $\mathscr{X}^{\mathrm{ev}}$ be the set of words $\mathbf{x} \in \mathscr{X}$ of even weight. Prove that either

(i) $\mathscr{X} = \mathscr{X}^{\mathrm{ev}}$ or

(ii) $\mathscr{X}^{\mathrm{ev}}$ is an $[N, k-1]$ linear subcode of $\mathscr{X}$.

Prove that if the generating matrix G of $\mathscr{X}$ has no zero column then the total weight $\sum_{\mathbf{x} \in \mathscr{X}} w(\mathbf{x})$ equals $N 2^{k-1}$.

[*Hint:* Consider the contribution from each column of G.]

Denote by $\mathscr{X}_{\mathrm{H},\ell}$ the binary Hamming code of length $N = 2^\ell - 1$ and by $\mathscr{X}_{\mathrm{H},\ell}^{\perp}$ the dual simplex code, $\ell = 3, 4, \ldots$. Is it always true that the N-vector $1 \ldots 1$ (with all digits one) is a codeword in $\mathscr{X}_{\mathrm{H},\ell}$? Let A_s and $A_s^{\perp}$ denote the number of words of weight s in $\mathscr{X}_{\mathrm{H},\ell}$ and $\mathscr{X}_{\mathrm{H},\ell}^{\perp}$, respectively, with $A_0 = A_0^{\perp} = 1$ and $A_1 = A_2 = 0$. Check that

$$
A_3 = N(N-1)/3!, \quad A_4 = N(N-1)(N-3)/4!,
$$

and

$$
A_5 = N(N-1)(N-3)(N-7)/5!.
$$

Prove that $A_{2^\ell - 1}^{\perp} = 2^\ell - 1$ (i.e. all non-zero words $\mathbf{x} \in \mathscr{X}_{\mathrm{H},\ell}^{\perp}$ have weight $2^{\ell-1}$). By using the last fact and the MacWilliams identity for binary codes, give a formula for A_s in terms of $K_s(2^{\ell-1})$, the value of the Kravchuk polynomial:

$$
K_s(2^{\ell-1}) = \sum_{j=0 \vee s + 2^{\ell-1} - 2^\ell + 1}^{s \wedge 2^{\ell-1}} \binom{2^{\ell-1}}{j} \binom{2^\ell - 1 - 2^{\ell-1}}{s-j} (-1)^j.
$$

Here $0 \vee s + 2^{\ell-1} - 2^\ell + 1 = \max [0, s + 2^{\ell-1} - 2^\ell + 1]$ and $s \wedge 2^{\ell-1} = \min [s, 2^{\ell-1}]$. Check that your formula gives the right answer for $s = N = 2^\ell - 1$.

Solution $\mathscr{X}^{\mathrm{ev}}$ is always a linear subcode of $\mathscr{X}$. In fact, for binary words $\mathbf{x}$ and $\mathbf{x}'$, $w(\mathbf{x}+\mathbf{x}') = w(\mathbf{x}) + w(\mathbf{x}') - 2w(\mathbf{x} \wedge \mathbf{x}')$, where digit $(\mathbf{x} \wedge \mathbf{x}')_j = x_j x'_j = \min[x_i, x'_i]$. If both $\mathbf{x}$ and $\mathbf{x}'$ are even words (have even weight) or odd (have odd weight) then $\mathbf{x}+\mathbf{x}'$ is even, and if $\mathbf{x}$ is even and $\mathbf{x}'$ odd then $\mathbf{x}+\mathbf{x}'$ is odd. So, if $\mathscr{X}^{\mathrm{ev}} \neq \mathscr{X}$ then $\mathscr{X}^{\mathrm{ev}}$ is a subgroup in $\mathscr{X}$ of index $[\mathscr{X}^{\mathrm{ev}} : \mathscr{X}]$ two. Thus, there are two cosets, and $\mathscr{X}^{\mathrm{ev}}$ is a half of $\mathscr{X}$. So, $\mathscr{X}^{\mathrm{ev}}$ is $[N, k-1]$ code.

Let $\mathbf{g}^{(j)} = (g_{1j}, \ldots, g_{kj})^{\mathrm{T}}$ be column j of G, the generating matrix of $\mathscr{X}$. Set $\mathscr{W}_j = \{i = 1, \ldots, k : g_{ij} = 1\}$, with $\sharp \mathscr{W}_j = w(\mathbf{g}^{(j)}) = w_j \geq 1, 1 \leq j \leq N$. The contribution into the sum $\sum_{\mathbf{x} \in \mathscr{X}} w(\mathbf{x})$ coming from $\mathbf{g}^{(j)}$ equals

$$2^{k-w_j} \times 2^{w_j-1} = 2^{k-1}.$$

Here 2^{k-w_j} represents the number of subsets of the complement $\{1, \ldots, k\} \setminus \mathscr{W}_j$ and 2^{w_j-1} the number of odd subsets of $\mathscr{W}_j$. Multiplying by N (the number of columns) gives $N2^{k-1}$.

If $H = H_{\mathrm{H}, \ell}$ is the parity-check matrix of the Hamming code $\mathscr{X}_{\mathrm{H}, \ell}$ then weight of row j of H equals the number of digits one in position j in the binary decomposition of numbers $1, \ldots, 2^l - 1, 1 \leq j \leq l$. So, $w(\mathbf{h}^{(j)}) = 2^{l-1}$ (a half of numbers $1, \ldots, 2^l - 1$ have zero in position j and a half one). Then for all $j = 1, \ldots, N$, the dot-product $\mathbf{1} \cdot \mathbf{h}^{(j)} = w(\mathbf{h}^{(j)})$ mod $2 = 0$, i.e. $1 \ldots 1 \in \mathscr{X}_{\mathrm{H}, \ell}$.

Now $A_3 = N(N-1)/3!$, the number of linearly dependent triples of columns of H (as the choice is made by fixing two distinct columns: the third is their sum). Next, $A_4 = N(N-1)(N-3)/4!$, the number of linearly dependent 4-ples of columns of H (as the choice is made by fixing: (a) two arbitrary distinct columns, (b) a third column that is not their sum, with (c) the fourth column being the sum of the first three), and similarly, $A_5 = N(N-1)(N-3)(N-7)/5!$, the number of linearly dependent 5-ples of columns of H. (Here $N-7$ indicates that while choosing the fourth column we should avoid any of $2^3 - 1 = 7$ linear combinations of the first three.)

In fact, any non-zero word $\mathbf{x}$ in the dual code $\mathscr{X}_{\mathrm{H}, \ell}^{\perp}$ has $w(\mathbf{x}) = 2^{l-1}$. To prove this note that the generating matrix of $\mathscr{X}_{\mathrm{H}, \ell}^{\perp}$ is H. So, write $\mathbf{x}$ as a sum of rows of H, and let $\mathscr{W}$ be the set of rows of H contributing into this sum, with $\sharp \mathscr{W} = w \leq \ell$. Then $w(\mathbf{x})$ equals the number of j among $1, 2, \ldots, 2^\ell - 1$ such that in the binary decomposition $j = 2^0 j_0 + 2^1 j_1 + \cdots + 2^{\ell-1} j_{l-1}$ the sum $\sum_{t \in \mathscr{W}} j_t$ mod 2 equals one. As before, this is equal to 2^{w-1} (the number of subsets of $\mathscr{W}$ of odd cardinality). So, $w(\mathbf{x}) = 2^{\ell-w+w-1} = 2^{\ell-1}$. Note that the rank of $\mathscr{X}_{H, l}^{\perp}$ is $2^\ell - 1 - (2^\ell - 1 - l) = l$ and the size $\sharp \mathscr{X}_{\mathrm{H}, \ell}^{\perp} = 2^\ell$.

The MacWilliams identity (in the reversed form) reads

$$A_s = \frac{1}{\sharp \mathscr{X}_{\mathrm{H}, \ell}^{\perp}} \sum_{i=1}^{N} A_i^{\perp} K_s(i) \tag{3.6.1}$$

where

$$K_s(i) = \sum_{j=0 \vee s+i-N}^{s \wedge i} \binom{i}{j}\binom{N-i}{s-j}(-1)^j, \tag{3.6.2}$$

with $0 \vee s+i-N = \max[0, s+i-N]$, $s \wedge i = \min[s, i]$. In our case, $A_0^{\perp} = 1, A_{2^\ell-1}^{\perp} = 2^\ell - 1$ (the number of the non-zero words in $\mathscr{X}_{\mathrm{H},\ell}^{\perp}$). Thus,

$$
\begin{aligned}
A_s &= \frac{1}{2^\ell}\left(1 + (2^\ell - 1)K_s(2^{\ell-1})\right) \\
&= \frac{1}{2^\ell}\left(1 + (2^\ell - 1)\sum_{j=0 \vee s+2^{\ell-1}-2^\ell+1}^{s \wedge 2^{\ell-1}} \binom{2^{\ell-1}}{j}\binom{2^\ell - 1 - 2^{\ell-1}}{2^\ell - 1 - j}(-1)^j\right).
\end{aligned}
$$

For $s = N = 2^\ell - 1$, $A_{2^\ell-1}$ can be either 1 (if the 2^ℓ-word $11\ldots1$ lies in $\mathscr{X}_{\mathrm{H},\ell}$) or 0 (if it doesn't). The last formula yields

$$
\begin{aligned}
A_{2^\ell-1} &= \frac{1}{2^\ell}\left(1 + (2^\ell - 1)\sum_{j=2^{\ell-1}}^{2^{\ell-1}} \binom{2^{\ell-1}}{j}\binom{2^\ell - 1 - 2^{\ell-1}}{2^\ell - 1 - j}\right) \\
&= \frac{1}{2^\ell}\left(1 + 2^\ell - 1\right) = 1,
\end{aligned}
$$

which agrees with the fact that $11\ldots1 \in \mathscr{X}_{\mathrm{H},\ell}$. $\qquad\square$

Problem 3.3 *Let ω be a root of $M(X) = X^5 + X^2 + 1$ in $\mathbb{F}_{32}$; given that $M(X)$ is a primitive polynomial for $\mathbb{F}_{32}$, ω is a primitive $(31, \mathbb{F}_{32})$-root of unity. Use elements $\omega, \omega^2, \omega^3, \omega^4$ to construct a binary narrow-sense primitive BCH code $\mathscr{X}$ of length 31 and designed distance 5. Identify the cyclotomic coset $\{i, 2i, \ldots, 2^{d-1}i\}$ for each of $\omega, \omega^2, \omega^3, \omega^4$. Check that ω and ω^3 suffice as defining zeros of $\mathscr{X}$ and that the actual minimum distance of $\mathscr{X}$ equals 5. Show that the generator polynomial $g(X)$ for $\mathscr{X}$ is the product*

$$
\begin{aligned}
(X^5 + X^2 + 1)&(X^5 + X^4 + X^3 + X^2 + 1) \\
&= X^{10} + X^9 + X^8 + X^6 + X^5 + X^3 + 1.
\end{aligned}
$$

Suppose you received a word $u(X) = X^{12} + X^{11} + X^9 + X^7 + X^6 + X^2 + 1$ from a sender who uses code $\mathscr{X}$. Check that $u(\omega) = \omega^3$ and $u(\omega^3) = \omega^9$, argue that $u(X)$ should be decoded as

$$c(X) = X^{12} + X^{11} + X^9 + X^7 + X^6 + X^3 + X^2 + 1$$

and verify that $c(X)$ is indeed a codeword in $\mathscr{X}$.

The field table for $\mathbb{F}_{32} = \mathbb{F}_2^5$ and the list of irreducible polynomials of degree 5 over $\mathbb{F}_2$ are also provided to help with your calculations.

The field table for $\mathbb{F}_{32} = \mathbb{F}_2^5$:

i	0	1	2	3	4	5	6	7
ω^i	00001	00010	00100	01000	10000	00101	01010	10100

i	8	9	10	11	12	13	14	15
ω^i	01101	11010	10001	00111	01110	11100	11101	11111

i	16	17	18	19	20	21	22	23
ω^i	11011	10011	00011	00110	01100	11000	10101	01111

i	24	25	26	27	28	29	30
ω^i	11110	11001	10111	01011	10110	01001	10010

The list of irreducible polynomials of degree 5 over $\mathbb{F}_2$:

$$X^5 + X^2 + 1, \ X^5 + X^3 + 1, \ X^5 + X^3 + X^2 + X + 1,$$

$$X^5 + X^4 + X^3 + X + 1, \ X^5 + X^4 + X^3 + X^2 + 1;$$

they all have order 31. Polynomial $X^5 + X^2 + 1$ *is primitive.*

Solution As $M(X) = X^5 + X^2 + 1$ is a primitive polynomial in $\mathbb{F}_2[X]$, any root ω of $M(X)$ is a primitive $(31, \mathbb{F}_2)$-root of unity, i.e. satisfies $\omega^{31} + 1 = 0$. Furthermore, $M(X)$ is the minimal polynomial for ω.

The BCH code $\mathscr{X}$ under construction is a cyclic code whose generator is a polynomial of smallest degree having $\omega, \omega^2, \omega^3, \omega^4$ among its roots (i.e. a cyclic code whose zeros form a minimal set including $\omega, \omega^2, \omega^3, \omega^4$). Thus, the generator polynomial $g(X)$ of X is the lcm of the minimal polynomials for $\omega, \omega^2, \omega^3, \omega^4$.

The cyclotomic coset for ω is $C = \{1, 2, 4, 8, 16\}$; hence

$$(X - \omega)(X - \omega^2)(X - \omega^4)(X - \omega^8)(X - \omega^{16}) = X^5 + X^2 + 1$$

is the minimal polynomial for ω, ω^2 and ω^4. The cyclotomic coset for ω^3 is $C = \{3, 6, 12, 24, 17\}$ and the minimal polynomial $M_{\omega^3}(X)$ for ω^3 equals

$$
\begin{aligned}
M_{\omega^3}(X) &= (X - \omega^3)(X - \omega^6)(X - \omega^{12})(X - \omega^{24})(X - \omega^{17}) \\
&= X^5 + (\omega^3 + \omega^6 + \omega^{12} + \omega^{24} + \omega^{17})X^4 + (\omega^9 + \omega^{15} + \omega^{27} \\
&\quad + \omega^{20} + \omega^{18} + \omega^{30} + \omega^{23} + \omega^{36} + \omega^{29} + \omega^{41})X^3 \\
&\quad + (\omega^{21} + \omega^{33} + \omega^{26} + \omega^{39} + \omega^{32} + \omega^{44} + \omega^{42} + \omega^{35} \\
&\quad + \omega^{47} + \omega^{53})X^2 + (\omega^{45} + \omega^{38} + \omega^{50} + \omega^{56} + \omega^{59})X + \omega^{62} \\
&= X^5 + X^4 + X^3 + X^2 + 1
\end{aligned}
$$

by a direct field-table calculation or by inspecting the list of irreducible polynomials over $\mathbb{F}_2$ of degree 5.

So, ω and ω^3 suffice as zeros, and the generating polynomial $g(X)$ equals

$$(X^5 + X^2 + 1)(X^5 + X^4 + X^3 + X^2 + 1)$$
$$= X^{10} + X^9 + X^8 + X^6 + X^5 + X^3 + 1,$$

as required. In other words:

$$\mathscr{X} = \{c(X) \in \mathbb{F}_2[X]/(X^{31} + 1) : c(\omega) = c(\omega^3) = 0\}$$

$$= \{c(X) \in \mathbb{F}_2[X]/(X^{31} + 1) : g(X)|c(X)\}.$$

The rank of $\mathscr{X}$ equals 21. The minimum distance of $\mathscr{X}$ equals 5, its designed distance. This follows from Theorem 3.3.20:

Let $N = 2^\ell - 1$. If $2^{\ell E} < \sum\limits_{0 \le i \le E+1} \binom{N}{i}$ then the binary narrow-sense primitive BCH code of designed distance $2E + 1$ has minimum distance $2E + 1$.

In fact, $N = 31 = 2^5 - 1$ with $\ell = 5$ and $E = 2$, i.e. $2E + 1 = 5$, and

$$1024 = 2^{10} < 1 + 31 + \frac{31 \times 30}{2} + \frac{31 \times 30 \times 29}{2 \times 3} = 4992.$$

Thus, $\mathscr{X}$ corrects two errors. The Berlekamp–Massey decoding procedure requires calculating the values of the received polynomial at the defining zeros. From the $\mathbb{F}_{32}$ field table we have

$$u(\omega) = \omega^{12} + \omega^{11} + \omega^9 + \omega^7 + \omega^6 + \omega^2 + 1 = \omega^3,$$
$$u(\omega^3) = \omega^{36} + \omega^{33} + \omega^{27} + \omega^{18} + \omega^6 + 1 = \omega^9.$$

So, $u(\omega^3) = u(\omega)^3$. As $u(\omega) = \omega^3$, we conclude that a single error occurred, at digit three, i.e. $u(X)$ is decoded by

$$c(X) = X^{12} + X^{11} + X^9 + X^7 + X^6 + X^3 + X^2 + 1$$

which is $(X^2 + 1)g(X)$ as required. $\qquad\square$

Problem 3.4   *Define the dual $\mathscr{X}^\perp$ of a linear $[N, k]$ code of length N and dimension k with alphabet $\mathbb{F}$. Prove or disprove that if $\mathscr{X}$ is a binary $[N, (N-1)/2]$ code with N odd then $\mathscr{X}^\perp$ is generated by a basis of $\mathscr{X}$ plus the word $1 \ldots 1$. Prove or disprove that if a binary code $\mathscr{X}$ is self-dual, $\mathscr{X} = \mathscr{X}^\perp$, then N is even and the word $1 \ldots 1$ belongs to $\mathscr{X}$.*

Prove that a binary self-dual linear $[N, N/2]$ code $\mathscr{X}$ exists for each even N. Conversely, prove that if a binary linear $[N, k]$ code $\mathscr{X}$ is self-dual then $k = N/2$.

Give an example of a non-binary linear self-dual code.

Solution   The dual $\mathscr{X}^\perp$ of the $[N, k]$ linear code $\mathscr{X}$ is given by

$$\mathscr{X}^\perp = \{\mathbf{x} = x_1 \ldots x_N \in \mathbb{F}^N : \mathbf{x} \cdot \mathbf{y} = 0 \quad \text{for all } \mathbf{y} \in \mathscr{X}\}$$

where $\mathbf{x} \cdot \mathbf{y} = x_1 y_1 + \cdots + x_N y_N$. Take $N = 5$, $k = (N-1)/2 = 2$,

$$\mathcal{X} = \begin{pmatrix} 1 & 0 & 0 & 0 & 0 \\ 0 & 1 & 0 & 0 & 0 \\ 1 & 1 & 0 & 0 & 0 \\ 0 & 0 & 0 & 0 & 0 \end{pmatrix}.$$

Then $\mathcal{X}^{\perp}$ is generated by

$$\begin{pmatrix} 0 & 0 & 1 & 0 & 0 \\ 0 & 0 & 0 & 1 & 0 \\ 0 & 0 & 0 & 0 & 1 \end{pmatrix}.$$

None of the vectors from $\mathcal{X}$ belongs to $\mathcal{X}^{\perp}$, so the claim is false.

Now take a self-dual code $\mathcal{X} = \mathcal{X}^{\perp}$. If the word $\mathbf{1} = 1\ldots1 \notin \mathcal{X}$ then there exists $\mathbf{x} \in \mathcal{X}$ such that $\mathbf{x} \cdot \mathbf{1} \neq 0$. But $\mathbf{x} \cdot \mathbf{1} = \sum x_i = w(\mathbf{x}) \bmod 2$. On the other hand, $\sum x_i = \mathbf{x} \cdot \mathbf{x}$, so $\mathbf{x} \cdot \mathbf{x} \neq 0$. But then $\mathbf{x} \notin \mathcal{X}^{\perp}$. Hence $\mathbf{1} \in \mathcal{X}$. But then $\mathbf{1} \cdot \mathbf{1} = 0$ which implies that N is even.

Now let $N = 2k$. Divide digits $1, \ldots, N$ into k disjoint pairs $(\alpha_1, \beta_1), \ldots, (\alpha_k, \beta_k)$, with $\alpha_i < \beta_i$. Then consider k binary words $\mathbf{x}^{(1)}, \ldots, \mathbf{x}^{(k)}$ of length N and weight 2, with the non-zero digits in the word $\mathbf{x}^{(i)}$ in positions (α_i, β_i). Then form the $[N, k]$ code generated by $\mathbf{x}^{(1)}, \ldots, \mathbf{x}^{(k)}$.

This code $\mathcal{X}$ is self-dual. In fact, $\mathbf{x}^{(i)} \cdot \mathbf{x}^{(i')} = 0$ for all i, i', hence $\mathcal{X} \subset \mathcal{X}^{\perp}$. Conversely, let $\mathbf{y} \in \mathcal{X}^{\perp}$. Then $\mathbf{y} \cdot \mathbf{x}^{(i)} = 0$ for all i. This means that for all i, $\mathbf{y}$ has either both 0 or both non-zero digits at positions (α_i, β_i). Then $\mathbf{y} \in \mathcal{X}$. So, $\mathcal{X} = \mathcal{X}^{\perp}$.

Now assume $\mathcal{X} = \mathcal{X}^{\perp}$. Then N is even. But the dimension must be k by the rank-nullity theorem.

The non-binary linear self-dual code is the ternary Golay $[12, 6]$ with a generating matrix

$$G = \begin{pmatrix} 1&0&0&0&0&0&0&1&1&1&1&1 \\ 0&1&0&0&0&0&1&0&1&2&2&1 \\ 0&0&1&0&0&0&1&1&0&1&2&2 \\ 0&0&0&1&0&0&1&2&1&0&1&2 \\ 0&0&0&0&1&0&1&2&2&1&0&1 \\ 0&0&0&0&0&1&1&1&2&2&1&0 \end{pmatrix}$$

Here rows of G are orthogonal (including self-orthogonal). Hence, $\mathcal{X} \subset \mathcal{X}^{\perp}$. But $\dim(\mathcal{X}) = \dim(\mathcal{X}^{\perp}) = 6$, so $\mathcal{X} = \mathcal{X}^{\perp}$. $\qquad\square$

Problem 3.5 Define a finite field $\mathbb{F}_q$ with q elements and prove that q must have the form $q = p^s$ where p is a prime integer and $s \geq 1$ a positive integer. Check that p is the characteristic of $\mathbb{F}_q$.

Prove that for any p and s as above there exists a finite field $\mathbb{F}_p^s$ with p^s elements, and this field is unique up to isomorphism.

Prove that the set $\mathbb{F}_{p^s}^*$ of the non-zero elements of $\mathbb{F}_{p^s}$ is a cyclic group $\mathbb{Z}_{p^s-1}$.

Write the field table for $\mathbb{F}_9$, identifying the powers ω^i of a primitive element $\omega \in \mathbb{F}_9$ as vectors over $\mathbb{F}_3$. Indicate all vectors α in this table such that $\alpha^4 = e$.

Solution A field $\mathbb{F}_q$ with q elements is a set of cardinality q with two commutative group operations, $+$ and $\cdot$, with standard distributivity rules. It is easy to check that $\mathrm{char}(\mathbb{F}_q) = p$ is a prime number. Then $\mathbb{F}_p \subset \mathbb{F}_q$ and $q = \sharp\mathbb{F}_q = p^s$ where $s = [\mathbb{F}_q : \mathbb{F}_p]$ is the dimension of $\mathbb{F}_q$ as a vector space over $\mathbb{F}_p$, a field of p elements.

Now, let $\mathbb{F}_q^*$, the multiplicative group of non-zero elements from $\mathbb{F}_q$, contain an element of order $q - 1 = \sharp\mathbb{F}_q^*$. In fact, every $b \in \mathbb{F}_q^*$ has a finite order $\mathrm{ord}(b) = r(b)$; set $r_0 = \max[r(b) : b \in \mathbb{F}_q^*]$. and fix $a \in \mathbb{F}_q^*$ with $r(a) = r_0$. Then $r(b)|r_0$ for all $b \in \mathbb{F}_q^*$. Next, pick γ, a prime factor of $r(b)$, and write $r(b) = \gamma^{s'}\omega, r_0 = \gamma^s\alpha$. Let us check that $s \geq s'$. Indeed, a^{γ^s} has order α, b^ω order $\gamma^{s'}$ and $a^{\gamma^s}b^\omega$ order $\gamma^{s'}\alpha$. Thus, if $s' > s$, we obtain an element of order $> r_0$. Hence, $s \geq s'$ which holds for any prime factor of $r(b)$, and $r(b)|r(a)$.

Then $b^{r(a)} = e$, for all $b \in \mathbb{F}_q^*$, i.e. the polynomial $X^{r_0} - e$ is divisible by $(X - b)$. It must then be the product $\prod_{b \in \mathbb{F}_q^*}(X - b)$. Then $r_0 = \sharp\mathbb{F}_q^* = q - 1$. Then $\mathbb{F}_q^*$ is a cyclic group with generator a.

For each prime p and positive integer s there exists at most one field $\mathbb{F}_q$ with $q = p^s$, up to isomorphism. Indeed, if $\mathbb{F}_q$ and $\mathbb{F}_q'$ are two such fields then they both are isomorphic to $\mathrm{Spl}(X^q - X)$, the splitting field of $X^q - X$ (over $\mathbb{F}_p$, the basic field).

The elements α of $\mathbb{F}_9 = \mathbb{F}_3 \times \mathbb{F}_3$ with $\alpha^4 = e$ are $e = 01$, $\omega^2 = 1 + 2\omega = 21$, $\omega^4 = 02$, $\omega^6 = 2 + \omega = 12$ where $\omega = 10$. $\qquad\square$

Problem 3.6 *Give the definition of a cyclic code of length N with alphabet $\mathbb{F}_q$. What are the defining zeros of a cyclic code and why are they always $(N, \mathbb{F}_q)$-roots of unity? Prove that the ternary Hamming $\left[\dfrac{3^s - 1}{2}, \dfrac{3^s - 1}{2} - s, 3\right]$ code is equivalent to a cyclic code and identify the defining zeros of this cyclic code.*

A sender uses the ternary $[13, 10, 3]$ Hamming code, with field alphabet $\mathbb{F}_3 = \{0, 1, 2\}$ and the parity-check matrix H of the form

$$\begin{pmatrix} 1 & 0 & 1 & 2 & 0 & 1 & 2 & 0 & 1 & 2 & 0 & 1 & 2 \\ 0 & 1 & 1 & 1 & 0 & 0 & 0 & 1 & 1 & 1 & 2 & 2 & 2 \\ 0 & 0 & 0 & 0 & 1 & 1 & 1 & 1 & 1 & 1 & 1 & 1 & 1 \end{pmatrix}.$$

The receiver receives the word $\mathbf{x} = 2\ 1\ 2\ 0\ 1\ 1\ 0\ 0\ 2\ 1\ 1\ 2\ 0$. How should he decode it?

Solution As $g(X)|(X^N - 1)$, all roots of $g(X)$ are Nth roots of unity. Let $\gcd(l, q - 1) = 1$. We prove that the Hamming $\left[\dfrac{q^l - 1}{q - 1}, \dfrac{q^l - 1}{q - 1} - l \right]$ code is equivalent to a cyclic code, with defining zero $\omega = \beta^{q-1}$ where β is the primitive $(q^l - 1)/(q-1)$-root of unity. Indeed, set $N = (q^l - 1)/(q-1)$. The splitting field $\mathrm{Spl}(X^N - 1) = \mathbb{F}_{q^r}$ where $r = \mathrm{ord}_N(q) = \min[s : N|(q^s - 1)]$. Then $r = l$ as $(q^l - 1)/(q - 1)|(q^l - 1)$ and l is the least such power. So, $\mathrm{Spl}(X^N - 1) = \mathbb{F}_{q^l}$.

If β is a primitive element is $\mathbb{F}_{q^l}$ then $\omega = \beta^{\frac{q^l-1}{N}} = \beta^{q-1}$ is a primitive Nth root of unity in $\mathbb{F}_{q^l}$. Write $\omega^0 = e, \omega, \omega^2, \ldots, \omega^{N-1}$ as column vectors in $\mathbb{F}_q \times \ldots \times \mathbb{F}_q$ and form an $l \times N$ check matrix H. We want to check that any two distinct columns of H are linearly independent. This is done exactly as in Theorem 3.3.14.

Then the code with parity-check matrix H has distance ≥ 3, rank $k \geq N - l$. The Hamming bound with $N = (q^l - 1)/(q - 1)$

$$ q^k \leq q^N \left(\sum_{0 \leq m \leq E} \binom{N}{m} (q-1)^m \right)^{-1}, \quad \text{with } E = \left\lfloor \frac{d-1}{2} \right\rfloor, \qquad (3.6.3) $$

shows that $d = 3$ and $k = N - l$. So, the cyclic code with the parity-check matrix H is equivalent to Hamming's.

To decode the code in question, calculate the syndrome $\mathbf{x}H^{\mathrm{T}} = 2\,0\,2 = 2 \cdot (1\,0\,1)$ indicating the error is in the 6th position. Hence, $\mathbf{x} - 2\mathbf{e}^{(6)} = \mathbf{y} + \mathbf{e}^{(6)}$ and the correct word is $\mathbf{c} = 2\,1\,2\,0\,1\,2\,0\,0\,2\,1\,1\,2\,0$. $\qquad \square$

Problem 3.7 *Compute the rank and minimum distance of the cyclic code with generator polynomial $g(X) = X^3 + X + 1$ and parity-check polynomial $h(X) = X^4 + X^2 + X + 1$. Now let ω be a root of $g(X)$ in the field $\mathbb{F}_8$. We receive the word $r(X) = X^5 + X^3 + X \,(\mathrm{mod}\, X^7 - 1)$. Verify that $r(\omega) = \omega^4$, and hence decode $r(X)$ using minimum-distance decoding.*

Solution A cyclic code $\mathscr{X}$ of length N has generator polynomial $g(X) \in \mathbb{F}_2[X]$ and parity-check polynomial $h(X) \in \mathbb{F}_2[X]$ with $g(X)h(X) = X^N - 1$. Recall that if $g(X)$ has degree k, i.e. $g(X) = a_0 + a_1 X + \cdots + a_k X^k$ where $a_k \neq 0$, then $g(X), Xg(X), \ldots, X^{n-k-1}g(X)$ form a basis for $\mathscr{X}$. In particular, the rank of $\mathscr{X}$ equals $N - k$. In this question, $k = 3$ and $\mathrm{rank}(\mathscr{X}) = 4$.

If $h(X) = b_0 + b_1 X + \cdots + b_{N-k} X^{N-k}$ then the parity-check matrix H of code $\mathscr{X}$ is

$$
\begin{pmatrix}
b_{N-k} & b_{N-k-1} & \cdots & b_1 & b_0 & 0 & \cdots & 0 & 0 \\
0 & b_{N-k} & b_{N-k-1} & \cdots & b_1 & b_0 & \cdots & 0 & 0 \\
0 & \ddots & \ddots & \ddots & \ddots & \ddots & \ddots & \\
0 & 0 & \cdots & 0 & b_{N-k} & b_{N-k-1} & \cdots & b_1 & b_0
\end{pmatrix}.
$$

$$\underbrace{\qquad\qquad\qquad\qquad\qquad\qquad\qquad\qquad\qquad\qquad}_{N}$$

The codewords of $\mathscr{X}$ are linear dependence relations between the columns of H. The minimum distance $d(\mathscr{X})$ of a linear code $\mathscr{X}$ is the minimum non-zero weight of a codeword.

In this question, $N = 7$, and

$$
H = \begin{pmatrix}
1 & 0 & 1 & 1 & 1 & 0 & 0 \\
0 & 1 & 0 & 1 & 1 & 1 & 0 \\
0 & 0 & 1 & 0 & 1 & 1 & 1
\end{pmatrix}.
$$

$$
\begin{array}{lcl}
\text{no zero column} & \Rightarrow & \text{no codewords of weight 1} \\
\text{no repeated column} & \Rightarrow & \text{no codewords of weight 2}
\end{array}
$$

Hence, $d(\mathscr{X}) = 3$. In fact, $\mathscr{X}$ is equivalent to Hamming's original $[7,4]$ code.

Since $g(X) \in \mathbb{F}_2[X]$ is irreducible, the code $\mathscr{X} \subset \mathbb{F}_2^7 = \mathbb{F}_2[X]/(X^7 - 1)$ is the cyclic code defined by ω. The multiplicative cyclic group $\mathbb{F}_8^* \simeq \mathbb{Z}_7^\times$ of non-zero elements of field $\mathbb{F}_8$ is

$$
\begin{aligned}
\omega^0 &= 1, \\
\omega, & \\
\omega^2, & \\
\omega^3 &= \omega + 1, \\
\omega^4 &= \omega^2 + \omega, \\
\omega^5 &= \omega^3 + \omega^2 = \omega^2 + \omega + 1, \\
\omega^6 &= \omega^3 + \omega^2 + \omega = \omega^2 + 1, \\
\omega^7 &= \omega^3 + \omega = 1.
\end{aligned}
$$

Next, the value $r(\omega)$ is

$$
\begin{aligned}
r(\omega) &= \omega + \omega^3 + \omega^5 \\
&= \omega + (\omega + 1) + (\omega^2 + \omega + 1) \\
&= \omega^2 + \omega \\
&= \omega^4,
\end{aligned}
$$

as required. Let $c(X) = r(X) + X^4 \bmod (X^7 - 1)$. Then $c(\omega) = 0$, i.e. $c(X)$ is a code-word. Since $d(\mathcal{X}) = 3$ the code is 1-error correcting. We just found a codeword $c(X)$ at distance 1 from $r(X)$. Then $r(X)$ is written as

$$c(X) = X + X^3 + X^4 + X^5 \bmod (X^7 - 1),$$

and should be decoded by $c(X)$ under minimum-distance decoding. □

Problem 3.8 If $\mathcal{X}$ is a linear $[N,k]$ code, define its weight enumeration polynomial $W_{\mathcal{X}}(s,t)$. Show that:

(a) $W_{\mathcal{X}}(1,1) = 2^k$,
(b) $W_{\mathcal{X}}(0,1) = 1$,
(c) $W_{\mathcal{X}}(1,0)$ has value 0 or 1,
(d) $W_{\mathcal{X}}(s,t) = W_{\mathcal{X}}(t,s)$ if and only if $W_{\mathcal{X}}(1,0) = 1$.

Solution If $\mathbf{x} \in \mathcal{X}$ the weight $w(\mathbf{x})$ of $\mathcal{X}$ is given by $w(\mathbf{x}) = \sharp\{x_i : x_i = 1\}$. Define the weight enumeration polynomial

$$W_{\mathcal{X}}(s,t) = \sum A_j s^j t^{N-j} \tag{3.6.4}$$

where $A_j = \sharp\{\mathbf{x} \in \mathcal{X} : w(\mathbf{x}) = j\}$. Then:
(a) $W_{\mathcal{X}}(1,1) = \sharp\{\mathbf{x} : \mathbf{x} \in \mathcal{X}\} = 2^{\dim \mathcal{X}} = 2^k$.
(b) $W_{\mathcal{X}}(0,1) = A_0 = \sharp\{\mathbf{0}\} = 1$; note $\mathbf{0} \in \mathcal{X}$ since $\mathcal{X}$ is a subspace.
(c) $W_{\mathcal{X}}(1,0) = 1 \Leftrightarrow A_N = 1$, i.e. $11\ldots1 \in \mathcal{X}$, $W_{\mathcal{X}}(1,0) = 0 \Leftrightarrow A_N = 0$, i.e. $11\ldots1 \notin \mathcal{X}$.
(d) $W_{\mathcal{X}}(s,t) = W_{\mathcal{X}}(t,s) \Rightarrow W(0,1) = W(1,0) \Rightarrow W_{\mathcal{X}}(1,0) = 1$ by (b).
So, if $W_{\mathcal{X}}(1,0) = 1$ then

$$\sharp\{\mathbf{x} \in \mathcal{X} : w(\mathbf{x}) = j\} = \sharp\{\mathbf{x} + 11\ldots1 : \mathbf{x} \in \mathcal{X}, w(\mathbf{x}) = j\}$$
$$= \sharp\{\mathbf{y} \in \mathcal{X} : w(\mathbf{y}) = N - j\}$$

and $W_{\mathcal{X}}(1,0) = 1$ implies $A_{N-j} = A_j$ for all j. Hence, $W_{\mathcal{X}}(s,t) = W_{\mathcal{X}}(t,s)$. □

Problem 3.9 State the MacWilliams identity, connecting the weight enumerator polynomials of a code $\mathcal{X}$ and its dual $\mathcal{X}^{\perp}$.

Prove that the weight enumerator of the binary Hamming code $\mathcal{X}_{H,l}$ of length $N = 2^l - 1$ equals

$$W_{\mathcal{X}_l^H}(z) = \frac{1}{2^l}\left[(1+z)^{2^l-1} + (2^l - 1)(1 - z^2)^{(2^l-2)/2}(1 - z)\right]. \tag{3.6.5}$$

Solution (The second part only) Let A_i be the number of codewords of weight i. Consider $i - 1$ columns of the parity-check matrix H. There are three possibilities:
(a) the sum of these columns is $\mathbf{0}$;

(b) the sum of these columns is one of the chosen columns;

(c) the sum of these columns is one of the remaining columns.

Possibility (a) occurs A_{i-1} times; possibility (c) occurs iA_i times as the selected combination of $i-1$ columns may be obtained from any word of weight i by dropping any of its non-zero components. Next, observe that possibility (b) occurs $(N-(i-2))A_{i-2}$ times. Indeed, this combination may be obtained from a codeword of weight $i-2$ by adding any of the $N-(i-2)$ remaining columns. However, we can choose $i-1$ columns in $\binom{N}{i-1}$ ways. Hence,

$$iA_i = \binom{N}{i-1} - A_{i-1} - (N-i+2)A_{i-2}, \tag{3.6.6}$$

which is trivially correct if $i > N+1$. If we multiply both sides by z^{i-1} and then sum over i we obtain an ODE

$$A'(z) = (1+z)^N - A(z) - NzA(z) + z^2 A'(z). \tag{3.6.7}$$

Since $A(0) = 1$, the unique solution of this ODE is

$$A(z) = \frac{1}{N+1}(1+z)^N + \frac{N}{N+1}(1+z)^{(N-1)/2}(1-z)^{(N+1)/2} \tag{3.6.8}$$

which is equivalent to (3.6.5). □

Problem 3.10   Let $\mathcal{X}$ *be a linear code over* $\mathbb{F}_2$ *of length N and rank k and let A_i be the number of words in $\mathcal{X}$ of weight i, $i = 0,\dots,N$. Define the weight enumerator polynomial of $\mathcal{X}$ as*

$$W(\mathcal{X},z) = \sum_{0 \le i \le N} A_i z^i.$$

Let $\mathcal{X}^\perp$ denote the dual code to $\mathcal{X}$. Show that

$$W\left(\mathcal{X}^\perp, z\right) = 2^{-k}(1+z)^N W\left(\mathcal{X}, \frac{1-z}{1+z}\right). \tag{3.6.9}$$

[*Hint:* Consider $g(\mathbf{u}) = \sum\limits_{\mathbf{v} \in \mathbb{F}_2^N} (-1)^{\mathbf{u} \cdot \mathbf{v}} z^{w(\mathbf{v})}$ *where $w(\mathbf{v})$ denotes the weight of the vector $\mathbf{v}$ and average over $\mathcal{X}$.*]

   Hence or otherwise show that if $\mathcal{X}$ corrects at least one error then the words of $\mathcal{X}^\perp$ have average weight $N/2$.

   Apply (3.6.9) to the enumeration polynomial of Hamming code,

$$W(\mathcal{X}_{Ham}, z) = \frac{1}{N+1}(1+z)^N + \frac{N}{N+1}(1+z)^{(N-1)/2}(1-z)^{(N+1)/2}, \tag{3.6.10}$$

to obtain the enumeration polynomial of the simplex code:

$$W(\mathcal{X}_{simp}, z) = 2^{-k}2^N/2^l + 2^{-k}(2^l-1)/2^l \times 2^N z^{2^{l-1}} = 1 + (2^l-1)z^{2^{l-1}}.$$

Solution The dual code $\mathscr{X}^\perp$, of a linear code $\mathscr{X}$ with the generating matrix G and the parity-check matrix H, is defined as a linear code with the generating matrix H. If $\mathscr{X}$ is an $[N,k]$ code, $\mathscr{X}^\perp$ is an $[N,N-k]$ code, and the parity-check matrix for $\mathscr{X}^\perp$ is G.

Equivalently, $\mathscr{X}^\perp$ is the code which is formed by the linear subspace in $\mathbb{F}_2^N$ orthogonal to $\mathscr{X}$ in the dot-product

$$\langle \mathbf{x}, \mathbf{y} \rangle = \sum_{1 \le i \le N} x_i y_i, \ \mathbf{x} = x_1 \dots x_N, \ \mathbf{y} = y_1 \dots y_N.$$

By definition,

$$W(\mathscr{X},z) = \sum_{\mathbf{u} \in \mathscr{X}} z^{w(\mathbf{u})}, \ W\left(\mathscr{X}^\perp, z\right) = \sum_{\mathbf{v} \in \mathscr{X}^\perp} z^{w(\mathbf{v})}.$$

Following the hint, consider the average

$$\frac{1}{\sharp \mathscr{X}} \sum_{\mathbf{u} \in \mathscr{X}} g(\mathbf{u}), \ \text{where} \ g(\mathbf{u}) = \sum_{\mathbf{v}} (-1)^{\langle \mathbf{u}, \mathbf{v} \rangle} z^{w(\mathbf{v})}. \tag{3.6.11}$$

Then write (3.6.11) as

$$\frac{1}{\sharp \mathscr{X}} \sum_{\mathbf{v}} z^{w(\mathbf{v})} \sum_{\mathbf{u} \in \mathscr{X}} (-1)^{\langle \mathbf{u}, \mathbf{v} \rangle}. \tag{3.6.12}$$

Note that when $\mathbf{v} \in \mathscr{X}^\perp$, the sum $\sum_{\mathbf{u} \in \mathscr{X}} (-1)^{\langle \mathbf{u}, \mathbf{v} \rangle} = \sharp \mathscr{X}$. On the other hand, when $\mathbf{v} \notin \mathscr{X}^\perp$ then there exists $\mathbf{u}_0 \in \mathscr{X}$ such that $\langle \mathbf{u}_0, \mathbf{v} \rangle \ne 0$ (i.e. $\langle \mathbf{u}_0, \mathbf{v} \rangle = 1$). Hence, if $\mathbf{v} \notin \mathscr{X}^\perp$, then, with the change of variables $\mathbf{u} \mapsto \mathbf{u} + \mathbf{u}_0$, we obtain

$$\sum_{\mathbf{u} \in \mathscr{X}} (-1)^{\langle \mathbf{u}, \mathbf{v} \rangle} = \sum_{\mathbf{u} \in \mathscr{X}} (-1)^{\langle \mathbf{u} + \mathbf{u}_0, \mathbf{v} \rangle}$$
$$= (-1)^{\langle \mathbf{u}_0, \mathbf{v} \rangle} \sum_{\mathbf{u} \in \mathscr{X}} (-1)^{\langle \mathbf{u}, \mathbf{v} \rangle} = - \sum_{\mathbf{u} \in \mathscr{X}} (-1)^{\langle \mathbf{u}, \mathbf{v} \rangle},$$

which yields that in this case $\sum_{\mathbf{u} \in \mathscr{X}} (-1)^{\langle \mathbf{u}, \mathbf{v} \rangle} = 0$. We conclude that the sum in (3.6.11) equals

$$\frac{1}{\sharp \mathscr{X}} \sum_{\mathbf{v} \in \mathscr{X}^\perp} z^{w(\mathbf{v})} \left(\sharp \mathscr{X}\right) = W\left(\mathscr{X}^\perp, z\right). \tag{3.6.13}$$

On the other hand, for $\mathbf{u} = u_1 \dots u_N$,

$$g(\mathbf{u}) = \sum_{v_1, \dots, v_N} \prod_{1 \le i \le N} z^{w(v_i)} (-1)^{u_i v_i}$$
$$= \prod_{1 \le i \le N} \sum_{a=0,1} z^{w(a)} (-1)^{a u_i}$$
$$= \prod_{1 \le i \le N} \left[1 + z(-1)^{u_i}\right]. \tag{3.6.14}$$

Here $w(a) = 0$ for $a = 0$ and $w(a) = 1$ for $a = 1$. The RHS of (3.6.14) equals

$$(1-z)^{w(\mathbf{u})}(1+z)^{N-w(\mathbf{u})}.$$

Hence, an alternative expression for (3.6.11) is

$$\frac{1}{\sharp\mathscr{X}}(1+z)^N \sum_{\mathbf{u}\in\mathscr{X}}\left(\frac{1-z}{1+z}\right)^{w(\mathbf{u})} = \frac{1}{\sharp\mathscr{X}}(1+z)^N W\left(\mathscr{X},\frac{1-z}{1+z}\right). \qquad (3.6.15)$$

Equating (3.6.13) and (3.6.15) yields

$$\frac{1}{\sharp\mathscr{X}}(1+z)^N W\left(\mathscr{X},\frac{1-z}{1+z}\right) = W\left(\mathscr{X}^\perp,z\right) \qquad (3.6.16)$$

which gives the required equation as $\sharp\mathscr{X} = 2^k$.

Next, differentiate (3.6.16) in z at $z = 1$. The RHS gives

$$\sum_{0\le i\le N} iA_i\left(\mathscr{X}^\perp\right) = \left(\sharp\mathscr{X}^\perp\right) \times \left(\text{the average weight in } \mathscr{X}^\perp\right).$$

On the other hand, in the LHS we have

$$\frac{\mathrm{d}}{\mathrm{d}z}\left(\frac{1}{\sharp\mathscr{X}}\sum_{0\le i\le N} A_i(\mathscr{X})(1-z)^i(1+z)^{N-i}\right)\Bigg|_{z=1}$$

$$= \frac{1}{\sharp\mathscr{X}}\left(N2^{N-1} - A_1(\mathscr{X})2^{N-1}\right) \text{ (only terms } i = 0,1 \text{ contribute)}$$

$$= \frac{2^N}{\sharp\mathscr{X}}\frac{N}{2} \quad (A_1(\mathscr{X}) = 0 \text{ as the code is at least 1-error correcting,}$$

$$\text{with distance } \ge 3).$$

Now take into account that

$$(\sharp\mathscr{X}) \times (\sharp\mathscr{X}^\perp) = 2^k \times 2^{N-k} = 2^N.$$

The equality

$$\text{the average weight in } \mathscr{X}^\perp = \frac{N}{2}$$

follows. The enumeration polynomial of the simplex code is obtained by substitution. In this case the average length is $(2^l - 1)/2$. □

Problem 3.11   *Describe the binary narrow-sense BCH code X of length 15 and the designed distance 5 and find the generator polynomial. Decode the message* 100000111000100.

Solution Take the binary narrow-sense BCH code X of length 15 and the designed distance 5. We have $\mathrm{Spl}(X^{15}-1) = \mathbb{F}_{2^4} = \mathbb{F}_{16}$. We know that X^4+X+1 is a primitive polynomial over $\mathbb{F}_{16}$. Let ω be a root of X^4+X+1. Then

$$M_1(X) = X^4+X+1, M_3(X) = X^4+X^3+X^2+X+1,$$

and the generator $g(X)$ for X is

$$g(X) = M_1(X)M_3(X) = X^8+X^7+X^6+X^4+1.$$

Take $g(X)$ as example of a codeword. Introduce 2 errors – at positions 4 and 12 – by taking

$$u(X) = X^{12}+X^8+X^7+X^6+1.$$

Using the field table for $\mathbb{F}_{16}$, obtain

$$u_1 = u(\omega) = \omega^{12}+\omega^8+\omega^7+\omega^6+1 = \omega^6$$

and

$$u_3 = u(\omega^3) = \omega^{36}+\omega^{24}+\omega^{18}+1 = \omega^9+\omega^3+1 = \omega^4.$$

As $u_1 \neq 0$ and $u_1^3 = \omega^{18} = \omega^3 \neq u_3$, deduce that ≥ 2 errors occurred. Calculate the locator polynomial

$$l(X) = 1+\omega^6X+(\omega^{13}+\omega^{12})X^2.$$

Substituting $1, \omega, \ldots, \omega^{14}$ into $l(X)$, check that ω^3 and ω^{11} are roots. This confirms that, if exactly 2 errors occurred their positions are 4 and 12 then the codeword sent was 100010111000000. □

Problem 3.12   For a word $\mathbf{x} = x_1\ldots x_N \in \mathbb{F}_2^N$ the weight $w(\mathbf{x})$ is the number of non-zero digits: $w(\mathbf{x}) = \sharp\{i : x_i \neq 0\}$. For a linear $[N,k]$ code X let A_i be the number of words in X of weight i $(0 \leq i \leq N)$. Define the weight enumerator polynomial $W(X,z) = \sum_{i=0}^{N} A_i z^i$. Show that if we use X on a binary symmetric channel with error-probability p, the probability of failing to detect an incorrect word is $(1-p)^N \left(W\left(X, \frac{p}{1-p}\right) - 1\right)$.

Solution Suppose we have sent the zero codeword $\mathbf{0}$. Then the error-probability

$$E = \sum_{\mathbf{x} \in X \setminus \mathbf{0}} \mathbf{P}(\mathbf{x} \mid \mathbf{0} \text{ sent}) = \sum_{i \geq 1} A_i p^i (1-p)^{N-i} =$$

$$(1-p)^N \left[\sum_{i \geq 0} A_i \left(\frac{p}{1-p}\right)^i - 1\right] = (1-p)^N \left(W\left(X, \frac{p}{1-p}\right) - 1\right).$$

□

Problem 3.13   Let $\mathcal{X}$ be a binary linear $[N,k,d]$ code, with the weight enumerator $W_{\mathcal{X}}(s)$. Find expressions, in terms of $W_{\mathcal{X}}(s)$, for the weight enumerators of:

(i) the subcode $\mathcal{X}^{\mathrm{ev}} \subseteq \mathcal{X}$ consisting of all codewords $\mathbf{x} \in \mathcal{X}$ of even weight,
(ii) the parity-check extension $\mathcal{X}^{\mathrm{pc}}$ of $\mathcal{X}$.

Prove that if d is even then there exists an $[N,k,d]$ code where each codeword has even weight.

Solution (i) All words with even weights from $\mathcal{X}$ belong to subcode $\mathcal{X}^{\mathrm{ev}}$. Hence

$$W_{\mathcal{X}}^{\mathrm{ev}}(s) = \frac{1}{2}\left[W_{\mathcal{X}}(s) + W_{\mathcal{X}}(-s)\right].$$

(ii) Clearly, all non-zero coefficients of weight enumeration polynomial for $\mathcal{X}^+$ corresponds to even powers of z, and $A_{2i}(\mathcal{X}^+) = A_{2i}(\mathcal{X}) + A_{2i-1}(\mathcal{X})$, $i = 1, 2, \ldots$. Hence,

$$W_{\mathcal{X}}^{\mathrm{pc}}(s) = \frac{1}{2}\left[(1+s)W_{\mathcal{X}}(s) + (1-s)W_{\mathcal{X}}(-s)\right].$$

If $\mathcal{X}$ is binary $[N,k,d]$ then you first truncate $\mathcal{X}$ to $\mathcal{X}^-$ then take the parity-check extension $(\mathcal{X}^-)^+$. This preserves k and d (if d is even) and makes all codewords of even weight.   □

Problem 3.14   Check that polynomials $X^4 + X^3 + X^2 + X + 1$ and $X^4 + X + 1$ are irreducible over $\mathbb{F}_2$. Are these polynomials primitive over $\mathbb{F}_2$? What about polynomials $X^3 + X + 1$, $X^3 + X^2 + 1$? $X^4 + X^3 + 1$?

Solution As both polynomials $X^4 + X^3 + X^2 + X + 1$ and $X^4 + X + 1$ do not vanish at $X = 0$ or $X = 1$, they are not divisible by X or $X + 1$. They are also not divisible by $X^2 + X + 1$, the only irreducible polynomial of degree 2, or by $X^3 + X + 1$ or $X^3 + X^2 + 1$, the only irreducible polynomials of degree 3. Hence, they are irreducible.
  The polynomial $X^4 + X^3 + X^2 + X + 1$ cannot be primitive polynomial as it divides $X^5 - 1$. Let us check that $X^4 + X + 1$ is primitive. Take $\mathbb{F}_2[X]/\langle X^4 + X + 1\rangle$ and use the $\mathbb{F}_2^4$ field table. The cyclotomic coset is $\{\omega, \omega^2, \omega^4, \omega^8\}$ (as $\omega^{16} = \omega$). The primitive polynomial $M_\omega(X)$ is then

$$(X - \omega)(X - \omega^2)(X - \omega^4)(X - \omega^8)$$
$$= X^4 - (\omega + \omega^2 + \omega^4 + \omega^8)X^2$$
$$+ (\omega\omega^2 + \omega\omega^4 + \omega\omega^8 + \omega^2\omega^4 + \omega^2\omega^8 + \omega^4\omega^8)X^2$$
$$- (\omega\omega^2\omega^4 + \omega\omega^2\omega^8 + \omega\omega^4\omega^8 + \omega^2\omega^4\omega^8)x + \omega\omega^2\omega^4\omega^8$$
$$= X^4 - (\omega + \omega^2 + \omega^4 + \omega^8)X^2 + (\omega^3\omega^5 + \omega^9 + \omega^6 + \omega^{10} + \omega^{12})X^2$$
$$- (\omega^7 + \omega^{11} + \omega^{13} + \omega^{14})X + \omega^{15} = X^4 + X + 1.$$

The order of $X^4 + X + 1$ is 15: other primitive polynomials of order 15 are $X^4 + X^3 + 1$ and $X^4 + X + 1$. Thus, the only primitive polynomial of degree 4 is $X^4 + X + 1$. Similarly, the only primitive polynomials of degree 3 are $X^3 + X + 1$ and $X^3 + X^2 + 1$, both of order 7.   □

Problem 3.15 *Suppose a binary narrow-sense BCH code is used, of length 15, designed distance 5, and the received word is $X^{10} + X^5 + X^4 + X + 1$. How is it decoded? If the received word is $X^{11} + X^{10} + X^6 + X^5 + X^4 + X + 1$, what is the number of errors?*

Solution Suppose the received word is

$$r(X) = X^{10} + X^5 + X^4 + X + 1,$$

and let ω be a primitive element in $\mathbb{F}_{16}$. Then

$$s_1 = r(\omega) = \omega^{10} + \omega^5 + \omega^4 + \omega + e$$
$$= 0111 + 0110 + 0011 + 0010 + 0001 = 0001 = e,$$
$$s_3 = r(\omega^3) = \omega^{30} + \omega^{15} + \omega^{12} + \omega^3 + e$$
$$= 0001 + 0001 + 1111 + 1000 + 0001 = 0110 = \omega^5.$$

See that $s_3 \neq s_1^3$: two errors. The error-locator polynomial

$$\sigma(X) = e + s_1 X + (s_3 s_1^{-1} + s_1^2)X^2 = e + X + (\omega^5 + e)X^2 = e + X + \omega^{10}X^2.$$

Checking for the roots, $\omega^0 = e, \omega^1, \omega^2, \omega^3, \omega^4, \omega^5, \omega^6$: no, ω^7: yes. Then divide:

$$(\omega^{10}X^2 + X + e)/(X + \omega^7) = \omega^{10}X + \omega^8 = \omega^{10}(X + \omega^{13}),$$

and identify the second root: ω^{13}. So, the errors occurred at positions $15 - 7 = 8$ and $15 - 13 = 2$. Decode:

$$r(X) \mapsto X^{10} + X^8 + X^5 + X^4 + X^2 + X + 1.$$

  □

Problem 3.16 *Prove that the binary code of length 23 generated by the polynomial $g(X) = 1 + X + X^5 + X^6 + X^7 + X^9 + X^{11}$ has minimal distance 7, and is perfect.*

[Hint: If $g^{\text{rev}}(X) = X^{11}g(1/X)$ is the reversal of $g(X)$ then

$$X^{23} + 1 \equiv (X+1)g(X)g^{\text{rev}}(X) \bmod 2.]$$

Solution First, show that the code is BCH, of designed distance 5. By the fresher's dream Lemma 3.1.5, if ω is a root of a polynomial $f(X) \in \mathbb{F}_2[X]$ then so is ω^2. Thus, if ω is a root of $g(X) = 1 + X + X^5 + X^6 + X^7 + X^9 + X^{11}$ then so are ω, ω^2, ω^4, ω^8, ω^{16}, ω^9, ω^{18}, ω^{13}, ω^3, ω^6, ω^{12}. This yields the design sequence

$\{\omega, \omega^2, \omega^3, \omega^4\}$. By the BCH bound (Theorem 2.5.39 and Theorem 3.2.9), the cyclic code $\mathscr{X}$ generated by $g(X)$ has distance ≥ 5.

Next, the parity-check extension, $\mathscr{X}^+$, is self-orthogonal. To check this, we need only to show that any two rows of the generating matrix of $\mathscr{X}^+$ are orthogonal. These are represented by the concatenated words

$$(X^i g(X)|1) \text{ and } (X^j g(X)|1).$$

Their dot-product equals

$$1 + (X^i g(X))(X^j g(X)) = 1 + \sum_r g_{i+r} g_{j+r}$$
$$= 1 + \sum_r g_{i+r} g^{\text{rev}}_{11-j-r}$$
$$= 1 + \text{coefficient of } X^{11+i-j} \text{ in } \underbrace{g(X) \times g^{\text{rev}}(X)}_{\substack{|| \\ 1 + \cdots + X^{22}}}$$
$$= 1 + 1 = 0.$$

We conclude that

any two words in $\mathscr{X}^+$ are dot-orthogonal.

Next, observe that all words in $\mathscr{X}^+$ have weights divisible by 4. Indeed, by inspection, all rows $(X^i g(X)|1)$ of the generating matrix of $\mathscr{X}^+$ have weight 8. Then, by induction on the number of rows involved in the sum, if $\mathbf{x} \in \mathscr{X}^+$ and $\mathbf{g}^{(i)} \sim (X^i g(X)|1)$ is a row of the generating matrix of $\mathscr{X}^+$ then

$$w(\mathbf{g}^{(i)} + \mathbf{x}) = w(\mathbf{g}^{(i)}) + w(\mathbf{x}) - 2w(\mathbf{g}^{(i)} \wedge \mathbf{x}), \qquad (3.6.17)$$

where $(\mathbf{g}^{(i)} \wedge \mathbf{x})_l = \min\ [(\mathbf{g}^{(i)})_l, x_l]$, $l = 1, \ldots, 24$. We know that 8 divides $w(\mathbf{g}^{(i)})$. Moreover, by the induction hypothesis, 4 divides $w(\mathbf{x})$. Next, by (3.6.17), $w(\mathbf{g}^{(i)} \wedge \mathbf{x})$ is even, so $2w(\mathbf{g}^{(i)} \wedge \mathbf{x})$ is divisible by 4. Then the LHS, $w(\mathbf{g}^{(i)} + \mathbf{x})$, is divisible by 4.

Therefore, the distance of $\mathscr{X}^+$ is 8, as it is ≥ 5 and is divisible by 4. (It is easy to see that it cannot be > 8 as then it would be 12.) Then the distance of the original code, $\mathscr{X}$, equals 7.

The code $\mathscr{X}$ is perfect 3-error correcting, since the volume of the 3-ball in $\mathbb{F}_2^{23}$ equals

$$\binom{23}{0} + \binom{23}{1} + \binom{23}{2} + \binom{23}{3} = 1 + 23 + 253 + 1771 = 2048 = 2^{11},$$

and

$$2^{11} \times 2^{12} = 2^{23}.$$

Here, obviously, 12 represents the rank and 23 the length.   □

Problem 3.17 Use the MacWilliams identity to prove that the weight distribution of a q-ary MDS code of distance d is

$$A_i = \binom{N}{i} \sum_{0 \le j \le i-d} (-1)^j \binom{i}{j} \left(q^{i-d+1-j} - 1 \right)$$

$$= \binom{N}{i} (q-1) \sum_{0 \le j \le i-d} (-1)^j \binom{i-1}{j} q^{i-d-j}, \quad d \le i \le N.$$

[*Hint:* To begin the solution,

(a) write the standard MacWilliams identity,
(b) swap $\mathscr{X}$ and $\mathscr{X}^{\perp}$,
(c) change $s \mapsto s^{-1}$,
(d) multiply by s^n and
(e) take the derivative d^r/ds^r, $0 \le r \le k$ (which equals $d(\mathscr{X}^{\perp}) - 1$).

Use the Leibniz rule

$$\frac{d^r}{ds^r} \left[f(s)g(s) \right] = \sum_{0 \le j \le r} \binom{r}{j} \left[\frac{d^j}{ds^j} f(s) \right] \left[\frac{d^{r-j}}{ds^{r-j}} g(s) \right]. \qquad (3.6.18)$$

Use the fact that $d(\mathscr{X}) = N - k + 1$ and $d(\mathscr{X}^{\perp}) = k + 1$ and obtain simplified equations involving $A_{N-k+1}, \ldots, A_{N-r}$ only. Subsequently, determine $A_{N-k+1}, \ldots, A_{N-r}$. Varying r, continue up to A_N.]

Solution The MacWilliams identity is

$$\sum_{1 \le i \le N} A_i^{\perp} s^i - \frac{1}{q^k} \sum_{1 \le i \le N} A_i (1-s)^i \left[1 + (q-1)s \right]^{N-i}.$$

Swap $\mathscr{X}$ and $\mathscr{X}^{\perp}$, change $s \mapsto s^{-1}$ and multiply by s^N. After this differentiate $r \le k$ times and substitute $s = 1$:

$$\frac{1}{q^k} \sum_{0 \le i \le N-r} \binom{N-i}{r} A_i = \frac{1}{q^r} \sum_{0 \le i \le r} A_i^{\perp} \binom{N-i}{N-r} \qquad (3.6.19)$$

(the Leibniz rule (3.6.18) is used here). Formula (3.6.19) is the starting point. For an MDS code, $A_0 = A_0^{\perp} = 1$, and

$$A_i = 0, \ 1 \le i \le N - k \ (= d - 1), \ A_i^{\perp} = 0, \ 1 \le i \le k \ (= d^{\perp} - 1).$$

Then

$$\binom{N}{r} \frac{1}{q^k} + \frac{1}{q^k} \sum_{i=N-k+1}^{N-r} \binom{N-i}{r} A_i = \frac{1}{q^r} \binom{N}{N-r} = \frac{1}{q^r} \binom{N}{r},$$

i.e.

$$\sum_{i=N-k+1}^{N-r} \binom{N-i}{r} A_i = \binom{N}{r}(q^{k-r}-1).$$

For $r = k$ we obtain $0 = 0$, for $r = k - 1$

$$A_{N-k+1} = \binom{N}{k-1}(q-1), \qquad (3.6.20)$$

for $r = k - 2$

$$\binom{k-1}{k-2} A_{N-k+1} + A_{N-k+2} = \binom{N}{k-2}(q^2-1),$$

etc. This is a triangular system of equations for $A_{N-k+1}, \ldots, A_{N-r}$. Varying r, we can get $A_{N-k+1}, \ldots, A_{N-1}$. The result is

$$
\begin{aligned}
A_i &= \binom{N}{i} \sum_{0 \le j \le i-d} (-1)^j \binom{i}{j}(q^{i-d+1-j}-1) \\
&= \binom{N}{i}\left[\sum_{0 \le j \le i-d}(-1)^j\binom{i-1}{j}(qq^{i-d-j}-1)\right. \\
&\qquad \left. - \sum_{1 \le j \le i-d+1}(-1)^{j-1}\binom{i-1}{j-1}(q^{i-d+1-j}-1)\right] \\
&= \binom{N}{i}(q-1)\sum_{0 \le j \le i-d}(-1)^j\binom{i-1}{j}q^{i-d-j}, d \le i \le N,
\end{aligned}
$$

as required.

In fact, (3.6.20) can be obtained without calculations: in an MDS code of rank k and distance d any $k = N - d + 1$ digits determine the codeword uniquely. Further, for any choice of $N - d$ positions there are exactly q codewords with digits 0 in these positions. One of them is the zero codeword, and the remaining $q - 1$ are of weight d. Hence,

$$A_{N-k+1} = A_d = \binom{N}{d}(q-1).$$

$\square$

Problem 3.18   *Prove the following properties of Kravchuk's polynomials $K_k(i)$.*

(a) For all q: $(q-1)^i \binom{N}{i} K_k(i) = (q-1)^k \binom{N}{k} K_i(k).$

(b) For $q = 2$: $K_k(i) = (-1)^k K_k(N-i).$

(c) For $q = 2$: $K_k(2i) = K_{N-k}(2i).$

Solution Write

$$K_k(i) = \sum_{0 \vee (i+k-N) \le j \le k \wedge i} \binom{i}{j} \binom{N-i}{k-j} (-1)^j (q-1)^{k-j}.$$

Next:

(a) The following straightforward equation holds true:

$$(q-1)^i \binom{N}{i} K_k(i) = (q-1)^k \binom{N}{k} K_i(k)$$

(as all summands become insensitive to swapping $i \leftrightarrow k$).

For $q = 2$ this yields $\binom{N}{i} K_k(i) = \binom{N}{k} K_i(k)$; in particular,

$$\binom{N}{i} K_0(i) = \binom{N}{0} K_i(0) = K_i(0).$$

(b) Also, for $q = 2$: $K_k(i) = (-1)^k K_k(N-i)$ (again straightforward, after swapping $i \leftrightarrow i - j$).

(c) Thus, still for $q = 2$: $\binom{N}{2i} K_k(2i) = \binom{N}{k} K_{2i}(k)$ which equals

$$(-1)^{2i} \binom{N}{N-k} K_{2i}(N-k) = \binom{N}{2i} K_{N-k}(2i). \text{ That is,}$$

$$K_k(2i) = K_{N-k}(2i).$$

□

Problem 3.19   What is an $(n, \mathbb{F}_q)$-root of unity? Show that the set $\mathbb{E}^{(n,q)}$ of the $(n, \mathbb{F}_q)$-roots of unity form a cyclic group. Check that the order of $\mathbb{E}^{(n,q)}$ equals n if n and q are co-prime. Find the minimal s such that $\mathbb{E}^{(n,q)} \subset \mathbb{F}_{q^s}$.

Define a primitive $(n, \mathbb{F}_q)$-root of unity. Determine the number of primitive $(n, \mathbb{F}_q)$-roots of unity when n and q are co-prime. If ω is a primitive $(n, \mathbb{F}_q)$-root of unity, find the minimal ℓ such that $\omega \in \mathbb{F}_{q^\ell}$.

Find representation of all elements of $\mathbb{F}_9$ as vectors over $\mathbb{F}_3$. Find all $(4, \mathbb{F}_9)$-roots of unity as vectors over $\mathbb{F}_3$.

Solution We know that any root of an irreducible polynomial of degree 2 over field $\mathbb{F}_3 = \{0, 1, 2\}$ belongs to $\mathbb{F}_9$. Take the polynomial $f(X) = X^2 + 1$ and denote its root by α (any of the two). Then all elements of $\mathbb{F}_9$ may be represented as $a_0 + a_1 \alpha$ where $a_0, a_1 \in \mathbb{F}_3$. In fact,

$$\mathbb{F}_9 = \{0, 1, \alpha, 1+\alpha, 2+\alpha, 2\alpha, 1+2\alpha, 2+2\alpha\}.$$

Another approach is as follows: we know that $X^8 - 1 = \prod_{1 \le i \le 8} (X - \zeta^i)$ in the field $\mathbb{F}_9$ where ζ is a primitive $(8, \mathbb{F}_9)$-root of unity. In terms of circular polynomials, $X^8 - 1 = Q_1(X)Q_2(X)Q_4(X)Q_8(X)$. Here $Q_n(x) = \prod_{s:\gcd(s,n)=1}(x - \omega^s)$ where ω is a primitive $(n, \mathbb{F}_9)$-root of unity. Write $X^8 - 1 = \prod_{d:d|8} Q_d(x)$. Next, compute

$$Q_1(X) = -1 + X, Q_2(X) = 1 + X, Q_4(X) = 1 + X^2,$$

$$Q_8(X) = (X^8 - 1)/(Q_1(X)Q_2(X)Q_4(X)) = (X^8 - 1)/(X^4 - 1) = X^4 + 1.$$

As $3^2 = 1 \bmod 8$, by Theorem 3.1.53 $Q_8(X)$ should be decomposed over $\mathbb{F}_3$ into product of $\phi(8)/2 = 2$ irreducible polynomials of degree 2. Indeed,

$$Q_8(X) = (X^2 + X + 2)(X^2 + 2X + 2).$$

Let ζ be a root of $X^2 + X + 2$, then it is a primitive root of degree 8 over $\mathbb{F}_3$ and $\mathbb{F}_9 = \mathbb{F}_3(\zeta)$. Hence, $\mathbb{F}_9 = \{0, \zeta, \zeta^2, \zeta^3, \zeta^4, \zeta^5, \zeta^6, \zeta^7, \zeta^8\}$, and $\zeta = 1 + \alpha$. Finally, we present the index table

$$\zeta = 1 + \alpha, \zeta^2 = 2\alpha, \zeta^3 = 1 + 2\alpha, \zeta^4 = 2,$$
$$\zeta^5 = 2 + 2\alpha, \zeta^6 = \alpha, \zeta^7 = 2 + \alpha, \zeta^8 = 1.$$

Hence, the roots of degree 4 are $\zeta^2, \zeta^4, \zeta^6, \zeta^8$. $\qquad\qquad\square$

Problem 3.20 *Define a cyclic code of length N over the field $\mathbb{F}_q$. Show that there is a bijection between the cyclic codes of length N, and the factors of $X^N - e$ in the polynomial ring $\mathbb{F}_q[X]$.*

Now consider binary cyclic codes. If N is an odd integer then we can find a finite extension K of $\mathbb{F}_2$ that contains a primitive Nth root of unity ω. Show that a cyclic code of length N with defining set $\{\omega, \omega^2, \ldots, \omega^{\delta-1}\}$ has minimum distance at least δ. Show that if $N = 2^\ell - 1$ and $\delta = 3$ then we obtain the Hamming $[2^\ell - 1, 2^\ell - 1 - \ell, 3]$ code.

Solution A linear code $\mathcal{X} \subset \mathbb{F}_q^{\times N}$ is a cyclic code if $x_1 \ldots x_N \in \mathcal{X}$ implies that $x_2, \ldots x_N x_1 \in \mathcal{X}$. Bijection of cyclic codes and factors of $X^N - 1$ can be established as in Corollary 3.3.3.

Passing to binary codes, consider, for brevity, $N = 7$. Factorising in $\mathbb{F}_2^7$ renders the decomposition

$$X^7 - 1 = (X - 1)(X^3 + X + 1)(X^3 + X^2 + 1) := (X - 1)f_1(X)f_2(X).$$

Suppose ω is a root of $f_1(X)$. Since $f_1(X)^2 = f_1(X^2)$ in $\mathbb{F}_2[X]$ we have

$$f_1(\omega) = f_1(\omega^2) = 0.$$

It follows that the cyclic code $\mathcal{X}$ with defining root ω has the generator polynomial $f_1(X)$ and the check polynomial $(X - 1)f_2(X) = X^4 + X^2 + X + 1$. This property characterises Hamming's original code (up to equivalence). The case where ω is a root of $f_2(X)$ is similar (in fact, we just reverse every codeword). For a general $N = 2^l - 1$, we take a primitive element $\omega \in \mathbb{F}_{2^l}$ and its minimal polynomial $M_\omega(X)$. The roots of $M_\omega(X)$ are $\omega, \omega^2, \ldots, \omega^{2^{l-1}}$, hence $\deg M_\omega(X) = l$. Thus, a code with defining root ω has rank $N - l = 2^l - 1 - l$, as in the Hamming $[2^l - 1, 2^l - l - 1]$ code. $\square$

Problem 3.21 *Write an essay comparing the decoding procedures for Hamming and two-error correcting BCH codes.*

Solution To clarify the ideas behind the BCH construction, we first return to the Hamming codes. The Hamming $[2^l - 1, 2^l - 1 - l]$ code is a perfect one-error correcting code of length $N = 2^l - 1$. The procedure of decoding the Hamming code is as follows. Having a word $\mathbf{y} = y_1 \ldots y_N$, $N = 2^l - 1$, form the syndrome $\mathbf{s} = \mathbf{y}H^{\mathrm{T}}$. If $\mathbf{s} = 0$, decode $\mathbf{y}$ by $\mathbf{y}$. If $\mathbf{s} \neq 0$ then $\mathbf{s}$ is among the columns of $H = H_{Ham}$. If this is column i, decode $\mathbf{y}$ by $\mathbf{x}_* = \mathbf{y} + \mathbf{e}_i$, where $\mathbf{e}_i = 0 \ldots 010 \ldots 0$ (1 in the ith position, 0 otherwise).

We can try the following idea to be able to correct more than one error (two to start with). Select $2l$ of the rows of the parity-check matrix in the form

$$\tilde{H} = \begin{pmatrix} H \\ \Pi H \end{pmatrix}. \tag{3.6.21}$$

Here ΠH_{Ham} is obtained by permuting the columns of H_{Ham} (Π is a permutation of degree $2^l - 1$). The new matrix $\tilde{H}$ contains $2l$ linearly independent rows: it then determines a $[2^l - 1, 2^l - 1 - 2l]$ linear code. The syndromes are now words of length $2l$ (or pairs of words of length l): $\mathbf{y}\tilde{H}^{\mathrm{T}} = (\mathbf{s}\mathbf{s}')$. A syndrome $(\mathbf{s}, \mathbf{s}')^{\mathrm{T}}$ may or may not be among the columns of $\tilde{H}$. Recall, we want the new code to be two-error correcting, and the decoding procedure to be similar to the one for the Hamming codes. Suppose two errors occur, i.e. $\mathbf{y}$ differs from a codeword $\mathbf{x}$ by two digits, say i and j. Then the syndrome is

$$\mathbf{y}\tilde{H}^{\mathrm{T}} = (\mathbf{s}_i + \mathbf{s}_j, \quad \mathbf{s}_{\Pi i} + \mathbf{s}_{\Pi j})$$

where $\mathbf{s}_k$ is the word representing column k in H. We organise our permutation so that, knowing vector $(\mathbf{s}_i + \mathbf{s}_j, \mathbf{s}_{\Pi i} + \mathbf{s}_{\Pi j})$, we can always find i and j (or equivalently, $\mathbf{s}_i$ and $\mathbf{s}_j$). In other words, we should be able to solve the equations

$$\mathbf{s}_i + \mathbf{s}_j = \mathbf{z}, \quad \mathbf{s}_{\Pi i} + \mathbf{s}_{\Pi j} = \mathbf{z}' \tag{3.6.22}$$

for any pair $(\mathbf{z}, \mathbf{z}')$ that may eventually occur as a syndrome under two errors.

A natural guess is to try a permutation Π that has some *algebraic* significance, e.g. $\mathbf{s}_{\Pi i} = \mathbf{s}_i \mathbf{s}_i = (\mathbf{s}_i)^2$ (a bad choice) or $\mathbf{s}_{\Pi i} = \mathbf{s}_i \mathbf{s}_i \mathbf{s}_i = (\mathbf{s}_i)^3$ (a good choice) or, generally, $\mathbf{s}_{\Pi i} = \mathbf{s}_i \mathbf{s}_i \cdots \mathbf{s}_i$ (k times). Say, one can try the multiplication mod $1 + X^N$; unfortunately, the multiplication does not lead to a field. The reason is that polynomial $1 + X^N$ is always *reducible*. So, suppose we organise the check matrix as

$$\widetilde{H}^{\mathrm{T}} = \begin{pmatrix} (1\dots00) & (1\dots00)^k \\ & \vdots \\ (1\dots11) & (1\dots11)^k \end{pmatrix}.$$

Then we have to deal with equations of the type

$$\mathbf{s}_i + \mathbf{s}_j = \mathbf{z}, \ \mathbf{s}_i^k + \mathbf{s}_j^k = \mathbf{z}'. \tag{3.6.23}$$

For solving (3.6.23), we need the *field structure* of the Hamming space, i.e. not only multiplication but also *division*. Any field structure on the Hamming space of length N is isomorphic to $\mathbb{F}_{2^N}$, and a concrete realisation of such a structure is $\mathbb{F}_2[X]/\langle c(X) \rangle$, a polynomial field modulo an irreducible polynomial $c(X)$ of degree N. Such a polynomial always exists: it is one of the primitive polynomials of degree N. In fact, the simplest consistent system of the form (3.6.23) is

$$\mathbf{s} + \mathbf{s}' = \mathbf{z}, \ \mathbf{s}^3 + \mathbf{s}'^3 = \mathbf{z}';$$

it is reduced to a single equation $\mathbf{z}\mathbf{s}^2 - \mathbf{z}^2\mathbf{s} + \mathbf{z}^3 - \mathbf{z}' = 0$, and our problem becomes to factorise the polynomial $zX^2 - z^2 X + z^3 - z'$.

For $N = 2^l - 1, l = 4$ we obtain $[15, 7, 5]$ code. The rank 7 is due to the linear independence of the columns of $\widetilde{H}$. The key point is to check that the code corrects up to two errors. First suppose we received a word $\mathbf{y} = y_1 \dots y_{15}$ in which two errors occurred in digits i and j that are unknown. In order to find these places, calculate the syndrome $\mathbf{y}\widetilde{H}^{\mathrm{T}} = (\mathbf{z}, \mathbf{z}')^{\mathrm{T}}$. Recall that $\mathbf{z}$ and $\mathbf{z}'$ are words of length 4; the total length of the syndrome is 8. Note that $\mathbf{z}' \neq \mathbf{z}^3$: if $\mathbf{z}' = \mathbf{z}^3$, precisely one error occurred. Write a pair of equations

$$\mathbf{s} + \mathbf{s}' = \mathbf{z}, \ \mathbf{s}^3 + \mathbf{s}'^3 = \mathbf{z}', \tag{3.6.24}$$

where $\mathbf{s}$ and $\mathbf{s}'$ are words of length 4 (or equivalently their polynomials), and the multiplication is mod $1 + X + X^4$. In the case of two errors it is guaranteed that there is exactly one pair of solutions to (3.6.24), one vector occupying position i and another position j, among the columns of the upper (Hamming) half of matrix $\widetilde{H}$. Moreover, (3.6.24) cannot have more than one pair of solutions because

$$\mathbf{z}' = \mathbf{s}^3 + \mathbf{s}'^3 = (\mathbf{s} + \mathbf{s}')(\mathbf{s}^2 + \mathbf{s}\mathbf{s}' + \mathbf{s}'^2) = \mathbf{z}(\mathbf{z}^2 + \mathbf{s}\mathbf{s}')$$

implies that

$$\mathbf{s}\mathbf{s}' = \mathbf{z}'\mathbf{z}^{-1} + \mathbf{z}^2. \tag{3.6.25}$$

Now (3.6.25) and the first equation in (3.6.24) give that $\mathbf{s}, \mathbf{s}'$ are precisely the roots of a quadratic equation

$$X^2 + \mathbf{z}X + \left(\mathbf{z}'\mathbf{z}^{-1} + \mathbf{z}^2\right) = 0 \qquad (3.6.26)$$

(with $\mathbf{z}'\mathbf{z}^{-1} + \mathbf{z}^2 \neq 0$). But the polynomial in the LHS of (3.6.26) cannot have more than two distinct roots (it could have no root or two coinciding roots, but it is excluded by the assumption that there are precisely two errors). In the case of a single error, we have $\mathbf{z}' = \mathbf{z}^3$; in this case $\mathbf{s} = \mathbf{z}$ is the only root and we just find the word $\mathbf{z}$ among the columns of the upper half of matrix $\widetilde{H}$.

Summarising, the decoding scheme, in the case of the above $[15, 7]$ code, is as follows: Upon receiving word $\mathbf{y}$, form a syndrome $\mathbf{y}\widetilde{H}^{\mathrm{T}} = (\mathbf{z}, \mathbf{z}')^{\mathrm{T}}$. Then

(i) If both $\mathbf{z}$ and $\mathbf{z}'$ are zero words, conclude that no error occurred and decode $\mathbf{y}$ by $\mathbf{y}$ itself.

(ii) If $\mathbf{z} \neq 0$ and $\mathbf{z}^3 = \mathbf{z}'$, conclude that a single error occurred and find the location of the error digit by identifying word $\mathbf{z}$ among the columns of the Hamming check matrix.

(iii) If $\mathbf{z} \neq 0$ and $\mathbf{z}^3 \neq \mathbf{z}'$, form the quadric (3.6.24), and if it has two distinct roots $\mathbf{s}$ and $\mathbf{s}'$, conclude that two errors occurred and locate the error digits by identifying words $\mathbf{s}$ and $\mathbf{s}'$ among the columns of the Hamming check matrix.

(iv) If $\mathbf{z} \neq 0$ and $\mathbf{z}^3 \neq \mathbf{z}'$ and quadric (3.6.26) has no roots, or if $\mathbf{z}$ is zero but $\mathbf{z}'$ is not, conclude that there are at least three errors.

Note that the case where $\mathbf{z} \neq 0$, $\mathbf{z}^3 \neq \mathbf{z}'$ and quadric (3.6.26) has a single root is impossible: if (3.6.26) has a root, $\mathbf{s}$ say, then either another root $\mathbf{s}' \neq \mathbf{s}$ or $\mathbf{z} = 0$ and a single error occurs.

The decoding procedure allows us to detect, in some cases, that more than three errors occurred. However, this procedure may lead to a wrong codeword when three or more errors occur. $\qquad\qquad\square$

4

Further Topics from Information Theory

In Chapter 4 it will be convenient to work in a general setting which covers both discrete and continuous-type probability distributions. To do this, we assume that probability distributions under considerations are given by their Radon–Nikodym derivatives with respect to underlying reference measures usually denoted by μ or ν. The role of a reference measure can be played by a counting measure supported by a discrete set or by the Lebesgue measure on $\mathbb{R}^d$; we need only that the reference measure is locally finite (i.e. it assigns finite values to compact sets). The Radon–Nikodym derivatives will be called probability mass functions (PMFs): they represent probabilities in the discrete case and probability density functions (PDFs) in the continuous case.

The initial setting of the channel capacity theory developed for discrete channels in Chapter 1 (see Section 1.4) goes almost unchanged for a continuously distributed noise by adopting the logical scheme:

a set $\mathscr{U}$ of messages, of cardinality $M = \lceil 2^{NR} \rceil$
$\rightarrow$ a codebook $\mathscr{X}$ of size M with codewords of length N
$\rightarrow$ reliable rate R of transmission through a noisy channel
$\rightarrow$ the capacity of the channel.

However, to simplify the exposition, we will assume from now on that encoding $\mathscr{U} \rightarrow \mathscr{X}$ is a one-to-one map and identify a code with its codebook.

4.1 Gaussian channels and beyond

Here we study channels with continuously distributed noise; they are the basic models in telecommunication, including both wireless and telephone transmission. The most popular model of such a channel is a memoryless additive Gaussian channel (MAGC) but other continuous-noise models are also useful. The case of an

366

MAGC is particularly attractive because it allows one to do some handy and far-reaching calculations with elegant answers.

However, Gaussian (and other continuously distributed) channels present a challenge that was absent in the case of finite alphabets considered in Chapter 1. Namely, because codewords (or, using a slightly more appropriate term, codevectors) can *a priori* take values from a Euclidean space (as well as noise vectors), the definition of the channel capacity has to be modified, by introducing a power constraint. More generally, the value of capacity for a channel will depend upon the so-called *regional constraints* which can generate analytic difficulties. In the case of MAGC, the way was shown by Shannon, but it took some years to make his analysis rigorous.

An input word of length N (designed to use the channel over N slots in succession) is identified with an input N-vector

$$\mathbf{x}(=\mathbf{x}^{(N)}) = \begin{pmatrix} x_1 \\ \vdots \\ x_N \end{pmatrix}.$$

We assume that $x_i \in \mathbb{R}$ and hence $\mathbf{x}^{(N)} \in \mathbb{R}^N$ (to make the notation shorter, the upper index (N) will be often omitted).

In an additive channels an input vector $\mathbf{x}$ is transformed to a random vector $\mathbf{Y}^{(N)} = \begin{pmatrix} Y_1 \\ \vdots \\ Y_N \end{pmatrix}$ where $\mathbf{Y} = \mathbf{x} + \mathbf{Z}$, or, component-wise,

$$Y_j = x_j + Z_j, \ 1 \le j \le N. \tag{4.1.1}$$

Here and below,

$$\mathbf{Z} = \begin{pmatrix} Z_1 \\ \vdots \\ Z_N \end{pmatrix}$$

is a noise vector composed of random variables $Z_1, \ldots, Z_N$. Thus, the noise can be characterised by a joint PDF $f^{no}(\underline{z}) \ge 0$ where

$$\underline{z} = \begin{pmatrix} z_1 \\ \vdots \\ z_N \end{pmatrix}$$

and the total integral $\int f^{\mathrm{no}}(\underline{z})dz_1 \ldots dz_N = 1$. The N-fold noise probability distribution is determined by integration over a given set of values for **Z**:

$$\mathbf{P}^{\mathrm{no}}(\mathbf{Z} \in A) = \int_A f^{\mathrm{no}}(\underline{z})dz_1 \ldots dz_N, \quad \text{for } A \subseteq \mathbb{R}^N.$$

Example 4.1.1   An additive channel is called *Gaussian* (an AGC, in short) if,

for each N, the noise vector $\begin{pmatrix} Z_1 \\ \vdots \\ Z_N \end{pmatrix}$ is a multivariate normal; cf. PSE I, p. 114.

We assume from now on that the mean value $\mathbb{E}Z_j = 0$. Recall that the multivariate normal distribution with the zero mean is completely determined by its covariance matrix. More precisely, the joint PDF $f^{\mathrm{no}}_{\mathbf{Z}^{(N)}}(\underline{z}^{(N)})$ for an AGC has the form

$$\frac{1}{(2\pi)^{N/2}\left(\det\Sigma\right)^{1/2}} \exp\left(-\frac{1}{2}\underline{z}^{\mathrm{T}}\Sigma^{-1}\underline{z}\right), \quad \underline{z} = \begin{pmatrix} z_1 \\ \vdots \\ z_N \end{pmatrix} \in \mathbb{R}^N. \tag{4.1.2}$$

Here Σ is an $N \times N$ matrix assumed to be real, symmetric and strictly positive definite, with entries $\Sigma_{jj'} = \mathbb{E}\left(Z_j Z_{j'}\right)$ representing the covariance of noise random variables Z_j and $Z_{j'}$, $1 \le j, j' \le N$. (Real strict positive definiteness means that Σ is of the form BB^{T} where B is an $N \times N$ real invertible matrix; if Σ is strictly positive definite then Σ has N mutually orthogonal eigenvectors, and all N eigenvalues of Σ are greater than 0.) In particular, each random variable Z_j is normal: $Z_j \sim \mathrm{N}(0, \sigma_j^2)$ where $\sigma_j^2 = \mathbb{E}Z_j^2$ coincides with the diagonal entry Σ_{jj}. (Due to strict positive definiteness, $\Sigma_{jj} > 0$ for all $j = 1, \ldots, N$.)

If in addition the random variables $Z_1, Z_2, \ldots$ are IID, the channel is called *memoryless* Gaussian (MGC) or a channel with (additive) Gaussian *white noise*. In this case matrix Σ is diagonal: $\Sigma_{ij} = 0$ when $i \ne j$ and $\Sigma_{ii} > 0$ when $i = j$. This is an important model example (both educationally and practically) since it admits some nice final formulas and serves as a basis for further generalisations.

Thus, an MGC has IID noise random variables $Z_i \sim \mathrm{N}(0, \sigma^2)$ where $\sigma^2 = \mathrm{Var}Z_i = \mathbb{E}Z_i^2$. For normal random variables, independence is equivalent to decorrelation. That is, the equality $\mathbb{E}\left(Z_j Z_{j'}\right) = 0$ for all $j, j' = 1, \ldots, N$ with $j \ne j'$ implies that the components $Z_1, \ldots, Z_N$ of the noise vector $\mathbf{Z}^{(N)}$ are mutually independent. This can be deduced from (4.1.2): if matrix Σ has $\Sigma_{jj'} = 0$ for $j \ne j'$ then Σ is diagonal, with $\det\Sigma = \prod_{1 \le j \le N} \Sigma_{jj}$, and the joint PDF in (4.1.2) decomposes into a product of N factors representing individual PDFs of components Z_j, $1 \le j \le N$:

$$\prod_{1 \le j \le N} \frac{1}{(2\pi\Sigma_{jj})^{1/2}} \exp\left(-\frac{z_j^2}{2\Sigma_{jj}}\right). \tag{4.1.3}$$

Moreover, under the IID assumption, with $\Sigma_{jj} \equiv \sigma^2 > 0$, all random variables $Z_j \sim$ N$(0, \sigma^2)$, and the noise distribution for an MGC is completely specified by a single parameter $\sigma > 0$. More precisely, the joint PDF from (4.1.3) is rewritten as

$$\left(\frac{1}{\sqrt{2\pi}\sigma}\right)^N \exp\left(-\frac{1}{2\sigma^2} \sum_{1 \le j \le N} z_j^2\right).$$

It is often convenient to think that an infinite random sequence $\mathbf{Z}_1^\infty = \{Z_1, Z_2, \ldots\}$ is given, and the above noise vector $\mathbf{Z}^{(N)}$ is formed by the first N members of this sequence. In the Gaussian case, $\mathbf{Z}_1^\infty$ is called a random Gaussian process; with $\mathbb{E}Z_j \equiv 0$, this process is determined, like before, by its covariance Σ, with $\Sigma_{ij} =$ Cov $(Z_i, Z_j) = \mathbb{E}(Z_i Z_j)$. The term 'white Gaussian noise' distinguishes this model from a more general model of a channel with 'coloured' noise; see below.

Channels with continuously distributed noise are analysed by using a scheme similar to the one adopted in the discrete case: in particular, if the channel is used for transmitting one of $M \sim 2^{RN}$, $R < 1$, encoded messages, we need a codebook that consists of M codewords of length N: $\mathbf{x}^{\mathrm{T}}(i) = (x_1(i), \ldots, x_N(i))$, $1 \le i \le M$:

$$\mathcal{X}_{M,N} = \left\{\mathbf{x}^{(N)}(1), \ldots, \mathbf{x}^{(N)}(M)\right\} = \left\{\begin{pmatrix} x_1(1) \\ \vdots \\ x_N(1) \end{pmatrix}, \ldots, \begin{pmatrix} x_1(M) \\ \vdots \\ x_N(M) \end{pmatrix}\right\}. \quad (4.1.4)$$

The codebook is, of course, presumed to be known to both the sender and the receiver. The transmission rate R is given by

$$R = \frac{\log_2 M}{N}. \quad (4.1.5)$$

Now suppose that a codevector $\mathbf{x}(i)$ had been sent. Then the received random vector $\mathbf{Y}(=\mathbf{Y}(i)) = \begin{pmatrix} x_1(i) + Z_1 \\ \vdots \\ x_N(i) + Z_N \end{pmatrix}$ is decoded by using a chosen decoder $d : \mathbf{y} \mapsto$ $d(\mathbf{y}) \in \mathcal{X}_{M,N}$. Geometrically, the decoder looks for the nearest codeword $\mathbf{x}(k)$, relative to a certain distance (adapted to the decoder); for instance, if we choose to use the Euclidean distance then vector $\mathbf{Y}$ is decoded by the codeword minimising the sum of squares:

$$d(\mathbf{Y}) = \arg\min\left[\sum_{1 \le j \le N}(Y_j(i) - x_j(l))^2 : \mathbf{x}(l) \in \mathcal{X}_{M,N}\right]; \quad (4.1.6)$$

when $d(\mathbf{y}) \ne \mathbf{x}(i)$ we have an error. Luckily, the choice of a decoder is conveniently resolved on the basis of the maximum-likelihood principle; see below.

There is an additional subtlety here: one assumes that, for an input word $\mathbf{x}$ to get a chance of successful decoding, it should belong to a certain 'transmittable' domain in $\mathbb{R}^N$. For example, working with an MAGC, one imposes the power constraint

$$\frac{1}{N} \sum_{1 \leq j \leq N} x_j^2 \leq \alpha \qquad (4.1.7)$$

where $\alpha > 0$ is a given constant. In the context of wireless transmission this means that the amplitude square power per signal in an N-long input vector should be bounded by α, otherwise the result of transmission is treated as 'undecodable'. Geometrically, in order to perform decoding, the input codeword $\mathbf{x}(i)$ constituting the codebook must lie inside the Euclidean ball $\mathbb{B}_{\ell_2}^N(\sqrt{\alpha N})$ of radius $r = \sqrt{\alpha N}$ centred at $\mathbf{0} \in \mathbb{R}^N$:

$$\mathbb{B}_{\ell_2}^{(N)}(r) = \left\{ \mathbf{x} = \begin{pmatrix} x_1 \\ \vdots \\ x_N \end{pmatrix} : \left(\sum_{1 \leq j \leq N} x_j^2 \right)^{1/2} \leq r \right\}.$$

The subscript ℓ_2 stresses that $\mathbb{R}^N$ with the standard Euclidean distance is viewed as a Hilbert ℓ_2-space.

In fact, it is not required that the whole codebook $\mathscr{X}_{M,N}$ lies in a decodable domain; the agreement is only that if a codeword $\mathbf{x}(i)$ falls outside then it is decoded wrongly with probability 1. Pictorially, the requirement is that 'most' of codewords lie within $\mathbb{B}_{\ell_2}^N((N\alpha)^{1/2})$ but not necessarily all of them. See Figure 4.1.

A reason for the 'regional' constraint (4.1.7) is that otherwise the codewords can be positioned in space at an arbitrarily large distance from each other, and, eventually, every transmission rate would become reliable. (This would mean that the capacity of the channel is infinite; although such channels should not be dismissed outright, in the context of an AGC the case of an infinite capacity seems impractical.)

Typically, the decodable region $\mathbb{D}^{(N)} \subset \mathbb{R}^N$ is represented by a ball in $\mathbb{R}^N$, centred at the origin, and specified relative to a particular distance in $\mathbb{R}^N$. Say, in the case of exponentially distributed noise it is natural to select

$$\mathbb{D}^{(N)} = \mathbb{B}_{\ell_1}^{(N)}(N\alpha) = \left\{ \mathbf{x} = \begin{pmatrix} x_1 \\ \vdots \\ x_N \end{pmatrix} : \sum_{1 \leq j \leq N} |x_j| \leq N\alpha \right\}$$

the ball in the ℓ_1-metric. When an output-signal vector falling within distance r from a codeword is decoded by this codeword, we have a correct decoding if (i) the output signal falls in exactly one sphere around a codeword, (ii) the codeword in question lies within $\mathbb{D}^{(N)}$, and (iii) this specific codeword was sent. We have possibly an error when more than one codeword falls into the sphere.

error in decoding
for the input signals
outside the domain

possible correct decoding
for the input signals
inside the domain

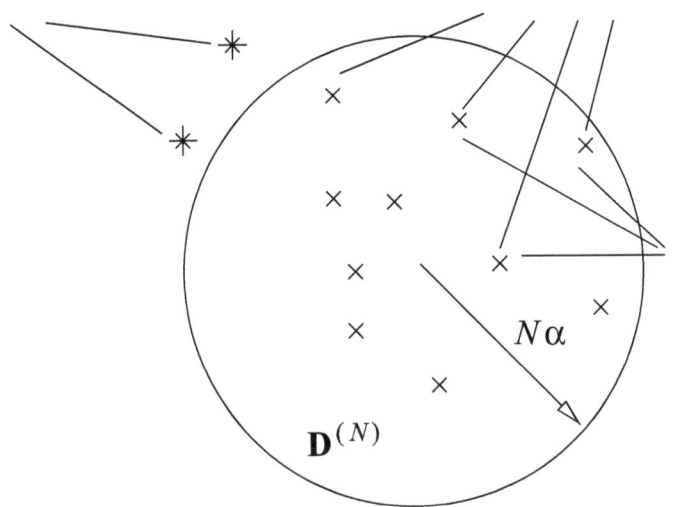

Figure 4.1

As in the discrete case, a more general channel is represented by a family of (conditional) probability distributions for received vectors of length N given that an input word $\mathbf{x}^{(N)} \in \mathbb{R}^N$ has been sent:

$$\mathbf{P}_{\text{ch}}^{(N)}(\,\cdot \mid \mathbf{x}^{(N)}) = \mathbf{P}_{\text{ch}}^{(N)}(\,\cdot \mid \text{word } \mathbf{x}^{(N)} \text{ sent}), \quad \mathbf{x} \in \mathbb{R}^N. \tag{4.1.8}$$

As before, $N = 1, 2, \ldots$ indicates how many slots of the channel were used for transmission, and we will consider the limit $N \to \infty$. Now assume that the distribution $\mathbf{P}_{\text{ch}}^{(N)}(\,\cdot \mid \mathbf{x}^{(N)})$ is determined by a PMF $f_{\text{ch}}^{(N)}(\mathbf{y}^{(N)} \mid \mathbf{x}^{(N)})$ relative to a fixed measure $\nu^{(N)}$ on $\mathbb{R}^N$:

$$\mathbf{P}_{\text{ch}}^{(N)}(\mathbf{Y}^{(N)} \in \mathbb{A} \mid \mathbf{x}^{(N)}) = \int_{\mathbb{A}} f_{\text{ch}}^{(N)}(\,\cdot \mid \mathbf{x}^{(N)}) d\nu^{(N)}(\mathbf{y}^{(N)}). \tag{4.1.9a}$$

A typical assumption is that $\nu^{(N)}$ is a product-measure of the form

$$\nu^{(N)} = \nu \times \cdots \times \nu \ (N \text{ times}); \tag{4.1.9b}$$

for instance, $\nu^{(N)}$ can be the Lebesgue measure on $\mathbb{R}^N$ which is the product of Lebesgue measures on $\mathbb{R}$: $d\mathbf{x}^{(N)} = dx_1 \times \cdots \times dx_N$. In the discrete case where digits x_i represent letters from an input channel alphabet $\mathcal{A}$ (say, binary, with $\mathcal{A} = \{0, 1\}$), ν is the counting measure on $\mathcal{A}$, assigning weight 1 to each symbol of the alphabet. Then $\nu^{(N)}$ is the counting measure on $\mathcal{A}^N$, the set of all input words of length N, assigning weight 1 to each such word.

Assuming the product-form reference measure $\nu^{(N)}$ (4.1.9b), we specify a memoryless channel by a product form PMF $f_{\mathrm{ch}}^{(N)}(\mathbf{y}^{(N)}\,|\,\mathbf{x}^{(N)})$:

$$f_{\mathrm{ch}}^{(N)}(\mathbf{y}^{(N)}\,|\,\mathbf{x}^{(N)}) = \prod_{1\leq j\leq N} f_{\mathrm{ch}}(y_j|x_j). \qquad (4.1.10)$$

Here $f_{\mathrm{ch}}(y|x)$ is the symbol-to-symbol channel PMF describing the impact of a single use of the channel. For an MGC, $f_{\mathrm{ch}}(y|x)$ is a normal $\mathrm{N}(x,\sigma^2)$. In other words, $f_{\mathrm{ch}}(y|x)$ gives the PDF of a random variable $Y = x + Z$ where $Z \sim \mathrm{N}(0,\sigma^2)$ represents the 'white noise' affecting an individual input value x.

Next, we turn to a codebook $\mathcal{X}_{M,N}$, the image of a one-to-one map $\mathcal{M} \to \mathbb{R}^N$ where $\mathcal{M}$ is a finite collection of messages (originally written in a message alphabet); cf. (4.1.4). As in the discrete case, the ML decoder d_{ML} decodes the received word $\mathbf{Y} = \mathbf{y}^{(N)}$ by maximising $f_{\mathrm{ch}}^{(N)}(\mathbf{y}\,|\,\mathbf{x})$ in the argument $\mathbf{x} = \mathbf{x}^{(N)} \in \mathcal{X}_{M,N}$:

$$d_{\mathrm{ML}}(\mathbf{y}) = \arg\max \left[f_{\mathrm{ch}}^{(N)}(\mathbf{y}\,|\,\mathbf{x}) : \mathbf{x} \in \mathcal{X}_{M,N} \right]. \qquad (4.1.11)$$

The case when maximiser is not unique will be treated as an error.

Another useful example is the joint typicality (JT) decoder $d_{\mathrm{JT}} = d_{\mathrm{JT}}^{(N),\varepsilon}$ (see below); it looks for the codeword $\mathbf{x}$ such that $\mathbf{x}$ and $\mathbf{y}$ lie in the ε-typical set T_ε^N:

$$d_{\mathrm{JT}}(\mathbf{y}) = \mathbf{x} \text{ if } \mathbf{x} \in \mathcal{X}_{M,N} \text{ and } (\mathbf{x},\mathbf{y}) \in T_\varepsilon^N. \qquad (4.1.12)$$

The JT decoder is designed – via a specific form of set T_ε^N – for codes generated as samples of a *random* code $\mathcal{X}_{M,N}$. Consequently, for given output vector $\mathbf{y}^N$ and a code $\mathcal{X}_{M,N}$, the decoded word $d_{\mathrm{JT}}(\mathbf{y}) \in \mathcal{X}_{M,N}$ may be not uniquely defined (or not defined at all), again leading to an error. A general decoder should be understood as a one-to-one map defined on a set $\mathbb{K}^{(N)} \subseteq \mathbb{R}^N$ taking points $\mathbf{y}^N \in \mathbb{K}^N$ to points $\mathbf{x} \in \mathcal{X}_{M,N}$; outside set $\mathbb{K}^{(N)}$ it may be not defined correctly. The decodable region $\mathbb{K}^{(N)}$ is a part of the specification of decoder $d^{(N)}$. In any case, we want to achieve

$$\mathbf{P}_{\mathrm{ch}}^{(N)}\left(d^{(N)}(\mathbf{Y}) \neq \mathbf{x}|\mathbf{x} \text{ sent}\right) = \mathbf{P}_{\mathrm{ch}}^{(N)}\left(\mathbf{Y} \notin \mathbb{K}^{(N)}|\mathbf{x} \text{ sent}\right)$$
$$+ \mathbf{P}_{\mathrm{ch}}^{(N)}\left(\mathbf{Y} \in \mathbb{K}^{(N)}, d(\mathbf{Y}) \neq \mathbf{x}|\mathbf{x} \text{ sent}\right) \to 0$$

as $N \to \infty$. In the case of an MGC, for any code $\mathcal{X}_{M,N}$, the ML decoder from (4.1.6) is defined uniquely almost everywhere in $\mathbb{R}^N$ (but does not necessarily give the right answer).

We also require that the input vector $\mathbf{x}^{(N)} \in \mathbb{D}^{(N)} \subset \mathbb{R}^N$ and when $\mathbf{x}^{(N)} \notin \mathbb{D}^{(N)}$, the result of transmission is rendered undecodable (regardless of the qualities of the decoder used). Then the average probability of error, while using codebook $\mathcal{X}_{M,N}$ and decoder $d^{(N)}$, is defined by

$$e^{\mathrm{av}}(\mathcal{X}_{M,N}, d^{(N)}, \mathbb{D}^{(N)}) = \frac{1}{M} \sum_{\mathbf{x} \in \mathcal{X}_{M,N}} e(\mathbf{x}, d^{(N)}, \mathbb{D}^{(N)}), \qquad (4.1.13a)$$

and the maximum probability of error by

$$e^{\max}(\mathscr{X}_{M,N}, d^{(N)}, \mathbb{D}^{(N)}) = \max \left[e(\mathbf{x}, d^{(N)}, \mathbb{D}^{(N)}) : \mathbf{x} \in \mathscr{X}_{M,N} \right]. \qquad (4.1.13b)$$

Here $e(\mathbf{x}, d^{(N)}, \mathbb{D}^{(N)})$ is the probability of error when codeword $\mathbf{x}$ had been transmitted:

$$e(\mathbf{x}, d^{(N)}, \mathbb{D}^{(N)}) = \begin{cases} 1, & \mathbf{x} \notin \mathbb{D}^{(N)}, \\ \mathbf{P}_{\mathrm{ch}}^{(N)}\left(d^{(N)}(\mathbf{Y}) \neq \mathbf{x} | \mathbf{x}\right), & \mathbf{x} \in \mathbb{D}^{(N)}. \end{cases} \qquad (4.1.14)$$

In (4.1.14) the order of the codewords in the codebook $\mathscr{X}_{M,N}$ does not matter; thus $\mathscr{X}_{M,N}$ may be regarded simply as a set of M points in the Euclidean space $\mathbb{R}^N$. Geometrically, we want the points of $\mathscr{X}_{M,N}$ to be positioned so as to maximise the chance of correct ML-decoding and lying, as a rule, within domain $\mathbb{D}^{(N)}$ (which again leads us to a sphere-packing problem).

To this end, suppose that a number $R > 0$ is fixed, the size of the codebook $\mathscr{X}_{M,N}$: $M = \lceil 2^{NR} \rceil$. We want to define a reliable transmission rate as $N \to \infty$ in a fashion similar to how it was done in Section 1.4.

Definition 4.1.2 Value $R > 0$ is called a *reliable transmission rate* with regional constraint $\mathbb{D}^{(N)}$ if, with $M = \lceil 2^{NR} \rceil$, there exist a sequence $\{\mathscr{X}_{M,N}\}$ of codebooks $\mathscr{X}_{M,N} \subset \mathbb{R}^N$ and a sequence $\{d^{(N)}\}$ of decoders $d^{(N)} : \mathbb{R}^N \to \mathbb{R}^N$ such that

$$\lim_{N \to \infty} e^{\mathrm{av}}(\mathscr{X}_{M,N}, d^{(N)}, \mathbb{D}^{(N)}) = 0. \qquad (4.1.15)$$

Remark 4.1.3 It is easy to verify that a transmission rate R reliable in the sense of average error-probability $e^{\mathrm{av}}(\mathscr{X}_{M,N}, d^{(N)}, \mathbb{D}^{(N)})$ is reliable for the maximum error-probability $e^{\max}(\mathscr{X}_{M,N}, d^{(N)}, \mathbb{D}^{(N)})$. In fact, assume that R is reliable in the sense of Definition 4.1.2, i.e. in the sense of the average error-probability. Take a sequence $\{\mathscr{X}_{M,N}\}$ of the corresponding codebooks with $M = \lceil 2^{RN} \rceil$ and a sequence $\{d_N\}$ of the corresponding decoding rules. Divide each code $\mathscr{X}_N$ into two halves, $\mathscr{X}_N^{(0)}$ and $\mathscr{X}_N^{(1)}$, by ordering the codewords in the non-decreasing order of their probabilities of erroneous decoding and listing the first $M^{(0)} = \lceil M/2 \rceil$ codewords in $\mathscr{X}_N^{(0)}$ and the rest, $M^{(1)} = M - M^{(0)}$, in $\mathscr{X}_N^{(1)}$. Then, for the sequence of codes $\{\mathscr{X}_{M,N}^{(0)}\}$:

(i) the information rate approaches the value R as $N \to \infty$ as

$$\frac{1}{N} \log M^{(0)} \geq R + O(N^{-1});$$

(ii) the maximum error-probability, while using the decoding rule d_N,

$$P_{\mathrm{e}}^{\max}\left(\mathscr{X}_N^{(0)}, d_N\right) \leq \frac{1}{M^{(1)}} \sum_{\mathbf{x}^{(N)} \in \mathscr{X}_N^{(1)}} P_{\mathrm{e}}(\mathbf{x}^{(N)}, d_N) \leq \frac{M}{M^{(1)}} P_{\mathrm{e}}^{\mathrm{av}}(\mathscr{X}_N, d_N).$$

Since $M/M^{(1)} \leq 2$, the RHS tends to 0 as $N \to \infty$. We conclude that R is a reliable transmission rate for the maximum error-probability. The converse assertion, that a reliable transmission rate R in the sense of the maximum error-probability is also reliable in the sense of the average error-probability, is obvious.

Next, the capacity of the channel is the supremum of reliable transmission rates:

$$C = \sup \left[R > 0 : R \text{ is reliable} \right]; \tag{4.1.16}$$

it varies from channel to channel and with the shape of constraining domains.

It turns out (cf. Theorem 4.1.9 below) that for the MGC, under the average power constraint threshold α (see (4.1.7)), the channel capacity $C(\alpha, \sigma^2)$ is given by the following elegant expression:

$$C(\alpha, \sigma^2) = \frac{1}{2} \log_2 \left(1 + \frac{\alpha}{\sigma^2} \right). \tag{4.1.17}$$

Furthermore, like in Section 1.4, the capacity $C(\alpha, \sigma^2)$ is achieved by a sequence of random codings where codeword $\mathbf{x}(i) = (X_1(i), \ldots, X_N(i))$ has IID components $X_j(i) \sim N(0, \alpha - \varepsilon_N)$, $j = 1, \ldots, N$, $i = 1, \ldots, M$, with $\varepsilon_N \to 0$ as $N \to \infty$. Although such random codings do not formally obey the constraint (4.1.7) for finite N, it is violated with a vanishing probability as $N \to \infty$ (since $\limsup_{N \to \infty} \mathbb{P}\left(\max \left[\frac{1}{N} \sum_{1 \leq j \leq N} X_j(i)^2 : 1 \leq i \leq M \right] \leq \alpha \right) = 1$ with a proper choice of ε_N). Consequently, the average error-probability (4.1.13a) goes to 0 (of course, for a random coding the error-probability becomes itself random).

Example 4.1.4 Next, we discuss an AGC with *coloured* Gaussian noise. Let a codevector $\mathbf{x} = (x_1, \ldots, x_N)$ have multi-dimensional entries

$$x_j = \begin{pmatrix} x_{j1} \\ \vdots \\ x_{jk} \end{pmatrix} \in \mathbb{R}^k, \quad 1 \leq j \leq N,$$

and the components Z_j of the noise vector

$$\mathbf{Z} = \begin{pmatrix} Z_1 \\ \vdots \\ Z_N \end{pmatrix}$$

are also random vectors of dimension k:

$$Z_j = \begin{pmatrix} Z_{j1} \\ \vdots \\ Z_{jk} \end{pmatrix}.$$

For instance, $Z_1, \ldots, Z_N$ may be IID $N(0, \Sigma)$ (with k-variate normal), where Σ is a given $k \times k$ covariance matrix.

The 'coloured' model arises when one uses a system of k scalar Gaussian channels in parallel. Here, a scalar signal x_{j1} is sent through channel 1, x_{j2} through channel 2, etc., at the jth use of the system. A reasonable assumption is that at each use the scalar channels produce jointly Gaussian noise; different channels may be independent (with matrix Σ being $k \times k$ diagonal) or dependent (when Σ is a general positive-definite $k \times k$ matrix).

Here a codebook, as before, is an (ordered or unordered) collection $\mathscr{X}_{M,N} = \left\{ \mathbf{x}(1), \ldots, \mathbf{x}(M) \right\}$ where each codeword $\mathbf{x}(i)$ is a 'multi-vector' $(x_1(i), \ldots, x_N(i))^{\mathrm{T}} \in \mathbb{R}^{k \times N} := \mathbb{R}^k \times \cdots \times \mathbb{R}^k$. Let Q be a positive-definite $k \times k$ matrix commuting with Σ: $Q\Sigma = \Sigma Q$. The power constraint is now

$$\frac{1}{N} \sum_{1 \le j \le N} \langle x_j(i), Q x_j(i) \rangle \le \alpha. \tag{4.1.18}$$

The formula for the capacity of an AGC with coloured noise is, not surprisingly, more complicated. As $\Sigma Q = Q\Sigma$, matrices Σ and Q may be simultaneously diagonalised. Let λ_i and γ_i, $i = 1, \ldots, k$, be the eigenvalues of Σ and Q, respectively (corresponding to the same eigenvectors). Then

$$C(\alpha, Q, \Sigma) = \frac{1}{2} \sum_{1 \le l \le k} \log_2 \left(1 + \frac{(v\gamma_l^{-1} - \lambda_l)_+}{\lambda_l} \right), \tag{4.1.19}$$

where $(v\gamma_l^{-1} - \lambda_l)_+ = \max \left(v\gamma_l^{-1} - \lambda_l, 0 \right)$. In other words, $(v\gamma_l^{-1} - \lambda_l)_+$ are the eigenvalues of the matrix $\left(vQ^{-1} - \Sigma \right)_+$ representing the positive-definite part of the Hermitian matrix $vQ^{-1} - \Sigma$. Next, $v = v(\alpha) > 0$ is determined from the condition

$$\mathrm{tr} \left[\left(v\mathbf{I} - Q\Sigma \right)_+ \right] = \alpha. \tag{4.1.20}$$

The positive-definite part $\left(v\mathbf{I} - Q\Sigma \right)_+$ is in turn defined by

$$\left(v\mathbf{I} - Q\Sigma \right)_+ = \Pi_+ \left(v\mathbf{I} - Q\Sigma \right) \Pi_+$$

where Π_+ is the orthoprojection (in $\mathbb{R}^k$) onto the subspace spanned by the eigenvectors of $Q\Sigma$ with eigenvalues $\gamma_l \lambda_l < v$. In (4.1.20) $\mathrm{tr} \left[\left(v\mathbf{I} - Q\Sigma \right)_+ \right] \ge 0$ (since $\mathrm{tr}\, AB \ge 0$ for all pair of positive-definite matrices), equals 0 for $v = 0$ (as

$\left(-Q\Sigma\right)_+ = \mathbf{0}$) and monotonically increases with v to $+\infty$. Therefore, for any given $\alpha > 0$, (4.1.20) determines the value of $v = v(\alpha)$ uniquely.

Though (4.1.19) looks much more involved than (4.1.17) both expressions are corollaries of two facts: (i) the capacity can be identified as the maximum of the mutual entropy between the (random) input and output signals, just as in the discrete case (cf. Sections 1.3 and 1.4), and (ii) the mutual information in the case of a Gaussian noise (white or coloured) is attained when the input signal is itself Gaussian whose covariance solves an auxiliary optimisation problem. In the case of (4.1.17) this optimisation problem is rather simple, while for (4.1.19) it is more complicated (but still has a transparent meaning).

Correspondingly, the random encoding achieving the capacity $C(\alpha; Q; \Sigma)$ is where signals $X_j(i)$, $1 \le j \le N$, $i = 1, \dots, M$, are IID, and $X_j(i) \sim \mathrm{N}(0, A - \varepsilon_N \mathbf{I})$ where A is the $k \times k$ positive-definite matrix maximising the determinant $\det(A + \Sigma)$ subject to the constraint $\operatorname{tr} QA = \alpha$; such a matrix turns out to be of the form $\left(vQ^{-1} - \Sigma\right)_+$. The random encoding provides a convenient tool for calculating the capacity in various models. We will discuss a number of such models in Worked Examples.

The notable difference emerging for channels with continuously distributed noise is that the entropy should be replaced – when appropriate – with the differential entropy. Recall the differential entropy introduced in Section 1.5. The mutual entropy between two random variables X and Y with the joint PMF $f_{X,Y}(x,y)$ relative to a reference measure $\mu \times v$ and marginal PMFs $f_X(x) = \int f_{X,Y}(x,y)v(\mathrm{d}y)$ and $f_Y(y) = \int f_{X,Y}(x,y)\mu(\mathrm{d}x)$ is

$$I(X:Y) = \mathbb{E}\, \log \frac{f_{X,Y}(X,Y)}{f_X(X)f_Y(Y)}$$
$$= \int f_{X,Y}(x,y) \log \frac{f_{X,Y}(x,y)}{f_X(x)f_Y(y)}\mu(\mathrm{d}x)v(\mathrm{d}y).$$

A similar definition works when X and Y are replaced by random vectors $\mathbf{X} = (X_1, \dots, X_N)$ and $\mathbf{Y} = (Y_1, \dots, Y_{N'})$ (or even multi-vectors where – as in Example 4.1.4 – components X_j and Y_j are vectors themselves):

$$I(\mathbf{X}^{(N)} : \mathbf{Y}^{(N')}) = \mathbb{E}\, \log \frac{f_{\mathbf{X}^{(N)}, \mathbf{Y}^{(N')}}(\mathbf{X}^{(N)}, \mathbf{Y}^{(N')})}{f_{\mathbf{X}^{(N)}}(\mathbf{X}^{(N)})f_{\mathbf{Y}^{(N')}}(\mathbf{Y}^{(N')})}. \qquad (4.1.21a)$$

Here $f_{\mathbf{X}^{(N)}}(\mathbf{x}^{(N)})$ and $f_{\mathbf{Y}^{(N')}}(\mathbf{y}^{(N')})$ are the marginal PMFs for $\mathbf{X}^{(N)}$ and $\mathbf{Y}^{(N')}$ (i.e. joint PMFs for components of these vectors).

Specifically, if $N = N'$, $\mathbf{X}^{(N)}$ represents a random input and $\mathbf{Y}^{(N)} = \mathbf{X}^{(N)} + \mathbf{Z}^{(N)}$ the corresponding random output of a channel with a (random) probability of error:

$$
E(\mathbf{x}^{(N)}, \mathbb{D}^{(N)}) =
\begin{cases}
1, & \mathbf{x}^{(N)} \notin \mathbb{D}^{(N)}, \\
\mathbf{P}_{ch}^{(N)}\left(d_{ML}(\mathbf{Y}^{(N)}) \neq \mathbf{x}^{(N)} | \mathbf{x}^{(N)}\right), & \mathbf{x}^{(N)} \in \mathbb{D}^{(N)};
\end{cases}
$$

cf. (4.1.14). Furthermore, we are interested in the expected value

$$
\mathscr{E}(\mathbf{P}_{\mathbf{X}^{(N)}}; \mathbb{D}^{(N)}) = \mathbb{E}\left[E(\mathbf{X}^{(N)}, \mathbb{D}^{(N)})\right]. \tag{4.1.21b}
$$

Next, given $\varepsilon > 0$, we can define the supremum of the mutual information per signal (i.e. per a single use of the channel), over all input probability distributions $P_{\mathbf{X}^{(N)}}$ with $\mathscr{E}(P_{\mathbf{X}^{(N)}}, \mathbb{D}^{(N)}) \leq \varepsilon$:

$$
\overline{C}_{\varepsilon,N} = \frac{1}{N} \sup \left[I(\mathbf{X}^{(N)} : \mathbf{Y}^{(N)}) : \mathscr{E}(P_{\mathbf{X}^{(N)}}, \mathbb{D}^{(N)}) \leq \varepsilon \right], \tag{4.1.22}
$$

$$
\overline{C}_{\varepsilon} = \limsup_{N \to \infty} \overline{C}_{\varepsilon,N} \quad \overline{C} = \liminf_{\varepsilon \to 0} \overline{C}_{\varepsilon}. \tag{4.1.23}
$$

We want to stress that the supremum in (4.1.22) should be taken over *all* probability distributions $P_{\mathbf{X}^{(N)}}$ of the input word $\mathbf{X}^{(N)}$ with the property that the expected error-probability is $\leq \varepsilon$, regardless of whether these distributions are discrete or continuous or mixed (contain both parts). This makes the correct evaluation of $\overline{C}_{N,\varepsilon}$ quite difficult. However, the limiting value $\overline{C}$ is more amenable, at least in some important examples.

We are now in a position to prove the converse part of the Shannon second coding theorem:

Theorem 4.1.5 (cf. Theorems 1.4.14 and 2.2.10.) *Consider a channel given by a sequence of probability distributions* $\mathbf{P}_{ch}\left(\cdot \mid \mathbf{x}^{(N)} \text{ sent} \right)$ *for the random output words* $\mathbf{Y}^{(N)}$ *and decodable domains* $\mathbb{D}^{(N)}$. *Then quantity* $\overline{C}$ *from (4.1.22), (4.1.23) gives an upper bound for the capacity:*

$$
C \leq \overline{C}. \tag{4.1.24}
$$

Proof Let R be a reliable transmission rate and $\{\mathscr{X}_{M,N}\}$ be a sequence of codebooks with $M = \sharp \, \mathscr{X}_{M,N} \sim 2^{NR}$ for which $\lim_{N \to \infty} e^{av}(\mathscr{X}_{M,N}, \mathbb{D}^{(N)}) = 0$. Consider the pair $(\mathbf{x}, d_{ML}(\mathbf{y}))$ where (i) $\mathbf{x} = \mathbf{x}_{eq}^{(N)}$ is the random input word equidistributed over $\mathscr{X}_{M,N}$, (ii) $\mathbf{Y} = \mathbf{Y}^{(N)}$ is the received word and (iii) $d_{ML}(\mathbf{y})$ is the codeword guessed while using the ML decoding rule d_{ML} after transmission. Words $\mathbf{x}$ and $d_{ML}(\mathbf{Y})$ run jointly over $\mathscr{X}_{M,N}$, i.e. have a discrete-type joint distribution. Then, by the generalised Fano inequality (1.2.23),

$$h_{\mathrm{discr}}(\mathbf{X}|d(\mathbf{Y})) \leq 1 + \log(M-1) \sum_{\mathbf{X} \in \mathscr{X}_{M,N}} \mathbb{P}(\mathbf{x} = \mathbf{x}, d_{\mathrm{ML}}(\mathbf{Y}) \neq \mathbf{x})$$

$$\leq 1 + \frac{NR}{M} \sum_{\mathbf{x} \in \mathscr{X}_{M,N}} \mathbf{P}_{\mathrm{ch}}(d_{\mathrm{ML}}(\mathbf{Y}) \neq \mathbf{x}|\mathbf{x} \text{ sent})$$

$$= 1 + NRe^{\mathrm{av}}(\mathscr{X}_{M,N}, \mathbb{D}^{(N)}) := N\theta_N,$$

where $\theta_N \to 0$ as $N \to \infty$. Next, with $h(\mathbf{X}_{\mathrm{eq}}^{(N)}) = \log M$, we have $NR - 1 \leq h(\mathbf{X}_{\mathrm{eq}}^{(N)})$. Therefore,

$$R \leq \frac{1 + h(\mathbf{X}_{\mathrm{eq}}^{(N)})}{N}$$

$$= \frac{1}{N}I(\mathbf{X}_{\mathrm{eq}}^{(N)} : d(\mathbf{Y}^{(N)})) + \frac{1}{N}h(\mathbf{X}_{\mathrm{eq}}^{(N)}|d(\mathbf{Y}^{(N)}))$$

$$+ \frac{1}{N} \leq \frac{1}{N}I(\mathbf{x}_{\mathrm{eq}}^{(N)} : \mathbf{Y}^{(N)}) + \theta_N.$$

For any given $\varepsilon > 0$, for N sufficiently large, the average error-probability will satisfy $e^{\mathrm{av}}(\mathscr{X}_{M,N}, \mathbb{D}^{(N)}) < \varepsilon$. Consequently, $R \leq \overline{C}_{\varepsilon,N}$, for N large enough. (Because the equidistribution over a codebook $\mathscr{X}_{M,N}$ with $e^{\mathrm{av}}(\mathscr{X}_{M,N}, \mathbb{D}^{(N)})$ gives a specific example of an input distribution $P_{\mathbf{X}^{(N)}}$ with $\mathscr{E}(P_{\mathbf{X}^{(N)}}, \mathbb{D}^{(N)}) \leq \varepsilon$.) Thus, for all $\varepsilon > 0$, $R \leq \overline{C}_\varepsilon$, implying that the transition rate $R \leq \overline{C}$. Therefore, $C \leq \overline{C}$, as claimed. $\square$

The bound $C \leq \overline{C}$ in (4.1.24) becomes exact (with $C = \overline{C}$) in many interesting situations. Moreover, the expression for $\overline{C}$ simplifies in some cases of interest. For example, for an MAGC instead of maximising the mutual information $I(\mathbf{X}^{(N)} : \mathbf{Y}^{(N)})$ for varying N it becomes possible to maximise $I(X : Y)$, the mutual information between single input and output signals subject to an appropriate constraint. Namely, for an MAGC,

$$C = \overline{C} = \sup\left[I(X : Y) : \mathbb{E}X^2 < \alpha\right]. \qquad (4.1.25\mathrm{a})$$

The quantity $\sup\left[I(X : Y) : \mathbb{E}X^2 \leq \alpha\right]$ is often called the information capacity of an MAGC, under the square-power constraint α. Moreover, for a general AGC,

$$C = \overline{C} = \lim_{N \to \infty} \frac{1}{N} \sup\left[I(\mathbf{X}^{(N)} : \mathbf{Y}^{(N)}) : \frac{1}{N} \sum_{1 \leq j \leq N} \mathbb{E}X_j^2 < \alpha\right]. \qquad (4.1.25\mathrm{b})$$

Example 4.1.6 Here we estimate the capacity $\overline{C}(\alpha, \sigma^2)$ of an MAGC with additive white Gaussian noise of variance σ^2, under the average power constraint (with $\mathbb{D}^{(N)} = \mathbb{B}^{(N)}((N\alpha)^{1/2})$ (cf. Example 4.1.1.), i.e. bound from above the right-hand side of (4.1.25b).

Given an input distribution $P_{\mathbf{X}^{(N)}}$, we write

$$
\begin{aligned}
I(\mathbf{X}^{(N)} : \mathbf{Y}^{(N)}) &= h(\mathbf{Y}^{(N)}) - h(\mathbf{Y}^{(N)}|\mathbf{X}^{(N)}) \\
&= h(\mathbf{Y}^{(N)}) - h(\mathbf{Z}^{(N)}) \\
&\leq \sum_{1 \leq j \leq N} h(Y_j) - h(\mathbf{Z}^{(N)}) \\
&= \sum_{1 \leq j \leq N} \left(h(Y_j) - h(Z_j) \right).
\end{aligned}
\tag{4.1.26}
$$

Denote by $\alpha_j^2 = \mathbb{E}X_j^2$ the second moment of the single-input random variable X_j, the jth entry of the random input vector $\mathbf{X}^{(N)}$. Then the corresponding random output random variable Y_j has

$$
\mathbb{E}Y_j^2 = \mathbb{E}(X_j + Z_j)^2 = \mathbb{E}X_j^2 + 2\mathbb{E}X_j Z_j + \mathbb{E}Z_j^2 = \alpha_j^2 + \sigma^2,
$$

as X_j and Z_j are independent and $\mathbb{E}Z_j = 0$.

Note that for a Gaussian channel, Y_j has a continuous distribution (with the PDF $f_{Y_j}(y)$ given by the convolution $\int \phi_{\sigma^2}(x - y) dF_{X_j}(x)$ where ϕ_{σ^2} is the PDF of $Z_j \sim N(0, \sigma^2)$). Consequently, the entropies figuring in (4.1.26) and – implicitly – in (4.1.25a,b) are the differential entropies. Recall that for a random variable Y_j with a PDF f_{Y_j}, under the condition $\mathbb{E}Y_j^2 \leq \alpha_j^2 + \sigma^2$, the maximum of the differential entropy $h(Y_j) \leq \dfrac{1}{2} \log_2[2\pi e(\alpha_j^2 + \sigma^2)]$. In fact, by Gibbs,

$$
\begin{aligned}
h(Y_j) &= - \int f_{Y_j}(y) \log_2 f_{Y_j}(y) dy \\
&< - \int f_{Y_j}(y) \log_2 \phi_{\alpha_j^2 + \sigma^2}(y) dy \\
&= \frac{1}{2} \log_2 \left[2\pi(\alpha_j^2 + \sigma^2) \right] + \frac{\log_2 e}{2(\alpha_j^2 + \sigma^2)} \mathbb{E}Y_j^2 \\
&\leq \frac{1}{2} \log_2 \left[2\pi e(\alpha_j^2 + \sigma^2) \right],
\end{aligned}
$$

and consequently,

$$
\begin{aligned}
I(X_j : Y_j) &= h(Y_j) - h(Z_j) \\
&\leq \log_2[2\pi e(\alpha_j^2 + \sigma^2)] - \log_2(2\pi e \sigma^2) \\
&= \log_2 \left(1 + \frac{\alpha_j^2}{\sigma^2} \right),
\end{aligned}
$$

with equality iff $Y_j \sim N(0, \alpha_j^2 + \sigma^2)$.

The bound $\sum_{1 \leq j \leq N} \mathbb{E}X_j^2 = \sum_{1 \leq j \leq N} \alpha_j^2 < N\alpha$ in (4.1.25b) implies, by the law of large numbers, that $\lim_{N \to \infty} P_{\mathbf{X}^{(N)}} \left(\mathbb{B}^{(N)}(\sqrt{N\alpha}) \right) = 1$. Moreover, for any input probability distribution $P_{\mathbf{X}^{(N)}}$ with $\mathbb{E}X_j^2 \leq \alpha_j^2$, $1 \leq j \leq N$, we have that

$$\frac{1}{N}I(\mathbf{X}^{(N)} : \mathbf{Y}^{(N)}) \leq \frac{1}{2N} \sum_{1 \leq j \leq N} \log_2 \left(1 + \frac{\alpha_j^2}{\sigma^2} \right).$$

The Jensen inequality, applied to the concave function $x \mapsto \log_2(1+x)$, implies

$$\frac{1}{2N} \sum_{1 \leq j \leq N} \log_2 \left(1 + \frac{\alpha_j^2}{\sigma^2} \right) \leq \frac{1}{2} \log_2 \left(1 + \frac{1}{N} \sum_{1 \leq j \leq N} \frac{\alpha_j^2}{\sigma^2} \right)$$

$$\leq \frac{1}{2} \log_2 \left(1 + \frac{\alpha}{\sigma^2} \right).$$

Therefore, in this example, the information capacity $\overline{C}$, taken as the RHS of (4.1.25b), obeys

$$\overline{C} \leq \frac{1}{2} \log_2 \left(1 + \frac{\alpha}{\sigma^2} \right). \tag{4.1.27}$$

After establishing Theorem 4.1.8, we will be able to deduce that the capacity $C(\alpha, \sigma^2)$ equals the RHS, confirming the answer in (4.1.17).

Example 4.1.7   For the coloured Gaussian noise the bound from (4.1.26) can be repeated:

$$I(\mathbf{X}^{(N)} : \mathbf{Y}^{(N)}) \leq \sum_{1 \leq j \leq N} [h(Y_j) - h(Z_j)].$$

Here we work with the mixed second-order moments for the random vectors of input and output signals X_j and $Y_j = X_j + Z_j$:

$$\alpha_j^2 = \mathbb{E}\langle X_j, QX_j \rangle, \quad \mathbb{E}\langle Y_j, QY_j \rangle = \alpha_j^2 + \mathrm{tr}\,(Q\Sigma), \quad \frac{1}{N} \sum_{1 \leq j \leq N} \alpha_j^2 \leq \alpha.$$

In this calculation we again made use of the fact that X_j and Z_j are independent and the expected value $\mathbb{E}Z_j = 0$.

Next, as in the scalar case, $\frac{1}{N}I(\mathbf{X}^{(N)} : \mathbf{Y}^{(N)})$ does not exceed the difference $h(Y) - h(Z)$ where $Z \sim \mathrm{N}(0, \Sigma)$ is the coloured noise vector and $Y = X + Z$ is a multivariate normal distribution maximising the differential entropy under the trace restriction. Formally:

$$\frac{1}{N}I(\mathbf{X}^{(N)} : \mathbf{Y}^{(N)}) \leq \overline{h}(\alpha, Q, \Sigma) - h(Z)$$

where K is the covariance matrix of a signal, and

$$\bar{h}(\alpha, Q, \Sigma) = \frac{1}{2} \max \left\{ \log \left[(2\pi)^k e \det(K + \Sigma) \right] : \right.$$
$$\left. K \text{ positive-definite } k \times k \text{ matrix with } \operatorname{tr}(QK) \leq \alpha \right\}.$$

Write Σ in the diagonal form $\Sigma = C\Lambda C^{\mathrm{T}}$ where C is an orthogonal and Λ the diagonal $k \times k$ matrix formed by the eigenvalues of Σ:

$$\Lambda = \begin{pmatrix} \lambda_1 & 0 & \cdots & 0 \\ 0 & \lambda_2 & \cdots & 0 \\ 0 & 0 & \ddots & 0 \\ 0 & 0 & \cdots & \lambda_k \end{pmatrix}.$$

Write $C^{\mathrm{T}}KC = B$ and maximise $\det(B + \Lambda)$ subject to the constraint

$$B \text{ positive-definite and } \operatorname{tr}(\Gamma B) \leq \alpha \text{ where } \Gamma = C^{\mathrm{T}}QC.$$

By the Hadamard inequality of Worked Example 1.5.10, $\det(B + \Lambda) \leq \prod_{1 \leq i \leq k} (B_{ii} + \lambda_i)$, with equality iff B is diagonal (i.e. matrices Σ and K have the same eigenbasis), and $B_{11} = \beta_1, \ldots, B_{kk} = \beta_k$ are the eigenvalues of K. As before, assume $Q\Sigma = \Sigma Q$, then $\operatorname{tr}(\Gamma B) = \sum_{1 \leq i \leq k} \gamma_i \beta_i$. So, we want to maximise the product $\prod_{1 \leq i \leq k} (\beta_i + \lambda_i)$, or equivalently, the sum

$$\sum_{1 \leq i \leq k} \log(\beta_i + \lambda_i), \text{ subject to } \beta_1, \ldots, \beta_k \geq 0 \text{ and } \sum_{1 \leq i \leq k} \gamma_i \beta_i \leq \alpha.$$

If we discard the regional constraints $\beta_1, \ldots, \beta_k \geq 0$, the Lagrangian

$$\mathscr{L}(\beta_1, \ldots, \beta_k; \kappa) = \sum_{1 \leq i \leq k} \log(\beta_i + \lambda_i) + \kappa \left(\alpha - \sum_{1 \leq i \leq k} \gamma_i \beta_i \right)$$

is maximised at

$$\frac{1}{\beta_i + \lambda_i} = \kappa \gamma_i, \text{ i.e. } \beta_i = \frac{1}{\kappa \gamma_i} - \lambda_i, \quad i = 1, \ldots, k.$$

To satisfy the regional constraint, we take

$$\beta_i = \left(\frac{1}{\kappa \gamma_i} - \lambda_i \right)_+, \quad i = 1, \ldots, k,$$

and adjust $\kappa > 0$ so that

$$\sum_{1 \leq i \leq k} \left(\frac{1}{\kappa} - \gamma_i \lambda_i \right)_+ = \alpha. \tag{4.1.28}$$

This yields that the information capacity $\overline{C}(\alpha, Q, \Sigma)$ obeys

$$\overline{C}(\alpha, Q, \Sigma) \leq \frac{1}{2} \sum_{1 \leq l \leq k} \log_2 \left(1 + \frac{(\nu \gamma_l^{-1} - \lambda_l)_+}{\lambda_l} \right), \tag{4.1.29}$$

where the RHS comes from (4.1.28) with $\nu = 1/\kappa$. Again, we will show that the capacity $C(\alpha, Q, \Sigma)$ equals the last expression, confirming the answer in (4.1.19).

We now pass to the direct part of the second Shannon coding theorem for general channels with regional restrictions. Although the statement of this theorem differs from that of Theorems 1.4.15 and 2.2.1 only in the assumption of constraints upon the codewords (and the proof below is a mere repetition of that of Theorem 1.4.15), it is useful to put it in the formal context.

Theorem 4.1.8   *Let a channel be specified by a sequence of conditional probabilities* $\mathbf{P}_{ch}^{(N)}(\cdot \, | \mathbf{x}^{(N)}$ *sent$)$ for the received word* $\mathbf{Y}^{(N)}$ *and a sequence of decoding constraints* $\mathbf{x}^{(N)} \in \mathbb{D}^{(N)}$ *for the input vector. Suppose that probability* $\mathbf{P}_{ch}^{(N)}(\cdot \, | \mathbf{x}^{(N)}$ *sent$)$ is given by a PMF* $f_{ch}(\mathbf{y}^{(N)} | \mathbf{x}^{(N)}$ *sent$)$ relative to a reference measure* $\nu^{(N)}$. *Given* $c > 0$, *suppose that there exists a sequence of input probability distributions* $P_{\mathbf{X}^{(N)}}$ *such that*

(i) $\lim\limits_{N \to \infty} P_{\mathbf{X}^{(N)}}(\mathbb{D}^{(N)}) = 1$,

(ii) *the distribution* $P_{\mathbf{X}^{(N)}}$ *is given by a PMF* $f_{\mathbf{X}^{(N)}}(\mathbf{x}^{(N)})$ *relative to a reference measure* $\mu^{(N)}$,

(iii) *the following convergence in probability holds true: for all* $\varepsilon > 0$,

$$\lim_{N \to \infty} \mathbb{P}_{\mathbf{X}^{(N)}, \mathbf{Y}^{(N)}} \left(T_\varepsilon^N \right) = 1,$$

$$T_\varepsilon^N = \left(\left| \frac{1}{N} \log_+ \frac{f_{\mathbf{X}^{(N)}, \mathbf{Y}^{(N)}} (\mathbf{x}^{(N)}, \mathbf{Y}^{(N)})}{f_{\mathbf{X}^{(N)}} (\mathbf{x}^{(N)}) f_{\mathbf{Y}^{(N)}} (\mathbf{Y}^{(N)})} - c \right| \leq \varepsilon \right), \tag{4.1.30a}$$

where

$$f_{\mathbf{X}^{(N)}, \mathbf{Y}^{(N)}} (\mathbf{x}^{(N)}, \mathbf{y}^{(N)}) = f_{\mathbf{X}^{(N)}} (\mathbf{x}^{(N)}) f_{ch}(\mathbf{y}^{(N)} | \mathbf{x}^{(N)} \text{ sent}),$$
$$f_{\mathbf{Y}^{(N)}} (\mathbf{y}^{(N)}) = \int f_{\mathbf{X}^{(N)}} (\widetilde{\mathbf{x}}^{(N)}) f_{ch}(\mathbf{y}^{(N)} | \widetilde{\mathbf{x}}^{(N)} \text{ sent}) \mu^{\times N} \left(d\widetilde{\mathbf{x}}^{(N)} \right). \tag{4.1.30b}$$

Then the capacity of the channel satisfies $C \geq c$.

Proof   Take $R < c$ and consider a random codebook $\{\mathbf{x}^N(1), \ldots, \mathbf{x}^N(M)\}$, with $M \sim 2^{NR}$, composed by IID codewords where each codeword $\mathbf{x}^N(j)$ is drawn according to $P^N = P_{\mathbf{x}^N}$. Suppose that a (random) codeword $\mathbf{x}^N(j)$ has been sent and a (random) word $\mathbf{Y}^N = \mathbf{Y}^N(j)$ received, with the joint PMF $f_{\mathbf{X}^{(N)}, \mathbf{Y}^{(N)}}$ as in (4.1.30b). We take $\varepsilon > 0$ and decode $\mathbf{Y}^N$ by using joint typicality:

$$d_{\mathrm{JT}}(\mathbf{Y}^N) = \mathbf{x}^N(i) \text{ when } \mathbf{x}^N(i) \text{ is the only vector among}$$
$$\mathbf{x}^N(1), \ldots, \mathbf{x}^N(M) \text{ such that } (\mathbf{x}^N(i), \mathbf{Y}^N) \in T_\varepsilon^N.$$

Here set T_ε^N is specified in (4.1.30a).

Suppose a random vector $\mathbf{x}^N(j)$ has been sent. It is assumed that an error occurs every time when

(i) $\mathbf{x}^N(j) \notin \mathbb{D}^{(N)}$, or
(ii) the pair $(\mathbf{x}^N(j), \mathbf{Y}^N) \notin T_\varepsilon^N$, or
(iii) $(\mathbf{x}^N(i), \mathbf{Y}^N) \in T_\varepsilon^N$ for some $i \neq j$.

These possibilities do not exclude each other but if none of them occurs then

(a) $\mathbf{x}^N(j) \in \mathbb{D}^{(N)}$ and
(b) $\mathbf{x}(j)$ is the only word among $\mathbf{x}^N(1), \ldots, \mathbf{x}^N(M)$ with $(\mathbf{x}^N(j), \mathbf{Y}^N) \in T_\varepsilon^N$.

Therefore, the JT decoder will return the correct result. Consider the average error-probability

$$\mathcal{E}_M(P^N) = \frac{1}{M} \sum_{1 \leq j \leq M} E(j, P^N)$$

where $E(j, P^N)$ is the probability that any of the above possibilities (i)–(iii) occurs:

$$E(j, P^N) = \mathbb{P}\left(\{\mathbf{x}^N(j) \notin \mathbb{D}^{(N)}\} \cup \{(\mathbf{x}^N(j), \mathbf{Y}^N) \notin T_\varepsilon^N\} \right.$$
$$\left. \cup \{(\mathbf{x}^N(i), \mathbf{Y}^N) \in T_\varepsilon^N \text{ for some } i \neq j\} \right)$$
$$= \mathbb{E}\mathbf{1}\left(\mathbf{x}^N(j) \notin \mathbb{D}^{(N)} \right)$$
$$+ \mathbb{E}\mathbf{1}\left(\mathbf{x}^N(j) \in \mathbb{D}^{(N)}, \ d_{\mathrm{JT}}(\mathbf{Y}^N) \neq \mathbf{x}^N(j) \right). \quad (4.1.31)$$

The symbols $\mathbb{P}$ and $\mathbb{E}$ in (4.1.31) refer to (1) a collection of IID input vectors $\mathbf{x}^N(1), \ldots, \mathbf{x}^N(M)$, and (2) the output vector $\mathbf{Y}^N$ related to $\mathbf{x}^N(j)$ by the action of the channel. Consequently, $\mathbf{Y}^N$ is independent of vectors $\mathbf{x}^N(i)$ with $i \neq j$. It is in-structive to represent the corresponding probability distribution $\mathbb{P}$ as the Cartesian product; e.g. for $j = 1$ we refer in (4.1.31) to

$$\mathbb{P} = P_{\mathbf{x}^N(1), \mathbf{Y}^N(1)} \times P_{\mathbf{x}^N(2)} \times \cdots \times P_{\mathbf{x}^N(M)}$$

where $P_{\mathbf{x}^N(1), \mathbf{Y}^N(1)}$ stands for the joint distribution of the input vector $\mathbf{x}^N(1)$ and the output vector $\mathbf{Y}^N(1)$, determined by the joint PMF

$$f_{\mathbf{x}^N(1), \mathbf{Y}^N(1)}(\mathbf{x}^N, \mathbf{y}^N) = f_{\mathbf{x}^N(1)}(\mathbf{x}^N) f_{\mathrm{ch}}(\mathbf{y}^N | \mathbf{x}^N \text{ sent}).$$

By symmetry, $E(j, P^N)$ does not depend on j, thus in the rest of the argument we can take $j = 1$. Next, probability $E(1, P^N)$ does not exceed the sum of probabilities

$$\mathbb{P}\left(\mathbf{x}^N(1) \notin \mathbb{D}^{(N)}\right) + \mathbb{P}\left((\mathbf{x}^N(1), \mathbf{Y}^N) \notin T_\varepsilon^N\right)$$
$$+ \sum_{i=2}^M \mathbb{P}\left((\mathbf{x}^N(i), \mathbf{Y}^N) \in T_\varepsilon^N\right).$$

Thanks to the condition that $\lim_{N \to \infty} P_{\mathbf{x}^{(N)}}(\mathbb{D}^{(N)}) = 1$, the first summand vanishes as $N \to \infty$. The second summand vanishes, again in the limit $N \to \infty$, because of (4.1.30a). It remains to estimate the sum $\sum_{i=2}^M \mathbb{P}\left((\mathbf{x}^N(i), \mathbf{Y}^N) \in T_\varepsilon^N\right)$.

First, note that, by symmetry, all summands are equal, so

$$\sum_{i=2}^M \mathbb{P}\left((\mathbf{x}^N(i), \mathbf{Y}^N) \in T_\varepsilon^N\right) = \left(2^{\lceil NR \rceil} - 1\right) \mathbb{P}\left((\mathbf{x}^N(2), \mathbf{Y}^N) \in T_\varepsilon^N\right).$$

Next, by Worked Example 4.2.3 (see (4.2.9) below)

$$\mathbb{P}\left((\mathbf{x}^N(2), \mathbf{Y}^N) \in T_\varepsilon^N\right) \leq 2^{-N(c - 3\varepsilon)}$$

and hence

$$\sum_{i=2}^m \mathbb{P}\left((\mathbf{x}^N(i), \mathbf{Y}^N) \in T_\varepsilon^N\right) \leq 2^{N(R - c + 3\varepsilon)}$$

which tends to 0 as $N \to \infty$ when $\varepsilon < (c - R)/3$.

Therefore, for $R < c$, $\lim_{N \to \infty} \mathscr{E}_M(P^N) = 0$. But $\mathscr{E}_M(P^N)$ admits the representation

$$\mathscr{E}_m(P^N) = \mathbb{E}_{P_{\mathbf{x}^N(1)} \times \cdots \times P_{\mathbf{x}^N(M)}} \left(\frac{1}{M} \sum_{1 \leq j \leq M} E(j)\right)$$

where quantity $E(j)$ represents the error-probability as defined in (4.1.14):

$$E(j) = \begin{cases} 1, & \mathbf{x}^N \notin \mathbb{D}^{(N)}, \\ P_{\mathrm{ch}}\left(d_{\mathrm{JT}}^{(N), \varepsilon}(\mathbf{Y}^N) \neq \mathbf{x}^N(j) | \mathbf{x}^N(j) \text{ sent}\right), & \mathbf{x}^N \in \mathbb{D}^{(N)}. \end{cases}$$

We conclude that there exists a sequence of sample codebooks $\mathscr{X}_{M,N}$ such that the average error-probability

$$\frac{1}{M} \sum_{\mathbf{x} \in \mathscr{X}_{M,N}} e(\mathbf{x}) \to 0$$

where $e(\mathbf{x}) = e(\mathbf{x}, \mathscr{X}_{M,N}, \mathbb{D}^{(N)}, d_{\mathrm{JT}}^{(N),\varepsilon})$ is the error-probability for the input word $\mathbf{x}$ in code $\mathscr{X}_{M,N}$, under the JT decoder and with regional constraint specified by $\mathbb{D}^{(N)}$:

$$e(\mathbf{x}) = \begin{cases} 1, & \mathbf{x}^N \notin \mathbb{D}^{(N)}, \\ \mathbf{P}_{\mathrm{ch}}\left(d_{\mathrm{JT}}^{(N),\varepsilon}(\mathbf{Y}^N) \neq \mathbf{x} | \mathbf{x} \text{ sent} \right), & \mathbf{x}^N \in \mathbb{D}^{(N)}. \end{cases}$$

Hence, R is a reliable transmission rate in the sense of Definition 4.1.2. This completes the proof of Theorem 4.1.8. $\qquad\square$

We also have proved in passing the following result.

Theorem 4.1.9 *Assume that the conditions of Theorem 4.1.5 hold true. Then, for all $R < C$, there exists a sequence of codes $\mathscr{X}_{M,N}$ of length N and size $M \sim 2^{RN}$ such that the maximum probability of error tends to 0 as $N \to \infty$.*

Example 4.1.10 Theorem 4.1.8 enables us to specify the expressions in (4.1.17) and (4.1.19) as the true values of the corresponding capacities (under the ML rule): for a scalar white noise of variance σ^2, under an average input power constraint $\sum\limits_{1 \leq j \leq N} x_j^2 \leq N\alpha$,

$$C(\alpha, \sigma^2) = \frac{1}{2} \log\left(1 + \frac{\alpha}{\sigma^2}\right),$$

for a vector white noise with variances $\underline{\sigma}^2 = (\sigma_1^2, \dots, \sigma_k^2)$, under the constraint $\sum\limits_{1 \leq j \leq N} x_j^{\mathrm{T}} x_j \leq N\alpha$,

$$C(\alpha, \underline{\sigma}^2) = \frac{1}{2} \sum_{1 \leq i \leq k} \log\left(1 + \frac{(v - \sigma_i^2)_+}{\sigma_i^2}\right), \text{ where } \sum_{1 \leq i \leq k} (v - \sigma_i^2)_+ = \alpha^2,$$

and for the coloured vector noise with a covariance matrix Σ, under the constraint $\sum\limits_{1 \leq j \leq N} x_j^{\mathrm{T}} Q x_j \leq N\alpha$,

$$C(\alpha, Q, \Sigma) = \frac{1}{2} \sum_{1 \leq i \leq k} \log\left(1 + \frac{(v \gamma_i^{-1} - \lambda_i)_+}{\lambda_i}\right),$$

where $\sum\limits_{1 \leq i \leq k} (v - \gamma_i \lambda_i)_+ = \alpha$.

Explicitly, for a scalar white noise we take the random coding where the signals $X_j(i)$, $1 \leq j \leq N$, $1 \leq i \leq M = \lceil 2^{NR} \rceil$, are IID $N(0, \alpha - \varepsilon)$. We have to check the conditions of Theorem 4.1.5 in this case: as $N \to \infty$,

(i) $\lim\limits_{N \to \infty} \mathbb{P}(\mathbf{x}^{(N)}(i) \in \mathbb{B}^{(N)}(\sqrt{N\alpha})$, for all $i = 1, \dots, M) = 1$;

(ii) $\lim_{\varepsilon \to 0} \lim_{N \to \infty} \theta_N = C(\alpha, \sigma^2)$ in probability where

$$\theta_N = \frac{1}{N} \sum_{1 \le j \le M} \log \frac{P(X,Y)}{P_X(X)P_Y(Y)}.$$

First, property (i):

$$\mathbb{P}\left(\mathbf{x}^{(N)}(i) \notin \mathbb{B}^{(N)}(\sqrt{N\alpha}), \text{ for some } i = 1, \ldots, M\right)$$

$$\le \mathbb{P}\left(\frac{1}{NM} \sum_{1 \le i \le M} \sum_{1 \le j \le N} X_j(i)^2 \ge \alpha\right)$$

$$= \mathbb{P}\left(\frac{1}{NM} \sum_{1 \le i \le M} \sum_{1 \le j \le N} (X_j(i)^2 - \sigma^2) \ge \varepsilon\right)$$

$$\le \mathbb{E}(X^2 - \sigma^2)^2 \left(\frac{1}{NM\varepsilon^2}\right) \to 0.$$

Next, (ii): since pairs (X_j, Y_j) are IID, we apply the law of large numbers and obtain that

$$\theta_N \to \mathbb{E} \log \frac{P(X,Y)}{P_X(X)P_Y(Y)} = I(X_1 : Y_1).$$

But

$$I(X_1 : Y_1) = h(Y_1) - h(Y_1|X_1)$$
$$= \frac{1}{2} \log \left[2\pi e(\alpha - \varepsilon + \sigma^2)\right] - \frac{1}{2} \log \left(2\pi e \sigma^2\right)$$
$$= \frac{1}{2} \log \left(1 + \frac{\alpha - \varepsilon}{\sigma^2}\right) \to C(\alpha, \sigma^2) \text{ as } \varepsilon \to 0.$$

Hence, the capacity equals $C(\alpha, \sigma^2)$, as claimed. The case of coloured noise is studied similarly.

Remark 4.1.11   Introducing a regional constraint described by a domain $\mathbb{D}$ does not mean one has to secure that the whole code $\mathcal{X}$ should lie in $\mathbb{D}$. To guarantee that the error-probability $P_e^{av}(\mathcal{X}) \to 0$ we only have to secure that the 'majority' of codewords $\mathbf{x}(i) \in \mathcal{X}$ belong to $\mathbb{D}$ when the codeword-length $N \to \infty$.

Example 4.1.12   Here we consider a non-Gaussian additive channel, where the noise vector

$$\mathbf{Z} = \begin{pmatrix} Z_1 \\ \vdots \\ Z_N \end{pmatrix}$$

has two-side exponential IID components $Z_j \sim (2) \operatorname{Exp}(\lambda)$, with the PDF

$$f_{Z_j}(z) = \frac{1}{2} \lambda e^{-\lambda |z|}, \quad -\infty < z < \infty,$$

where Exp denotes the exponential distribution, $\lambda > 0$ and $\mathbb{E}|Z_j| = 1/\lambda$ (see PSE I, Appendix). Again we will calculate the capacity under the ML rule and with a regional constraint $\mathbf{x}^{(N)} \in \text{Ł}(N\alpha)$ where

$$\text{Ł}(N\alpha) = \left\{ \mathbf{x}^{(N)} \in \mathbb{R}^N : \sum_{1 \le j \le N} |x_j| \le N\alpha \right\}.$$

First, observe that if the random variable X has $\mathbb{E}|X| \le \alpha$ and the random variable Z has $\mathbb{E}|Z| \le \zeta$ then $\mathbb{E}|X+Z| \le \alpha + \zeta$. Next, we use the fact that a random variable Y with PDF f_Y and $\mathbb{E}|Y| \le \eta$ has the differential entropy

$$h(Y) \le 2 + \log_2 \eta; \quad \text{with equality iff } Y \sim (2) \operatorname{Exp}(1/\eta).$$

In fact, as before, by Gibbs

$$
\begin{aligned}
h(Y) &= -\int f_Y(y) \log f_Y(y) \mathrm{d}y \\
&\le -\int f_Y(y) \log \phi^{(2) \operatorname{Exp}(1/\eta)}(y) \mathrm{d}y \\
&= 1 + \frac{1}{\eta} \int f_Y(y) |y| \mathrm{d}y + \log \eta \\
&= 1 + \log \eta \le 2 + \log \eta \\
&\quad + \frac{1}{\eta} \mathbb{E}|Y| \\
&= -\int \phi^{(2) \operatorname{Exp}(1/\eta)}(y) \log \phi^{(2) \operatorname{Exp}(1/\eta)}(y) \mathrm{d}y,
\end{aligned}
$$

and the equalities are achieved only when $f_Y = \phi^{(2) \operatorname{Exp}(1/\eta)}$.

Then, by the converse part of the SSCT,

$$
\begin{aligned}
\frac{1}{N} I(\mathbf{x}^{(N)} : \mathbf{Y}^{(N)}) &= \frac{1}{N} \sum h(Y_j) - h(Z_j) \\
&\le \frac{1}{N} \sum [2 + \log_2(\alpha_j + \lambda^{-1}) - 2 + \log_2(\lambda)] \\
&= \frac{1}{N} \sum \log_2 (1 + \alpha_j \lambda)] \\
&\le \log_2 (1 + \alpha \lambda).
\end{aligned}
$$

The same arguments as before establish that the RHS gives the capacity of the channel.

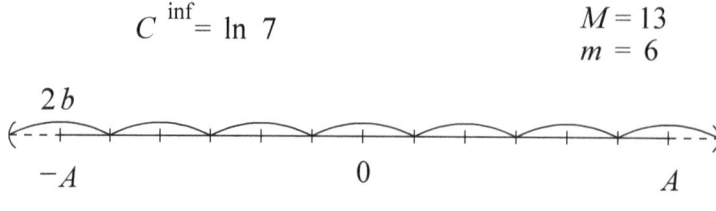

Figure 4.2

Worked Example 4.1.13 *Next, we consider a channel with an additive uniform noise, where the noise random variable $Z \sim \mathrm{U}(-b,b)$ with $b > 0$ representing the limit for the noise amplitude. Let us choose the region constraint for the input signal as a finite set $\mathscr{A} \subset \mathbb{R}$ (an input 'alphabet') of the form $\mathscr{A} = \{a, a+b, \ldots, a+(M-1)b\}$. Compute the information capacity of the channel:*

$$C^{\mathrm{inf}} = \sup \left[I(X:Y) : \ p_X(\mathscr{A}) = 1, Y = X + Z \right].$$

Solution Because of the shift-invariance, we can assume that $a = -A$ and $a + Mb = A$ where $2A = Mb$ is the 'width' of the input signal set. The formula $I(X:Y) = h(Y) - h(Y|X)$ where $h(Y|X) = h(Z) = \ln(2b)$ (in nats) shows that we must maximise the output signal entropy $h(Y)$. The limits for Y are $-A - b \leq Y \leq A + b$, so the distribution P_Y must be as close to uniform $\mathrm{U}(-A-b, A+b)$ as possible.

First, suppose M is odd: $\sharp\mathscr{A} = 2m+1$, with

$$\mathscr{A} = \{0, \pm A/m, \pm 2A/m, \ldots, \pm A\} \ \text{and} \ b = A/m.$$

That is, the points of $\mathscr{A}$ partition the interval $[-A, A]$ into $2m$ intervals of length A/m; the 'extended' interval $[-A-b, A+b]$ contains $2(m+1)$ such intervals. The maximising probability distribution P_X can be spotted without calculations: it assigns equal probabilities $1/(m+1)$ to $m+1$ points

$$-A, -A+2b, \ldots, A-2b, A.$$

In other words, we 'cross off' every second 'letter' from $\mathscr{A}$ and use the remaining letters with equal probabilities.

In fact, with $P_X(-A) = P_X(-A+2b) = \cdots = P_X(A)$, the output signal PDF f_Y assigns the value $[2b(m+1)]^{-1}$ to every point $y \in [-A-b, A+b]$. In other words, $Y \sim \mathrm{U}(-A-b, A+b)$ as required. The information capacity C^{inf} in this case is equal (in nats) to

$$\ln(2A+2b) - \ln 2b = \ln(1+m). \tag{4.1.32}$$

Say, for $M = 3$ (three input signals, at $-A$, 0, A, and $b = A$), $C^{\mathrm{inf}} = \ln 2$. For $M = 5$ (five input signals, at $-A$, $-A/2$, 0, $A/2$, A, and $b = A/2$), $C^{\mathrm{inf}} = \ln 3$. See Figure 4.2 for $M = 13$. $\qquad\square$

Remark 4.1.14 It can be proved that (4.1.32) gives the maximum mutual information $I(X:Y)$ between the input and output signals X and $Y = X + Z$ when (i) the noise random variable $Z \sim \mathrm{U}(-b,b)$ is independent of X and (ii) X has a general distribution supported on the interval $[-A,A]$ with $b = A/m$. Here, the mutual information $I(X:Y)$ is defined according to Kolmogorov:

$$I(X:Y) = \sup_{\xi,\eta} I(X_\xi : Y_\eta) \qquad (4.1.33)$$

where the supremum is taken over all finite partitions ξ and η of intervals $[-A,A]$ and $[-A-b,A+b]$, and X_ξ and Y_η stand for the quantised versions of random variables X and Y, respectively.

In other words, the input-signal distribution P_X with

$$P_X(-A) = P_X(-A+2b) = \cdots = P_X(A-2b) = P_X(A) = \frac{1}{m+1} \qquad (4.1.34)$$

maximises $I(X:Y)$ under assumptions (i) and (ii). We denote this distribution by $P_X^{(A,A/m)}$, or, equivalently, $P_X^{(bm,b)}$.

However, if $M = 2m$, i.e. the number $\sharp\mathscr{A}$ of allowed signals is even, the calculation becomes more involved. Here, clearly, the uniform distribution $\mathrm{U}(-A - b, A + b)$ for the output signal Y cannot be achieved. We have to maximise $h(Y) = h(X + Z)$ within the class of piece-wise constant PDFs f_Y on $[-A - b, A + b]$; see below.

Equal spacing in $[-A,A]$ is generated by points $\pm A/(2m - 1)$, $\pm 3A/(2m - 1),\ldots,\pm A$; they are described by the formula $\pm(2k - 1)A/(2m - 1)$ for $k = 1,\ldots,m$. These points divide the interval $[-A,A]$ into $(2m - 1)$ intervals of length $2A/(2m - 1)$. With $Z \sim \mathrm{U}(-b,b)$ and $A = b(m - 1/2)$, we again have the output-signal PDF $f_Y(y)$ supported in $[-A - b, A + b]$:

$$f_Y(y) = \begin{cases} p_m/(2b), & \text{if } b(m - 1/2) \leq y \leq b(m + 1/2), \\ (p_k + p_{k+1})/(2b), & \text{if } b(k - 1/2) \leq y \leq b(k + 1/2) \\ & \qquad \text{for } k = 1,\ldots,m - 1, \\ (p_{-1} + p_1)/(2b), & \text{if } -b/2 \leq y \leq b/2, \\ (p_k + p_{k+1})/(2b), & \text{if } b(k - 1/2) \leq y \leq b(k + 1/2) \\ & \qquad \text{for } k = -1,\ldots,-m + 1, \\ p_{-m}/(2b), & \text{if } -b(m + 1/2) \leq y \leq -b(m - 1/2), \end{cases}$$

where

$$p_{\pm k} = p_X\left(\pm b\left(k - \frac{1}{2}\right)\right) = \mathbb{P}\left(X = \pm\frac{(2k-1)A}{2m-1}\right), \ k = 1,\dots,m,$$

stand for the input-signal probabilities. The entropy $h(Y) = h(X + Z)$ is written as

$$-\frac{p_m}{2}\ln\frac{p_m}{2b} - \sum_{1\le k<m}\frac{p_k + p_{k+1}}{2}\ln\frac{p_k + p_{k+1}}{2b} - \frac{p_{-1} + p_1}{2}\ln\frac{p_{-1} + p_1}{2b}$$
$$- \sum_{-m<k\le-1}\frac{p_k + p_{k+1}}{2}\ln\frac{p_k + p_{k+1}}{2b} - \frac{p_{-m}}{2}\ln\frac{p_{-m}}{2b}.$$

It turns out that the maximising distribution P_X has $p_{-k} = p_k$, for $k = 1,\dots,m$. Thus, we face an optimisation problem:

$$\text{maximise} \ \ G(\underline{p}) = -p_m\ln\frac{p_m}{2b} - \sum_{1\le k<m}(p_k + p_{k+1})\ln\frac{p_k + p_{k+1}}{2b} - p_1\ln\frac{p_1}{b}$$

$$(4.1.35)$$

subject to the probabilistic constraints $p_k \ge 0$ and $2\sum_{1\le k\le m}p_k = 1$. The Lagrangian $\mathscr{L}(P_X;\lambda)$ reads

$$\mathscr{L}(P_X;\lambda) = G(\underline{p}) + \lambda(2p_1 + \cdots + 2p_m - 1)$$

and is maximised when

$$\frac{\partial}{\partial p_k}\mathscr{L}(P_X;\lambda) = 0, \ \ k = 1,\dots,m.$$

Thus, we have m equations, with the same RHS:

$$-\ln\frac{p_m(p_{m-1} + p_m)}{4b^2} - 2 + 2\lambda = 0, \ \text{(implies)} \ p_m(p_{m-1} + p_m) = 4b^2 e^{2\lambda - 2},$$
$$-\ln\frac{(p_{k-1} + p_k)(p_k + p_{k+1})}{4b^2} - 2 + 2\lambda = 0,$$
$$\text{(implies)} \ (p_{k-1} + p_k)(p_k + p_{k+1}) = 4b^2 e^{2\lambda - 2}, \ 1 < k < m,$$
$$-\ln\frac{2p_1(p_1 + p_2)}{4b^2} - 2 + 2\lambda = 0 \ \text{(implies)} \ 2p_1(p_1 + p_2) = 4b^2 e^{2\lambda - 2}.$$

This yields

$$\left.\begin{aligned}p_m = p_{m-1} + p_{m-2} = \cdots = p_3 + p_2 = 2p_1,\\ p_m + p_{m-1} = p_{m-2} + p_{m-3} = \cdots = p_2 + p_1,\end{aligned}\right\} \text{for } m \text{ even,}$$

and

$$p_m = p_{m-1} + p_{m-2} = \cdots = p_2 + p_1,$$
$$p_m + p_{m-1} = p_{m-2} + p_{m-3} = \cdots = p_3 + p_2 = 2p_1, \quad \bigg\} \text{ for } m \text{ odd.}$$

For small values of $M = 2m$ the solution is straightforward. Viz., for $M = 2$ (two input signals at $\pm A$ with $b = 2A$): $p_1 = 1/2$ and the maximising output-signal PDF is

$$f_Y(y) = \begin{cases} 1/(4b), & A \le y \le 3A, \\ 1/(2b), & -A \le y \le A, \\ 1/(4b), & -3A \le y \le -A, \end{cases} \quad \text{yielding } C^{\text{inf}} = (\ln 2)/2.$$

For $M = 4$ (four input signals at $-A$, $-A/3$, $A/3$, A, with $b = 2A/3$): $p_1 = 1/6$, $p_2 = 1/3$, and the maximising output-signal PDF is

$$f_Y(y) = \begin{cases} 1/(6b), & A \le y \le 5A/3 \text{ and } -5A/3 \le y \le -A, \\ 1/(4b), & 2A/3 \le y \le A \text{ and } -A \le y \le -2A/3, \\ 1/(6b), & -2A/3 \le y \le 2A/3, \end{cases}$$

which yields $C^{\text{inf}} = \ln(6^{1/2}4^{1/3}/2)$.

For $M = 6$ (six input signals at $-A$, $-3A/5$, $-A/5$, $A/5$, $3A/5$, A, with $b = 2A/5$): $p_1 = 1/6$, $p_2 = 1/12$, $p_3 = 1/4$. Similarly, for $M = 8$ (eight input signals at $-A$, $-5A/7$, $-3A/7$, $-A/7$, $A/7$, $3A/7$, $5A/7$, A, with $b = 2A/7$): $p_1 = 1/10$, $p_2 = 3/20$, $p_3 = 1/20$, $p_4 = 1/5$.

In general, we can write all probabilities in terms of p_1. Viz., for m even:

$$p_m = 2p_1,$$
$$p_{m-1} = p_2 - p_1,$$
$$p_{m-2} = 3p_1 - p_2,$$
$$p_{m-3} = 2(p_2 - p_1),$$
$$p_{m-4} = 4p_1 - 2p_2,$$
$$\vdots$$
$$p_3 = \left(\frac{m}{2} - 1\right)(p_2 - p_1),$$
$$p_2 = \frac{m+2}{m}p_1,$$

whence

$$p_2 = \frac{m+2}{m} p_1,$$

$$p_3 = \frac{m-2}{m} p_1,$$

$$p_4 = \frac{m+4}{m} p_1,$$

$$p_5 = \frac{m-4}{m} p_1,$$

$$\vdots$$

$$p_{m-2} = \frac{2m-2}{m},$$

$$p_{m-1} = \frac{2}{m},$$

$$p_m = 2p_1,$$

$$\text{with } p_1 = \frac{1}{2(m+1)}. \qquad (4.1.36)$$

The corresponding PDF f_Y gives the value

$$h(Y) = -\frac{1}{2}\ln\frac{1}{4m(m+1)b^2} \quad\text{and}\quad C_{\mathscr{A}}^{\text{inf}} = -\frac{1}{2}\ln\frac{1}{4m(m+1)} - \ln 2. \quad (4.1.37)$$

On the other hand, for a general odd m, the maximising input-signal distribution P_X has

$$p_1 = \frac{m+1}{2m(m+1)},$$

$$p_2 = \frac{m-1}{2m(m+1)},$$

$$p_3 = \frac{m+3}{2m(m+1)},$$

$$p_4 = \frac{m-3}{2m(m+1)},$$

$$\vdots$$

$$p_{m-1} = \frac{1}{2m(m+1)},$$

$$p_m = \frac{m}{m(m+1)}. \qquad (4.1.38)$$

This yields the same answer for the maximum entropy and the restricted capacity:

$$h(Y) = -\frac{1}{2}\ln\frac{1}{4m(m+1)b^2} \quad \text{and} \quad C_{\mathscr{A}}^{\text{inf}} = -\frac{1}{2}\ln\frac{1}{4m(m+1)} - \ln 2. \quad (4.1.39)$$

In future, we will refer to the input-signal distributions specified in (4.1.36) and (4.1.38) as $\widetilde{P}_X^{(A,2A/(2m-1))}$.

Remark 4.1.15 It is natural to suggest that the above formulas give the maximum mutual information $I(X:Y)$ when (i) the noise random variable $Z \sim U(-b,b)$ is independent of X and (ii) the input-signal distribution P_X is confined to $[-A,A]$ with $b = 2A/(2m-1)$, but otherwise is arbitrary (with $I(X:Y)$ again defined as in (4.1.33)). A further-reaching (and more speculative) conjecture is about the maximiser under the above assumptions (i) and (ii) but for arbitrary $A > b > 0$, not necessarily with A/b being integer or half-integer. Here number $M = 2A/b+1$ will not be integer either, but remains worth keeping as a value of reference.

So when b decays from A/m to $A/(m+1)$ (or, equivalently, A grows from bm to $b(m+1)$ and, respectively, M increases from $2m+1$ to $2m+3$), the maximiser $P_X^{(A,b)}$ evolves from $P_X^{(bm,b)}$ to $P_X^{(b(m+1),b)}$; at $A = b(m+1/2)$ (when $M = 2(m+1)$) distribution $P_X^{(A,b)}$ may or may not coincide with the distribution $\widetilde{P}_X^{(A,b)}$ from (4.1.36), (4.1.38).

To (partially) clarify the issue, consider the case where $A/2 \le b \le A$ (i.e. $3 \le M \le 5$) and assume that the input-signal distribution P_X has

$$P_X(-A) = P_X(A) = p \quad \text{and} \quad P_X(0) = 1 - 2p \quad \text{where } 0 \le p \le \frac{1}{2}. \quad (4.1.40)$$

Then

$$h_Y(Y) = -\frac{1}{b}\left[Ap\ln\frac{p}{2b} + (2b-A)(1-p)\ln\frac{1-p}{2b} + (A-b)(1-2p)\ln\frac{1-2p}{2b}\right], \quad (4.1.41)$$

and the equation $dh(Y)/dp = 0$ is equivalent to

$$p^A = (1-p)^{2b-A}(1-2p)^{2(A-b)}. \quad (4.1.42)$$

For $b = A/2$ this yields $p^A = (1-2p)^A$, i.e. $p = 1 - 2p$ whence $p = 1/3$; similarly, for $b = A$, $p = 1/2$. These coincide with previously obtained results. For $b = 2A/3$ we have that

$$p^A = (1-p)^{A/3}(1-2p)^{2A/3};$$

i.e.

$$p^3 = (1-p)(1-2p)^2. \quad (4.1.43a)$$

We are interested in the solution lying in $(0, 1/2)$ (in fact, in $(1/3, 1/2)$). For $b = 3A/4$, the equation becomes

$$p^A = (1-p)^{A/2}(1-2p)^{A/2},$$

i.e.

$$p^2 = (1-p)(1-2p), \qquad\qquad (4.1.43b)$$

whence $p = (3 - \sqrt{5})/2$.

Example 4.1.16 It is useful to look at the example where the noise random variable Z has two components: discrete and continuous. To start with, one could try the case where

$$f_Z(z) = q\delta_0 + (1-q)\phi(z; \sigma^2),$$

i.e. $Z = 0$ with probability q and $Z \sim N(0, \sigma^2)$ with probability $1 - q \in (0, 1)$. (So, $1 - q$ gives the total probability of error.) Here, we consider the case

$$f_Z = q\delta_0 + (1-q)\frac{1}{2b}\mathbf{1}(|z| \le b),$$

and study the input-signal PMF of the form

$$P_X(-A) = p_{-1}, \ P_X(0) = p_0, \ P_X(A) = p_1, \qquad\qquad (4.1.44a)$$

where

$$p_{-1}, p_0, p_1 \ge 0, \ p_{-1} + p_0 + p_1 = 1, \qquad\qquad (4.1.44b)$$

with $b = A$ and $M = 3$ (three signal levels in $(-A, A)$). The input-signal entropy is

$$h(X) = h(p_{-1}, p_0, p_1) = -p_{-1}\ln p_{-1} - p_0 \ln p_0 - p_1 \ln p_1.$$

The output-signal PMF has the form

$$f_Y(y) = q\left(p_{-1}\delta_{-A} + p_0\delta_0 + p_1\delta_A\right) + (1-q)\frac{1}{2b}$$
$$\times \left[p_{-1}\mathbf{1}(-2A \le y \le 0) + p_0\mathbf{1}(-A \le y \le A) + p_1\mathbf{1}(0 \le y \le 2A)\right]$$

and its entropy $h(Y)$ (calculated relative to the reference measure μ on $\mathbb{R}$, whose absolutely continuous component coincides with the Lebesgue and discrete component assigns value 1 to points $-A$, 0 and A) is given by

$$h(Y) = -q\ln q - (1-q)\ln(1-q) - qh(p_{-1}, p_0, p_1)$$
$$- (1-q)A\left[p_{-1}\ln\frac{p_{-1}}{2A} + (p_{-1} + p_0)\ln\frac{p_{-1} + p_0}{2A}\right.$$
$$\left. + (p_0 + p_1)\ln\frac{p_0 + p_1}{2A} + p_1\ln\frac{p_1}{2A}\right].$$

By symmetry, $h(Y)$ is maximised when $p_{-1} = p_1 = p$, $p_0 = 1 - 2p$, and we have to maximise, in $q \in (0, 1)$, the expression

$$h(Y) = h(q, 1 - q) - qh(p, p, 1 - 2p) - (1 - q)A\left[2p\ln\frac{p}{2A} + (1 - 2p)\ln\frac{1 - 2p}{2A}\right],$$

for a given $q \in (0, 1)$.

Differentiating yields

$$\frac{d}{dp}h(Y) = 0 \leftrightarrow \frac{p}{1 - 2p} = \left(\frac{p}{1 - p}\right)^{-(1-q)A/q}.$$

If $(1 - q)A/q > 1$ this equation yields a unique solution which defines an optimal input-signal distribution P_X of the form (4.1.44a)–(4.1.44b).

If we wish to see what value of q yields the maximum of $h(Y)$ (and hence, the maximum information capacity), we differentiate in q as well:

$$\frac{d}{dq}h(Y) = 0 \leftrightarrow \log\frac{q}{1 - q} = (A - 1)h(p, p, 1 - 2p) - 2A\ln 2A.$$

If we wish to consider a continuously distributed input signal on $[-A, A]$, with a PDF $f_X(x)$, then the output random variable $Y = X + Z$ has the PDF given by the convolution:

$$f_Y(y) = \frac{1}{2b}\int_{(y-b)\vee(-A)}^{(y+b)\wedge A} f_X(x)dx.$$

The differential entropy $h(Y) = -\int f_Y(y)\ln f_Y(y)dy$, in terms of f_X, takes the form

$$h(X + Z) = -\frac{1}{2b}\int_{-A}^{A} f_X(x)\int_{-b}^{b}\ln\left[\frac{1}{2b}\int_{(x+z-b)\vee(-A)}^{(x+z+b)\wedge A} f_X(x')dx'\right]dzdx.$$

The PDF f_X minimising the differential entropy $h(X + Z)$ yields a solution to

$$0 = \int_{-b}^{b}\left(\ln\left[\frac{1}{2b}\int_{(x+z-b)\vee(-A)}^{(x+z+b)\wedge A} f_X(x')dx'\right] + f_X(x)\right.$$
$$\left.\times\left[\int_{(x+z-b)\vee(-A)}^{(x+z+b)\wedge A} f_X(x')dx'\right]^{-1}\left[f_X(x + z + b) - f_X(x + z - b)\right]\right)dz.$$

An interesting question emerges when we think of a two-time-per-signal use of a channel with a uniform noise. Suppose an input signal is represented by a point $\mathbf{x} = (x_1, x_2)$ in a plane $\mathbb{R}^2$ and assume as before that $Z \sim U(-b, b)$, independently of the input signal. Then the square $S_b(\mathbf{x}) = (x_1 - b, x_1 + b) \times (x_2 - b, x_2 + b)$, with the uniform PDF $1/(4b^2)$, outlines the possible positions of the output signal Y given that $X = (x_1, x_2)$. Suppose that we have to deal with a finite input alphabet $\mathscr{A} \subset \mathbb{R}^2$; then the output-signal domain is the finite union $\mathscr{B} = \cup_{\mathbf{x}\in\mathscr{A}}S(\mathbf{x})$. The above argument shows that if we can find a subset $\mathscr{A}' \subseteq \mathscr{A}$ such that squares $S_b(\mathbf{x})$

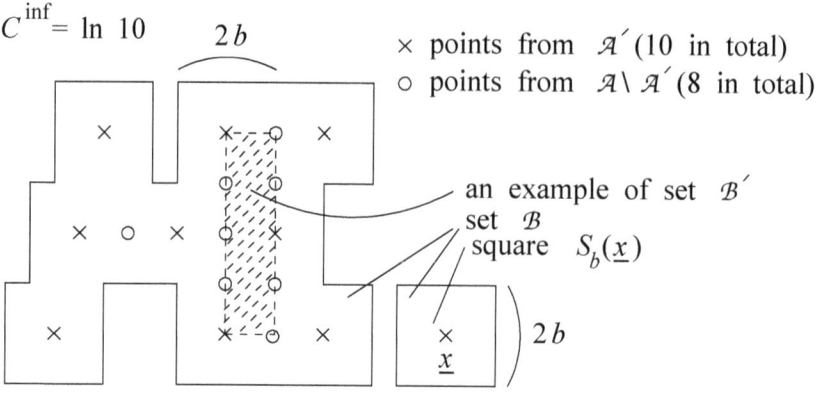

Figure 4.3

with $\mathbf{x} \in \mathscr{A}'$ partition domain $\mathscr{B}$ (i.e. cover $\mathscr{B}$ but do not intersect each other) then, for the input PMF $P_\mathbf{x}$ with $P_\mathbf{x}(\mathbf{x}) = 1/(\sharp\mathscr{A}')$ (a uniform distribution over $\mathscr{A}'$), the output-vector-signal PDF f_Y is uniform on $\mathscr{B}$ (that is, $f_Y(y) = 1/(\text{area of } \mathscr{B})$). Consequently, the output-signal entropy $h(Y) = \ln(\text{area of } \mathscr{B})$ is attaining the maximum over all input-signal PMFs $P_\mathbf{x}$ with $P_\mathbf{x}(\mathscr{A}) = 1$ (and even attaining the maximum over all input-signal PMFs $P_\mathbf{x}$ with $P_\mathbf{x}(\mathscr{B}') = 1$ where $\mathscr{B}' \subset \mathscr{B}$ is an arbitrary subset with the property that $\cup_{\mathbf{x}' \in \mathscr{B}'} S(\mathbf{x}')$ lies within $\mathscr{B}$). Finally, the information capacity for the channel under consideration,

$$C^{\text{inf}} = \frac{1}{2} \ln \frac{\text{area of } \mathscr{B}}{4b^2} \quad \text{nats/(scalar input signal)}.$$

See Figure 4.3.

To put it differently, any bounded set $\mathbb{D}_2 \subset \mathbb{R}^2$ that can be partitioned into disjoint squares of length $2b$ yields the information capacity

$$C_2^{\text{inf}} = \frac{1}{2} \ln \frac{\text{area of } \mathbb{D}_2}{4b^2} \quad \text{nats/(scalar input signal)},$$

of an additive channel with a uniform noise over $(-b, b)$, when the channel is used two times per scalar input signal and the random vector input $\mathbf{x} = (X_1, X_2)$ is subject to the regional constraint $\mathbf{x} \in \mathbb{D}_2$. The maximising input-vector PMF assigns equal probabilities to the centres of squares forming the partition.

A similar conclusion holds in $\mathbb{R}^3$ when the channel is used three times for every input signal, i.e. the input signal is a three-dimensional vector $\mathbf{x} = (x_1, x_2, x_3)$, and so on. In general, when we use a K-dimensional input signal $\mathbf{x} = (x_1, \ldots, x_k) \in \mathbb{R}^K$, and the regional constraint is $\mathbf{x} \in \mathbb{D}_K \subset \mathbb{R}^K$ where $\mathbb{D}_K$ is a bounded domain that can be partitioned into disjoint cubes of length $2b$, the information capacity

$$C_K^{\text{inf}} = \frac{1}{K} \ln \frac{\text{volume of } \mathbb{D}_K}{(2b)^K} \quad \text{nats/(scalar input signal)}$$

is achieved at the input-vector-signal PMF $P_{\mathbf{x}}$ assigning equal masses to the centres of the cubes forming the partition.

As $K \to \infty$, the quantity C_K may converge to a limit C_∞^{inf} yielding the capacity per scalar input signal under the sequence of regional constraint domains $\mathbb{D}_K$. A trivial example of such a situation is where $\mathbb{D}_K$ is a K-dimensional cube

$$S_b^K = (-2bm, 2bm)^{\times K};$$

then $C_K^{\text{inf}} = \ln(1+m)$ does not depend on K (and the channel is memoryless).

4.2 The asymptotic equipartition property in the continuous time setting

> *The errors of a wise man make your rule,*
> *Rather than perfections of a fool.*
> William Blake (1757–1821), English poet

This section provides a missing step in the proof of Theorem 4.1.8 and additional Worked Examples. We begin with a series of assertions illustrating the asymptotic equipartition property in various forms. The central facts are based on the Shannon–McMillan–Breiman (SMB) theorem which is considered a cornerstone of information theory. This theorem gives the information rate of a stationary ergodic process $\mathbf{X} = (X_n)$. Recall that a transformation of a probability space T is called ergodic if every set A such that $TA = A$ almost everywhere, satisfies $\mathbb{P}(A) = 0$ or 1. For a stationary ergodic source with a finite expected value, Birkhoff's ergodic theorem states the law of large numbers (with probability 1):

$$\frac{1}{n} \sum_{i=1}^{n} X_i \to \mathbb{E}X. \tag{4.2.1}$$

Typically, for a measurable function $f(X_t)$ of ergodic process,

$$\frac{1}{n} \sum_{i=1}^{n} f(X_i) \to \mathbb{E}f(X). \tag{4.2.2}$$

Theorem 4.2.1 (Shannon–McMillan–Breiman) *For any stationary ergodic process* $\mathbf{X}$ *with finitely many values the information rate* $R = h$, *i.e. the limit in* (4.2.3) *exists in the sense of the a.s. convergence and equals to entropy*

$$- \lim_{n \to \infty} \frac{1}{n} \log p_{X_0^{n-1}} \left(X_0^{n-1} \right) = h \text{ a.s.} \tag{4.2.3}$$

The proof of Theorem 4.2.1 requires some auxiliary lemmas and is given at the end of the section.

Worked Example 4.2.2   (A general asymptotic equipartition property) *Given a sequence of random variables* $X_1, X_2, \ldots$, *for all* $N = 1, 2, \ldots$, *the distribution of the random vector* $\mathbf{x}_1^N = \begin{pmatrix} X_1 \\ \vdots \\ X_N \end{pmatrix}$ *is determined by a PMF* $f_{\mathbf{x}_1^N}(\mathbf{x}_1^N)$ *with respect to measure* $\mu^{(N)} = \mu \times \cdots \times \mu$ *(N factors). Suppose that the statement of the Shannon–McMillan–Breiman theorem holds true:*

$$- \frac{1}{N} \log f_{\mathbf{x}_1^N}(\mathbf{x}_1^N) \to h \text{ in probability,}$$

where $h > 0$ *is a constant (typically,* $h = \lim_{i \to \infty} h(X_i)$*). Given* $\varepsilon > 0$, *consider the typical set*

$$S_\varepsilon^N = \left\{ \mathbf{x}_1^N = \begin{pmatrix} x_1 \\ \vdots \\ x_N \end{pmatrix} : -\varepsilon \le \frac{1}{N} \log f_{\mathbf{x}_1^N}(\mathbf{x}_1^N) + h \le \varepsilon \right\}.$$

The volume $\mu^{(N)}(S_\varepsilon^N) = \int_{S_\varepsilon^N} \mu(dx_1) \ldots \mu(dx_N)$ *of set* S_ε^N *has the following properties:*

$$\mu^{(N)}(S_\varepsilon^N) \le 2^{N(h+\varepsilon)}, \text{ for all } \varepsilon \text{ and } N, \tag{4.2.4}$$

and, for $0 < \varepsilon < h$ *and for all* $\delta > 0$,

$$\mu^{(N)}(S_\varepsilon^N) \ge (1-\delta)2^{N(h-\varepsilon)}, \text{ for } N \text{ large enough, depending on } \delta. \tag{4.2.5}$$

Solution Since $\mathbb{P}(\mathbb{R}^N) = \int_{\mathbb{R}^N} f_{\mathbf{x}_1^N}(\mathbf{x}_1^N) \prod_{1 \le j \le N} \mu(dx_j) = 1$, we have that

$$1 = \int_{\mathbb{R}^N} f_{\mathbf{x}_1^N}(\mathbf{x}_1^N) \prod_{1 \le j \le N} \mu(dx_j)$$

$$\ge \int_{S_\varepsilon^N} f_{\mathbf{x}_1^N}(\mathbf{x}_1^N) \prod_{1 \le j \le N} \mu(dx_j)$$

$$\ge 2^{-N(h+\varepsilon)} \int_{S_\varepsilon^N} \prod_{1 \le j \le N} \mu(dx_j) = 2^{-N(h+\varepsilon)} \mu^{(N)}(S_\varepsilon^N),$$

giving the upper bound (4.2.4). On the other hand, given $\delta > 0$, we can take N large so that $\mathbb{P}(S_\varepsilon^N) \geq 1 - \delta$, in which case, for $0 < \varepsilon < h$,

$$
\begin{aligned}
1 - \delta &\leq \mathbb{P}(S_\varepsilon^N) \\
&= \int_{S_\varepsilon^N} f_{\mathbf{X}_1^N}(\mathbf{x}_1^N) \prod_{1 \leq j \leq N} \mu(\mathrm{d}x_j) \\
&\leq 2^{-N(h-\varepsilon)} \int_{S_\varepsilon^N} \prod_{1 \leq j \leq N} \mu(\mathrm{d}x_j) = 2^{-N(h-\varepsilon)} \mu^{(N)}(S_\varepsilon^N).
\end{aligned}
$$

This yields the lower bound (4.2.5). $\qquad \square$

The next step is to extend the asymptotic equipartition property to joint distributions of pairs $\mathbf{X}_1^N$, $\mathbf{Y}_1^N$ (in applications, $\mathbf{X}_1^N$ will play a role of an input and $\mathbf{Y}_1^N$ of an output of a channel). Formally, given two sequences of random variables, $X_1, X_2, \ldots$ and $Y_1, Y_2, \ldots$, for all $N = 1, 2, \ldots$, consider the joint distribution of the random vectors $\mathbf{X}_1^N = \begin{pmatrix} X_1 \\ \vdots \\ X_N \end{pmatrix}$ and $\mathbf{Y}_1^N = \begin{pmatrix} Y_1 \\ \vdots \\ Y_N \end{pmatrix}$ which is determined by a (joint) PMF $f_{\mathbf{X}_1^N, \mathbf{Y}_1^N}$ with respect to measure $\mu^{(N)} \times \nu^{(N)}$ where $\mu^{(N)} = \mu \times \cdots \times \mu$ and $\nu^{(N)} = \nu \times \cdots \times \nu$ (N factors in both products). Let $f_{\mathbf{X}_1^N}$ and $f_{\mathbf{Y}_1^N}$ stand for the (joint) PMFs of vectors $\mathbf{X}_1^N$ and $\mathbf{Y}_1^N$, respectively.

As in Worked Example 4.2.2, we suppose that the statements of the Shannon–McMillan–Breiman theorem hold true, this time for the pair $(\mathbf{X}_1^N, \mathbf{Y}_1^N)$ and each of $\mathbf{X}_1^N$ and $\mathbf{Y}_1^N$: as $N \to \infty$,

$$
\begin{aligned}
-\frac{1}{N} \log f_{\mathbf{X}_1^N}(\mathbf{X}_1^N) \to h_1, \quad &-\frac{1}{N} \log f_{\mathbf{Y}_1^N}(\mathbf{Y}_1^N) \to h_2, \\
-\frac{1}{N} \log f_{\mathbf{X}_1^N, \mathbf{Y}_1^N}(\mathbf{X}_1^N, \mathbf{Y}_1^N) &\to h,
\end{aligned} \quad \text{in probability,}
$$

where h_1, h_2 and h are positive constants, with

$$
h_1 + h_2 \geq h; \tag{4.2.6}
$$

typically, $h_1 = \lim_{i \to \infty} h(X_i)$, $h_2 = \lim_{i \to \infty} h(Y_i)$, $h = \lim_{i \to \infty} h(X_i, Y_i)$ and $h_1 + h_2 - h = \lim_{i \to \infty} I(X_i : Y_i)$. Given $\varepsilon > 0$, consider the typical set formed by sample pairs $(\mathbf{x}_1^N, \mathbf{y}_1^N)$ where

$$
\mathbf{x}_1^N = \begin{pmatrix} x_1 \\ \vdots \\ x_N \end{pmatrix}
$$

and

$$\mathbf{y}_1^N = \begin{pmatrix} y_1 \\ \vdots \\ y_N \end{pmatrix}.$$

Formally,

$$T_\varepsilon^N = \Big\{ (\mathbf{x}_1^N, \mathbf{y}_1^N): \ -\varepsilon \le \frac{1}{N} \log f_{\mathbf{x}_1^N}(\mathbf{x}_1^N) + h_1 \le \varepsilon,$$

$$-\varepsilon \le \frac{1}{N} \log f_{\mathbf{Y}_1^N}(\mathbf{y}_1^N) + h_2 \le \varepsilon,$$

$$-\varepsilon \le \frac{1}{N} \log f_{\mathbf{x}_1^N, \mathbf{Y}_1^N}(\mathbf{x}_1^N, \mathbf{y}_1^N) + h \le \varepsilon \Big\}; \quad (4.2.7)$$

by the above assumption we have that $\lim\limits_{N \to \infty} \mathbb{P}(T_\varepsilon^N) = 1$ for all $\varepsilon > 0$. Next, define the volume of set T_ε^N:

$$\mu^{(N)} \times \nu^{(N)}(T_\varepsilon^N) = \int_{T_\varepsilon^N} \mu^{(N)}(\mathbf{dx}_1^N) \nu^{(N)}(\mathbf{dy}_1^N).$$

Finally, consider an independent pair $\left(\widetilde{\mathbf{X}}_1^N, \widetilde{\mathbf{Y}}_1^N \right)$ where component $\widetilde{\mathbf{X}}_1^N$ has the same PMF as $\mathbf{X}_1^N$ and $\widetilde{\mathbf{Y}}_1^N$ the same PMF as $\mathbf{Y}_1^N$. That is, the joint PMF for $\widetilde{\mathbf{X}}_1^N$ and $\widetilde{\mathbf{Y}}_1^N$ has the form

$$f_{\widetilde{\mathbf{X}}_1^N, \widetilde{\mathbf{Y}}_1^N}(\mathbf{x}_1^N, \mathbf{y}_1^N) = f_{\mathbf{X}_1^N}(\mathbf{x}_1^N) f_{\mathbf{Y}_1^N}(\mathbf{y}_1^N). \quad (4.2.8)$$

Next, we assess the volume of set T_ε^N and then the probability that $\left(\widetilde{\mathbf{x}}_1^N, \widetilde{\mathbf{Y}}_1^N \right) \in T_\varepsilon^N$.

Worked Example 4.2.3 (A general joint asymptotic equipartition property)
(I) *The volume of the typical set has the following properties:*

$$\mu^{(N)} \times \nu^{(N)}(T_\varepsilon^N) \le 2^{N(h+\varepsilon)}, \text{for all } \varepsilon \text{ and } N, \quad (4.2.9)$$

and, for all $\delta > 0$ and $0 < \varepsilon < h$, for N large enough, depending on δ,

$$\mu^{(N)} \times \nu^{(N)}(T_\varepsilon^N) \ge (1 - \delta) 2^{N(h-\varepsilon)}. \quad (4.2.10)$$

(II) *For the independent pair $\left(\widetilde{\mathbf{X}}_1^N, \widetilde{\mathbf{Y}}_1^N \right)$,*

$$\mathbb{P}\left(\left(\widetilde{\mathbf{X}}_1^N, \widetilde{\mathbf{Y}}_1^N \right) \in T_\varepsilon^N \right) \le 2^{-N(h_1+h_2-h-3\varepsilon)}, \text{ for all } \varepsilon \text{ and } N, \quad (4.2.11)$$

and, for all $\delta > 0$, for N large enough, depending on δ,

$$\mathbb{P}\left(\left(\widetilde{\mathbf{X}}_1^N, \widetilde{\mathbf{Y}}_1^N \right) \in T_\varepsilon^N \right) \ge (1 - \delta) 2^{-N(h_1+h_2-h+3\varepsilon)}, \quad \text{for all } \varepsilon. \quad (4.2.12)$$

Solution (I) Completely follows the proofs of (4.2.4) and (4.2.5) with integration of $f_{\mathbf{X}_1^N, \mathbf{Y}_1^N}$.

(II) For the probability $\mathbb{P}\left(\left(\tilde{\mathbf{X}}_1^N, \tilde{\mathbf{Y}}_1^N \right) \in T_{\varepsilon}^N \right)$ we obtain (4.2.11) as follows:

$$\mathbb{P}\left(\left(\tilde{\mathbf{X}}_1^N, \tilde{\mathbf{Y}}_1^N \right) \in T_{\varepsilon}^N \right) = \int_{T_{\varepsilon}^N} f_{\tilde{\mathbf{X}}_1^N, \tilde{\mathbf{Y}}_1^N} \mu(\mathrm{d}\mathbf{x}_1^N) \nu(\mathrm{d}\mathbf{y}_1^N)$$

by definition

$$= \int_{T_{\varepsilon}^N} f_{\mathbf{X}_1^N}(\mathbf{x}_1^N) f_{\mathbf{Y}_1^N}(\mathbf{y}_1^N) \mu(\mathrm{d}\mathbf{x}_1^N) \nu(\mathrm{d}\mathbf{y}_1^N)$$

substituting (4.2.8)

$$\leq 2^{-N(h_1-\varepsilon)} 2^{-N(h_2-\varepsilon)} \int_{T_{\varepsilon}^N} \mu(\mathrm{d}\mathbf{x}_1^N) \nu(\mathrm{d}\mathbf{y}_1^N)$$

according to (4.2.7)

$$\leq 2^{-N(h_1-\varepsilon)} 2^{-N(h_2-\varepsilon)} 2^{N(h+\varepsilon)} = 2^{-N(h_1+h_2-h-3\varepsilon)}$$

because of bound (4.2.9).

Finally, by reversing the inequalities in the last two lines, we can cast them as

$$\geq 2^{-N(h_1+\varepsilon)} 2^{-N(h_2+\varepsilon)} \int_{T_{\varepsilon}^N} \mu(\mathrm{d}\mathbf{x}_1^N) \nu(\mathrm{d}\mathbf{y}_1^N)$$

according to (4.2.7)

$$\geq (1-\delta) 2^{-N(h_1+\varepsilon)} 2^{-N(h_2+\varepsilon)} 2^{N(h-\varepsilon)} = (1-\delta) 2^{-N(h_1+h_2-h+3\varepsilon)}$$

because of bound (4.2.10).

Formally, we assumed here that $0 < \varepsilon < h$ (since it was assumed in (4.2.10)), but increasing ε only makes the factor $2^{-N(h_1+h_2-h+3\varepsilon)}$ smaller. This proves bound (4.2.12). □

A more convenient (and formally a broader) extension of the asymptotic equipartition property is where we suppose that the statements of the Shannon–McMillan–Breiman theorem hold true directly for the ratio $f_{\mathbf{X}_1^N, \mathbf{Y}_1^N}(\mathbf{X}_1^N, \mathbf{Y}_1^N) / [f_{\mathbf{X}_1^N}(\mathbf{X}_1^N) f_{\mathbf{Y}_1^N}(\mathbf{Y}_1^N))]$. That is,

$$\frac{1}{N} \log \frac{f_{\mathbf{X}_1^N, \mathbf{Y}_1^N}(\mathbf{X}_1^N, \mathbf{Y}_1^N)}{f_{\mathbf{X}_1^N}(\mathbf{X}_1^N) f_{\mathbf{Y}_1^N}(\mathbf{Y}_1^N)} \to c \text{ in probability,} \tag{4.2.13}$$

where $c > 0$ is a constant. Recall that $f_{\mathbf{X}_1^N,\mathbf{Y}_1^N}$ represents the joint PMF while $f_{\mathbf{X}_1^N}$ and $f_{\mathbf{x}_1^N}$ individual PMFs for the random input and output vectors $\mathbf{x}^N$ and $\mathbf{Y}^N$, with respect to reference measures $\mu^{(N)}$ and $v^{(N)}$:

$$f_{\mathbf{X}_1^N,\mathbf{Y}_1^N}(\mathbf{x}_1^N,\mathbf{y}_1^N) = f_{\mathbf{X}_1^N}(\mathbf{x}_1^N)f_{\text{ch}}(\mathbf{y}_1^N|\mathbf{x}_1^N \text{ sent }),$$
$$f_{\mathbf{Y}_1^N}(\mathbf{Y}_1^N) = \int f_{\mathbf{X}_1^N,\mathbf{Y}_1^N}(\mathbf{x}_1^N,\mathbf{y}_1^N)\mu^{(N)}(\mathrm{d}\mathbf{x}_1^N).$$

Here, for $\varepsilon > 0$, we consider the typical set

$$T_\varepsilon^N = \left\{(\mathbf{X}_1^N,\mathbf{y}_1^N): -\varepsilon \le \frac{1}{N}\log\frac{f_{\mathbf{X}_1^N,\mathbf{Y}_1^N}(\mathbf{x}_1^N,\mathbf{y}_1^N)}{f_{\mathbf{X}_1^N}(\mathbf{x}_1^N)f_{\mathbf{Y}_1^N}(\mathbf{y}_1^N)} - c \le \varepsilon\right\}; \qquad (4.2.14)$$

by assumption (4.2.13) we have that $\lim_{N\to\infty}\mathbb{P}\left((\mathbf{X}_1^N,\mathbf{Y}_1^N) \in T_\varepsilon^N\right) = 1$ for all $\varepsilon > 0$.

Again, we will consider an independent pair $\left(\widetilde{\mathbf{X}}_1^N,\widetilde{\mathbf{Y}}_1^N\right)$ where component $\widetilde{\mathbf{X}}_1^N$ has the same PMF as $\mathbf{X}_1^N$ and $\widetilde{\mathbf{Y}}_1^N$ the same PMF as $\mathbf{Y}_1^N$.

Theorem 4.2.4 (Deviation from the joint asymptotic equipartition property) *Assume that property (4.2.13) holds true. For an independent pair $\left(\widetilde{\mathbf{X}}_1^N,\widetilde{\mathbf{Y}}_1^N\right)$, the probability that $\left(\widetilde{\mathbf{X}}_1^N,\widetilde{\mathbf{Y}}_1^N\right) \in T_\varepsilon^N$ obeys*

$$\mathbb{P}\left(\left(\widetilde{\mathbf{X}}_1^N,\widetilde{\mathbf{Y}}_1^N\right) \in T_\varepsilon^N\right) \le 2^{-N(c-\varepsilon)}, \quad \text{for all } \varepsilon \text{ and } N, \qquad (4.2.15)$$

and, for all $\delta > 0$, for N large enough, depending on δ,

$$\mathbb{P}\left(\left(\widetilde{\mathbf{X}}_1^N,\widetilde{\mathbf{Y}}_1^N\right) \in T_\varepsilon^N\right) \ge (1-\delta)2^{-N(c+\varepsilon)}, \quad \text{for all } \varepsilon. \qquad (4.2.16)$$

Proof Again, we obtain (4.2.15) as follows:

$$\mathbb{P}\left(\left(\widetilde{\mathbf{X}}_1^N,\widetilde{\mathbf{Y}}_1^N\right) \in T_\varepsilon^N\right) = \int_{T_\varepsilon^N} f_{\widetilde{\mathbf{X}}_1^N,\widetilde{\mathbf{Y}}_1^N}\,\mu^{\times N}(\mathrm{d}\mathbf{X}_1^N)v^{\times N}(\mathrm{d}\mathbf{y}_1^N)$$
$$= \int_{T_\varepsilon^N} f_{\mathbf{X}_1^N}(\mathbf{x}_1^N)f_{\mathbf{Y}_1^N}(\mathbf{y}_1^N)\mu(\mathrm{d}\mathbf{X}_1^N)v(\mathrm{d}\mathbf{y}_1^N)$$
$$= \int_{T_\varepsilon^N}\exp\left[-\frac{f_{\mathbf{X}_1^N,\mathbf{Y}_1^N}(\mathbf{x}_1^N,\mathbf{y}_1^N)}{f_{\mathbf{X}_1^N}(\mathbf{x}_1^N)f_{\mathbf{Y}_1^N}(\mathbf{y}_1^N)}\right]$$
$$\times f_{\mathbf{X}_1^N,\mathbf{Y}_1^N}(\mathbf{x}_1^N,\mathbf{y}_1^N)\mu^{\times N}(\mathrm{d}\mathbf{x}_1^N)v^{\times N}(\mathrm{d}\mathbf{y}_1^N)$$
$$\le 2^{-N(c-\varepsilon)}\int_{T_\varepsilon^N} f_{\mathbf{X}_1^N,\mathbf{Y}_1^N}(\mathbf{x}_1^N,\mathbf{y}_1^N)\mu(\mathrm{d}\mathbf{x}_1^N)v(\mathrm{d}\mathbf{y}_1^N)$$
$$= 2^{-N(c-\varepsilon)}\mathbb{P}\left((\mathbf{X}_1^N,\mathbf{Y}_1^N) \in T_\varepsilon^N\right)$$
$$\le 2^{-N(c-\varepsilon)}.$$

The first equality is by definition, the second step follows by substituting (4.2.8), the third is by direct calculation, and the fourth because of the bound (4.2.14).

Finally, by reversing the inequalities in the last two lines, we obtain the bound (4.2.16):

$$\geq 2^{-N(c+\varepsilon)} \int_{T_{\varepsilon}^N} f_{\mathbf{X}_1^N, \mathbf{Y}_1^N}(\mathbf{x}_1^N, \mathbf{y}_1^N) \mu(d\mathbf{x}_1^N) \nu(d\mathbf{y}_1^N)$$

$$= 2^{-N(c+\varepsilon)} \mathbb{P}\left((\mathbf{X}_1^N, \mathbf{Y}_1^N) \in T_{\varepsilon}^N\right) \geq 2^{-N(c+\varepsilon)}(1 - \delta),$$

the first inequality following because of (4.2.14).   ☐

Worked Example 4.2.5   *Let* $\mathbf{x} = \{X(1), \ldots, X(n)\}^{\mathrm{T}}$ *be a given vector/collection of random variables. Let us write* $\mathbf{x}(C)$ *for subcollection* $\{X(i) : i \in C\}$ *where C is a non-empty subset in the index set* $\{1, \ldots, n\}$. *Assume that the joint distribution for any subcollection* $\mathbf{x}(C)$ *with* $\sharp C = k$, $1 \leq k \leq n$, *is given by a joint PMF* $f_{\mathbf{x}(C)}$ *relative to measure* $\mu \times \cdots \times \mu$ *(k factors, each corresponding to a random variable*

$X(i)$ *with* $i \in C$*). Similarly, given a vector* $\mathbf{x} = \begin{pmatrix} x(1) \\ \vdots \\ x(n) \end{pmatrix}$ *of values for* $\mathbf{x}$*, denote by*

$\mathbf{x}(C)$ *the argument* $\{x(i) : i \in C\}$ *(the sub-column in* $\mathbf{x}$ *extracted by picking the rows with* $i \in C$*). By the Gibbs inequality, for all partitions* $\{C_1, \ldots, C_s\}$ *of set* $\{1, \ldots, n\}$ *into non-empty disjoint subsets* $C_1, \ldots, C_s$ *(with* $1 \leq s \leq n$*), the integral*

$$\int f_{\mathbf{x}}(\mathbf{x}) \log \frac{f_{\mathbf{x}_1^n}(\mathbf{x})}{f_{\mathbf{x}(C_1)}(\mathbf{x}(C_1)) \ldots f_{\mathbf{x}(C_s)}(\mathbf{x}(C_s))} \prod_{1 \leq j \leq n} \mu(dx(j)) \geq 0. \qquad (4.2.17)$$

What is the partition for which the integral in (4.2.17) attains its maximum?

Solution   The partition in question has $s = n$ subsets, each consisting of a single point. In fact, consider the partition of set $\{1, \ldots, n\}$ into single points; the corresponding integral equals

$$\int f_{\mathbf{x}}(\mathbf{x}) \log \frac{f_{\mathbf{x}_1^n}(\mathbf{x})}{\prod_{1 \leq i \leq n} f_{X_i}(x_i)} \prod_{1 \leq j \leq n} \mu(dx(j)). \qquad (4.2.18)$$

Let $\{C_1, \ldots, C_s\}$ be any partition of $\{1, \ldots, n\}$. Multiply and divide the fraction under the log by the product of joint PMFs $\prod_{1 \leq i \leq s} f_{\mathbf{x}(C_i)}(\mathbf{x}(C_i))$. Then the integral (4.2.18) is represented as the sum

$$\int f_{\mathbf{x}}(\mathbf{x}) \log \frac{f_{\mathbf{x}_1^n}(\mathbf{x})}{\prod_{1 \leq i \leq s} f_{\mathbf{x}(C_i)}(\mathbf{x}(C_i))} \prod_{1 \leq j \leq n} \mu(dx(j)) + \text{terms} \geq 0.$$

The answer follows.   ☐

Worked Example 4.2.6   Let $\mathbf{x} = \{X(1),\ldots,X(n)\}$ be a collection of random variables as in Worked Example 4.2.5, and let Y be another random variable. Suppose that there exists a joint PMF $f_{\mathbf{x},Y}$, relative to a measure $\mu^{(n)} \times \nu$ where $\mu^{(n)} = \mu \times \cdots \times \mu$ (n times). Given a subset $C \subseteq \{1,\ldots,n\}$, consider the sum

$$I(\mathbf{x}(C):Y) + \mathbb{E}\big[I(\mathbf{x}(\overline{C}:Y)|\mathbf{x}(C)\big].$$

Here $\mathbf{x}(C) = \{X(i) : i \in C\}$, $\mathbf{x}(\overline{C}) = \{X(i) : i \notin C\}$, and $\mathbb{E}\big[I(\mathbf{x}(\overline{C}:Y)|\mathbf{x}(C)\big]$ stands for the expectation of $I(\mathbf{x}(\overline{C}:Y)$ conditional on the value of $\mathbf{x}(C)$. Prove that this sum does not depend on the choice of set C.

Solution Check that the expression in question equals $I(\mathbf{x}:Y)$.   □

In Section 4.3 we need the following facts about parallel (or product) channels.

Worked Example 4.2.7   (Lemma A in [173]; see also [174]) *Show that the capacity of the product of r time-discrete Gaussian channels with parameters $(\alpha_j, p^{(j)}, \sigma_j^2)$ equals*

$$C = \sum_{1 \le j \le r} \frac{\alpha_j}{2} \ln\left(1 + \frac{p^{(j)}}{\alpha_j \sigma_j^2}\right). \tag{4.2.19}$$

Moreover, (4.2.19) holds when some of the α_js equal $+\infty$: in this case the corresponding summand takes the form $p^{(j)}/\sigma_j^2$.

Solution Suppose that multi-vector data $\mathbf{x} = \{x_1,\ldots,x_r\}$ are transmitted via r parallel channels of capacities $C_1,\ldots,C_r$, where each vector $x_j = \begin{pmatrix} x_{j1} \\ \vdots \\ x_{jn_j} \end{pmatrix} \in \mathbb{R}^{n_j}$. It is convenient to set $n_j = \lceil \alpha_j \tau \rceil$ where $\tau \to \infty$. It is claimed that the capacity for this product-channel equals the sum $\sum_{1 \le i \le r} C_i$. By induction, it is sufficient to consider the case $r = 2$. For the direct part, assume that $R < C_1 + C_2$ and $\varepsilon > 0$ are given. For τ sufficiently large we must find a code for the product channel with $M = e^{R\tau}$ codewords and $P_e < \varepsilon$. Set $\eta = (C_1 + C_2 - R)/2$. Let $\mathscr{X}^1$ and $\mathscr{X}^2$ be codes for channels 1 and 2 respectively with $M_1 \sim e^{(C_1-\eta)\tau}$ and $M_2 \sim e^{(C_2-\eta)\tau}$ and error-probabilities $P_e^{\mathscr{X}^1}, P_e^{\mathscr{X}^2} \le \varepsilon/2$. Construct a concatenation code $\mathscr{X}$ with codewords $x = x_k^1 x_l^2$ where $x_\bullet^i \in \mathscr{X}^i$, $i = 1,2$. Then, for the product-channel under consideration, with codes $\mathscr{X}^1$ and $\mathscr{X}^2$, the error-probability $P_e^{\mathscr{X}^1,\mathscr{X}^2}$ is decomposed as follows:

$$P_e^{\mathscr{X}^1,\mathscr{X}^2} = \frac{1}{M_1 M_2} \sum_{1 \le k \le M_1, 1 \le l \le M_2} \mathbb{P}\big(\text{error in channel 1 or 2} \big| x_k^1 x_l^2 \text{ sent}\big).$$

By independence of the channels, $P_e^{\mathscr{X}^1,\mathscr{X}^2} \le P_e^{\mathscr{X}^1} + P_e^{\mathscr{X}^2} \le \varepsilon$ which yields the direct part.

The proof of the inverse is more involved and we present only a sketch, referring the interested reader to [174]. The idea is to apply the so-called list decoding: suppose we have a code $\mathscr{Y}$ of size M and a decoding rule $d = d^{\mathscr{Y}}$. Next, given that a vector y has been received at the output port of a channel, a list of L possible code-vectors from $\mathscr{Y}$ has to be produced, by using a decoding rule $\tilde{d} = \tilde{d}^{\mathscr{Y}}_{\text{list}}$, and the decoding (based on rule $\tilde{d}$) is successful if the correct word is in the list. Then, for the average error-probability $P_e = P_e^{\mathscr{Y}}(d)$ over code $\mathscr{Y}$, the following inequality is satisfied:

$$P_e \geq P_e(\tilde{d}) P_e^{\text{AV}}(L,d) \tag{4.2.20}$$

where the error-probability $P_e(\tilde{d}) = P_e^{\mathscr{Y}}(\tilde{d})$ refers to list decoding and $P_e^{\text{AV}}(L,d) = P_e^{\text{AV}}(\mathscr{Y},L,d)$ stands for the error-probability under decoding rule d averaged over all subcodes in $\mathscr{Y}$ of size L.

Now, going back to the product-channel with marginal capacities C_1 and C_2, choose $R > C_1 + C_2$, set $\eta = (R - C_1 - C_2)/2$ and let the list size be $L = e^{R_L \tau}$, with $R_L = C_2 + \eta$. Suppose we use a code $\mathscr{Y}$ of size $e^{R\tau}$ with a decoding rule d and a list decoder $\tilde{d}$ with the list-size L. By using (4.2.20), write

$$P_e \geq P_e(\tilde{d}) P_e^{\text{AV}}(e^{R_L \tau}, d) \tag{4.2.21}$$

and use the facts that $R_L > C_2$ and the value $P_e^{\text{AV}}(e^{R_L \tau}, d)$ is bounded away from zero. The assertion of the inverse part follows from the following observation discussed in Worked Example 4.2.8. Take $R_2 < R - R_L$ and consider subcodes $\mathscr{L} \subset \mathscr{Y}$ of size $\sharp \mathscr{L} = e^{R_2 \tau}$. Suppose we choose subcode $\mathscr{L}$ at random, with equal probabilities. Let $M_2 = e^{R_2 \tau}$ and $P_e^{\mathscr{Y}, M_2}(d)$ stand for the mean error-probability averaged over all subcodes $\mathscr{L} \subset \mathscr{Y}$ of size $\sharp \mathscr{L} = e^{R_2 \tau}$. Then

$$P_e(\tilde{d}) \geq P_e^{\mathscr{Y}, M_2}(d) + \varepsilon(\tau) \tag{4.2.22}$$

where $\varepsilon(\tau) \to 0$ as $\tau \to \infty$. □

Worked Example 4.2.8   Let $L = e^{R_L \tau}$ and $M = e^{R\tau}$. We aim to show that if $R_2 < R - R_L$ and $M_2 = e^{R_2 \tau}$ then the following holds. Given a code $\mathscr{X}$ of size M, a decoding rule d and a list decoder $\tilde{d}$ with list-size L, consider the mean error-probability $P_e^{\mathscr{X}, M_2}(d)$ averaged over the equidistributed subcodes $\mathscr{S} \subset \mathscr{X}$ of size $\sharp \mathscr{S} = M_2$. Then $P_e^{\mathscr{X}, M_2}(d)$ and the list-error-probability $P_e^{\mathscr{X}}(\tilde{d})$ satisfy

$$P_e^{\mathscr{X}}(\tilde{d}) \geq P_e^{\mathscr{X}, M_2}(d) + \varepsilon(\tau) \tag{4.2.23}$$

where $\varepsilon(\tau) \to 0$ as $\tau \to \infty$.

Solution Let $\mathscr{X}$, $\mathscr{S}$ and d be as above and suppose we use a list decoder $\tilde{d}$ with list-length L.

Given a subcode $\mathscr{S} \subset \mathscr{X}$ with M_2 codewords, we will use the following decoding. Let $\mathscr{L}$ be the output of decoder $\tilde{d}$. If exactly one element $x_j \in \mathscr{S}$ belongs to $\mathscr{L}$, the decoder for $\mathscr{S}$ will declare x_j. Otherwise, it will pronounce an error. Denote the decoder for $\mathscr{S}$ by $d^{\mathscr{S}}$. Thus, given that $x_k \in \mathscr{S}$ was transmitted, the resulting error-probability, under the above decoding rule, takes the form

$$P_{ek} = \sum_{\mathscr{L}} p(\mathscr{L}|x_k) E_{\mathscr{S}}(\mathscr{L}|x_k)$$

where $p(\mathscr{L}|x_k)$ is a probability of obtaining the output $\mathscr{L}$ after transmitting x_k under the rule $d^{\mathscr{X}}$ and $E_{\mathscr{S}}(\mathscr{L}|x_k)$ is the error-probability for $d^{\mathscr{S}}$. Next, split $E_{\mathscr{S}}(\mathscr{L}|x_k) = E^1_{\mathscr{S}}(\mathscr{L}|x_k) + E^2_{\mathscr{S}}(\mathscr{L}|x_k)$ where $E^1_{\mathscr{S}}(\mathscr{L}|x_k)$ stands for the probability that $x_k \notin \mathscr{L}$ and $E^2_{\mathscr{S}}(\mathscr{L}|x_k)$ for the probability that word $x_k \in \mathscr{L}$ was decoded by a wrong code-vector from $\mathscr{S}$ (both probabilities conditional upon sending x_k). Further, $E^2_{\mathscr{S}}(\mathscr{L}|x_k)$ is split into a sum of (conditional) probabilities $E_{\mathscr{S}}(\mathscr{L}, x_j|x_k)$ that the decoder returned vector $x_j \in \mathscr{L}$ with $j \neq k$.

Let $P_e^{\mathscr{S}}(d) = P_e^{\mathscr{S},\text{AV}}(d)$ denote the average error-probability for subcode $\mathscr{S}$. The above construction yields

$$P_e^{\mathscr{S}}(d) \leq \frac{1}{M_2} \sum_{k:\, x_k \in \mathscr{S}} \sum_{\mathscr{L}} p(\mathscr{L}|x_k) \left[E^1_{\mathscr{S}}(\mathscr{L}|x_k) + \sum_{j \neq k} E_{\mathscr{S}}(\mathscr{L}, x_j|x_k) \right]. \quad (4.2.24)$$

Inequality (4.2.24) is valid for any subcode $\mathscr{S}$. We now select $\mathscr{S}$ at random from $\mathscr{X}$ choosing each subcode of size M_2 with equal probability. After averaging over all such subcodes we obtain a bound for the averaged error-probability $P_e^{\mathscr{X},M_2} = P_e^{\mathscr{X},M_2}(d)$:

$$P_e^{\mathscr{X},M_2} \leq P_e^{\mathscr{X}}(\tilde{d}) + \frac{1}{M_2} \sum_{k=1}^{M_2} \sum_{\mathscr{L}} \sum_{j \neq k} \left\langle p(\mathscr{L}|x_k) E_{\bullet}(\mathscr{L}, x_j|x_k) \right\rangle^{\mathscr{X},M_2} \quad (4.2.25)$$

where $\langle \ \rangle^{\mathscr{X},M_2}$ means the average over all selections of subcodes. As x_j and x_k are chosen independently,

$$\left\langle p(\mathscr{L}|x_k) E^2_{\bullet}(\mathscr{L}, x_j) \right\rangle^{\mathscr{X},M_2} = \left\langle p(\mathscr{L}|x_k) \right\rangle^{\mathscr{X},M_2} \left\langle E^2_{\bullet}(\mathscr{L}, x_j) \right\rangle^{\mathscr{X},M_2}.$$

Next,

$$\left\langle p(\mathscr{L}|x_k) \right\rangle^{\mathscr{X},M_2} = \sum_{x \in \mathscr{X}} \frac{1}{M} p(\mathscr{L}|x), \quad \left\langle E^2_{\bullet}(\mathscr{L}, x_j|x_k) \right\rangle^{\mathscr{X},M_2} = \frac{L}{M},$$

and we obtain

$$P_e^{\mathscr{X},M_2} \leq P_e^{\mathscr{X}}(\tilde{d}) + \frac{1}{M_2} \sum_{k=1}^{M_2} \sum_{\mathscr{L}} \left(\sum_{x \in \mathscr{X}} \frac{1}{M} p(\mathscr{L}|x) \right) \left(\sum_{j \neq k} \frac{L}{M} \right)$$

which implies

$$P_e^{\mathscr{X},M_2} \leq P_e^{\mathscr{X}}(\tilde{d}) + \frac{M_2 L}{M}. \tag{4.2.26}$$

Since $M_2 L/M = e^{R_2 \tau} e^{-(R-R_L)\tau} \to 0$ when $\tau \to \infty$ as $R_2 < R - R_L$, inequality (4.2.23) is proved. $\square$

We now give the proof of Theorem 4.2.1. Consider the sequence of kth-order Markov approximations of a process $\mathbf{X}$, by setting

$$p^{(k)}(X_0^{n-1}) = p_{X_0^{k-1}}(X_0^{k-1}) \prod_{i=k}^{n-1} p(X_i | X_{i-k}^{i-1}). \tag{4.2.27}$$

Set also

$$H^{(k)} = \mathbb{E}\big[-\log p\,(X_0 | X_{-k}^{-1})\big] = h(X_0 | X_{-k}^{-1}) \tag{4.2.28}$$

and

$$\overline{H} = \mathbb{E}\big[-\log p\,(X_0 | X_{-\infty}^{-1})\big] = h(X_0 | X_{-\infty}^{-1}). \tag{4.2.29}$$

The proof is based on the following three results: Lemma 4.2.9 (the sandwich lemma), Lemma 4.2.10 (a Markov approximation lemma) and Lemma 4.2.11 (a no-gap lemma).

Lemma 4.2.9   *For any stationary process* $\mathbf{X}$,

$$\limsup_{n \to \infty} \frac{1}{n} \log \frac{p^{(k)}(X_0^{n-1})}{p(X_0^{n-1})} \leq 0 \text{ a.s.}, \tag{4.2.30}$$

$$\limsup_{n \to \infty} \frac{1}{n} \log \frac{p(X_0^{n-1})}{p(X_0^{n-1} | X_{-\infty}^{-1})} \leq 0 \text{ a.s.} \tag{4.2.31}$$

Proof   If A_n is a support event for $p_{X_0^{n-1}}$ (i.e. $\mathbb{P}(X_0^{n-1} \in A_n) = 1$), write

$$\mathbb{E}\frac{p^{(k)}(X_0^{n-1})}{p(X_0^{n-1})} = \sum_{x_0^{n-1} \in A_n} p(x_0^{n-1}) \frac{p^{(k)}(x_0^{n-1})}{p(x_0^{n-1})}$$

$$= \sum_{x_0^{n-1} \in A_n} p^{(k)}(x_0^{n-1})$$

$$= p^{(k)}(A) \leq 1.$$

Similarly, if $B_n = B_n(X_{-\infty}^{-1})$ is a support event for $p_{X_0^{n-1}|X_{-\infty}^{-1}}$ (i.e. $\mathbb{P}(X_0^{n-1} \in B_n|X_{-\infty}^{-1}) = 1$), write

$$\mathbb{E}\frac{p(X_0^{n-1})}{p(X_0^{n-1}|X_{-\infty}^{-1})} = \mathbb{E}_{X_{-\infty}^{-1}} \sum_{x_0^{n-1} \in B_n} p(x_0^{n-1}|X_{-\infty}^{-1})\frac{p(x_0^{n-1})}{p(x_0^{n-1}|X_{-\infty}^{-1})}$$

$$= \mathbb{E}_{X_{-\infty}^{-1}} \sum_{x_0^{n-1} \in B_n} p(x_0^{n-1})$$

$$= \mathbb{E}_{X_{-\infty}^{-1}}\mathbb{P}(B_n) \leq 1.$$

By the Markov inequality,

$$\mathbb{P}\left(\frac{p^{(k)}(X_0^{n-1})}{p(X_0^{n-1})} \geq t_n\right) = \mathbb{P}\left(\frac{1}{n}\log\frac{p^{(k)}(X_0^{n-1})}{p(X_0^{n-1})} \geq \frac{1}{n}\log t_n\right) \leq \frac{1}{t_n},$$

and similarly for $\mathbb{P}\left(\dfrac{p(X_0^{n-1})}{p(X_0^{n-1}|X_{-\infty}^{-1})} \geq t_n\right)$. Letting $t_n = n^2$ so that $\sum\limits_n 1/t_n < \infty$ and using the Borel–Cantelli lemma completes the proof. $\qquad\square$

Lemma 4.2.10 *For a stationary ergodic process* **X**,

$$-\frac{1}{n}\log p^{(k)}(X_0^{n-1}) \overset{\text{a.s.}}{\Rightarrow} H^{(k)}, \tag{4.2.32}$$

$$-\frac{1}{n}\log p(X_0^{n-1}|X_{-\infty}^{-1}) \overset{\text{a.s.}}{\Rightarrow} \overline{H}. \tag{4.2.33}$$

Proof Substituting $f = -\log\ p(X_0|X_{-k}^{-1})$ and $f = -\log\ p(X_0|X_{-\infty}^{-1})$ into Birkhoff's ergodic theorem (see for example Theorem 9.1 from [36]) yields

$$-\frac{1}{n}\log\ p^{(k)}(X_0^{n-1}) = -\frac{1}{n}\log\ p(X_0^{k-1}) - \frac{1}{n}\sum_{i=k}^{n-1}\log\ p^{(k)}(X_i|X_{i-k}^{i-1}) \overset{\text{a.s.}}{\Rightarrow} 0 + H^{(k)} \tag{4.2.34}$$

and

$$-\frac{1}{n}\log\ p(X_0^{n-1}|X_{-\infty}^{-1}) = -\frac{1}{n}\sum_{i=0}^{n-1}\log\ p(X_i|X_{-\infty}^{i-1}) \overset{\text{a.s.}}{\Rightarrow} \overline{H}, \tag{4.2.35}$$

respectively.

So, by Lemmas 4.2.9 and 4.2.10,

$$\limsup_{n\to\infty}\frac{1}{n}\log\ \frac{1}{p(X_0^{n-1})} \leq \lim_{n\to\infty}\frac{1}{n}\log\ \frac{1}{p^{(k)}(X_0^{n-1})} = H^{(k)}, \tag{4.2.36}$$

and

$$\liminf_{n\to\infty} \frac{1}{n} \log \frac{1}{p(X_0^{n-1})} \geq \lim_{n\to\infty} \frac{1}{n} \log \frac{1}{p(X_0^{n-1}|X_{-\infty}^{-1})} = \overline{H},)$$

which we rewrite as

$$\overline{H} \leq \liminf_{n\to\infty} -\frac{1}{n} \log p(X_0^{n-1}) \leq \limsup_{n\to\infty} -\frac{1}{n} \log p(X_0^{n-1}) \leq H^{(k)}. \qquad (4.2.37)$$

$\square$

Lemma 4.2.11 *For any stationary process* **X**, $H^{(k)} \searrow \overline{H} = H$.

Proof The convergence $H^{(k)} \searrow H$ follows by stationarity and by conditioning not to increase entropy. It remains to show that $H^{(k)} \searrow \overline{H}$, so that $\overline{H} = H$. The Doob–Lévy martingale convergence theorem for conditional probabilities yields

$$\overset{\text{a.s.}}{p\left(X_0 = x_0|X_{-k}^{-1}\right) \Rightarrow p\left(X_0 = x_0|X_{-\infty}^{-1}\right), \quad k \to \infty.} \qquad (4.2.38)$$

As the set of values I is supposed to be finite, and the function $p \in [0, 1] \mapsto -p \log p$ is bounded, the bounded convergence theorem gives that as $k \to \infty$,

$$H^{(k)} = \mathbb{E} - \sum_{x_0 \in I} p(X_0 = x_0|X_{-k}^{-1}) \log p(X_0 = x_0|X_{-k}^{-1})$$

$$\to \mathbb{E}\left[-\sum_{x_0 \in I} p(X_0 = x_0|X_{-\infty}^{-1}) \log p(X_0 = x_0|X_{-\infty}^{-1}) \right] = \overline{H}.$$

$\square$

4.3 The Nyquist–Shannon formula

In this section we give a rigorous derivation of the famous Nyquist–Shannon formula[1] for the capacity of a continuous-time channel with the power constraint and a finite bandwidth, the result broadly considered an ultimate fact of information theory. Our exposition follows (with minor deviations) the paper [173]. Because it is quite long, we divide the section into subsections, each of which features a particular step of the construction.

Harry Nyquist (1889–1976) is considered a pioneer of information theory whose works, together with those of Ralph Hartley (1888–1970), helped to create the concept of the channel capacity.

---

[1] Some authors speak in this context of a Shannon–Hartley theorem.

The setting is as follows. Fix numbers $\tau, \alpha, p > 0$ and assume that every τ seconds a coder produces a real code-vector

$$\mathbf{x} = \begin{pmatrix} x_1 \\ \vdots \\ x_n \end{pmatrix}$$

where $n = \lceil \alpha \tau \rceil$. All vectors $\mathbf{x}$ generated by the coder lie in a finite set $\mathcal{X} = \mathcal{X}_n \subset \mathbb{R}^n$ of cardinality $M \sim 2^{R_b \tau} = e^{R_n \tau}$ (a codebook); sometimes we write, as before, $\mathcal{X}_{M,n}$ to stress the role of M and n. It is also convenient to list the code-vectors from $\mathcal{X}$ as $\mathbf{x}(1), \ldots, \mathbf{x}(M)$ (in an arbitrary order) where

$$\mathbf{x}(i) = \begin{pmatrix} x_1(i) \\ \vdots \\ x_n(i) \end{pmatrix}, \quad 1 \leq i \leq M.$$

Code-vector $\mathbf{x}$ is then converted into a continuous-time signal

$$x(t) = \sum_{i=1}^{n} x_i \phi_i(t), \quad \text{where } 0 \leq t \leq \tau, \tag{4.3.1}$$

by using an orthonormal basis in $\mathsf{L}_2[0, \tau]$ formed by functions $\phi_i(t), i = 1, 2, \ldots$ (with $\int_0^\tau \phi_i(t) \overline{\phi_j(t)} \mathrm{d}t = \delta_{ij}$). Then the entry x_i can be recovered by integration:

$$x_i = \int_0^\tau x(t) \overline{\phi_i(t)} \mathrm{d}t. \tag{4.3.2}$$

The instantaneous signal power at time t is associated with $|x(t)|^2$; then the square-norm $||\mathbf{x}||^2 = \int_0^\tau |x(t)|^2 \mathrm{d}t = \sum_{1 \leq i \leq n} |x_i|^2$ will represent the full energy of the signal in the interval $[0, \tau]$. The upper bound on the total energy spent on transmission takes the form

$$||\mathbf{x}||^2 \leq p\tau, \quad \text{or } \mathbf{x} \in \mathbb{B}_n(\sqrt{p\tau}). \tag{4.3.3}$$

(In the theory of waveguides, the dimension n is called the Nyquist number and the value $W = n/(2\tau) \sim \alpha/2$ the bandwidth of the channel.)

The code-vector $\mathbf{x}(i)$ is sent through an additive channel, where the receiver gets the (random) vector

$$\mathbf{Y} = \begin{pmatrix} Y_1 \\ \vdots \\ Y_n \end{pmatrix} \quad \text{where } Y_k = x_k(i) + Z_k, \ 1 \leq k \leq n. \tag{4.3.4}$$

The assumption we will adopt is that

$$
\mathbf{Z} = \begin{pmatrix} Z_1 \\ \vdots \\ Z_n \end{pmatrix}
$$

is a vector with IID entries $Z_k \sim \mathrm{N}(0, \sigma^2)$. (In applications, engineers use the representation $Z_i = \int_0^\tau Z(t)\overline{\phi_i(t)}dt$, in terms of a 'white noise' process $Z(t)$.)

From the start we declare that if $\mathbf{x}(i) \in \mathscr{X} \setminus \mathbb{B}_n(\sqrt{p\tau})$, i.e. $||\mathbf{x}(i)||^2 > p\tau$, the output signal vector $\mathbf{Y}$ is rendered 'non-decodable'. In other words, the probability of correctly decoding the output vector $\mathbf{Y} = \mathbf{x}(i) + \mathbf{Z}$ with $||\mathbf{x}(i)||^2 > p\tau$ is taken to be zero (regardless of the fact that the noise vector $\mathbf{Z}$ can be small and the output vector $\mathbf{Y}$ close to $\mathbf{x}(i)$, with a positive probability).

Otherwise, i.e. when $||\mathbf{x}(i)||^2 \le p\tau$, the receiver applies, to the output vector $\mathbf{Y}$, a decoding rule $d(= d_{n,\mathscr{X}})$, i.e. a map $\mathbf{y} \in \mathbb{K} \mapsto d(\mathbf{y}) \in \mathscr{X}$ where $\mathbb{K} \subset \mathbb{R}^n$ is a 'decodable domain' (where map d had been defined). In other words, if $\mathbf{Y} \in \mathbb{K}$ then vector $\mathbf{Y}$ is decoded as $d(\mathbf{Y}) \in \mathscr{X}$. Here, an error arises either if $\mathbf{Y} \notin \mathbb{K}$ or if $d(\mathbf{Y}) \ne \mathbf{x}(i)$ given that $\mathbf{x}(i)$ was sent. This leads to the following formula for the probability of erroneously decoding the input code-vector $\mathbf{x}(i)$:

$$
P_{\mathrm{e}}(i,d) = \begin{cases} 1, & ||\mathbf{x}(i)||^2 > p\tau, \\ \mathbf{P}_{\mathrm{ch}}\left(\mathbf{Y} \notin \mathbb{K} \text{ or } d(\mathbf{Y}) \ne \mathbf{x}(i) | \mathbf{x}(i) \text{ sent}\right), & ||\mathbf{x}(i)||^2 \le p\tau. \end{cases} \tag{4.3.5}
$$

The average error-probability $P_{\mathrm{e}} = P_{\mathrm{e}}^{\mathscr{X},\mathrm{av}}(d)$ for the code $\mathscr{X}$ is then defined by

$$
P_{\mathrm{e}} = \frac{1}{M} \sum_{1 \le i \le M} P_{\mathrm{e}}(i,d). \tag{4.3.6}
$$

Furthermore, we say that R_{bit} (or R_{nat}) is a reliable transmission rate (for given α and p) if for all $\varepsilon > 0$ we can specify $\tau_0(\varepsilon) > 0$ such that for all $\tau > \tau_0(\varepsilon)$ there exists a codebook $\mathscr{X}$ of size $\sharp \mathscr{X} \sim e^{R_{\mathrm{nat}}\tau}$ and a decoding rule d such that $P_{\mathrm{e}} = P_{\mathrm{e}}^{\mathscr{X},\mathrm{av}}(d) < \varepsilon$. The channel capacity C is then defined as the supremum of all reliable transmission rates, and the argument from Section 4.1 yields

$$
C = \frac{\alpha}{2} \ln\left(1 + \frac{p}{\alpha\sigma^2}\right) \text{ (in nats)}; \tag{4.3.7}
$$

cf. (4.1.17). Note that when $\alpha \to \infty$, the RHS in (4.3.7) tends to $p/(2\sigma^2)$.

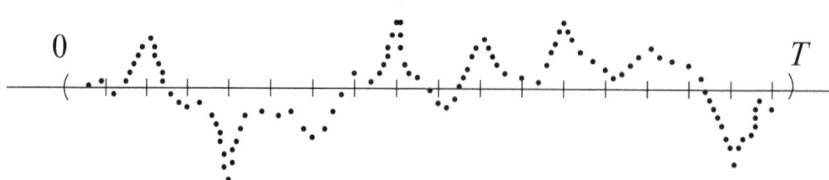

Figure 4.4

In the time-continuous set-up, Shannon (and Nyquist before him) discussed an application of formula (4.3.7) to *band-limited* signals. More precisely, set $W = \alpha/2$; then the formula

$$C = W \ln \left(1 + \frac{p}{2\sigma_0^2 W} \right) \qquad (4.3.8)$$

should give the capacity of the time-continuous additive channel with white noise of variance $\sigma^2 = \sigma_0^2 W$, for a band-limited signal $x(t)$ with the spectrum in $[-W, W]$ and of energy per unit time $\leq p$.

This sentence, perfectly clear to a qualified engineer, became a stumbling point for mathematicians and required a technically involved argument for justifying its validity. In engineers' language, an 'ideal' orthonormal system on $[0, \tau]$ to be used in (4.3.1) would be a collection of $n \sim 2W\tau$ equally spaced δ-functions. In other words, it would have been very convenient to represent the code-vector $\mathbf{x}(i) = (x_1(i), \ldots, x_n(i))$ by a function $f_i(t)$, of the time argument $t \in [0, \tau]$, given by the sum

$$f_i(t) = \sum_{1 \leq k \leq n} x_k(i)\delta \left(t - \frac{k}{2W} \right) \qquad (4.3.9)$$

where $n = \lceil 2W\tau \rceil$ (and $\alpha = 2W$). Here $\delta(t)$ represents a 'unit impulse' appearing near time 0 and graphically visualised as an 'acute unit peak' around point $t = 0$. Then the shifted function $\delta(t - k/(2W))$ yields a peak concentrated near $t = k/(2W)$, and the graph of function $f_i(t)$ is shown in Figure 4.4.

We may think that our coder produces functions $x_i(t)$ every τ seconds, and each such function is the result of encoding a message i. Moreover, within each time interval of length τ, the peaks $x_k(i)\delta(t - k/(2W))$ appear at time-step $1/(2W)$. Here $\delta(t - k/(2W))$ is the time-shifted Dirac delta-function.

The problem is that $\delta(t)$ is a so-called 'generalised function', and $\delta \notin \mathbb{L}_2$. A way to sort out this difficulty is to pass the signal through a low-frequency filter. This produces, instead of $f_i(t)$, the function $\tilde{f}_i(t)(= \tilde{f}_{W,i}(t))$ given by

$$\tilde{f}_i(t) = \sum_{1 \le k \le n} x_k(i) \operatorname{sinc}(2Wt - k). \tag{4.3.10}$$

Here

$$\operatorname{sinc}(2Wt - k) = \frac{\sin(\pi(2Wt - k))}{\pi(2Wt - k)} \tag{4.3.11}$$

is the value of the shifted and rescaled (normalised) *sinc function*:

$$\operatorname{sinc}(s) = \begin{cases} \dfrac{\sin(\pi s)}{\pi s}, & s \ne 0, \\ 1, & s = 0, \end{cases} \quad s \in \mathbb{R}, \tag{4.3.12}$$

featured in Figure 4.5.

The procedure of removing high-frequency harmonics (or, more generally, high-resolution components) and replacing the signal $f_i(t)$ with its (approximate) lower-resolution version $\tilde{f}_i(t)$ is widely used in modern computer graphics and other areas of digital processing.

Example 4.3.1   (The Fourier transform in $\mathbb{L}_2$) Recall that the Fourier transform $\phi \mapsto \mathbf{F}\phi$ of an integrable function ϕ (i.e. a function with $\int |\phi(x)| dx < +\infty$) is defined by

$$[\mathbf{F}\phi](\omega) = \int \phi(x) e^{i\omega x} dx, \quad \omega \in \mathbb{R}. \tag{4.3.13}$$

The inverse Fourier transform can be written as an inverse map:

$$[\mathbf{F}^{-1}\phi](x) - \frac{1}{2\pi} \int \phi(\omega) e^{-i\omega x} d\omega. \tag{4.3.14}$$

A profound fact is that (4.3.13) and (4.3.14) can be extended to square-integrable functions $\phi \in \mathbb{L}_2(\mathbb{R})$ (with $\|\phi\|^2 = \int |\phi(x)|^2 dx < +\infty$). We have no room here to go into detail; the enthusiastic reader is referred to [127]. Moreover, the Fourier-transform techniques turn out to be extremely useful in numerous applications. For instance, denoting $\mathbf{F}\phi = \hat{\phi}$ and writing $\mathbf{F}^{-1}\hat{\phi} = \phi$, we obtain from (4.3.13), (4.3.14) that

$$\phi(x) = \frac{1}{2\pi} \int \hat{\phi}(\omega) e^{-ix\omega} d\omega. \tag{4.3.15}$$

In addition, for any two square-integrable functions $\phi_1, \phi_2 \in \mathbb{L}_2(\mathbb{R})$,

$$2\pi \int \phi_1(x) \overline{\phi_2(x)} dx = \int \hat{\phi}_1(\omega) \overline{\hat{\phi}_2(\omega)} d\omega. \tag{4.3.16}$$

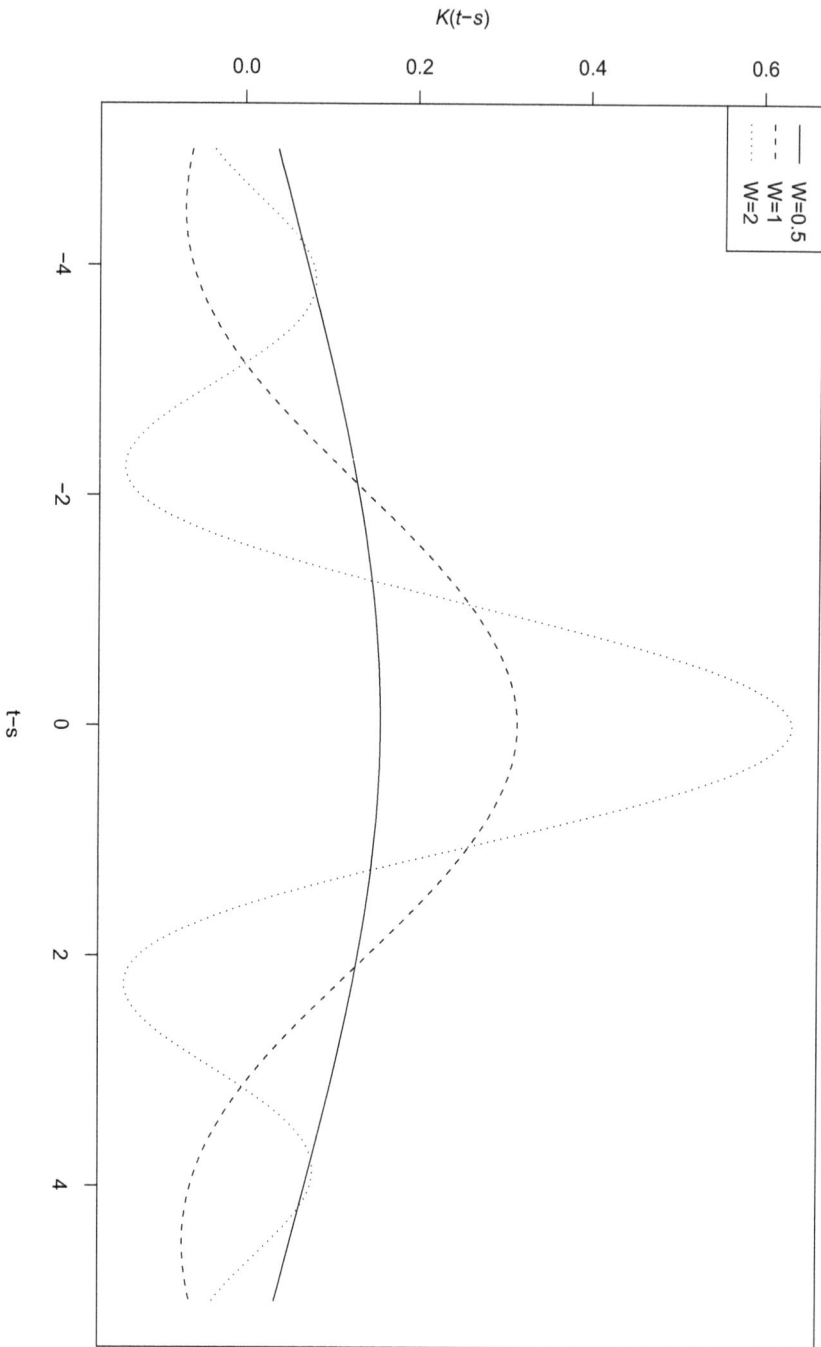

Figure 4.5

Furthermore, the Fourier transform can be defined for generalised functions too; see again [127]. In particular, the equations similar to (4.3.13)–(4.3.14) for the delta-function look like this:

$$\delta(t) = \frac{1}{2\pi} \int e^{-i\omega t} d\omega, \quad 1 = \int \delta(t) e^{it\omega} dt, \tag{4.3.17}$$

implying that the Fourier transform of the Dirac delta is $\widehat{\delta}(\omega) \equiv 1$. For the shifted delta-function we obtain

$$\delta\left(t - \frac{k}{2W}\right) = \frac{1}{2\pi} \int e^{ik\omega/(2W)} e^{-i\omega t} d\omega. \tag{4.3.18}$$

The Shannon–Nyquist formula is established for a device where the channel is preceded by a 'filter' that 'cuts off' all harmonics $e^{\pm it\omega}$ with frequencies ω outside the interval $[-2\pi W, 2\pi W]$. In other words, a (shifted) unit impulse $\delta(t - k/(2W))$ in (4.3.18) is replaced by its cut-off version which emerges after the filter cuts off the harmonics $e^{-it\omega}$ with $|\omega| > 2\pi W$.

The sinc function (a famous object in applied mathematics) is a classical function arising when we reduce the integral in ω in (4.3.17) to the interval $[-\pi, \pi]$:

$$\mathrm{sinc}(t) = \frac{1}{2\pi} \int_{-\pi}^{\pi} e^{-i\omega t} d\omega, \quad \mathbf{1}_{[-\pi,\pi]}(\omega) = \int \mathrm{sinc}(t) e^{it\omega} dt, \quad t, \omega \in \mathbb{R}^1 \tag{4.3.19}$$

(symbolically, function $\mathrm{sinc} = \mathbf{F}^{-1}\mathbf{1}_{[-\pi,\pi]}$). In our context, the function $t \mapsto A\,\mathrm{sinc}(At)$ can be considered, for large values of parameter $A > 0$, as a convenient approximation for $\delta(t)$. A customary caution is that $\mathrm{sinc}(t)$ is not an integrable function on the whole axis $\mathbb{R}$ (due to the $1/t$ factor), although it is square-integrable: $\int \left(\mathrm{sinc}\,t\right)^2 dt < \infty$. Thus, the right equation in (4.3.19) should be understood in an $Ł_2$-sense.

However, it does not make the mathematical and physical aspects of the theory less tricky (as well as engineering ones). Indeed, an ideal filter producing a clear cut of unwanted harmonics is considered, rightly, as 'physically unrealisable'. Moreover, assuming that such a perfect device is available, we obtain a signal $\widetilde{f}_i(t)$ that is no longer confined to the time interval $[0, \tau]$ but is widely spread in the whole time axis. To overcome this obstacle, one needs to introduce further technical approximations.

Worked Example 4.3.2   *Verify that the functions*

$$t \mapsto \left(2\sqrt{\pi W}\right) \mathrm{sinc}\,(2Wt - k), \quad k = 1, \ldots, n, \tag{4.3.20}$$

are orthonormal in the space $Ł_2(\mathbb{R}^1)$:

$$(4\pi W) \int \left[\mathrm{sinc}\,(2Wt - k)\right] \left[\mathrm{sinc}\,(2Wt - k')\right] dt = \delta_{kk'}.$$

Solution The shortest way to see this is to write the Fourier-decomposition (in $Ł_2(\mathbb{R})$) implied by (4.3.19):

$$2\sqrt{\pi W}\, \text{sinc}\,(2Wt - k) = \frac{1}{2\sqrt{\pi W}} \int_{-2\pi W}^{2\pi W} e^{ik\omega/(2W)} e^{-it\omega} d\omega \qquad (4.3.21)$$

and check that the functions representing the Fourier-transforms

$$\frac{1}{2\sqrt{\pi W}} \mathbf{1}\,(|\omega| \le 2\pi W) e^{ik\omega/(2W)}, \quad k = 1,\ldots,n,$$

are orthonormal. That is,

$$\frac{1}{4\pi W} \int_{-2\pi W}^{2\pi W} e^{i(k-k')\omega/(2W)} d\omega = \delta_{kk'} \qquad (4.3.22)$$

where

$$\delta_{kk'} = \begin{cases} 1, & k = k', \\ 0, & k \ne k', \end{cases}$$

is the Kronecker symbol. But (4.3.22) can be verified by a standard integration. □

Since functions in (4.3.20) are orthonormal, we obtain that

$$\|\mathbf{x}(i)\|^2 = (4\pi W)\|\tilde{f_i}\|^2, \text{ where } \|\tilde{f_i}\|^2 = \int |\tilde{f_i}(t)|^2 dt, \qquad (4.3.23)$$

and functions $\tilde{f_i}$ have been introduced in (4.3.10). Thus, the power constraint can be written as

$$\|\tilde{f_i}\|^2 \le p\tau/4\pi W = p_0. \qquad (4.3.24)$$

In fact, the coefficients $x_k(i)$ coincide with the values $\tilde{f_i}(k/(2W))$ of function $\tilde{f_i}$ calculated at time points $k/(2W)$, $k = 1,\ldots,n$; these points can be referred to as 'sampling instances'.

Thus, the input signal $\tilde{f_i}(t)$ develops in continuous time although it is completely specified by its values $\tilde{f_i}(k/(2W)) = x_k(i)$. Thus, if we think that different signals are generated in disjoint time intervals $(0, \tau), (\tau, 2\tau),\ldots$, then, despite interference caused by infinite tails of the function $\text{sinc}(t)$, these signals are clearly identifiable through their values at sampling instances.

The Nyquist–Shannon assumption is that signal $\tilde{f_i}(t)$ is transformed in the channel into

$$g(t) = \tilde{f_i}(t) + \tilde{Z}(t). \qquad (4.3.25)$$

Here $\tilde{Z}(t)$ is a stationary continuous-time Gaussian process with the zero mean ($\mathbb{E}\tilde{Z}(t) \equiv 0$) and the (auto-)correlation function

$$\mathbb{E}\big[\tilde{Z}(s)\tilde{Z}(t+s)\big] = 2\sigma_0^2 W \operatorname{sinc}(2Wt), \quad t,s \in \mathbb{R}. \tag{4.3.26}$$

In particular, when t is a multiple of π/W (i.e. point t coincides with a sampling instance), the random variables $\tilde{Z}(s)$ and $\tilde{Z}(t+s)$ are independent. An equivalent form of this condition is that the spectral density

$$\Phi(\omega) := \int e^{it\omega} \mathbb{E}\big[\tilde{Z}(0)\tilde{Z}(t)\big]\, dt = \sigma_0^2 \mathbf{1}(|\omega| < 2\pi W). \tag{4.3.27}$$

We see that the received continuous-time signal $y(t)$ can be identified through its values $y_k = y\left(\dfrac{k}{2W}\right)$ via equations

$$y_k = x_k(i) + Z_k \text{ where } Z_k = \tilde{Z}\left(\frac{k}{2W}\right) \text{ are IID } N(0, 2\sigma_0^2 W).$$

This corresponds to the system considered in Section 4.1 with $p = 2Wp_0$ and $\sigma^2 = 2\sigma_0^2 W$. It has been generally believed in the engineering community that the capacity C of the current system is given by (4.3.8), i.e. the transmission rates below this value of C are reliable and above it they are not.

However, a number of problems are to be addressed, in order to understand formula (4.3.8) rigorously. One is that, as was noted above, a 'shar' filter band-limiting the signal to a particular frequency interval is an idealised device. Another is that the output signal $g(t)$ in (4.3.25) can be reconstructed after it has been recorded over a small time interval because any sample function of the form

$$t \in \mathbb{R} \mapsto \sum_{1 \le k \le n} (x_k(i) + z_k) \operatorname{sinc}(2Wt - k) \tag{4.3.28}$$

is analytic in t. Therefore, the notion of *rate* should be properly re-defined.

The simplest solution (proposed in [173]) is to introduce a class of functions $\mathscr{A}(\tau, W, p_0)$ which are

(i) approximately band-limited to W cycles per a unit of time (say, a second),
(ii) supported by a time interval of length τ (it will be convenient to specify this interval as $[-\tau/2, \tau/2]$),
(iii) have the total energy (the $Ł_2(\mathbb{R})$-norm) not exceeding $p_0 \tau$.

These restrictions determine the regional constraints upon the system.

Thus, consider a code $\mathscr{X}$ of size $M \sim e^{R\tau}$, i.e. a collection of functions $\tilde{f}_1(t), \ldots, \tilde{f}_M(t)$, of a time variable t. If a given code-function $\tilde{f}_i \notin \mathscr{A}(\tau, W, p_0)$, it is declared non-decodable: it generates an error with probability 1. Otherwise, the signal $\tilde{f}_i \in \mathscr{A}(\tau, W, p_0)$ is subject to the additive Gaussian noise $\tilde{Z}(t)$ with mean $\mathbb{E}\tilde{Z}(t) \equiv 0$ characterised by (4.3.27) and is transformed to $g(t) = \tilde{f}_i(t) + \tilde{Z}(t)$, the signal at the output port of the channel (cf. (4.3.25)). The receiver uses a decoding rule, i.e. a map $d : \mathbb{K} \to \mathscr{X}$ where $\mathbb{K}$ is, as earlier, the domain of definition of d, i.e. some given class of functions where map d is defined. (As before, the decoder d may vary with the code, prompting the notation $d = d^{\mathscr{X}}$.) Again, if $g \notin \mathbb{K}$, the transmission is considered as erroneous. Finally, if $g \in \mathbb{K}$ then the received signal g is decoded by the code-function $d^{\mathscr{X}}(g)(t) \in \mathscr{X}$. The probability of error for code $\mathscr{X}$ when the code-signal generated by the coder was $\tilde{f}_i \in \mathscr{X}$ is set to be

$$P_e(i) = \begin{cases} 1, & \tilde{f}_i \notin \mathscr{A}(\tau, W, p_0), \\ \mathbf{P}_{ch}\left(\mathbb{K}^c \cup \{g : d^{\mathscr{X}}(g) \neq \tilde{f}_i\}\right), & \tilde{f}_i \in \mathscr{A}(\tau, W, p_0). \end{cases} \qquad (4.3.29)$$

The average error-probability $P_e = P_e^{\mathscr{X}, \mathrm{av}}(d)$ for code $\mathscr{X}$ (and decoder d) equals

$$P_e = \frac{1}{M} \sum_{1 \leq i \leq M} P_e(i, d). \qquad (4.3.30)$$

Value $R(= R_{\mathrm{nat}})$ is called a reliable transmission rate if, for all $\varepsilon > 0$, there exists τ and a code $\mathscr{X}$ of size $M \sim e^{R\tau}$ such that $P_e < \varepsilon$.

Now fix a value $\eta \in (0, 1)$. The class $\mathscr{A}(\tau, W, p_0) = \mathscr{A}(\tau, W, p_0, \eta)$ is defined as the set of functions $f^\circ(t)$ such that

(i) $f^\circ = D_\tau f$ where

$$D_\tau f(t) = f(t)\mathbf{1}(|t| < \tau/2), \quad t \in \mathbb{R},$$

and $f(t)$ has the Fourier transform $\int e^{it\omega} f(t)\,dt$ vanishing for $|\omega| > 2\pi W$;

(ii) the ratio

$$\frac{||f^\circ||^2}{||f||^2} \geq 1 - \eta;$$

and

(iii) the norm $||f^\circ||^2 \leq p_0 \tau$.

In other words, the 'transmittable' signals $f^\circ \in \mathscr{A}(\tau, W, p_0, \eta)$ are 'sharply localised' in time and 'nearly band-limited' in frequency.

The Nyquist–Shannon formula can be obtained as a limiting case from several assertions; the simplest one is Theorem 4.3.3 below. An alternative approach will be presented later in Theorem 4.3.7.

Theorem 4.3.3   *The capacity $C = C(\eta)$ of the above channel with constraint domain $\mathscr{A}(\tau, W, p_0, \eta)$ described in conditions (i)–(iii) above is given by*

$$C = W \ln \left(1 + \frac{p_0}{2\sigma_0^2 W} \right) + \frac{\eta}{1 - \eta} \frac{p_0}{\sigma_0^2}. \tag{4.3.31}$$

As $\eta \to 0$,

$$C(\eta) \to W \ln \left(1 + \frac{p_0}{2\sigma_0^2 W} \right) \tag{4.3.32}$$

which yields the Nyquist–Shannon formula (4.3.8).

Before going to (quite involved) technical detail, we will discuss some facts relevant to the product, or parallel combination, of r time-discrete Gaussian channels. (In essence, this model was discussed at the end of Section 4.2.) Here, every τ time units, the input signal is generated, which is an ordered collection of vectors

$$\{\mathbf{x}^{(1)}, \dots, \mathbf{x}^{(r)}\} \text{ where } \mathbf{x}^{(j)} = \begin{pmatrix} x_1^{(j)} \\ \vdots \\ x_{n_j}^{(j)} \end{pmatrix} \in \mathbb{R}^{n_j}, \ 1 \le j \le r, \tag{4.3.33}$$

and $n_j = \lceil \alpha_j \tau \rceil$ with α_j being a given value (the speed of the digital production from coder j). For each vector $\mathbf{x}^{(j)}$ we consider a specific power constraint:

$$\left\| \mathbf{x}^{(j)} \right\|^2 \le p^{(j)} \tau, \ 1 \le j \le r. \tag{4.3.34}$$

The output signal is a collection of (random) vectors

$$\{\mathbf{Y}^{(1)}, \dots, \mathbf{Y}^{(r)}\} \text{ where } \mathbf{Y}^{(j)} = \begin{pmatrix} Y_1^{(j)} \\ \vdots \\ Y_{n_j}^{(j)} \end{pmatrix} \text{ and } Y_k^{(j)} = x_k^{(j)} + Z_k^{(j)}, \tag{4.3.35}$$

with $Z_k^{(j)}$ being IID random variables, $Z_k^{(j)} \sim \mathrm{N}\left(0, \sigma^{(j)2}\right)$, $1 \le k \le n_j$, $1 \le j \le r$.

A codebook $\mathscr{X}$ with information rate R, for the product-channel under consideration, is an array of M input signals,

$$\left\{\begin{array}{c} \left(\mathbf{x}^{(1)}(1),\ldots,\mathbf{x}^{(r)}(1)\right) \\ \left(\mathbf{x}^{(1)}(2),\ldots,\mathbf{x}^{(r)}(2)\right) \\ \cdots \quad \cdots \quad \cdots \\ \left(\mathbf{x}^{(1)}(M),\ldots,\mathbf{x}^{(r)}(M)\right) \end{array}\right\}, \tag{4.3.36}$$

each of which has the same structure as in (4.3.33). As before, a decoder d is a map acting on a given set $\mathbb{K}$ of sample output signals $\left\{\mathbf{y}^{(1)},\ldots,\mathbf{y}^{(r)}\right\}$ and taking these signals to $\mathscr{X}$.

As above, for $i = 1,\ldots,M$, we define the error-probability $P_e(i,d)$ for code $\mathscr{X}$ when sending an input signal $\left(\mathbf{x}^{(1)}(i),\ldots,\mathbf{x}^{(r)}(i)\right)$:

$$P_e(i,d) = 1, \quad \text{if } \left|\mathbf{x}^{(j)}(i)\right|^2 \geq p^{(j)}\tau \text{ for some } j = 1,\ldots,r,$$

and

$$P_e(i,d) = \mathbf{P}_{\mathrm{ch}}\Big(\left\{\mathbf{Y}^{(1)},\ldots,\mathbf{Y}^{(r)}\right\} \notin \mathbb{K} \text{ or}$$
$$d(\left\{\mathbf{Y}^{(1)},\ldots,\mathbf{Y}^{(r)}\right\}) \neq \left\{\mathbf{x}^{(1)}(i),\ldots,\mathbf{x}^{(r)}(i)\right\}$$
$$\big|\left\{\mathbf{x}^{(1)}(i),\ldots,\mathbf{x}^{(r)}(i)\right\} \text{ sent}\Big),$$
$$\text{if } \left|\mathbf{x}^{(j)}(i)\right|^2 < p^{(j)}\tau \quad \text{for all } j = 1,\ldots,r.$$

The average error-probability $P_e = P_e^{\mathscr{X},\mathrm{av}}(d)$ for code $\mathscr{X}$ (while using decoder d) is then again given by

$$P_e = \frac{1}{M}\sum_{1 \leq i \leq M} P_e(i,d).$$

As usual, R is said to be a reliable transmission rate if for all $\varepsilon > 0$ there exists a $\tau_0 > 0$ such that for all $\tau > \tau_0$ there exists a code $\mathscr{X}$ of cardinality $M \sim e^{R\tau}$ and a decoding rule d such that $P_e < \varepsilon$. The capacity of the combined channel is again defined as the supremum of all reliable transmission rates. In Worked Example 4.2.7 the following fact has been established (cf. Lemma A in [173]; see also [174]).

Lemma 4.3.4 *The capacity of the product-channel under consideration equals*

$$C = \sum_{1 \leq j \leq r} \frac{\alpha_j}{2} \ln\left(1 + \frac{p^{(j)}}{\alpha_j \sigma_j^2}\right). \tag{4.3.37}$$

Moreover, (4.3.37) holds when some of the α_j equal $+\infty$: in this case the corresponding summand takes the form $p^{(j)}/2\sigma_j^2$.

Our next step is to consider jointly constrained products of time-discrete Gaussian channels. We discuss the following types of joint constraints.

Case I. Take $r = 2$, assume $\sigma_1^2 = \sigma_2^2 = \sigma_0^2$ and replace condition (4.3.34) with

$$\|\mathbf{x}^{(1)}\|^2 + \|\mathbf{x}^{(2)}\|^2 < p_0 \tau. \qquad (4.3.38a)$$

In addition, if $\alpha_1 \leq \alpha_2$, we introduce $\beta \in (0,1)$ and require that

$$\|\mathbf{x}^{(2)}\|^2 \leq \beta \left(\|\mathbf{x}^{(1)}\|^2 + \|\mathbf{x}^{(2)}\|^2 \right). \qquad (4.3.38b)$$

Otherwise, i.e. if $\alpha_2 \leq \alpha_1$, formula (4.3.38b) is replaced by

$$\|\mathbf{x}^{(1)}\|^2 \leq \beta \left(\|\mathbf{x}^{(1)}\|^2 + \|\mathbf{x}^{(2)}\|^2 \right). \qquad (4.3.38c)$$

Case II. Here we take $r = 3$ and assume that $\sigma_1^2 = \sigma_2^2 \geq \sigma_3^2$ and $\alpha_3 = +\infty$. The requirements are now that

$$\sum_{1 \leq j \leq 3} \|\mathbf{x}^{(j)}\|^2 < p_0 \tau \qquad (4.3.39a)$$

and

$$\|\mathbf{x}^{(3)}\|^2 \leq \beta \sum_{1 \leq j \leq 3} \|\mathbf{x}^{(j)}\|^2. \qquad (4.3.39b)$$

Case III. As in Case I, take $r = 2$ and assume $\sigma_1^2 = \sigma_2^2 = \sigma_0^2$. Further, let $\alpha_2 = +\infty$. The constraints now are

$$\|\mathbf{x}^{(1)}\|^2 < p_0 \tau \qquad (4.3.40a)$$

and

$$\|\mathbf{x}^{(2)}\|^2 < \beta \left(\|\mathbf{x}^{(1)}\|^2 + \|\mathbf{x}^{(2)}\|^2 \right). \qquad (4.3.40b)$$

Worked Example 4.3.5   (cf. Theorem 1 in [173]). *We want to prove that the capacities of the above combined parallel channels of types I–III are as follows.*

Case I, $\alpha_1 \leq \alpha_2$:

$$C = \frac{\alpha_1}{2} \ln \left(1 + \frac{(1 - \zeta) p_0}{\alpha_1 \sigma_0^2} \right) + \frac{\alpha_2}{2} \ln \left(1 + \frac{\zeta p_0}{\alpha_2 \sigma_0^2} \right) \qquad (4.3.41a)$$

where

$$\zeta = \min \left[\beta, \frac{\alpha_2}{\alpha_1 + \alpha_2} \right]. \qquad (4.3.41b)$$

If $\alpha_2 \leq \alpha_1$, subscripts 1 and 2 should replace each other in these equations. Further, when $\alpha_i = +\infty$, one uses the limiting expression $\lim_{\alpha \to \infty} (\alpha/2)\ln(1 + v/\alpha) = v/2$. In particular, if $\alpha_1 < \alpha_2 = +\infty$ then $\beta = \zeta$, and the capacity becomes

$$C = \frac{\alpha_1}{2} \ln \left(1 + \frac{(1-\beta)p_0}{\alpha_1 \sigma_0^2} \right) + \beta \frac{p_0}{2\sigma_0^2}. \tag{4.3.41c}$$

This means that the best transmission rate is attained when one puts as much 'energy' into channel 2 as is allowed by (4.3.38b).

Case II:

$$C = \frac{\alpha_1}{2} \ln \left(1 + \frac{(1-\beta)p_0}{(\alpha_1 + \alpha_2)\sigma_1^2} \right)$$
$$+ \frac{\alpha_2}{2} \ln \left(1 + \frac{(1-\beta)p_0}{(\alpha_1 + \alpha_2)\sigma_1^2} \right) + \frac{\beta p}{2\sigma_3^2}. \tag{4.3.42}$$

Case III:

$$C = \frac{\alpha_1}{2} \ln \left(1 + \frac{p_0}{\alpha_1 \sigma_0^2} \right) + \frac{\beta p_0}{2(1-\beta)\sigma_0^2}. \tag{4.3.43}$$

Solution We present the proof for Case I only. For definiteness, assume that $\alpha_1 < \alpha_2 \leq \infty$. First, the direct part. With $p_1 = (1-\zeta)p_0$, $p_2 = \zeta p_0$, consider the parallel combination of two channels, with individual power constraints on the input signals $\mathbf{x}^{(1)}$ and $\mathbf{x}^{(2)}$:

$$\left\| \mathbf{x}^{(1)} \right\|^2 \leq p_1 \tau, \quad \left\| \mathbf{x}^{(2)} \right\|^2 \leq p_2 \tau. \tag{4.3.44a}$$

Of course, (4.3.44a) implies (4.3.38a). Next, with $\zeta \leq \beta$, condition (4.3.38b) also holds true. Then, according to the direct part of Lemma 4.3.4, any rate R with $R < C_1(p_1) + C_2(p_2)$ is reliable. Here and below,

$$C_i(q) = \frac{\alpha_i}{2} \ln \left(1 + \frac{q}{\alpha_i \sigma_0^2} \right), \quad i = 1, 2. \tag{4.3.44b}$$

This implies the direct part.

A longer argument is needed to prove the inverse. Set $C^* = C_1(p_1) + C_2(p_2)$. The aim is to show that any rate $R > C^*$ is not reliable. Assume the opposite: there exists such a reliable $R = C^* + \varepsilon$; let us recall what it formally means. There exists a sequence of values $\tau^{(l)} \to \infty$ and (a) a sequence of codes

$$\mathscr{X}^{(l)} = \left\{ \mathbf{x}(i) = \left\{ \mathbf{x}^{(1)}(i), \mathbf{x}^{(2)}(i) \right\}, \ 1 \leq i \leq M^{(l)} \right\}$$

of size $M^{(l)} \sim e^{R\tau^{(l)}}$ composed of 'combined' code-vectors $\mathbf{x}(i) = \{\mathbf{x}^{(1)}(i), \mathbf{x}^{(2)}(i)\}$ with square-norms $\|\mathbf{x}(i)\|^2 = \|\mathbf{x}^{(1)}(i)\|^2 + \|\mathbf{x}^{(2)}(i)\|^2$, and (b) a sequence of decoding maps $d^{(l)} : \mathbf{y} \in \mathbb{K}^{(l)} \mapsto d^{(l)}(\mathbf{y}) \in \mathscr{X}^{(l)}$ such that $P_\mathrm{e} \to 0$. Here, as before, $P_\mathrm{e} = P_\mathrm{e}^{\mathscr{X}^{(l)}, \mathrm{av}}(d^{(l)})$ stands for the average error-probability:

$$P_\mathrm{e} = \frac{1}{M^{(l)}} \sum_{1 \le i \le M^{(l)}} P_\mathrm{e}(i, d^{(l)})$$

calculated from individual error-probabilities $P_\mathrm{e}(i, d^{(l)})$:

$$P_\mathrm{e}(i, d^{(l)}) = \begin{cases} 1, & \text{if } \|\mathbf{x}(i)\|^2 > p_0 \tau^{(l)} \text{ or } \|\mathbf{x}^{(2)}(i)\|^2 > \beta \|\mathbf{x}(i)\|^2, \\ \mathbf{P}_\mathrm{ch}\left(\mathbf{Y} \notin \mathbb{K}^{(l)} \text{ or } d^{(l)}(\mathbf{Y}) \ne \mathbf{x}(i) \mid \mathbf{x}(i) \text{ sent}\right), \\ \qquad \text{if } \|\mathbf{x}(i)\|^2 \le p_0 \tau^{(l)} \text{ and } \|\mathbf{x}^{(2)}(i)\|^2 \le \beta \|\mathbf{x}(i)\|^2. \end{cases}$$

The component vectors

$$\mathbf{x}^{(1)}(i) = \begin{pmatrix} x_1^{(1)}(i) \\ \vdots \\ x_{\lceil \alpha_1 \tau^{(l)} \rceil}^{(1)}(i) \end{pmatrix} \in \mathbb{R}^{\lceil \alpha_1 \tau^{(l)} \rceil} \quad \text{and} \quad \mathbf{x}^{(2)}(i) = \begin{pmatrix} x_1^{(1)}(i) \\ \vdots \\ x_{\lceil \alpha_2 \tau^{(l)} \rceil}^{(2)}(i) \end{pmatrix} \in \mathbb{R}^{\lceil \alpha_2 \tau^{(l)} \rceil}$$

are sent through their respective parts of the parallel-channel combination, which results in output vectors

$$\mathbf{Y}^{(1)} = \begin{pmatrix} Y_1^{(1)}(i) \\ \vdots \\ Y_{\lceil \alpha_1 \tau^{(l)} \rceil}^{(1)}(i) \end{pmatrix} \in \mathbb{R}^{\lceil \alpha_1 \tau^{(l)} \rceil}, \quad \mathbf{Y}^{(2)} = \begin{pmatrix} Y_1^{(2)}(i) \\ \vdots \\ Y_{\lceil \alpha_2 \tau^{(l)} \rceil}^{(2)}(i) \end{pmatrix} \in \mathbb{R}^{\lceil \alpha_2 \tau^{(l)} \rceil}$$

forming the combined output signal $\mathbf{Y} = \{\mathbf{Y}^{(1)}, \mathbf{Y}^{(2)}\}$. The entries of vectors $\mathbf{Y}^{(1)}$ and $\mathbf{Y}^{(2)}$ are sums

$$Y_j^{(1)} = x_j^{(1)}(i) + Z_j^{(1)}, \quad Y_k^{(2)} = x_k^{(2)}(i) + Z_k^{(2)},$$

where $Z_j^{(1)}$ and $Z_k^{(2)}$ are IID, $\mathrm{N}(0, \sigma_0^2)$ random variables. Correspondingly, $\mathbf{P}_\mathrm{ch}$ refers to the joint distribution of the random variables $Y_j^{(1)}$ and $Y_k^{(2)}$, $1 \le j \le \lceil \alpha_1 \tau^{(l)} \rceil$, $1 \le k \le \lceil \alpha_2 \tau^{(l)} \rceil$.

Observe that function $q \mapsto C_1(q)$ is uniformly continuous in q on $[0, p_0]$. Hence, we can find an integer J_0 large enough such that

$$\left| C_1(q) - C_1\left(q - \frac{\zeta p_0}{J_0}\right) \right| < \frac{\varepsilon}{2}, \quad \text{for all } q \in (0, \zeta p_0).$$

Then we partition the code $\mathscr{X}^{(l)}$ into J_0 classes (subcodes) $\mathscr{X}_j^{(l)}$, $j = 1, \ldots, J_0$: a code-vector $\left(\mathbf{x}^{(1)}(i), \mathbf{x}^{(2)}(i)\right)$ falls in class $\mathscr{X}_j^{(l)}$ if

$$(j-1)\frac{\zeta p_0 \tau}{J_0} < \sum_{1 \le k \le \lceil \alpha_2 \tau^{(l)} \rceil} (x_k^{(2)})^2 \le j\frac{\zeta p_0 \tau}{J_0}. \tag{4.3.45a}$$

Since a transmittable code-vector $\mathbf{x}$ has a component $\mathbf{x}^{(2)}$ with $\|\mathbf{x}^{(2)}\|^2 \le \zeta \|\mathbf{x}\|^2$, each such $\mathbf{x}$ lies in one and only one class. (We make an agreement that zero code-vectors belong to $\mathscr{X}_1^{(l)}$.) The class $\mathscr{X}_j^{(l)}$ containing the most code-vectors is denoted by $\mathscr{X}_*^{(l)}$. Then, obviously, the cardinality $\sharp \mathscr{X}_*^{(l)} \ge M^{(l)}/J_0$, and the transmission rate R_* of code $\mathscr{X}_*^{(l)}$ satisfies

$$R_* \ge R - \frac{1}{\tau^{(l)}} \ln J_0. \tag{4.3.45b}$$

On the other hand, the maximum error-probability for subcode $\mathscr{X}_*^{(l)}$ is not larger than for the whole code $\mathscr{X}^{(l)}$ (when using the same decoder $d^{(l)}$); consequently, the error-probability $P_e^{\mathscr{X}_*^{(l)}, \mathrm{av}}\left(d^{(l)}\right) \le P_e^{(l)} \to 0$.

Having a fixed number J_0 of classes in the partition of $\mathscr{X}^{(l)}$, we can find at least one $j_0 \in \{1, \ldots, J_0\}$ such that, for infinitely many l, the most numerous class $\mathscr{X}_*^{(l)}$ coincides with $\mathscr{X}_j^{(l)}$. Reducing our argument to those l, we may assume that $\mathscr{X}_*^{(l)} = \mathscr{X}_{j_0}^{(l)}$ for all l. Then, for all $\left(\mathbf{x}^{(1)}, \mathbf{x}^{(2)}\right) \in \mathscr{X}_*^{(l)}$, with

$$\mathbf{x}^{(i)} = \begin{pmatrix} x_1^{(i)} \\ \vdots \\ x_{n_i}^{(i)} \end{pmatrix}, \quad i = 1, 2,$$

using (4.3.38a) and (4.3.45a)

$$\left\|\mathbf{x}^{(1)}\right\|^2 \le \left(1 - \frac{(j_0 - 1)\zeta}{J_0}\right) p_0 \tau^{(l)}, \quad \left\|\mathbf{x}^{(2)}\right\|^2 \le \frac{j_0 \zeta}{J_0} p_0 \tau^{(l)}.$$

That is, $\left\{\left(\mathscr{X}_*^{(l)}, d^{(l)}\right)\right\}$ is a coder/decoder sequence for the 'standard' parallel-channel combination (cf. (4.3.34)), with

$$p_1 = \left[1 - \frac{(j_0 - 1)\zeta}{J_0}\right] p_0 \text{ and } p_2 = \frac{j_0 \zeta}{J_0} p_0.$$

As the error-probability $P_e^{\mathscr{X}_*^{(l)}, \mathrm{av}}\left(d^{(l)}\right) \to 0$, rate R is reliable for this combination of channels. Hence, this rate does not surpass the capacity:

$$R^* \le C_1 \left(\left(1 - \frac{(j_0 - 1)\zeta}{J_0}\right) p_0\right) + C_2 \left(\frac{j_0 \zeta}{J_0} p_0\right).$$

Here and below we refer to the definition of $C_i(u)$ given in (4.3.44b), i.e.

$$R^* \leq C_1((1-\delta)p_0) + C_2(\delta p_0) + \frac{\varepsilon}{2} \qquad (4.3.46)$$

where $\delta = j_0\zeta/J_0$.

Now note that, for $\alpha_2 \geq \alpha_1$, the function

$$\delta \mapsto C_1((1-\delta)p_0) + C_2(\delta p_0)$$

increases in δ when $\delta < \alpha_2/(\alpha_1 + \alpha_2)$ and decreases when $\delta > \alpha_2/(\alpha_1 + \alpha_2)$. Consequently, as $\delta = j_0\zeta/J_0 \leq \zeta$, we obtain that, with $\zeta = \min[\beta, \alpha_2/(\alpha_1 + \alpha_2)]$,

$$C_1((1-\delta)p_0) + C_2(\delta p_0) \leq C_1(p_1) + C_2(p_2) = C^*. \qquad (4.3.47)$$

In turn, this implies, owing to (4.3.45b), (4.3.46) and (4.3.47), that

$$R \leq C^* + \frac{\varepsilon}{2} + \frac{1}{\tau^{(l)}} \ln J_0, \text{ or } R \leq C^* + \frac{\varepsilon}{2} \text{ when } \tau^{(l)} \to \infty.$$

The contradiction to $R = C^* + \varepsilon$ yields the inverse. □

Example 4.3.6 (Prolate spheroidal wave functions (PSWFs); see [146], [90], [91]) For any given $\tau, W > 0$ there exists a sequence of real functions $\psi_1(t), \psi_2(t), \ldots$, of a variable $t \in \mathbb{R}$, belonging to the Hilbert space $Ł_2(\mathbb{R})$ (i.e. with $\int \psi_n(t)^2 dt < \infty$), called prolate spheroidal wave functions (PSWFs), such that

(a) The Fourier transforms $\widehat{\psi}_n(\omega) = \int \psi(t)e^{it\omega}dt$ vanish for $|\omega| > 2\pi W$; moreover, the functions $\psi_n(t)$ form an orthonormal basis in the Hilbert subspace formed by functions from $Ł_2(\mathbb{R})$ with this property.

(b) The functions $\psi_n^{\circ}(t) := \psi_n(t)\mathbf{1}(|t| < \tau/2)$ (the restrictions of $\psi_n(t)$ to $(-\tau/2, \tau/2)$) are pairwise orthogonal:

$$\int \psi_n^{\circ}(t)\psi_{n'}^{\circ}(t)dt = \int_{-\tau/2}^{\tau/2} \psi_n(t)\psi_{n'}(t)dt = 0 \text{ when } n \neq n'. \qquad (4.3.48a)$$

Furthermore, functions ψ_n° form a complete system in $Ł_2(-\tau/2, \tau/2)$: if a function $\varphi \in Ł_2(-\tau/2, \tau/2)$ has $\int_{-\tau/2}^{\tau/2} \varphi(t)\psi_n(t)dt = 0$ for all $n \geq 1$ then $\varphi(t) = 0$ in $Ł_2(-\tau/2, \tau/2)$.

(c) The functions $\psi_n(t)$ satisfy, for all $n \geq 1$ and $t \in \mathbb{R}$, the equations

$$\lambda_n \psi_n(t) = 2W \int_{-\tau/2}^{\tau/2} \psi_n(s) \operatorname{sinc}\left(2W\pi(t-s)\right)ds. \qquad (4.3.48b)$$

That is, functions $\psi_n(t)$ are the eigenfunctions, with the eigenvalues λ_n, of the integral operator $\varphi \mapsto \int \varphi(s) K(\cdot, s) \, ds$ with the integral kernel

$$K(t,s) = \mathbf{1}(|s| < \tau/2)(2W)\operatorname{sinc}\left(2W(t-s)\right)$$
$$= \mathbf{1}(|s| < \tau/2)\frac{\sin(2\pi W(t-s))}{\pi(t-s)}, \quad -\tau/2 \le s; t \le \tau/2.$$

(d) The eigenvalues λ_n satisfy the condition

$$\lambda_n = \int_{-\tau/2}^{\tau/2} \psi_n(t)^2 dt \text{ with } 1 > \lambda_1 > \lambda_2 > \cdots > 0.$$

An equivalent formulation can be given in terms involving the Fourier transforms $[\mathbf{F}\psi_n^\circ](\omega) = \int \psi_n^\circ(t) e^{it\omega} dt$:

$$\frac{1}{2\pi} \int_{-2\pi W}^{2\pi W} |[\mathbf{F}\psi_n^\circ](\omega)|^2 \, d\omega \Big/ \int_{-\tau/2}^{\tau/2} |\psi_n(t)|^2 dt = \lambda_n,$$

which means that λ_n gives a 'frequency concentration' for the truncated function ψ_n°.

(e) It can be checked that functions $\psi_n(t)$ (and hence numbers λ_n) depend on W and τ through the product $W\tau$ only. Moreover, for all $\theta \in (0,1)$, as $W\tau \to \infty$,

$$\lambda_{\lceil 2W\tau(1-\theta)\rceil} \to 1, \quad \text{and} \quad \lambda_{\lceil 2W\tau(1+\theta)\rceil} \to 0. \tag{4.3.48c}$$

That is, for τ large, nearly $2W\tau$ of values λ_n are close to 1 and the rest are close to 0.

An important part of the argument that is currently developing is the Karhunen–Loève decomposition. Suppose $Z(t)$ is a Gaussian random process with spectral density $\Phi(\omega)$ given by (4.3.27). The Karhunen–Loève decomposition states that for all $t \in (-\tau/2, \tau/2)$, the random variable $Z(t)$ can be written as a convergent (in the mean-square sense) series

$$Z(t) = \sum_{n \ge 1} A_n \psi_n(t), \tag{4.3.49}$$

where $\psi_1(t), \psi_2(t), \ldots$ are the PSWFs discussed in Worked Example 4.3.9 below and $A_1, A_2, \ldots$ are IID random variables with $A_n \sim \mathrm{N}(0, \lambda_n)$ where λ_n are the corresponding eigenvalues. Equivalently, one writes $Z(t) = \sum_{n \ge 1} \sqrt{\lambda_n} \xi_n \psi_n(t)$ where $\xi_n \sim \mathrm{N}(0,1)$ IID random variables.

The proof of this fact goes beyond the scope of this book, and the interested reader is referred to [38] or [103], p. 144.

The idea of the proof of Theorem 4.3.3 is as follows. Given W and τ, an input signal $s^\circ(t)$ from $\mathscr{A}(\tau, W, p_0, \eta)$ is written as a Fourier series in the PSWFs ψ_n. In this series, the first $2W\tau$ summands represent the part of the signal confined between the frequency band-limits $\pm 2\pi W$ and the time-limits $\pm \tau/2$. Similarly, the noise realisation $Z(t)$ is decomposed in a series in functions ψ_n. The action of the continuous-time channel is then represented in terms of a parallel combination of two jointly constrained discrete-time Gaussian channels. Channel 1 deals with the first $2W\tau$ PSWFs in the signal decomposition and has $\alpha_1 = 2W$. Channel 2 receives the rest of the expansion and has $\alpha_2 = +\infty$. The power constraint $\|s\|^2 \le p_0\tau$ leads to a joint constraint, as in (4.3.38a). In addition, a requirement emerges that the energy allocated outside the frequency band-limits $\pm 2\pi W$ or time-limits $\pm \tau/2$ is small: this results in another power constraint, as in (4.3.38b). Applying Worked Example 4.3.5 for Case I results in the assertion of Theorem 4.3.3.

To make these ideas precise, we first derive Theorem 4.3.7 which gives an alternative approach to the Nyquist–Shannon formula (more complex in formulation but somewhat simpler in the (still quite lengthy) proof).

Theorem 4.3.7 *Consider the following modification of the model from Theorem 4.3.3. The set of allowable signals $\mathscr{A}_2(\tau, W, p_0, \eta)$ consists of functions $t \in \mathbb{R} \mapsto s(t)$ such that*

(1) $\|s\|^2 = \int |s(t)|^2 dt \le p_0\tau$,

(2) *the Fourier transform $[\mathbf{F}s](\omega) = \int s(t)e^{it\omega}dt$ vanishes when $|\omega| > 2\pi W$, and*

(3) *the ratio $\displaystyle\int_{-\tau/2}^{\tau/2} |s(t)|^2 dt \Big/ \|s\|^2 > 1 - \eta$. That is, the functions $s \in \mathscr{A}(\tau, W, p_0, \eta)$ are 'sharply band-limited' in frequency and 'nearly localised' in time.*

The noise process is Gaussian, with the spectral density vanishing when $|\omega| > 2\pi W$ and equal to σ_0^2 for $|\omega| \le 2\pi W$.

Then the capacity of such a channel is given by

$$C = C_\eta = W \ln\left(1 + (1 - \eta)\frac{p_0}{2\sigma_0^2 W}\right) + \frac{\eta p_0}{2\sigma_0^2}. \qquad (4.3.50)$$

As $\eta \to 0$,

$$C_\eta \to W \ln\left(1 + \frac{p_0}{2\sigma_0^2 W}\right)$$

yielding the Nyquist–Shannon formula (4.3.8).

Proof of Theorem 4.3.7 First, we establish the direct half. Take

$$R < W \ln \left(1 + \frac{(1-\eta)p_0}{2\sigma_0^2 W} \right) + \frac{\eta p_0}{2\sigma_0^2} \qquad (4.3.51)$$

and take $\delta \in (0,1)$ and $\xi \in (0, \min[\eta, 1-\eta])$ such that R is still less than

$$C^* = W(1-\delta) \ln \left(1 + \frac{(1-\eta+\xi)p_0}{2\sigma_0^2 W(1-\delta)} \right) + \frac{(\eta-\xi)p_0}{2\sigma_0^2}. \qquad (4.3.52)$$

According to Worked Example 4.3.5, C^* is the capacity of a jointly constrained discrete-time pair of parallel channels as in Case I, with

$$\alpha_1 = 2W(1-\delta), \quad \alpha_2 = +\infty, \quad \beta = \eta - \xi, \quad p = p_0, \quad \sigma^2 = \sigma_0^2; \qquad (4.3.53)$$

cf. (4.3.41a). We want to construct codes and decoding rules for the time-continuous version of the channel, yielding asymptotically vanishing probability of error as $\tau \to \infty$. Assume $\left(\mathbf{x}^{(1)}, \mathbf{x}^{(2)} \right)$ is an allowable input signal for the parallel pair of discrete-time channels with parameters given in (4.3.53). The input for the time-continuous channel is the following series of (W, τ) PSWFs:

$$s(t) = \sum_{1 \le k \le \lceil \alpha_1 \tau \rceil} x_k^{(1)} \psi_k(t) + \sum_{1 \le k < \infty} x_k^{(2)} \psi_{k+\lceil \alpha_1 \tau \rceil}(t). \qquad (4.3.54)$$

The first fact to verify is that the signal in (4.3.54) belongs to $\mathscr{A}_2(\tau, W, p_0, \eta)$, i.e. satisfies conditions (1)–(3) of Theorem 4.3.7.

To check property (1), write

$$\|s\|^2 = \sum_{1 \le k \le \lceil \alpha_1 \tau \rceil} \left(x_k^{(1)} \right)^2 + \sum_{1 \le k < \infty} \left(x_k^{(2)} \right)^2 = \left\| \mathbf{x}^{(1)} \right\|^2 + \left\| \mathbf{x}^{(2)} \right\|^2 \le p_0 \tau.$$

Next, the signal $s(t)$ is band-limited, inheriting this property from the PSWFs $\psi_k(t)$. Thus, (2) holds true.

A more involved argument is needed to establish property (3). Because the PSWFs $\psi_k(t)$ are orthogonal in $\mathsf{L}_2[-\tau/2, \tau/2]$ (cf. (4.3.48a)), and using the monotonicity of the values λ_n (cf. (4.3.48b)), we have that

$$1 - \int_{-\tau/2}^{\tau/2} |s(t)|^2 \mathrm{d}t \bigg/ \|s\|^2 = \frac{\|(1-D_\tau)s\|^2}{\|s\|^2}$$

$$= \sum_{1 \le k \le \lceil \alpha_1 \tau \rceil} \frac{(1-\lambda_k)\left(x_k^{(1)} \right)^2}{\left\| \mathbf{x}^{(1)} \right\|^2 + \left\| \mathbf{x}^{(2)} \right\|^2} + \sum_{1 \le k < \infty} \frac{(1-\lambda_{k+\lceil \alpha_1 \tau \rceil})\left(x_k^{(2)} \right)^2}{\left\| \mathbf{x}^{(1)} \right\|^2 + \left\| \mathbf{x}^{(2)} \right\|^2}$$

$$\le \left(1 - \lambda_{\lceil \alpha_1 \tau \rceil} \right) \frac{\left\| \mathbf{x}^{(1)} \right\|^2}{\left\| \mathbf{x}^{(1)} \right\|^2 + \left\| \mathbf{x}^{(2)} \right\|^2} + \frac{\left\| \mathbf{x}^{(2)} \right\|^2}{\left\| \mathbf{x}^{(1)} \right\|^2 + \left\| \mathbf{x}^{(2)} \right\|^2}.$$

Now, as $\tau \to \infty$, the value $\lambda_{\lceil \alpha_1 \tau \rceil} \to 1$ (see (4.3.48c)). With the ratio $\|\mathbf{x}^{(1)}\|^2 / \left(\|\mathbf{x}^{(1)}\|^2 + \|\mathbf{x}^{(2)}\|^2 \right) \leq 1$, we have that for τ large enough,

$$\left(1 - \lambda_{\lceil \alpha_1 \tau \rceil} \right) \frac{\|\mathbf{x}^{(1)}\|^2}{\|\mathbf{x}^{(1)}\|^2 + \|\mathbf{x}^{(2)}\|^2} \leq \xi.$$

Next, the ratio $\|\mathbf{x}^{(2)}\|^2 / \left(\|\mathbf{x}^{(1)}\|^2 + \|\mathbf{x}^{(2)}\|^2 \right) \leq \eta - \xi$ (referring to (4.3.38b)). This finally yields

$$1 - \int_{-\tau/2}^{\tau/2} |s(t)|^2 dt \Big/ \|s\|^2 = \frac{\|(1 - D_\tau)s\|^2}{\|s\|^2} \leq \xi + \eta - \xi = \eta,$$

i.e. property (3).

Further, the noise can be expanded in accordance with Karhunen–Loève:

$$Z(t) = \sum_{1 \leq k \leq \lceil \alpha_1 \tau \rceil} Z_k^{(1)} \psi_k(t) + \sum_{1 \leq k < \infty} Z_k^{(2)} \psi_{k+\lceil \alpha_1 \tau \rceil}(t). \tag{4.3.55}$$

Here again, $\psi_k(t)$ are the PSWFs and IID random variables $Z_k^{(j)} \sim N(0, \lambda_k)$. Correspondingly, the output signal is written as

$$Y(t) = \sum_{1 \leq k \leq \lceil \alpha_1 \tau \rceil} Y_k^{(1)} \psi_k(t) + \sum_{1 \leq k < \infty} Y_k^{(2)} \psi_{k+\lceil \alpha_1 \tau \rceil}(t) \tag{4.3.56}$$

where

$$Y_k^{(j)} = x_k^{(j)} + Z_k^{(j)}, \quad j = 1, 2, \ k \geq 1. \tag{4.3.57}$$

So, the continuous-time channel is equivalent to a jointly constrained parallel combination. As we checked, the capacity equals C^* specified in (4.3.52). Thus, for $R < C^*$ we can construct codes of rate R and decoding rules such that the error-probability tends to 0.

For the converse, assume that there exists a sequence $\tau^{(l)} \to \infty$, a sequence of transmissible domains $\mathscr{A}_2^{(l)}(\tau^{(l)}, W, p_0, \eta^{(l)})$ described in (1)–(3) and a sequence of codes $\mathscr{X}^{(l)}$ of size $M = \lceil e^{R\tau^{(l)}} \rceil$ where

$$R > W \ln \left(1 + \frac{(1 - \eta)p_0}{2W\sigma_0^2} \right) + \frac{\eta p_0}{\sigma_0^2}.$$

As usual, we want to show that the error-probability $P_e^{\mathscr{X}^{(l)}, \mathrm{av}}(d^{(l)})$ does not tend to 0.

As before, we take $\delta > 0$ and $\xi \in (0, 1 - \eta)$ to ensure that $R > C^*$ where

$$C^* = W(1 + \delta) \ln \left[1 + \frac{(1 - \eta - \xi)}{(1 - \xi)} \frac{p_0}{2W\sigma_0^2(1 + \delta)} \right] + \frac{\eta p_0}{(1 - \xi)\sigma_0^2}.$$

Then, as in the argument on the direct half, C^* is the capacity of the type I jointly constrained parallel combination of channels with

$$\beta = \frac{\eta}{1-\xi}, \quad \sigma^2 = \sigma_0^2, \quad p = p_0, \quad \alpha_1 = 2W(1+\delta), \quad \alpha_2 = +\infty. \qquad (4.3.58)$$

Let $s(t) \in \mathscr{X}^{(l)} \cap \mathscr{A}_2^{(l)}(\tau^{(l)}, W, p_0, \eta^{(l)})$ be a continuous-time code-function. Since the PSWFs $\psi_k(t)$ form an ortho-basis in $Ł_2(\mathbb{R})$, we can decompose

$$s(t) = \sum_{1\leq k \leq \lceil \alpha_1 \tau^{(l)} \rceil} x_k^{(1)} \psi_k(t) + \sum_{1\leq k < \infty} x_k^{(2)} \psi_{k+\lceil \alpha_1 \tau^{(l)} \rceil}(t), t \in \mathbb{R}. \qquad (4.3.59)$$

We want to show that the discrete-time signal $\mathbf{x} = (\mathbf{x}^{(1)}, \mathbf{x}^{(2)})$ represents an allowable input to the type I jointly constrained parallel combination specified in (4.3.38a–c). By orthogonality of PSWFs $\psi_k(t)$ in $Ł_2(\mathbb{R})$ we can write

$$||\mathbf{x}||^2 = ||s||^2 \leq p_0 \tau^{(l)}$$

ensuring that condition (4.3.38a) is satisfied. Further, using orthogonality of PSW functions $\psi_k(t)$ in $Ł_2(-\tau/2, \tau/2)$ and the fact that the eigenvalues λ_k decrease monotonically, we obtain that

$$1 - \int_{-\tau^{(l)}/2}^{\tau^{(l)}/2} |s(t)|^2 dt \,\Big/\, ||s||^2 = \frac{||(1 - D_{\tau^{(l)}})s||^2}{||s||^2}$$

$$= \sum_{1\leq k \leq \lceil \alpha_1 \tau^{(l)} \rceil} \frac{(1 - \lambda_k)\left(x_k^{(1)}\right)^2}{||\mathbf{x}||^2} + \sum_{1\leq k < \infty} \frac{\left(1 - \lambda_{k+\lceil \alpha_1 \tau^{(l)} \rceil}\right)\left(x_k^{(2)}\right)^2}{||\mathbf{x}||^2}$$

$$\geq \left(1 - \lambda_{\lceil \alpha_1 \tau^{(l)} \rceil}\right) \frac{||\mathbf{x}^{(2)}||^2}{||\mathbf{x}||^2}.$$

By virtue of (4.3.48c), $\lambda_{\lceil \alpha_1 \tau^{(l)} \rceil} \leq \xi$ for l large enough. Moreover, since $1 - \int_{-\tau^{(l)}/2}^{\tau^{(l)}/2} |s(t)|^2 dt \,\Big/\, ||s||^2 \leq \eta$, we can write

$$\frac{||\mathbf{x}^{(2)}||^2}{||\mathbf{x}||^2} \leq \frac{\eta}{1-\xi}$$

and deduce property (4.3.38b).

Next, as in the direct half, we again use the Karhunen–Loève decomposition of noise $Z(t)$ to deduce that for each code for the continuous-time channel there corresponds a code for the jointly constrained parallel combination of discrete-time channels, with the same rate and error-probability. Since R is $> C^*$, the capacity of the discrete-time channel, the error-probability $P_e^{\mathscr{X}^{(l)},\mathrm{av}}(d^{(l)})$ remains bounded away from 0 as $l \to \infty$. This yields the converse. $\qquad\square$

Proof of Theorem 4.3.3 (Sketch) The formal argument proceeds as in Theorem 4.3.7: we have to prove the direct and converse parts of the theorem. Recall that the direct part states that the capacity is $\geq C$, the value indicated in (4.3.31), while the converse/inverse that it is $\leq C$. For the direct part, the channel is decomposed into the product of two parallel channels, as in Case III, with

$$\alpha_1 = 2W(1-\theta), \quad \alpha_2 = +\infty, \quad p = p_0, \quad \sigma^2 = \sigma_0^2, \quad \beta = \eta - \xi, \tag{4.3.60}$$

where $\theta \in (0,1)$ (cf. property (e) of PSWFs in Example 4.3.6) and $\xi \in (0,\eta)$ are auxiliary values.

For the converse half we use the decomposition into two parallel channels, again as in Case III, with

$$\alpha_1 = 2W(1+\theta), \quad \alpha_2 = +\infty, \quad p = p_0, \quad \sigma^2 = \sigma_0^2, \quad \beta = \frac{\eta}{1-\xi}. \tag{4.3.61}$$

Here, as before, value $\theta \in (0,1)$ emerges from property (e) of PSWFs, whereas value $\xi \in (0,1)$. $\qquad\square$

Summing up our previous observations we obtain the famous

Lemma 4.3.8 (The Nyquist–Shannon–Kotelnikov–Whittaker sampling lemma) *Let f be a function $t \in \mathbb{R} \mapsto f(t) \in \mathbb{R}$ with $\int |f(t)|dt < +\infty$. Suppose that the Fourier transform*

$$[\mathbf{F}f](\omega) = \int e^{it\omega} f(t)dt$$

vanishes for $|\omega| > 2\pi W$. Then, for all $x \in \mathbb{R}$, function f can be uniquely reconstructed from its values $f(x + n/(2W))$ calculated at points $x + n/(2W)$, where $n = 0, \pm 1, \pm 2$. More precisely, for all $t \in \mathbb{R}$,

$$f(t) = \sum_{n \in \mathbb{Z}^1} f\left(\frac{n}{2W}\right) \frac{\sin\left[2\pi(Wt - n)\right]}{2\pi(Wt - n)}. \tag{4.3.62}$$

Worked Example 4.3.9 *By the famous uncertainty principle of quantum physics, a function and its Fourier transform cannot be localised simultaneously in finite intervals $[-\tau, \tau]$ and $[-2\pi W, 2\pi W]$. What could be said about the case when both function and its Fourier transform are nearly localised? How can we quantify the uncertainty in this case?*

Solution Assume the function $f \in \mathbb{L}_2(\mathbb{R})$ and let $\widehat{f} = \mathbf{F}f \in L_2(\mathbb{R})$ be the Fourier transform of f. (Recall that space $\mathbb{L}_2(\mathbb{R})$ consists of functions f on $\mathbb{R}$ with $||f||^2 =$

$\int |f(t)|^2 dt < +\infty$ and that for all $f, g \in Ł_2(\mathbb{R})$, the inner product $\int f(t)\overline{g}(t)dt$ is finite.) We shall see that if

$$\int_{t_0-\tau/2}^{t_0+\tau/2} |f(t)|^2 dt \Big/ \int_{-\infty}^{\infty} |f(t)|^2 dt = \alpha^2 \qquad (4.3.63)$$

and

$$\int_{-2\pi W}^{2\pi W} |\mathbf{F}f(\omega)|^2 d\omega \Big/ \int_{-\infty}^{\infty} |\mathbf{F}f(\omega)|^2 d\omega = \beta^2 \qquad (4.3.64)$$

then $W\tau \geq \eta$, where $\eta = \eta(\alpha, \beta)$ will be found explicitly. (The inequality will be sharp, and functions yielding equality will be specified.)

Consider the linear operators $f \in Ł_2(\mathbb{R}) \mapsto Df \in Ł_2(\mathbb{R})$ and $f \in Ł_2(\mathbb{R}) \mapsto Bf \in Ł_2(\mathbb{R})$ given by

$$Df(t) = f(t)\mathbf{1}(|t| \leq \tau/2) \qquad (4.3.65)$$

and

$$Bf(t) = \frac{1}{2\pi} \int_{-2\pi W}^{2\pi W} \mathbf{F}f(\omega)e^{-i\omega t} d\omega = \frac{1}{\pi} \int_{-\infty}^{\infty} f(s)\frac{\sin 2\pi W(t-s)}{t-s} ds. \qquad (4.3.66)$$

We are interested in the product of these operators, $A = BD$:

$$Af(t) = \frac{1}{\pi} \int_{-\tau/2}^{\tau/2} f(s)\frac{\sin 2\pi W(t-s)}{t-s} ds; \qquad (4.3.67)$$

see Example 4.3.6. The eigenvalues λ_n of A obey $1 > \lambda_0 > \lambda_1 > \cdots$ and tend to zero as $n \to \infty$; see [91]. We are interested in the eigenvalue λ_0: it can be shown that λ_0 is a function of the product $W\tau$. In fact, the eigenfunctions (ψ_j) of (4.3.67) yield an orthonormal basis in $Ł_2(\mathbb{R})$; at the same time these functions form an orthogonal basis in $Ł_2[-\tau/2, \tau/2]$:

$$\int_{-\tau/2}^{\tau/2} \psi_j(t)\psi_i(t)dt = \lambda_i \delta_{ij}.$$

As usual, the angle between f and g in Hilbert space $Ł_2(\mathbb{R})$ is determined by

$$\theta(f, g) = \cos^{-1}\left(\frac{1}{||f|| \, ||g||} \text{Re} \int f(t)\overline{g}(t)dt\right). \qquad (4.3.68)$$

The angle between two subspaces is the minimal angle between vectors in these subspaces. We will show that there exists a positive angle $\theta(\mathscr{B}, \mathscr{D})$ between the subspaces $\mathscr{B}$ and $\mathscr{D}$, the image spaces of operators B and D. That is, $\mathscr{B}$ is the linear subspace of all band-limited functions while $\mathscr{D}$ is that of all time-limited functions. Moreover,

$$\theta(\mathscr{B}, \mathscr{D}) = \cos^{-1}\sqrt{\lambda_0} \qquad (4.3.69)$$

and $\inf_{f \in \mathscr{B}, g \in \mathscr{D}} \theta(f,g)$ is achieved when $f = \psi_0, g = D\psi_0$ where ψ_0 is the (unique) eigenfunction with the eigenvalue λ_0.

To this end, we verify that for any $f \in \mathscr{B}$

$$\min_{g \in \mathscr{D}} \theta(f,g) = \cos^{-1} \frac{||Df||}{||f||}. \tag{4.3.70}$$

Indeed, expand $f = f - Df + Df$ and observe that the integral $\int [f(t) - Df(t)]g(t)dt = 0$ (since the supports of g and $f - Df$ are disjoint). This implies that

$$\left| \mathrm{Re} \int f(t)\overline{g}(t)dt \right| \leq \left| \int f(t)\overline{g}(t)dt \right| = \left| \int Df(t)\overline{g}(t)dt \right|.$$

Hence,

$$\frac{1}{||f|| ||g||} \mathrm{Re} \int f(t)\overline{g}(t)dt \leq \frac{||Df||}{||f||}$$

which implies (4.3.70), by picking $g = Df$.

Next, we expand $f = \sum_{n=0}^{\infty} a_n \psi_n$, relative to the eigenfunctions of A. This yields the formula

$$\cos^{-1} \frac{||Df||}{||f||} = \cos^{-1} \left(\frac{\sum_n |a_n|^2 \lambda_n}{\sum_n |a_n|^2} \right)^{1/2}. \tag{4.3.71}$$

The supremum of the RHS in f is achieved when $a_n = 0$ for $n \geq 1$, and $f = \psi_0$. We conclude that there exists the minimal angle between subspaces $\mathscr{B}$ and $\mathscr{D}$, and this angle is achieved on the pair $f = \psi_0$, $g = D\psi_0$, as required. $\square$

Next, we establish

Lemma 4.3.10   *There exists a function $f \in \mathcal{L}_2$ such that $||f|| = 1$, $||Df|| = \alpha$ and $||Bf|| = \beta$ if and only if α and β fall in one of the following cases (a)–(d):*

(a) $\alpha = 0$ and $0 \leq \beta < 1$;
(b) $0 < \alpha < \sqrt{\lambda_0} < 1$ and $0 \leq \beta \leq 1$;
(c) $\sqrt{\lambda_0} \leq \alpha < 1$ and $\cos^{-1}\alpha + \cos^{-1}\beta \geq \cos^{-1}\sqrt{\lambda_0}$;
(d) $\alpha = 1$ and $0 < \beta \leq \sqrt{\lambda_0}$.

Proof   Given $\alpha \in [0,1]$, let $\mathscr{G}(\alpha)$ be the family of functions $f \in L^2$ with norms $||f|| = 1$ and $||Df|| = \alpha$. Next, determine $\beta^*(\alpha) := \sup_{f \in \mathscr{G}(\alpha)} ||Bf||$.

(a) If $\alpha = 0$, the family $\mathscr{G}(0)$ can contain no function with $\beta = ||Bf|| = 1$. Furthermore, if $||Df|| = 0$ and $||Bf|| = 1$ for $f \in \mathscr{B}$ then f is analytic and $f(t) = 0$ for $|t| < \tau/2$, implying $f \equiv 0$. To show that $\mathscr{G}(0)$ contains functions with all values of $\beta \in [0,1)$, we set $\tilde{f}_n = \dfrac{\psi_n - D\psi_n}{\sqrt{1 - \lambda_n}}$. Then the norm $||B\tilde{f}_n|| = \sqrt{1 - \lambda_n}$. Since there exist eigenvalues λ_n arbitrarily close to zero, $||B\widetilde{f}_n||$ becomes arbitrarily close to 1. By considering the functions $e^{ipt}\widetilde{f}(t)$ we can obtain all values of β between points $\sqrt{1-\lambda_n}$ since

$$||Be^{ipt}\widetilde{f}|| = \left(\int_{-p-\pi W}^{-p+\pi W} |F_n(\omega)|^2 d\omega \right)^{1/2}.$$

The norm $||Be^{ipt}\widetilde{f}||$ is continuous in p and approaches 0 as $p \to \infty$. This completes the analysis of case (a).

(b) When $0 < \alpha < \sqrt{\lambda_0} < 1$, we set

$$\widetilde{f} = \frac{\sqrt{\alpha^2 - \lambda_n}\psi_0 - \sqrt{\lambda_0 - \alpha^2}\psi_n}{\sqrt{\lambda_0 - \lambda_n}},$$

for n large when the eigenvalue λ_n is close to 0. We have that $\widetilde{f} \in \mathscr{B}$, $||\widetilde{f}|| = ||B\widetilde{f}|| = 1$, while a simple computation shows that $||D\widetilde{f}|| = \alpha$. This includes the case $\beta = 1$ as, by choosing $e^{ipt}\widetilde{f}(t)$ appropriately, we can obtain any $0 < \beta < 1$.

(c) and (d) If $\sqrt{\lambda_0} \leq \alpha < 1$ we decompose $f \in \mathscr{G}(\alpha)$ as follows:

$$f = a_1 Df + a_2 Bf + g \tag{4.3.72}$$

with g orthogonal to both Df and Bf. Taking the inner product of the sum in the RHS of (4.3.72), subsequently, with f, Df, Bf and g we obtain four equations:

$$1 = a_1\alpha^2 + a_2\beta^2 + \int g(t)\overline{f}(t)dt,$$

$$\alpha^2 = a_1\alpha^2 + a_2 \int Bf(t)\overline{Dg}(t)dt,$$

$$\beta^2 = a_1 \int Df(t)\overline{Bf}(t)dt + a_2\beta^2,$$

$$\int f(t)\overline{g}(t)dt = ||g||^2.$$

These equations imply

$$\alpha^2 + \beta^2 - 1 + ||g||^2 = a_1 \int Df(t)\overline{Bf}(t)dt + a_2 \int Bf(t)\overline{Df}(t)dt.$$

By eliminating $\int g(t)\overline{f}(t)dt$, a_1 and a_2 we find, for $\alpha\beta \neq 0$,

$$\beta^2 = \frac{1-\alpha^2-||g||^2}{(\beta^2 - \int Bf(t)\overline{Df}(t)dt)}\beta^2$$

$$+\left[1 - \frac{1-\alpha^2-||g||^2}{\alpha^2(\beta^2 - \int Bf(t)\overline{Df}(t)dt)}\int Bf(t)\overline{Df}(t)dt\right]$$

$$\times \int Df(t)\overline{Bf}(t)dt$$

which is equivalent to

$$\beta^2 - 2\mathrm{Re}\int Df(t)\overline{Bf}(t)dt$$

$$\leq -\alpha^2 + \left((1-\frac{1}{\alpha^2\beta^2}\left|\int Df(t)\overline{Bf}(t)dt\right|^2\right) \qquad (4.3.73)$$

$$-||g||^2\left(1-\frac{1}{\alpha^2\beta^2}\left|\int Df(t)\overline{Bf}(t)dt\right|^2\right). \qquad (4.3.74)$$

In terms of the angle θ, we can write

$$\alpha\beta\cos\theta = \mathrm{Re}\int Df(t)\overline{Bf}(t)dt \leq \left|\int Df(t)\overline{Bf}(t)dt\right| \leq \alpha\beta.$$

Substituting into (4.3.74) and completing the square we obtain

$$(\beta - \alpha\cos\theta)^2 \leq (1-\alpha^2)\sin^2\theta \qquad (4.3.75)$$

with equality if and only if $g=0$ and the integral $\int Df(t)\overline{B}f(t)dt$ is real. Since $\theta \geq \cos^{-1}\sqrt{\lambda_0}$, (4.3.75) implies that

$$\cos^{-1}\alpha + \cos^{-1}\beta \geq \cos^{-1}\sqrt{\lambda_0}. \qquad (4.3.76)$$

The locus of points (α,β) satisfying (4.3.76) is up and to the right of the curve where

$$\cos^{-1}\alpha + \cos^{-1}\beta = \cos^{-1}\sqrt{\lambda_0}. \qquad (4.3.77)$$

See Figure 4.6.

Equation (4.3.77) holds for the function $\tilde{f} = b_1\psi_0 + b_2D\psi_0$ with

$$b_1 = \sqrt{\frac{1-\alpha^2}{1-\lambda_0}} \quad \text{and} \quad b_2 = \frac{\alpha}{\sqrt{\lambda_0}} - \sqrt{\frac{1-\alpha^2}{1-\lambda_0}}.$$

All intermediate values of β are again attained by employing $e^{ipt}\tilde{f}$. $\square$

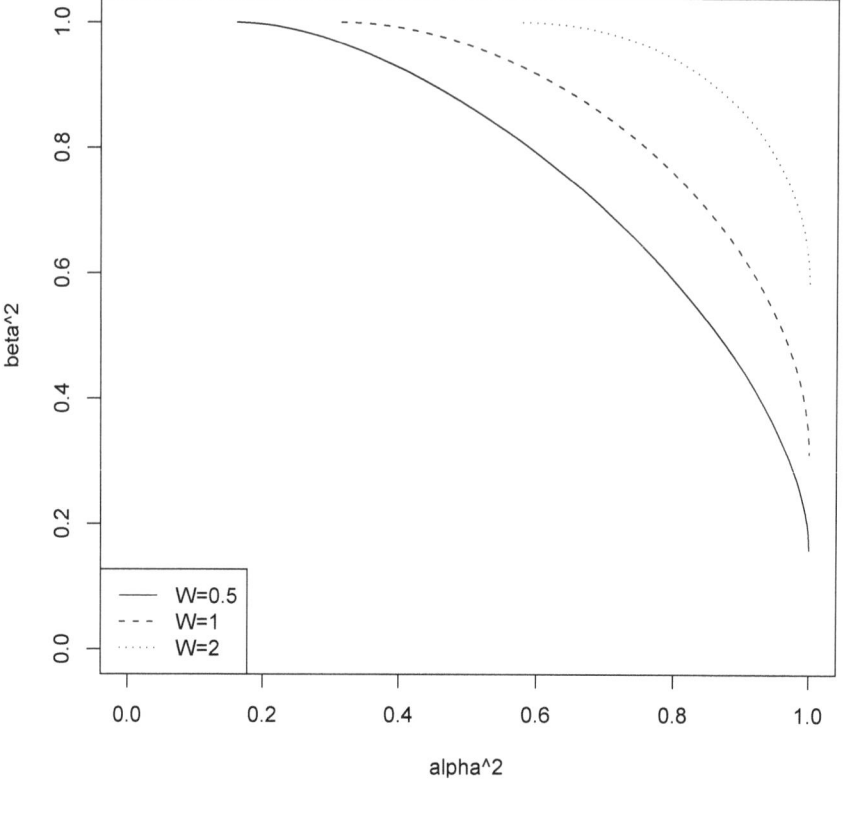

Figure 4.6

4.4 Spatial point processes and network information theory

For a discussion of capacity of distributed systems and construction of random
codebooks based on point processes we need some background. Here we study the
spatial Poisson process in $\mathbb{R}^d$, and some more advanced models of point process
are introduced with a good code distance. This section could be read independently
of PSE II, although some knowledge of its material may be very useful.

Definition 4.4.1   (cf. PSE II, p. 211) Let μ be a measure on $\mathbb{R}$ with values $\mu(A)$
for measurable subsets $A \subseteq \mathbb{R}$. Assume that μ is (i) *non-atomic* and (ii) *σ-finite*,
i.e. (i) $\mu(A) = 0$ for all countable sets $A \subset \mathbb{R}$ and (ii) there exists a partition $\mathbb{R} =$
$\cup_j J_j$ of $\mathbb{R}$ into pairwise disjoint intervals $J_1, J_2, \ldots$ such that $\mu(J_j) < \infty$. We say
that a random counting measure M defines a *Poisson random measure* (PRM, for
short) with mean, or intensity, measure μ if for all collection of pairwise disjoint
intervals $I_1, \ldots, I_n$ on $\mathbb{R}$, the values $M(I_k)$, $k = 1, \ldots, n$, are independent, and each
$M(I_k) \sim \mathrm{Po}(\mu(I_k))$.

We will state several facts, without proof, about the existence and properties of the Poisson random measure introduced in Definition 4.4.1.

Theorem 4.4.2   *For any non-atomic and σ-finite measure μ on $\mathbb{R}_+$ there exists a unique PRM satisfying Definition 4.4.1. If measure μ has the form $\mu(dt) = \lambda dt$ where $\lambda > 0$ is a constant (called the intensity of μ), this PRM is a Poisson process $PP(\lambda)$. If the measure μ has the form $\mu(dt) = \lambda(t)dt$ where $\lambda(t)$ is a given function, this PRM gives an inhomogeneous Poisson process $PP(\lambda(t))$.*

Theorem 4.4.3   *(The mapping theorem) Let μ be a non-atomic and σ-finite measure on $\mathbb{R}$ such that for all $t \geq 0$ and $h > 0$, the measure $\mu(t, t+h)$ of the interval $(t, t+h)$ is positive and finite (i.e. the value $\mu(t, t+h) \in (0, \infty)$), with $\lim_{h \to 0} \mu(0, h) = 0$ and $\mu(\mathbb{R}_+) = \lim_{u \to +\infty} \mu(0, u) = +\infty$. Consider the function*

$$f : u \in \mathbb{R}_+ \mapsto \mu(0, u),$$

and let f^{-1} be the inverse function of f. (It exists because $f(u) = \mu(0, u)$ is strictly monotone in u.) Let M be the $\mathrm{PRM}(\mu)$. Define a random measure f^*M by

$$(f^*M)(I) = M(\mu(f^{-1}I)) = M(\mu(f^{-1}(a), f^{-1}(b))), \qquad (4.4.1)$$

for interval $I = (a, b) \subset \mathbb{R}_+$, and continue it on $\mathbb{R}$. Then $f^*M \sim PP(1)$, i.e. f^*M yields a Poisson process of the unit rate.

We illustrate the above approach in a couple of examples.

Worked Example 4.4.4   *Let the rate function of a Poisson process $\Pi = PP(\lambda(x))$ on the interval $S = (-1, 1)$ be*

$$\lambda(x) = (1+x)^{-2}(1-x)^{-3}.$$

Show that Π has, with probability 1, infinitely many points in S, and that they can be labelled in ascending order as

$$\cdots X_{-2} < X_{-1} < X_0 < X_1 < X_2 < \cdots$$

with $X_0 < 0 < X_1$.

Show that there is an increasing function $f : S \to \mathbb{R}$ with $f(0) = 0$ such that the points $f(X)(X \in \Pi)$ form a Poisson process of unit rate on $\mathbb{R}$, and use the strong law of large numbers to show that, with probability 1,

$$\lim_{n \to +\infty} (2n)^{1/2}(1 - X_n) = \frac{1}{2}. \qquad (4.4.2)$$

Find a corresponding result as $n \to -\infty$.

Solution Since

$$\int_{-1}^{1} \lambda(x)dx = \infty,$$

there are with probability 1 infinitely many points of Π in $(-1,1)$. On the other hand,

$$\int_{-1+\delta}^{1-\delta} \lambda(x)dx < \infty$$

for every $\delta > 0$, so that $\Pi(-1+\delta, 1-\delta)$ is finite with probability 1. This is enough to label uniquely in ascending order the points of Π. Let

$$f(x) = \int_{0}^{x} \lambda(y)dy.$$

As $f : S \to \mathbb{R}$ is increasing, f maps Π into a Poisson process whose mean measure μ is given by

$$\mu(a,b) = \int_{f^{-1}(a)}^{f^{-1}(b)} \lambda(x)dx = b - a.$$

With this choice of f, the points $(f(X_n))$ form a Poisson process of unit rate on $\mathbb{R}$. The strong law of large numbers shows that, with probability 1, as $n \to \infty$,

$$n^{-1}f(X_n) \to 1, \text{ and } n^{-1}f(X_{-n}) \to -1.$$

Now, observe that

$$\lambda(x) \sim \frac{1}{4}(1-x)^{-3} \text{ and } f(x) \sim \frac{1}{8}(1-x)^{-2}, \text{ as } x \to 1.$$

Thus, as $n \to \infty$, with probability 1,

$$n^{-1}\frac{1}{8}(1-X_n)^{-2} \to 1,$$

which is equivalent to (4.4.2). Similarly,

$$\lambda(x) \sim \frac{1}{8}(1+x)^{-2} \text{ and } f(x) \sim \frac{1}{8}(1+x)^{-1}, \text{ as } x \to -1,$$

implying that with probability 1, as $n \to \infty$,

$$n^{-1}\frac{1}{8}(1+X_{-n})^{-1} \to 1.$$

Hence, with probability 1

$$\lim_{n\to\infty} n(1+X_{-n}) = \frac{1}{8}.$$

□

Worked Example 4.4.5 *Show that, if $Y_1 < Y_2 < Y_3 < \cdots$ are points of a Poisson process on $(0, \infty)$ with constant rate function λ, then*

$$\lim_{n \to \infty} Y_n / n = \lambda$$

with probability 1. Let the rate function of a Poisson process $\Pi = PP(\lambda(x))$ on $(0, 1)$ be

$$\lambda(x) = x^{-2}(1 - x)^{-1}.$$

Show that the points of Π can be labelled as

$$\cdots < X_{-2} < X_{-1} < \frac{1}{2} < X_0 < X_1 < \cdots$$

and that

$$\lim_{n \to -\infty} X_n = 0, \quad \lim_{n \to \infty} X_n = 1.$$

Prove that

$$\lim_{n \to \infty} n X_{-n} = 1$$

with probability 1. What is the limiting behaviour of X_n as $n \to +\infty$?

Solution The first part again follows from the strong law of large numbers. For the second part we set

$$f(x) = \int_{1/2}^{x} \lambda(\xi) d\xi,$$

and use the fact that f maps Π into a PP of constant rate on $(f(0), f(1))$: $f(\Pi) = PP(1)$. In our case, $f(0) = -\infty$ and $f(1) = \infty$, and so $f(\Pi)$ is a PP on $\mathbb{R}$. Its points may be labelled

$$\cdots < Y_{-2} < Y_{-1} < 0 < Y_0 < Y_1 < \cdots$$

with

$$\lim_{n \to -\infty} Y_n = -\infty, \quad \lim_{n \to +\infty} Y_n = +\infty.$$

Then $X_n = f^{-1}(Y_n)$ has the required properties.

The strong law of large numbers applied to Y_{-n} gives

$$\lim_{n \to -\infty} \frac{f(X_n)}{n} = \lim_{n \to -\infty} \frac{Y_n}{n} = 1, \quad \text{a.s.}$$

Now, as $x \to 0$,

$$f(x) = -\int_{x}^{1/2} \xi^{-2}(1 - \xi)^{-1} d\xi \sim -\int_{x}^{1/2} \xi^{-2} d\xi \sim -x^{-1},$$

implying that

$$\lim_{n \to \infty} \frac{X_{-n}^{-1}}{n} = 1, \ \text{i.e.} \ \lim_{n \to \infty} nX_{-n} = 1, \ \text{a.s.}$$

Similarly,

$$\lim_{n \to +\infty} \frac{f(X_n)}{n} = 1, \ \text{a.s.},$$

and as $x \to 1$,

$$f(x) \sim \int_{1/2}^{x} (1 - \xi)^{-1} d\xi \sim - \ln(1 - x).$$

This implies that

$$\lim_{n \to \infty} - \frac{\ln(1 - X_n)}{n} = 1, \ \text{a.s.}$$

$\square$

Next, we discuss the concept of a Poisson random measure (PRM) on a general set E. Formally, we assume that E had been endowed with a σ-algebra $\mathscr{E}$ of subsets, and a measure μ assigning to every $A \in \mathscr{E}$ a value $\mu(A)$, so that if $A_1, A_2, \ldots$ are pairwise disjoint sets from $\mathscr{E}$ then

$$\mu\left(\cup_n A_n\right) = \sum_n \mu(A_n).$$

The value $\mu(E)$ can be finite or infinite. Our aim is to define a random counting measure $M = (M(A), A \in \mathscr{E})$, with the following properties:
(a) The random variable $M(A)$ takes non-negative integer values (including, possibly, $+\infty$). Furthermore,

$$M(A) \begin{cases} \sim \text{Po}(\lambda \mu(A)), & \text{if } \mu(A) < \infty, \\ = +\infty \ \text{with probability } 1, & \text{if } \mu(A) = \infty. \end{cases} \tag{4.4.3}$$

(b) If $A_1, A_2, \ldots \in \mathscr{E}$ are disjoint sets then

$$M\left(\cup_i A_i\right) = \sum_i M(A_i). \tag{4.4.4}$$

(c) The random variables $M(A_1), M(A_2), \ldots$ are independent if sets $A_1, A_2, \ldots \in \mathscr{E}$ are disjoint. That is, for all finite collections of disjoint sets $A_1, \ldots, A_n \in \mathscr{E}$ and non-negative integers $k_1, \ldots, k_n$

$$\mathbb{P}\left(M(A_i) = k_i, \ 1 \le i \le n\right) = \prod_{1 \le i \le n} \mathbb{P}\left(M(A_i) = k_i\right). \tag{4.4.5}$$

First assume that $\mu(E) < \infty$ (if not, split E into subsets of finite measure). Fix a random variable $M(E) \sim \text{Po}(\lambda\mu(E))$. Consider a sequence $X_1, X_2, \ldots$ of IID random points in E, with $X_i \sim \mu/\mu(E)$, independently of $M(E)$. It means that for all $n \geq 1$ and sets $A_1, \ldots, A_n \in \mathscr{E}$ (not necessarily disjoint)

$$\mathbb{P}(M(E) = n, \ X_1 \in A_1, \ldots, X_n \in A_n) = e^{-\lambda\mu(E)} \frac{(\lambda\mu(E))^n}{n!} \prod_{i=1}^{n} \frac{\mu(A_i)}{\mu(E)}, \quad (4.4.6)$$

and conditionally,

$$\mathbb{P}(X_1 \in A_1, \ldots, X_n \in A_n | M(E) = n) = \prod_{i=1}^{n} \frac{\mu(A_i)}{\mu(E)}. \quad (4.4.7)$$

Then set

$$M(A) = \sum_{i=1}^{M(E)} \mathbf{1}(X_i \in A), \ A \in \mathscr{E}. \quad (4.4.8)$$

Theorem 4.4.6 *If $\mu(E) < \infty$, equation (4.4.8) defines a random measure M on E satisfying properties (a)–(c) above.*

Worked Example 4.4.7 *Let M be a Poisson random measure of intensity λ on the plane $\mathbb{R}^2$. Denote by $C(r)$ the circle $\{\mathbf{x} \in \mathbb{R}^2 : |\mathbf{x}| < r\}$ of radius r in $\mathbb{R}^2$ centred at the origin and let R_k be the largest radius such that $C(R_k)$ contains precisely k points of M. [Thus $C(R_0)$ is the largest circle about the origin containing no points of M, $C(R_1)$ is the largest circle about the origin containing a single point of M, and so on.] Calculate $\mathbb{E}R_0$, $\mathbb{E}R_1$ and $\mathbb{E}R_2$.*

Solution Clearly,

$$\mathbb{P}(R_0 > r) = \mathbb{P}(C(r) \text{ contains no point of } M) = e^{-\lambda\pi r^2}, \quad r > 0,$$

and

$$\mathbb{P}(R_1 > r) = \mathbb{P}(C(r) \text{ contains at most one point of } M)$$
$$= (1 + \lambda\pi r^2)e^{-\lambda\pi r^2}, \quad r > 0.$$

Similarly,

$$\mathbb{P}(R_2 > r) = \left[1 + \lambda\pi r^2 + \frac{1}{2}(\lambda\pi r^2)^2\right] e^{-\lambda\pi r^2}, \quad r > 0.$$

Then

$$\mathbb{E}R_0 = \int_0^\infty \mathbb{P}(R_0 > r)\mathrm{d}r = \frac{1}{\sqrt{2\pi\lambda}} \int_0^\infty e^{-\pi\lambda r^2}\mathrm{d}(\sqrt{2\pi\lambda}r) = \frac{1}{2\sqrt{\lambda}},$$

$$\mathbb{E}R_1 = \int_0^\infty \mathbb{P}(R_1 > r)\mathrm{d}r$$

$$= \frac{1}{2\sqrt{\lambda}} + \int_0^\infty e^{-\pi\lambda r^2}\left(\lambda\pi r^2\right)\mathrm{d}r$$

$$= \frac{1}{2\sqrt{\lambda}} + \frac{1}{2\sqrt{2\pi\lambda}} \int_0^\infty (2\pi\lambda r^2)e^{-\pi\lambda r^2}\mathrm{d}(\sqrt{2\pi\lambda}r)$$

$$= \frac{3}{4\sqrt{\lambda}},$$

$$\mathbb{E}R_2 = \frac{3}{4\sqrt{\lambda}} + \int_0^\infty \frac{\left(\lambda\pi r^2\right)^2}{2}e^{-\pi\lambda r^2}\mathrm{d}r$$

$$= \frac{3}{4\sqrt{\lambda}} + \frac{1}{8\sqrt{2\pi\lambda}} \int_0^\infty (2\lambda\pi r^2)^2 e^{-\pi\lambda r^2}\mathrm{d}(\sqrt{2\lambda\pi}r)$$

$$= \frac{3}{4\sqrt{\lambda}} + \frac{3}{16\sqrt{\lambda}}$$

$$= \frac{15}{16\sqrt{\lambda}}.$$

$\square$

We shall use for the PRM M on the phase space E with intensity measure μ constructed in Theorem 4.4.6 the notation $\mathrm{PRM}(E,\mu)$. Next, we extend the definition of the PRM to integral sums: for all functions $g: E \to \mathbb{R}_+$ define

$$M(g) = \sum_{i=1}^{M(E)} g(X_i) := \int g(y)\mathrm{d}M(y); \qquad (4.4.9)$$

summation is taken over all points $X_i \in E$, and $M(E)$ is the total number of such points. Next, for a general $g: E \to \mathbb{R}$ we set

$$M(g) = M(g_+) - M(-g_-),$$

with the standard agreement that $+\infty - a = +\infty$ and $a - \infty = -\infty$ for all $a \in (0,\infty)$. [When both $M(g_+)$ and $M(-g_-)$ equal ∞, the value $M(g)$ is declared not defined.] Then

Theorem 4.4.8   (Campbell theorem) *For all $\theta \in \mathbb{R}$ and for all functions $g: E \to \mathbb{R}$ such that $e^{\theta g(y)} - 1$ is μ-integrable*

$$\mathbb{E}e^{\theta M(g)} = \exp\left[\lambda \int_E \left(e^{\theta g(y)} - 1\right)\mathrm{d}\mu(y)\right]. \qquad (4.4.10)$$

Proof Write

$$\mathbb{E}e^{\theta M(g)} = \mathbb{E}\left[\mathbb{E}\left(e^{\theta M(g)}|M(E)\right)\right]$$
$$= \sum_k \mathbb{P}(M(E)=k)\mathbb{E}\left(\exp\left[\theta\sum_{i=1}^k g(X_i)\right]|M(E)=k\right).$$

Owing to conditional independence (4.4.7),

$$\mathbb{E}\left(\exp\left[\theta\sum_{i=1}^k g(X_i)\right]|M(E)=k\right) = \prod_{i=1}^k \mathbb{E}e^{\theta g(X_i)} = \left(\mathbb{E}e^{\theta g(X_1)}\right)^k$$
$$= \left(\frac{1}{\mu(E)}\int_E e^{\theta g(x)}d\mu(x)\right)^k,$$

and

$$\mathbb{E}e^{\theta M(g)} = \sum_k e^{-\lambda\mu(E)}\frac{(\lambda\mu(E))^k}{k!}\frac{1}{(\mu(E))^k}\left(\int_E e^{\theta g(x)}d\mu(x)\right)^k$$
$$= e^{-\lambda\mu(E)}\exp\left[\lambda\int_E e^{\theta g(x)}d\mu(x)\right]$$
$$= \exp\left[\lambda\int_E \left(e^{\theta g(x)}-1\right)d\mu(x)\right].$$

$\square$

Corollary 4.4.9 *The expected value of $M(g)$ is given by*

$$\mathbb{E}M(g) = \lambda\int_E g(y)d\mu(y);$$

it exists if and only if the integral on the RHS is well defined.

Proof The proof follows by differentiation of the MGF at $\theta=0$. $\square$

Example 4.4.10 Suppose that the wireless transmitters are located at the points of Poisson process Π on $\mathbf{R}^2$ of rate λ. Let r_i be the distance from transmitter i to the central receiver at 0, and the minimal distance to a transmitter is r_0. Suppose that the power of the received signal is $Y = \sum_{X_i\in\Pi}\frac{P}{r_i^\alpha}$ for some $\alpha > 2$. Then

$$\mathbb{E}e^{\theta Y} = \exp\left[2\lambda\pi\int_{r_0}^\infty \left(e^{\theta g(r)}-1\right)r dr\right], \qquad (4.4.11)$$

where $g(r) = \frac{P}{r^\alpha}$ where P is the transmitter power.

A popular model in application is the so-called *marked point process* with the space of marks D. This is simply a random measure on $\mathbf{R}^d \times D$ or on its subset. We will need the following product property proved below in the simplest set-up.

Theorem 4.4.11 (The product theorem) *Suppose that a Poisson process with the constant rate λ is given on $\mathbb{R}$, and marks Y_i are IID with distribution ν. Define a random measure M on $\mathbf{R}_+ \times D$ by*

$$M(A) = \sum_{n=1}^{\infty} I\big((T_n, Y_n) \in A\big), \; A \subseteq \mathbb{R}_+ \times D. \tag{4.4.12}$$

This measure is a PRM on $\mathbb{R}_+ \times D$, with the intensity measure $\lambda m \times \nu$ where m is a Lebesgue measure.

Proof First, consider a set $A \subseteq [0,t) \times D$ where $t > 0$. Then

$$M(A) = \sum_{n=1}^{N_t} \mathbf{1}\big((T_n, Y_n) \in A\big).$$

Consider the MGF $\mathbb{E}e^{\theta M(A)}$ and use standard conditioning

$$\mathbb{E}e^{\theta M(A)} = \mathbb{E}\left[\mathbb{E}\left(e^{\theta M(A)} | N_t\right)\right]$$

$$= \sum_{k=0}^{\infty} \mathbb{P}(N_t = k) \mathbb{E}\left(e^{\theta M(A)} | N_t = k\right).$$

We know that $N_t \sim \mathrm{Po}(\lambda t)$. Further, given that $N_t = k$, the jump points $T_1, \ldots, T_k$ have the conditional joint PDF $f_{T_1, \ldots, T_k}(\cdot | N_t = k)$ given by (4.4.7). Then, by using further conditioning, by $T_1, \ldots, T_k$, in view of the independence of the Y_n, we have

$$\mathbb{E}\left(e^{\theta M(A)} | N_t = k\right)$$

$$= \mathbb{E}\left[\mathbb{E}\left(e^{\theta M(A)} | N_t = k; T_1, \ldots, T_k\right)\right]$$

$$= \int_0^t \cdots \int_0^t dx_k \ldots dx_1 f_{T_1, \ldots, T_k}(x_1, \ldots, x_k | N = k)$$

$$\times \mathbb{E}\left(\exp \theta \left(\sum_{i=1}^k I\big((x_i, Y_i) \in A\big)\right) \Big| N_t = k; T_1 = x_1, \ldots, T_k = x_k\right)$$

$$= \frac{1}{t^k} \left(\int_0^t \int_D e^{\theta I_A(x,y)} d\nu(y) dx\right)^k.$$

Then

$$\mathbb{E}e^{\theta M(A)} = e^{-\lambda t} \sum_{k=0}^{\infty} \frac{(\lambda t)^k}{k!} \frac{1}{t^k} \left(\int_0^t \int_D e^{\theta I_A(x,y)} \mathrm{d}v(y)\mathrm{d}x \right)^k$$

$$= \exp\left[\lambda \int_0^t \int_D \left(e^{\theta I_A(x,y)} - 1 \right) \mathrm{d}v(y)\mathrm{d}x \right].$$

The expression $e^{\theta I_A(x,y)} - 1$ takes value $e^{\theta} - 1$ for $(x,y) \in A$ and 0 for $(x,y) \notin A$. Hence,

$$\mathbb{E}e^{\theta M(A)} = \exp\left[(e^{\theta} - 1)\lambda \int_A \mathrm{d}v(y)\mathrm{d}x \right], \quad \theta \in \mathbb{R}. \qquad (4.4.13)$$

Therefore, $M(A) \sim \mathrm{Po}(\lambda m \times v(A))$.

Moreover, if $A_1, \dots, A_n$ are disjoint subsets of $[0,t) \times D$ then the random variables $M(A_1), \dots, M(A_n)$ are independent. To see this, note first that, by definition, M is additive: $M(A) = M(A_1) + \cdots + M(A_n)$ where $A = A_1 \cup \cdots \cup A_n$. From (4.4.13)

$$\mathbb{E}e^{\theta M(A)} = \exp\left[(e^{\theta} - 1)\lambda \sum_{i=1}^{n} \int_{A_i} \mathrm{d}v(y)\mathrm{d}x \right] = \prod_{i=1}^{n} \mathbb{E}e^{\theta M(A_i)}, \quad \theta \in \mathbb{R},$$

which implies independence.

So, the restriction of M to $\overline{E}_n = [0,n) \times D$ is an $(\overline{E}_n, \lambda \mathrm{d}m_n \times v)$ PRM, where $m_n = m|_{[0,n)}$. Then, by the extension property, M is an $(\mathbb{R}_+ \times D, \lambda m \times v)$ PRM. □

Worked Example 4.4.12 *Use the product and Campbell's theorems to solve the following problem. Stars are scattered over three-dimensional space $\mathbb{R}^3$ in a Poisson process Π with density $v(X)$ $(X \in \mathbb{R}^3)$. Masses of the stars are IID random variables; the mass m_X of a star at X has PDF $\rho(X, \mathrm{d}m)$. The gravitational potential at the origin is given by*

$$F = \sum_{X \in \Pi} \frac{Gm_X}{|X|},$$

where G is a constant. Find the MGF $\mathbb{E}e^{\theta F}$.

A galaxy occupies a sphere of radius R centred at the origin. The density of stars is $v(\mathbf{x}) = 1/|\mathbf{x}|$ for points $\mathbf{x}$ inside the sphere; the mass of each star has the exponential distribution with mean M. Calculate the expected potential due to the galaxy at the origin. Let C be a positive constant. Find the distribution of the distance from the origin to the nearest star whose contribution to the potential F is at least C.

Solution Campbell's theorem says that if M is a Poisson random measure on the space E with intensity measure v and $a : E \to \mathbb{R}$ is a bounded measurable function then

$$\mathbb{E}e^{\theta\Sigma} = \exp\left(\int_E \left(e^{\theta a(y)} - 1\right) v(dy)\right),$$

where

$$\Sigma = \int_E a(y)M(dy) = \sum_{X \in \Pi} a(X).$$

By the product theorem, pairs (X, m_X) (position, mass) form a PRM on $\mathbb{R}^3 \times \mathbb{R}_+$, with intensity measure $\mu(dx \times dm) = v(x)dx\rho(x, dm)$. Then by Campbell's theorem:

$$\mathbb{E}e^{\theta F} = \exp\left(\int_{\mathbb{R}^3} \int_0^\infty \mu(dx \times dm)\left(e^{\theta Gm/|x|} - 1\right)\right).$$

The expected potential at the origin is $\mathbb{E}F = \dfrac{d\mathbb{E}e^{\theta F}}{d\theta}\Big|_{\theta=0}$ and equals

$$\int_{\mathbb{R}^3} v(x)dx \int_0^\infty \rho(x, dm)\frac{Gm}{|x|} = GM \int_{\mathbb{R}^3} dx\frac{1}{|x|^2}\mathbf{1}(|x| \leq R).$$

In the spherical coordinates,

$$\int_{\mathbb{R}^3} dx\frac{1}{|x|^2}\mathbf{1}(|x| \leq R) = \int_0^R dr\frac{1}{r^2}r^2 \int d\vartheta \cos \vartheta \int d\phi = 4\pi R$$

which yields

$$\mathbb{E}F = 4\pi GMR.$$

Finally, let D be the distance to the nearest star contributing to F at least C. Then, by the product theorem,

$$\mathbb{P}(D \geq d) = \mathbb{P}(\text{no points in } A) = \exp\left(-\mu(A)\right).$$

Here

$$A = \left\{(x, m) \in \mathbb{R}^3 \times \mathbb{R}_+ : |x| \leq d, \ \frac{Gm}{|x|} \geq C\right\},$$

and $\mu(A) = \int_A \mu(\mathrm{d}x \times \mathrm{d}m)$ is represented as

$$\int_0^d \mathrm{d}r \frac{1}{r} r^2 \int \mathrm{d}\vartheta \cos \vartheta \int \mathrm{d}\phi M^{-1} \int_{Cr/G}^{\infty} \mathrm{d}m e^{-m/M}$$

$$= 4\pi \int_0^d \mathrm{d}r r e^{-Cr/(GM)}$$

$$= 4\pi \left(\frac{GM}{C}\right)^2 \left(1 - e^{-Cd/(GM)} - \frac{Cd}{GM} e^{-Cd/(GM)}\right).$$

This determines the distribution of D on $[0, R]$. □

In distributed systems of transmitters and receivers like wireless networks of mobile phones the admissible communication rate between pairs of nodes in the wireless network depends on their random positions and their transmission strategies. Usually, the transmission is performed along the chain of transmitters from the source to destination. So, the new interesting direction in information theory has emerged; some experts even coined the term 'network information theory'. This field of research has many connections with probability theory, in particular, percolation and spatial point processes. We do not attempt here to give even a glimpse of this rapidly developing field, but no presentation of information theory nowadays can completely avoid network aspects. Here we touch slightly a few topics and refer the interested reader to [48] and the literature cited therein.

Example 4.4.13 Suppose that the receiver is located at point y and the transmitters are scattered on the plane $\mathbf{R}^2$ at the points of $x_i \in \Pi$ of Poisson process of rate λ. Then the simplest model for the power of the received signal is

$$Y = \sum_{x_i \in \Pi} P\ell(|x_i - y|) \tag{4.4.14}$$

where P is the emitted signal power and the function ℓ describes the fading of the signal. In the case of so-called Rayleigh fading $\ell(|x|) = e^{-\beta|x|}$, and in the case of the power fading $\ell(|x|) = |x|^{-\alpha}, \alpha > 2$. By the Campbell theorem

$$\phi(\theta) = \mathbb{E}[e^{\theta Y}] = \exp\left(2\lambda\pi \int_0^{\infty} r(e^{\theta P\ell(r)} - 1)\mathrm{d}r\right). \tag{4.4.15}$$

A more realistic model of the wireless network may be described as follows. Suppose that receivers are located at points $y_j, j = 1, \ldots, J$, and transmitters are scattered on the plane $\mathbf{R}^2$ at the points of $x_i \in \Pi$ of Poisson process of rate λ.

Assuming that the signal S_k from the point x_k is amplified by the coefficient $\sqrt{P}$ we write the signal

$$Y_j = \sum_{x_k \in \Pi} h_{jk} S_k + Z_j, j = 1, \ldots, J. \qquad (4.4.16)$$

Here the simplest model of the transmission function is

$$h_{jk} = \sqrt{P} \frac{e^{2\pi i r_{jk}/\nu}}{r_{jk}^{\alpha/2}}, \qquad (4.4.17)$$

where ν is the transmission wavelength and $r_{jk} = |y_j - x_k|$. The noise random variables Z_j are assumed to be IID $N(0, \sigma_0^2)$. A similar formula could be written for Rayleigh fading. We know that in the case of $J = 1$ and a single transmitter $K = 1$, by the Nyquist–Shannon theorem of Section 4.3, the capacity of the continuous time, additive white Gaussian noise channel $Y(t) = X(t)\ell(x, y) + Z(t)$ with attenuation factor $\ell(x, y)$, subject to the power constraint $\int_{-\tau/2}^{\tau/2} X^2(t) dt < P\tau$, bandwidth W, and noise power spectral density σ_0^2, is

$$C = W \log \left(1 + \frac{P\ell^2(x, y)}{2W \sigma_0^2} \right). \qquad (4.4.18)$$

Next, consider the case of finite numbers K of transmitters and J of receivers

$$y_j(t) = \sum_{i=1}^{K} \ell(x_i, y_j) x_i(t) + z_j(t), j = 1, \ldots, J, \qquad (4.4.19)$$

with a power constraint P_k for transmitter $k = 1, \ldots, K$. Using Worked Example 4.3.5 for the capacity of parallel channels it can be proved (cf. [48]) that the capacity of the channel is

$$C = \sum_{k=1}^{K} W \log \left(1 + \frac{P_k s_k^2}{2W \sigma_0^2} \right) \qquad (4.4.20)$$

where s_k is the kth largest singular value of the matrix $L = \left(\ell(|y_j - x_k|) \right)$. Next, we assume that the bandwidth $W = 1$. It is also interesting to describe the capacity region of a distributed system with K transmitters and J receivers under constant, on average, power of transmission $K^{-1} \sum_k P_k \leq P$. Again, the interested reader is referred to [48] where the following capacity domain for allowable rates R_{kj} is established:

$$\sum_{k=1}^{K} \sum_{j=1}^{J} R_{kj} \leq \max_{P_k \geq 0, \sum_k P_k \leq KP} \sum_{k=1}^{K} \log \left(1 + \frac{P_k s_k^2}{2\sigma_0^2} \right). \qquad (4.4.21)$$

Theorem 4.4.14 *Consider an arbitrary configuration S of $2n$ nodes placed inside the box B_n of area n (i.e. size $\sqrt{n}$); partition them into two sets S_1 and S_2, so* *that $S_1 \cap S_2 = \emptyset$, $S_1 \cup S_2 = S, \sharp S_1 = \sharp S_2 = n$. The sum $C_n = \sum_{k=1}^{n} \sum_{j=1}^{n} R_{kj}$ of reliable transmission rates in model (4.4.19) from the transmitters $x_k \in S_1$ to the receivers $y_i \in S_2$ is bounded from above:*

$$C_n = \sum_{k=1}^{n} \sum_{j=1}^{n} R_{kj} \leq \max_{P_k \geq 0, \Sigma P_k \leq nP} \sum_{k=1}^{n} \log \left(1 + \frac{P_k s_k^2}{2\sigma_0^2} \right),$$

where s_k is the kth largest singular value of the matrix $L = (\ell(x_k, y_j))$, σ_0^2 is the noise power spectral density, and the bandwidth $W = 1$.

This result allows us to find the asymptotic of capacity as $n \to \infty$. In the most interesting case of Rayleigh fading $R(n) = C_n/n \sim O(\frac{(\log n)^2}{\sqrt{n}})$; in the case of power $\alpha > 2$ fading, $R(n) \sim O(\frac{n^{1/\alpha}(\log n)^2}{\sqrt{n}})$: see again [48].

Next we discuss the interference limited networks. Let Π be a Poisson process of rate λ on the plane $\mathbb{R}^2$. Let the function $\ell : \mathbb{R}^2 \times \mathbb{R}^2 \to \mathbb{R}_+$ describing an attenuation factor of signal emitted from x at point y be symmetric: $\ell(x, y) = \ell(y, x), x, y \in \mathbb{R}^2$. The most popular examples are $\ell(x, y) = Pe^{-\beta |x-y|}$ and $\ell(x, y) = \frac{P}{|x-y|^\alpha}, \alpha > 2$. A general theory is developed under the following assumptions:
(i) $\ell(x, y) = \ell(|x - y|)$, $\int_{r_0}^{\infty} r\ell(r)dr < \infty$ for some $r_0 > 0$.
(ii) $l(0) > k\sigma_0^2/P$, $\ell(x) \leq 1$ for all $x > 0$ where $k > 0$ is an admissible level of interference.
(iii) ℓ is continuous and strictly decreasing where it is non-zero.
For each pair of points $x_i, x_j \in \Pi$ define the signal/noise ratio

$$\text{SNR}(x_i \to x_j) = \frac{P\ell^2(x_i, x_j)}{\sigma_0^2 + \gamma \Sigma_{k \neq i,j} P\ell^2(x_k, x_j)} \tag{4.4.22}$$

where $P, \sigma_0^2, k > 0$ and $0 \leq \gamma < \frac{1}{k}$. We say that a transmitter located at x_i can send a message to receiver located at x_j if $\text{SNR}(x_i \to x_j) \geq k$. For any $k > 0$ and $0 < \kappa < 1$, let $A_n(k, \kappa)$ be an event that there exists a set S_n of at least κn points of Π such that for any two points $s, d \in S_n$, $\text{SNR}(s, d) > k$. It can be proved (see [48]) that for all $\kappa \in (0, 1)$ there exists $k = k(\kappa)$ such that

$$\lim_{n \to \infty} \mathbb{P}\left(A_n(k(\kappa), \kappa)\right) = 1. \tag{4.4.23}$$

Then we say that the network is supercritical at interference level $k(\kappa)$; it means that the number of other points the given transmitter (say, located at the origin 0) could communicate to, by using re-transmission at intermediate points, is infinite with a positive probability.

First, we note that any given transmitter may be directly connected to at most $1 + (\gamma k)^{-1}$ receivers. Indeed, suppose that n_x nodes are connected to the node x. Denote by x_1 the node connected to x and such that

$$\ell(|x_1 - x|) \leq \ell(|x_i - x|), i = 2, \ldots, n_x. \tag{4.4.24}$$

Since x_1 is connected to x we have

$$\frac{P\ell(|x_1 - x|)}{\sigma_0^2 + \gamma \sum\limits_{i=2}^{\infty} P\ell(|x_i - x|)} \geq k$$

which implies

$$P\ell(|x_1 - x|) \geq k\sigma_0^2 + k\gamma \sum_{i \geq 2} P\ell(|x_i - x|)$$

$$\geq k\sigma_0^2 + k\gamma(n_x - 1)P\ell(|x_1 - x|) + k\gamma \sum_{i \geq n_x + 1} P\ell(|x_i - x|)$$

$$\geq k\gamma(n_x - 1)P\ell(|x_1 - x|). \tag{4.4.25}$$

We conclude from (4.4.25) that $n_x \leq 1 + (k\gamma)^{-1}$. However, the network *percolates* for some values of parameters in view of (4.4.23). This means that with positive probability a given transmitter may be connected to an infinite number of others with re-transmissions. In particular, the model percolates for $\gamma = 0$, above the critical rate for percolation of Poisson flow λ_{cr}. It may be demonstrated that for $\lambda > \lambda_{cr}$ the critical value of $\gamma^*(\lambda)$ first increases with λ but then starts to decay because the interference becomes too strong. The proof of the following result may be found in [48].

Theorem 4.4.15   Let λ_{cr} be the critical node density for $\gamma = 0$. For any node density $\lambda > \lambda_{cr}$, there exists $\gamma^*(\lambda) > 0$ such that for $\gamma \leq \gamma^*(\lambda)$, the interference model percolates. For $\lambda \to \infty$ we have that

$$\gamma^*(\lambda) = O(\lambda^{-1}). \tag{4.4.26}$$

Another interesting connection with the theory of spatial point processes in $\mathbb{R}^N$ is in using realisations of point processes for producing random codebooks. An alternative (and rather efficient) way to generate a random coding attaining the value $C(\alpha)$ in (4.1.17) is as follows. Take a Poisson process $\Pi^{(N)}$ in $\mathbb{R}^N$, of rate $\lambda_N = e^{NR_N}$ where $R_N \to R$ as $N \to \infty$. Here $R < \frac{1}{2} \log \frac{1}{2\pi e \sigma_0^2}$ where σ_0^2 be the variance of additive Gaussian noise in a channel. Enlist in the codebook $\mathscr{X}_{M,N}$ the random points $X(i)$ from process $\xi^{(N)}$ lying inside the Euclidean ball $\mathbb{B}^{(N)}(\sqrt{N\alpha})$ and surviving the following 'purge'. Fix $r > 0$ (the minimal distance of the random code) and for any point X_j of a Poisson process $\Pi^{(N)}$ generate an IID random variable $T_j \sim \mathrm{U}([0,1])$ (a random mark). Next, for every point X_j of the original Poisson process examine the ball $\mathbb{B}^{(N)}(X_j, r)$ of radius r centred at X_j. The point X_j will survive only if its mark T_j is strictly smaller than the marks of all other points from $\Pi^{(N)}$ lying in $\mathbb{B}^{(N)}(X_j, r)$. The resulting point process $\xi^{(N)}$ is known as the Matérn process; it is an example of a more general construction discussed in the recent paper [1].

The main parameter of a random codebook with codewords $\mathbf{x}^{(N)}$ of length N is the induced distribution of the distance between codewords. In the case of code-books generated by stationary point processes it is convenient to introduce a func-tion $K(t)$ such that $\lambda^2 K(t)$ gives the expected number of ordered pairs of distinct points in a unit volume less than distance t apart. In other words, $\lambda K(t)$ is the ex-pected number of further points within t of an arbitrary point of a process. Say, for Poisson process on $\mathbb{R}^2$ of rate λ, $K(t) = \pi t^2$. In random codebooks we are in-terested in models where $K(t)$ is much smaller for small and moderate t. Hence, random codewords appear on a small distance from one another much more rarely than in a Poisson process. It is convenient to introduce the so-called *product density*

$$\rho(t) = \frac{\lambda^2}{c(t)} \frac{dK(t)}{dt}, \tag{4.4.27}$$

where $c(t)$ depends on the state space of the point process. Say, $c(t) = 2\pi t$ on $\mathbf{R}^1$, $c(t) = 2\pi t^2$ on $\mathbf{R}^2$, $c(t) = 2\pi \sin t$ on the unit sphere, etc.

Some convenient models of this type have been introduced by B. Matérn. Here we discuss two rather intuitive models of point processes on $\mathbb{R}^N$. The first is ob-tained by sampling a Poisson process of rate λ and deleting any point which is within $2R$ of any other whether or not this point has already been deleted. The rate of this process for $N = 2$ is

$$\lambda_{M,1} = \lambda e^{-4\pi\lambda R^2}. \tag{4.4.28}$$

The product density $k(t) = 0$ for $t < 2R$, and

$$\rho(t) = \lambda^2 e^{-2U(t)}, t > 2R,$$

where

$$U(t) = \text{meas}[B((0,0), 2R) \cup B((t,0), 2R)]. \tag{4.4.29}$$

Here $B((0,0), 2R)$ is the ball with centre $(0,0)$ of radius $2R$, and $B((t,0), 2R)$ is the ball with centre $(t,0)$ of radius $2R$. For varying λ this model has the maximum rate of $(4\pi e R^2)^{-1}$ and so cannot model densely packed codes. This is 10% of the theoretical bound $(\sqrt{12R^2})^{-1}$ which is attained by the triangular lattice packing, cf. [1].

The second Matérn model is an example of the so-called marked point process. The points of a Poisson process of rate λ are independently marked by IID random variables with distribution $U([0, 1])$. A point is deleted if there is another point of the process within distance $2R$ which has a bigger mark whether or not this point has already been deleted. The rate of this process for $N = 2$ is

$$\lambda_{M,2} = (1 - e^{-\lambda c})/c, c = U(0) = 4\pi R^2. \tag{4.4.30}$$

The product density $\rho(t) = 0$ for $t < 2R$, and

$$\rho(t) = \frac{2U(t)\left(1 - e^{-4\pi R^2 \lambda}\right) - 2c\left(1 - e^{-\lambda U(t)}\right)}{cU(t)(U(t) - c)}, t > 2R. \tag{4.4.31}$$

An equivalent definition is as follows. Given two points X and Y of the primary Poisson process on the distance $t = |X - Y|$ define the probability $k(t) = \rho(t)/\lambda^2$ that both of them are retained in the secondary process. Then $k(t) = 0$ for $t < 2R$, and

$$k(t) = \frac{2U(t)(1 - e^{-4\pi R^2 \lambda}) - 8\pi R^2 (1 - e^{-\lambda U(t)})}{4\lambda^2 \pi R^2 U(t)(U(t) - 4\pi R^2)}, t > 2R.$$

Example 4.4.16 (Outage probability in a wireless network) Suppose a receiver is located at the origin and transmitters are distributed according to the Matérn hard-core process with the inner radius r_0. We suppose that no transmitters are closer to each other than r_0 and the coverage distance is a. The sum of received powers at the central receiver from the signals from all wireless network is written as

$$X_{r_0} = \sum_{J_{r_0,a}} \frac{P}{r_i^\alpha} \tag{4.4.32}$$

where $J_{r_0,a}$ denotes the set of interfering transmitters such that $r_0 \le r_i < a$. Let λ_P be the rate of Poisson process producing a Matérn process after thinning. The rate of thinned process is

$$\lambda = \frac{1 - \exp\left(-\lambda_P \pi r_0^2\right)}{\pi r_0^2}.$$

Using the Campbell theorem we compute the MGF of X_{r_0}:

$$\phi(\theta) = \mathbb{E}\left(e^{\theta X_{r_0}}\right)$$

$$= \exp\left(\lambda_P \pi (a^2 - r_0^2) \int_0^1 q(t) dt \left[\int_{r_0}^a \frac{2r}{(a^2 - r_0^2)} e^{\theta g(r)} dr - 1\right]\right). \tag{4.4.33}$$

Here $g(r) = \dfrac{P}{r^\alpha}$ and $q(t) = \exp\left(-\lambda_P \pi r_0^2 t\right)$ is the retaining probability of a point

of mark t. Since $\displaystyle\int_0^1 q(t)\mathrm{d}t = \dfrac{\lambda}{\lambda_P}$, we obtain

$$\phi(\theta) = \exp\left(\lambda \pi (a^2 - r_0^2) \left[\int_{r_0}^a \frac{2r}{(a^2 - r_0^2)} e^{\theta g(r)} \mathrm{d}r - 1\right]\right). \tag{4.4.34}$$

Now we can compute all absolute moments of the interfering signal:

$$\mu_k = \lambda \pi \int_{r_0}^a 2r(g(r))^k \mathrm{d}r = \frac{2\lambda \pi}{k\alpha - 2}\left(\frac{P^k}{r_0^{k\alpha - 2}} - \frac{P^k}{a^{k\alpha - 2}}\right). \tag{4.4.35}$$

Engineers say that outage happens at the central receiver, i.e. the interference prevents one from reading a signal obtained from a sender at distance r_s, if

$$\frac{P/r_s^\alpha}{\sigma_0^2 + \Sigma_{J_{r_0,a}} P/r_i^\alpha} \leq k.$$

Here, σ_0^2 is the noise power, r_s is the distance to sender and k is the minimal SIR (signal/noise ratio) required for successful reception. Different approximations of outage probability based on the moments computed in (4.4.35) are developed. Typically, the distribution of X_{r_0} is close to log-normal; see, e.g., [113].

4.5 Selected examples and problems from cryptography

Cryptography, commonly defined as 'the practice and study of hiding information', became a part of many courses and classes on coding; in our exposition we mainly follow the traditions of the Cambridge course *Coding and Cryptography*. We keep the theoretical explanations to the bare minimum and refer the reader to specialised books for details. Cryptography has a long and at times fascinating history where mathematics is interleaved with other sciences and even non-sciences. It has inspired countless fiction and half-fiction books, films, and broadcast programmes; its popularity does not seem to be waning.

A popular method of producing encrypted digit sequences is through the so-called feedback shift registers. We will restrict ourselves to the binary case, working with string spaces $\mathcal{H}_{n,2} = \{0,1\}^n = \mathbb{F}_2^{\times n}$.

Definition 4.5.1   A (general) binary *feedback shift register* of length d is a map $\{0,1\}^d \to \{0,1\}^d$ of the form

$$(x_0,\ldots,x_{d-1}) \mapsto (x_1,\ldots,x_{d-1},f(x_0,\ldots,x_{d-1}))$$

for some function $f : \{0,1\}^d \to \{0,1\}$ (a feedback function). The initial string $(x_0, \ldots, x_{d-1})$ is called an *initial fill*; it produces an *output stream* $(x_n)_{n \geq 0}$ satisfying the recurrence equation

$$x_{n+d} = f(x_n, \ldots, x_{n+d-1}), \quad \text{for all } n \geq 0. \tag{4.5.1}$$

A feedback shift register is said to be *linear* (an LFSR, for short) if function f is linear and $c_0 = 1$:

$$f(x_0, \ldots, x_{d-1}) = \sum_{i=0}^{d-1} c_i x_i, \quad \text{where } c_i = 0, 1, \ c_0 = 1; \tag{4.5.2}$$

in this case the recurrence equation is linear:

$$x_{n+d} = \sum_{i=0}^{d-1} c_i x_{n+i} \text{ for all } n \geq 0. \tag{4.5.3}$$

It is convenient to write (4.5.3) in the matrix form

$$\mathbf{x}_{n+1}^{n+d} = \mathbf{V} \mathbf{x}_n^{n+d-1} \tag{4.5.4}$$

where

$$\mathbf{V} = \begin{pmatrix} 0 & 1 & 0 & \cdots & 0 & 0 \\ 0 & 0 & 1 & \cdots & 0 & 0 \\ \vdots & \vdots & \vdots & \ddots & \vdots & \vdots \\ 0 & 0 & 0 & \cdots & 0 & 1 \\ c_0 & c_1 & c_2 & \cdots & c_{d-2} & c_{d-1} \end{pmatrix}, \quad \mathbf{x}_n^{n+d-1} = \begin{pmatrix} x_n \\ x_{n+1} \\ \vdots \\ x_{n+d-2} \\ x_{n+d-1} \end{pmatrix}. \tag{4.5.5}$$

By the expansion of the determinant along the first column one can see that $\det \mathbf{V} = 1 \bmod 2$: the cofactor for the $(n,1)$ entry c_0 is the matrix $\mathbf{I}_{d-1}$. Hence,

$$\det \mathbf{V} = c_0 \det \mathbf{I}_{d-1} = c_0 = 1, \text{ and the matrix } \mathbf{V} \text{ is invertible.} \tag{4.5.6}$$

A useful concept is the *auxiliary*, or *feedback*, polynomial of an LFSR from (4.5.3):

$$C(X) = c_0 + c_1 X + \cdots + c_{d-1} X^{d-1} + X^d. \tag{4.5.7}$$

Observe that general feedback shift registers, after an initial run, become periodic:

Theorem 4.5.2   *The output stream (x_n) of a general feedback shift register of length d has the property that there exists integer r, $0 \leq r < 2^d$, and integer D, $1 \leq D < 2^d - r$, such that $x_{k+D} = x_k$ for all $k \geq r$.*

Proof A segment $x_M \ldots x_{M+d-1}$ determines uniquely the rest of the output stream in (4.5.1), i.e. $(x_n, n \geq M + d - 1)$. We see that if such a segment is reproduced in the stream, it will be repeated. There are 2^d different possibilities for a string of d subsequent digits. Hence, by the pigeonhole principle, there exists $0 \leq r < R < 2^d$ such that the two segments of length d of the output stream, from positions r and R onwards, will be the same: $x_{r+j} = x_{R+j}$, $0 \leq r < R < d$. Then, as was noted, $x_{r+j} = x_{R+j}$ for all $j \geq 0$, and the assertion holds true with $D = R - r$. $\qquad \square$

In the linear case (LFSR), we can repeat the above argument, with the zero string discarded. This allows us to reduce 2^d to $2^d - 1$. However, an LFSR is periodic in a 'proper sense':

Theorem 4.5.3 *An LFSR* (x_n) *is periodic, i.e. there exists* $D \leq 2^d - 1$ *such that* $x_{n+D} = x_n$ *for all n. The smallest D with this property is called the* period *of the LFSR.*

Proof Indeed, the column vectors $\mathbf{x}_n^{n+d-1}$, $n \geq 0$, are related by the equation $\mathbf{x}_{n+1} = \mathbf{V}\mathbf{x}_n = \mathbf{V}^{n+1}\mathbf{x}_0$, $n \geq 0$, where matrix $\mathbf{V}$ was defined in (4.5.5). We noted that $\det \mathbf{V} = c_0 \neq 0$ and hence $\mathbf{V}$ is invertible. As was said before, we may discard the zero initial fill. For each vector $\mathbf{x}_n \in \{0,1\}^d$ there are only $2^d - 1$ non-zero possibilities. Therefore, as in the proof of Theorem 4.5.2, among the initial $2^d - 1$ vectors $\mathbf{x}_n$, $0 \leq n \leq 2^d - 2$, either there will be repeats, or there will be a zero vector. The second possibility can be again discarded, as it leads to the zero initial fill. Thus, suppose that the first repeat was for j and $D + j$: $\mathbf{x}_j = \mathbf{x}_{j+D}$, i.e. $\mathbf{V}^{j+D}\mathbf{x}_0 = \mathbf{V}^j\mathbf{x}_0$. If $j \neq 0$, we multiply by $\mathbf{V}^{-1}$ and arrive at an earlier repeat. So: $j = 0$, $D \leq 2^d - 1$ and $\mathbf{V}^D\mathbf{x}_0 = \mathbf{x}_0$. Then, obviously, $\mathbf{x}_{n+D} = \mathbf{V}^{n+D}\mathbf{x}_0 = \mathbf{V}^n\mathbf{x}_0 = \mathbf{x}_n$. $\qquad \square$

Worked Example 4.5.4 *Give an example of a general feedback register with output k_j, and initial fill $(k_0, k_1, \ldots, k_N)$, such that*

$$(k_n, k_{n+1}, \ldots, k_{n+N}) \neq (k_0, k_1, \ldots, k_N) \text{ for all } n \geq 1.$$

Solution Take $f \colon \{0,1\}^2 \to \{0,1\}^2$ with $f(x_1, x_2) = x_2 1$. The initial fill 00 yields 00111111111$\ldots$. Here, $k_{n+1} \neq 0 = k_1$ for all $n \geq 1$. $\qquad \square$

Worked Example 4.5.5 *Let matrix $\mathbf{V}$ be defined by (4.5.5), for the linear recursion (4.5.3). Define and compute the characteristic and minimal polynomials for $\mathbf{V}$.*

Solution The characteristic polynomial of matrix $\mathbf{V}$ is $h_{\mathbf{V}}(X) \in \mathbb{F}_2[X] = X \mapsto \det(X\mathbf{I} - \mathbf{V})$:

$$h_{\mathbf{V}}(X) = \det \begin{pmatrix} X & 1 & 0 & \cdots & 0 & 0 \\ 0 & X & 1 & \cdots & 0 & 0 \\ \vdots & \vdots & \vdots & \ddots & \vdots & \vdots \\ 0 & 0 & 0 & \cdots & X & 1 \\ c_0 & c_1 & c_2 & \cdots & c_{d-2} & (c_{d-1}+X) \end{pmatrix} \qquad (4.5.8)$$

(recall, entries 1 and c_i are considered in $\mathbb{F}_2$). Expanding along the bottom row, the polynomial $h_{\mathbf{V}}(t)$ is written as a linear combination of determinants of size $(d-1) \times (d-1)$ (co-factors):

$$c_0 \det \begin{pmatrix} 1 & 0 & \cdots & 0 & 0 \\ X & 1 & \cdots & 0 & 0 \\ \vdots & \vdots & \ddots & \vdots & \vdots \\ 0 & 0 & \cdots & X & 1 \end{pmatrix} + c_1 \det \begin{pmatrix} X & 0 & \cdots & 0 & 0 \\ X & 1 & \cdots & 0 & 0 \\ \vdots & \vdots & \ddots & \vdots & \vdots \\ 0 & 0 & \cdots & X & 1 \end{pmatrix}$$

$$+ \cdots + c_{d-2} \det \begin{pmatrix} X & 1 & \cdots & 0 & 0 \\ 0 & X & \cdots & 0 & 0 \\ \vdots & \vdots & \ddots & \vdots & \vdots \\ 0 & 0 & \cdots & 0 & 1 \end{pmatrix}$$

$$+ (c_{d-1}+X) \det \begin{pmatrix} X & 1 & \cdots & 0 & 0 \\ 0 & X & \cdots & 0 & 0 \\ \vdots & \vdots & \ddots & \vdots & \vdots \\ 0 & 0 & \cdots & 0 & X \end{pmatrix}$$

$$= c_0 + c_1 X + \cdots + c_{d-2} X^{d-2} + (c_{d-1}+X)X^{d-1}$$
$$= \sum_{0 \le i \le d-1} c_i X^i + X^d,$$

which gives the characteristic polynomial $C(X)$ of the recursion.

By the Cayley–Hamilton theorem,

$$h_{\mathbf{V}}(\mathbf{V}) = c_0 \mathbf{I} + c_1 \mathbf{V} + \cdots + c_{d-1} \mathbf{V}^{d-1} + \mathbf{V}^d = \mathbf{O}.$$

The minimal polynomial, $m_{\mathbf{V}}(X)$, of matrix $\mathbf{V}$ is the polynomial of minimal degree such that $m_{\mathbf{V}}(\mathbf{V}) = \mathbf{O}$. It is a divisor of $h_{\mathbf{V}}(X)$, and every root of $h_{\mathbf{V}}(X)$ is a root of $m_{\mathbf{V}}(X)$. The difference between $m_{\mathbf{V}}(X)$ and $h_{\mathbf{V}}(X)$ is in multiplicities: the multiplicity of a root μ of $m_{\mathbf{V}}(X)$ equals the maximal size of the Jordan cell of $\mathbf{V}$ corresponding to μ whereas for $h_{\mathbf{V}}(X)$ it is the sum of the sizes of all Jordan cells in $\mathbf{V}$ corresponding to μ.

To calculate $m_{\mathbf{V}}(X)$, we:

(i) take a basis $\mathbf{e}_1, \ldots, \mathbf{e}_d$ (in $\mathbb{F}_2^{\times d}$);
(ii) then for any vector $\mathbf{e}_j$ we find the minimal number d_j such that vectors $\mathbf{e}_j$, $\mathbf{V}\mathbf{e}_j, \ldots, \mathbf{V}^{d_j}\mathbf{e}_j, \mathbf{V}^{d_j+1}\mathbf{e}_j$ are linearly dependent;
(iii) identify the corresponding linear combination

$$a_0^{(j)}\mathbf{e}_j + a_1^{(j)}\mathbf{V}\mathbf{e}_j + \cdots + a_{d_j}^{(j)}\mathbf{V}^{d_j}\mathbf{e}_j + \mathbf{V}^{d_j+1}\mathbf{e}_j = 0.$$

(iv) Further, we form the corresponding polynomial

$$m_{\mathbf{V}}^{(j)}(X) = \sum_{0 \le i \le d_j} a_i^{(j)} X^i + X^{d_j+1}.$$

(v) Then,

$$m_{\mathbf{V}}(X) = \text{lcm}\left[m_{\mathbf{V}}^{(1)}(X), \ldots, m_{\mathbf{V}}^{(d)}(X)\right].$$

In our case, it is convenient to take

$$\mathbf{e}_j = \begin{pmatrix} 0 \\ \vdots \\ 1 \\ \vdots \\ 0 \end{pmatrix} \begin{matrix} \vdots \\ \sim j. \\ \vdots \end{matrix}$$

Then $\mathbf{V}^j\mathbf{e}_1 = \mathbf{e}_j$, and we obtain that $d_1 = d$, and

$$m_{\mathbf{V}}^{(1)}(X) = \sum_{0 \le i \le d-1} c_i X^i + X^d = h_{\mathbf{V}}(X).$$

$\square$

We see that the feedback polynomial $C(X)$ of the recursion coincides with the characteristic and the minimal polynomial for $\mathbf{V}$. Observe that at $X = 0$ we obtain

$$h_{\mathbf{V}}(0) = C(0) = c_0 = 1 = \det \mathbf{V}. \tag{4.5.9}$$

Any polynomial can be identified through its roots; we saw that such a description may be extremely useful. In the case of an LFSR, the following example is instructive.

Theorem 4.5.6 *Consider the binary linear recurrence in (4.5.3) and the corresponding auxiliary polynomial $C(X)$ from (4.5.7).*

(a) *Suppose $\mathbb{K}$ is a field containing $\mathbb{F}_2$ such that polynomial $C(X)$ has a root α of multiplicity m in $\mathbb{K}$. Then, for all $k = 0, 1, \ldots, m-1$,*

$$x_n = A(n,k)\alpha^n, \quad n = 0, 1, \ldots, \tag{4.5.10}$$

is a solution to (4.5.3) in $\mathbb{K}$, *where*

$$A(n,k) = \begin{cases} 1, & k=0, \\ \left[\prod_{0 \le l \le k-1} (n-l)_+ \right] \bmod 2, & k \ge 1. \end{cases} \quad (4.5.11)$$

Here, and below, $(a)_+$ stands for $\max[a,0]$. In other words, sequence $\mathbf{x}^{(k)} = (x_n)$, where x_n is given by (4.5.10), is an output of the LFSR with auxiliary polynomial $C(X)$.

(b) Suppose $\mathbb{K}$ is a field containing $\mathbb{F}_2$ such that $C(X)$ factorises in $\mathbb{K}$ into linear factors. Let $\alpha_1, \ldots, \alpha_r \in \mathbb{K}$ be distinct roots of $C(X)$ of multiplicities $m_1, \ldots, m_r$, with $\sum_{1 \le i \le r} m_i = d$. Then the general solution of (4.5.3) in $\mathbb{K}$ is

$$x_n = \sum_{1 \le i \le r} \sum_{0 \le k \le m_i - 1} b_{i,k} A(n,k) \alpha_i^n \quad (4.5.12)$$

for some $b_{u,v} \in \mathbb{K}$. In other words, sequences $\mathbf{x}^{(i,k)} = (x_n)$, where $x_n = A(n,k)\alpha_i^n$ and $A(n,k)$ is given by (4.5.11), span the set of all output streams of the LFSR with auxiliary polynomial $C(X)$.

Proof (a) If $C(X)$ has a root $\alpha \in K$ of multiplicity m then $C(X) = (X-\alpha)^m \widetilde{C}(X)$ where $\widetilde{C}(X)$ is a polynomial of degree $d-m$ (with coefficients from a field $\mathbb{K}' \subseteq \mathbb{K}$). Then, for all $k=0,\ldots,m-1$, and for all $n \ge d$, the polynomial

$$D_{k,n}(X) := X^k \frac{\mathrm{d}^k}{\mathrm{d}X^k} \left[X^{n-d} C(X) \right]$$

(with coefficients taken mod 2) vanishes at $X = \alpha$ (in field $\mathbb{K}$):

$$D_{k,n}(\alpha) = \sum_{0 \le i \le d-1} c_i A(n-d+i,k) \alpha^{n-d+i} + A(n,k)\alpha^n.$$

This yields

$$A(n,k)\alpha^n = \sum_{0 \le i \le d-1} c_i A(n-d+i,k)\alpha^{n-d+i}.$$

Thus, stream $\mathbf{x}^{(k)} = (x_n)$ with x_n as in (4.5.10) solves the recursion $x_n = \sum_{0 \le i \le d-1} c_i x_{n-d+i}$ in $\mathbb{K}$. The number of such solutions equals m, the multiplicity of root α.

(b) First, observe that the set of output streams $(x_n)_{n \ge 0}$ forms a linear space $\mathbb{W}$ over $\mathbb{K}$ (in the set of all sequences with entries from $\mathbb{K}$). The dimension of $\mathbb{W}$ equals d, as every stream is uniquely defined by a seed (initial fill) $x_0 x_1 \ldots x_{d-1} \in \mathbb{K}^d$. On the other hand, $d = \sum_{1 \le i \le r} m_i$, the total number of sequences $\mathbf{x}^{(i,k)} = (x_n^{(i,k)})$ with entries

$$x_n^{(i,k)} = A(n,k)\alpha_i^n, \quad n = 0,1,\ldots.$$

Thus, it suffices to check that the streams $\mathbf{x}^{(i,k)}$, where $i = 1,\ldots,r$, $k = 0, 1 \ldots, m_i - 1$, are linearly independent over $\mathbb{K}$.

To this end, take a linear combination $\sum\limits_{1 \leq i \leq r} \sum\limits_{0 \leq k \leq m_i - 1} b_{i,k}\mathbf{x}^{(i,k)}$ and assume it gives $\mathbf{0}$. Let us also agree that sequence $\mathbf{x}^{(i,k)} = \mathbf{0}$ for $k < 0$. It is convenient to introduce a shift operator $\mathbf{x} = (x_n) \mapsto \mathbf{S}\mathbf{x}$ where sequence $\mathbf{S}\mathbf{x} = (x'_n)$ has entries $x'_n = x_{n+1}$, $n = 0, 1, \ldots$. The key observation is as follows. Let $\mathbf{I}$ stand for the identity transformation. Then for all $\beta \in \mathbb{K}$,

$$(\mathbf{S} - \beta\mathbf{I})\mathbf{x}^{(i,k)} = (\alpha_i - \beta)\mathbf{x}^{(i,k)} + k\alpha_i\mathbf{x}^{(i,k-1)}.$$

In fact, the nth entry of the sequence $(\mathbf{S} - \beta\mathbf{I})\mathbf{x}^{(i,k)}$ equals

$$A(n+1,k)\alpha_i^{n+1} - \beta A(k,n)\alpha_i^n$$
$$= [A(n,k) + kA(n,k-1)]\alpha_i^{n+1} - \beta A(n,k)\alpha_i^n$$
$$= (\alpha_i - \beta)A(n,k)\alpha_i^n + k\alpha_i A(n,k-1)\alpha_i^n,$$

in agreement with the above equation for sequences. We have used here the elementary equation

$$A(n+1,k) = A(n,k) + kA(n,k-1).$$

Then, iterating, we obtain

$$(\mathbf{S} - \beta_1\mathbf{I})(\mathbf{S} - \beta_2\mathbf{I})\mathbf{x}^{(i,k)} = (\alpha_i - \beta_1)(\alpha_i - \beta_2)\mathbf{x}^{(i,k)}$$
$$+ k\alpha_i(\alpha_i - \beta_1 + \alpha_i - \beta_2)\mathbf{x}^{(i,k-1)} + k^2\alpha_i^2\mathbf{x}^{(i,k-2)}$$
$$= (\mathbf{S} - \beta_2\mathbf{I})(\mathbf{S} - \beta_1\mathbf{I})\mathbf{x}^{(i,k)},$$

and so on (all operations with coefficients are performed in field $\mathbb{K}$). In particular, with $\beta = \alpha_i$:

$$(\mathbf{S} - \alpha_i\mathbf{I})^l\mathbf{x}^{(i,k)} = \begin{cases} (k\alpha_i)^l\mathbf{x}^{(i,k-l)}, & 1 \leq l \leq k, \\ 0, & l > k. \end{cases}$$

Now consider the product of operators $\prod\limits_{1 \leq i < r} (\mathbf{S} - \alpha_i\mathbf{I})^{m_i}(\mathbf{S} - \alpha_r\mathbf{I})^{m_r - 1}$ applied to our vanishing linear combination $\sum\limits_{1 \leq i \leq r} \sum\limits_{0 \leq k \leq m_i - 1} b_{i,k}\mathbf{x}^{(i,k)}$. The only term that survives comes from the summand $b_{r,m_r-1}\mathbf{x}^{(r,m_r-1)}$. This gives

$$b_{r,m_r-1} \prod\limits_{1 \leq i < r} (\alpha_i - \alpha_r)^{m_i}[(m_r - 1)\alpha_r]^{m_r-1}\mathbf{x}^{(i,0)} = \mathbf{0}.$$

Hence, $b_{r,m_r-1} = 0$. Next, we apply $\prod\limits_{1 \leq i < r} (\mathbf{S} - \alpha_i\mathbf{I})^{m_i}(\mathbf{S} - \alpha_r\mathbf{I})^{m_r-2}$ to obtain that $b_{r,m_r-2} = 0$. Continuing in a similar fashion, we can guarantee that each coefficient $b_{i,k} = 0$. $\square$

Upon seeing a stream of digits $(x_n)_{n\geq 0}$, an observer may wish to determine whether it was produced by an LFSR. This can be done by using the so-called Berlekamp–Massey (BM) algorithm, solving a system of linear equations. If a sequence (x_n) comes from an LFSR with feedback polynomial $C(X) = \sum\limits_{i=0}^{d-1} c_i X^i + X^d$ then the recurrence $x_{n+d} = \sum\limits_{i=0}^{d-1} c_i x_{n+i}$ for $n = 0, \ldots, d$ can be written in a vector-matrix form $\mathbf{A}_d \mathbf{c}_d = \mathbf{0}$ where

$$\mathbf{A}_d = \begin{pmatrix} x_0 & x_1 & x_2 & \cdots & x_d \\ x_1 & x_2 & x_3 & \cdots & x_{d+1} \\ \vdots & \vdots & \vdots & \ddots & \vdots \\ x_d & x_{d+1} & x_{d+2} & \cdots & x_{2d} \end{pmatrix}, \quad \mathbf{c}_d = \begin{pmatrix} c_0 \\ c_1 \\ \vdots \\ c_{d-1} \\ 1 \end{pmatrix}. \tag{4.5.13}$$

Consequently, the $(d+1) \times (d+1)$ matrix $\mathbf{A}_d$ must have determinant 0, and the $(d+1)$-dimensional vector $\mathbf{c}_d$ must lie in the null-space $\ker \mathbf{A}_d$.

The algorithm begins with an inspection of matrix $\mathbf{A}_r$ for a small value of r (known to be $\leq d$):

$$\mathbf{A}_r = \begin{pmatrix} x_0 & x_1 & x_2 & \cdots & x_r \\ x_1 & x_2 & x_3 & \cdots & x_{r+1} \\ \vdots & \vdots & \vdots & \ddots & \vdots \\ x_r & x_{r+1} & x_{r+2} & \cdots & x_{2r} \end{pmatrix}.$$

We calculate $\det \mathbf{A}_r$: if $\det \mathbf{A}_r \neq 0$, we conclude that $d \neq r$ and increase r by 1. If $\det \mathbf{A}_r = 0$ then we solve the equation $\mathbf{A}_r \mathbf{a}_r = 0$, i.e. try $d = r$:

$$\mathbf{A}_d \begin{pmatrix} a_0 \\ a_1 \\ \vdots \\ a_{d-1} \\ 1 \end{pmatrix} = 0, \quad \text{where } \mathbf{A}_d = \begin{pmatrix} x_0 & x_1 & \cdots & x_d \\ x_1 & x_2 & \cdots & x_{d+1} \\ \vdots & \vdots & \ddots & \vdots \\ x_d & x_{d+1} & \cdots & x_{2d} \end{pmatrix}$$

(e.g. by Gaussian elimination) and test sequence (x_n) for the recursion $x_{n+d} = \sum\limits_{0 \leq i \leq d-1} a_i x_{n+i}$. If we discover a discrepancy, we choose a different vector $\mathbf{c}_r \in \ker \mathbf{A}_r$ or – if it fails – increase r.

The BM algorithm can be stated in an elegant algebraic form. Given a sequence (x_n), consider a formal power series in X: $\sum\limits_{j=0}^{\infty} x_j X^j$. The fact that (x_n) is produced by the LFSR with a feedback polynomial $C(X)$ is equivalent to the fact that the above series is obtained by dividing a polynomial $A(X) = \sum\limits_{i=0}^{d} a_i X^i$ by $C(X)$:

$$\sum_{j=0}^{\infty} x_j X^j = \frac{A(X)}{C(X)}. \qquad (4.5.14)$$

Indeed, as $c_0 = 1$, $A(X) = C(X) \sum\limits_{j=0}^{\infty} x_j X^j$ is equivalent to

$$a_n = \sum_{i=1}^{n} c_i x_{n-i}, \quad n = 1, \ldots, \qquad (4.5.15)$$

or

$$x_n = \begin{cases} a_n - \sum\limits_{i=1}^{n-1} c_i x_{n-i}, & n = 0, 1, \ldots, d, \\ -\sum\limits_{i=0}^{n-1} c_i x_{n-i}, & n > d. \end{cases} \qquad (4.5.16)$$

In other words, $A(X)$ takes part in specifying the initial fill, and $C(X)$ acts as the feedback polynomial.

Worked Example 4.5.7   *What is a linear feedback shift register? Explain the Berlekamp–Massey method for recovering the feedback polynomial of a linear feedback shift register from its output. Illustrate in the case when we observe outputs*

$$1\,0\,1\,0\,1\,1\,0\,0\,1\,0\,0\,0\,\ldots,$$

$$0\,1\,0\,1\,1\,1\,1\,0\,0\,0\,1\,0\,\ldots$$

and

$$1\,1\,0\,0\,1\,0\,1\,1.$$

Solution An initial fill $x_0 \ldots x_{d-1}$ produces an output stream $(x_n)_{n \geq 0}$ satisfying the recurrence equation

$$x_{n+d} = \sum_{i=0}^{d-1} c_i x_{n+i} \text{ for all } n \geq 0.$$

The feedback polynomial

$$C(X) = c_0 + c_1 X + \cdots + c_{d-1} X^{d-1} + X^d$$

is the characteristic polynomial for this recurrence equation determining its solutions. We will assume that coefficient $c_0 \neq 0$; otherwise value x_n has no impact on x_{n+d} and the register can be treated as the one of length $d - 1$.

The Berlekamp–Massey algorithm begins with an inspection of matrix

$$\mathbf{A}_1 = \begin{pmatrix} 1 & 0 \\ 0 & 1 \end{pmatrix}, \quad \text{with } \det \mathbf{A}_1 \neq 0,$$

but

$$\mathbf{A}_2 = \begin{pmatrix} 1 & 0 & 1 \\ 0 & 1 & 0 \\ 1 & 0 & 1 \end{pmatrix}, \quad \text{with } \det \mathbf{A}_2 = 0,$$

and $\mathbf{A}_2 \begin{pmatrix} c_0 \\ c_1 \\ 1 \end{pmatrix} = 0$ has the solution $c_0 = 1, c_1 = 0$. This gives the recursion

$$x_{n+2} = x_n,$$

which does not fit the remaining digits. So, we move to $\mathbf{A}_3$:

$$\mathbf{A}_3 = \begin{pmatrix} 1 & 0 & 1 & 0 \\ 0 & 1 & 0 & 1 \\ 1 & 0 & 1 & 1 \\ 0 & 1 & 1 & 0 \end{pmatrix}, \quad \text{with } \det \mathbf{A}_3 \neq 0,$$

and then to $\mathbf{A}_4$:

$$\mathbf{A}_4 = \begin{pmatrix} 1 & 0 & 1 & 0 & 1 \\ 0 & 1 & 0 & 1 & 1 \\ 1 & 0 & 1 & 1 & 0 \\ 0 & 1 & 1 & 0 & 0 \\ 1 & 1 & 0 & 0 & 1 \end{pmatrix}, \quad \text{with } \det \mathbf{A}_4 = 0.$$

The equation $\mathbf{A}_4 \mathbf{c}_4 = 0$ is solved by $\mathbf{c}_4 = \begin{pmatrix} 1 \\ 0 \\ 0 \\ 0 \\ 1 \end{pmatrix}$. This yields

$$x_{n+4} = x_n + x_{n+3},$$

which fits the rest of the string. In the second example we have:

$$\det \begin{pmatrix} 0 & 1 \\ 1 & 0 \end{pmatrix} \neq 0, \det \begin{pmatrix} 0 & 1 & 0 \\ 1 & 0 & 1 \\ 0 & 1 & 1 \end{pmatrix} \neq 0, \det \begin{pmatrix} 0 & 1 & 0 & 1 \\ 1 & 0 & 1 & 1 \\ 0 & 1 & 1 & 1 \\ 1 & 1 & 1 & 1 \end{pmatrix} \neq 0$$

and

$$\begin{pmatrix} 0 & 1 & 0 & 1 & 1 \\ 1 & 0 & 1 & 1 & 1 \\ 0 & 1 & 1 & 1 & 1 \\ 1 & 1 & 1 & 1 & 0 \\ 1 & 1 & 1 & 0 & 0 \end{pmatrix} \begin{pmatrix} 1 \\ 1 \\ 0 \\ 0 \\ 1 \end{pmatrix} = 0.$$

This yields the solution: $d = 4$, $x_{n+4} = x_n + x_{n+1}$. The linear recurrence relation is satisfied by every term of the output sequence given. The feedback polynomial is then $X^4 + X + 1$.

In the third example the recursion is $x_{n+3} = x_n + x_{n+1}$. $\qquad\qquad\qquad$ □

LFSRs are used for producing *additive stream ciphers*. Additive stream ciphers were invented in 1917 by Gilbert Vernam, at the time an engineer with the AT&T Bell Labs. Here, the sending party uses an output stream from an LFSR (k_n) to encrypt a plain text (p_n) by (z_n) where

$$z_n = p_n + k_n \bmod 2, \ \ n \geq 0. \qquad\qquad (4.5.17)$$

The recipient would decrypt it by

$$p_n = z_n + k_n \bmod 2, \ \ n \geq 0, \qquad\qquad (4.5.18)$$

but of course he must know the initial fill $k_0 \ldots k_{d-1}$ and the string $c_0 \ldots c_{d-1}$. The main deficiency of the stream cipher is its periodicity. Indeed, if the generating LFSR has period D then it is enough for an 'attacker' to have in his possession a cipher text $z_0 z_1 \ldots z_{2D-1}$ and the corresponding plain text $p_0 p_1 \ldots p_{2D-1}$, of length $2D$. (Not an unachievable task for a modern-day Sherlock Holmes.) If by some luck the attacker knows the value of the period D then he only needs $z_0 z_1 \ldots z_{D-1}$ and $p_0 p_1 \ldots p_{D-1}$. This will allow the attacker to break the cipher, i.e. to decrypt the whole plain text, however long.

Clearly, short-period LFSRs are easier to break when they are used repeatedly. The history of World War II and the subsequent Cold War has a number of spectacular examples (German code-breakers succeeding in part in reading British Navy codes, British and American code-breakers succeeding in breaking German codes, the American project 'Venona' deciphering Soviet codes) achieved because of intensive message traffic. However, even ultra-long periods cannot guarantee safety.

As far as this section of the book is concerned, the period of an LFSR can be increased by combining several LFSRs.

Theorem 4.5.8 *Suppose a stream (x_n) is produced by an LFSR of length d_1, period D_1 and with an auxiliary polynomial $C_1(X)$, and a stream (y_n) by an LFSR* of length d_2, period D_2 and with an auxiliary polynomial $C_2(X)$. Let $\alpha_1, \ldots, \alpha_{r_1}$ and $\beta_1, \ldots, \beta_{r_2}$ be the distinct roots of $C_1(X)$ and $C_2(X)$, respectively, lying in some field $\mathbb{K} \supset \mathbb{F}_2$. Let m_i be the multiplicity of root α_i and m'_j be the multiplicity of root β_j, with $d_1 = \sum_{1 \leq i \leq r_1} m_i$ and $d_2 = \sum_{1 \leq i \leq r_2} m'_i$. Then

(a) Stream $(x_n + y_n)$ is produced by an LFSR with the auxiliary polynomial $\mathrm{lcm}(C_1(X), C_2(X))$.

(b) Stream $(x_n y_n)$ is produced by an LFSR with the auxiliary polynomial $C(X) = \prod_{1 \leq i \leq r_1} \prod_{1 \leq j \leq r_2} (X - \alpha_i \beta_j)^{m_i + m'_j - 1}$.

In particular, the period of the resulting LFSR is in both cases divisible by $\mathrm{lcm}(D_1, D_2)$.

Proof According to Theorem 4.5.6, the output streams (x_n) and (y_n) for the LFSRs in question have the following form in field $\mathbb{K}$:

$$x_n = \sum_{1 \leq i \leq r_1} \sum_{0 \leq k \leq m_i - 1} a_{i,k} A(n,k) \alpha_i^n, \quad y_n = \sum_{1 \leq j \leq r_2} \sum_{0 \leq l \leq m'_j - 1} b_{j,l} A(n,l) \beta_j^n, \quad (4.5.19)$$

for some $a_{i,k}, b_{j,l} \in \mathbb{K}$.

(a) Writing $x_n + y_n$ as the sum of the expressions from (4.5.19) and grouping similar terms leads to the statement (a).

(b) For the product $x_n y_n$ we have the expression

$$\sum_{i,j} \sum_{k,l} a_{i,k} b_{j,l} A(n,k) A(n,l) (\alpha_i \beta_j)^n.$$

The product $a_{i,k} b_{j,l} A(n,k) A(n,l)$ can be written as a sum $\sum_{k \wedge l \leq t \leq k+l-1} A(n,t) u_t(a_{i,k}, b_{j,l})$ where coefficients $u_t(a_{i,k}, b_{j,l}) \in \mathbb{K}$. This gives the following representation of $x_n y_n$:

$$\sum_{1 \leq i \leq r_1} \sum_{1 \leq j \leq r_2} \sum_{0 \leq t \leq m_i + m'_j - 2} A(n,t) \sum_{k,l: k \wedge l \leq t \leq k+l-1} u_t(a_{i,k}, b_{j,l}) (\alpha_i \beta_j)^n$$

which in turn can be written as

$$x_n y_n = \sum_{1 \leq i \leq r_1} \sum_{1 \leq j \leq r_2} \sum_{0 \leq t \leq m_i + m'_j - 2} A(n,t) v_{i,j;t} (\alpha_i \beta_j)^n$$

corresponding to the generic form of the output stream for the LFSR with the auxiliary polynomial $C(X)$ in statement (b). □

Despite serious drawbacks, LFSRs remain in use in a variety of situations: they allow simple enciphering and deciphering without 'lookahead' and display a 'local' effect of an error, be it encoding, transmission or decoding. More generally, non-linear LFSRs often offer only marginal advantages while bringing serious disadvantages, in particular with deciphering.

> ... *an error by the same example*
> *Will rush into the state.*
> William Shakespeare (1564–1616), English playwright and poet
> from *Merchant of Venice*

Worked Example 4.5.9 (a) *Let (x_n), (y_n), (z_n) be three streams produced by LFSRs. Set*

$$k_n = x_n \text{ if } y_n = z_n,$$

$$k_n = y_n \text{ if } y_n \neq z_n.$$

Show that k_n is also a stream produced by a linear feedback register.

(b) *A cipher stream is given by a linear feedback register of known length d. Show that, given plain text and ciphered text of length $2d$, we can find the cipher stream.*

Solution (a) For three streams (x_n), (y_n), (z_n) produced by LFSRs we set

$$k_n = x_n + (x_n + y_n)(y_n + z_n) \quad \text{(in } \mathbb{F}_2\text{)}.$$

So it suffices to note that (pointwise) sums and products of streams produced by LFSRs also yield some streams produced by LFSRs.

(b) Suppose the plain text is $y_1 \, y_2 \ldots y_{2d}$, and the ciphered text is $x_1 + y_1 \, x_2 + y_2 \ldots x_{2d} + y_{2d}$. Then we can recover $x_1 \ldots x_{2d}$. We know that $c_1 \ldots c_d$ must satisfy d simultaneous linear equations

$$x_{d+j} = \sum_{i=1}^{d} c_i x_{j+i-1}, \quad \text{for } j = 1, 2, \ldots, d.$$

Solve these to find $c_1, c_2, \ldots, c_d$ and hence the cipher stream. $\square$

Worked Example 4.5.10 *A binary non-linear feedback register of length 4 has defining relation*

$$x_{n+1} = x_{n-1} + x_n x_{n-2} + x_{n-3}.$$

Show that the state space contains cycles of lengths 1, 4, 9 and 2.

Solution There are $2^4 = 16$ initial binary strings. By inspection,

$$0000 \mapsto 0000 \qquad\qquad\qquad \text{(a cycle of length 1),}$$

$$0001 \mapsto 0010 \mapsto 0100 \mapsto 1000 \mapsto 0001 \qquad \text{(a cycle of length 4),}$$

$$0011 \mapsto 0111 \mapsto 1111 \mapsto 1110 \mapsto 1101$$

$$\mapsto 1011 \mapsto 0110 \mapsto 1100 \mapsto 1001 \mapsto 0011 \quad \text{(a cycle of length 9),}$$

$$0101 \mapsto 1010 \mapsto 0101 \qquad\qquad\qquad \text{(a cycle of length 2).}$$

All 16 initial fills have appeared in the list, so the analysis is complete. $\qquad\square$

Worked Example 4.5.11 *Describe how an additive stream cipher operates. What is a one-time pad? Explain briefly why a one-time pad is safe if used only once but becomes unsafe if used many times. A one-time pad is used to send the message $x_1x_2x_3x_4x_5x_6y_7$ which is encoded as 0101011. By mistake, it is reused to send the message $y_0x_1x_2x_3x_4x_5x_6$ which is encoded as 0100010. Show that $x_1x_2x_3x_4x_5x_6$ is one of two possible messages, and find the two possibilities.*

Solution A one-time pad is an example of a cipher based on a *random key* and proposed by Gilbert Vernam and Joseph Mauborgne (the Chief of the USA Signal Corps during World War II). The cipher uses a random number generator producing a sequence $k_1k_2k_3\ldots$ from the alphabet J of size q. More precisely, each letter is uniformly distributed over J and different letters are independent. A message $m = a_1a_2\ldots a_n$ is encrypted as $c = c_1c_2\ldots c_n$ where

$$c_i = a_i + k_i \pmod{q}.$$

To show that the one-time pad achieves perfect secrecy, write

$$\mathbb{P}(M = m, C = c) = \mathbb{P}(M = m, K = c - m)$$

$$= \mathbb{P}(M = m)\mathbb{P}(K = c - m) = \mathbb{P}(M = m)\frac{1}{q^n};$$

here the subtraction $c - m$ is digit-wise and mod q. Hence, the conditional probability

$$\mathbb{P}(C = c | M = m) = \frac{\mathbb{P}(M = m, C = c)}{\mathbb{P}(M = m)} = \frac{1}{q^n}$$

does not depend on m. Hence, M and C are independent.

Working in $\mathbb{F}_2$, consider a cipher key stream $k_1\ k_2\ k_3\ldots$. The plain (input) text stream $p_1\ p_2\ p_3\ldots$ is encrypted as the cipher text stream $c_1\ c_2\ c_3\ldots$, where $c_j = p_j + k_j$. If the k_j are IID random numbers and the cipher key stream is only used once (which happens in practice) then we have a one-time pad. (It is assumed that the cipher key stream is known only to the sender and the recipient.) In the example, we have

$$x_1 x_2 x_3 x_4 x_5 x_6 y_7 \mapsto 0101011,$$
$$y_0 x_1 x_2 x_3 x_4 x_5 x_6 \mapsto 0100010.$$

Suppose $x_1 = 0$. Then

$$k_0 = 0, k_1 = 1, x_2 = 0, k_2 = 0, x_3 = 0, k_3 = 0, x_4 = 1, k_1 = 0,$$
$$x_5 = 1, k_5 = 0, x_6 = 1, k_6 = 1.$$

Thus,

$$k = 0100101, \quad x = 000111.$$

If $x_1 = 1$, every digit changes, so

$$k = 1011010, \quad x = 111000.$$

Alternatively, set $x_0 = y_0$ and $x_7 = y_7$. If the first cipher is $q_1 q_2 \ldots$, the second is $p_1 p_2 \ldots$ and the one-time pad is $k_1, k_2, \ldots$, then

$$q_j = x_{j+1} + k_j, \quad p_j = x_j + k_j.$$

So,

$$x_j + x_{j+1} = q_j + p_j,$$

and

$$x_1 + x_2 = 0, \quad x_2 + x_3 = 0,$$
$$x_3 + x_4 = 1, \quad x_4 + x_5 = 0, \quad x_5 + x_6 = 0.$$

This yields

$$x_1 = x_2 = x_3, \quad x_4 = x_5 = x_6, \quad x_4 = x_3 + 1.$$

The message is 000111 or 111000. □

Worked Example 4.5.12   (a) *Let $\theta : \mathbb{Z}_+ \to \{0, 1\}$ be given by $\theta(n) = 1$ if n is odd, $\theta(n) = 0$ if n is even. Consider the following recurrence relation over $\mathbb{F}_2$:*

$$u_{n+3} + u_{n+2} + u_{n+1} + u_n = 0. \tag{4.5.20}$$

Is it true that the general solution of (4.5.20) is $u_n = A + B\theta(n) + C\theta(n^2)$? If it is true, prove it. If not, explain why it is false and state and prove the correct result.

(b) *Solve the recurrence relation $u_{n+2} + u_n = 1$ over $\mathbb{F}_2$, subject to $u_0 = 1, u_1 = 0$, expressing the solution in terms of θ and n.*

(c) *Four streams w_n, x_n, y_n, z_n are produced by linear feedback registers. If we set*

$$k_n = \begin{cases} x_n + y_n + z_n & \text{if } z_n + w_n = 1, \\ x_n + w_n & \text{if } z_n + w_n = 0, \end{cases}$$

show that k_n is also a stream produced by a linear feedback register.

Solution (a) Observe that $\theta(n^2) = \theta(n)$, so the suggested sum contains only two arbitrary constants. Now consider $g(n) = \theta\left(n(n-1)/2\right)$. Then

$$g(n+3) + g(n+2) + g(n+1) + g(n)$$
$$= \theta\left(\frac{(n+3)(n+2)}{2}\right) + \theta\left(\frac{(n+2)(n+1)}{2}\right)$$
$$+ \theta\left(\frac{(n+1)n}{2}\right) + \theta\left(\frac{n(n-1)}{2}\right)$$
$$= \theta\left((n+2)^2 + n^2\right) = 0,$$

and $g(0) = g(1) = 0, g(2) = 1$. Then we substitute $n = 0$ and $n = 1$ into the relation $a\theta(n) + b + cg(n) = 0$, and observe that $a = b = c = 0$. So, $\theta(n), 1, g(n)$ are independent. Thus, $A\theta(n) + B + Cg(n)$ is a general solution of the third-order difference equation.

(b) First try to solve the recurrence relation $u_{n+2} + u_n = 1$ without additional conditions

$$g(n) + g(n+2) = \theta\left(\frac{n(n-1)}{2} + \frac{(n+2)(n+1)}{2}\right)$$
$$= \theta\left(\frac{n^2 - n + n^2 + 3n + 2}{2}\right)$$
$$= \theta\left(n^2 + n + 1\right) = 1.$$

Now substitute $n = 0$ and $n = 1$ into relation $u_n = A + B\theta(n) + g(n)$ to get $A = B = 1$. Thus, $u_n = 1 + \theta(n) + g(n)$.

(c) The sequence k_n is produced by the linear register

$$k_n = x_n + w_n + (z_n + w_n)(y_n + z_n + w_n).$$

$\square$

In the next part of this section, we discuss properties of a class of cryptosystems used in modern practice and called *public-key ciphers*, focusing in particular on the RSA and the bit commitment cryptosystems.

Definition 4.5.13   We say that a formal *cryptosystem* is given, if we can identify:

(a) a set $\mathscr{P}$ of *plaintexts* (source messages in the language of Chapter 1);

(b) a set $\mathscr{C}$ of *ciphertexts* (codewords in the language of Chapter 1);

c) a set $\mathscr{K}$ of *keys* that label the encoding maps;

(d) the set $\mathscr{E}$ of *encryptic functions* (encoding maps) where each function E_k takes $P \in \mathscr{P} \mapsto E_k(P) \in \mathscr{C}$ and is labelled by an element $k \in \mathscr{K}$;

(e) the set $\mathscr{D}$ of *decryptic functions* (decoding maps) where each function D_k takes $C \in \mathscr{C} \mapsto D_k(C) \in \mathscr{P}$ and is again labelled by an element $k \in \mathscr{K}$;

such that

(f) for all key $e \in \mathcal{K}$ there is a key $d \in \mathcal{K}$, with the property that $D_d(E_e(P)) = P$ for all plaintext $P \in \mathcal{P}$.

Example 4.5.14 Suppose that two parties, Bob and Alice, intend to have a two-side private communication. They want to exchange their keys, E_A and E_B, by using an insecure binary channel. An obvious protocol is as follows. Alice encrypts a plain-text m as $E_A(m)$ and sends it to Bob. He encrypts it as $E_B(E_A(m))$ and returns it to Alice. Now we make a crucial assumption that E_A and E_B commute for any plaintext m' : $E_A \circ E_B(m') = E_B \circ E_A(m')$. In this case Alice can decrypt this message as $D_A(E_A(E_B(m))) = E_B(m)$ and send this to Bob, who then calculates $D_B(E_B(m)) = m$. Under this protocol, at no time during the transaction is an unencrypted message transmitted.

However, a further thought shows that this is no solution at all. Indeed, suppose that Alice uses a one-time pad k_A and Bob uses a one-time pad k_B. Then any single interception provides no information about plaintext m. However, if all three transmissions are intercepted, it is enough to take the sum

$$(m + k_A) + (m + k_A + k_B) + (m + k_B) = m$$

to obtain the plaintext m. So, more sophisticated protocols should be developed: this is where public key cryptosystems are helpful.

Another popular example is a network of investors and brokers dealing in a market and using an open access cryptosystem such as RSA. An investor's concern is that a broker will buy shares without her authorisation and, in the case of a loss, claim that he had a written request from the client. Indeed, it is easy for a broker to generate a coded order requesting to buy the stocks as the encoding key is in the public domain. On the other hand, a broker may be concerned that if he buys the shares by the investor's request and the market goes down, the investor may claim that she never ordered this transaction and that her coded request is a fake.

However, it is easy to develop a protocol which addresses these concerns. An investor Alice sends to a broker Bob, together with her request p to buy shares, her 'electronic signature' $f_B f_A^{-1}(p)$. After receiving this message Bob sends a receipt r encoded as $f_A f_B^{-1}(r)$. If a conflict emerges, both sides can provide a third party (say, a court) with these coded messages and the keys. Since no-one but Alice could generate the message coded by $f_B f_A^{-1}$ and no-one but Bob could generate the message coded by $f_A f_B^{-1}$, no doubts would remain. This is the gist of *bit commitment*. The above-mentioned RSA (Rivest–Shamir–Adelman) scheme is a prime example of a public key cryptosystem. Here, a recipient user (Bob, possibly a collective entity) sets

$$N = pq, \text{ where } p \text{ and } q \text{ are two large primes, kept secret.} \quad (4.5.21)$$

Number N is often called the RSA modulus (and made public). The value of the Euler totient function is

$$\phi(N) = (p-1)(q-1), \text{ kept secret.}$$

Next, the recipient user chooses (or is given by the key centre) an integer l such that

$$1 < l < \phi(N) \text{ and } \gcd(\phi(N), l) = 1. \quad (4.5.22)$$

Finally, an integer d is computed (again, by Bob or on his behalf) such that

$$1 < d < \phi(N) \text{ and } ld = 1 \mod \phi(N). \quad (4.5.23)$$

[The value of d can be computed via the extended Euclid's algorithm.] The public key e_B used for encryption is the pair (N, l) (listed in the public directory). The sender (Alice), when communicating to Bob, understands that Bob's plaintext and ciphertext sets are $\mathscr{P} = \mathscr{C} = \{1, \ldots, N-1\}$. She then encrypts her chosen plaintext $m = 1, \ldots, N-1$ as the ciphertext

$$E_{N,l}(m) = c \text{ where } c = m^l \mod N. \quad (4.5.24)$$

Bob's private key d_B is the pair (N, d) (or simply number d): it is kept secret from public but made known to Bob. The recipient decrypts ciphertext c as

$$D_d(c) = c^d \mod N. \quad (4.5.25)$$

In the literature, l is often called the encryption and d the decryption exponent. Theorem 4.5.15 below guarantees that

$$D_d(c) = m^{dl} = m \mod N, \quad (4.5.26)$$

i.e. the ciphertext c is decrypted correctly. More precisely,

Theorem 4.5.15 *For all integers $m = 0, \ldots, N-1$, the equation (4.5.26) holds true, where l and d satisfy (4.5.22) and (4.5.23) and N is as in (4.5.21).*

Proof By virtue of (4.5.23),

$$ld = 1 + b(p-1)(q-1)$$

where b is an integer. Then

$$(m^l)^d = m^{ld} = m^{1+b(p-1)(q-1)} = m\left(m^{(p-1)}\right)^{(q-1)b}.$$

Recall the Euler–Fermat theorem: *If* $\gcd(m, p) = 1$ *then* $m^{\phi(p)} = 1 \bmod p$. *We deduce that if m is not divisible by p then*

$$(m^l)^d = m \bmod p. \tag{4.5.27}$$

Otherwise, i.e. when $p|m$, (4.5.27) still holds as m and $(m^l)^d$ are both equal to $0 \bmod p$. By a similar argument,

$$(m^l)^d = m \bmod q. \tag{4.5.28}$$

By the Chinese remainder theorem (CRT) – [28], [114] – (4.5.27) and (4.5.28) imply (4.5.26). □

Example 4.5.16   Suppose Bob has chosen $p = 29$, $q = 31$, with $N = 899$ and $\phi(N) = 840$. The smallest possible value of e with $\gcd(l, \phi(N)) = 1$ is $l = 11$, after that 13 followed by 17, and so on. The (extended) Euclid algorithm yields $d = 611$ for $l = 11$, $d = 517$ for $l = 13$, and so on. In the first case, the encrypting key $E_{899,11}$ is

$$m \mapsto m^{11} \bmod 899, \text{ that is, } E_{899,11}(2) = 250.$$

The ciphertext 250 is decoded by

$$D_{611}(250) = 250^{611} \bmod 899 = 2,$$

with the help of the computer. [The computer is needed even after the simplification rendered by the use of the CRT. For instance, the command in *Mathematica* is PowerMod[250,611,899].]

Worked Example 4.5.17   (a) *Referring to the RSA cryptosystem with public key* (N, l) *and private key* $(\phi(N), d)$, *discuss possible advantages or disadvantages of taking* (i) $l = 2^{32} + 1$ *or* (ii) $d = 2^{32} + 1$.

(b) *Let a (large) number N be given, and we know that N is a product of two distinct prime numbers,* $N = pq$, *but we do not know the numbers p and q. Assume that another positive integer, m, is given, which is a multiple of* $\phi(N)$. *Explain how to find p and q.*

(c) *Describe how to solve the bit commitment problem by means of the RSA.*

Solution Using $l = 2^{32} + 1$ provides fast encryption (you need just 33 multiplications using repeated squaring). With $d = 2^{32} + 1$ one can decrypt messages quickly (but an attacker can easily guess it).

(b) Next, we show that if we know a multiple m of $\phi(N)$ then it is 'easy' to factor N. Given positive integers $y > 1$ and $M > 1$, denote by $\mathrm{ord}_M(y)$ the order of y relative to M:

$$\mathrm{ord}_M(y) = \min\left[s = 1, 2, \ldots : y^s = 1 \bmod M\right].$$

Assume that $m = 2^a b$ where $a \geq 0$ and b is odd. Set

$$\mathbb{X} = \left\{x = 1, 2, \ldots, N : \mathrm{ord}_p(x^b) \neq \mathrm{ord}_q(x^b)\right\}. \tag{4.5.29}$$

Given N, l and d, we put $m = dl - 1$. As $\phi(N)|dl - 1$ we can use Lemma 4.5.18 below to factor N. We select $x < N$. Suppose $\gcd(x, N) = 1$; otherwise the search is already successful. The probability of finding a non-trivial factor is $1/2$, so the probability of failure after r random choices of $x \in \mathbb{X}$ is $1/2^r$.

(c) The bit commitment problem arises in the following case: Alice sends a message to Bob in such a way that

 (i) Bob cannot read the message until Alice sends further information;
(ii) Alice cannot change the message.

A solution is to use the electronic signature: Bob cannot read the message until Alice (later) reveals her private key. This does not violate conditions (i), (ii) and makes it (legally) impossible for Alice to refuse acknowledging her authorship.

$\square$

Lemma 4.5.18 (i) *Let $N = pq$, m be as before, i.e. $\phi(N)|m$, and define the set $\mathbb{X}$ as in (4.5.29). If $x \in \mathbb{X}$ then there exists $0 \leq t < a$ such that $\gcd\left(x^{2^t b} - 1, N\right) > 1$ is a non-trivial factor of $N = pq$.*
(ii) *The cardinality $\sharp\mathbb{X} \geq \phi(N)/2$.*

Proof (i) Put $y = x^b \bmod N$. The Euler–Fermat theorem implies that $x^{\phi(N)} \equiv 1 \bmod N$ and hence $y^{2^a} \equiv 1 \bmod N$. Then

$$\mathrm{ord}_p(x^b) \text{ and } \mathrm{ord}_q(x^b) \text{ are powers of } 2.$$

As we know, $\mathrm{ord}_p(x^b) \neq \mathrm{ord}_q(x^b)$; say, $\mathrm{ord}_p(x^b) < \mathrm{ord}_q(x^b)$. Then there exists $0 \leq t < a$ such that

$$y^{2^t} \equiv 1 \bmod p, \quad y^{2^t} \not\equiv 1 \bmod q.$$

So, $\gcd\left(y^{2^t} - 1, N\right) = p$, as required.

(ii) By the CRT, there is a bijection

$$x \in \{1, \ldots, N\} \quad \leftrightarrow \quad (x \bmod p, \ x \bmod q) \in \{1, \ldots, p\} \times \{1, \ldots, q\},$$

with the agreement that $N \leftrightarrow (p, q)$. Then it suffices to show that if we partition set $\{1, \ldots, p\}$ into subsets according to the value of $\mathrm{ord}_p(x^b)$, $x \in \mathbb{X}$, then each subset has size $\leq (p-1)/2$. We will do this by exhibiting such a subset of size $(p-1)/2$. Note that

$$\phi(N)|2^a b \text{ implies that there exists } \gamma \in \{1,\ldots,p-1\}$$
$$\text{such that } \text{ord}_p(\gamma^b) \text{ is a power of 2.}$$

In turn, the latter statement implies that

$$\text{ord}_p(\gamma^{\delta b}) \begin{cases} = \text{ord}_p(\gamma^b), & \delta \text{ odd}, \\ < \text{ord}_p(\gamma^b), & \delta \text{ even}. \end{cases}$$

Therefore, $\{\gamma^{\delta b} \bmod p : \delta \text{ odd}\}$ is the required subset. □

Our next example of a cipher is the *Rabin*, or *Rabin–Williams* cryptosystem. Here, again, one uses the factoring problem to provide security. For this system, the relation with the factoring problem has been proved to be mutual: knowing the solution to the factoring problem breaks the cryptosystem, and the ability of breaking the cryptosystem leads to factoring. [That is not so in the case with the RSA: it is not known whether breaking the RSA enables one to solve the factoring problem.]

In the Rabin system the recipient user (Alice) chooses at random two large primes, p and q, with

$$p = q = 3 \bmod 4. \tag{4.5.30}$$

Furthermore:

Alice's public key is $N = pq$; her secret key is the pair (p,q);
Alice's plaintext and ciphertext are numbers $m = 0, 1 \ldots, N-1$, $\quad$ (4.5.31)
and her encryption rule is $E_N(m) = c$ where $c = m^2 \bmod N$.

To decrypt a ciphertext c addressed to her, Alice computes

$$m_p = c^{(p+1)/4} \bmod p \text{ and } m_q = c^{(q+1)/4} \bmod q. \tag{4.5.32}$$

Then

$$\pm m_p = c^{1/2} \bmod p \text{ and } \pm m_q = c^{1/2} \bmod q,$$

i.e. $\pm m_p$ and $\pm m_q$ are the square roots of $c \bmod p$ and $\bmod q$, respectively. In fact,

$$\left(\pm m_p\right)^2 = c^{(p+1)/2} = c^{(p-1)/2} c = \left(\pm m_p\right)^{p-1} c = c \bmod p;$$

at the last step the Euler–Fermat theorem has been used. The argument for $\pm m_q$ is similar. Then Alice computes, via Euclid's algorithm, integers $u(p)$ and $v(q)$ such that

$$u(p)p + v(q)q = 1.$$

Finally, Alice computes

$$\pm r = \pm\big[u(p)pm_q + v(q)qm_p\big] \bmod N$$

and

$$\pm s = \pm\big[u(p)pm_q - v(q)qm_p\big] \bmod N.$$

These are four square roots of $c \bmod N$. The plaintext m is one of them. To secure that she can identify the original plaintext, Alice may reduce the plaintext space $\mathscr{P}$, allowing only plaintexts with some special features (like the property that their first 32 and last 32 digits are repetitions of each other), so that it becomes unlikely that more than one square root has this feature. However, such a measure may result in a reduced difficulty of breaking the cipher as it will be not always true that the 'reduced' problem is equivalent to factoring.

> *I have often admired the mystical way of Pythagoras*
> *and the secret magic of numbers.*
> Thomas Browne (1605–1682), English author who wrote on
> medicine, religion, science and the esoteric

Example 4.5.19   Alice uses prime numbers $p = 11$ and $q = 23$. Then $N = 253$. Bob encrypts the message $m = 164$, with

$$c = m^2 \bmod N = 78.$$

Alice calculates $m_p = 1, m_q = 3, u(p) = -2, v(q) = 1$. Then Alice computes

$$r = \pm[u(p)pm_q + v(q)qm_p] \bmod N = 210 \text{ and } 43$$

$$s = \pm[u(p)pm_q - v(q)qm_p] \bmod N = 164 \text{ and } 89$$

and finds out the message $m = 164$ among the solutions: $164^2 = 78 \bmod 253$.

We continue with the *Diffie–Hellman key exchange scheme*. Diffie and Hellman proposed a protocol enabling a pair of users to exchange secret keys via insecure channels. The Diffie–Hellman scheme is not a public-key cryptosystem but its importance has been widely recognised since it forms a basis for the ElGamal signature cryptosystem.

The Diffie–Hellman protocol is related to the discrete logarithm problem (DLP): we are given a prime number p, field $\mathbb{F}_p$ with the multiplicative group $\mathbb{F}_p^* \simeq \mathbb{Z}_{p-1}$ and a generator γ of $\mathbb{F}_p^*$ (i.e. a primitive element in $\mathbb{F}_p^*$). Then, for all $b \in \mathbb{F}_p^*$, there exists a unique $\alpha \in \{0, 1, \ldots, p-2\}$ such that

$$b = \gamma^\alpha \bmod p. \tag{4.5.33}$$

Then α is called the *discrete logarithm*, mod p, of b to base γ; some authors write $\alpha = \mathrm{dlog}_\gamma b$ mod p. Computing discrete logarithms is considered a difficult problem: no efficient (polynomial) algorithm is known, although there is no proof that it is indeed a non-polynomial problem. [In an additive cyclic group $\mathbb{Z}/(n\mathbb{Z})$, the DLP becomes $b = \gamma\alpha$ mod n and is solved by the Euclid algorithm.]

The Diffie–Hellman protocol allows Alice and Bob to establish a common secret key using field tables for $\mathbb{F}_p$, for a sufficient quantity of prime numbers p. That is, they know a primitive element γ in each of these fields. They agree to fix a large prime number p and a primitive element $\gamma \in \mathbb{F}_p$. The pair (p, γ) may be publicly known: Alice and Bob can fix p and γ through the insecure channel.

Next, Alice chooses $a \in \{0, 1, \ldots, p-2\}$ at random, computes

$$A = \gamma^a \bmod p$$

and sends A to Bob, keeping a secret. Symmetrically, Bob chooses $b \in \{0, 1, \ldots, p-2\}$ at random, computes

$$B = \gamma^b \bmod p$$

and sends B to Alice keeping b secret. Then

Alice computes B^a mod p and Bob computes A^b mod p,

and their secret key is the common value

$$K = \gamma^{ab} = B^a = A^b \bmod p.$$

The attacker may intercept p, γ, A and B but knows

neither $a = \mathrm{dlog}_\gamma A$ mod p nor $b = \mathrm{dlog}_\gamma B$ mod p.

If the attacker can find discrete logarithms mod p then he can break the secret key: this is the only known way to do so. The opposite question – solving the discrete logarithm problem if he is able to break the protocol – remains open (it is considered an important problem in public key cryptography).

However, like previously discussed schemes, the Diffie–Hellman protocol has a particular weak point: it is vulnerable to the *man in the middle* attack. Here, the attacker uses the fact that neither Alice nor Bob can verify that a given message really comes from the opposite party and not from a third party. Suppose the attacker can intercept all messages between Alice and Bob. Suppose he can impersonate Bob and exchange keys with Alice pretending to be Bob and at the same time impersonate Alice and exchange keys with Bob pretending to be Alice. It is necessary to use electronic signatures to distinguish this forgery.

We conclude Section 4.5 with the *ElGamal cryptosystem* based on *electronic signatures*. The ElGamal cipher can be considered a development of the Diffie–Hellman protocol. Both schemes are based on the difficulty of the discrete logarithm problem (DLP). In the ElGamal system, a recipient user, Alice, selects a prime number p and a primitive element $\gamma \in \mathbb{F}_p$. Next, she chooses, at random, an exponent $a \in \{0, \ldots, p-2\}$, computes

$$A = \gamma^a \bmod p$$

and announces/broadcasts

the triple (p, γ, A), her public key.

At the same time, she keeps in secret

exponent a, her private key.

Alice's plaintext set $\mathscr{P}$ is numbers $0, 1, \ldots, p-1$.

Another user Bob, wishing to send a message to Alice and knowing triple (p, γ, A), chooses, again at random, an exponent $b \in \{0, 1, \ldots, p-2\}$, and computes

$$B = \gamma^b \bmod p.$$

Then Bob lets Alice know B (which he can do by broadcasting value B). The value B will play the role of Bob's 'signature'. In contrast, the value b of Bob's exponent is kept secret.

Now, to send to Alice a message $m \in \{0, 1, \ldots, p-1\}$, Bob encrypts m by the pair

$$E_b(m) = (B, c) \quad \text{where} \quad c = A^b m \bmod p.$$

That is, Bob's ciphertext consists of two components: the encrypted message c and his *signature B*.

Clearly, values A and B are parts of the Diffie–Hellman protocol; in this sense the latter can be considered as a part of the ElGamal cipher. Further, the encrypted message c is the product of m by A^b, the factor combining part A of Alice's public key and Bob's exponent b.

When Alice receives the ciphertext (B, c) she uses her secret key a. Namely, she divides c by $B^a \bmod p$. A convenient way is to calculate $x = p - 1 - a$: as $1 \le a \le p-2$, the value x also satisfies $1 \le x \le p-2$. Then Alice decrypts c by $B^x c \bmod p$. This yields the original message m, since

$$B^x c = \gamma^{b(p-1-a)} A^b m = \left(\gamma^{p-1}\right)^b \left(\gamma^a\right)^{-b} A^b m = A^{-b} A^b m = m \bmod p.$$

Example 4.5.20 With $p = 37$, $\gamma = 2$ and $a = 12$ we have

$$A = \gamma^a \bmod p = 26$$

and Alice's public key is $(p = 37, \gamma = 2, A = 26)$, her plaintexts are $0, 1, \ldots, 36$ and private key $a = 12$. Assume Bob has chosen $b = 32$; then

$$B = 2^{32} \bmod 37 = 4.$$

Suppose Bob wants to send $m = 31$. He encrypts m by

$$c = A^b m \bmod p = (26)^{32} m \bmod 37 = 10 \times 31 \bmod 37 = 14.$$

Alice decodes this message as $2^{32} = 7$ and $7^{24} = 26 \bmod 37$,

$$14 \times 2^{32(37-12-1)} \bmod 37 = 14 \times 7^{24} = 14 \times 26 \bmod 37 = 31.$$

Worked Example 4.5.21 *Suppose that Alice wants to send the message 'today' to Bob using the ElGamal encryption. Describe how she does this using the prime $p = 15485863$, $\gamma = 6$ a primitive root mod p, and her choice of $b = 69$. Assume that Bob has private key $a = 5$. How does Bob recover the message using the Mathematica program?*

Solution Bob has public key $(15485863, 6, 7776)$, which Alice obtains. She converts the English plaintext using the alphabet order to the numerical equivalent: $19, 14, 3, 0, 24$. Since $26^5 < p < 26^6$, she can represent the plaintext message as a single 5-digit base 26 integer:

$$m = 19 \times 26^4 + 14 \times 26^3 + 3 \times 26^2 + 0 \times 26 + 24 = 8930660.$$

Now she computes $\gamma^b = 6^{69} = 13733130 \bmod 15485863$, then

$$m\gamma^{ab} = 8930660 \times 7776^{69} = 4578170 \bmod 15485863.$$

Alice sends $c = (13733130, 4578170)$ to Bob. He uses his private key to compute

$$(\gamma^b)^{p-1-a} = 13733130^{15485863-1-5} = 2620662 \bmod 15485863$$

and

$$(\gamma)^{-a} m\gamma^{ab} = 2620662 \times 4578170 = 8930660 \bmod 15485863,$$

and converts the message back to the English plaintext. □

Worked Example 4.5.22 (a) *Describe the Rabin–Williams scheme for coding a message x as x^2 modulo a certain N. Show that, if N is chosen appropriately, breaking this code is equivalent to factorising the product of two primes.*

(b) *Describe the RSA system associated with a public key e, a private key d and the product N of two large primes.*

Give a simple example of how the system is vulnerable to a homomorphism attack. Explain how a signature system prevents such an attack. Explain how to factorise N when e, d and N are known.

Solution (a) Fix two large primes $p, q \equiv -1 \bmod 4$ which forms a private key; the broadcasted public key is the product $N = pq$. The properties used are:

(i) If p is a prime, the congruence $a^2 \equiv d \bmod p$ has at most two solutions.

(ii) For a prime $p = -1 \bmod 4$, i.e. $p = 4k - 1$, if the congruence $a^2 \equiv c \bmod p$ has a solution then $a \equiv c^{(p+1)/4} \bmod p$ is one solution and $a \equiv -c^{(p+1)/4} \bmod p$ is another solution. [Indeed, if $c \equiv a^2 \bmod p$ then, by the Euler–Fermat theorem, $c^{2k} = a^{4k} = a^{(p-1)+2} = a^2 \bmod p$, implying $c^k = \pm a$.]

The message is a number m from $\mathcal{M} = \{0, 1, \ldots, N-1\}$. The encrypter (Bob) sends (broadcasts) $\tilde{m} = m^2 \bmod N$. The decrypter (Alice) uses property (ii) to recover the two possible values of $m \bmod p$ and two possible values of $m \bmod q$. The CRT then yields four possible values for m: three of them would be incorrect and one correct.

So, if one can factorise N then the code would be broken. Conversely, suppose that we can break the code. Then we can find all four distinct square roots $u_1, u_2, u_3, u_4 \bmod N$ for a general u. (The CRT plus property (i) shows that u has zero or four square roots unless it is a multiple of p and q.) Then $u_j u^{-1}$ (calculable via Euclid's algorithm) gives rise to the four square roots, $1, -1, \varepsilon_1$ and ε_2, of $1 \bmod N$, with

$$\varepsilon_1 \equiv 1 \bmod p, \quad \varepsilon_1 \equiv -1 \bmod q$$

and

$$\varepsilon_2 \equiv -1 \bmod p, \quad \varepsilon_2 \equiv 1 \bmod q.$$

By interchanging p and q, if necessary, we may suppose we know ε_1. As $\varepsilon_1 - 1$ is divisible by p and not by q, the $\gcd(\varepsilon_1 - 1, N) = p$; that is, p can be found by Euclid's algorithm. Then q can also be identified.

In practice, it can be done as follows. Assuming that we can find square roots mod N, we pick x at random and solve the congruence $x^2 \equiv y^2 \bmod N$. With probability $1/2$, we have $x \not\equiv \pm y \bmod N$. Then $\gcd(x - y, N)$ is a non-trivial factor of N. We repeat the procedure until we identify a factor; after k trials the probability of success is $1 - 2^{-k}$.

(b) To define the RSA cryptosystem let us randomly choose large primes p and q. By Fermat's little theorem,

$$x^{p-1} \equiv 1 \bmod p, \quad x^{q-1} \equiv 1 \bmod q.$$

Thus, by writing $N = pq$ and $\lambda(N) = \mathrm{lcm}(p-1, q-1)$, we have

$$x^{\lambda(N)} \equiv 1 \bmod N,$$

for all integers x coprime to N.

Next, we choose e randomly. Either Euclid's algorithm will reveal that e is not co-prime to $\lambda(N)$ or we can use Euclid's algorithm to find d such that

$$de \equiv 1 \bmod \lambda(N).$$

With a very high probability a few trials will give appropriate d and e.

We now give out the value e of the public key and the value of N but keep secret the private key d. Given a message m with $1 \le m \le N-1$, it is encoded as the integer c with

$$1 \le c \le N-1 \text{ and } c \equiv m^e \bmod N.$$

Unless m is not co-prime to N (an event of negligible probability), we can decode by observing that

$$m \equiv m^{de} \equiv c^d \bmod N.$$

As an example of a homomorphism attack, suppose the system is used to transmit a number m (dollars to be paid) and someone knowing this replaces the coded message c by c^2. Then

$$(c^2)^d \equiv m^{2de} \equiv m^2$$

and the recipient of the (falsified) message believes that m^2 dollars are to be paid.

Suppose that a signature $B(m)$ is also encoded and transmitted, where B is a many-to-one function with no simple algebraic properties. Then the attack above will produce a message and signature which do not correspond, and the recipient will know that the message was tampered with.

Suppose e, d and N are known. Since

$$de - 1 \equiv 0 \pmod{\lambda(N)}$$

and $\lambda(N)$ is even, $de-1$ is even. Thus $de-1 = 2^a b$ with b odd and $a \ge 1$.

Choose x at random. Set $z \equiv x^b \bmod N$. By the CRT, z is a square root of 1 mod $N = pq$ if and only if it is a square root of 1 mod p and q. As $\mathbb{F}_2$ is a field,

$$x^2 \equiv 1 \bmod p \Leftrightarrow (x-1)(x+1) \equiv 0 \bmod p$$
$$\Leftrightarrow (x-1) \equiv 0 \bmod p \text{ or } (x+1) \equiv 0 \bmod p.$$

Thus 1 has four square roots $w \bmod N$ satisfying $w \equiv \pm 1 \bmod p$ and $w \equiv \pm 1 \bmod q$. In other words,

$w \equiv 1 \bmod N$, $w \equiv -1 \bmod N$,
$w \equiv w_1 \bmod N$ with $w_1 \equiv 1 \bmod p$ and $w_1 \equiv -1 \bmod q$
or
$w \equiv w_2 \pmod N$ with $w_2 \equiv -1 \bmod p$ and $w_1 \equiv 1 \bmod q$.

Now z (the square root of $1 \bmod N$) cannot satisfy $z \equiv 1 \bmod N$. If $w \equiv -1 \bmod N$, we are unlucky and try again. Otherwise we know that $z+1$ is not congruent to $0 \bmod N$ but is divisible by one of the two prime factors of N. Applying Euclid's algorithm yields the common factor. Having found one prime factor, we can find the other one by division or by looking at $z-1$.

Since square roots of 1 are algebraically indistinguishable, the probability of this method's failure tends to 0 rapidly with the number of trials. $\qquad\square$

4.6 Additional problems for Chapter 4

Problem 4.1 (a) *Let $(N_t)_{t\geq 0}$ be a Poisson process of rate $\lambda > 0$ and $p \in (0,1)$. Suppose that each jump in (N_t) is counted as type one with probability p and type two with probability $1-p$, independently for different jumps and independently of the Poisson process. Let $M_t^{(1)}$ be the number of type-one jumps and $M_t^{(2)} = N_t - M_t^{(1)}$ the number of type-two jumps by time t. What is the joint distribution of the pair of processes $(M_t^{(1)})_{t\geq 0}$ and $(M_t^{(2)})_{t\geq 0}$? What if we fix probabilities $p_1,\ldots,p_m$ with $p_1 + \cdots + p_m = 1$ and consider m types instead of two?*

(b) *A person collects coupons one at a time, at jump times of a Poisson process $(N_t)_{t\geq 0}$ of rate λ. There are m types of coupons, and each time a coupon of type j is obtained with probability p_j, independently of the previously collected coupons and independently of the Poisson process. Let T be the first time when a complete set of coupon types is collected. Show that*

$$\mathbb{P}(T < t) = \prod_{j=1}^{m}\left(1 - e^{-p_j\lambda t}\right). \qquad (4.6.1)$$

Let $L = N_T$ be the total number of coupons collected by the time the complete set of coupon types is obtained. Show that $\lambda\,\mathbb{E}T = \mathbb{E}L$. Hence, or otherwise, deduce that $\mathbb{E}L$ does not depend on λ.

Solution Part (a) directly follows from the definition of a Poisson process.

(b) Let T_j be the time of the first collection of a type j coupon. Then $T_j \sim \mathrm{Exp}(p_j\lambda)$, independently for different j. We have

$$T = \max\left[T_1,\ldots,T_m\right],$$

and hence

$$\mathbb{P}(T < t) = \mathbb{P}\left(\max\left[T_1,\ldots,T_m\right] < t\right) = \prod_{j=1}^{m}\mathbb{P}(T_j < t) = \prod_{j=1}^{m}\left(1 - e^{-p_j\lambda t}\right).$$

Next, observe that the random variable L counts the jumps in the original Poisson process (N_t) until the time of collecting a complete set of coupon types. That is:

$$T = \sum_{i=1}^{L} S_i,$$

where $S_1, S_2, \ldots$ are the holding times in (N_t), with $S_j \sim \text{Exp}(\lambda)$, independently for different j. Then

$$\mathbb{E}(T|L = n) = n\mathbb{E}S_1 = n\lambda^{-1}.$$

Moreover, L is independent of the random variables $S_1, S_2, \ldots$. Thus,

$$\mathbb{E}T = \sum_{n \geq m} \mathbb{P}(L = n)\mathbb{E}(T|L = n) = \mathbb{E}S_1 \sum_{n \geq m} n\mathbb{P}(L = n) = \lambda^{-1}\mathbb{E}L.$$

But

$$\lambda\mathbb{E}T = \lambda \int_0^\infty \mathbb{P}(T > t)dt$$

$$= \lambda \int_0^\infty \left[1 - \prod_{j=1}^{m}\left(1 - e^{-p_j\lambda t}\right)\right]dt$$

$$= \int_0^\infty \left[1 - \prod_{j=1}^{m}\left(1 - e^{-p_j t}\right)\right]dt,$$

and the RHS does not depend on λ.

Equivalently, L is identified as the number of collections needed for collecting a complete set of coupons when collections occur at positive integer times $t = 1, 2, \ldots$, with probability p_j of obtaining a coupon of type j, regardless of the results of previous collections. In this construction, λ does not figure, so the mean $\mathbb{E}L$ does not depend on λ (as, in fact, the whole distribution of L). $\qquad\qquad\square$

Problem 4.2 *Queuing systems are discussed in detail in PSE II. We refer to this topic occasionally as they provide a rich source of examples in point processes. Consider a system of k queues in series, each with infinitely many servers, in which, for $i = 1, \ldots, k-1$, customers leaving the ith queue immediately arrive at the $(i+1)$th queue. Arrivals to the first queue form a Poisson process of rate λ. Service times at the ith queue are all independent with distribution F, and independent of service times at other queues, for all i. Assume that initially the system is empty and write $V_i(t)$ for the number of customers at queue i at time $t \geq 0$. Show that $V_1(t), \ldots, V_k(t)$ are independent Poisson random variables.*

In the case $F(t) = 1 - e^{-\mu t}$ show that

$$\mathbb{E}V_i(t) = \frac{\lambda}{\mu}\mathbb{P}(N_t \geq i), \quad t \geq 0, \quad i = 1, \ldots, k, \tag{4.6.2}$$

where $(N_t)_{t \geq 0}$ is a Poisson process of rate μ.

Suppose now that arrivals to the first queue stop at time T. Determine the mean number of customers at the ith queue at each time $t \geq T$.

Solution We apply the product theorem to the Poisson process of arrivals with random vectors $Y_n = (S_n^1, \ldots, S_n^k)$ where S_n^i is the service time of the nth customer at the ith queue. Then

$$V_i(t) = \text{the number of customers in the } i\text{th queue at time t}$$

$$= \sum_{n=1}^{\infty} \mathbf{1}\Big(\text{the } n\text{th customer arrived in the first queue at}$$

$$\text{time } J_n \text{ is in the } i\text{th queue at time } t\Big)$$

$$= \sum_{n=1}^{\infty} \mathbf{1}\Big(J_n > 0, \ S_n^1, \ldots, S_n^k \geq 0,$$

$$J_n + S_n^1 + \cdots + S_n^{i-1} < t < J_n + S_n^1 + \cdots + S_n^i\Big)$$

$$= \sum_{n=1}^{\infty} \mathbf{1}\Big[(J_n, (S_n^1, \ldots, S_n^k)) \in A_i(t)\Big] = M(A_i(t)).$$

Here $(J_n : n \in \mathbb{N})$ denote the jump times of a Poisson process of rate λ, and the measures M and v on $(0, \infty) \times \mathbf{R}_+^k$ are defined by

$$M(A) = \sum_{n=1}^{\infty} \mathbf{1}\big((J_n, Y_n) \in A\big), \ \ A \subset (0, \infty) \times \mathbf{R}_+^k$$

and

$$v\big((0, t] \times B\big) = \lambda t \mu(B).$$

The product theorem states that M is a Poisson random measure on $(0, \infty) \times \mathbf{R}_+^k$ with intensity measure v. Next, the set $A_i(t) \subset (0, \infty) \times \mathbb{R}_+^k$ is defined by

$$A_i(t) = \big\{(\tau, s^1, \ldots, s^k) : 0 < \tau < t, \ s^1, \ldots, s^k \geq 0$$

$$\text{and } \tau + s^1 + \cdots + s^{i-1} \leq t < \tau + s^1 + \cdots + s^i\big\}$$

$$= \Big\{(\tau, s^1, \ldots, s^k) : 0 < \tau < t, \ s^1, \ldots, s^k \geq 0$$

$$\text{and } \sum_{l=1}^{i-1} s^l \leq t - \tau < \sum_{l=1}^{i} s^l\Big\}.$$

Sets $A_i(t)$ are pairwise disjoint for $i = 1, \ldots, k$ (as $t - \tau$ can fall between subsequent partial sums $\sum_{l=1}^{i-1} s^l$ and $\sum_{l=1}^{i} s^l$ only once). So, the random variables $V_i(t)$ are independent Poisson.

A direct verification is through the joint MGF. Namely, let $N_t \sim \mathrm{Po}(\lambda t)$ be the number of arrivals at the first queue by time t. Then write

$$M_{V_1(t),\ldots,V_k(t)}(\theta_1,\ldots,\theta_k) = \mathbb{E}\exp\left(\theta_1 V_1(t) + \cdots + \theta_k V_k(t)\right)$$

$$= \mathbb{E}\left[\mathbb{E}\exp\left(\sum_{i=1}^{k}\theta_i V_i(t)\,\Big|\,N_t;\, J_1,\ldots,J_{N_t}\right)\right].$$

In turn, given $n = 1, 2, \ldots$ and points $0 < \tau_1 < \cdots < \tau_n < t$, the conditional expectations is

$$\mathbb{E}\exp\left(\sum_{i=1}^{k}\theta_i V_i(t)\,\Big|\,N_t = n;\, J_1 = \tau_1,\ldots,J_n = \tau_n\right)$$

$$= \mathbb{E}\exp\left(\sum_{i=1}^{k}\theta_i\sum_{j=1}^{n}\mathbf{1}\left[(\tau_j,(S_j^1,\ldots,S_j^k)) \in A_i(t)\right]\right)$$

$$= \mathbb{E}\exp\left(\sum_{j=1}^{n}\sum_{i=1}^{k}\theta_i\mathbf{1}\left[(\tau_j,(S_j^1,\ldots,S_j^k)) \in A_i(t)\right]\right)$$

$$= \prod_{j=1}^{n}\mathbb{E}\exp\left(\sum_{i=1}^{k}\theta_i\mathbf{1}\left[(\tau_j,(S_j^1,\ldots,S_j^k)) \in A_i(t)\right]\right).$$

Next, perform summation over n and integration over $\tau_1,\ldots,\tau_n$:

$$\mathbb{E}\left[\mathbb{E}\exp\left(\sum_{i=1}^{k}\theta_i V_i(t)\,\Big|\,N_t;\, J_1,\ldots,J_{N_t}\right)\right] = \sum_{n=1}^{\infty}\lambda^n e^{-\lambda t}\int_0^t\int_0^{\tau_n}\cdots\int_0^{\tau_2}$$

$$\times\prod_{j=1}^{n}\mathbb{E}\exp\left(\sum_{i=1}^{k}\theta_i\mathbf{1}\left[(\tau_j,(S_j^1,\ldots,S_j^k)) \in A_i(t)\right]\right)d\tau_1\cdots d\tau_{n-1}d\tau_n$$

$$= \sum_{n=1}^{\infty}\frac{\lambda^n}{n!}e^{-\lambda t}\left(\int_0^t\mathbb{E}\exp\left(\sum_{i=1}^{k}\theta_i\mathbf{1}\left[(\tau,(S^1,\ldots,S^k)) \in A_i(t)\right]\right)d\tau\right)^n$$

$$= \exp\left(\lambda\int_0^t\left[\mathbb{E}\exp\left(\sum_{i=1}^{k}\theta_i\mathbf{1}\left[(\tau,(S^1,\ldots,S^k)) \in A_i(t)\right]\right) - 1\right]d\tau\right)$$

$$= \exp\left[\lambda\int_0^t\sum_{i=1}^{k}\mathbb{P}\left((\tau,(S^1,\ldots,S^k)) \in A_i(t)\right)(e^{\theta_i} - 1)d\tau\right]$$

$$= \prod_{i=1}^{k}\exp\left[(e^{\theta_i} - 1)\lambda\int_0^t\mathbb{P}\left(\sum_{l=1}^{i-1}S^l < t - \tau < \sum_{l=1}^{i}S^l\right)d\tau\right].$$

By the uniqueness of a random variable with a given MGF, this implies that

$$V_i(t) \sim \mathrm{Po}\left(\lambda\int_0^t\mathbb{P}\left(\sum_{l=1}^{i-1}S^l < t - \tau < \sum_{l=1}^{i}S^l\right)d\tau\right), \quad \text{independently.}$$

If $F(t) = 1 - e^{-\mu t}$ then partial sums $S_1, S_1 + S_2, \ldots$ mark the subsequent points of a Poisson process $(\tilde{N}_s)$ of rate μ. In this case, $\mathbb{E}V_i(t) = v(A_i(t))$ equals

$$\lambda \int_0^t \mathbb{P}\left(\sum_{l=1}^{i-1} S^l \leq t - \tau < \sum_{l=1}^{i} S^l \right) d\tau = \lambda \int_0^t \mathbb{P}(\tilde{N}_{t-\tau} = i - 1) d\tau$$

$$= \lambda \mathbb{E} \int_0^t \mathbf{1}(\tilde{N}_s = i - 1) ds = \frac{\lambda}{\mu} \mathbb{P}(\tilde{N}_t \geq i).$$

Finally, write $V_i(t, T)$ for the number of customers in queue i at time t after closing the entrance at time T. Then

$$\mathbb{E}V_i(t, T) = \lambda \int_0^T \mathbb{P}(\tilde{N}_{t-\tau} = i - 1) d\tau = \lambda \mathbb{E} \int_{t-T}^t \mathbf{1}(\tilde{N}_s = i - 1) ds$$

$$= \frac{\lambda}{\mu} \big[\mathbb{P}(\tilde{N}_t \geq i) - \mathbb{P}(\tilde{N}_{t-T} \geq i) \big].$$

$\square$

Problem 4.3   The arrival times of customers at a supermarket form a Poisson process of rate λ. Each customer spends a random length of time, S, collecting items to buy, where S has PDF $(f(s,t)\colon s \geq 0)$ for a customer arriving at time t. Customers behave independently of one another. At a checkout it takes time $g(S)$ to buy the items collected. The supermarket has a policy that nobody should wait at the checkout, so more tills are made available as required. Find

 (i) the probability that the first customer has left before the second has arrived,
(ii) the distribution of the number of checkouts in use at time T.

Solution (i) If J_1 is the arrival time of the first customer then $J_1 + S_1$ is the time he enters the checkout till and $J_1 + S_1 + g(S_1)$ the time he leaves. Let J_2 be the time of arrival of the second customer. Then $J_1, J_2 - J_1 \sim \mathrm{Exp}(\lambda)$, independently.
   Then

$$\mathbb{P}(S_1 + g(S_1) < J_2 - J_1) = \int_0^\infty dt_1 \lambda e^{-\lambda t_1} \int_0^\infty dt_2 \lambda e^{-\lambda t_2} \int_0^{t_2} ds_1 f(s_1, t_1) \mathbf{1}(s_1 + g(s_1) < t_2)$$

$$= \int_0^\infty dt_1 \lambda e^{-\lambda t_1} \int_0^\infty ds_1 f(s_1, t_1) \int_{s_1 + g(s_1)}^\infty dt_2 \lambda e^{-\lambda t_2}$$

$$= \int_0^\infty dt_1 \lambda e^{-\lambda t_1} \int_0^\infty ds_1 f(s_1, t_1) e^{-\lambda(s_1 + g(s_1))}.$$

(ii) Let N_T^{ch} be the number of checkouts used at time T. By the product theorem, 4.4.11, $N_T^{\text{ch}} \sim \text{Po}(\Lambda(T))$ where

$$\Lambda(T) = \lambda \int_0^T du \int_0^\infty ds\, f(s, u)\mathbf{1}(u + s < T,\ u + s + g(s) > T)$$

$$= \lambda \int_0^T du \int_0^\infty ds\, f(s, u)\mathbf{1}(T - g(s) < u + s < T).$$

In fact, if $N_T^{\text{arr}} \sim \text{Po}(\lambda T)$ is the number of arrivals by time T, then

$$N_T^{\text{ch}} = \sum_{i=1}^{N_T^{\text{arr}}} \mathbf{1}(J_i + S_i < T < J_i + S_i + g(S_i)),$$

and the MGF

$$\mathbb{E} \exp\left(\theta N_T^{\text{ch}}\right) = \mathbb{E}\left[\mathbb{E}\left(\exp\left(\theta N_T^{\text{ch}}\right) \mid N_T^{\text{arr}};\ J_1, \ldots, J_{N_T^{\text{arr}}}\right)\right]$$

$$= e^{-\lambda T} \sum_{k=0}^\infty \lambda^k \int_0^T \int_0^{t_k} \cdots \int_0^{t_2} \prod_{i=1}^k \mathbb{E}\exp\left[\theta\mathbf{1}(t_i + S_i < T\right.$$
$$\left. < t_i + S_i + g(S_i))\right] dt_1 \cdots dt_k$$

$$= e^{-\lambda T} \sum_{k=0}^\infty \frac{\lambda^k}{k!} \int_0^T \cdots \int_0^T \prod_{i=1}^k \mathbb{E}\exp\left[\theta\mathbf{1}(t_i + S_i < T\right.$$
$$\left. < t_i + S_i + g(S_i))\right] dt_1 \cdots dt_k$$

$$= e^{-\lambda T} \sum_{k=0}^\infty \frac{\lambda^k}{k!} \left(\int_0^T \mathbb{E}\exp\left[\theta\mathbf{1}(t + S < T < t + S + g(S))\right]\right)^k$$

$$= \exp\left[\lambda \int_0^T \mathbb{E}\left(\exp\left[\theta\mathbf{1}(t + S < T < t + S + g(S))\right] - 1\right) dt\right]$$

$$= \exp\left[\lambda(e^\theta - 1) \int_0^T \mathbb{P}(t + S < T < t + S + g(S)) dt\right]$$

$$= \exp\left[(e^\theta - 1)\lambda \int_0^T \int_0^\infty f(s, u)\mathbf{1}(u + s < T < u + s + g(s)) ds\, du\right],$$

which verifies the claim. $\qquad \square$

Problem 4.4 *A library is open from 9am to 5pm. No student may enter after 5pm; a student already in the library may remain after 5pm. Students arrive at the library in the period from 9am to 5pm in the manner of a Poisson process of rate λ. Each student spends in the library a random amount of time, H hours, where*

$0 \le H \le 8$ is a random variable with PDF h and $\mathbf{E}[H] = 1$. The periods of stay of different students are IID random variables.

(a) Find the distribution of the number of students who leave the library between 3pm and 4pm.

(b) Prove that the mean number of students who leave between 3pm and 4pm is $\mathbf{E}[\min(1,(7-H)_+)]$, where w_+ denotes $\max[w,0]$.

(c) What is the number of students still in the library at closing time?

Solution The library is open from 9am to 5pm. Students arrive as a PP(λ). The problem is equivalent to an $M/GI/\infty$ queue (until 5pm, when the restriction of no more arrivals applies, but for problems involving earlier times this is unimportant).

Denote by J_n the arrival time of the nth student using the 24 hour clock.

Denote by H_n the time the nth student spends in the library.

Again use the product theorem, 4.4.11, for the random measure on $(0,8) \times (0,8)$ with atoms (J_n, Y_n), where $(J_n : n \in \mathbf{N})$ are the arrival times and $(Y_n : n \in \mathbf{N})$ are periods of time that students stay in the library. Define measures on $(0,\infty) \times \mathbf{R}_+$ by $\mu((0,t) \times B) = \lambda t \mu(B)$, $N(A) = \sum_n \mathbf{1}_{((J_n,H_n)\in A)}$. Then N is a Poisson random measure with intensity $v([0,t] \times [0,y]) = \lambda t F(y)$, where $F(y) = \int_0^y h(x)dx$ (the time $t = 0$ corresponds to 9am).

(a) Now, the number of students leaving the library between 3pm and 4pm (i.e. $6 \le t \le 7$) has a Poisson distribution Po($v(A)$) where $A = \{(r,s) : s \in [0,7], r \in [6-s, 7-s] \text{ if } s \le 6; r \in [0, 7-s] \text{ if } s > 6\}$. Here

$$v(A) = \int_0^8 \lambda dF(r) \int_{(6-r)_+}^{(7-r)_+} ds = \int_0^8 \lambda \left[(7-r)_+ - (6-r)_+\right] dF(r).$$

So, the distribution of students leaving the library between 3pm and 4pm is Poisson with rate $= \lambda \int_0^7 [(7-y)_+ - (6-y)_+]dF(r)$.

(b)

$$(7-y)_+ - (6-y)_+ = \begin{cases} 0, & \text{if } y \ge 7, \\ 7-y, & \text{if } 6 \le y \le 7, \\ 1, & \text{if } y \le 6. \end{cases}$$

The mean number of students leaving the library between 3pm and 4pm is

$$v(A) = \int_0^8 \lambda\,[\min(1,(7-r)_+]\mathrm{d}F(r) = \lambda\mathbf{E}[\min(1,(7-H)_+)]$$

as required.

(c) For students still to be there at closing time we require $J + H \ge 8$, as H ranges over $[0,8]$, and J ranges over $[8-H,8]$. Let

$$B = \{(t,x) : t \in [0,8], x \in [8-t,8]\}.$$

So,

$$v(B) = \lambda\int_0^8 \mathrm{d}t \int_{8-t}^8 \mathrm{d}F(x) = \lambda\int_0^8 \mathrm{d}F(x)\int_0^{8-x}\mathrm{d}t$$

$$= \lambda\int_0^8(8-x)\mathrm{d}F(x) = 8\lambda\int_0^8\mathrm{d}F(x) - \lambda\int_0^8 x\mathrm{d}F(x),$$

but $\int_0^8\mathrm{d}F(x) = 1$ and $\int_0^8 x\mathrm{d}F(x) = \mathbf{E}[H] = 1$ imply $\lambda\mathbf{E}[H] = \lambda$. Hence, the expected number of students in the library at closing time is 7λ. $\qquad\square$

Problem 4.5 (i) *Prove Campbell's theorem, i.e. show that if M is a Poisson random measure on the state space E with intensity measure μ and $a : E \to \mathbf{R}$ is a bounded measurable function, then*

$$\mathbf{E}[e^{\theta X}] = \exp\left[\int_E (e^{\theta a(y)} - 1)\mu(\mathrm{d}y)\right], \qquad (4.6.3)$$

where $X = \int_E a(y)M(\mathrm{d}y)$ (assume that $\lambda = \mu(E) < \infty$).

(ii) *Shots are heard at jump times $J_1, J_2, \ldots$ of a Poisson process with rate λ. The initial amplitudes of the gunshots $A_1, A_2, \ldots \sim \mathrm{Exp}(2)$ are IID exponentially distributed with parameter 2, and the amplitudes decay linearly at rate α. Compute the MGF of the total amplitude X_t at time t:*

$$X_t = \sum_n A_n(1 - \alpha(t-J_n)_+)\mathbf{1}_{(J_n \le t)};$$

$x_+ = x$ *if $x \ge 0$ and 0 otherwise.*

Solution (i) Conditioned on $M(E) = n$, the atoms of M form a random sample $Y_1, \ldots, Y_n$ with distribution $\dfrac{1}{\lambda}\mu$, so

$$\mathbf{E}[e^{\theta X} \mid M(E) = n] = \mathbf{E}\Big[e^{\theta \sum\limits_{k=1}^{n} a(Y_k)}\Big]$$

$$= \left(\int_E e^{\theta a(y)} \mu(\mathrm{d}y)/\lambda\right)^n.$$

Hence,

$$\mathbf{E}[e^{\theta X}] = \sum_n \mathbf{E}[e^{\theta X} \mid M(E) = n]\mathbf{P}(M(E) = n)$$

$$= \sum_n \left(\int_E e^{\theta a(y)} \mu(\mathrm{d}y)/\lambda\right)^n \frac{e^{-\lambda}\lambda^n}{n!}$$

$$= \exp\left(\int_E (e^{\theta a(y)} - 1)\mu(\mathrm{d}y)\right).$$

(ii) Fix t and let $E = [0, t] \times \mathbf{R}^+$ and v and M be such that $v(\mathrm{d}s, \mathrm{d}x) = 2\lambda e^{-2x}\mathrm{d}s\mathrm{d}x$, $M(B) = \sum_n \mathbf{1}_{\{(J_n, A_n) \in B\}}$. By the product theorem M is a Poisson random measure with intensity measure v. Set $a_t(s, x) = x(1 - \alpha(t - s))_+$, then $X_t = \int_E a_t(s, x)M(\mathrm{d}s, \mathrm{d}x)$. So, by Campbell's theorem, for $\theta < 2$,

$$\mathbf{E}[e^{\theta X_t}] = \exp\left(\int_E (e^{\theta a_t(s,x)} - 1)v(\mathrm{d}s, \mathrm{d}x)\right)$$

$$= e^{-\lambda t} \exp\left(2\lambda \int_0^t \int_0^\infty e^{-x(2 - \theta(1 - \alpha(t-s))_+)}\mathrm{d}x\mathrm{d}s\right)$$

$$= e^{-\lambda t} \exp\left(2\lambda \int_0^t \mathrm{d}s \frac{1}{2 - \theta(1 - \alpha(t - s))_+}\right)$$

$$= e^{-\lambda \min[t, 1/\alpha]} \left(\frac{2 - \theta + \theta\alpha \min[t, 1/\alpha]}{2 - \theta}\right)^{\frac{2\lambda}{\theta\alpha}}$$

by splitting integral $\int_0^t = \int_0^{t - \frac{1}{\alpha}} + \int_{t - \frac{1}{\alpha}}^t$ in the case $t > \frac{1}{\alpha}$. $\square$

Problem 4.6 *Seeds are planted in a field $S \subset \mathbf{R}^2$. The random way they are sown means that they form a Poisson process on S with density $\lambda(x, y)$. The seeds grow into plants that are later harvested as a crop, and the weight of the plant at (x, y) has* mean $m(x,y)$ and variance $v(x,y)$. The weights of different plants are independent random variables. Show that the total weight W of all the plants is a random variable with finite mean

$$I_1 = \iint_S m(x,y)\lambda(x,y)\,dxdy$$

and variance

$$I_2 = \iint_S \{m(x,y)^2 + v(x,y)\}\,\lambda(x,y)\,dxdy,$$

so long as these integrals are finite.

Solution Suppose first that

$$\mu = \int_S \lambda(x,y)\,dxdy$$

is finite. Then the number N of plants is finite and has the distribution $\mathrm{Po}(\mu)$. Conditional on N, their positions may be taken as independent random variables (X_n, Y_n), $n = 1,\ldots,N$, with density λ/μ on S. The weights of the plants are then independent, with

$$\mathbf{E}W = \int_S m(x,y)\lambda(x,y)\mu^{-1}\,dxdy = \mu^{-1}I_1$$

and

$$\mathbf{E}W^2 = \int_S [m(x,y)^2 + v(x,y)]\lambda(x,y)\mu^{-1}\,dxdy = \mu^{-1}I_2,$$

where I_1 and I_2 are finite. Hence,

$$\mathbf{E}(W|N) = \sum_{n=1}^{N} \mu^{-1}I_1 = N\mu^{-1}I_1$$

and

$$\mathrm{Var}\,(W|N) = \sum_{n=1}^{N} \left(\mu^{-1}I_2 - \mu^{-2}I_1^2\right) = N\left(\mu^{-1}I_2 - \mu^{-2}I_1^2\right).$$

Then

$$\mathbf{E}W = \mathbf{E}N\mu^{-1}I_1 = I_1$$

and

$$\mathrm{Var}\,W = \mathbf{E}\left[\mathrm{Var}\,(W|N)\right] + \mathrm{Var}\left[\mathbf{E}(W|N)\right]$$
$$= \mu\left(\mu^{-1}I_2 - \mu^{-2}I_1^2\right) + \left(\mathrm{Var}\,N\right)\mu^{-2}I_1^2 = I_2,$$

as required.

If $\mu = \infty$, we divide S into disjoint S_k on which λ is integrable, then write $W = \sum_k W_{(k)}$ where the harvests $W_{(k)}$ on S_k are independent, and use

$$\mathbb{E}W = \sum_k \mathbb{E}W_{(k)} = \sum_k \int_{S_k} m(x,y)\lambda(x,y)\mathrm{d}x\mathrm{d}y$$
$$= \int_S m(x,y)\lambda(x,y)\mathrm{d}x\mathrm{d}y$$

and similarly for $\mathrm{Var}\,W$.   □

Problem 4.7   *A line L in $\mathbb{R}^2$ not passing through the origin O can be defined by its perpendicular distance $p > 0$ from O and the angle $\theta \in [0, 2\pi)$ that the perpendicular from O to L makes with the x-axis. Explain carefully what is meant by a Poisson process of such lines L.*

A Poisson process Π of lines L has mean measure μ given by

$$\mu(B) = \iint_B \mathrm{d}p\,\mathrm{d}\theta \tag{4.6.4}$$

for $B \subseteq (0,\infty) \times [0,2\pi)$. A random countable set $\Phi \subset \mathbb{R}^2$ is defined to consist of all intersections of pairs of lines in Π. Show that the probability that there is at least one point of Φ inside the circle with centre O and radius r is less than

$$1 - (1 + 2\pi r)e^{-2\pi r}.$$

Is Φ a Poisson process?

Solution Suppose that μ is a measure on the space $\mathscr{L}$ of lines in $\mathbb{R}^2$ not passing through 0. A Poisson process with mean measure μ is a random countable subset Π of $\mathscr{L}$ such that

(1) the number $N(A)$ of points of Π in a measurable subset A of $\mathscr{L}$ has distribution $\mathrm{Po}(\mu(A))$, and
(2) for disjoint $A_1,\ldots,A_n$, the $N(A_j)$ are independent.

In the problem, the number N of lines which meet the disc D of centre 0 and radius r equals the number of lines with $p < r$. It is Poisson with mean

$$\int_0^r \int_0^{2\pi} \mathrm{d}p\mathrm{d}\theta = 2\pi r.$$

If there is at least one point of Φ in D then there must be at least two lines of Π meeting D, and this has probability

$$\sum_{n\geq 2} \frac{(2\pi r)^n}{n!}e^{-2\pi r} = 1 - (1+2\pi r)e^{-2\pi r}.$$

The probability of a point of Φ lying in D is strictly less than this, because there may be two lines meeting D whose intersection lies outside D.

Finally, Φ is *not* a Poisson process, since it has with positive probability collinear points. □

Problem 4.8 *Particular cases of the Poisson–Dirichlet distribution for the random sequence $(p_1, p_2, p_3, \ldots)$ with parameter θ appeared in PSE II the definition is given below. Show that, for any polynomial ϕ with $\phi(0) = 0$,*

$$\mathbb{E}\left\{\sum_{n=1}^{\infty} \phi(p_n)\right\} = \theta \int_0^1 \phi(x)x^{-1}(1-x)^{\theta-1}dx. \tag{4.6.5}$$

What does this tell you about the distribution of p_1?

Solution The simplest way to introduce the Poisson–Dirichlet distribution is to say that $\mathbf{p} = (p_1, p_2, \ldots)$ has the same distribution as (ξ_n/σ), where $\{\xi_n, n = 1, 2, \ldots\}$ are the points in descending order of a Poisson process on $(0, \infty)$ with rate $\theta x^{-1}e^{-x}$, and $\sigma = \sum_{n\geq 1} \xi_n$. By Campbell's theorem, σ is a.s. finite and has distribution Gam(θ) (where $\theta > 0$ can be arbitrary) and is independent from the vector $\mathbf{p} = (p_1, p_2, \ldots)$ with

$$p_1 \geq p_2 \geq \cdots, \quad \sum_{n\geq 1} p_n = 1, \quad \text{with probability } 1.$$

Here Gam stands for the Gamma distribution; see PSE I, Appendix.

To prove (4.6.5), we can take $p_n = \xi_n/\sigma$ and use the fact that σ and $\mathbf{p}$ are independent. For $k \geq 1$,

$$\mathbb{E}\left(\sum_{n\geq 1} \xi_n^k\right) = \int_0^{\infty} x^k \theta x^{-1} e^{-x} dx = \theta\Gamma(k).$$

The left side equals

$$\mathbb{E}\left(\sigma^k \sum_{n\geq 1} p_n^k\right) = \Gamma(\theta+k)\Gamma(\theta)^{-1}\mathbb{E}\left(\sum_{n\geq 1} p_n^k\right).$$

Thus,

$$\mathbb{E}\left(\sum_{n\geq 1} p_n^k\right) = \frac{\theta\Gamma(k)\Gamma(\theta)}{\Gamma(k+\theta)} = \theta \int_0^1 x^{k-1}(1-x)^{\theta-1}dx.$$

We see that the identity (4.6.5) holds for $\phi(x) = x^k$ (with $k \geq 1$) and hence by linearity for all polynomials with $\phi(0) = 0$.

Approximating step functions by polynomials shows that the mean number of p_n in an interval (a,b) (with $0 < a < b < 1$) equals

$$\theta \int_a^b x^{-1}(1-x)^{\theta-1} dx.$$

If $a > 1/2$, there can be at most one such p_n, so that p_1 has the PDF

$$\theta x^{-1}(1-x)^{\theta-1} \text{ on } (1/2,1).$$

But this fails on $(0, 1/2)$, and the identity (4.6.5) does not determine the distribution of p_1 on this interval. □

Problem 4.9   *The positions of trees in a large forest can be modelled as a Poisson process Π of constant rate λ on $\mathbb{R}^2$. Each tree produces a random number of seeds having a Poisson distribution with mean μ. Each seed falls to earth at a point uniformly distributed over the circle of radius r whose centre is the tree. The positions of the different seeds relative to their parent tree, and the numbers of seeds produced by a given tree, are independent of each other and of Π. Prove that, conditional on Π, the seeds form a Poisson process Π^* whose mean measure depends on Π. Is the unconditional distribution of Π^* that of a Poisson process?*

Solution By a direct calculation, the seeds from a tree at X form a Poisson process with rate

$$\rho_X(x) = \begin{cases} \pi^{-1}r^{-2}, & |x-X| < r, \\ 0, & \text{otherwise.} \end{cases}$$

Superposing these independent Poisson processes gives a Poisson process with rate

$$\Lambda_\Pi(x) = \sum_{X \in \Pi} \rho_X(x);$$

it clearly depends on Π. The unrealistic assumption of a circular uniform distribution is chosen to create no doubt about this dependence – in this case Π can be reconstructed from the contours of Λ_Π.

Here we meet for the first time the doubly stochastic (Cox) processes, i.e. Poisson process with random intensity. The number of seeds in a bounded set Λ has mean

$$\mathbb{E}N(A) = \mathbb{E}\mathbb{E}[N(A)|\Pi] = \mathbb{E}\int_A \Lambda_\Pi(x) dx$$

and variance

$$\operatorname{Var} N(A) = \mathbb{E}\left(\operatorname{Var}\left[N(A)|\Pi\right]\right) + \operatorname{Var}\left(\mathbb{E}\left[N(A)|\Pi\right]\right)$$
$$= \mathbb{E}N(A) + \operatorname{Var}\left[\int \Lambda_{\Pi}(x)dx\right]$$
$$> \mathbb{E}N(A).$$

Hence, Π^* is *not* a Poisson process. $\qquad\square$

Problem 4.10 *A uniform Poisson process* Π *in the unit ball of* $\mathbb{R}^3$ *is one whose mean measure is Lebesgue measure (volume) on*

$$\mathbf{B} = \{(x,y,z) \in \mathbb{R}^3 \ : \ r^2 = x^2 + y^2 + z^2 \leqslant 1\}.$$

Show that

$$\Pi_1 = \{r \ : \ (x,y,z) \in \Pi\}$$

is a Poisson process on $[0,1]$ *and find its mean measure. Show that*

$$\Pi_2 = \{(x/r, y/r, z/r) \ : \ (x,y,z) \in \Pi\}$$

is a Poisson process on the boundary of $\mathbf{B}$, *whose mean measure is a multiple of surface area. Are* Π_1 *and* Π_2 *independent processes?*

Solution By the mapping theorem, Π_1 is Poisson, with expected number of points in (a,b) equal to $\lambda \times$ (the volume of the shell with radii a and b), i.e.

$$\lambda\left(\frac{4}{3}\pi b^3 - \frac{4}{3}\pi a^3\right).$$

Thus, the mean measure of Π_1 has the PDF

$$4\lambda\pi r^2 \ (0 < r < 1).$$

Similarly, the expected number of points of Π_2 in $A \subseteq \partial\mathbf{B}$ equals

$$\lambda \times (\text{the conic volume from } 0 \text{ to } A) = \frac{1}{3}\lambda \times (\text{the surface area of } A).$$

Finally, Π_1 and Π_2 are *not* independent since they have the same number of points. $\qquad\square$

Problem 4.11 *The points of* Π *are coloured randomly either red or green, the probability of any point being red being* r, $0 < r < 1$, *and the colours of different points being independent. Show that the red and the green points form independent Poisson processes.*

Solution If $A \subseteq S$ has $\mu(A) < \infty$ then write

$$N(A) = N_1(A) + N_2(A)$$

where N_1 and N_2 are the numbers of red and green points. Conditional on $N(A) = n$, $N_1(A)$ has the binomial distribution $\mathrm{Bin}(n, r)$. Thus,

$$\mathbb{P}\big(N_1(A) = k, \, N_2(A) = l\big)$$

$$= \mathbb{P}\big(N(A) = k + l\big)\mathbb{P}\big(N_1(A) = k | N(A) = k + l\big)$$

$$= \frac{\mu(A)^{k+l} e^{-\mu(A)}}{(k+l)!} \binom{k+l}{k} r^k (1-r)^l$$

$$= \frac{[r\mu(A)]^k e^{-r\mu(A)}}{k!} \frac{[(1-r)\mu(A)]^l e^{-(1-r)\mu(A)}}{l!}.$$

Hence, $N_1(A)$ and $N_2(A)$ are independent Poisson random variables with means $r\mu(A)$ and $(1-r)\mu(A)$, respectively.

If $A_1, A_2, \ldots$ are disjoint sets then the pairs

$$(N_1(A_1), N_2(A_1)), \; (N_1(A_2), N_2(A_2)), \ldots$$

are independent, and hence

$$\big(N_1(A_1), N_1(A_2), \ldots\big) \text{ and } \big(N_2(A_1), N_2(A_2), \ldots\big)$$

are two independent sequences of independent random variables. If $\mu(A) = \infty$ then $N(A) = \infty$ a.s., and since $r > 0$ and $1 - r > 0$, there are a.s. infinitely many red and green points in A. $\qquad\square$

Problem 4.12   *A model of a rainstorm falling on a level surface (taken to be the plane $\mathbb{R}^2$) describes each raindrop by a triple (X, T, V), where $X \in \mathbb{R}^2$ is the horizontal position of the centre of the drop, T is the instant at which the drop hits the plane, and V is the volume of water in the drop. The points (X, T, V) are assumed to form a Poisson process on $\mathbb{R}^4$ with a given rate $\lambda(x, t, v)$. The drop forms a wet circular patch on the surface, with centre X and a radius that increases with time, the radius at time $(T + t)$ being a given function $r(t, V)$. Find the probability that a point $\xi \in \mathbb{R}^2$ is dry at time τ, and show that the total rainfall in the storm has expectation*

$$\int_{\mathbb{R}^4} v\lambda(x, t, v)\,dx\,dt\,dv$$

if this integral converges.

Solution Thus, $\xi \in \mathbb{R}^2$ is wet iff there is a point of Π with $t < \tau$ and

$$||X - \xi|| < r(\tau - t, V)$$

(there no problem about whether or not the inequality is strict since the difference involves events of zero probability). The number of points of Π satisfying these two inequalities is Poisson, with mean

$$\mu = \int \lambda(x,t,v)\mathbf{1}(t < \tau, ||x - \xi|| < r(\tau - t, v))\mathrm{d}x\mathrm{d}t\mathrm{d}v.$$

Hence, the probability that ξ is dry is $e^{-\mu}$ (or 0 if $\mu = +\infty$). Finally, the formula for the expected total rainfall,

$$\sum_{(X,T,V)\in\Pi} V,$$

is a direct application of Campbell's theorem.   □

Problem 4.13   *Let M be a Poisson random measure on $E = \mathbb{R} \times [0, \pi)$ with constant intensity λ. For $(x, \theta) \in E$, denote by $l(x, \theta)$ the line in $\mathbb{R}^2$ obtained by rotating the line $\{(x,y) : y \in \mathbb{R}\}$ through an angle θ about the origin.*
  Consider the line process $L = M \circ l^{-1}$.

  (i) *What is the distribution of the number of lines intersecting the disk $D_a = \{z \in \mathbb{R}^2 : |z| \le a\}$?*
 (ii) *What is the distribution of the distance from the origin to the nearest line?*
(iii) *What is the distribution of the distance from the origin to the kth nearest line?*

Solution (i) A line intersects the disk $D_a = \{z \in \mathbb{R}^2 : |z| \le a\}$ if and only if its representative point (x, θ) lies in $(-a, a) \times [0, \pi)$. Hence,

$$\sharp \text{ of lines intersecting } D_a \sim \mathrm{Po}(2a\pi\lambda).$$

(ii) Let Y be the distance from the origin to the nearest line. Then

$$\mathbb{P}(Y \ge a) = \mathbb{P}(M((-a,a) \times [0,\pi)) = 0) = \exp(-2a\lambda\pi),$$

i.e. $Y \sim \mathrm{Exp}(2\pi\lambda)$.

(iii) Let $Y_1, Y_2, \ldots$ be the distances from the origin to the nearest line, the second nearest line, and so on. Then the Y_i are the atoms of the PRM N on $\mathbb{R}_+$ which is obtained from M by the projection $(x, \theta) \mapsto |x|$. By the mapping theorem, N is the Poisson process on $\mathbb{R}_+$ of rate $2\pi\lambda$. Hence, $Y_k \sim \mathrm{Gam}(k, 2\lambda\pi)$, as $Y_k = S_1 + \cdots + S_k$ where $S_i \sim \mathrm{Exp}(2\pi\lambda)$, independently.   □

Problem 4.14   *One wishes to transmit one of M equiprobable distinct messages through a noisy channel. The jth message is encoded by the sequence of scalars a_{jt} ($t = 1, 2, \ldots, n$) which, after transmission, is received as $a_{jt} + \varepsilon_t$ ($t = 1, 2, \ldots, n$). Here the noise random variables ε_t are independent and normally distributed, with zero mean and with time-dependent variance* Var $\varepsilon_t = v_t$.

Find an inference rule at the receiver for which the average probability that the message value is incorrectly inferred has the upper bound

$$\mathbb{P}(\text{error}) \leq \frac{1}{M} \sum_{1 \leq j \neq k \leq M} \exp(-d_{jk}/8), \qquad (4.6.6)$$

where

$$d_{jk} = \sum_{1 \leq t \leq n} (a_{jt} - a_{kt})^2 \Big/ v_t.$$

Suppose that M = 2 and that the transmitted waveforms are subject to the power constraint $\sum_{1 \leq t \leq n} a_{jt}^2 \leq K$, $j = 1, 2$. *Which of the two waveforms minimises the probability of error?*

[*Hint*: You may assume validity of the bound $\mathbb{P}(Z \geq a) \leq \exp(-a^2/2)$, where Z is a standard $N(0, 1)$ random variable.]

Solution   Let $f_j = f_{\text{ch}}(\mathbf{y}|X = \mathbf{a}_j)$ be the PDF of receiving a vector $\mathbf{y}$ given that a 'waveform' $A_j = (a_{jt})$ was transmitted. Then

$$\mathbb{P}(\text{error}) \leq \frac{1}{M} \sum_j \sum_{k:k \neq j} \mathbb{P}(\{\mathbf{y} : f_k(\mathbf{y}) \geq f_j(\mathbf{y})|X = A_j\}).$$

Let V be the diagonal matrix with the diagonal elements v_j. In the present case,

$$f_j = C \exp\left[-\frac{1}{2} \sum_{t=1}^{n} (y_t - a_{jt})^2 / v_t\right]$$

$$= C \exp\left(-\frac{1}{2}(Y - A_j)^{\mathsf{T}} V^{-1} (Y - A_j)\right).$$

Then if $X = A_j$ and $Y = A_j + \varepsilon$ we have

$$\log f_k - \log f_j = -\frac{1}{2}(A_j - A_k + \varepsilon)^{\mathsf{T}} V^{-1} (A_j - A_k + \varepsilon) + \frac{1}{2} \varepsilon^{\mathsf{T}} V^{-1} \varepsilon$$

$$= -\frac{1}{2} d_{jk} - (A_j - (A_k)^{\mathsf{T}} V^{-1} \varepsilon)$$

$$= -\frac{1}{2} d_{jk} + \sqrt{d_{jk}} Z$$

where $Z \sim N(0,1)$. Thus, by the hint, (4.6.6) follows:

$$\mathbb{P}(f_k \geq f_j) = \mathbb{P}(Z > \sqrt{d_{jk}}/2) \leq e^{-d_{jk}/8}.$$

In the case $M = 2$ we have to maximise

$$d_{12} = (A_1 - A_2)^{\mathrm{T}} V^{-1}(A_1 - A_2) = \sum_{1 \leq t \leq n} (a_{1t} - a_{2t})^2 / v_t$$

subject to

$$\sum_t a_{jt}^2 \leq K \text{ or } (\mathbf{a})_j^{\mathrm{T}}(\mathbf{a})_j \leq K, j = 1, 2.$$

By Cauchy–Schwarz,

$$(A_1 - A_2)^{\mathrm{T}} V^{-1}(A_1 - A_2) \leq \left(\sqrt{A_1^{\mathrm{T}} V^{-1} A_1} + \sqrt{A_2^{\mathrm{T}} V^{-1} A_2} \right)^2 \tag{4.6.7}$$

with equality holding when $A_1 = \text{const} A_2$. Further, in our case V is diagonal, and (4.6.7) is maximised when $A_j^{\mathrm{T}} A_j = K$, $j = 1, 2, \ldots$ We conclude that

$$a_{1t} = -a_{2t} = b_t$$

with b_t non-zero only for t such that v_t is minimal, and $\sum_t b_t^2 = K$.   □

Problem 4.15 *A random variable Y is distributed on the non-negative integers. Show that the maximum entropy of Y, subject to $\mathbb{E}Y \leq M$, is*

$$-M \log M + (M+1) \log(M+1)$$

attained by a geometric distribution with mean M.

A memoryless channel produces outputs Y from non-negative integer-valued inputs X by

$$Y = X + \varepsilon,$$

where ε is independent of X, $\mathbb{P}(\varepsilon = 1) = p$, $\mathbb{P}(\varepsilon = 0) = 1 - p = q$ and inputs X are constrained by $\mathbb{E}X \leq q$. Show that, provided $p \leq 1/3$, the optimal input distribution is

$$\mathbb{P}(X = r) = (1+p)^{-1} \left[\frac{1}{2^{r+1}} - \left(\frac{-p}{q} \right)^{r+1} \right], \quad r = 0, 1, 2, \ldots,$$

and determine the capacity of the channel.

Describe, very briefly, the problem of determining the channel capacity if $p > 1/3$.

Solution First, consider the problem

$$\text{maximise } h(Y) = -\sum_{y \geq 0} p_y \log p_y \text{ subject to } \begin{cases} p_y \geq 0, \\ \sum_y p_y = 1, \\ \sum_y y p_y = M. \end{cases}$$

The solution, found by using Lagrangian multipliers, is

$$p_y = (1-\lambda)\lambda^y, \ y = 0, 1, \ldots, \text{ with } M = \frac{\lambda}{1-\lambda}, \text{ or } \lambda = \frac{M}{M+1},$$

with the optimal value

$$h(Y) = (M+1)\log(M+1) - M\log M.$$

Next, for $g(m) = (m+1)\log(m+1) - m\log m$,

$$g'(m) = \log(m+1) - \log m > 0,$$

implying that $h(Y) \nearrow$ when $M \nearrow$. Therefore, the maximiser and the optimal value are the same for $\mathbb{E}Y \leq M$, as required.

Now, the capacity $C = \sup[h(Y) - h(Y|X)] = h(Y) - h(\varepsilon)$, and the condition $\mathbb{E}X \leq q$ implies that $\mathbb{E}Y \leq q + \mathbb{E}\varepsilon = q + p = 1$. With $h(\varepsilon) = -p\log p - q\log q$, we want Y geometric, with $M = 1$, $\lambda = 1/2$, yielding

$$C = 2\log 2 + p\log p + q\log q = \log\left(4p^p q^q\right).$$

Then

$$\begin{aligned}
\mathbb{E}z^X &= \frac{\mathbb{E}z^Y}{\mathbb{E}z^\varepsilon} = \left(\frac{1-\lambda}{1-\lambda z}\right) \Big/ (pz+q) = \frac{1}{(2-z)(q+pz)} \\
&= \frac{(2-z)^{-1} + p(q+pz)^{-1}}{1+p} \\
&= \frac{1}{1+p}\left(\sum (1/2)^{1+r} z^r + (p/q)\sum(-p/q)^r z^r\right).
\end{aligned}$$

If $p > 1/3$ then $p/q > 1/2$ and the alternate probabilities become negative, which means that there is no distribution for X giving an optimum for Y. Then we would have to maximise

$$-\sum_y p_y \log p_y, \text{ subject to } p_y = p\pi_{y-1} + q\pi_y,$$

where $\pi_y \geq 0$, $\sum_y \pi_y = 1$ and $\sum_y y\pi_y \leq q$.   □

Problem 4.16 *Assuming the bounds on channel capacity asserted by the second coding theorem, deduce the capacity of a memoryless Gaussian channel.*

A channel consists of r independent memoryless Gaussian channels, the noise in the ith channel having variance v_i, $i = 1, 2, \ldots, n$. The compound channel is subject to an overall power constraint $\mathbb{E}\left(\sum_i x_{it}^2\right) \le p$, for each t, where x_{it} is the input of channel i at time t. Determine the capacity of the compound channel.

Solution For the first part see Section 4.3.

If the power in the ith channel is reduced to p_i, we would have capacity

$$C' = \frac{1}{2}\sum_i \log\left(1 + \frac{p_i}{v_i}\right).$$

The actual capacity is given by $C = \max C'$ subject to $p_1, \ldots, p_r \ge 0$, $\sum_i p_i = p$. Thus, we have to maximise the Lagrangian

$$\mathcal{L} = \frac{1}{2}\sum_i \log\left(1 + \frac{p_i}{v_i}\right) - \lambda \sum_i p_i,$$

with

$$\frac{\partial}{\partial p_i}\mathcal{L} = \frac{1}{2}(v_i + p_i)^{-1} - \lambda, i = 1, \ldots, r$$

and the maximum at

$$p_i = \max\left[0, \frac{1}{2\lambda} - v_i\right] = \left(\frac{1}{2\lambda} - v_i\right)_+.$$

To adjust the constraint, choose $\lambda = \lambda^*$ where λ^* is determined from

$$\sum_i \left(\frac{1}{2\lambda^*} - v_i\right)_+ = p.$$

The existence and uniqueness of λ^* follows since the LHS monotonically decreases from $+\infty$ to 0. Thus,

$$C = \frac{1}{2}\sum_i \log\left(\frac{1}{2\lambda^* v_i}\right).$$

□

Problem 4.17 *Here we consider random variables taking values in a given set $\mathbb{A}$ (finite, countable or uncountable) whose distributions are determined by PMFs with respect to a given reference measure μ. Let ψ be a real function and β a real number. Prove that the maximum $h^{max}(X)$ of the entropy $h(X) = -\int f_X(X)\log f_X(x)\mu(dx)$ subject to the constraint $\mathbb{E}\psi(X) = \beta$ is achieved at the random variable X^* with the PMF*

$$f_{X^*}(x) = \frac{1}{\Xi}\exp\left[-\gamma\psi(x)\right] \tag{4.6.8a}$$

where $\Xi = \Xi(\gamma) = \int \exp\left[-\gamma\psi(x)\right]\mu(dx)$ is the normalising constant and γ is chosen so that

$$\mathbb{E}\psi(X^*) = \int \frac{\psi(x)}{\Xi}\exp\left[-\gamma\psi(x)\right]\mu(dx) = \beta. \qquad (4.6.8b)$$

Assume that the value γ with the property $\int \frac{\psi(x)}{\Xi}\exp\left[-\gamma\psi(x)\right]\mu(dx) = \beta$ exists.

Show that if, in addition, function ψ is non-negative, then, for any given $\beta > 0$, the PMF f_{X^*} from (4.6.8a), (4.6.8b) maximises the entropy $h(X)$ under a wider constraint $\mathbb{E}\psi(X) \leq \beta$.

Consequently, calculate the maximal value of $h(X)$ subject to $\mathbb{E}\psi(X) \leq \beta$, in the following cases: (i) when $\mathbb{A}$ is a finite set, μ is a positive measure on $\mathbb{A}$ (with $\mu_i = \mu(\{i\}) = 1/\mu(\mathbb{A})$ where $\mu(\mathbb{A}) = \sum_{j\in\mathbb{A}} \mu_j$) and $\psi(x) \equiv 1$, $x \in \mathbb{A}$; (ii) when $\mathbb{A}$ is an arbitrary set, μ is a positive measure on $\mathbb{A}$ with $\mu(\mathbb{A}) < \infty$ and $\psi(x) \equiv 1$, $x \in \mathbb{A}$; (iii) when $\mathbb{A} = \mathbb{R}$ is a real line, μ is the Lebesgue measure and $\psi(x) = |x|$; (iv) when $\mathbb{A} = \mathbb{R}^d$, μ is a d-dimensional Lebesgue measure and $\psi(x) = \sum_{1|leqj\leq d} K_{ij}x_ix_j$, where $K = (K_{ij})$ is a $d \times d$ positive definite real matrix.

Solution With $\ln f_X^*(x) = -\gamma\psi(x) - \ln\Xi$, we use the Gibbs inequality:

$$h(X) = -\int f_X(x)\ln f_X(x)\mu(dx) \leq \int f_X(x)\left[\gamma\psi(x)+\ln\Xi\right]\mu(dx)$$

$$= \int f_{X^*}(x)\left[\gamma\psi(x)+\ln\Xi\right]\mu(dx) = h(X^*)$$

with equality if and only if $X \sim X^*$. This proves the first assertion.

If $\psi \geq 0$, the expected value $\mathbb{E}\psi(X) \geq 0$, and γ is minimal when the constraint is satisfied. $\qquad \square$

Bibliography

[1] V. Anantharam, F. Baccelli. A Palm theory approach to error exponents. In *Proceedings of the 2008 IEEE Symposium on Information Theory*, Toronto, pp. 1768–1772, 2008.

[2] J. Adámek. *Foundations of Coding: Theory and Applications of Error-Correcting Codes, with an Introduction to Cryptography and Information Theory*. Chichester: Wiley, 1991.

[3] D. Applebaum. *Probability and Information: An Integrated Approach*. Cambridge: Cambridge University Press, 1996.

[4] R.B. Ash. *Information Theory*. New York: Interscience, 1965.

[5] E.F. Assmus, Jr., J.D. Key. *Designs and their Codes*. Cambridge: Cambridge University Press, 1992.

[6] K.A. Arwini, C.T.J. Dodson. *Information Geometry: Near Randomness and Near Independence*. Lecture notes in mathematics, 1953. Berlin: Springer, 2008.

[7] D. Augot, M. Stepanov. A note on the generalisation of the Guruswami–Sudan list decoding algorithm to Reed–Muller codes. In *Gröbner Bases, Coding, and Cryptography*. RISC Book Series. Springer, Heidelberg, 2009.

[8] R.U. Ayres. *Manufacturing and Human Labor as Information Processes*. Laxenburg: International Institute for Applied System Analysis, 1987.

[9] A.V. Balakrishnan. *Communication Theory* (with contributions by J.W. Carlyle et al.). New York: McGraw-Hill, 1968.

[10] J. Baylis. *Error-Correcting Codes: A Mathematical Introduction*. London: Chapman & Hall, 1998.

[11] A. Betten et al. *Error-Correcting Linear Codes Classification by Isometry and Applications*. Berlin: Springer, 2006.

[12] T. Berger. *Rate Distortion Theory: A Mathematical Basis for Data Compression*. Englewood Cliffs, NJ: Prentice-Hall, 1971.

[13] E.R. Berlekamp. *A Survey of Algebraic Coding Theory*. Wien: Springer, 1972.

[14] E.R. Berlekamp. *Algebraic Coding Theory*. New York: McGraw-Hill, 1968.

[15] J. Berstel, D. Perrin. *Theory of Codes*. Orlando, FL: Academic Press, 1985.

[16] J. Bierbrauer. *Introduction to Coding Theory*. Boca Raton, FL: Chapman & Hall/CRC, 2005.

[17] P. Billingsley. *Ergodic Theory and Information*. New York: Wiley, 1965.

[18] R.E. Blahut. *Principles and Practice of Information Theory*. Reading, MA: Addison-Wesley, 1987.

[19] R.E. Blahut. *Theory and Practice of Error Control Codes*. Reading, MA: Addison-Wesley, 1983. See also *Algebraic Codes for Data Transmission*. Cambridge: Cambridge University Press, 2003.

[20] R.E. Blahut. *Algebraic Codes on Lines, Planes, and Curves*. Cambridge: Cambridge University Press, 2008.

[21] I.F. Blake, R.C. Mullin. *The Mathematical Theory of Coding*. New York: Academic Press, 1975.

[22] I.F. Blake, R.C. Mullin. *An Introduction to Algebraic and Combinatorial Coding Theory*. New York: Academic Press, 1976.

[23] I.F. Blake (ed). *Algebraic Coding Theory: History and Development*. Stroudsburg, PA: Dowden, Hutchinson & Ross, 1973.

[24] N. Blachman. *Noise and its Effect on Communication*. New York: McGraw-Hill, 1966.

[25] R.C. Bose, D.K. Ray-Chaudhuri. On a class of errors, correcting binary group codes. *Information and Control*, **3**(1), 68–79, 1960.

[26] W. Bradley, Y.M. Suhov. The entropy of famous reals: some empirical results. *Random and Computational Dynamics*, **5**, 349–359, 1997.

[27] A.A. Bruen, M.A. Forcinito. *Cryptography, Information Theory, and Error-Correction: A Handbook for the 21st Century*. Hoboken, NJ: Wiley-Interscience, 2005.

[28] J.A. Buchmann. *Introduction to Cryptography*. New York: Springer-Verlag, 2002.

[29] P.J. Cameron, J.H. van Lint. *Designs, Graphs, Codes and their Links*. Cambridge: Cambridge University Press, 1991.

[30] J. Castiñeira Moreira, P.G. Farrell. *Essentials of Error-Control Coding*. Chichester: Wiley, 2006.

[31] W.G. Chambers. *Basics of Communications and Coding*. Oxford: Clarendon, 1985.

[32] G.J. Chaitin. *The Limits of Mathematics: A Course on Information Theory and the Limits of Formal Reasoning*. Singapore: Springer, 1998.

[33] G. Chaitin. *Information-Theoretic Incompleteness*. Singapore: World Scientific, 1992.

[34] G. Chaitin. *Algorithmic Information Theory*. Cambridge: Cambridge University Press, 1987.

[35] F. Conway, J. Siegelman. *Dark Hero of the Information Age: In Search of Norbert Wiener, the Father of Cybernetics*. New York: Basic Books, 2005.

[36] T.M. Cover, J.M. Thomas. *Elements of Information Theory*. New York: Wiley, 2006.

[37] I. Csiszár, J. Körner. *Information Theory: Coding Theorems for Discrete Memoryless Systems*. New York: Academic Press, 1981; Budapest: Akadémiai Kiadó, 1981.

[38] W.B. Davenport, W.L. Root. *Random Signals and Noise*. New York: McGraw Hill, 1958.

[39] A. Dembo, T. M. Cover, J. A. Thomas. Information theoretic inequalities. *IEEE Transactions on Information Theory*, **37**, (6), 1501–1518, 1991.

[40] R.L. Dobrushin. Taking the limit of the argument of entropy and information functions. *Teoriya Veroyatn. Primen.*, **5**, (1), 29–37, 1960; English translation: *Theory of Probability and its Applications*, **5**, 25–32, 1960.

[41] F. Dyson. The Tragic Tale of a Genius. *New York Review of Books*, July 14, 2005.

[42] W. Ebeling. *Lattices and Codes: A Course Partially Based on Lectures by F. Hirzebruch*. Braunschweig/Wiesbaden: Vieweg, 1994.

[43] N. Elkies. Excellent codes from modular curves. *STOC'01: Proceedings of the 33rd Annual Symposium on Theory of Computing* (Hersonissos, Crete, Greece), pp. 200–208, NY: ACM, 2001.

[44] S. Engelberg. *Random Signals and Noise: A Mathematical Introduction*. Boca Raton, FL: CRC/Taylor & Francis, 2007.

[45] R.M. Fano. *Transmission of Information: A Statistical Theory of Communication*. New York: Wiley, 1961.

[46] A. Feinstein. *Foundations of Information Theory*. New York: McGraw-Hill, 1958.

[47] G.D. Forney. *Concatenated Codes*. Cambridge, MA: MIT Press, 1966.

[48] M. Franceschetti, R. Meester. *Random Networks for Communication. From Statistical Physics to Information Science*. Cambridge: Cambridge University Press, 2007.

[49] R. Gallager. *Information Theory and Reliable Communications*. New York: Wiley, 1968.

[50] A. Gofman, M. Kelbert, Un upper bound for Kullback–Leibler divergence with a small number of outliers. *Mathematical Communications*, **18**, (1), 75–78, 2013.

[51] S. Goldman. *Information Theory*. Englewood Cliffs, NJ: Prentice-Hall, 1953.

[52] C.M. Goldie, R.G.E. Pinch. *Communication Theory*. Cambridge: Cambridge University Press, 1991.

[53] O. Goldreich. *Foundations of Cryptography*, Vols 1, 2. Cambridge: Cambridge University Press, 2001, 2004.

[54] V.D. Goppa. *Geometry and Codes*. Dordrecht: Kluwer, 1988.

[55] S. Gravano. *Introduction to Error Control Codes*. Oxford: Oxford University Press, 2001.

[56] R.M. Gray. *Source Coding Theory*. Boston: Kluwer, 1990.

[57] R.M. Gray. *Entropy and Information Theory*. New York: Springer-Verlag, 1990.

[58] R.M. Gray, L.D. Davisson (eds). *Ergodic and Information Theory*. Stroudsburg, CA: Dowden, Hutchinson & Ross, 1977 .

[59] V. Guruswami, M. Sudan. Improved decoding of Reed–Solomon codes and algebraic geometry codes. *IEEE Trans. Inform. Theory*, **45**, (6), 1757–1767, 1999.

[60] R.W. Hamming. *Coding and Information Theory*. 2nd ed. Englewood Cliffs, NJ: Prentice-Hall, 1986.

[61] T.S. Han. *Information-Spectrum Methods in Information Theory*. New York: Springer-Verlag, 2002.

[62] D.R. Hankerson, G.A. Harris, P.D. Johnson, Jr. *Introduction to Information Theory and Data Compression*. 2nd ed. Boca Raton, FL: Chapman & Hall/CRC, 2003.

[63] D.R. Hankerson et al. *Coding Theory and Cryptography: The Essentials*. 2nd ed. New York: M. Dekker, 2000. (Earlier version: D. G. Hoffman et al. *Coding Theory: The Essentials*. New York: M. Dekker, 1991.)

[64] W.E. Hartnett. *Foundations of Coding Theory*. Dordrecht: Reidel, 1974.

[65] S.J. Heims. *John von Neumann and Norbert Wiener: From Mathematics to the Technologies of Life and Death.* Cambridge, MA: MIT Press, 1980.

[66] C. Helstrom. *Statistical Theory of Signal Detection.* 2nd ed. Oxford: Pergamon Press, 1968.

[67] C.W. Helstrom. *Elements of Signal Detection and Estimation.* Englewood Cliffs, NJ: Prentice-Hall, 1995.

[68] R. Hill. *A First Course in Coding Theory.* Oxford: Oxford University Press, 1986.

[69] T. Ho, D.S. Lun. *Network Coding: An Introduction.* Cambridge: Cambridge University Press, 2008.

[70] A. Hocquenghem. Codes correcteurs d'erreurs. *Chiffres,* **2**, 147–156, 1959.

[71] W.C. Huffman, V. Pless. *Fundamentals of Error-Correcting Codes.* Cambridge: Cambridge University Press, 2003.

[72] J.F. Humphreys, M.Y. Prest. *Numbers, Groups, and Codes.* 2nd ed. Cambridge: Cambridge University Press, 2004.

[73] S. Ihara. *Information Theory for Continuous Systems.* Singapore: World Scientific, 1993.

[74] F.M. Ingels. *Information and Coding Theory.* Scranton: Intext Educational Publishers, 1971.

[75] I.M. James. *Remarkable Mathematicians. From Euler to von Neumann.* Cambridge: Cambridge University Press, 2009.

[76] E.T. Jaynes. *Papers on Probability, Statistics and Statistical Physics.* Dordrecht: Reidel, 1982.

[77] F. Jelinek. *Probabilistic Information Theory.* New York: McGraw-Hill, 1968.

[78] G.A. Jones, J.M. Jones. *Information and Coding Theory.* London: Springer, 2000.

[79] D.S. Jones. *Elementary Information Theory.* Oxford: Clarendon Press, 1979.

[80] O. Johnson. *Information Theory and the Central Limit Theorem.* London: Imperial College Press, 2004.

[81] J. Justensen. A class of constructive asymptotically good algebraic codes. *IEEE Transactions Information Theory,* **18**(5), 652–656, 1972.

[82] M. Kelbert, Y. Suhov. Continuity of mutual entropy in the large signal-to-noise ratio limit. In *Stochastic Analysis 2010,* pp. 281–299, 2010. Berlin: Springer.

[83] N. Khalatnikov. *Dau, Centaurus and Others.* Moscow: Fizmatlit, 2007.

[84] A.Y. Khintchin. *Mathematical Foundations of Information Theory.* New York: Dover, 1957.

[85] T. Klove. *Codes for Error Detection.* Singapore: World Scientific, 2007.

[86] N. Koblitz. *A Course in Number Theory and Cryptography.* New York: Springer, 1993.

[87] H. Krishna. *Computational Complexity of Bilinear Forms: Algebraic Coding Theory and Applications of Digital Communication Systems.* Lecture notes in control and information sciences, Vol. 94. Berlin: Springer-Verlag, 1987.

[88] S. Kullback. *Information Theory and Statistics.* New York: Wiley, 1959.

[89] S. Kullback, J.C. Keegel, J.H. Kullback. *Topics in Statistical Information Theory.* Berlin: Springer, 1987.

[90] H.J. Landau, H.O. Pollak. Prolate spheroidal wave functions, Fourier analysis and uncertainty, II. *Bell System Technical Journal,* 64–84, 1961.

[91] H.J. Landau, H.O. Pollak. Prolate spheroidal wave functions, Fourier analysis and uncertainty, III. The dimension of the space of essentially time- and band-limited signals. *Bell System Technical Journal*, 1295–1336, 1962.

[92] R. Lidl, H. Niederreiter. *Finite Fields*. Cambridge: Cambridge University Press, 1997.

[93] R. Lidl, G. Pilz. *Applied Abstract Algebra*. 2nd ed. New York: Wiley, 1999.

[94] E.H. Lieb. Proof of entropy conjecture of Wehrl. *Commun. Math. Phys.*, **62**, (1), 35–41, 1978.

[95] S. Lin. *An Introduction to Error-Correcting Codes*. Englewood Cliffs, NJ; London: Prentice-Hall, 1970.

[96] S. Lin, D.J. Costello. *Error Control Coding: Fundamentals and Applications*. Englewood Cliffs, NJ: Prentice-Hall, 1983.

[97] S. Ling, C. Xing. *Coding Theory*. Cambridge: Cambridge University Press, 2004.

[98] J.H. van Lint. *Introduction to Coding Theory*. 3rd ed. Berlin: Springer, 1999.

[99] J.H. van Lint, G. van der Geer. *Introduction to Coding Theory and Algebraic Geometry*. Basel: Birkhäuser, 1988.

[100] J.C.A. van der Lubbe. *Information Theory*. Cambridge: Cambridge University Press, 1997.

[101] R.E. Lewand. *Cryptological Mathematics*. Washington, DC: Mathematical Association of America, 2000.

[102] J.A. Llewellyn. *Information and Coding*. Bromley: Chartwell-Bratt; Lund: Studentlitteratur, 1987.

[103] M. Loève. *Probability Theory*. Princeton, NJ: van Nostrand, 1955.

[104] D.G. Luenberger. *Information Science*. Princeton, NJ: Princeton University Press, 2006.

[105] D.J.C. Mackay. *Information Theory, Inference and Learning Algorithms*. Cambridge: Cambridge University Press, 2003.

[106] H.B. Mann (ed). *Error-Correcting Codes*. New York: Wiley, 1969 .

[107] M. Marcus. Dark Hero of the Information Age: In Search of Norbert Wiener, the Father of Cybernetics. *Notices of the AMS* **53**, (5), 574–579, 2005.

[108] A. Marshall, I. Olkin. *Inequalities: Theory of Majorization and its Applications*. New York: Academic Press, 1979 .

[109] V.P. Maslov, A.S. Chernyi. On the minimization and maximization of entropy in various disciplines. *Theory Probab. Appl.* **48**, (3), 447–464, 2004.

[110] F.J. MacWilliams, N.J.A. Sloane. *The Theory of Error-Correcting Codes*, Vols I, II. Amsterdam: North-Holland, 1977.

[111] R.J. McEliece. *The Theory of Information and Coding*. Reading, MA: Addison-Wesley, 1977. 2nd ed. Cambridge: Cambridge University Press, 2002.

[112] R. McEliece. *The Theory of Information and Coding*. Student ed. Cambridge: Cambridge University Press, 2004.

[113] A. Menon, R.M. Buecher, J.H. Read. Impact of exclusion region and spreading in spectrum-sharing ad hoc networks. ACM 1-59593-510-X/06/08, 2006 .

[114] R.A. Mollin. *RSA and Public-Key Cryptography*. New York: Chapman & Hall, 2003.

[115] R.H. Morelos-Zaragoza. *The Art of Error-Correcting Coding*. 2nd ed. Chichester: Wiley, 2006.

[116] G.L. Mullen, C. Mummert. *Finite Fields and Applications*. Providence, RI: American Mathematical Society, 2007.

[117] A. Myasnikov, V. Shpilrain, A. Ushakov. *Group-Based Cryptography*. Basel: Birkhäuser, 2008.

[118] G. Nebe, E.M. Rains, N.J.A. Sloane. *Self-Dual Codes and Invariant Theory*. New York: Springer, 2006.

[119] H. Niederreiter, C. Xing. *Rational Points on Curves over Finite Fields: Theory and Applications*. Cambridge: Cambridge University Press, 2001.

[120] W.W. Peterson, E.J. Weldon. *Error-Correcting Codes*. 2nd ed. Cambridge, MA: MIT Press, 1972. (Previous ed. W.W. Peterson. *Error-Correcting Codes*. Cambridge, MA: MIT Press, 1961.)

[121] M.S. Pinsker. *Information and Information Stability of Random Variables and Processes*. San Francisco: Holden-Day, 1964.

[122] V. Pless. *Introduction to the Theory of Error-Correcting Codes*. 2nd ed. New York: Wiley, 1989.

[123] V.S. Pless, W.C. Huffman (eds). *Handbook of Coding Theory*, Vols 1, 2. Amsterdam: Elsevier, 1998.

[124] P. Piret. *Convolutional Codes: An Algebraic Approach*. Cambridge, MA: MIT Press, 1988.

[125] O. Pretzel. *Error-Correcting Codes and Finite Fields*. Oxford: Clarendon Press, 1992; Student ed. 1996.

[126] T.R.N. Rao. *Error Coding for Arithmetic Processors*. New York: Academic Press, 1974.

[127] M. Reed, B. Simon. *Methods of Modern Mathematical Physics*, Vol. II. Fourier analysis, self-adjointness. New York: Academic Press, 1975.

[128] A. Rényi. *A Diary on Information Theory*. Chichester: Wiley, 1987; initially published Budapest: Akad'emiai Kiadó, 1984.

[129] F.M. Reza. *An Introduction to Information Theory*. New York: Constable, 1994.

[130] S. Roman. *Coding and Information Theory*. New York: Springer, 1992.

[131] S. Roman. *Field Theory*. 2nd ed. New York: Springer, 2006.

[132] T. Richardson, R. Urbanke. *Modern Coding Theory*. Cambridge: Cambridge University Press, 2008.

[133] R.M. Roth. *Introduction to Coding Theory*. Cambridge: Cambridge University Press, 2006.

[134] B. Ryabko, A. Fionov. *Basics of Contemporary Cryptography for IT Practitioners*. Singapore: World Scientific, 2005.

[135] W.E. Ryan, S. Lin. *Channel Codes: Classical and Modern*. Cambridge: Cambridge University Press, 2009.

[136] T. Schürmann, P. Grassberger. Entropy estimation of symbol sequences. *Chaos*, **6**, (3), 414–427, 1996.

[137] P. Seibt. *Algorithmic Information Theory: Mathematics of Digital Information Processing*. Berlin: Springer, 2006.

[138] C.E. Shannon. A mathematical theory of cryptography. *Bell Lab. Tech. Memo.*, 1945.

[139] C.E. Shannon. A mathematical theory of communication. *Bell System Technical Journal*, **27**, July, October, 379–423, 623–658, 1948.

[140] C.E. Shannon: *Collected Papers*. N.J.A. Sloane, A.D. Wyner (eds). New York: IEEE Press, 1993.

[141] C.E. Shannon, W. Weaver. *The Mathematical Theory of Communication*. Urbana, IL: University of Illinois Press, 1949.

[142] P.C. Shields. *The Ergodic Theory of Discrete Sample Paths*. Providence, RI: American Mathematical Society, 1996.

[143] M.S. Shrikhande, S.S. Sane. *Quasi-Symmetric Designs*. Cambridge: Cambridge University Press, 1991.

[144] S. Simic. Best possible global bounds for Jensen functionals. *Proc. AMS*, **138**, (7), 2457–2462, 2010.

[145] A. Sinkov. *Elementary Cryptanalysis: A Mathematical Approach*. 2nd ed. revised and updated by T. Feil. Washington, DC: Mathematical Association of America, 2009.

[146] D. Slepian, H.O. Pollak. Prolate spheroidal wave functions, Fourier analysis and uncertainty, Vol. I. *Bell System Technical Journal*, 43–64, 1961 .

[147] W. Stallings. *Cryptography and Network Security: Principles and Practice*. 5th ed. Boston, MA: Prentice Hall; London: Pearson Education, 2011.

[148] H. Stichtenoth. *Algebraic Function Fields and Codes*. Berlin: Springer, 1993.

[149] D.R. Stinson. *Cryptography: Theory and Practice*. 2nd ed. Boca Raton, FL; London: Chapman & Hall/CRC, 2002.

[150] D. Stoyan, W.S. Kendall. J. Mecke. *Stochastic Geometry and its Applications*. Berlin: Academie-Verlag, 1987 .

[151] C. Schlegel, L. Perez. *Trellis and Turbo Coding*. New York: Wiley, 2004.

[152] Š. Šujan. *Ergodic Theory, Entropy and Coding Problems of Information Theory*. Praha: Academia, 1983.

[153] P. Sweeney. *Error Control Coding: An Introduction*. New York: Prentice Hall, 1991.

[154] Te Sun Han, K. Kobayashi. *Mathematics of Information and Coding*. Providence, RI: American Mathematical Society, 2002.

[155] T.M. Thompson. *From Error-Correcting Codes through Sphere Packings to Simple Groups*. Washington, DC: Mathematical Association of America, 1983.

[156] R. Togneri, C.J.S. deSilva. *Fundamentals of Information Theory and Coding Design*. Boca Raton, FL: Chapman & Hall/CRC, 2002.

[157] W. Trappe, L.C. Washington. *Introduction to Cryptography: With Coding Theory*. 2nd ed. Upper Saddle River, NJ: Pearson Prentice Hall, 2006.

[158] M.A. Tsfasman, S.G. Vlădut. *Algebraic-Geometric Codes*. Dordrecht: Kluwer Academic, 1991.

[159] M. Tsfasman, S. Vlădut, T. Zink. Modular curves, Shimura curves and Goppa codes, better than Varshamov–Gilbert bound. *Mathematics Nachrichten*, **109**, 21–28, 1982.

[160] M. Tsfasman, S. Vlădut, D. Nogin. *Algebraic Geometric Codes: Basic Notions*. Providence, RI: American Mathematical Society, 2007.

[161] M.J. Usher. *Information Theory for Information Technologists*. London: Macmillan, 1984.

[162] M.J. Usher, C.G. Guy. *Information and Communication for Engineers*. Basingstoke: Macmillan, 1997

[163] I. Vajda. *Theory of Statistical Inference and Information*. Dordrecht: Kluwer, 1989.

[164] S. Verdú. *Multiuser Detection*. New York: Cambridge University Press, 1998.

[165] S. Verdú, D. Guo. A simple proof of the entropy–power inequality. *IEEE Trans. Inform. Theory*, **52**, (5), 2165–2166, 2006.

[166] L.R. Vermani. *Elements of Algebraic Coding Theory*. London: Chapman & Hall, 1996.

[167] B. Vucetic, J. Yuan. *Turbo Codes: Principles and Applications*. Norwell, MA: Kluwer, 2000.

[168] G. Wade. *Coding Techniques: An Introduction to Compression and Error Control*. Basingstoke: Palgrave, 2000.

[169] J.L. Walker. *Codes and Curves*. Providence, RI: American Mathematical Society, 2000.

[170] D. Welsh. *Codes and Cryptography*. Oxford, Oxford University Press, 1988.

[171] N. Wiener. *Cybernetics or Control and Communication in Animal and Machine*. Cambridge, MA: MIT Press, 1948; 2nd ed: 1961, 1962.

[172] J. Wolfowitz. *Coding Theorems of Information Theory*. Berlin: Springer, 1961; 3rd ed: 1978.

[173] A.D. Wyner. The capacity of the band-limited Gaussian channel. *Bell System Technical Journal*, 359–395, 1996 .

[174] A.D. Wyner. The capacity of the product of channels. *Information and Control*, 423–433, 1966.

[175] C. Xing. Nonlinear codes from algebraic curves beating the Tsfasman–Vlǎdut–Zink bound. *IEEE Transactions Information Theory*, **49**, 1653–1657, 2003.

[176] A.M. Yaglom, I.M. Yaglom. *Probability and Information*. Dordrecht, Holland: Reidel, 1983.

[177] R. Yeung. *A First Course in Information Theory*. Boston: Kluwer Academic, 1992; 2nd ed. New York: Kluwer, 2002.

Index

For EU product safety concerns, contact us at Calle de José Abascal, 56–1°,
28003 Madrid, Spain or eugpsr@cambridge.org.

www.ingramcontent.com/pod-product-compliance
Ingram Content Group UK Ltd.
Pitfield, Milton Keynes, MK11 3LW, UK
UKHW051007240426
470322UK00018B/550